The Subnuclear Series • Volume 44

Proceedings of the International School of Subnuclear Physics

THE LOGIC OF NATURE, COMPLEXITY AND NEW PHYSICS:

FROM QUARK-GLUON PLASMA TO SUPERSTRINGS, QUANTUM GRAVITY AND BEYOND

THE SUBNUCLEAR SERIES

Series Editor: ANTONINO ZICHICHI, *European Physical Society, Geneva, Switzerland*

1. 1963 STRONG, ELECTROMAGNETIC, AND WEAK INTERACTIONS
2. 1964 SYMMETRIES IN ELEMENTARY PARTICLE PHYSICS
3. 1965 RECENT DEVELOPMENTS IN PARTICLE SYMMETRIES
4. 1966 STRONG AND WEAK INTERACTIONS
5. 1967 HADRONS AND THEIR INTERACTIONS
6. 1968 THEORY AND PHENOMENOLOGY IN PARTICLE PHYSICS
7. 1969 SUBNUCLEAR PHENOMENA
8. 1970 ELEMENTARY PROCESSES AT HIGH ENERGY
9. 1971 PROPERTIES OF THE FUNDAMENTAL INTERACTIONS
10. 1972 HIGHLIGHTS IN PARTICLE PHYSICS
11. 1973 LAWS OF HADRONIC MATTER
12. 1974 LEPTON AND HADRON STRUCTURE
13. 1975 NEW PHENOMENA IN SUBNUCLEAR PHYSICS
14. 1976 UNDERSTANDING THE FUNDAMENTAL CONSTITUENTS OF MATTER
15. 1977 THE WHYS OF SUBNUCLEAR PHYSICS
16. 1978 THE NEW ASPECTS OF SUBNUCLEAR PHYSICS
17. 1979 POINTLIKE STRUCTURES INSIDE AND OUTSIDE HADRONS
18. 1980 THE HIGH-ENERGY LIMIT
19. 1981 THE UNITY OF THE FUNDAMENTAL INTERACTIONS
20. 1982 GAUGE INTERACTIONS: Theory and Experiment
21. 1983 HOW FAR ARE WE FROM THE GAUGE FORCES?
22. 1984 QUARKS, LEPTONS, AND THEIR CONSTITUENTS
23. 1985 OLD AND NEW FORCES OF NATURE
24. 1986 THE SUPERWORLD I
25. 1987 THE SUPERWORLD II
26. 1988 THE SUPERWORLD III
27. 1989 THE CHALLENGING QUESTIONS
28. 1990 PHYSICS UP TO 200 TeV
29. 1991 PHYSICS AT THE HIGHEST ENERGY AND LUMINOSITY: To Understand the Origin of Mass
30. 1992 FROM SUPERSTRINGS TO THE REAL SUPERWORLD
31. 1993 FROM SUPERSYMMETRY TO THE ORIGIN OF SPACE–TIME
32. 1994 FROM SUPERSTRING TO PRESENT-DAY PHYSICS
33. 1995 VACUUM AND VACUA: The Physics of Nothing
34. 1996 EFFECTIVE THEORIES AND FUNDAMENTAL INTERACTIONS
35. 1997 HIGHLIGHTS OF SUBNUCLEAR PHYSICS: 50 Years Later
36. 1998 FROM THE PLANCK LENGTH TO THE HUBBLE RADIUS
37. 1999 BASICS AND HIGHLIGHTS IN FUNDAMENTAL PHYSICS
38. 2000 THEORY AND EXPERIMENT HEADING FOR NEW PHYSICS
39. 2001 NEW FIELDS AND STRINGS IN SUBNUCLEAR PHYSICS
40. 2002 FROM QUARKS AND GLUONS TO QUANTUM GRAVITY
41. 2003 FROM QUARKS TO BLACK HOLES: PROGRESS IN UNDERSTANDING THE LOGIC OF NATURE
42. 2004 HOW AND WHERE TO GO BEYOND THE STANDARD MODEL

Volume 1 was published by W. A. Benjamin, Inc., New York; 2–8 and 11–12 by Academic Press, New York and London; 9–10 by Editrice Compositori, Bologna; 13–29 by Plenum Press, New York and London; 30–44 by World Scientific, Singapore.

43. 2005 TOWARDS NEW MILESTONES IN OUR QUEST TO GO BEYOND THE STANDARD MODEL

44. 2006 THE LOGIC OF NATURE, COMPLEXITY AND NEW PHYSICS: FROM QUARK-GLUON PLASMA TO SUPERSTRINGS, QUANTUM GRAVITY AND BEYOND

The Subnuclear Series • Volume 44

Proceedings of the International School of Subnuclear Physics

THE LOGIC OF NATURE, COMPLEXITY AND NEW PHYSICS:

FROM QUARK-GLUON PLASMA TO SUPERSTRINGS, QUANTUM GRAVITY AND BEYOND

Edited by

Antonino Zichichi

European Physical Society
Geneva, Switzerland

World Scientific

NEW JERSEY • LONDON • SINGAPORE • BEIJING • SHANGHAI • HONG KONG • TAIPEI • CHENNAI

Published by

World Scientific Publishing Co. Pte. Ltd.
5 Toh Tuck Link, Singapore 596224
USA office: 27 Warren Street, Suite 401-402, Hackensack, NJ 07601
UK office: 57 Shelton Street, Covent Garden, London WC2H 9HE

British Library Cataloguing-in-Publication Data
A catalogue record for this book is available from the British Library.

The Subnuclear Series — Vol. 44
THE LOGIC OF NATURE, COMPLEXITY AND NEW PHYSICS
From Quark-Gluon Plasma to Superstrings, Quantum Gravity and Beyond
Proceedings of the International School of Subnuclear Physics

ISBN-13 978-981-283-245-0
ISBN-10 981-283-245-9

Printed in Singapore by World Scientific Printers

Richard Henry Dalitz (Erice, 1994)

This volume is dedicated to the memory of Richard Henry Dalitz, whose contributions to the Subnuclear Physics School activity during many decades, since its foundation, have been outstanding as testified by his superb series of lectures, which have contributed to the School's reputation worldwide.

PREFACE

During August/September 2006, a group of 96 physicists from 57 laboratories in 18 countries met in Erice to participate in the 44th Course of the International School of Subnuclear Physics. The countries represented by the participants were: Belgium, Brazil, Canada, France, Germany, Greece, Hungary, India, Israel, Italy, Japan, the Netherlands, Poland, Romania, Russia, Spain, Ukraine, United Kingdom and the United States of America.

The School was sponsored by the Academies of Sciences of Estonia, Georgia, Lithuania, Russia and Ukraine; the Chinese Academy of Sciences; the Commission of the European Communities; the European Physical Society; the Italian Ministry of Education, University and Scientific Research; the Sicilian Regional Government; the Weizmann Institute of Science; the World Federation of Scientists and the World Laboratory.

Each lecture of the School was dedicated to the memory of Richard H. Dalitz. The School's purpose was to focus attention on the phenomenological and theoretical developments in Gauge Theories, as well as in global and local Supersymmetry, including all the other sectors of Subnuclear Physics. Moreover, special emphasis was given to investigate if and how Complexity could shed new light in the greatest achievement of Reductionism, i.e. the Standard Model and its extension which predicts GUT, the existence of the Superworld and the resolution of the quantum-gravity problem via the powerful theoretical structure of Relativistic Quantum String Theory. Finally, experimental highlights from the most relevant sources of new data were presented and discussed, including the latest news concerning future facilities.

An original feature of the School, introduced in 1996, is a series of special sessions devoted to "New Talents", with the aim of encouraging and promoting young physicists to achieve recognition at an international level. This is a serious problem in Experimental Physics, where collaborations count several hundreds of participants and it is almost impossible for young fellows to be known. Even if with much less emphasis, the problem exists also in Theoretical Physics. So we decided to offer 25 young fellows the possibility to give an open presentation of the results of their studies, followed by a discussion. Among them, 15 best "New Talents" have been selected for publication in this volume, and five were given an award. In particular, one award for an original work in theoretical physics, one for valuable work in experimental physics, one for an original work in data analysis, one for the best theoretical presentation and one for the best experimental presentation were assigned.

During the organization and the running of this year's Course, I enjoyed the collaboration of my colleague and friend, Gerardus 't Hooft, who shared with me the Directorship of the Course. I would like to thank him, together with the group of invited scientists and all the people who contributed to the success of this year's Course.

I hope the reader will enjoy the book as much as the students attending the lectures and the discussion sessions, which, as every year, have been the focal point of the School's activity. Thanks to the work of the Scientific Secretaries, the discussions have been reproduced as faithfully as possible. At various stages of my work I have enjoyed the collaboration of many friends whose contributions have been extremely important for the School and are highly appreciated. I thank them most warmly. A final acknowledgement to all those in Erice, Bologna and Geneva, who have helped me on so many occasions and to whom I feel very indebted.

Antonino Zichichi
Geneva, October 2006

CONTENTS

Complexity of chaotic fields and standard model parameters

Christian Beck
School of Mathematical Sciences, Queen Mary, University of London, Mile End Road, London E1 4NS, UK

Abstract

In order to understand the parameters of the standard model of electroweak and strong interactions (coupling constants, masses, mixing angles) one needs to embed the standard model into some larger theory that accounts for the observed values. This means some additional sector is needed that fixes and stabilizes the values of the fundamental constants of nature. In these lecture notes we describe in non-technical terms how such a sector can be constructed. Our additional sector is based on rapidly fluctuating scalar fields that, although completely deterministic, evolve in the strongest possible chaotic way and exhibit complex behaviour. These chaotic fields generate potentials for moduli fields, which ultimately fix the fundamental parameters. The chaotic dynamics can be physically interpreted in terms of vacuum fluctuations. These vacuum fluctuations are different from those of QED and QCD but coupled with the same moduli fields as QED and QCD are. The vacuum energy generated by the chaotic fields underlies the currently observed dark energy of the universe. Our theory correctly predicts the numerical values of the electroweak and strong coupling constants using a simple principle, the minimization of vacuum energy. Implementing some additional discrete symmetry assumptions one also obtains predictions for fermion masses, as well as a Higgs mass prediction of 154 GeV.

1 Introduction

String theories predict an enourmous amount of possible vacua after compactification, of the order 10^{120}. In each of these vacua the cosmological constant as well as the fundamental constants of nature can have different values. One is led to the so-called 'landscape' picture [1]. The landscape in a sense represents the set of all possible blueprints of the universe. To select the right vacuum, i.e. the one we observe around us right now, often an anthropic point of view is chosen. But is an anthropic principle really the last word and the ultimate answer to all our questions?

A natural idea to avoid an anthropic selection principle would be that the universe will not be left alone with its choice of 10^{120} vacua but that it gets some help. This help should have the form of an additional sector, a theory as yet not included in the ordinary standard model, neither in ordinary string theories. The additional sector should yield some general principle to select, fix and stabilize the standard model parameters in the way we do observe them. In fact, it should do much more: It should choose the right gauge groups, the correct amount of supersymmetry, the correct number of flavour families, it should create an excess of matter over anti-matter, and so on. The additional sector should fix the most relevant information about the future universe at a very early stage, similar as a DNA string fixes the most important information of a human being already when the first cells are formed.

We are still far away from such a theory. But some numerical observations have recently been made [2, 3, 4] that seem to give us a hint how this additional sector could look like. Our aim here is to explain the relevant concepts in a non-technical way.

In practice standard model parameters can be thought of as being fixed by so-called moduli fields. A varying standard model parameter (e.g. the fine structure constant) can be essentially regarded as (a simple function of) such a moduli field. These moduli fields evolve to minima of their potentials. So if we know the correct moduli potentials describing the world around us, we also know the correct standard model parameters. So what could be a theory to construct these moduli potentials? In principle the potentials should follow from the embedding theory (e.g. M theory + compactification + supersymmetry breaking), but little is known in practice due to the enourmous complexity inherent in the above theories. But as with any unknown theory we can be guided by first trying to find an empirical theory that does the correct thing, i.e. reproduces the observed value of the fine structure constant and other fundamental constants, and then later try to embed it into a greater context. The interesting thing is that such an empirical theory is possible [2, 3, 4]: There is a class of highly nonlinear chaotic dynamical system that seem to reproduce the 'correct' standard model parameters by a simple selection mechanism, the minimization of vacuum energy.

Physically the above chaotic dynamics can be regarded as describing rapidly fluctuating scalar fields associated with vacuum fluctuations. These are vacuum fluctuations different from those of QED or QCD, so what would be the physical embedding? The most natural embedding is to associate the above chaotic vacuum fluctuations with the currently observed dark energy in the universe [4, 5]. Since nobody really knows what dark energy is there is a lot of freedom in the dark energy sector and enough 'space' to embed new things! The chaotic fields, living in the dark energy sector, generate effective potentials for moduli fields — just the same moduli fields that are responsible for the fundamental constants of the standard model of electroweak and strong interactions. The moduli fields then move to the minima of the potentials generated by the chaotic fields, and fix the fundamental constants of nature. The chaotic sector appears to provide a possible answer to the question why we do observe certain numerical values of standard model parameters (such as the fine structure constant or the strong coupling at W-mass scale) in nature, others not. It can be used to avoid anthropic considerations for fundamental constants. Moreover, it generates a small cosmological constant in a rather natural way [4].

These lecture notes are organized as follows: In section 2 we provide some background information on moduli fields, variable fine structure constants and related topics. In section 3 recall the method of stochastic quantization introduced by Parisi and Wu [6, 7]. We then introduce the chaotic fields in section 4, and show how they can generate potentials for moduli fields in section 5. Finally, in section 6 we provide numerical evidence that there are various local minima of the potentials that do reproduce known standard model parameters with high precision. We obtain excellent agreement with measured values seen in collider experiments. We will also make a few predictions for unknown parameters of the standard model, such as the Higgs mass, based on the chaotic field dynamics and some additional symmetry assumptions.

2 Moduli fields and variable fine structure constant

The action of the standard model is well known [8], we don't have to work that out here. The term of the action involving the electromagnetic field tensor $F_{\mu\nu}$ is given by

$$S_F = \int d^4x\sqrt{-g}\left(-\frac{1}{4}F_{\mu\nu}F^{\mu\nu}\right). \tag{1}$$

In models with a variable fine structure constant $\alpha(t)$ [9, 10, 11, 12, 13] the above action is modified as follows:

$$S_F = \int d^4x\sqrt{-g}\left(-\frac{1}{4}B_F(\chi/M^*)F_{\mu\nu}F^{\mu\nu}\right). \tag{2}$$

Here $\chi(t)$ is a homogeneous scalar field, henceforth called the *moduli field* (a related field in string theory is the dilaton field). M^* is a mass of the order of magnitude of the reduced Planck mass. B_F is a function that is in principle determined by the embedding theory (e.g. string theory or M theory). In Bekenstein's model [12],

$$B_F(\chi/M^*) = e^{-2(\chi-\bar{\chi})/M^*}, \tag{3}$$

but other choices are possible as well. Changes in the field $\chi(t)$ imply changes in the fine structure constant $\alpha(t)$. The relation is

$$\frac{\alpha(t)}{\bar{\alpha}} = \frac{1}{B_F(\chi(t)/M^*)} \tag{4}$$

where $B_F(\bar{\chi}/M^*) = 1$. Here $\bar{\alpha}$ is the stationary value of $\alpha(t)$, and $\bar{\chi}$ the stationary value of $\chi(t)$.

The interpretation of the above generalized action (2) is quite obvious: Normally, the electromagnetic field tensor would contribute with the term $-\frac{1}{4}F_{\mu\nu}F^{\mu\nu}$ in the action. This relation is now only satisfied in the stationary case. Otherwise the strength of this term is given by a prefactor B_F that depends on the value of the field $\chi(t)$ at a given time t. The fine structure constant $\alpha(t)$ is a simple function of the moduli field $\chi(t)$ according to eq. (4).

For small displacements of the field $\chi(t)$ from its stationary value we may write

$$B_F(\chi(t)/M^*) \approx B_F(\bar{\chi}/M^*) + \frac{\chi(t)-\bar{\chi}}{M^*}B_F'(\bar{\chi}/M^*), \tag{5}$$

where B'_F is the derivative of the function B_F. We thus obtain in leading order from eq. (4)

$$\alpha(t) = \bar{\alpha}\left(1 - B'_F(\bar{\chi}/M^*)\frac{\chi(t) - \bar{\chi}}{M^*}\right). \tag{6}$$

In Bekenstein's model, $B'_F(\bar{\chi}/M^*) = -2$. Clearly, the fine structure constant approaches its stationary value $\bar{\alpha}$ if the moduli field χ approaches its stationary value $\bar{\chi}$. Using $B'_F(\bar{\chi}/M^*) = -2$ we may also write eq. (6) as

$$\frac{1}{2}M^*\frac{\alpha(t) - \bar{\alpha}}{\bar{\alpha}} = \chi(t) - \bar{\chi} \tag{7}$$

which shows that the fine structure constant and the moduli field are basically the same.

To introduce a dynamics for the moduli field χ, we need to know its potential $V(\chi)$. In a Robertson-Walker metric the dynamics is then given by

$$\ddot{\chi} + 3H\dot{\chi} + \frac{\partial V}{\partial \chi} = -\xi_m \frac{\rho_m}{M^*}, \tag{8}$$

where H is the Hubble parameter and ξ_m is the coupling of the field χ to matter (in particular dark matter). ρ_m is the matter density of the universe at a given time. In typical models studied in the literature [9], the coupling ξ_m is very small.

Whereas in general the potential V is unknown, near to the equilibrium point $\bar{\chi}$ we may expand it as

$$V(\chi) = \frac{1}{2}m^2(\chi - \bar{\chi})^2 + const \tag{9}$$

so that

$$\frac{\partial V}{\partial \chi} = m^2(\chi - \bar{\chi}). \tag{10}$$

In [9] m is chosen as an extremely small mass parameter, of the order of the current value H_0 of the Hubble parameter:

$$m \sim H_0 \tag{11}$$

That is to say, one considers an ultralight scalar field χ with a mass of the order 10^{-33} eV. There is motivation for the existence of such ultralight fields from extended supergravity theories [14]. In terms of the fine structure constant $\alpha(t)$, one obtains near the stationary point

$$V(\chi) = \frac{1}{2} m^2 M^{*2} \frac{1}{(B'_F(\bar{\chi}/M^*))^2} \frac{(\alpha(t) - \bar{\alpha})^2}{\bar{\alpha}^2} + const \tag{12}$$

By construction, the energy density $m^2 M^{*2}$ associated with this potential is of the same order of magnitude as the dark energy density of the universe at the present time.

Of course, in a similar way one can introduce further moduli fields corresponding to other standard model coupling constants, masses and mixing parameters as well. The fine structure constant was just one example. In fact, for each relevant standard model parameter there should be a corresponding moduli field. So we need about 20 such fields.

3 Stochastic quantization

Let us now proceed to a 2nd-quantized theory. An elegant method to do 2nd quantization is via the so-called stochastic quantization method. In the Parisi-Wu approach of stochastic quantization one considers a stochastic differential equation evolving in a fictitious time variable s, the drift term being given by the classical field equation [6, 7]. Quantum mechanical expectations correspond to expectations with respect to the generated stochastic processes in the limit $s \to \infty$. The fictitious time s is different from the physical time t, it is just a helpful fifth coordinate to do 2nd quantization. Neglecting spatial gradients the field under consideration is a function of physical time t and fictitious time s. The stochastically quantized equation of motion of a homogeneous scalar field φ in Robertson-Walker metric is

$$\frac{\partial}{\partial s} \varphi = \ddot{\varphi} + 3H\dot{\varphi} + V'(\varphi) + L(s,t), \tag{13}$$

where H is the Hubble parameter, V is the potential under consideration and $L(s,t)$ is Gaussian white noise, δ-correlated both in s and t. For e.g. a numerical simulation we may discretize eq. (13) using

$$s = n\tau \tag{14}$$

$$t = i\delta, \tag{15}$$

where n and i are integers and τ is a fictitious time lattice constant, δ is a physical time lattice constant. The continuum limit requires $\tau \to 0$, $\delta \to 0$, but since quantum field theory is expected to break down at the Planck scale, it can make physical sense to keep a small lattice constant of that size as an effective cutoff:

$$\tau \sim \frac{1}{m_{Pl}^2}, \delta \sim \frac{1}{m_{Pl}}. \tag{16}$$

Eq. (13) yields the discrete dynamics

$$\frac{\varphi_{n+1}^i - \varphi_n^i}{\tau} = \frac{1}{\delta^2}(\varphi_n^{i+1} - 2\varphi_n^i + \varphi_n^{i-1}) + 3\frac{H}{\delta}(\varphi_n^i - \varphi_n^{i-1}) + V'(\varphi_n^i) + noise. \tag{17}$$

This can be written as the following recurrence relation for the field φ_n^i

$$\varphi_{n+1}^i = (1-a)\left\{\varphi_n^i + \frac{\tau}{1-a}V'(\varphi_n^i)\right\} + 3\frac{H\tau}{\delta}(\varphi_n^i - \varphi_n^{i-1}) + \frac{a}{2}(\varphi_n^{i+1} + \varphi_n^{i-1}) + \tau \cdot noise, \tag{18}$$

where a dimensionless coupling constant a is introduced as

$$a := \frac{2\tau}{\delta^2}. \tag{19}$$

This coupling a is a free parameter of our stochastically quantized theory. A priori it can take on any value, reminescent of a moduli field.

We may also introduce a dimensionless field variable Φ_n^i by writing $\varphi_n^i = \Phi_n^i p_{max}$, where p_{max} is some (so far) arbitrary energy scale. The above scalar field dynamics is equivalent to a spatially extended dynamical system (a coupled map lattice) of the form

$$\Phi_{n+1}^i = (1-a)T(\Phi_n^i) + \frac{3}{2}H\delta a(\Phi_n^i - \Phi_n^{i-1}) + \frac{a}{2}(\Phi_n^{i+1} + \Phi_n^{i-1}) + \tau \cdot noise, \tag{20}$$

where the local map T is given by

$$T(\Phi) = \Phi + \frac{\tau}{p_{max}(1-a)}V'(p_{max}\Phi). \tag{21}$$

Here the prime means

$$' = \frac{\partial}{\partial \varphi} = \frac{1}{p_{max}}\frac{\partial}{\partial \Phi}. \tag{22}$$

A symmetric dynamics of the form

$$\Phi_{n+1}^i = (1-a)T(\Phi_n^i) + \frac{a}{2}(\Phi_n^{i+1} + \Phi_n^{i-1}) + \tau \cdot noise \tag{23}$$

is obtained if $H\delta << 1$, equivalent to

$$\delta << H^{-1}. \tag{24}$$

This approximation is valid if the universe is much older than the physical time lattice constant δ. In this case the term proportional to H in eq. (20) can be neglected. The local map T depends on the potential V under consideration. Since we restrict ourselves to real scalar fields φ, T is a 1-dimensional map.

Let us summarize the main result of this section: Iterating a recurrence relation of the form (23) is *equivalent* to considering a stochastically quantized scalar field. The relation between the map T and the potential V is

$$V'(\varphi) = \frac{1-a}{\tau}\left\{-\varphi + p_{max}T\left(\frac{\varphi}{p_{max}}\right)\right\}. \tag{25}$$

Integration yields

$$V(\varphi) = \frac{1-a}{\tau}\left\{-\frac{1}{2}\varphi^2 + p_{max}\int d\varphi\, T\left(\frac{\varphi}{p_{max}}\right)\right\} + const, \tag{26}$$

which in terms of the dimensionless field Φ this can be written as

$$V(\varphi) = \frac{1-a}{\tau}p_{max}^2\left\{-\frac{1}{2}\Phi^2 + \int d\Phi T(\Phi)\right\} + const. \tag{27}$$

4 Introducing chaotic fields

We now come to the crucial point, namely what type of dynamics is generated by (23) for various types of scalar field theories. Take the example of an ultralight moduli field χ. For these types of fields $p_{max} \sim M^* \sim m_{Pl}$ is large, whereas $V'(\chi) \sim m^2(\chi - \bar{\chi})$ is extremely small. This means the mapping T in eq. (21) is extremely close to an identity and the field moves very slowly. Effectively this means that our discrete dynamics (23) approximates very well a smooth continuum evolution of the field χ. The field χ smoothly approaches a minimum of the potential and stays there, apart from some small fluctuations induced by the noise term.

Once again, for a moduli field χ the energy scale associated with the field variable is very large, of the order of the Planck mass $M^* \sim m_{Pl}$, whereas the potential V contains an extremely small mass parameter m, of the order of the present Hubble constant H_0. We could ask whether for symmmetry reasons maybe another scalar field φ exists that has just the opposite properties, i.e. the energy scale of that field variable φ is extremely small (of the order H_0) but its potential V contains a mass term of order M^*?

Indeed, such a field has been studied in [4]. Due to the fact that the forcing is now very strong, this field exhibits strongly chaotic behaviour if it is stochastically quantized. This can immediately be seen from eq. (21): With p_{max} being small, the map T is now far away from the identity. Chaotic behaviour is possible. The small noise term in eq. (23) can actually be neglected in the chaotic case.

We may go one step further in our symmetry considerations. The moduli field is as slow and regular as a field can be. What would be the other extreme? What would be a field that is as rapidly fluctuating and irregular as possible? In other words, which scalar field dynamics would create the strongest possible chaotic behaviour?

The above question has been solved in [15, 16] and is well understood from a nonlinear dynamics point of view. One knows that the maps with strongest chaotic properties (being smooth and deterministic at the same time) are those conjugated to a Bernoulli shift [17] (a shift of integer symbols in suitable coordinates). Among those, certain deterministic maps are even more 'random' than others, in the sense of having least higher-order correlations: These are the so-called Tchebyscheff maps T_N of N-th order, defined as

$$T_2(\Phi) = 2\Phi^2 - 1 \quad (28)$$

$$T_3(\Phi) = 4\Phi^3 - 3\Phi \quad (29)$$

$$\ldots = \ldots \quad (30)$$

$$T_N(\Phi) = \cos(N \arccos \Phi) \quad (31)$$

They can arise out of the above dynamics (23) for suitable potentials V.

The most important scalar field potential in particle physics is of course a double-well potential. So let us consider the distinguished example of a φ^4-theory generating strongest possible chaotic behaviour. Take the potential

$$V_{-3}(\varphi) = \frac{1-a}{\tau}\left\{\varphi^2 - \frac{1}{p_{max}^2}\varphi^4\right\} + const, \quad (32)$$

or, in terms of the dimensionless field Φ,

$$V_{-3}(\varphi) = \frac{1-a}{\tau} p_{max}^2 (\Phi^2 - \Phi^4) + const. \tag{33}$$

Applying our formalism of the previous section, we end up with the following local map:

$$\Phi_{n+1} = T_{-3}(\Phi_n) = -4\Phi_n^3 + 3\Phi_n \tag{34}$$

T_{-3} is the (negative) third-order Tchebyscheff map. It is conjugated to a Bernoulli shift of 3 symbols, and generates the strongest possible stochastic behaviour possible for a smooth low-dimensional deterministic dynamical system.

Apparently, starting from the potential (32) we obtain by second quantization a field φ that rapidly fluctuates in fictitious time on some finite interval, provided that initially $\varphi_0 \in [-p_{max}, p_{max}]$. The small noise term in eq. (23) can be neglected as compared to the deterministic chaotic fluctuations of the field. We physically interprete these rapid changes of the field φ as representing vacuum fluctuations. Of course these are vacuum fluctuations different from those of QED or QCD. Since the expectation of the vacuum energy associated with the chaotic field is $\langle V_{-3}(\varphi)\rangle \sim p_{max}^2/\tau \sim H_0^2 m_{Pl}^2$, such a chaotic field yields the correct order of magnitude of vacuum energy density in order to account for a small cosmological constant. Hence we assume that the chaotic fields underly dark energy.

The above example of a chaotic φ^4-theory can be generalized. There are various discrete degrees of freedom to introduce a deterministic chaotic field dynamics that is as random as possible. Consider a 1-dimensional lattice, the lattice sites are labelled by integers i. At each lattice site i we have a dimensionless field variable Φ_n^i which evolves in discrete time n. In the uncoupled case, the dynamics is given by

$$\Phi_{n+1}^i = \pm T_N(\Phi_n^i), \tag{35}$$

where $\pm T_N$ is either the positive of the negative Tchebyscheff map. The initial value Φ_0 is chosen on the interval $[-1, 1]$, the iterates Φ_n^i then stay in this interval, but evolve in a deterministic chaotic way. The dynamics is conjugated to a Bernoulli shift of N symbols, which means that in suitable coordinates the iteration process is like shifting symbols in a random symbol sequence. There are further discrete degrees of freedom to do the coupling to the nearest neighbours. Instead of coupling to the variables $\Phi_n^{i\pm1}$ at the neighboured sites as in eq. (23), we could also use the updated variables $\pm T_N(\Phi_n^{i\pm 1})$. Since n describes fictitious time, it's really not clear which choice is the correct one. For sure both degrees of freedom exist. All these discrete degrees of freedom can be written in a compact form as follows:

$$\Phi_{n+1}^i = (1-a)T_N(\Phi_n^i) + s\frac{a}{2}(T_N^b(\Phi_n^{i-1}) + T_N^b(\Phi_n^{i+1})), \qquad (36)$$

where s is a sign variable taking on the values ± 1. The choice $s = +1$ is called 'diffusive coupling', but for symmetry reasons it also makes sense sense to study the choice $s = -1$, which we call 'anti-diffusive coupling'. The integer b distinguishes between the forward and backward coupling form, $b = 1$ corresponds to forward coupling ($T_N^1(\Phi) := T_N(\Phi)$), $b = 0$ to backward coupling ($T_N^0(\Phi) := \Phi$). We consider random initial conditions, periodic boundary conditions and large lattices of size i_{max}.

One can easily check that for odd N the choice of s is irrelevant (since odd Tchebyscheff maps satisfy $T_N(-\Phi) = -T_N(\Phi)$), whereas for even N the sign of s is relevant and a different dynamics arises. Hence, restricting ourselves to $N = 2$ and $N = 3$, in total 6 relevant chaotic scalar field theories arise, characterized by $(N, b, s) = (2,1,+1), (2,0,+1), (2,1,-1), (2,0,-1)$ and $(N, b) = (3,1), (3,0)$. For easier notation, in the following we will label these chaotic field theories as $2A, 2B, 2A^-, 2B^-, 3A, 3B$, respectively.

The important thing to remember from this section is that eq. (36) just describes a degenerated stochastically quantized scalar field with strongest possible chaotic properties. For somebody from the nonlinear dynamics and complexity community, the dynamics (36) represents the most natural thing in the world: It's just a coupled map lattice, a standard example of a spatio-temporal dynamical system [18]. For somebody from the elementary particle physics or string theory community, however, eq. (36) may look somewhat unusual and unfamiliar at first sight. But it's sometimes worth to learn new things!

The dynamics (36) is a dissipative deterministic dynamical system. It exhibits chaotic behaviour and produces information, measured by a positive KS entropy [17]. Dissipative deterministic systems as a model of quantum gravity have also been suggested in [19].

5 Generating potentials for moduli fields

Clearly, if the coupling a in the above chaotic field dynamics (36) is chosen as $a = 0$, then there are no correlations between neighboured lattice sites:

$$\langle \Phi_n^i \Phi_n^{i+1} \rangle = 0 \tag{37}$$

Here the notation $\langle \cdots \rangle$ denotes the expectation value, which can be numerically calculated by iterating the maps for random initial conditions and doing a time average. Note that the index i denotes physical time (in units of the lattice constant δ). If we physically interpret the chaotic fluctuations of the field Φ_n^i as some sort of vacuum fluctuations underlying the dark energy of the universe (see [4] for more details), then the above condition (37) looks physically very reasonable: We want subsequent vacuum fluctuations to be uncorrelated in physical time, because otherwise they wouldn't really describe spontaneous fluctuations.

We may, however, insist on an interacting theory, i.e. $a \neq 0$. Are there still some distinguished couplings $a^* \neq 0$ where we can keep the above condition of a vanishing correlation function in physical time? Of course such a state would again be distinguished as being as random as possible, where this concept is now extended from fictitious time to physical time. There are also a couple of other reasons why one wants a vanishing correlation function, see [3] for more details.

In fact, numerical investigations show that the above states with vanishing correlation exist and distinguish certain coupling constants a^*. These are numerically observed to coincide with known standard model coupling strengths (see next section). In general one can show [2, 3] that the quantity

$$W(a) = \frac{1}{2} \langle \Phi_n^i \Phi_n^{i+1} \rangle \tag{38}$$

can be physically interpreted as the interaction energy of the chaotic field theory under consideration (in suitable units). Hence states of strongest random properties, described by a vanishing correlation function, have vanishing interaction energy.

Besides the interaction energy, there is also another relevant vacuum energy associated with the chaotic fields. This is the self energy $V(a)$, given by

$$V^{(2)}(a) = \langle \Phi \rangle - \frac{2}{3} \langle \Phi^3 \rangle \quad (N = 2), \tag{39}$$

respectively

$$V^{(3)}(a) = \frac{3}{2}\langle\Phi^2\rangle - \langle\Phi^4\rangle \quad (N = 3). \tag{40}$$

For a derivation, see [3]. The self energy describes the vacuum energy associated with the potentials that generate the chaotic dynamics in fictitious time (the additive constant is fixed by some symmetry considerations).

In the following, we will use both the interaction energy and the self energy to generate suitable potentials for moduli fields. Recall that classically the moduli field χ obeys

$$\ddot{\chi} + 3H\dot{\chi} + V'(\chi) = 0, \tag{41}$$

where we neglected possible interactions with dark matter. In the vicinity of the stationary state $\bar{\chi}$, one has $V'(\chi) = m^2(\chi - \bar{\chi})$ and the moduli field is essentially the same as the standard model coupling constant α. According to eq. (7) we have

$$\chi(t) = \frac{M^*}{2\bar{\alpha}}\alpha(t) + \bar{\chi} - \frac{1}{2}M^*. \tag{42}$$

Putting eq. (42) into (41), we obtain an equation for α,

$$\ddot{\alpha} + 3H\dot{\alpha} + V'(\alpha) = 0, \tag{43}$$

where locally the potential is given by

$$V(\alpha) = \frac{1}{2}m^2(\alpha - \bar{\alpha})^2. \tag{44}$$

Eq. (43) is just the equation of a damped harmonic oscillator, provided α is in the vicinity of $\bar{\alpha}$.

The crucial point is now that we assume that *the same* moduli fields are responsible for the coupling constants of the standard model of electroweak and strong interactions and those of the dark sector described by the chaotic fields. This means $a = \alpha$. The chaotic field wants to find a state of strongest possible random properties described by a vanishing correlation function $\langle\Phi_n^i\Phi_n^{i+1}\rangle$. Hence the moduli field χ given by eq. (42) with $\alpha = a$ needs to adjust its value. This can be achieved by choosing in eq. (43) the formal potential[1]

$$V(\alpha) = -m^2 \int_0^\alpha d\alpha' \langle\Phi_n^i\Phi_n^{i+1}\rangle. \tag{45}$$

---

[1]Our sign conventions relate to moduli potentials generated by *positive* Tchebyscheff maps.

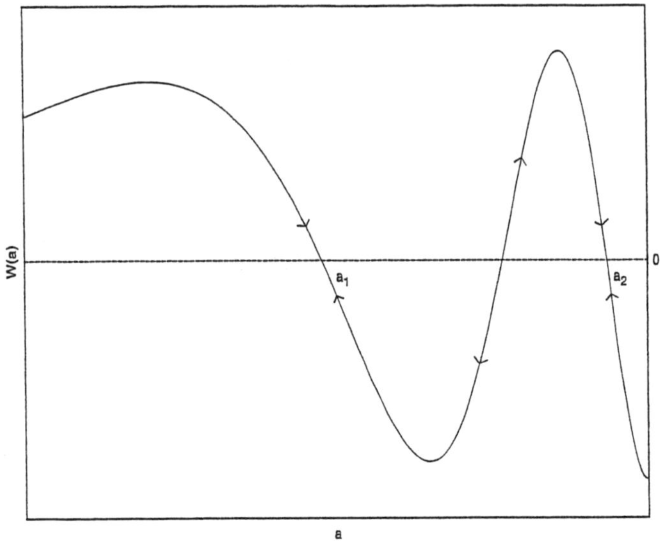

Figure 1: Basic idea underlying this paper. A priori all kinds of standard model couplings $\alpha = a$ (values of moduli fields) are possible. These are then driven into stable zeros of the interaction energy $W(a)$ of the chaotic fields. In the above schematic picture, the stable zeros are denoted by a_1 and a_2.

By construction, local minima of this potential are stable stationary points corresponding to stable (attracting) zeros of the forcing $V'(\alpha) = -m^2\langle\Phi_n^i\Phi_n^{i+1}\rangle$ (see Fig. 1). If the damping H in eq. (43) is large as compared to m, and if the initial displacement is not too large, then the stationary state $\bar{\alpha}$ is rapidly approached.

For other types of moduli fields, we may choose in eq. (43) the self energy potential $V(\alpha) = m^2V^{(N)}(\alpha)$ as given in eq. (39) or (40). These types of moduli fields then evolve to local minima of the self energy.

It is well known that standard model interaction strengths actually depend on the relevant energy scale E. We have the running electroweak and strong coupling constants. For example, the fine structure constant $\alpha_{el}(E)$ slightly increases with E, and the strong coupling α_s rapidly decreases with E.

What should we now take for the energy (or temperature) E of the moduli fields near to their stationary state? In other words, if a standard model coupling $\bar{\alpha}$ is fixed as a minimum of the potential $V(\alpha)$, at which energy scale

E is this standard model coupling fixed? *A priori* this is unknown. However, there is extensive numerical evidence [2, 3] that the distinguished couplings $\bar{\alpha} = a^*$ which correspond to minima of the potentials numerically coincide with running standard model coupling constants $\alpha(E)$ with an energy E being given by

$$E = \frac{1}{2}N(m_B + m_{f_1} + m_{\bar{f}_2}). \tag{46}$$

Here N is the index of the Tchebyscheff map considered, and $m_B, m_{f_1}, m_{\bar{f}_2}$ denote the masses of a boson B and a fermion f_1 and anti-fermion $\bar{f}_2$ —not just some exotic bosons and fermions but precisely those that we know from the standard model. One typically observes particle combinations $B, f_1, \bar{f}_2$ that describe possible interaction states in the standard model, for example a decay of the form $B \to f_1 + \bar{f}_2$ or a reaction $f_1 + \bar{f}_2 \to B$. Detailed evidence will be given in the next section. Formula (46) formally reminds us of the energy levels $E_N = \frac{N}{2}\hbar\omega$ of a quantum mechanical harmonic oscillator, with low-energy levels ($N = 2, 3$) given by the masses of the standard model particles.

6 Numerical results

We now present our numerical results (much more details can be found in [2, 3]). In particular, we will show that stable zeros of the correlation functions (states of strongest random properties for chaotic fields, which are stationary states for moduli fields) reproduce known standard model coupling constants and masses with high precision. Our results are obtained by long-term iteration of the chaotic dynamics. One numerically calculates, for a given coupling a, the interaction energy as the time average

$$W(a) = \frac{1}{2}\langle \Phi_n^i \Phi_n^{i+1} \rangle = \frac{1}{2} \lim_{n\to\infty} \lim_{J\to\infty} \frac{1}{MJ} \sum_{n=1}^{M} \sum_{i=1}^{J} \Phi_n^i \Phi_n^{i+1}, \tag{47}$$

where the Φ_n^i evolve according to (36) with random initial conditions. In practice, we used a finite lattice of size $J = 10000$ with periodic boundary conditions, and iteration numbers corresponding to several weeks of CPU time. Everybody is welcome to reproduce and verify the numerical results described below—the recurrence relation (36) can be very easily installed on any computer. What one observes is that the stable zeros of $W(a)$ coincide

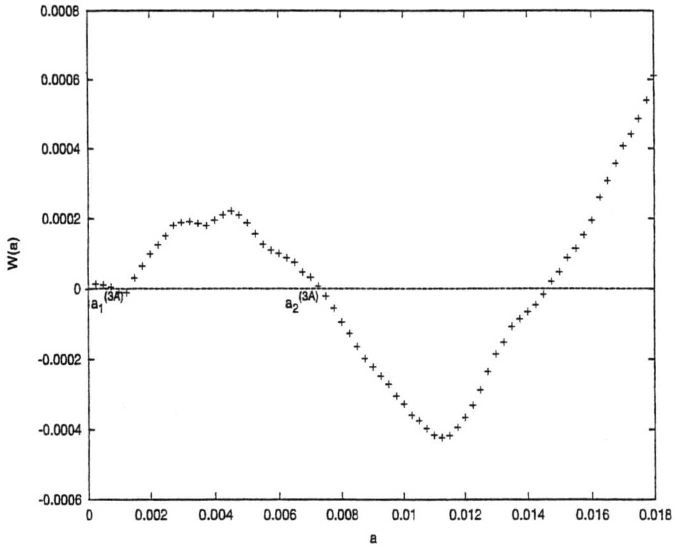

Figure 2: Interaction energy of the 3A dynamics in the small-coupling region.

with known standard model interaction strengths, thus indicating the physical relevance of the theory presented presented in sections 1–5. The chaotic fields appear to determine the moduli potentials in precisely the way we need them to be for a physically realistic vacuum. They can 'help' the universe to find the 'right' vacuum out of an incredibly large number of choices. Anthropic considerations can be avoided in this context.

6.1 The 3A dynamics—electric interaction strengths of electrons and d-quarks

Fig. 2 shows the interaction energy $W(a) = \frac{1}{2}\langle \Phi_n^i \Phi_n^{i+1} \rangle$ of the chaotic $3A$ dynamics in the small-coupling region. We observe two stable zeros of the interaction energy in the low-coupling region:

$$
\begin{aligned}
a_1^{(3A)} &= 0.0008164(8) \\
a_2^{(3A)} &= 0.0073038(17)
\end{aligned}
$$

Previously these stable zeros were denoted by a^* or $\bar{a}$. A stable zero satisfies $V'(a^*) = 0$ and $V''(a^*) > 0$, with V given by eq. (45).

Remarkably, the zero $a_2^{(3A)}$ appears to numerically coincide with the fine structure constant $\alpha_{el} \approx 1/137$. To construct a suitable physical interpretation in the sense of eq. (46), let us choose $B = \gamma$, $f_1 = e^-, \bar{f}_2 = e^+$. The relevant energy scale underlying this moduli field is then given by $E = (3/2)(m_\gamma + 2m_e) = 3m_e$, according to eq. (46). Hence our standard model interpretation of the stationary moduli state described by $a_2^{(3A)}$ suggests the numerical identity

$$a_2^{(3A)} = \alpha_{el}(3m_e). \tag{48}$$

For a precise numerical comparison let us estimate the running electromagnetic coupling at this energy scale. We may use the 1st-order QED formula

$$\alpha_{el}(E) = \alpha_{el}(0)\left\{1 + \frac{2\alpha_{el}(0)}{\pi}\sum_i f_i \int_0^1 dx\, x(1-x)\log\left(1 + \frac{E^2}{m_i^2}x(1-x)\right)\right\}. \tag{49}$$

The sum is over all charged elementary particles, m_i denotes their (free) masses, and f_i are charge factors given by 1 for e, μ, τ-leptons, $\frac{4}{3}$ for u, c, t-quarks and $\frac{1}{3}$ for d, s, b-quarks. Using this formula, we get $\alpha_{el}(3m_e) = 0.007303$, to be compared with $a_2^{(3A)} = 0.0073038(17)$. There is excellent agreement. Inverting the argument, the above zero of the interaction energy of the chaotic field can be used to predict the numerical value of the fine structure constant from first principles.

Next, we notice that the other zero $a_1^{(3A)}$ has approximately the value $\frac{1}{9}\alpha_{el}$. This could mean that the chaotic 3A field also has a mode that provides evidence for electrically interacting d-quarks. Our interpretation is

$$a_1^{(3A)} = \alpha_{el}^d(3m_d) = \frac{1}{9}\alpha_{el}(3m_d), \tag{50}$$

where $\alpha_{el}^d = \frac{1}{9}\alpha_{el}$ denotes the electromagnetic interaction strength of d-quarks. In the harmonic oscillator interpretation (46), we may choose $B = \gamma$, $f_1 = d, f_2 = \bar{d}$. Formula (49), as an estimate, yields for $m_d \approx 9$ MeV the value $\alpha_{el}(3m_d) = 0.007349$, which coincides very well with $9a_1^{(3A)} = 0.007348(7)$. The value $9a_1^{(3A)}$ actually translates to the energy scale $E_d = (26.0 \pm 6.4)$ MeV. This yields $m_d = \frac{1}{3}E_d = (8.7 \pm 2.1)$ MeV, which coincides with estimates of the $\overline{MS}$ current quark mass of the d quark at the proton mass renormalization scale.

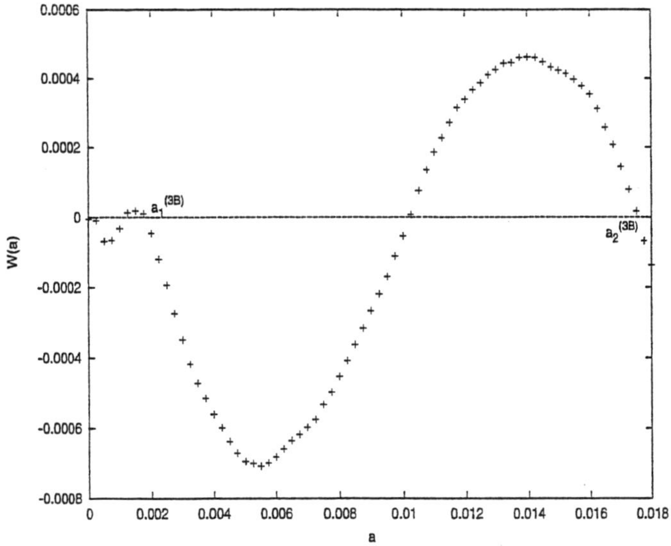

Figure 3: Interaction energy of the 3B dynamics in the low-coupling region.

6.2 The 3B dynamics —weak interaction strengths of neutrinos and u-quarks

The interaction energy $W(a)$ of the 3B field is plotted in Fig. 3. In the low-coupling region $a \in [0, 0.018]$ we observe the following stable zeros of $W(a)$:

$$\begin{aligned} a_1^{(3B)} &= 0.0018012(4) \\ a_2^{(3B)} &= 0.017550(1) \end{aligned}$$

If our approach is consistent, we should be able to find an interpretation of $a_1^{(3B)}$ and $a_2^{(3B)}$ in terms of moduli fields fixing the standard model coupling strengths of u-quarks and neutrinos.

Let us start with $a_2^{(3B)}$. For left-handed neutrinos, the weak coupling due to the exchange of Z^0-bosons is given by

$$\alpha_{weak}^{\nu_L} = \alpha_{el} \frac{1}{4 \sin^2 \theta_W \cos^2 \theta_W}. \tag{51}$$

Here θ_W is the weak mixing angle. In the following we will treat $\sin^2 \theta_W$ as an effective constant, and regard α_{el} as the running electromagnetic coupling. Other renormalization schemes are also possible, but yield only minor numerical differences. Experimentally, the effective weak mixing angle is measured as $\sin^2\theta_W = \bar{s}_l^2 \approx 0.2315$ [20]. Assuming that in addition to the left-handed neutrino interacting weakly there is an electron interacting electrically, the two interaction processes can add up independently if the electron is right-handed, since right-handed electrons cannot interact with left-handed neutrinos. Hence a possible standard model interpretation of the zero $a_2^{(3B)}$ would be

$$a_2^{(3B)} = \alpha_{el}(3m_e) + \alpha_{weak}^{\nu_L}(3m_{\nu_e}) = a_2^{(3A)} + \alpha_{el}(3m_{\nu_e})\frac{1}{4\sin^2\theta_W\cos^2\theta_W} \quad (52)$$

In the harmonic oscillator interpretation of eq. (46), we choose B massless, $f_1 = \nu_L$, $\bar{f}_2 = \bar{\nu_L}$ in addition to the process already described by $a_2^{(3A)}$. Putting in the experimentally measured value of $\sin^2\theta_W = 0.2315$, we obtain for the right-hand side of eq. (52) the value 0.01756, which coincides to 4 digits with the observed stationary value of the moduli field $a_2^{(3B)} = 0.01755$.

Next, let us interpret $a_1^{(3B)}$. In analogy to the joint appearance of ν and e, we should also expect to find evidence for a weakly interacting u-quark, together with a d-quark interacting electrically. Clearly, the u-quark could also interact electrically, but for symmetry reasons we expect the pair (u, d) to interact in a similar way as (ν, e). A right-handed u-quark interacts weakly with the coupling

$$\alpha_{weak}^{u_R} = \frac{4}{9}\alpha_{el}\frac{\sin^2\theta_W}{\cos^2\theta_W}. \quad (53)$$

Adding up the electrical interaction strength of a d-quark, a natural interpretation, quite similar to that of the zero $a_2^{(3B)}$, is

$$a_1^{(3B)} = \alpha_{el}^d(3m_d) + \alpha_{weak}^{u_R}(3m_u) = a_1^{(3A)} + \frac{4}{9}\alpha_{el}(3m_u)\frac{\sin^2\theta_W}{\cos^2\theta_W} \quad (54)$$

The harmonic oscillator interpretation of this moduli state is B massless, $f_1 = u_R$, $f_2 = \bar{u_R}$ in addition to the process underlying $a_1^{(3A)}$. Numerically, taking $\sin^2\theta_W = 0.2315$ and evaluating the running α_{el} using $m_u \approx 5$ MeV, we obtain for the right-hand side of eq. (54) 0.001800, which should be compared with $a_2^{(3B)} = 0.001801$. Again we have perfect agreement within the first 4 digits. It is remarkable that the same universal effective value $\sin^2\theta_W = 0.2315$ can be used consistently for both leptons (couplings $a_2^{(3A)}$,

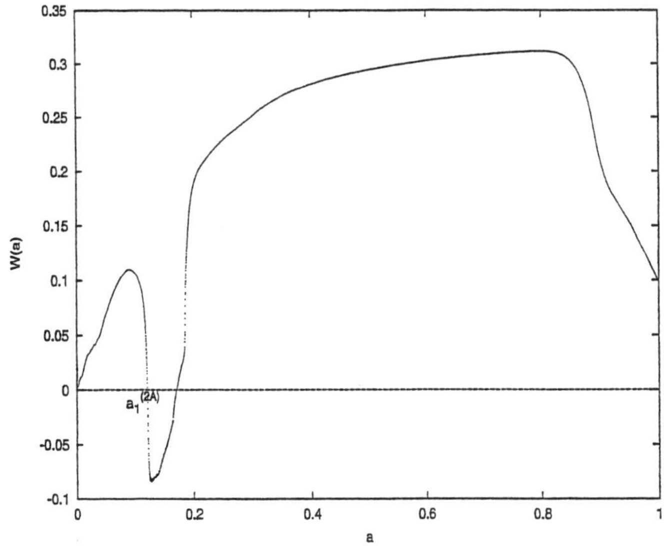

Figure 4: Interaction energy of the 2A dynamics.

$a_2^{(3B)}$) and quarks (couplings $a_1^{(3A)}$, $a_1^{(3B)}$). Inverting the formulas, the observed stationary values of the moduli fields can be used to predict that the weak mixing angle is $\bar{s}_l^2 \approx 0.2315$.

Note that generally the backward coupling form of the $N = 3$ chaotic fields seems to describe a spinless state (formed by e_R and ν_L, respectively d_L and u_R), whereas the forward coupling form just describes one particle species with non-zero spin (e or d). A similar statement will turn out to hold for the $N = 2$ theories, replacing fermions by bosons.

6.3 The 2A dynamics —strong interaction strength at the W-mass scale

If electroweak coupling strengths are fixed by suitable moduli potentials generated by chaotic fields, then something similar should also be the case for the strong coupling strength α_s. Let us now look at chaotic fields with $N = 2$. Fig. 4 shows the interaction energy $W(a)$ of the 2A dynamics. Only one stable zero is observed:

$$a_1^{(2A)} = 0.120093(3) \tag{55}$$

We notice that it numerically seems to coincide with the strong coupling constant α_s at the W- or Z mass scale, which is experimentally measured as $\alpha_s(m_Z) \approx 0.118$.

For symmetry reasons, it seems plausible that if the $N = 3$ dynamics fixes the electroweak couplings at the smallest fermionic mass scales, then the $N = 2$ dynamics could fix the strong couplings at the smallest bosonic mass scales. The lightest massive gauge boson is indeed the $W^\pm$. Hence our physical interpretation associated with $a_1^{(2A)}$ would be $B = W^\pm$, $f_1 = u$, $\bar{f}_2 = \bar{d}$ (respectively $f_1 = d$, $\bar{f}_2 = \bar{u}$), and since $N = 2$ formula (46) implies

$$a_1^{(2A)} = \alpha_s(m_W + m_u + m_d) \approx \alpha_s(m_W). \tag{56}$$

Since the W-mass is known with high precision, eq. (56) and (55) yield quite a precise prediction for the strong coupling α_s. We can evolve it with high precision to arbitrary energy scales, using the well known perturbative formulas from QCD, obtaining

$$\alpha_s(m_{Z^0}) = 0.117804(12). \tag{57}$$

This prediction of α_s from the zero of the chaotic 2A dynamics is clearly consistent with the experimentally measured value 0.118 and in fact much more precise than current experiments can verify.

6.4 The 2B dynamics—the lightest scalar glueball

The interaction energy of the 2B dynamics is shown in Fig. 5. $W(a)$ has only one non-trivial zero

$$a_1^{(2B)} = 0.3145(1). \tag{58}$$

It is a stable zero, so it should describe an observable stable stationary state of a moduli field. One possibility is to interpret this as a strong coupling at the lightest glueball mass scale. The lightest scalar glueball has spin $J^{PC} = 0^{++}$ and is denoted by gg^{0++} in the following. In our oscillator interpretation we take $B = gg^{0++}$, $f_1 = u$, $\bar{f}_2 = \bar{u}$, thus

$$a_1^{(2B)} = \alpha_s(m_{gg^{0++}} + 2m_u) \approx \alpha_s(m_{gg^{0++}}). \tag{59}$$

The 2B dynamics then describes two bosons at the same time (two gluons forming a glueball), similar to the 3B dynamics, which described two fermions at the same time (left and right handed). In both cases a spin 0 state is

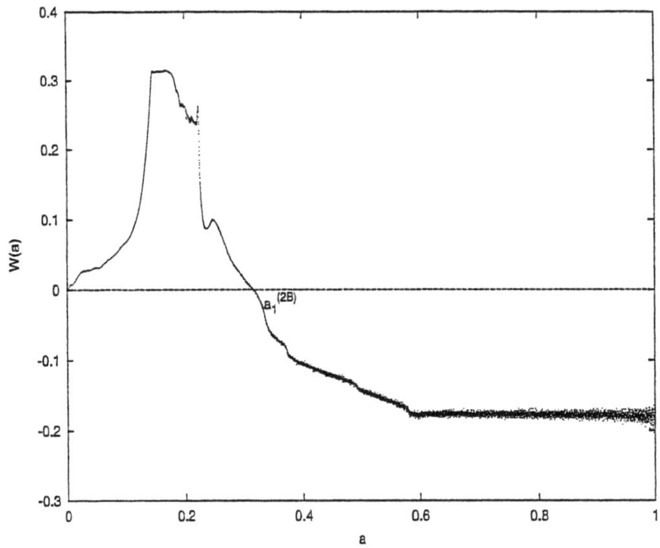

Figure 5: Interaction energy of the 2B dynamics.

formed in total. In lattice gauge calculations including dynamical fermions the smallest scalar glueball mass is estimated as $m_{gg^{0++}} = (1.74 \pm 0.07)$ GeV [21] and at this energy the running strong coupling constant is experimentally measured to be $\alpha_s \approx 0.32$. This clearly is consistent with the observed value of $a_1^{(2B)}$.

6.5 The 2A$^-$ and 2B$^-$ dynamics — towards a Higgs mass prediction

Two chaotic field theories are still remaining, namely those with $N = 2$ and antidiffusive coupling. The interaction energies $W(a)$ of the 2A$^-$ and 2B$^-$ chaotic fields are shown in Fig. 6 and 7. Let us now try to find a suitable physical interpretation for the observed smallest stable zeros $a_1^{(2A^-)} = 0.1758(1)$ and $a_1^{(2B^-)} = 0.095370(1)$. Again let us be guided by discrete symmetry considerations when attributing the various stationary moduli states to standard model particles. We saw that the smallest stable zero of the 2A dynamics describes a boson with non-zero spin (the lightest massive gauge boson $W^\pm$) and the smallest stable zero of the 2B dynamics a boson without spin (the

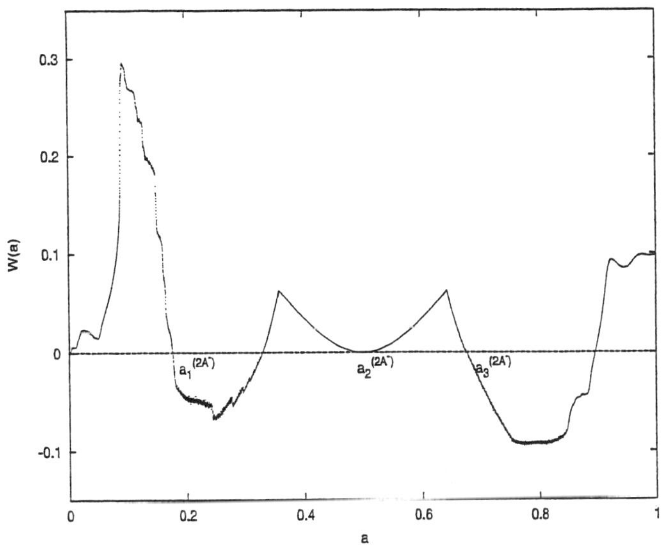

Figure 6: Interaction energy of the $2A^-$ dynamics.

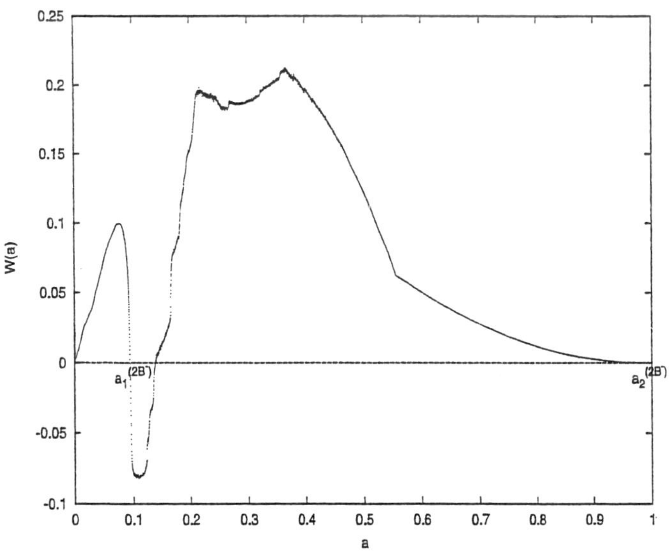

Figure 7: Interaction energy of the $2B^-$ dynamics.

lightest scalar glueball). Thus it seems reasonable to assume that the smallest stable zero of the 2A$^-$ gives us information on yet another bosonic particle with non-zero spin, possibly the lightest glueball with spin $J^{PC} = 2^{++}$, and the smallest zero of the 2B$^-$ dynamics on yet another bosonic particle with spin 0, possibly the lightest Higgs boson.

Let us start with $a_1^{(2A^-)}$. Our physical interpretation of this stationary moduli field state in the sense of eq. (46) is $B = gg^{2++}, f_1 = q, \bar{f}_2 = \bar{q}$, where $q, \bar{q}$ are suitable quarks. From the strong coupling interpretation

$$a_1^{(2A^-)} = \alpha_s(E_1) \tag{60}$$

the energy $E_1 = m_{gg^{2++}} + 2m_q$ can again be determined from the usual QCD formula, evolving our previously determined $\alpha_s(m_W)$ to lower energies. One obtains $E_1 \approx 10.45$ GeV. In lattice gauge calculations the mass of the lightest 2^{++} glueball is estimated as being roughly 2 GeV. We thus get the correct order of magnitude of the 2^{++} glueball mass if we assume that the quarks in eq. (46) are bottom quarks. With this interpretation a glueball mass $m_{gg2^{++}} = (10.45 - 2 \cdot 4.23)$ GeV =2.0 GeV is predicted, using $m_b = 4.23$ GeV.

Next, let us consider $a_1^{(2B^-)}$. Having already obtained information on the W boson in section 6.3, it is likely that there are also moduli fields that encode the Higgs boson. Our physical interpretation is $B = H, f_1 = q, \bar{f}_2 = \bar{q}$, where H is the lightest Higgs boson and $q, \bar{q}$ are suitable quarks. The strong coupling interpretation

$$a_1^{(2B^-)} = \alpha_s(E_2) \tag{61}$$

yields $E_2 = 483.4(3)$ GeV $=m_H + 2m_q$. However, experimental and theoretical arguments imply that the Higgs mass should be in the region 100...200 GeV. Hence we only obtain a consistent value for the Higgs mass if we assume that the quarks involved are t quarks. This is similar to the zero $a_1^{(2A^-)}$, where the quarks involved were also heavy quarks. Generally, chaotic fields with antidiffusive couplings seem to encode information on heavy quarks rather than light ones.

From the self energy of the 2B dynamics, one can obtain quite a precise prediction of the free top mass, namely $m_t = 164.5(2)$ GeV, corresponding to a top pole mass of 174.4(3) GeV (see [2, 3] for more details). With this value the zero $a_1^{(2B^-)}$ yields a Higgs mass prediction of

$$m_H = E_2 - 2m_t = 154.4(5) GeV. \tag{62}$$

This is a very precise prediction, the statistical error is very small. But of course the main source of uncertainty is a theoretical uncertainty, namely whether our harmonic oscillator interpretation (in the sense of eq. (46)) of the zero $a_1^{(2B^-)}$ in terms of a Higgs boson and two top quarks is correct. For example, assuming that a supersymmetric extension of the standard model is correct, then the zero could also describe other particle states whose masses add up to 483.4 GeV.

7 New physics?

A few stable zeros of interaction energies remain that cannot be interpreted in the standard model context. The $3A$ dynamics has yet another stable zero $a_3^{(3A)} = 0.07318$. Interpreted as a running strong coupling, this would correspond to an energy $E = \frac{3}{2}(m_B + m_{f_1} + m_{\bar{f}_2}) \approx 7.85$ TeV. One could speculate that this might describe a (rather large) supersymmetric particle scale. There are a few further stable zeros in the large coupling region: $a = 0.9141$ and $a = 1$ ($3A$), $a = 0.3496$ ($3B$), $a = 0.675$ ($2A^-$) and $a = 1$ ($2B^-$). Possibly these stationary moduli states could describe gravitational couplings at the Planck scale.

So far we only talked about the interaction energy of chaotic fields. But one can also look at the self energy as given in eq. (39) or (40). Again a large number of local minima are observed that can be associated with known standard model interaction strengths, the energy E being again given by eq. (46). We don't have the space to describe all the details here, but refer to [2, 3]. From local minima of the self energy one can get some rather precise predictions of fermion masses, in particular for the heavy fermions t, b, c, τ. The 'landscape' generated by the self energy has also minima that can be identified with Yukawa couplings and gravitational couplings.

The lighter fermions are much more difficult to deal with in this context. Since for very small couplings the self energy exhibits scaling behaviour with log-periodic oscillations [22], one is only able to give predictions of light fermion masses modulo 2, and there are also some other theoretical ambiguities on how to associate the various minima with the light particles. For an early attempt to predict neutrino masses, see [2]. Clearly more work is needed to eliminate the ambiguities for light fermions.

Proceeding to higher energies (by investigating the chaotic dynamics for coupling constants a that coincide with running standard model coupling constants couplings $\alpha_1, \alpha_2, \alpha_3$ at large energies), and using again the self energy as the relevant observable, one can also look for local minima that potentially could describe supersymmetric particles with masses in the TeV region. No such minima are found [3]. Generally the evidence for supersymmetric particles in the TeV region from our chaotic fields is pretty meagre, to say the least. One would expect to see lots of local minima in the relevant coupling region if the conventional ideas on supersymmetry breaking at the TeV scale are correct. The self energies of the chaotic fields do not provide any evidence for such particles: There are simply no minima in the relevant coupling region, the only exception being perhaps the zero $a_3^{(3A)}$ of the 3A interaction energy. However, minima *are* found at the energy scale 10^{16} GeV and at the Planck scale. See [3] for more details. If future accelerator experiments do not find any evidence for supersymmetric particles, a theoretical reason could be that they do not fit into the 'landscape' generated by the chaotic fields.

Talking about new physics, new physics is of course represented by the existence of the chaotic fields themselves. For small couplings a, their equation of state is close to $w = -1$ [4], hence they can account for dark energy in the universe. But for large couplings they can have an equation of state close to $w \approx 0$ [4]. Thus, in principle at least, chaotic fields could account for both, dark energy *and* dark matter in the universe.

8 Conclusion

We have introduced chaotic scalar fields of as a model of vacuum fluctuations in the dark energy sector. These chaotic fields were used to generate potentials for moduli fields. We numerically observe that minima of the potentials lead to realistic standard model coupling strengths. The values of the fine structure constant, of the weak mixing angle, and of the strong coupling at the W mass scale are obtained with high precision and correspond to stable stationary values of moduli fields. Based on additional discrete symmetry assumptions, a value of the Higgs mass of 154 GeV is predicted from the chaotic field dynamics. This prediction can be experimentally tested in the near future.

The theory described in this paper may be regarded as a 'tip of an iceberg'. It still needs to be embedded into a greater context. In particular, its possible relation to supersymmetry breaking mechanisms has to be clarified. However, one statement can be made without any doubt: The nonlinear dynamics given by eq. (36) appears to distinguish certain numerical values of coupling constants a that do coincide with known standard model coupling constants with very high precision. A random coincidence can really be excluded. In this way the chaotic fields can help to select the 'correct' vacuum out of an enormous number of possibilities, shaping the world around us in precisely the way we know it.

References

[1] L. Susskind, hep-th/0302219

[2] C. Beck, Physica D **171**, 72 (2002) [hep-th/0105152]

[3] C. Beck, *Spatio-temporal chaos and vacuum fluctuations of quantized fields*, World Scientific, Singapore (2002)

[4] C. Beck, Phys. Rev. D **69**, 123515 (2004) [astro-ph/0310479]

[5] D.N. Spergel et al., astro-ph/0603449

[6] G. Parisi and Y.S. Wu, Sci. Sin. **24**, 483 (1981)

[7] P.H. Damgaard and H. Hüffel, eds., *Stochastic Quantization*, World Scientific, Singapore (1988)

[8] G. Kane, *Modern Elementary Particle Physics*, Addison-Wesley, Redwood City (1987)

[9] C.L. Gardner, Phys. Rev. D **68**, 043513 (2003) [astro-ph/0305080]

[10] T. Damour and A.M. Polyakov, Nucl. Phys. B **423**, 532 (1994) [hep-th/9401069]

[11] H.B. Sandvik, J.D. Barrow, and J. Magueijo, Phys. Rev. Lett. **88**, 031301 (2002) [astro-ph/0107512]

[12] J.D. Bekenstein, Phys. Rev. D **25**, 1527 (1982)

[13] K.A. Olive and M. Pospelov, Phys. Rev. D **65**, 085044 (2002) [hep-ph/0110377]

[14] R. Kallosh, hep-th/0205315

[15] C. Beck, Nonlinearity **4**, 1131 (1991)

[16] A. Hilgers and C. Beck, Physica D **156**, 1 (2001)

[17] C. Beck and F. Schlögl, *Thermodynamics of Chaotic Systems*, Cambridge University Press, Cambridge (1993)

[18] K. Kaneko (ed.), *Theory and Applications of coupled map lattices*, Wiley, New York (1993)

[19] G. 't Hooft, Class. Quant. Grav. **16**, 3263 (1999) [gr-qc/9903084]

[20] Particle Data Group, http://pdg.lbl.gov

[21] M Boglione, MR Pennington, Phys. Rev. Lett. **97**, 1998 (1997)

[22] S. Groote and C. Beck, Phys. Rev. E **74**, 046216 (2006) [nlin.CD/0603067]

CHAIRMAN: C. BECK

Scientific Secretaries: V. Parihar, V. Begun

DISCUSSION I

-Machado:

Does your chaotic quantization formalism accommodate asymptotic freedom e.g. in case the coupling constants run asymptotically free on a microscopic scale?

-Beck:

I look at ordinary quantized field theories for the standard model as it is. I use the technique of stochastic quantization, on a large scale at least. Parisi & Wu have shown for Euclidean metric and some assumptions that it is equivalent to the second quantizing standard model. So I use this technique where the chaotic noise can enter very easily to make in a sense a more general path integral. So phenomena like confinement are not discussed here. If the stochastic quantization method is correct as Parisi & Wu have shown then it must contain those phenomena and to my knowledge there are some lattice QCD calculations that have been done which use algorithms just based on stochastic quantization and reproduce the correct thing. My aim is to extend that case where the Gaussian random fields are just Gaussian noise uncorrelated structures, embed that in a bigger theory where you have a very small scale chaotic structure. What is not investigated in these numerical calculations is coupling to particle sector of the standard model. Chaotic noise on its own is discussed e.g. when we investigate a sector on a computer, we investigate that sector on its own and translate the coupling to some other field. The chaotic noise sector is investigated and it is seen that the vacuum energy has some certain interesting features.

-Hisamatsu:

As told by you that the presence of gap between the Standard Model parameters as predicted in the chaotic string theory and those measured by the experiments, there is an indication of large vacuum energy. Please elaborate. According to your article in 2001, the top mass predicted from this formalism of yours, seems to be already inconsistent with that measured in the experiment.

-Beck:

There is no contradiction in the top quark masses as measured experimentally and that in the chaotic strings. There are two top masses, one is the free one and the other one is the so called the pole top masses. The paper referred to cites that the free top mass comes out as 164 GeV and I use the well known QCD formulas which gives the relation between the pole masses and these free masses, using which the mass is 174 GeV. One has to be careful in comparing the type of masses. What I presume is that my system fixes the free masses which are real input parameters in the Standard Model but the experimental mass is the pole mass and there is an article cited in the paper, which gives the formula to translate from one to the other.

-Lopez Mateos:

How do you make the connection between the highly mathematical constructions that you have introduced with the actual physics?

-Beck:

That's a legitimate question. Basically, so far, only the equations were shown to you that show some interesting behavior. Some ideas in this regard, will be presented tomorrow e.g. a Dark energy model which is based on these chaotically evolving fields which could be embedded into the dark energy sector and then the sense of dark energy could possibly be to fix and stabilize the observed Standard Model parameters. There may be other embeddings into a larger theory viz. String theory.

-Didenko:

To what extent, your equations based on Parisi-Wu correspond to usual second order quantization procedure?

-Beck:

There are rigorous theorems that prove that if you sum up iterates of the type considered in my model for very small coupling $\boldsymbol{a}$, and rescale the sum of the iterates by the number of iterates then that sum converges to the stochastic process called the Wiener process which is just the integral over Gaussian white noise. This kind of Limit Theorem tells you that if you look at these scaled maps from far away, on Standard Model scales, they would just be like simulation of path integrals which hides all these complicated structures at microscopic scale hence it must reproduce ordinary quantum field theory. Only at the Planck scale it has a complicated behavior whereby you could claim to have a modified field theory.

-Karliner:

Your paper hep-th/0105152 contains a prediction for the strange quark mass m_S = 164 MeV which agrees very well with the average value given in the 2000 edition of Review of Particle Properties, m_S = 167 MeV. But the recent RPP edition gives the reanalyzed value to be $m_S = 103 \pm 20$ MeV. So it seems that your prediction fits too well the old data which has become obsolete.

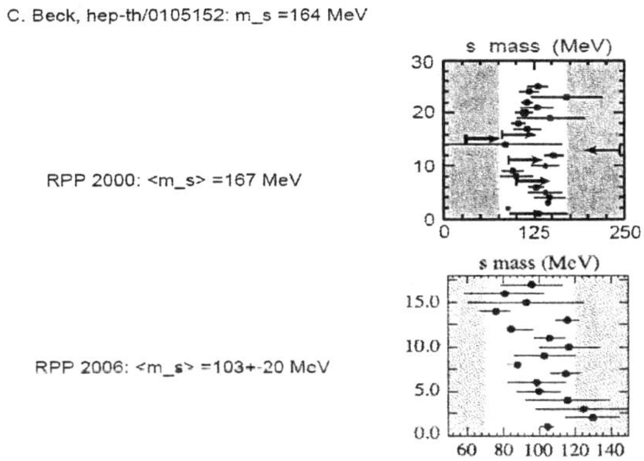

-Beck:

What have been studied are the minima of the landscapes considered in my model which may not be correctly predicting the above mentioned quark masses.

-Pinfold:

Why does your model seem to predict current quark masses rather than the constituent quark masses?

-Beck:

I am not studying this in the framework of QCD, whereby one would have a completely different dynamics. This is definitely a valid question, on why the choice of current quark masses is preferred over the constituent ones. But in the present scenario, the current quark masses are turning out to be a good choice.

-Bourjaily:

When determining masses and couplings, do you couple your computer code to the Standard Model Lagrangian with Standard Model masses and couplings?

-Beck:

My work corresponds to a certain dynamics which could be called ‘chaotic strings’, but it is not dependent on any QED input. It’s just the dynamics that is iterated and one can calculate the interaction energy. You find that the second zero with negative slope has the numerical value of the fine structure constant. One finds a coincidence, when a corresponding calculation is done for the running coupling in QED (where the energy scale is three times the electron mass). I have a feeling that the physical reason behind this coincidence could be that these chaotic maps might have existed long before the space-time structure was created. In the beginning there was just a shift of information and the chaotic strings somehow encode the information on the Standard Model parameters because they have minimum vacuum energy. Then these maps would have spilled the information to the space time structure of the universe that was created later.

-Bourjaily:

Calculation of zeroes for various values of a were calculated without coupling to the Standard Model?

-Beck:

Yes, without coupling to the Standard Model. It's very natural in this case because originally, you have Gaussian white noise (does not depend on Standard Model fields). As in the case of classical statistical mechanics, the Langevin noise is defined as rapidly evolving that has nothing to do with the drift term. In my case, I replace this Gaussian white noise with something chaotic keeping the property that is not influenced by the Standard Model in the drift term.

-Goldstein:

What sets the mass scale in this picture? There are no masses in input.

-Beck:

This is dimensionless dynamics. In order to set a scale one has to decide e.g. or the size of the vacuum energy, which so far was calculated in dimensionless units. This is equivalent to asking, how to embed it in a greater theory. One could choose the anthropic principle to determine the size of vacuum energy, which will be discussed later in the lectures.

-Tsallis:

If I generalize your model by allowing long-range interactions through a coupling which decays as $1/|i-j|^{\alpha}$ ($\alpha \geq 0$), your model would then be the $\alpha \rightarrow \infty$ limit. One would then have intriguing curves instead of your interesting numbers. Could you imagine, even very speculatively, any physical interpretation for those curves?

-Beck:

One could try out the suggested model but I would doubt that it would yield any of the observables that are obtained in my simplistic model. The basic idea is to correlate the results obtained to ones that are observed in the real world.

-Susskind:

How would you claim that in this scheme, what you produce is Unitary?

-Beck:

I don't know that.

-Zichichi:

How would a 1 dimensional chaotic field care about the Standard Model parameters? Are there any inputs at the beginning from the Standard Model?

-Beck:

It is indeed a surprise that just using a one dimensional chaotic field with nearest neighbors coupling (running between 0 and 1) one can produce results in the Standard model. But one can not get the value of fine structure constant just from QED equations.

One has to add something which could be string theory, M-theory or 'chaotic quantization'. This is just a numerical discovery which indicates the fact that the Standard Model fields should be amended by a 1-dimensional theory that evolves as a Gaussian noise on large scales.

CHAIRMAN: C. BECK

Scientific Secretaries: V. Parihar, V. Begun

DISCUSSION II

-Bourjaily:

You quoted a theorem that chaotic quantization for a=0 is asymptotically stochastic quantization (because the noise is asymptotically Gaussian white noise). So a=0 does become standard QFT in an exact sense. But there were plots of interaction energy as a function of "a" (ranging from 0 to 1), shown by you. Do you believe "a" is 'small enough' so QFT is asymptotically unchanged or do you propose that QFT must be modified even asymptotically?

-Beck:

That's an excellent question. What this Chebycheff maps scenario means is that if one assumes that there exist larger couplings of relevance that one cannot absorb into Gaussian noise of chaotic quantization then one would need to embed the dynamics somewhere else. That's the reason the Dark energy model was introduced by me so that these new physics fields (bigger a) could be introduced into dark energy sector and gotten rid off. So yes, we need something in addition to the normal Gaussian like quantum field theory.

-Antonio:

When you evolve your coupled map lattices, does that correspond to evolution of the chaotic string in real time?

-Beck:

Time in the Parisi -Wu approach is the fictitious time. In stochastic quantization the fictitious time is what you use to calculate the second quantized expectation values. In the dark energy model that I talked about this fictitious time is taken in the infinite limit. In statistical mechanics sense, one would then talk about the systems in equilibrium. Technically, the computer time to let these fields evolve is about a week or so.

-Antonio:

Does your theory predict 3 heavy neutrinos and do they agree with the light neutrinos, according to the see-saw mechanism?

-Beck:

In my 2001 paper, in fact I have three minima which would most plausibly correspond to three massive neutrinos but I would not argue which minimum corresponds to which neutrino. As I have said, from a theoretic point of view there is some ambiguity for the light fermions, the reason for that is that I am in the scaling region of these strings and there every pattern repeats with a factor of 2. So a prediction of state with mass m could also imply existence of states with mass 2 m or 0.5 m.

-Machado:

I was wondering as to what would be the connection between the Standard Model couplings and the chaotic field couplings and what happens if the couplings run? Could a renormalization flow be implemented?

-Beck:

What I have done till this point is like a tip of an iceberg and the theory is not complete. I would regard this as a renormalization flow and the argument is that the couplings in Standard Model viz. the fine structure constants are not fixed a-priori by the QED equations. I would imagine that even before the space time structure of the universe came into existence, any coupling constant between 0 and 1 could have been true and then there was probably just a shift of information rather than say matter, radiation or vacuum energy, just as represented in these Chebycheff maps. You can say the pre-Standard Model couplings change according to the potential generated by the interaction energy of these chaotic strings. One can intuitively think that if any coupling in the Standard Model viz. the fine structure constant, tries to become large then it is immediately driven back into the stable zeroes of the landscapes that I have shown. Hence in a way this can be regarded as a renormalization flow but I would stress that the picture is not yet complete.

-Lopez Mateos:

Your prediction of Higgs mass is at what energy scale and why?

-Beck:

If the Higgs mass is depending very strongly on the energy scale then I can't say, which energy scale does my theory represents. All I can say that I have evidence from N=2 chaotic strings for a massive particle which has precisely the mass of the W boson. Also, I have evidence for a massive boson of around 154 GeV which could be Higgs.

-Krotov:

Suppose we add the interaction, $L = \gamma\, \varphi^6$ to the Lagrangian of the Standard Model. Now, if we redo the whole procedure which you had described, would the correct prediction be $\gamma = 0$?

-Beck:

At present I am looking at noise on its own which is interesting because noise is supposed to be completely independent of any drift term. It would be an interesting task to couple to Standard Model and see what could come out.

-Haidt:

Suppose preons existed, how would you see them in your scheme?

-Beck:

Then you should see fraction of charges. No, I do not see this. It does not exclude preons but there is no evidence for them in my model as it stands now.

-Korthals-Altes:

Can you measure the θ parameter and other parameters in the Standard Model?

-Beck:

I have not done it yet. There are much more parameters that should be measured. I am just at the very beginning in my study.

-Cifarelli:

How do you introduce the dimensions of mass?

-Beck:

I introduce a dimensionless dynamics that may have existed before any mass and any space-time. This dimensionless dynamics depends on the dimensionless coupling constant "a". Then I compare this dimensionless coupling constant with running Standard Model coupling constants and translate my dimensionless coupling into a mass scale.

-Bourjaily:

Your analysis for the string related to α_{QCD} shows zeros near α=0.12. This is approaching the values where you expect non-Gaussian behavior. Do you see any phase transitions or confinement in the dynamics of modified QFT?

-Beck:

It depends on what string you look at. For the string you mention at the strong coupling 0.12 it is still very chaotic, so I am sure I still get a Gaussian there, but if I go to 0.6 or 0.7 then it is known that the system can loose ergodicity. It means you have a phase transition, but keep in mind it is not QCD, it is some addition, some picture that resemble some properties of QCD.

-Loehr:

You said that in your results you do not see any results of low energy SUSY. What in your formalism sets the scale up to which you think you can make statements?

-Beck:

I do not know what is that scale, but I assume that my formalism is valid up to the Planck scale. When I evolve the couplings with supersymmetric β-functions, then I do not see any minimum at the grand unified coupling $\alpha_1 = \alpha_2 = \alpha_3$. However, if I evolve with the ordinary non-supersymmetric β-functions, then first α_2 and α_3 meet, and I observe a minimum there in the landscape and then all couplings meet at the Planck scale.

QCD at low energy: the simplicity of complex non-perturbative phenomena

Gilberto Colangelo

Institut für Theoretische Physik, Universität Bern
Sidlerstr. 5, 3012 Bern, Switzerland

May 2, 2007

Abstract

These lectures provide a short introduction to chiral perturbation theory and to the effective field theory method in general. The focus is mainly on the foundations of the method and only at the end I discuss two recent applications. In a school about complexity, effective field theories provide a nice example of how one can describe complicated nonperturbative phenomena in simple terms.

1 Introduction

Effective field theories are an ubiquitous tool in physics. Indeed very rarely does one hope to describe a physical system at all length scales (or energies), and if one concentrates on a certain range and considers only the modes which get excited at those energies, then one is doing an effective theory. Although the atom contains quarks and gluons (inside the nucleus, that is), we do not need to be able to describe the strong interactions if we want to understand atomic spectra or how several atoms interact and form bound states.

In particle physics we do have a theory that could in principle be valid at all length scales, QCD. In the real world quarks and gluons do have also other types of interactions and interact with other forms of matter too, but one could imagine a world which is completely governed by QCD and where

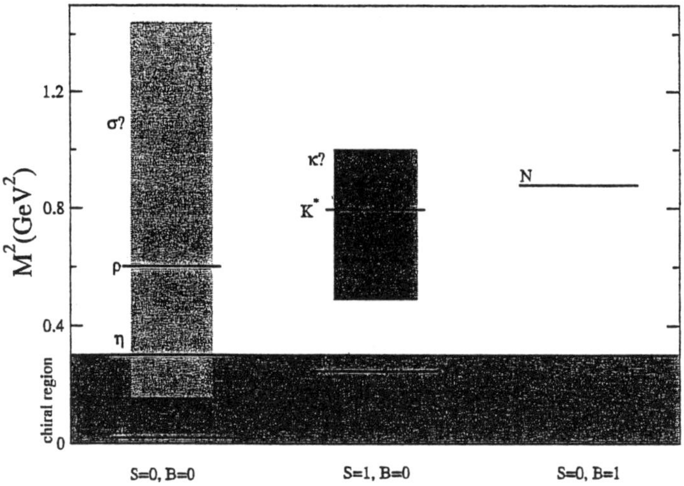

Figure 1: The lowest-lying part of the QCD spectrum in the sectors of strangeness and baryon number 0 and 1. The bands for the σ and the κ indicate the uncertainty in the mass as given by the PDG.

nothing else happens. This imaginary world is realized by the people who do lattice QCD calculations. Of course they do not do this to see what happens at very small length scales, but rather to test how well this theory describes the non-perturbative phenomena which occur at large length scales. At this school Richard Kenway [1] has explained how lattice calculations are performed and what the most recent developments in the field are. He showed that in recent years there have been a number of important developments which make a direct comparison to the phenomenology finally possible. As always, however, the progress is still rather slow, despite the steady increase in computer power, improvement in algorithms and clever new ideas by physicists. The difficulty of this enterprise lies in the fact that one wants to see new degrees of freedom emerge at low energy out of the local interactions among quarks and gluons. We lack analytical tools to do that and the numerical route is understandably not easy.

At very low-energy, however, the complex nonperturbative phenomena which generate hadrons out of quarks and gluons and rule their interactions, become simple, again. This happens because QCD undergoes a spontaneous symmetry breaking: although in the chiral limit the Lagrangian is invariant under $SU(N_f)_L \times SU(N_f)_R$ (with N_f the number of light flavours) transformations, the vacuum is invariant only under the vector part of this large group. As predicted by the Goldstone theorem massless particles appear in the spectrum, and their interaction becomes weak at low energy. There are no massless hadrons in nature, but an octet of light ones (as illustrated in Fig. 1), which have exactly the quantum numbers predicted by the Goldstone theorem. Their small mass is readily understood in terms of a small mass term in the QCD Lagrangian – this explicit chiral symmetry breaking generates a small mass for the Goldstone bosons. Since the interactions among these Goldstone bosons vanish in the limit of zero quark masses and zero momenta, one can do a perturbation theory around this limit and expand in small masses and momenta. This is the main idea of chiral perturbation theory (CHPT) which I will illustrate in some more details in what follows. As shown in Fig. 1 the theory can be applied in the energy region where only the Goldstone bosons are excited.

Since the method is completely general and can be applied to any system which undergoes spontaneous symmetry breaking, I will put an emphasis on this and describe its foundations without making specific reference to QCD as long as this is possible (even if for convenience I may sometimes call the Goldstone bosons "pions"). The plan of the paper is as follows. In Sect. 2 I will discuss the Goldstone theorem and how one can construct an effective field theory on its basis. In Sect. 3 I will specifically concentrate on the quantum character of the effective field theory and discuss in detail the role and meaning of loop diagrams and how one can carry out the renormalization program. The latter question is particularly relevant in view of the fact that this kind of effective field theories is intrinsically non-renormalizable. In Sect. 4 I will discuss a few recent results in CHPT and finally give an outlook and draw a few conclusions in Sect. 5.

2 Effective field theories for systems with spontaneous symmetry breaking

2.1 Goldstone theorem

The Goldstone theorem is one of the few physics results which are absolutely general and most relevant in very many different fields. If the Hamiltonian of a system is invariant under a group G of global transformations, while the vacuum only under a subgroup H, then the spectrum of the system contains massless excitations. The number of these depends only on the rank n_G of the group G and that of the subgroup H, n_H, and is given by the difference of the two: $n_{\mathrm{GB}} = n_G - n_H$. These massless excitations are usually referred to as Goldstone bosons (which explains the subscript GB above). We indicate with Q_i, $i = 1, \ldots, n_G$ the generators of the group G, with H_i, $i = 1, \ldots n_H$ those of H, and with X_i, $i = 1, \ldots, n_{\mathrm{GB}}$, the broken ones – the states obtained by letting these act on the vacuum do not give back the vacuum:

$$X_i|0\rangle \neq |0\rangle \; . \tag{1}$$

On the other hand, since G is a symmetry of the Hamiltonian $\mathcal{H}$, the latter commutes with all the generators, which implies that all the states $Q_i|0\rangle$ are degenerate with the vacuum:

$$\mathcal{H}X_i|0\rangle = X_i\mathcal{H}|0\rangle = 0 \; . \tag{2}$$

The states $X_i|0\rangle$ are the Goldstone bosons. The generators X_i do not form an algebra, and therefore there is no subgroup of G which is generated by them – they generate the coset space G/H, and there is a one–to–one mapping between the elements of this space and the Goldstone boson fields. The existence of this mapping dictates the transformation properties of the Goldstone boson fields under the group G [2]. Without entering the details, we just stress that these are nonlinear.

There is a second part to the Goldstone theorem, which is equally important as the first, best known one. If we denote the one-particle Goldstone boson states by $|\pi^a\rangle$ and consider the matrix elements of the Noether currents J_i^μ of the symmetry group G between the vacuum and these states:

$$\langle 0|J_i^\mu(0)|\pi^a(p)\rangle = iF_i^a p^\mu \tag{3}$$

these define a $n_{\mathrm{GB}} \times n_G$ matrix F_i^a. The Goldstone theorem says that this matrix is of maximal rank, *i.e.* of rank n_{GB}. In other words, in the proper basis, the Noether currents of the broken generators have nonzero matrix elements with the corresponding pion fields: $F_b^a = \delta_{ab}F_a$. Together with the fact that the Noether currents are conserved, this seemingly uninteresting statement implies that the Goldstone bosons interact weakly at low energy: more precisely, the strength of their interaction vanishes with the square of the momenta.

The construction of the effective Lagrangian for systems with spontaneous symmetry breaking is based on the Goldstone theorem. The first part implies that there are massless particles in the spectrum – if these are the only ones, then we can construct an effective Lagrangian containing only these degrees of freedom. Moreover, the symmetry transformation properties of the Goldstone bosons imply that the Lagrangian has to be nonlinear in the corresponding fields – therefore, it is necessarily nonrenormalizable. This would seem to destroy the usefulness of such an approach, because nonrenormalizability is often connected to lack of predictivity. Here is where the second part of the theorem plays an important role: since the interaction vanishes with the square of the momenta, in the Lagrangian one must discover that terms without derivatives are not allowed (by symmetry), and moreover this provides an organizing principle, a counting scheme, in the infinite number of terms of the nonrenormalizable Lagrangian.

We now consider the case of QCD and briefly describe the construction of the effective Lagrangian in the chiral limit. If we consider N_f massless flavours (and disregard heavy quarks) the Lagrangian of QCD is

$$\mathcal{L}^{(0)}_{\rm QCD} = \bar{q}_L i \not{D} q_L + \bar{q}_R i \not{D} q_R - \frac{1}{4} G^a_{\mu\nu} G^{a\mu\nu} \qquad q = \begin{pmatrix} q_1 \\ \vdots \\ q_{N_f} \end{pmatrix} \tag{4}$$

and is symmetric under a large global group of transformations:

$$\tilde{G} = SU(N_f)_L \times SU(N_f)_R \times U(1)_V \times U(1)_A \ . \tag{5}$$

The two $U(1)$ factors are not relevant for the present discussion for two opposite reasons: $U(1)_V$ remains conserved and the baryon number is the corresponding conserved quantity, while $U(1)_A$ is broken by quantum effects (is anomalous). The group G we have been referring to in the previous discussion is, in the case of QCD:

$$G = SU(N_f)_L \times SU(N_f)_R \ . \tag{6}$$

The vacuum, on the other hand, is only invariant under the vector transformations, those for which $g_L = g_R$,where $(g_L, g_R) \in SU(N_f)_L \times SU(N_f)_R$. QCD in the chiral limit of N_f flavours is therefore expected to have $N_f^2 - 1$ massless particles with the quantum numbers of the broken generators, the axial ones, *i.e.* pseudoscalar particles.

The construction of the effective Lagrangian proceeds as follows:

1. The pions are collected into a matrix–valued field U belonging to the coset space G/H. Since $SU(N_f)_L \times SU(N_f)_R/SU(N_f)_V$ is isomorph to $SU(N_f)$ the matrix U is a $SU(N_f)$ matrix. The fact that it is an element of the coset is seen in its transformation properties under $SU(N_f)_L \times SU(N_f)_R$:
$$U' = g_R U g_L^\dagger \ . \tag{7}$$
The pion fields can be identified with the coefficients of the generators of $SU(N_f)$, if we write U as an exponential
$$U = \exp[i\phi/F] \ , \qquad \phi = \sum_{a=1}^{N_f^2-1} \phi_a \lambda_a \ , \tag{8}$$
where λ_a are the generators (the Gell-Mann matrices for $N_f = 3$ or the Pauli matrices for $N_f = 2$), and F is a constant with dimensions of an energy, necessary if we want to have the pion fields canonically normalized.

2. The effective Lagrangian must contain only the fields which are active in the energy range of interest (in our case the pions) and must respect the symmetries of the system. In this case this means constructing a Lagrangian containing the U matrix which is invariant under (7). Renormalizability is not an issue, and we can therefore have as many derivatives as we want. The possible terms with up to two derivatives are
$$\mathcal{L}_{\text{eff}} = f_1(U) + f_2(U)\langle U^\dagger \Box U\rangle + f_3(U)\langle \partial_\mu U^\dagger \partial^\mu U\rangle + O(\partial^4) \ , \tag{9}$$
where $\langle A\rangle$ stands for the trace of the matrix A.

3. After having listed all possible terms compatible with the symmetries of our system, we can look for redundancies and simplify our Lagrangian. In this case we have two observations: the Lagrangian (9) can only be invariant under $SU(N_f)_L \times SU(N_f)_R$ if the functions f_i do not depend at all from U – given the transformation property (7), any matrix element of U can be transformed at will, and the f_i remain invariant only if they do not depend on any of them. The term f_1 is therefore just a constant which we can drop. Under the integral sign the term proportional to f_2 can be transformed into the third one by

partial integration. In conclusion we are left with one single term if we consider terms with up to two derivatives:

$$\mathcal{L}_{\text{eff}} = \mathcal{L}_2 + \mathcal{L}_4 + \mathcal{L}_6 + \dots , \qquad \mathcal{L}_2 = \frac{F^2}{4}\langle \partial_\mu U^\dagger \partial^\mu U \rangle \ , \tag{10}$$

where the constant factor has been chosen such that the kinetic term for the pion fields has canonical form (the generators of $SU(N_f)$ are normalized such that $\langle \lambda_a \lambda_b \rangle = 2\delta_{ab}$):

$$\mathcal{L}_2 = \frac{1}{2}\partial_\mu \phi_a \partial^\mu \phi_a + O(\phi^4) \ . \tag{11}$$

The constant F drops out in the kinetic term but appears in all the interaction terms. The effective Lagrangian becomes predictive once we can fix this constant. In order to do this we can look at the Noether currents of the symmetry of the system. At low energy they are represented by

$$V_i^\mu = i\frac{F^2}{4}\langle \lambda_i [\partial^\mu U, U^+] \rangle \qquad A_i^\mu = i\frac{F^2}{4}\langle \lambda_i \{\partial^\mu U, U^+\} \ . \tag{12}$$

The matrix elements of the axial currents between the vacuum and the pion fields are given by

$$\langle 0 | A_i^\mu | \pi^k(p) \rangle = i p^\mu \delta_{ik} F \ . \tag{13}$$

The constant F is the pion decay constant in the chiral limit. If we want to work to leading-order accuracy we can stop here and start calculating.

The effective Lagrangian contains only one term at this level, but since this is written with the matrix U, which is an exponential of the pion fields, it contains local interaction terms among as many pion fields as one wants. The interaction vanishes if we consider static fields, as implied by the Goldstone theorem. Here we already see an advantage of the effective Lagrangian method: to show that the Goldstone theorem implies a vanishing interaction among pions of zero momentum requires quite some labour. If we instead use the effective Lagrangian, we immediately find out that symmetry forbids the interaction among static pions. This is true for any other consequence of symmetry: with the effective Lagrangian one can derive it without "thinking", but just doing the usual quantum field theory calculations.

2.2 Explicit symmetry breaking

In the real world there are no massless hadrons – the lightest hadrons, however, are pseudoscalars which do form an octet (triplet) under $SU(3)_V$ ($SU(2)_V$) transformations, and so have the right quantum numbers for being the Goldstone bosons of chiral symmetry breaking in QCD with $N_f = 3$ ($N_f = 2$). The presence of a small mass for the candidate Goldstone bosons can be traced back to the presence of a small explicit chiral symmetry breaking term in the QCD Lagrangian – the only possible such term (if we want to consider a renormalizable theory) is a quark mass term.

$$\mathcal{L}^{\rm QCD} = \mathcal{L}_0^{\rm QCD} - \bar{q}\mathcal{M}q \tag{14}$$

where $\mathcal{M} = \mathrm{diag}(m_u, m_d, m_s)$ is the quark mass matrix. As long as this symmetry breaking term is small, we can consider the limit in which the latter vanishes and make an expansion around it. This amounts to an expansion in the quark masses. Whatever quantity we consider, we will express it as its value in the chiral limit plus corrections which are suppressed by powers of the quark masses. These corrections are given by matrix elements of the symmetry breaking term among appropriate external states – calculating these is difficult, as any matrix element in QCD, but since the external states are the unperturbed ones (in the chiral limit), and the symmetry breaking term has well defined transformation properties under chiral transformations, we can at least keep track of symmetry constraints on the correction terms.

Let us consider the mass of a generic hadron h, as an example:

$$M_h^2 = M_{h,0}^2 + \langle h|\bar{q}\mathcal{M}q|h\rangle + O(m_q^2) \ . \tag{15}$$

The corrections to the chiral limit value of the mass are determined by the matrix elements $\langle h|\bar{q}q|h\rangle$. For example, if we consider the pions, these are massless in the chiral limit, $M_{\pi,0}^2 = 0$ and the expansion reads [3]:

$$M_\pi^2 = -(m_u + m_d)\frac{1}{F_\pi^2}\langle 0|\bar{q}q|0\rangle + O(m_q^2) \tag{16}$$

where we have used a symmetry relation among matrix elements of $\bar{q}q$, a chiral Ward identity:

$$\langle \pi|\bar{q}q|\pi\rangle = -\frac{1}{F_\pi^2}\langle 0|\bar{q}q|0\rangle =: B_0 \ . \tag{17}$$

The quark condensate $\langle 0|\bar{q}q|0\rangle$ which determines the size of the pion mass squared to leading order in the quark mass expansion, is an order parameter for the spontaneous chiral symmetry breaking.

The effective Lagrangian method is a useful tool also in this circumstance, as it allows one to derive the symmetry constraints on the perturbative corrections in a systematic way. The implementation proceeds as follows: if one considers the quark mass matrix as an external, matrix-valued hermitian scalar field s, and assigns to it a specific transformation property under chiral transformations:

$$s' = g_R s g_L^\dagger \tag{18}$$

then one makes the QCD Lagrangian formally chiral invariant even in the presence of a mass term. Explicit symmetry breaking occurs because the "vacuum" around which one expands is not at $s = 0$ but rather at $s = \mathcal{M}$. The corresponding effective Lagrangian must also contain this external field s, it must still be chiral invariant, and explicit symmetry breaking is obtained by taking for s the constant value $s = \mathcal{M}$. The latter is considered as a perturbation and we expand in powers of s. The minimal number is one, and the only possible term in the effective Lagrangian is:

$$C\langle U^\dagger s + s^\dagger U\rangle = C\langle s(U^\dagger + U)\rangle \ , \tag{19}$$

where C is dimensionful constant. Considering directly the limit $s = \mathcal{M}$ we obtain the leading order effective Lagrangian:

$$\mathcal{L}_2 = \frac{F^2}{4}\left[\langle \partial_\mu U^+ \partial^\mu U\rangle + \langle 2B_0\mathcal{M}\left(U + U^\dagger\right)\rangle\right] \ , \tag{20}$$

where we have fixed the constant $C = B_0F^2/2$ by calculating the value of the quark condensate in the effective theory.

With this effective Lagrangian we can make calculations and even compare these to measurements. For example we can evaluate the masses of the pseudo Goldstone bosons, and obtain:

$$\begin{aligned} M_\pi^2 &= (m_u + m_d)B_0 + O(m_q^2) \\ M_{K^+}^2 &= (m_u + m_s)B_0 + O(m_q^2) \\ M_{K^0}^2 &= (m_d + m_s)B_0 + O(m_q^2) \\ M_\eta^2 &= \frac{1}{3}(m_u + m_d + 4m_s)B_0 + O(m_q^2) \end{aligned} \tag{21}$$

These leading order expressions tell us a number of things. First of all they relate pseudoscalar mass ratios to quark mass ratios [4]:

$$\begin{aligned} M_K^2/M_\pi^2 &= (m_s+\hat{m})/2\hat{m} & \Rightarrow m_s/\hat{m} = 25.9 \\ M_\eta^2/M_\pi^2 &= (2m_s+\hat{m})/3\hat{m} & \Rightarrow m_s/\hat{m} = 24.3 \end{aligned} \tag{22}$$

and moreover they imply a prediction for the eta mass in terms of the masses of the pion and kaon:

$$3M_\eta^2 = 4M_K^2 - M_\pi^2 \ , \tag{23}$$

a relation first obtained by Gell-Mann and Okubo [5].

Besides masses, we can of course also calculate decay and scattering amplitudes. For example, the $\pi\pi$ scattering amplitude to leading order is given by [6]:

$$\begin{aligned} \langle \pi^c(p_3)\pi^d(p_4) \text{ out} \,|\, \text{in } \pi^a(p_1)\pi^b(p_2)\rangle &= \delta^{ab}\delta^{cd}A(s,t,u) + \delta^{ac}\delta^{bd}A(t,u,s) \\ &\quad + \delta^{ad}\delta^{bc}A(u,s,t) \ , \\ A(s,t,u) &= \frac{s-M_\pi^2}{F_\pi^2} \ , \end{aligned} \tag{24}$$

where the leading order expressions have been substituted with the physical pion mass and decay constant. Out of the isospin invariant amplitude $A(s,t,u)$ we can construct amplitudes of a given isospin in a given channel. Evaluating the amplitudes at threshold gives us (modulo a normalization factor) the (S wave) scattering lengths. The $I=0$ and 2 scattering lengths are given by

$$a_0^0 = \frac{7M_\pi^2}{32\pi F_\pi^2} = 0.16 \ , \quad a_0^2 = -\frac{M_\pi^2}{16\pi F_\pi^2} = -0.045 \ . \tag{25}$$

Later we will see that these quantities are now known up to next-to-next-to-leading order, with very small uncertainties, and that there are a number of experiments running, which can test these precise predictions at a similar level of accuracy.

2.3 External fields

The introduction of a scalar field coupled to the scalar quark bilinear has served the purpose of dealing with explicit symmetry breaking. We have also mentioned, in passing, that with the help of this scalar field we can evaluate in the effective theory matrix elements of $\bar{q}q$ – more precisely, we can derive automatically and systematically the symmetry relations respected by the matrix elements of this quark bilinear. For example, we are not able to calculate the quark condensate, but can relate the matrix element of $\bar{q}q$ between two one-pion states to the condensate. The same trick can be repeated with other operators, like the pseudoscalar bilinear $\bar{q}\gamma_5 q$, or the Noether currents $\bar{q}\lambda_a\gamma_\mu q$ and $\bar{q}\lambda_a\gamma_\mu\gamma_5 q$. This is the approach adopted by Gasser and Leutwyler [7, 8], who introduced the necessary external fields in QCD:

$$\mathcal{L} = \mathcal{L}_0^{\rm QCD} + \bar{q}\gamma^\mu(v_\mu + \gamma_5 a_\mu)q - \bar{q}(s - i\gamma_5 p)q \ , \tag{26}$$

then considered the generating functional, defined as the vacuum-to-vacuum amplitude with this action:

$$e^{iZ[v,a,s,p]} = \langle 0|Te^{i\int d^4x\mathcal{L}}|0\rangle \ , \tag{27}$$

and stated that a low energy representation of the latter is conveniently obtained by evaluating the path integral over the pion fields with the action given by the effective Lagrangian constructed according to the principles we have discussed so far:

$$e^{iZ[v,a,s,p]} = \mathcal{N}^{-1}\int [dU]e^{i\int d^4x\mathcal{L}_{\rm eff}} \ . \tag{28}$$

The transformation property of the pseudoscalar external field is identical to that of the scalar one (18). Those of the external vector and axial-vector fields, could also be defined analogously. It is useful, however, to promote the chiral symmetry from global to local – while in this case the transformation properties of the scalar and pseudoscalar fields remain unchanged, those of the vector and axial-vector fields must be changed as follows

$$\begin{aligned} r_\mu = v_\mu + a_\mu \ &\stackrel{G}{\longrightarrow}\ r'_\mu = g_R r_\mu g_R^{-1} + i g_R \partial_\mu g_R^{-1} \ , \\ l_\mu = v_\mu - a_\mu \ &\stackrel{G}{\longrightarrow}\ l'_\mu = g_L l_\mu g_L^{-1} + i g_L \partial_\mu g_L^{-1} \ . \end{aligned} \tag{29}$$

The addition of the derivative term compensates the one generated by the kinetic term of the quark fields and makes QCD *locally* invariant under chiral transformations. Since the action itself is invariant, so is also the generating functional (not quite, because there is the chiral anomaly – we forget this for a moment and come back to it below):

$$Z[v', a', s', p'] = Z[v, a, s, p] \ . \tag{30}$$

The generating functional contains all possible Green functions of the quark bilinears to which the external fields are coupled. The invariance of the generating functional translates into relations among the Green functions – symmetry relations, *i.e.* Ward identities. The introduction of the derivative terms in the transformation of the vector and axial-vector fields has the consequence that the Ward identities contained in Eq. (30) also explicitly imply conservation of the axial and vector currents. In summary, Eq. (30) is a very compact way to write all possible Ward identities following from chiral invariance.

2.4 Chiral perturbation theory

With the effective Lagrangian method one aims at giving an explicit representation of the generating functional which respects the invariance property (30). This is obtained through a perturbative calculation of the path integral, evaluated with an effective Lagrangian containing only the low-energy degrees of freedom, *i.e.* the Goldstone bosons:

$$e^{iZ[v,a,s,p]} = \int [dU] e^{i\int dx \mathcal{L}_{\rm eff}(U,v,a,s,p)} \ , \tag{31}$$

where the effective action $S_{\rm eff}[U,v,a,s,p] = \int dx \mathcal{L}_{\rm eff}(U,v,a,s,p)$ must contain all possible terms which are invariant under local chiral transformations – the latter is the only requirement we have to make if we want to be completely general. The effective Lagrangian which is invariant under local chiral transformations is obtained from Eq. (20) by transforming the normal derivatives into covariant ones and adding the pseudoscalar to the scalar external field – of course Eq. (20) represents only the leading order whereas the full effective Lagrangian contains terms with any number of derivatives and external fields:

$$\mathcal{L}_{\rm eff} = \mathcal{L}_2 + \mathcal{L}_4 + \mathcal{L}_6 + \ldots \quad \mathcal{L}_2 = \frac{F^2}{4} \langle D_\mu U^+ D^\mu U + \chi^\dagger U + U^\dagger \chi \rangle \ , \tag{32}$$

where $D_\mu U = \partial_\mu U - i r_\mu U + i U l_\mu$ and $\chi = 2B_0(s + ip)$.

The reasoning we just stated is absolutely general and for theories without anomalies this would be the end of the story. In QCD, however, the chiral symmetry is anomalous, and the generating functional is correspondingly noninvariant. It is possible, however, to evaluate exactly the noninvariant term in the transformation – in this case the effective Lagrangian has to be constructed such that it reproduces the anomalous noninvariance of the generating functional. Wess and Zumino showed that in QCD this is indeed possible [9, 10, 11]. In the chiral counting the noninvariant term is first encountered at order p^4 – if we stop at order p^2 the Lagrangian in (32) is all there is.

The advantage of this formulation is that one directly deals with Green functions of quark bilinears, and these transform linearly under chiral transformations. Formulating symmetry properties working directly with the pions would be a lot more difficult, because these transform nonlinearly – moreover the exact transformation properties of the pions vary, depending on the representation one chooses for the pion fields. Indeed in this formulation of the effective Lagrangian method the pions move to the background and play the role only of integration variables in the path integral – in the foreground is the invariance property of the path integral. The latter is constructed in such a way that the Green functions it describes contain poles at the proper places – poles due to the exchange of pions. The residues of these poles give the matrix elements of pions. This is exactly what one would do if one were able to calculate directly the path integral of QCD (and it is exactly what one does with lattice calculations).

Another practical advantage of the external fields is that the W and Z bosons as well as the photons, couple to quarks exactly like the v and a fields we introduced. If we want to evaluate matrix elements involving quarks and these bosonic fields we just have to substitute the appropriate combinations of a and v with the W, Z, or photon fields and take the matrix element we are interested in.

3 Higher orders

In Eq. (32) we have specified only the leading order effective Lagrangian and just listed the next-to-leading and next-to-next-to leading term. If one wants to evaluate a physical quantity to a higher accuracy one needs to take these into account as well. But not only: as shown in Eq. (31), the generating functional (any Green function) of QCD at low energy is obtained by evaluating the path integral with the action specified by the effective Lagrangian – if we want to increase the accuracy of a calculation we have to evaluate also loop diagrams. The counting of loop diagrams in CHPT has been discussed first by Weinberg [12] and can be summarized by the statement that every loop counts like order p^2: a complete calculation at order p^4 must necessarily include one-loop diagrams with vertices from the $\mathcal{L}_2$ Lagrangian as well as tree diagrams with the $\mathcal{L}_4$ Lagrangian. If we want to go up to order p^6 we must take into account two-loop diagrams with vertices only from $\mathcal{L}_2$ and one-loop diagrams with only one vertex from $\mathcal{L}_4$ and tree diagrams from $\mathcal{L}_6$.

These are the rules of the game – if one wants to play it one may stop reading here and start calculating. A student who has learned to do loop calculations within a renormalizable quantum field theory, can do it also using an effective Lagrangian. It is true, however, that the use of an effective Lagrangian for calculating loops poses some conceptual problems – a good student at this point will certainly ask himself: "when I calculate loops I integrate over all possible momenta, and if I use vertices and propagators from the effective Lagrangian, I am using these well beyond their range of validity. How can I trust such a calculation? What is its meaning?"

In this section I will provide answers to these questions by analyzing the loop calculation of one simple quantity, the scalar form factor of the pion, which is defined as

$$\langle \pi^i(p_1)\pi^j(p_2)|\hat{m}(\bar{u}u+\bar{d}d)|0\rangle =: \delta^{ij}\Gamma(t) \ , \quad t=(p_1+p_2)^2 \ , \tag{33}$$

where $\hat{m} = (m_s + m_d)/2$. This matrix element is relevant to the decay $h \to \pi\pi$, which would have been the main decay mode for a very light Higgs (of course this scenario is now experimentally excluded). The tree-level calculation of this matrix element is simple and leads to

$$\Gamma(t) = 2\hat{m}B = M_\pi^2 + \mathcal{O}(p^4) \ . \tag{34}$$

This result, which we worked out from the Lagrangian, is actually a consequence of a general theorem, due to Feynman and Hellman.[13] This states that the expectation value of the perturbation in an eigenstate of the total Hamiltonian determines the derivative of the energy level with respect to the strength of the perturbation:

$$\hat{m}\frac{\partial M_\pi^2}{\partial \hat{m}} = \langle \pi|\hat{m}\bar{q}q|\pi\rangle = \Gamma(0) \ . \tag{35}$$

The value of the form factor at zero momentum transfer is fixed by this theorem, and a simple power counting implies that at leading order the scalar form factor is a constant – at order p^2 the theorem completely fixes the form factor. On the other hand no general principle forbids a dependence of the form factor on t (to the contrary, they imply it, as we will see), and to generate this we necessarily have to go beyond leading order.

Before starting the loop calculation, let us have a look at what happens to the form factor once we include tree-level contributions from higher-order terms in the Lagrangian. We now present the Lagrangian at order p^4 for the case in which one expands around $m_u = m_d = 0$ and keeping m_s at its physical value. In this case the chiral symmetry is $SU(2)_L \times SU(2)_R$. The Lagrangian of order p^2 remains unchanged – the only change is that the field U, and its logarithm ϕ are now 2×2 matrices. At order p^4 the Lagrangian is simpler, because we can use more trace identities to reduce the number of independent terms. For two light flavours this is a sum of seven terms[1]

$$\begin{aligned}\mathcal{L}_4 &= \ell_1\frac{1}{4}\langle u_\mu u^\mu\rangle^2 + \ell_2\frac{1}{4}\langle u_\mu u_\nu\rangle\langle u^\mu u^\nu\rangle + \ell_3\frac{1}{16}\langle\chi_+\rangle^2 + \ell_4\frac{i}{4}\langle u_\mu \chi_-^\mu\rangle \\ &- \ell_5\frac{1}{2}\langle f_{-\mu\nu}f_-^{\mu\nu}\rangle + \ell_6\frac{i}{4}\langle[u_\mu,u_\nu]f_+^{\mu\nu}\rangle - \ell_7\frac{1}{16}\langle\chi_-\rangle^2 ,\end{aligned} \quad (36)$$

where we have used the compact notation (with $U = u^2$):

$$\begin{aligned} u_\mu &= iu^\dagger D_\mu U u^\dagger = u_\mu^\dagger \\ \chi_\pm &= u^\dagger \chi u^\dagger \pm u\chi^\dagger u \\ \chi_\pm^\mu &= u^\dagger D^\mu \chi u^\dagger \pm u D^\mu \chi^\dagger u \\ f_\pm^{\mu\nu} &= u F_L^{\mu\nu} u^\dagger \pm u^\dagger F_R^{\mu\nu} u \ , \end{aligned} \quad (37)$$

and $F^{\mu\nu}_{R,L}$ are the field strengths of r_μ and l_μ . Only two of these seven terms contribute to the scalar form factor, those proportional to ℓ_3 and ℓ_4:

$$\Gamma_{[\ell_3,\ell_4]}(t) = \frac{M^2}{F^2}\left[-4M^2\ell_3 + t\ell_4\right] \ , \quad (38)$$

the calculation is recommended as an easy exercise, as is the calculation of the contribution of ℓ_3 to the pion mass. Once these two calculations are completed, one can then check that the Feynman–Hellman theorem is respected also in this case.

[1] If we disregard contact terms, i.e. those that depend only on the external fields. Notice also that in the theory with $N_f = 2$ and no electromagnetic interactions there is no chiral anomaly.

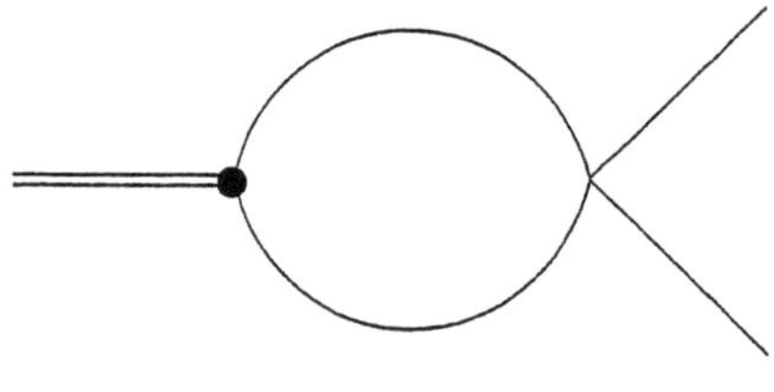

Figure 2: One-loop diagram for the scalar form factor of the pion. The double line stands for the scalar source, whereas single lines for pions.

The result in Eq. (38), a tree-level calculation with the next-to-leading order Lagrangian, is merely a statement about how the symmetry constrains this particular matrix element: at this order the scalar form factor can have at most a term linear in t. No symmetry relation exists between the constant term and the coefficient of the linear term, hence we have two different constants. The constant term is related to the derivative of the pion mass with respect to the strength of the symmetry-breaking term in the Lagrangian, whereas the coefficient of the linear term is related to the correction to the pion decay constant (again, it is a very good exercise to calculate the latter with the Lagrangian in Eq. (36)).

3.1 Loops and renormalization

Besides the less interesting tadpole graphs and the wave function renormalization there is only one graph for this process, the one shown in Fig. 2. Its structure is

$$\int \frac{d^4l}{(2\pi)^4} \frac{\{M^2, p^2, p\cdot l, l^2\}}{(l^2-M^2)((p-l)^2-M^2)} \Rightarrow xT(M^2) + yJ(t) \tag{39}$$

$$T(M^2) = \int \frac{d^4l}{i(2\pi)^4} \frac{1}{(M^2-l^2)}\,, \; J(t) = \int \frac{d^4l}{i(2\pi)^4} \frac{1}{(M^2-l^2)(M^2-(p-l)^2)}\,,$$

where $p = p_1 + p_2$. We have indicated, in the first integral, all the terms that can appear in the numerator, and, after the arrow, the two possible structures to which the various terms can be reduced. The momenta and masses in the numerator come from the four-pion vertex on the right-hand side of the diagram. The power counting for this integral shows that it represents a correction of order p^2 to the leading-order term (the integration measure, which is of order p^4, is compensated by the two propagators) in agreement with the general counting rule stated by Weinberg [12].

If we expand the tadpole integral $T(M^2)$ and the loop integral $J(t)$ in a Taylor series in their respective arguments:

$$T(M^2) = a + bM^2 + \bar{T}(M^2) \ , \quad J(t) = J(0) + \bar{J}(t) \ , \tag{40}$$

only the first terms in the expansion are divergent, whereas both $\bar{T}(M^2)$ and $\bar{J}(t)$ are finite – this is easily seen by taking a sufficient number of derivatives on the loop integrals in Eq. (39). It is left as an exercise to the reader to show that the one-loop diagrams that we have neglected can only produce terms like $T(M^2)$. Therefore the contribution of the loop diagrams to the scalar form factor has the following structure:

$$\Gamma(t) \sim \frac{M^2}{F^2}\left[x_1 bM^2 + x_2 tJ(0) + x_1\bar{T}(M^2) + x_2\bar{J}(t)\right] \ . \tag{41}$$

The divergent part of the loop diagrams has exactly the same structure as the counterterm contribution calculated above: to remove it we simply need to define the counterterms properly (in this case the constants ℓ_3 and ℓ_4).

The principles we have followed in the construction of the effective Lagrangian only appealed to symmetry arguments, and therefore allowed for an infinite number of terms. Once this is accepted, there is no problem of principle in carrying through the renormalization program, because all counterterms necessary to remove the divergences must already be present: as anticipated, the difference between renormalizable and non-renormalizable theories is, in a sense, a technical detail.

To renormalize the form factor, the two counterterms ℓ_3 and ℓ_4 have to be defined in the following way

$$\ell_3 = \ell_3^r(\mu) - \frac{1}{2}\lambda \ , \quad \ell_4 = \ell_4^r(\mu) + 2\lambda \ , \tag{42}$$

where λ is divergent and dependent on the regularization method (in dimensional regularization, for example, it is defined by $\lambda = (c\mu)^{d-4}/(d-4)$, with c an arbitrary constant that defines the regularization scheme).

If we look at the definition of the operators in front of ℓ_3 and ℓ_4 we see that (as usual) they contain an infinite number of pion fields. For example they both contain a term with one scalar source and 6 pion fields. Does it mean that if we had calculated that matrix element to one loop we would have found the same divergent part for ℓ_3 and ℓ_4? Or, to put it differently, that once we calculate the divergent part of the scalar form factor then we know the divergent part of all other matrix elements with a higher number of pion fields?

The answer is yes. Chiral symmetry puts a strong constraint on the divergences: they have to be chiral-invariant terms. This conclusion is originally due to Weinberg [12] on the basis of a highly plausible, but still heuristic argument. It is now put on a solid basis by the work of Leutwyler [18]: he proved that to calculate hadronic Green functions with an effective Lagrangian, such that they respect the Ward identities implied by the chiral symmetry, one necessarily has to start from a chiral-invariant effective Lagrangian. The non-trivial part in this statement is that it takes into account also quantum effects: anomalous symmetries show that it is not always true that a symmetry that exists at the classical level survives the quantum corrections – or vice versa, that to have a symmetrical quantum theory one necessarily has to start from a symmetrical classical Lagrangian.

The theorem applies also to the divergent part of the quantum corrections: they have to be chiral-invariant. A general theorem of quantum field theory states that the divergent part of a loop graph is a polynomial in the external masses and momenta. These two theorems lead to the conclusion that the divergences, order by order, can be reabsorbed by the chiral-invariant counterterms.

3.2 Chiral logarithms

If we expand the form factor in a Taylor series in t, we can write it in the following form:

$$\Gamma(t) = \Gamma(0)\left[1 + \frac{1}{6}\langle r^2\rangle_S^\pi t + O(t^2)\right] \quad . \tag{43}$$

The coefficient of the linear term, properly normalized, is called the scalar radius of the pion. Its size is a measure of the spatial extension of the pion when probed with a scalar source. We have stated above that the coupling constant that appears in this quantity, ℓ_4, also determines the first correction of the pion decay constant around the chiral limit. There is another piece of information on the scalar radius, which we can already gather from the simple sketch of the loop calculation given above:

$$\langle r^2\rangle_S^\pi \sim J(0) = \int \frac{d^4l}{i(2\pi)^4} \frac{1}{(l^2 - M^2)^2} \sim \ln\frac{M^2}{\Lambda^2} \ , \tag{44}$$

namely that the scalar radius contains an infrared divergence. In the chiral limit this quantity diverges. This divergence should not be removed and does not represent a problem because it has a physical meaning, in the sense that when the pion becomes massless the cloud of pions surrounding any hadron (and therefore also the pion itself) extends to an infinite range. A quantity that measures the spatial extension of this cloud should indeed become infinite in the chiral limit. Notice that the scalar form factor is finite and remains finite also in the chiral limit – it is only the first derivative in t, calculated at $t = 0$ that diverges when the pion mass goes to zero.

These infrared divergences are present everywhere in pion physics, and in many cases they are among the most important physical effects (less so in the case of kaons). Their relevance was first pointed out by Li and Pagels[14]. The effective Lagrangian method provides a systematic way to calculate these effects. It is also interesting to observe that the coefficients of these logs are given by renormalization group equations – despite several attempts, and a thorough understanding of the relation between these logs and the renormalization group equations for nonrenormalizable theories, however, a way to resum the chiral logs has not yet been found [12, 15].

As we have seen in the above example, the chiral logs arise from the infrared region in the loop integrals – precisely the region where we should fully trust the vertices of our effective Lagrangian. One may be less at ease with the ultraviolet region of the loop integrals: there one has no justification for the use of the effective Lagrangian. On the other hand, through the process of renormalization, that part of the integrals is completely removed and substituted with unknown constants, the counterterms. As is sometimes said, these parametrize our ignorance of the physics that lies above the range of applicability of the effective Lagrangian.

3.3 Loops and unitarity

Until now the only part of the loop integrals that we have not analysed is the finite, analytically non-trivial part of the loop integral, the function $\bar{J}(t)$. Its presence is a consequence of unitarity, as we are now going to explain. According to the Watson theorem [16], above threshold but below the inelasticity threshold, the phase of the scalar form factor is equal to the S-wave $\pi\pi$ phase shift with isospin $I = 0$, δ_0^0. As is well known, this theorem is a consequence of unitarity:

$$\text{Im}\bar{\Gamma}(t) = \sigma(t)\bar{\Gamma}(t)t_0^{0*}(t) = \bar{\Gamma}(t)e^{-i\delta_0^0}\sin\delta_0^0 = |\bar{\Gamma}(t)|\sin\delta_0^0 \ , \tag{45}$$

where $\sigma(t) = [1 - 4M_\pi^2/t]^{1/2}$ and $\bar{\Gamma}(t) = \Gamma(t)/\Gamma(0)$, and t_0^0 is the $I = 0$ S-wave of $\pi\pi$ scattering. The unitarity relation (45) shows that the leading-order expression $\bar{\Gamma}(t) = 1$ cannot be the whole story: if we want an accurate description of the form factor away from $t = 0$, we need to include higher orders and, in particular, loops – imaginary parts can only be generated by loop graphs.

Notice that since at leading order $\bar{\Gamma}(t)$ is $\mathcal{O}(1)$, and the phase δ_0^0 is $\mathcal{O}(p^2)$, the $\mathcal{O}(p^2)$ imaginary part (which is a next-to-leading order correction) is completely fixed by leading-order quantities:

$$\text{Im}\bar{\Gamma}^{(2)}(t) = \delta_0^{0(2)}(t) = \sigma(t)\frac{2t - M_\pi^2}{32\pi F_\pi^2} \ . \tag{46}$$

The use of the effective Lagrangian method to calculate the form factor guarantees that this relation is satisfied. The complete expression for the one-loop scalar form factor reads as follows:

$$\bar{\Gamma}(t) = 1 + \frac{t}{16\pi^2 F_\pi^2}(\bar{l}_4 - 1) + \frac{2t - M_\pi^2}{2F_\pi^2}\bar{J}(t) \ , \tag{47}$$

where $\bar{J}(t)$ is the subtracted one-loop integral (40). Its explicit expression reads:

$$\bar{J}(t) = \frac{1}{16\pi^2}\left[\sigma(t)\ln\frac{\sigma(t) - 1}{\sigma(t) + 1} + 2\right] \ . \tag{48}$$

The reader can now easily verify that the imaginary part of the form factor at this order indeed satisfies (46).

3.4 Dispersion relation for the scalar form factor

A real analytic function must be real on the real axis: the scalar form factor is non-analytic from threshold ($4M_\pi^2$) up to infinity. On the basis of very general arguments, which mainly use the causality principle, one can prove that as a function of t, the scalar form factor must be analytic everywhere else (see

Ref.[17] for a general discussion of the analyticity properties of amplitudes and Green functions). The non-analyticity of the form factor on the real axis can be further characterized, and described as a discontinuity:

$$\bar{\Gamma}(t+i\varepsilon) = \bar{\Gamma}^*(t-i\varepsilon) = |\bar{\Gamma}(t)|e^{i\delta_0^0(t)} \ . \tag{49}$$

Given these analytic properties, we can write the following dispersion relation:

$$\bar{\Gamma}(t) = 1 + bt + \frac{t^2}{\pi}\int_{4M_\pi^2}^{\infty}\frac{dt'}{t'^2}\frac{\mathrm{Im}\bar{\Gamma}(t')}{t'-t} \ , \tag{50}$$

where, for later convenience, we have subtracted the dispersive integral twice – we will come back to the issue of how many subtractions are necessary for the dispersive integral to converge. The dispersion relation shows that, if we know the subtraction constants (in this case only one, b) and the imaginary part on the real axis, we can reconstruct the scalar form factor everywhere on the complex plane.

It is no surprise that any perturbative calculation in a quantum field theory produces amplitudes and Green functions with the correct analytic properties. Using an effective field theory makes no difference: the form factor calculated to one loop in CHPT must have the correct analytic properties, and must satisfy (at the perturbative level) the dispersion relation (50). To convince ourselves that this is actually the case, let us first apply the chiral counting to the dispersion relation:

$$\bar{\Gamma}^{(0)} = 1 \ , \quad b \sim \mathcal{O}(1) \ , \quad \mathrm{Im}\bar{\Gamma}(t') = \mathcal{O}(p^2) \ . \tag{51}$$

As we have seen the $\mathcal{O}(p^2)$ imaginary part is fully fixed by leading-order quantities, (46), and apart from an unconstrained polynomial term, the real part must be given by the dispersive integral over this known imaginary part. We leave it as an exercise to prove that this is true. For this it is useful to know the dispersive representation of the loop integral

$$\bar{J}(t) = \frac{t}{16\pi^2}\int_{4M_\pi^2}^{\infty}\frac{dt'}{t'}\frac{\sigma(t')}{t'-t} \ . \tag{52}$$

In the previous section we showed that the renormalization procedure removes the contributions to the loop integrals where the momentum squared of the pions is large. This was reassuring because we cannot hope that our effective Lagrangian describes highly virtual pions well. In the present section we are dealing with the contribution to the loop integrals from real pions: the dispersive integrals. As we have seen above, these extend all the way up to infinity, as required by analyticity. In the perturbative expansion that we are considering, the imaginary part of the form factor, which is in the integrand, is evaluated only to leading non-trivial order. This description of the imaginary part can be reasonably accurate only in the low energy region: still, in the integral, it is used all the way up to infinity. How can we trust the dispersive integral?

In fact, we do not. At least not for the contribution that comes from the high-energy region. Suppose we decide to remove the part of the dispersive integral from $\Lambda = 1$ GeV to infinity. We should then subtract from the form factor a term like:

$$\frac{t^2}{\pi}\int_{\Lambda^2}^{\infty}\frac{\delta_0^{0(2)}(t')}{t'^2(t'-t)} = \frac{t^2}{\pi}\int_{\Lambda^2}^{\infty}\frac{\delta_0^{0(2)}(t')}{t'^3}\left(1+\frac{t}{t'}+O(t^2)\right) \ . \tag{53}$$

The Taylor expansion inside the integral can be safely performed because the CHPT calculation of the form factor is valid only for $t \ll \Lambda^2$. In the chiral language, this part of the dispersive integral can be represented as a polynomial series starting at order p^4, i.e. at an order which is beyond the accuracy at which we are presently working. This shows that worrying about these contributions to the dispersive integrals is futile: the only sensible way to improve the evaluation of the dispersive integral is to go one order higher in the chiral expansion. This would automatically give a representation of the form factor that contains the dispersive integral over the imaginary part correct up to order p^4. For those who are interested in the numerics, the first term in the Taylor expansion of the integral (53) is equal to:

$$\frac{t^2}{\pi}\int_{\Lambda^2}^{\infty}\frac{\delta_0^{0(2)}(t')}{t'^3} = 0.7\left[t(\,\mathrm{GeV}^2)^2\right] \ , \tag{54}$$

which means, for $t = (0.5\,\mathrm{GeV})^2$ (which is about the upper limit of validity of the chiral expansion), a 4% correction to the leading-order result. Also numerically everything is well under control.

4 Applications

There are plenty of applications of the effective Lagrangian method and they are not restricted to particle physics, although the latter is the field where the applications have reached the highest level of sophistication. If one wants to understand the phenomenology of strongly interacting particles at low energy, CHPT provides the proper framework to do it – not that there are no puzzles or experimental results which are not quite explained or understood, but even in these cases, the use of CHPT allows us to learn something from the puzzles. In this field the most interesting development of recent years is that experiments are reaching a high level of precision, and that the theory, stimulated by the perspective of having very precise data, has anticipated that and provided some remarkably precise predictions. This is obtained mainly in pion physics – the expansion in the mass of the strange quarks inevitably leads to a slowly converging series. In the following I will discuss two recent applications in pion physics which illustrate how, through a combined use of CHPT and dispersion relations one can reach a high level of precision, even for quantities which are in principle outside the region of application of CHPT, like the mass of resonances.

4.1 $\pi\pi$ scattering

In Sect. 2.2 we have discussed the leading order calculation of the $\pi\pi$ scattering amplitude and the numerical prediction (which is now 40 years old!) for the S-wave scattering lengths (25). In the meanwhile these have been calculated to NLO and NNLO. The formula of the S-wave scattering lengths to NLO is quite instructive [7]:

$$
\begin{aligned}
a_0^0 &= \frac{7M_\pi^2}{32\pi F_\pi^2}\left[1+\frac{M_\pi^2}{3}\langle r^2\rangle_S^\pi+\frac{200\pi F_\pi^2 M_\pi^2}{7}(a_2^0+2a_2^2)\right.\\
&\qquad\left.-\frac{M_\pi^2}{672\pi^2 F_\pi^2}(15\bar{\ell}_3-353)\right]=0.16\cdot 1.25=0.20\\
2a_0^0-5a_0^2 &= \frac{3M_\pi^2}{4\pi F_\pi^2}\left[1+\frac{M_\pi^2}{3}\langle r^2\rangle_S^\pi+\frac{41M_\pi^2}{192\pi^2 F_\pi^2}\right]=0.624\ , \qquad (55)
\end{aligned}
$$

as it shows explicitly that the core of the effective Lagrangian method is simply in establishing symmetry relations among different quantities. At leading order everything can be expressed in terms of the pion mass and decay constant – these fully specify the Lagrangian. At NLO we have seven new constants in the game, cf. Eq. (36), and they have a less direct physical meaning. Still, whenever they appear one can reexpress them in terms of other physical quantities and write a NLO chiral expression in a way which makes the symmetry relations explicit. In Eq. (55) the two S-wave scattering lengths are expressed in terms of the pion scalar radius and the two D-wave scattering lengths ($a_2^{I=0,2}$) – plus the low-energy constant ℓ_3 which we have already encountered in Sec. 3, and which is the coefficient of the quadratic term in the quark mass expansion of the pion mass squared. A particular combination of the S-scattering lengths has a very simple relation to the scalar radius, as shown in Eq. (55). If one can pin down these quantities, one can work out a more accurate numerical prediction for the S-wave scattering lengths, as was originally done by Gasser and Leutwyler [7]. The remarkable outcome of the numerical estimate is that with respect to leading order, the NLO prediction for a_0^0 is 25% higher. For an expansion in $\hat{m}/1$ GeV $\sim 5\%$ this is quite a large correction. The reason for such a large correction is to

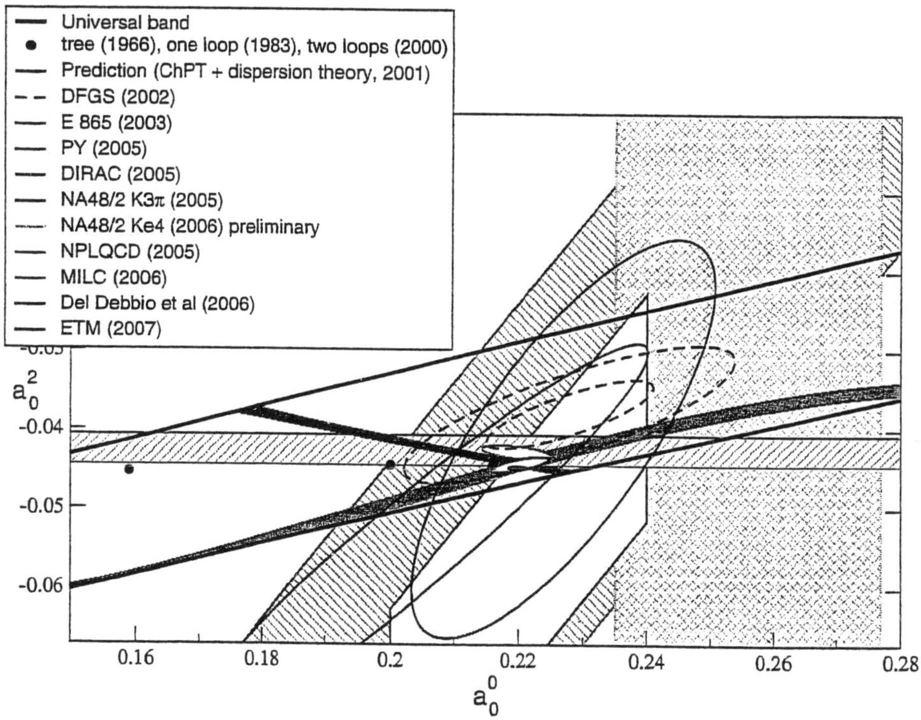

Figure 3: Theory predictions, lattice calculations and measurement of a_0^0 and a_0^2. See the text for more details.

be found in the presence of large chiral logs, also present in the scalar radius discussed in Sect. 3.1. The chiral logs in the scattering lengths are

$$a_0^0 = \frac{7M_\pi^2}{32\pi F_\pi^2}\left[1+\frac{9}{2}\ell_\chi+\ldots\right] \qquad a_0^2 = -\frac{M_\pi^2}{16\pi F_\pi^2}\left[1-\frac{3}{2}\ell_\chi+\ldots\right] , \tag{56}$$

where

$$\ell_\chi = \frac{M_\pi^2}{16\pi^2 F_\pi^2}\ln\frac{\mu^2}{M_\pi^2} , \tag{57}$$

and indeed the coefficient in a_0^0 happens to be quite large.

The $\pi\pi$ scattering amplitude has been later calculated even to NNLO [19], and an accurate estimate of the numerical prediction and the corresponding uncertainties has been worked out on the basis of a matching with dispersion relations [20]:

$$a_0^0 = 0.220 \pm 0.005, \qquad a_0^2 = -0.0444 \pm 0.0010 . \tag{58}$$

These remarkably precise predictions are currently being tested by a number of experiments. A summary of the current situation is shown in Fig. 3 and is as follows

$$\begin{aligned}
a_0^0 &= 0.26 \pm 0.05 && [21]\\
a_0^0 &= 0.216 \pm 0.013 \pm 0.003 && [22]\\
|a_0^0 - a_0^2| &= 0.264\ {}^{+0.033}_{-0.020} && [23]\\
a_0^0 - a_0^2 &= 0.268 \pm 0.010 \pm 0.013 && [24]\\
a_0^0 &= 0.256 \pm 0.011 \qquad \text{— PRELIMINARY —} && [25] .
\end{aligned} \tag{59}$$

After the old Geneva-Saclay experiment [21], the most recent ones have confirmed the CHPT prediction, although not yet reaching the same level of accuracy. The most recent number from the NA48/2 experiment [25], a measurement of the phase shift difference $\delta_0^0 - \delta_1^1$ in K_{e4} decays, and indirectly of the scattering length a_0^0, although preliminary, does show some tension with the prediction (58), and with the other K_{e4}-decay measurement of the E865 Collaboration from Brookhaven [22]. A completely different method to measure the difference $|a_0^0 - a_0^2|$ (through a measurement of the lifetime of pionium) is used by the DIRAC collaboration [23]. The same combination is measured by NA48/2 [24] in a cusp effect in $K \to 3\pi$ decays [26]. In Fig. 3 also the phenomenological analyses of Descotes-Genon *et al.* (entry DFGS

(2002) in the legenda, see Ref. [27]) and of Peláez and Ynduráin (entry PY (2005) in the legenda, see Ref. [28]) are shown.

In the same figure, besides the experimental numbers just discussed, also some lattice data have been plotted. One of them, from the NPLQCD collaboration [29] is a direct calculation on the lattice of the $I = 2$ S-wave scattering length and agrees well with the chiral prediction. The other three [30, 31, 32], are calculations of the constants $\ell_{3,4}$ which are two of the crucial inputs in the chiral calculation. The points on the plane corresponding to these calculations are obtained by using these numbers together with those of [20] for the remaining constants. All three agree with the chiral numbers. We stress, however, that the MILC numbers are the only ones obtained with three light sea (staggered) quark flavours – the other two simulations are done without strange quarks in the sea.

4.2 The σ meson

As we have discussed in Sec. 2.2, the lowest part of the QCD spectrum can be well understood within the effective Lagrangian framework: the lightest hadrons would be massless in the chiral limit and the pattern of their small masses is well explained by leading order formulae. It is of course an essential test of our understanding of strong interactions to be able to explain the whole hadronic spectrum, including resonances. One of the problems of this enterprise is to obtain reliable experimental information on these resonances. This appears to be particularly difficult for resonances with the quantum numbers of the vacuum: a look at the PDG [33] shows that, despite many relevant experimental results, the lightest resonance with isospin and spin zero – commonly known as the σ – is not known very well, cf. also Fig. 1. According to the PDG its mass lies between 400 and 1200 MeV and its half-width between 250 and 500 MeV. In this case, even if one were able to calculate the σ resonance parameters from first principles, it would hardly be a significant test to compare these to the "experimental" numbers. The main problem, in this case, is that in order to translate data into resonance parameters, one needs a theoretical description of the data – for most experimental situations it is difficult to have a model–independent parametrization, and as a result the same data can be interpreted in terms of wildly differing resonance parameters.

In a recent work [34] we have shown two things: first, that dispersion relations for scattering amplitudes do provide a model-independent way to

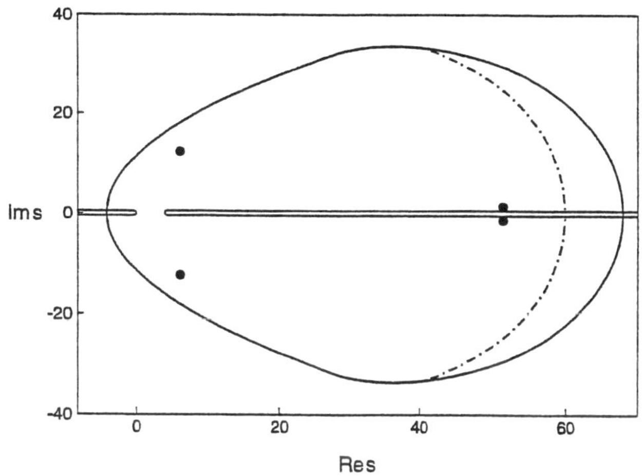

Figure 4: Domain of validity of the Roy equations for the partial waves $t_\ell^I(s)$ in the complex plane of the variable s, here expressed in pion mass squared units.

translate data into resonance parameters, and second, that in the case of the σ, the crucial input is represented by the two S-wave scattering lengths. The argument goes as follows: the Roy equations provide a dispersive representation for the partial waves of the $\pi\pi$ scattering amplitude which reads

$$\begin{aligned} t_0^0(s) &= a + (s - 4M_\pi^2)\, b + \int_{4M_\pi^2}^{\Lambda^2} ds' \left\{ K_0(s,s')\, \mathrm{Im}\, t_0^0(s') \right. \\ &\left. + K_1(s,s')\, \mathrm{Im}\, t_1^1(s') + K_2(s,s')\, \mathrm{Im}\, t_0^2(s') \right\} + d_0^0(s)\,. \end{aligned} \tag{60}$$

This is a twice subtracted dispersion relation which takes into account crossing symmetry in order to

1. write the contribution of the left-hand cut in terms of the imaginary parts of other partial waves in the physical region;

2. write the two subtraction constants in terms of the two S-wave scattering lengths $a = a_0^0$, $b = (2\,a_0^0 - 5\,a_0^2)/(12M_\pi^2)$.

The so-called driving term $d_0^0(s)$ collects the dispersion integrals over the higher partial waves ($\ell \geq 2$), as well as the high energy end of the integral over the S- and P-waves. The kernels $K_i(s,s')$ are known functions. The dispersive representation (60) is valid even for complex values of s, if these are inside the ellipse shown in Fig. 4, which is the region inside which the Roy equations have been rigorously proved to be valid. A direct evaluation of the representation (60) gives, however, the partial wave on the first Riemann sheet – the poles corresponding to resonances are to be found on the second sheet.

In order to reach the second sheet one can use the S matrix element $S_0^0(s) = 1 - 2\sqrt{4\,M_\pi^2/s - 1}\; t_0^0(s)$ and its unitarity property, which implies the relation

$$S_0^0(s)^{II} = 1/S_0^0(s)^{I} \; . \tag{61}$$

This shows that the amplitude contains a pole on the second sheet if and only if $S_0^0(s)$ has a zero on the physical sheet. So, all we need to do is numerically evaluate Eq. (60) for complex values of s in the domain where it has been shown to hold and find out whether or not $S_0^0(s)$ has zeros there. The analysis presented in Ref. [34] yields the two pairs of zeros shown in Fig. 4. The one corresponding to the σ (the other is the $f_0(980)$) has parameters

$$M_\sigma = 441\ ^{+16}_{-8}\ \mathrm{MeV}\,, \quad \Gamma_\sigma = 544\ ^{+18}_{-25}\ \mathrm{MeV}\ . \tag{62}$$

An analogous analysis has been recently applied to the κ [35] with similarly precise results.

We stress that the precision in the final numbers (62) is a direct consequence of the precise chiral prediction for the scattering lengths (58). The rest of the experimental input – the imaginary parts in the dispersive integrals in Eq. (60) – plays only a relatively minor role so that, despite relatively large experimental uncertainties, the corresponding contribution to the error in Eq. (62) is limited. One could, however, completely give up any use of theory in the calculation, and use only experimental input, even for the scattering lengths, cf. (59). This would increase the final error in (62) and shift somewhat the central values. The outcome, however, would be much more precise than what is still listed in the PDG.

5 Conclusions and outlook

In these lectures I have provided an introduction to effective field theories for systems with spontaneous symmetry breaking. I have mainly concentrated on the foundations of the method and have emphasized the systematic and rigorous aspects of it: a proper application of effective field theories is strictly equivalent to a full exploitation of the symmetry properties of the underlying theory and of some general principles, like analyticity and unitarity, and *nothing else.* At the end I have discussed two recent applications, both related to $\pi\pi$ scattering. The two S-wave scattering lengths are two very interesting physical quantities which can be

1. predicted with very high accuracy within chiral perturbation theory (combined with Roy equations);

2. measured with very high accuracy and with different methods in running experiments;

3. calculated from first principles in lattice QCD.

Comparing the numbers obtained in these three manners represents a thorough test of our understanding of the strong interactions – the precision of this test is a rare exception in hadronic physics and illustrates well the power of the effective field theory method.

If these lectures have sparkled some interest in the reader, she will certainly want to read more. A thorough and detailed introduction to chiral perturbation theory can be found in Ref. [36], where the reader can also find an extensive bibliography and suggestions for further readings.

The very last word I want to spend to emphasize once more the universal character of effective field theories: applications cover all aspects of low–energy hadronic physics (in particular as far as the connections to lattice calculations and the related necessary extrapolations are concerned), but also extend to other very different physical systems, like antiferromagnets [37], or gravitational interactions [38].

Acknowledgment

It is a pleasure to thank Antonino Zichichi for the kind invitation and warm hospitality. Many thanks also to the team of organizers who made the school run very smoothly and especially to the students for creating a stimulating atmosphere. I am grateful to Heiri Leutwyler for a careful reading of the manuscript and for preparing Fig. 3.

References

[1] R. Kenway, these proceedings.

[2] S. R. Coleman, J. Wess and B. Zumino, Phys. Rev. **177** (1969) 2239. C. G. . Callan, S. R. Coleman, J. Wess and B. Zumino, Phys. Rev. **177** (1969) 2247.

[3] M. Gell-Mann, R. J. Oakes and B. Renner, Phys. Rev. **175** (1968) 2195.

[4] J. Gasser and H. Leutwyler, Nucl. Phys. B **94** (1975) 269, S. Weinberg, in *A Festschrift for I. I. Rabi*, ed. L. Motz, Trans. New York Acad. Sci.Ser. II 38 (1977) 185. For a review of the early literature, see J. Gasser and H. Leutwyler, Phys. Rept. **87** (1982) 77.

[5] M. Gell-Mann, Phys. Rev. **106**(1957) 1296, S. Okubo, Prog. Theor. Phys. **27**(1962) 949.

[6] S. Weinberg, Phys. Rev. Lett. **17** (1966) 616.

[7] J. Gasser and H. Leutwyler, Annals Phys. **158** (1984) 142.

[8] J. Gasser and H. Leutwyler, Nucl. Phys. B **250** (1985) 465.

[9] W. A. Bardeen, Phys. Rev. **184** (1969) 1848.

[10] J. Wess and B. Zumino, Phys. Lett. B **37** (1971) 95.

[11] E. Witten, Nucl. Phys. B **223** (1983) 422.

[12] S. Weinberg, Physica **A 96** (1979) 327.

[13] R. P. Feynman, Phys. Rev. **56** (1939) 340; H. Hellman, *Einführung in die Quantenchemie*, Deuticke Verlag (Leipzig, 1937); S. T. Epstein, *Amer. J. Phys.* **22** (1954) 613.

[14] L. F. Li and H. Pagels, Phys. Rev. Lett. **26** (1971) 1204.

[15] M. Buchler and G. Colangelo, Eur. Phys. J. C **32** (2003) 427 [arXiv:hep-ph/0309049]. M. Bissegger and A. Fuhrer, Phys. Lett. B **646** (2007) 72 [arXiv:hep-ph/0612096], and arXiv:hep-ph/0701132.

[16] K. M. Watson, Phys. Rev. **88** (1952) 1163.

[17] S. Weinberg, *Quantum field theory*, Cambridge University Press (Cambridge, 1995).

[18] H. Leutwyler, Annals Phys. **235** (1994) 165 [arXiv:hep-ph/9311274].

[19] J. Bijnens, G. Colangelo, G. Ecker, J. Gasser and M. E. Sainio, Phys. Lett. B **374** (1996) 210 [arXiv:hep-ph/9511397],
and Nucl. Phys. B **508** (1997) 263 [Erratum-ibid. B **517** (1998) 639] [arXiv:hep-ph/9707291].

[20] G. Colangelo, J. Gasser and H. Leutwyler, Nucl. Phys. B **603** (2001) 125 [arXiv:hep-ph/0103088].

[21] L. Rosselet *et al.*, Phys. Rev. D **15** (1977) 574.

[22] S. Pislak *et al.* [BNL-E865 Collaboration], Phys. Rev. Lett. **87** (2001) 221801 [arXiv:hep-ex/0106071],
and Phys. Rev. D **67** (2003) 072004 [arXiv:hep-ex/0301040].

[23] B. Adeva *et al.* [DIRAC Collaboration], Phys. Lett. B **619** (2005) 50 [arXiv:hep-ex/0504044].

[24] J. R. Batley *et al.* [NA48/2 Collaboration], Phys. Lett. B **633** (2006) 173 [arXiv:hep-ex/0511056].

[25] L. Masetti, arXiv:0704.1307 [hep-ex].

[26] N. Cabibbo, Phys. Rev. Lett. **93** (2004) 121801 [arXiv:hep-ph/0405001].

[27] S. Descotes-Genon, N. H. Fuchs, L. Girlanda and J. Stern, Eur. Phys. J. C **24** (2002) 469 [arXiv:hep-ph/0112088].

[28] J. R. Pelaez and F. J. Yndurain, Phys. Rev. D **71** (2005) 074016 [arXiv:hep-ph/0411334].

[29] S. R. Beane, P. F. Bedaque, K. Orginos and M. J. Savage [NPLQCD Collaboration], Phys. Rev. D **73** (2006) 054503 [arXiv:hep-lat/0506013].

[30] C. Bernard *et al.* [MILC Collaboration], arXiv:hep-lat/0609053.

[31] L. Del Debbio, L. Giusti, M. Luscher, R. Petronzio and N. Tantalo, JHEP **0702** (2007) 056 [arXiv:hep-lat/0610059].

[32] Ph. Boucaud *et al.* [ETM Collaboration], arXiv:hep-lat/0701012.

[33] W. M. Yao *et al.* [Particle Data Group], J. Phys. G **33** (2006) 1.

[34] I. Caprini, G. Colangelo and H. Leutwyler, Phys. Rev. Lett. **96** (2006) 132001 [arXiv:hep-ph/0512364].

[35] S. Descotes-Genon and B. Moussallam, Eur. Phys. J. C **48** (2006) 553 [arXiv:hep-ph/0607133].

[36] S. Scherer, Adv. Nucl. Phys. **27** (2003) 277 [arXiv:hep-ph/0210398].

[37] C. Brugger, C. P. Hofmann, F. Kampfer, M. Moser, M. Pepe and U. J. Wiese, arXiv:cond-mat/0612363.

[38] J. F. Donoghue, Phys. Rev. D **50** (1994) 3874 [arXiv:gr-qc/9405057].

CHAIRMAN: G. COLANGELO

Scientific Secretaries: E.P. Borges, S. Pacetti

DISCUSSION I

– *Csanád:*

What do you think, what kind of ways there are to calculate experimentally checkable, accessible signals of chiral symmetry restoration in high density and temperature matter ?

– *Colangelo:*

So, you would like to have experimentally accessible signal of a kind of symmetry restoration.

– *Csanád:*

Yes, in high energy heavy ion collisions.

– *Colangelo:*

I think I don't know.

– *Csanád:*

Like some kind of cross sections or no calculations are done yet on this subject.

– *Colangelo:*

Well, we can do theoretical calculations like evaluate the $q\bar{q}$ condensate in the theory. We can do that also at low, non-zero temperature, and you see that as you increase the temperature, the condensate starts to disappear, it melts away. On the other hand, within this framework you can not reach the point where the condensate completely disappears because at that point you don't have the Goldstone bosons, then the theory breaks down. How you can see this experimentally, I don't know. Maybe somebody in the audience would like to answer this. The question is: how chiral symmetry restoration can be experimentally seen?

– *Karliner:*

There is a recent activity having to do with the effect of possible restoration of chiral symmetry at high energies. The idea is that high excited states in QCD should not care about the fact that the vacuum breaks chiral symmetry. So the spectrum of highly excited states should have parity doublets, etc. The relevant papers are by Glozman (hep-ph/0606038), Shifman (hep-ph 0507246) Jaffe et al. PRL 96 (2006) 121601.

– *Krotov:*

At what scale the relation between $\langle 0|q\bar{q}|0\rangle$ and the masses at the pseudo-scalar octet is defined ?

– *Colangelo:*

Both the quark masses and $\langle q\bar{q} \rangle$ depend on μ, but their product is μ-invariant. Well, the masses and the $q\bar{q}$ operator renormalize in a contravariant way.

– *Vartanov:*

When you go to the loop expansion, do the operators from the Lagrangian with dimension 6 give you finite contribution to the Green functions and observables from the part of the Lagrangian with dimension 4 ?

– *Colangelo:*

So, let me see if I understand: you're saying, the operators at order p^6, they will correct those at order p^4. I mean they do give corrections to observables. So if you look at any observable you will have contributions of order p^2, of order p^4, p^6 and so on. However, the transformation property of each of these operators is different, so that they never mix and you can distinguish precisely the constant in front of one of them from the constant in front of the other. It is not like in a renormalizable theory, where you have to redefine what you mean by your coupling constant as you go to higher and higher orders. In this case the constants which appear in front of each operator contribute in different ways if you look to different observables.

– *Vartanov*:

In other words, it means that the chiral symmetry is preserved.

– *Colangelo:*

Well, I mean you can choose to make your life miserable and use a regularization which does not respect chiral symmetry or introduce an extra mass scale and then you have also mixing, so when you do loop calculations, you may also correct the constants which you had an order before, but this is an unnecessary complication: you can do things in such a way that every loop correction is one order higher and once you have done correction at a certain order, you have renormalized your constants at a certain order, this renormalization does not need to be updated anymore. You said something about using the Lagrangian at tree level. I mean, you certainly have to do loops, tomorrow's lecture will be devoted to explain the meaning and technically how the evaluation of loop corrections works.

– *Sonnenschein:*

Can the coefficients of the higher order terms in the chiral Lagrangian be determined? There is a string inspired model, which has a closed formula to all higher order terms.

– *Colangelo:*

The main idea in this approach is that we want to find a low energy effective theory of QCD. The form of the Lagrangian we can derive, but then, in order for that to be the low-energy effective theory of QCD, we must have exact values for each of these constants. We cannot yet calculate all of them. In principle we can do it. If you have powerful enough computers, we could determine all of them on the lattice. If you say: "I have a theory which has exactly the same symmetries of QCD", the low energy effective theory will have exactly the same form and in this theory I can even calculate the constants, it is certainly possible, there is nothing wrong with that. If you compare the two things, what you are doing is a test of your fundamental theory.

– *Sonnenschein:*

I am talking about a theory that clearly have the same symmetries that so I wonder whether you can determine the coefficients just from comparing to data and then one can expand this formula and check whether they match or not.

– *Colangelo:*

If you have a fundamental theory in which you can calculate all these coefficients, then you can compare to experiment, it is a test that can be done, tomorrow I will discuss some numerics for these constants.

– *Tsallis:*

The cumulant expansion in statistical mechanics preserves the N-dependence term by term. Is your expansion precisely that or analogous to that ?

– *Colangelo:*

I don't know enough about cumulant expansion to say whether it is analogous or not.

– *'t Hooft:*

I think it is. When you do statistical mechanics, the cumulant expansions are very similar to the irreducible diagram expansion field theories. Basically the effective models are generated by irreducible diagrams, so it should be the same thing.

– *Goldstein:*

In your last slides you introduced external fields *s, p, a, v*. Is *p* the same field as the pion field ?

– *Colangelo:*

I mean, these are external fields, so they do not have any kinetic term, in the QCD Lagrangian they are coupled to $\bar{q}q$ bilinear. It means that the quantity: $e^{iZ[v,a,s,p]}$ is a compact way of writing all possible Green functions which involve four quark bilinears. So, the final aim of the effective field theory method is to provide an explicit representation of this object which in QCD, in principle, is a very complicated thing, is something which we can not calculate. In this case we can calculate it. You do it by calculating a path integral in perturbation theory. Once you have done it, for example you take two derivatives of the pseudoscalar with respect to pseudoscalar field then you set all other fields to zero, this is the Green function of $\bar{q}\gamma^5 q$, at a certain point times $\bar{q}\gamma^5 q$ at another point. You take a Fourier transform and you will find out this has a pole. The residue of this pole is the coupling constant between $\bar{q}q$ and pion field. If you now take four derivatives with respect to *p*, you generate a four-point Green function. The maximal divergence is four poles, and if you take the residue of these four poles, this is the $\pi\pi$ scattering amplitude. So, this object here is really all we want to know about QCD, and the statement is that you can provide an explicit representation at low energy. It is off-shell, it is the only sensible way which you have in QCD to define off-shell quantities, you can do it only starting from quark fields. It does not make sense to have pion fields off-shell, there is no way, no unique way of defining that. This is a very powerful way to formulate this effective Lagrangian method.

– *Kenway:*

In lattice QCD we make heavy use of chiral perturbation theory to extrapolate our results to the chiral limit. How far should we trust these expressions ?

– *Colangelo:*

The experience in recent years has shown that lattice results are a lot more optimistic about the convergence of this series than we could be. Until now it has been very difficult to see anything beyond leading order: the meson masses seem to depend linearly with the quark masses, and this certainly for values of quark masses where I would say: "no, you should not trust the chiral expansion". So, the way I view this is that, this is a puzzle which I do not understand. It would certainly be no good to extract a low energy constant from such data, even if they show that they are in agreement with the chiral expansion, if they are at too high masses. The rule of thumb is you are doing expansions in momenta or masses with respect to the typical scale of QCD, i.e.: *1 Ge*V. In phenomenology one can see that there are sectors where you are lucky and the scale of QCD is a relatively high one, and other sectors where there are phenomena which go beyond the pure Goldstone boson dynamics, which happen at the lower scale, where the convergence seems to be worse. And I think we have quite some experience in the phenomenology to be able to judge them case by case, also, if you look at lattice quantities, where you should expect this convergence to happen. One thing that is quite striking, for example, tomorrow you will see more, is the computation of the scattering lengths. The isospin 0 phase is very difficult to calculate on the lattice, while for the isospin 2 phase there are several calculations. Both have been calculated up to two loops, order p^6, so you can judge within this framework rather well whether the series converges or not. The isospin 0 phase has a very bad convergence, so, for pions of the order of *250 MeV*, you would have already corrections of the order 100%, but for the isospin 2 case, that series behaves much better. So the rule of thumb is easy, but then there can be very different behaviours, depending on the individual cases.

– *Noferini:*

If the chiral symmetry is restored, for example in the first stages of heavy ion collisions, the mass of mesons should be lower. Is it possible to observe a similar effect ?

– *Colangelo:*

If the chiral condensate melts away, you have the impression that the mass of the pions goes to zero. On the other hand, there is nothing anymore to protect the pion for getting a mass which is vanishing in the chiral limit and in that case the pion should be degenerate with scalar partner. So it's not necessarily so, that the mass will become lower, the pion, for example, might increase its mass.

– *Krotov:*

I have a question about this reconstruction of the mass term of the Gasser Leutwyler model. You said that, to restore the invariance of the Lagrangian, you need to rotate masses. The statement is clear. The question is why don't we do the same in ordinary QCD, to restore, for example, chiral invariance ? You see my point ? Suppose we have zero masses in ordinary QCD. In that case we have $SU(3)\times SU(3)\times U(1)\times U(1)$.

Then, let us have non zero masses. In this case we can still rotate these masses and restore the invariance of the Lagrangian. Is it true ? It seems that it should be true.

– *Colangelo:*

Let me try to answer looking at, again, this expression with external fields. So, this object here (Z to something) is chiral invariant. Because we have to define the external fields to transform in such a way that when we do chiral transformations, this generating functional remains invariant. Now, the only difference, or, the way you can look at explicitly chiral symmetry breaking in this case is that, instead of expanding around $s=0$, we expand around a finite value of the scalar external field. So if you want, you have a theory which is symmetric but then your vacuum, the point around which you are doing your expansion, the point which defines the vacuum state, this is non invariant, because your have chosen s to have a particular value, and at that point, once you have chosen that vacuum with a constant value, you can not restore anymore the symmetry by transformation.

– *Krotov:*

Let me rephrase, what you have been saying in order to understand: is it correct that you are saying that the masses in Lagrangian, you should consider like some vacuum expectation values ?

– *Colangelo:*

No. I mean, the arbitrary values of the fields around which you chose to expand, there is nothing spontaneous, it is a choice.

– *Krotov:*

I really do not understand your point. Why can you rotate the masses in this model and you can not rotate in QCD, what is the difference ?

– *Colangelo:*

What I am trying to do when I say, "I rotate the masses and I make the Lagrangian invariant", is not to remove the explicitly symmetry breaking. I do not want to remove it. The theory has explicit symmetry breaking. I am doing this trick to be able to use still the effective Lagrangian method, even in the presence of such an explicit symmetry breaking.

– *Krotov:*

But why is it possible ? Why you can use it ?

– *Colangelo:*

Well, the crucial point is if this explicit symmetry breaking is large or small, it is not a matter of principle. In principle the symmetry is broken. There is no way of removing this. But if it is small, you can deal with explicit symmetry breaking as a perturbation. That is the old idea behind this construction. And, assigning this transformation property, defining the mass term as a spurion, as an external field, is only a way to be able to derive an effective Lagrangian systematically and rigorously, but nothing beyond that.

CHAIRMAN: G. COLANGELO

Scientific Secretaries : D. Derkach, A. Sizanov

DISCUSSION II

- *'t Hooft*

Probably my question might require another lecture which I don't want you to give but you did not mention anything of the instanton interactions which cause the major effects as you know a little bit for higher energy but particularly for the $\eta - \pi$ splitting and things like that.

- *Colangelo*

This contribution goes into the splitting of the η' from the rest. It's one of the inputs in the construction of the theory because you don't do $U(3)\otimes U(3)$ but $SU(3)\otimes SU(3)$ and you simply forget the η'. Now also in connection to what professor Zichichi asked me privately: you can do a $U(3)\otimes U(3)$ theory if at the same time you do a $\frac{1}{N_c}$ expansion. And there are people who pursue this approach. Then it is a consistent scheme in that case. You only have to make the assumption that the splitting vanishes like $\frac{1}{N_c}$.

- *'t Hooft*

Yes, but in real QCD the splitting is rather big. Such a thing can very nicely be added to your chiral lagrangians and effective lagrangians you have done. You can just add the instanton term and go on with the computation of these things and effective part of other interactions.

- *Colangelo*

I mean the problem is that you have a mass, a scale which does not vanish in the chiral limit. Then the scheme which you have does not have the counting, this nice counting scheme property. So you need something which suppresses it, you must have a limit where this splitting vanishes and this is why $\frac{1}{N_c}$ is useful.

- *Susskind*

The thing that you think to do (...) was actually done by Weinberg a number of years ago.

- *Colangelo*

I mean there is recent work in this direction which also addresses questions like mixing angles. I wanted also to suggest a question. In the poster of the school there is "why composite $q\bar{q}$ and $l\bar{l}$ scalar particles had never been clearly established?" and this picture shows that indeed they have been never clearly established. These are the σ parameters that you find in the PDG put over plane: real part of the mass and imaginary part of the mass. The points are all over the plane and this is the region which the PDG suggests. And well something I had not time to discuss is that we add a point here with very small error bars: the statement is the σ-pole is here with these uncertainties which are conservatively estimated.

- *Susskind*

How broad is it?

- *Colangelo*

Well, you see the imaginary part is half of the width. So it's about 500.

- *Zichichi*

...with the uncertainty is incredibly small.

- *Colangelo*

So if anybody wants to know about this I'd be happy to explain something.

- *Machado*

I am Pedro Machado from Utrecht. I am actually asking a bit far question. I just wonder if you could comment further on the application of this particular perturbative approach to General Relativity like you said something in your closing slides.

- *Colangelo*

As everybody knows there is a big problem with interpreting General Relativity as a quantum field theory. But if you take the philosophy which I have discussed today you can just say General Relativity (Einstein-Hilbert action) is simply the lowest possible term which you have and you can calculate loops, then you will find that you need counterterms. You can calculate contributions to observables from these loops and from the counterterms and see what happens. John Donoghue from US has done work in this direction. I think the conclusion was that the observable effects from these higher order contributions would be extremely small and so very difficult to observe.

- *Pacetti*

First of all I'd like to remind you that there is a very well established scalar which is the f_0 but unfortunately it seems to be a tetraquark instead of usual $q\bar{q}$. I have two questions. About the σ. Which is in this field the role played by $\pi\pi$ scattering amplitude (Adler zero). The second question. You have used a dispersive approach when you wrote a dispersion relation with only a right-hand cut which is only a cut in a time-like region. What about the left-hand cut which is $(-\infty,0)$.

- *Colangelo*

I'll answer first the second question. What leads to this very precise determination of the σ-parameter is the use of such a dispersion relation. This was derived by Roy in 1972. This is what you are referring to. The amplitude is written as a subtraction polynomial and a dispersive integral. And then the dispersive integral seems to be only on the right-hand side from $4M_\pi^2$ to infinity (the cut-off is just for convenience). But it is not the case because if this was so then you would find here only the imaginary part of the same wave. But you have also the imaginary part from the other waves which means you've taken into account here the left-hand cut (contributions from other channels). In fact if you take one of these kernels you see the standard pole term which gives you the standard dispersive integral. But here you have also this logarithm which gives you an imaginary part if s is negative which means you have the left-hand cut here. And this representation you can use also for complex values of s. Of course you would be however on the first sheet whereas to find resonances you have to look for poles on the second sheet. However unitarity of the S-matrix implies that the S-matrix on the second sheet is equal to the inverse of the S-matrix on the first sheet. So poles on the second sheet correspond to the zeros on the first sheet. And in the first sheet you can use this representation. So you simply look at where you find poles with this representation. The reason why people find the scatter of points is that they are trying to do an analytic extrapolation from the real axis off into the complex plane. And in doing this they use some parameterization. This is an unstable procedure. Whereas with this dispersion relation you only use data on the real axis. And to go off the real axis you don't have to do anything special at all. And the reason why this final determination is so precise is that this is a low energy resonance and if you look at the various contribution to this dispersion relation at low energy you see this green line as Weinberg calculated in 1966. The subtraction term in the dispersion relation is the red line, not very far from here. And the true thing is this grey band. So at low energy you really understand the amplitude in terms of these subtraction constants. Since chiral symmetry fixes them very precisely then you are able to determine this resonance very precisely. I mean if you want to believe theory you determine it really very precisely. However you can also use the data on these subtraction constants. So I think one should agree and conclude that the determination of the σ -pole is really certain.

You have mentioned that the other zero (it is here) is a very important feature at low energy. It is a matter of interpretation whether this is what drives the σ -pole, but certainly the effect of the subtraction polynomial is crucial.

F. DENEF, M.R. DOUGLAS

"Complexity and Landscape in String Theory"

The Content of the Lectures can be found in:

- *"Flux Compactification"*, M.R. Douglas and S. Kachru, Rev. Mod. Phys. 79:733-796, 2007;
- *"Physics of String Flux Compactifications"*, F. Denef, M.R. Douglas and S. Kachru, Ann. Rev. Nucl. Part. Sci. 57:119-144, 2007;
- *"Computational Complexity of the Landscape I."*, F. Denef, M.R. Douglas, Annals Phys. 322:1096-1142, 2007.

CHAIRMAN: M.R. DOUGLAS

Scientific Secretaries: A. Marrani, I. Rubinskiy

DISCUSSION I

- Didenko:

What are the conditions that lead to a finite number of Calabi-Yau (CY) manifolds?

- Douglas:

First, we should make the question a bit more precise. The way the mathematicians would ask it, is are there finitely many moduli spaces of complex structures of CY threefolds.

It is similar to asking if there are finitely many topologies, but in principle one topology might admit several complex structures. Having said this, well, mathematicians don't know whether the number is finite. Most think it is, but there is not really any convincing argument, and there a few such as Miles Reid who say they think the number is infinite. Most of what people know comes from a systematic construction called a hypersurface in a toric variety. It is a broad generalization of the quintic hypersurface, which is the manifold of solutions to a single equation in P^4. Physically, it amounts to all the CY's you can get from Witten's linear sigma model, with an abelian gauge group, and a single F-term equation. A toric variety which can be used to define a CY corresponds to a certain type of polytope called reflexive, and classifying these is a combinatorial problem. Kreuzer and Skarke, two string theorists, did this by computer, and there are about 400 million reflexive polytopes. Since several polytopes can correspond to the same CY, this is an upper bound on the number of toric hypersurfaces. One gets a lower bound around 30 thousand by listing the different Betti numbers which arise. But there also exist CYs which do not belong to this class, and nobody knows how many. Some mathematicians as well as mathematically-minded string theorists are trying to classify CYs with a 1-parameter moduli space. There are some arguments that only 14 1-parameter CY threefolds exist.

It should also be remarked that, even if the number of CYs were infinite, Bobby Acharya and I showed that as one goes down the list of CY's, the size of the extra dimensions must become bigger and bigger, eventually reaching the macroscopic regime in which Kaluza-Klein modes would show up in 4-dimensional physics, say by modifying the inverse square law of gravity.

- Susskind:

Thus, by assuming some conditions of reasonability on the 4-dimensional physics resulting from the compactification, does one obtain that the total number of compact internal CY threefolds cannot be infinite?

- Douglas:

Yes, this is right. Such conditions are given by the constraint to fit the known experimental data, for example the inverse square law of gravity, known down to 10 microns or so. Such cuts in the analysis imply that only a finite set of CY's could ever agree with data. However, this argument leads to a very large upper bound.

- *Krotov:*

Suppose you compactify your starting higher-dimensional theory on two CYs which are topologically the same but with different radii, yielding different masses of a certain Kaluza-Klein mode. Would you consider such two CY compactifications distinct or not?

- *Douglas:*

Physically, they are distinct vacua, since they give different masses of a particle which in principle is observable. By definition, they are related by varying the moduli (free parameters) of their Ricci-flat metric; consequently, the important issue here is the study of the possible stabilization mechanisms of such moduli.

- *Korthals-Altes:*

In order to make the discussion more concrete, are there one or two phrase definitions of what a Calabi-Yau threefold is?

- *Douglas:*

Yes, it is a complex Kaehler manifold, which is Ricci-flat, i.e. that has a vanishing Ricci tensor. The ones usually used in string compactification are 6 real dimensional, they exist in any even real dimension, starting with 1 (the two-torus) and 2 (the four-torus and the K3 manifold). Physically, the definition is motivated by requiring N=1 supersymmetry in the low energy 4-dimensional theory. This requires a globally defined covariantly constant spinor, a very special type of solution of the massless Dirac equation. One can show that this is equivalent to the mathematical conditions, and then Yau's theorem is used to argue for the existence of the Ricci-flat metric. We need this theorem because we cannot solve for these metrics explicitly.

- *Pinfold:*

You said that the constancy of α in the early Universe is a "prediction" of string theory, but isn't it rather a result of effective field theory?

- *Douglas:*

Well, while the argument essentially comes from naturalness in effective field theory, one does not really know in general that effective field theories must be natural; nothing prevents one from tuning to obtain a very flat potential in the α direction. This is a major loophole , as is shown by the large number of physics papers which take the possibility of varying α seriously. Indeed, the loophole would still be there if there were too many string theory vacua, more than 10^1000 or so; thus, it is a non trivial claim. But if the number is not so large, and people see varying α, then I for one would say that the landscape, and perhaps all of string theory, is falsified. There are even claims for varying α based on quasar spectral lines. These are not generally accepted, but the point is, this is a testable prediction.

- *Sonnenschein:*

I think it is a little bit premature to count vacua, since we do not know how to quantize M-theory nor string theory in the RR background in the NSR formulation. I guess your counting of vacua is consistently performed in an effective classical field theory, but, when you quantize, the situation may drastically change.Let us take the example of a non-Abelian gauge theory. At the classical level, every Lie group in any representation might be chosen as the gauge one. However, after quantizing in most cases anomalies would arise, and therefore most cases would be ruled out. By adding the additional condition of asymptotic freedom, only a very few cases would survive at the quantum level. Might this happen also in the string theory framework?

- *Douglas:*

My answer to such an important question is that, first, we have been studying and constructing string vacua for twenty years now, and seem to know quite a lot about the problem. There are many consistency conditions on the vacua, having to do with anomalies, tadpoles, and the like, and people think they have identified them. We can add to this that most of the currently investigated string vacua models are weakly coupled. While non-perturbative corrections are important, they can be estimated semiclassically, to the accuracy needed to establish the existence of the vacua. Therefore, we are reasonably confident about our results. Finally, if we were making a big mistake, I would hope that this would show up in many contradictions in the final results.

- *Karliner*:

Can you say a few words on how long it takes to reach these various vacuum states?
We have experience that in simple systems, such as spin glasses, it can take a very long time to get out from a metastable state. Is the lifetime of Universe long enough to allow for the Universe itself to reach such vacua?

- *Douglas:*

Well, the issue of dynamics of string vacua is not currently fully understood, and extremely confusing to discuss. The dynamics of spin glass systems is complicated, and here I would just say that string vacua are more complicated than spin glasses.

- *Susskind:*

Let me try to answer the question. The Universe, as we know it and as we usually think of it, began at the point where a transition took place, from whatever existed before to the vacuum in which we currently are. So, there is actually no question whether our Universe had enough time to reach such a vacuum state.

- *Brandt:*

Are CY threefolds the simplest typology of internal manifolds on which one can compactify string theory?

- *Douglas:*

Yes, if one wants some of the 10-dimensional supersymmetry to be preserved in 4 dimensions. Clearly, this is not the unique choice, just the simplest one, and there is no reason that Nature must be so simple.

- *Brandt:*

So, do we actually live in an unique CY threefold?

- *Douglas:*

Yes, according to the way we currently understand string theory. Transitions between different internal compact manifolds might have been important in the very early stages of our Universe, or also in the so-called pre Big Bang era. However, at present time we live in only one internal manifold.

- *Brandt:*

Thus, might we be able to single out such an unique internal manifold by doing physics up to the Planck scale, supposing that we can do it?

- *Douglas*:

In principle yes, but unfortunately we are unable to perform experiments up to such an energy range.

- *Krotov:*

I would like you to spend few more words on the Bousso-Polchinski model. You mentioned that in such a framework there is a cancellation between vacuum energy and the energy of the fluxes. I appreciated if you could explain how such a cancellation takes place, since it seems to me a kind of fine-tuning.

- *Douglas:*

Well, it is a fine tuning in the sense that in most vacua the cancellation does not hold. However, we live in a vacuum for which the cancellation holds, either because it is the only type of vacuum we can live in, or for some other reason external to the model.

- *Krotov:*

What is the spectrum of the Bousso-Polchinski model?

- *Douglas:*

This model deals with the vacuum state of the Universe, it does not concern the particles and their mass spectrum. One could say that its "spectrum" is given by the possible vacuum energies of the Universe.

- *Aschieri:*

Why are you treating CY compactifications, rather than compactifications on generalized CY threefolds or G(2) holonomy manifolds?

- *Douglas:*

Well, we can compactify on more general manifolds, but for now I personally prefer to work with CY threefolds, since they are by far the best understood manifolds, both at a mathematical level and at the level of compactification. But the other manifolds are studied too, for example G(2) by Acharya and collaborators, and generalized CY by many people, and this is important.

- *Krotov:*

If we relax the assumptions of Ricci flatness and vanishing of the first Chern class, or also of Kaehler structure of the complex internal threefold, and we compactify on generic 6-dimensional internal manifolds, is it true that we generally get an infinite number of string vacua?

- *Douglas:*

Well, it is a too far-reaching question. People have made claims of this sort, but so far only for models which we do not understand in enough detail to really believe the claims. But it is important to check. In favour of the finiteness conjecture, besides the many results we have for better understood aspects of the problem, like type IIB fluxes and brane configurations, bundles in heterotic compactification, and so on,there are various oversimplified but suggestive arguments on this point. For example, considering geometrical compactifications of string theory, one gets a fundamental trichotomy between positive, zero and negative curvature cases, which generalizes the classification of Riemann surfaces as genus 0,1, and genus 2 or greater.

A finite number of string vacua is obtained in the positive and zero curvature cases. Actually, the positive curvature or Freund-Rubin case is under better control than the zero curvature Calabi-Yau case. One can even get finiteness results for negatively curved manifolds, but always if one assumes some lower bound on the Ricci tensor. If one does not have that, the number of compactifications is obviously infinite. It is good that, as seen in Freund and Rubin's original work, turning on flux in the internal manifold tends to force the internal curvature to be positive. If it turned out that one could compactify on big classes of negatively curved manifolds, the finiteness conjecture would likely be wrong.

CHAIRMAN: F. DENEF (for Prof. M.R. DOUGLAS)

Scientific Secreteries: P. Machado, P. Reska

DISCUSSION II

- *Aschieri:*

I would like to understand better the computational complexity, the computational aspect. I give you these two analogies: I would like to understand better the relation between the computational problems and other ways of figuring out - of calculating - things without a computer. I give you these two examples to clarify my question. If you want to calculate the volume of a very strange closed surface, that may be quite a challenging problem - on the other hand, if you fill it up with water and then you weight it immediately. Or if you want to figure out which is the minimum surface that touches a series of matches disposed arbitrarily in space. That, again, may be complicated - you put it in bubble soap and you have the answer. So, I hope I made it clear. Thank you.

- *Denef:*

Right, so that's a very interesting question, but it seems to be - and I'll say more about that in my lecture tomorrow - that no matter how hard you try there is no way to come up with such a natural way of solving things, that is, no way you can come up with such a thing to solve an actual NP hard problem. So, your example of soap bubbles or minimal surfaces is instructive - people have indeed suggested that and they have even suggested this as a proof of the fact that NP equal P, thereby claiming the Clay Prize and so on: so, exactly like you said, you can do this experiment, like for example you take - so, the actual NP hard problem is given, say, a number of dots on the plane, find the tree going between these dots that has minimal length. You could try to solve that along the lines that you suggest, so you take two glass plates, you put the number of dot-things in between them and you put that in soap and what you see is that the soap will form a minimal surface connecting these dots. However, the actual NP hard problem is not to find some local minimum, but to find the absolute minimum, so what this soap construction will typically do is get stuck in some metastable local minimum, but it will fail to find the actual absolute minimum. This is something you see in many other physical systems, like, for example, finding the ground state of a spin glass - this is an NP hard problem - and indeed, when you try to make nature find these ground states by taking your spin glass and cooling it down, it will take absurdly long relaxation times to actually get there. So what is hard for is on paper is also hard for nature, and that's actually a quite remarkable thing. I'll say more about this tomorrow, like, you can in fact deduce many of the fundamental principles of quantum mechanics - for example, linearity, unitarity, the fact that probability goes like psi-squared - just by requiring that you cannot use quantum mechanics to solve NP complete problems in polynomial time.

- *Bourjaily:*

I've been thinking a bit about computability and I have a question to ask about - it's a lot more naive than P versus NP, but maybe more relevant to the universe, and that's just whether or not you can solve a problem. I mean, surely you can solve NP problems if you wait long enough - maybe. That's not exactly true if the universe is de Sitter, for example, in which case you'll eventually freeze out and you won't be able to do any computations any more. So it's like our universe, as long as we're not in some quintessence system. So, I guess, is there any way to find the ideal computer? Is it a bit per Planck unit of time and a unit in every Planck length cubed, or, is there a sense in which you can actually define a problem as being unsolvable in our universe with the cosmological constant?

- *Denef:*

Yeah, we think so, because of the reasons you mentioned - but it doesn't necessarily mean that it's unsolvable according to the fundamental principles of physics. Because, for example - well, as you say -, if you live in dS space eventually, due to the expansion, you'll die and everything is cold and computations will end physically; but that's not really a fundamental upper bound because you can always imagine universes with smaller and smaller cosmological constants - or even zero cosmological constant, where this problem will not arise. On the other hand, indeed, there are bounds of apparently truly fundamental nature, related to finiteness of, for example, the speed of light and finiteness of the Planck constant, and those indeed do put upper bounds on what you can compute within, then, a certain area or time or this and that. What the exact upper bound is I don't know.

- *Lockyer:*

This morning, Michael said that after the inflationary period that the universe goes through, it would have to decide which of the solutions it was going to choose, and then he went into a long discussion about how big a computer you would need in order to find the right solution. I was questioning whether that was even the right way to think about the problem - I want to know whether it was just an interesting question and so, then, I can entertain it, or is it really the way you think about things happening?

- *Denef:*

Or the way I think about those things happening?

- *Lockyer:*

Well, you're answering for Michael.

- *Denef:*

Yeah, exactly, a sort of awkward position... Well, these are amusing questions and it's interesting, I think, to try to make sense of them. But I cannot say that I have a clean vision on what those questions really physically mean, or if there's any really deep significance. You see, the problem with early universe cosmology is that it's such a swamp of confusing ideas, it's not even clear how to really set up the problem; of course, you could make some assumptions: so you say, Let's say we start in some general landscape in some metastable vacuum with positive cosmological constant, so that's the initial state, and then you say, Well, could if be now that there's some dynamical selection mechanism that takes you right from there, guaranteed or almost certainly guaranteed, to the universe with the smallest positive cosmological constant. Then, from a computational point of view, such a thing seems highly unlikely, or impossible, because this, when generalized to there existing some such dynamical selection mechanism, seems to solve, or be able to solve, NP hard problems in polynomial time - if you really could do this jumping. So you would say Ok, that's not going to happen. Is this interesting? Not really, because you know physically, already from the start, those things don't happen and you have a very complicated potential with a huge number of minima - and so no laws of quantum or classical mechanics are miraculously going to find this absolute minimum, for example. This is compatible with computational complexity arguments, but to say that it's a triumph of computational complexity would be to go too far. Nevertheless, there's probably something there to be much better thought through, I guess.

- *Susskind:*

My question is about very very closely related things, but let me phrase it in a particularly graphic, colorful way that, maybe, helps crystallize it. Evolution, as I said, is another example of trying to work your way through a very very complicated landscape and find something interesting, and, computationally, it seems like it could be very difficult, but we could ask two separate, different questions - you sort of just asked them. We could ask how long does it take to create the perfect, best human being, the human being who is, ideally, best in every possible way suited to whatever it happens to be - Gerard, for example. I think the answer would be a very very long time, probably exponential in all sorts of things. But we could ask another question: how long does it take to make a human being of any kind that can do some thinking, in other words, that's good enough, within some reasonable bounds of what we would call evolutionary success. My impression is that that's a lot easier and probably is polynomial. The analog for cosmology is, if you're looking for some trick to get through to the lowest possible positive cosmological constant - the most perfect universe, in some sense -, the timescale is probably incredibly long; but if all you're trying to do is to get through to a universe which is livable, that might just be polynomial - particularly, if you have all kinds of branching processes allowed, as in evolution. Would you agree with that?

- *Denef:*

Well, not quite, or only up to a certain extent. The problem is you have to, of course, be more precise about what you mean by acceptable, and the issue depends on how you define that. So, to put the problem slightly more abstractly and more mathematically: take the Bousso-Polchisnki problem, for example. The first version of your problem would be what is the smallest positive cosmological constant vacuum? The second version would be, is there a vacuum which has cosmological constant between zero and epsilon? Now, if you're allowed to vary everything, to choose epsilon freely, and the number of fluxes and those things - if all those things are given parameters -, then also the second problem is still NP hard, so you'd still need exponential time. Of course, things start to become different if you're no longer allowed to choose this epsilon yourself, if it's just set to one, or the such, and you're only allowed to vary the rest, in that case your problem may become polynomial - but I'm not sure, I'd have to think about that...

When you talk about NP completeness and NP hardness and so on, these are mathematical asymptotic notions, so you have to be very precise about what you keep constant and what you scale to infinity and those things. However, they do also have practical implications for finite-sized problems. Again, you could perhaps take the analogy with spin-glass: it's true that a spin-glass will rapidly find a relatively low energy state, but then there is actually a narrow band in which it'll take an extremely long time before it gets there, and the absolute minimum it may never find. But even if you have just a narrow band above the absolute minimum, then that's something that's going to be very hard. So it's a matter of relative size - but ten to the minus one-hundred and twenty, as the size of the cosmological constant we know versus the fundamental scale tends to be, of course, tiny, so...

- *Susskind:*

But, somehow, what's left out of the discussion, in the case of evolution, is the branching nature of evolution. It's not just one sequence of things looking for something, but it's a branching bush that grows out in every direction and finds everything, and it doesn't take very long for it to find everything - or, at least, to find many things...

- *Denef*:

Yes, but also that can actually be formalized in a way (although it doesn't really take away the mistery of evolution, I mean, it's still remarkable that it works so efficiently, I would say), or better, another way of defining what a problem in NP is as follows. The mathematical formalization of a standard computer is that of a so-called deterministic Turing machine, some idealized computer that does things step by step. Now there's also a thing called a non-deterministic Turing machine, that is the same as an ordinary computer but with the remarkable difference that it's allowed at every computation step to multiply itself. So, whereas an ordinary computer, an ordinary Turing machine, follows just one computation chain, a non-deterministic Turing machine follows a computation tree, because it keeps on splitting and branching and branching and branching. That concept turns out to be very useful to define various computational complexity classes, and, for example, the class NP can now be defined as problems that can be solved with a non-deterministic Turing machine in a polynomial number of steps, that is, problems that are such that a non-deterministic Turing machine that halts after a finite number of steps has at least one computation branch that answers Yes if the correct answer is Yes and, in the case the answer is No, all the branches should say no. So, in a way, evolution works like a non-deterministic Turing machine, because it keeps on branching off and branching off - and of course, at a certain point, the parallel fails, given finiteness of resources and so on; but, also, evolution does it very efficiently, as you once pointed out, because it recycles, you don't need this exponentially growing amount of resources, as organisms die and grow and remain, and so on - so, in a way, it works like that.

- *Beck*:

I have a question. Suppose there was some computer investigation of these ten to the one-hundred and twenty vacua, and suddenly you're lucky and you find one that has the right cosmological constant. Does that guarantee in any way that it also does everything else you wan to achieve, that it reproduces the Standard Model physics and things like that, or are these completely different aspects?

- *Denef*:

Yes, absolutely not - it certainly doesn't guarantee that. There may be several vacua, many vacua, even, that have such a small cosmological constant and they may be completely unrelated to each other in all other aspects.

- *Beck*:

So is it kind of wishful thinking that there is one which does everything you want to achieve?

- *Denef*:

No, no, there may be one that - well, we believe there is one, the one in which we live, that achieves everything, but if you're just selecting the cosmological constants, that is, trying to find a vacuum that matches the cosmological constant, that doesn't mean that everything else will be ok, automatically, unfortunately. So that may be a motivation to not start buying CPUs to actually try to solve that problem.

- *Bourjaily*:

I remember reading a story a few years now about two people in India who discovered a polynomial algorithm for answering whether a certain number p was prime, and that was a really big result and very special. But one of the things the article mentioned and made strong point of was that it didn't really change computing time of computing whether or not p was prime, because there were stochastic algorithms that existed so that the expected time of answering whether p was prime was going to beat this polynomial algorithm by leaps and bounds - it just wasn't a deterministic algorithm, because in worst case scenario it could take np time.

- *Denef*:

That's right, so in practice, probabilistic polynomial time algorithms are as good, or sometimes even better than, deterministic polynomial time ones. One way of looking at it is, of course, that even what we think of as deterministic polynomial time algorithms in computers are in a way also probabilistic, because there could always be some cosmic ray hitting your computer and screwing up the computation. But with these probabilistic polynomial time algorithms, or bounded probabilistic polynomial time algorithms, that is, these algorithms that do the computation with some randomized algorithm, you can actually make the probability of error as small as you want - by just repeating things, for example. So you can easily make the probability of error much smaller than the probability of a cosmic ray hitting your computer - thus, from that point of view, they're definitely not worse. In terms of more abstract notions, people believe, even though it's not proven, that the class of problems that can be solved probabilistically in polynomial time is actually identical to the class of problems that can be solved deterministically in polynomial time.

- *Bourjaily:*

Right, I was going to ask if NP problems could be mapped into to P-stochastic problems...

- *Denef*:

No. Well, of course if NP turns out to be equal to P, then yes, but it's widely believed that this probabilistic class, called BPP, is actually close to P or the same as P and then there is NP up there, and there's this huge gap. The same hold for quantum computers, by the way: quantum computers are believed not to be able to solve NP hard problems.

CHAIRMAN: F. DENEF

Scientific Secretaries: P. Goldin, M. Csanád

DISCUSSION III

- *Krotov:*

I have a question on this picture, the Bousso-Polchinski model when you have a small region for cancellation of the cosmological constant. This certain picture is valid only when the metric g is delta symbol, right? But it could be arbitrarily complicated? And it could be even not in the plane but at some high dimensional space in three dimensions?

- Denef:

Yes, that's right. It is in a very high dimensional space. In two dimensions, three dimensions, the problem is very easy. But it becomes exponentially harder with the number of dimensions of the lattice. This specific problem is already NP-hard for a general diagonal metric with different components. There are related problems such as the shortest lattice factor problem: given a lattice, what is the shortest vector in the lattice. For a diagonal metric that would not be NP-complete because it's always going to be one of the basis vectors, you just have to go through the list of n of them if the lattice is n-dimensional and you already know the answer. But in the case of general, off-diagonal matrix again you have a problem that people think is NP-complete, although in fact no one has been able to prove that it is strictly NP-complete in that case.

Krotov:

And from what information do they extract the metric g_{ij}?

- *Denef*:

Well, ok, so this is just a model. In general, if you take one step closer to a real string theory realization it would be determined essentially by the kinetic term of the fluxes which is determined by the geometry, by the metric that exist on the Calabi-Yau. So when you deform the Calabi-Yau this g_{ij} will also change.

- *Krotov:*

But is it possible to write some formula?

- *Denef*:

Yes, it is possible. Well, in the simplest cases, how should I say, the prettiest cases maybe, you really can. So for example if you have three-form fluxes on a Calabi-Yau three-fold then you can write completely explicit expressions of this metric in terms of the so-called period-matrix of the holomorphic three-form. The Calabi-Yau has some holomorphic three-form, a unique one, and you can take the integral of that holomorphic three-form over a bases of independent three-cycles, so you get some list of numbers which depends on the deformations of the Calabi-Yau. From that and derivatives of those things and combinations of those you get a completely explicit formula for that metric. You still have to compute these periods, but again, there are techniques developed for that, and essentially they are given by generalized hypergeometric functions.

- *Krotov*:

So it depends on periods?

- *Denef*:

Yes, that's right.

- *Krotov*:

Thanks.

- *Bourjaily*:

This is a short question, maybe a follow-up. In classical physics, is a variational minimization of a Lagrangian, or just variational calculus, is that an NP problem or a P problem?

- *Denef*:

Well, you have to be more specific what the input is, what the problem size is.

- *Bourjaily*:

Lets say I throw a ball in a gravitational field, we have the in-point and the out-point. We know we can solve it. But if you want to just fed to a computer for a generic Lagrangian?

- *Denef*:

Well, the extremum would be given by some differential equations. Solving this differential equation is in fact a polynomial problem in the input size, it can be still a problem very hard task to do on a computer but it is polynomial because you really start with the initial conditions and you just go step by step, it's not that you had to try various different configurations.

- *Bourjaily*:

Is computing the amplitude in a quantum field problem an NP problem?

- *Denef*:

Again, you should be more specific, but I think, for example lattice QCD calculations and so on, as far as I know (I'm not an expert, but there are experts here), they are polynomial for example in the lattice size. But you see, that can still take a lot of time. But these things tend to be polynomial. A general rough rule, it's combinatorial problems for which the general dumb algorithm is that you just try out all possibilities, and the number of possibilities is exponential in the input size, which tend to be NP problems. Not problems you can solve constructively step-by-step.

- *Susskind*:

It is equivalent to solving the Schrödinger-equations, which is truly polynomial.

- *Denef*:

Exactly.

- *Goldstein*:

I have a little of formulating the question but there is in fact the title of this whole school, the logic of Nature, complexity and the new physics, and there is something about this new approach to figuring out what our Universe looks like that seems to be a different kind of logic of discovery than what I'm familiar with. It's saying there is some theory, they could state string theory, and that theory has almost uncountable, well countable, you counted, number of solutions, and we don't know what most of those solutions look like, but they have gauge groups, they have certain topological structures. How will we ever discover whether that theory is correct or not given this multiplicity of solutions without putting many more constraints in the system. Do we even know, well, we know that heterotic strings in E_6 and group structures that barely resemble what we see in the Standard Model and GUTs, but we don't really know if there is contained in that multiverse of solutions anything that resembles the Standard Model at this point. So I'm just confused by this approach and wonder if people were satisfied with it? Clearly, Susskind is satisfied with it...

- *Susskind*:

I am not!

- *Denef*:

This is a lot of questions. First of all, the theory is there, right, it's mathematically consistent, it incorporates gravity, so we better find a way to figure out if it's a valid theory or not. We cannot say like, well, man, this looks difficult, let's just throw it away and try to find a different theory. It's there, it is mathematically correct, so we have to do something. Question two was sort of, like, don't we have to bring more constraints in, because the landscape looks so vast... Well, I believe, that both from the experimental and theoretical side we will be able to narrow down things further which will enable us to make progress in this. But you know, this takes time of course. Thirdly, you said, that we don't even know if there is anything even remotely looking like the Standard Model in this landscape, but I don't agree with that, we actually do know of many models that do look remotely like the Standard Model. It's going too far to say that there is a model with a positive cosmological constant, everything under control, exactly the Standard Model gauge group, and so on, but there are models with those ingredients separately. For example if you say, let's forget about moduli stabilization, let's just try to get the spectrum – people have done that. By spectrum, I mean, the correct gauge group, the correct light matter and so on. I don't mean like the masses exactly. But there are plenty proofs of principle that these constructions are possible within the framework of string theory. Sp presumably somewhere in this landscape exactly our world can be realized, so we better put in some more effort and try to find this.

- *Goldstein*:

You have been also talking about NP-complete problems and told, that even if I gave you what I claim is a solution, it might take a very long time to show that it is.

- *Denef*:

No. by definition, if you come up with a candidate-solution for an NP-problem, this can be verified in polynomial time. So if you somehow could come up with a solution, with some vacuum or some configuration of string theory that is supposed to satisfy all this things, then in principle you could verify it. But the problem is finding it, finding the right ingredients. But with the right intuition maybe that can be done. Of course then you'd run into the problems whether this is the only possibility or not. Anyway this whole thing about NP-completeness was not meant to show we should give up or to shoot my fellow string theorists in the foot or something like that, it just says, if these are the assumptions we make and these are the models then we will be running into this problem. But problems are there to be solved.

- *Susskind*:

Since my name was mentioned, can I make a comment? I hope I have not given the impression that I think everything is such in wonderful satisfactory order. I don't. It's not that I think the situation is good. It's just this is what the situation is, as best as I can tell, that's what I can say. It's not exactly what I would have chosen the whole thing to be like. It's a mess, it's a big mess, and it's going to be very hard to sort out things, and no, I'm not satisfied by it.

- *Csanad*:

Maybe I'm wrong or it is a stupid question, but I would like to ask this. But a very simple minimizing problem can be a P-problem, solvable in polynomial time. But if we start to complicate the function it can evolve to an NP-problem, if it has many local minima or very complicated structure. Basically, finding these vacuums is also a minimizing problem.

My question is if there is some kind of transition between there.

- *Denef*:

You see, if people say something is an NP problem, what they actually mean is a family of problems. You also need to be able to scale up the problem because the asymptotic behavior of the computation time is what is needed. So more precisely we mean that we have a family of problems, and there will never be an algorithm that is able to solve the problem for every member of the family in polynomial time. But of course there always can be easy examples like minimizing a simple function. You can consider some family or some class of functions, of course some of them will be trivial to minimize. You can just roll down and that would be it.

- *Csanad*:

So there can be sub-families of problems. And there is no theory of these families and sub-family of problems?

- *Denef*:

No, no, there are, one can wonder if one has a specific NP-hard problem to solve, whether is it really true that it is generically hard to solve or it is just one particular instance that is hard to solve. Then of course you have to define what generic means; you have to define a measure on the space of possibilities. Once you do that, you can figure out if this problem is indeed hard to solve on average or just hard to solve in some perverted cases. It's a whole industry, because clearly the stakes are high, since these are real life problems.

- *Goldstein*:

One of your statement was hard is hard. Does it mean that quantum computers will not help in solving NP problems? Or will it help in some particular cases?

- *Denef*:

There is no consensus about that among computational complexity theorists. Most people believe that indeed a quantum computer will not be able to solving NP complete problems in polynomial time either. The problem is really to prove things like that. This is clearly very hard, because people don't even know whether NP equals P. If NP equals P, you will be definitely able to solve it with a quantum computer as well… But then, under certain assumptions, you can actually obtain evidence that it is not going to happen. Naively it seems that a quantum computer has, exponential resources. If you have N qbits, the general state is specified by 2 to the N numbers. It looks like an exponential amount of resources. Indeed it is not so hard to design systems such that somewhere in this wave-function there is a solution, but it is difficult to extract this solution… If you have nonlinearity, than it would be relativity easy, because you can be smart and design some systems where the correct solution which is somewhere in this wave-function starts suppressing everything else and it emerges as the only one. But that's not the case. So the answer to your question would be no. For example the problem of factorization is not NP complete. For problems, which have structure like prime-factorization, quantum computers can obviously dramatically speed up computation time.

- *Susskind*:

Let's consider biology. There were this ten to billion possibilities. The way biology was "solved" was not by going through ten to the billion possibilities until we found the right one, the way it was solved was by a bunch of smart experimentalist by measuring the genome and it took about three years. The corresponding thing would be here to measure the fluxes. Just to measure the fluxes and when you measured the fluxes, however you do it, you'll then check. Right? Do we know if there is any reason to believe, that building the technology to measure the fluxes, building the Super Planckotron, whatever it is going to be, it is necessarily non polynomial in time. Then a billion years from now, one billion years from now, those experimentalists will not able to build an accelerator as big as galaxy.

- *Denef*:

Actually the time you need to build such an accelerator is independent of the problem size, I would say. You just need to build an accelerator, which has enough energy to probe the extra dimensions. So this is just a fixed number, which is independent on how many fluxes there are and those things. And we may even be lucky, the extra dimensions could still turn out to be micron-size or something, you know.

- *Susskind*:

So the future could be in experimental physics and not in mathematical study of complexity but simply measuring the fluxes.

- *Denef*:

The future will be in experimental physics, that is clear. But yes, the problem could be solved by measuring the fluxes – exactly.

- *Zichichi*:

This is not going to be, it is experimental. No theorist can tell us if the superworld exists. We need to make the experiment. Nothing of these arguments can apply to solve our problems. This is given the problem, you are going to tell us, if it is NP complete, the theoretical description as they have done with Polchinski model, but it is experimental. I wonder if the same kind of arguments, if this school was in 1945 before the discovery, the three fundamental discoveries which opened the new world. The same conclusions would apply, but nothing was known of what we know now.

- *Aschieri*:

Consider a gas of particles with a short-range interaction, and the problem of finding the minimum energy of such gas particles comparing it with free particles of gas. It can be a computational problem quite complex, but on the other hand using asymptotic methods, the school of Leeb has proved and given an exact answer.

The question is, on the one hand, does this mean that this has nothing to do with and NP problem. Can you think about some problem than can be solved analytically but computationally with machine extremely complicated and it is not NP.

- *Denef*:

If you have some formula that gives analytic solution for every input, it is obviously not NP complete problem. Because you know, a formula... what is a formula? You put in the input, you compute the thing, and you get the output. If such a formula exists and can be computed in polynomial time, you have a polynomial time algorithm. So if your problem is of that kind, that for any inputs size you get an explicit answer, the problem is certainly not NP complete or NP hard. But nevertheless these things can be sometimes interesting because not every single problem is going to be NP complete that we encounter in string theory. So, this definitely may be useful, but it definitely means it is not NP complete.

Black holes, qubits and the Fano Plane

M. J. Duff [1]† and S. Ferrara[2]‡

† *The Blackett Laboratory, Imperial College London,*
Prince Consort Road, London SW7 2AZ
and
‡*Physics Department, Theory Unit, CERN, CH1211, Geneva23, Switzerland and*
INFN-Laboratori Nazionale di Frascati, Via E. Fermi 40, 00044 Frascati, Italy

ABSTRACT

We review some recently established connections between the mathematics of black hole entropy in string theory and that of multipartite entanglement in quantum information theory. In the case of $N = 2$ black holes and the entanglement of three qubits, the quartic $[SL(2)]^3$ invariant, Cayley's hyperdeterminant, provides both the black hole entropy and the measure of tripartite entanglement. In the case of $N = 8$ black holes and the entanglement of seven qubits, the quartic E_7 invariant of Cartan provides both the black hole entropy and the measure of a particular tripartite entanglement encoded in the Fano plane.

[1] m.duff@imperial.ac.uk
[2] Sergio.Ferrara@cern.ch

Contents

1 Black holes and qubits

It sometimes happens that two very different areas of theoretical physics share the same mathematics. This may eventually lead to the realisation that they are, in fact, dual descriptions of the same physical phenomena, or it may not. Either way, it frequently leads to new insights in both areas. In this paper the two areas in question are black hole entropy in string theory and qubit entanglement in quantum information theory. Going one way, we shall learn that the entropy of the so-called STU $N = 2$ black hole is given by the "hyperdeterminant", a quantity first introduced by Cayley in 1845 and which describes the tripartite entanglement of three qubits [1, 2, 3]. Going the other way, we discover that the exceptional group E_7, the U-duality group of $N = 8$ supergravity, plays a part in the tripartite entanglement of seven qubits [4, 5].

We begin in section 2 with an interesting subsector of string compactification to four dimensions which is provided by the STU model whose low energy limit is described by $N = 2$ supergravity coupled to three vector multiplets. One may regard it as a truncation of an $N = 4$ theory obtained by compactifying the heterotic string on T^6 where S, T, U correspond to the dilaton/axion, complex Kahler form and complex structure fields respectively. It exhibits an $SL(2, Z)_S$ strong/weak coupling duality and an $SL(2, Z)_T \times SL(2, Z)_U$ target space duality. By string/string duality, this is equivalent to a Type IIA string on $K3 \times T^2$ with S and T exchanging roles [6, 7, 8]. Moreover, by mirror symmetry this is in turn equivalent to a Type IIB string on the mirror manifold with T and U exchanging roles. Another way to obtain this model is by truncation of the N=8 theory that results from T^7 compactification of M-theory. Either way, the truncated theory has a combined $[SL(2, Z)]^3$ duality and complete $S - T - U$ triality symmetry [9]. Alternatively, one may simply start with this $N = 2$ theory directly as an interesting four-dimensional supergravity in its own right, as described in section 2.

The model admits extremal black holes solutions carrying four electric and magnetic charges and we organize these 8 charges into the $2 \times 2 \times 2$ *hypermatrix*, a_{ABD}, and display the $S - T - U$ symmetric Bogomolnyi mass formula [9]. Associated with this hypermatrix is a *hyperdeterminant*, Det a_{ABD}, discussed in section 3, first introduced by Cayley in 1845 [10]. The black hole entropy, first calculated in [11], is quartic in the charges and must be invariant under $[SL(2, Z)]^3$ and under triality. The main result of section 4, is to show [1] that this entropy given by the square root of Cayley's hyperdeterminant:

$$S = \pi\sqrt{|\text{Det } a_{ABD}|}. \tag{1.1}$$

The hyperdeterminant also makes it appearance in quantum information theory [13]. Let the three qubit system ABD (Alice, Bob amd Daisy) be in a pure state $|\Psi\rangle$, and let the components of $|\Psi\rangle$ in the standard basis be a_{ABD}:

$$|\Psi\rangle = a_{ABD}|ABD\rangle \tag{1.2}$$

or

$$\begin{aligned}|\Psi\rangle &= a_{000}|000\rangle + a_{001}|001\rangle + a_{010}|010\rangle + a_{011}|011\rangle \\ &+ a_{100}|100\rangle + a_{101}|101\rangle + a_{110}|110\rangle + a_{111}|111\rangle\end{aligned} \tag{1.3}$$

Then the three way entanglement of the three qubits A, B and D is given by the *3-tangle* [12]

$$\tau_3(ABD) = 4|\text{Det } a_{ABD}|. \tag{1.4}$$

The 3-tangle is maximal for the GHZ state $|000\rangle + |111\rangle$ [28] and vanishes for the states $p|100\rangle + q|010\rangle + r|001\rangle$. The relation between three qubit quantum entanglement and the Cayley hyperdeterminant was pointed out by Miyake and Wadati [13].

As far as we can tell [1], the appearance of the Cayley hyperdeterminant in these two different contexts of stringy black hole entropy (where the a_{ABD} are integers and the symmetry is $[SL(2,Z)]^3$) and three-qubit quantum entanglement (where the a_{ABD} are complex numbers and the symmetry is $[SL(2,C]^3)$ is a purely mathematical coincidence. Nevertheless, it has already provided fascinating new insights [1, 2, 3, 4, 5] into the connections between strings, black holes, and quantum information[3].

In section 6 we extend the argument to the $N = 8$ case and, noting that

$$E_{7(7)}(Z) \supset [SL(2,Z)]^7 \tag{1.5}$$

and

$$E_7(C) \supset [SL(2,C)]^7, \tag{1.6}$$

show that the corresponding system in quantum information theory is that of seven qubits (Alice, Bob, Charlie, Daisy, Emma, Fred and George). However, the larger symmetry requires that they undergo at most tripartite entanglement of a very specific kind. As discussed in section 8, the entanglement measure will be given by the quartic Cartan $E_7(C)$ invariant [16, 17, 18, 19]. The entanglement may be represented by the Fano plane [15] which also provides the multiplication table of the octonions. See also the interesting paper by Levay [5] who noted independently the connection to the Fano plane.

2 The N=2 STU model

Consider the three complex scalars axion/dilaton field S, the complex Kahler form field T and the complex structure field U

$$\begin{aligned} S &= S_1 + iS_2 \\ T &= T_1 + iT_2 \\ U &= U_1 + iU_2 \,. \end{aligned} \tag{2.1}$$

This complex parameterization allows for a natural transformation under the various $SL(2,Z)$ symmetries. The action of $SL(2,Z)_S$ is given by

$$S \to \frac{aS+b}{cS+d}, \tag{2.2}$$

[3]A third application [14], not considered in this paper, is the Nambu-Goto string whose action is also given by $\sqrt{|\text{Det } a_{ABD}|}$ in spacetime signature $(2,2)$.

where a, b, c, d are integers satisfying $ad - bc = 1$, with similar expressions for $SL(2, Z)_T$ and $SL(2, Z)_U$. Defining the matrices $\mathcal{M}_S$, $\mathcal{M}_T$ and $\mathcal{M}_U$ via

$$\mathcal{M}_S = \frac{1}{S_2}\begin{pmatrix} 1 & S_1 \\ S_1 & |S|^2 \end{pmatrix} , \tag{2.3}$$

the action of $SL(2, Z)_S$ now takes the form

$$\mathcal{M}_S \to {\omega_S}^T \mathcal{M}_S \omega_S , \tag{2.4}$$

where

$$\omega_S = \begin{pmatrix} d & b \\ c & a \end{pmatrix} , \tag{2.5}$$

with similar expressions for $\mathcal{M}_T$ and $\mathcal{M}_U$. We also define the $SL(2, Z)$ invariant tensors

$$\epsilon_S = \epsilon_T = \epsilon_U = \begin{pmatrix} 0 & 1 \\ -1 & 0 \end{pmatrix} . \tag{2.6}$$

Starting from the heterotic string,the bosonic action for the graviton $g_{\mu\nu}$, dilaton η, two-form $B_{\mu\nu}$ four $U(1)$ gauge fields A_S^a and two complex scalars T and U is [9]

$$\begin{aligned} I_{STU} &= \frac{1}{16\pi G}\int d^4x\sqrt{-g}e^{-\eta}\Big[R_g + g^{\mu\nu}\partial_\mu\eta\partial_\nu\eta - \frac{1}{12}g^{\mu\lambda}g^{\nu\tau}g^{\rho\sigma}H_{\mu\nu\rho}H_{\lambda\tau\sigma} \\ &\quad + \frac{1}{4}\mathrm{Tr}(\partial{\mathcal{M}_T}^{-1}\partial\mathcal{M}_T) + \frac{1}{4}\mathrm{Tr}(\partial{\mathcal{M}_U}^{-1}\partial\mathcal{M}_U) \\ &\quad - \frac{1}{4}{F_{S\mu\nu}}^T(\mathcal{M}_T \times \mathcal{M}_U){F_S}^{\mu\nu}\Big] . \end{aligned} \tag{2.7}$$

where the metric $g_{\mu\nu}$ is related to the four-dimensional canonical Einstein metric $g^c_{\mu\nu}$ by $g_{\mu\nu} = e^\eta {g^c}_{\mu\nu}$ and where

$$H_{\mu\nu\rho} = 3(\partial_{[\mu}B_{\nu\rho]} - \tfrac{1}{2}{A_{S[\mu}}^T(\epsilon_T \times \epsilon_U)F_{S\nu\rho]}). \tag{2.8}$$

This action is manifestly invariant under T-duality and U-duality, with

$$F_{S\mu\nu} \to ({\omega_T}^{-1} \times {\omega_U}^{-1})F_{S\mu\nu} , \qquad \mathcal{M}_{T/U} \to \omega^T_{T/U}\,\mathcal{M}_{T/U}\,\omega_{T/U} , \tag{2.9}$$

and with η, $g_{\mu\nu}$ and $B_{\mu\nu}$ inert. Its equations of motion and Bianchi identities (but not the action itself) are also invariant under S-duality (2.2), with T and ${g^c}_{\mu\nu}$ inert and with

$$\begin{pmatrix} {F_{S\mu\nu}}^a \\ {\tilde{F}_{S\mu\nu}}{}^a \end{pmatrix} \to \omega_S^{-1}\begin{pmatrix} {F_{S\mu\nu}}^a \\ {\tilde{F}_{S\mu\nu}}{}^a \end{pmatrix} , \tag{2.10}$$

where

$${\tilde{F}_{S\mu\nu}}{}^a = -S_2[({\mathcal{M}_T}^{-1} \times {\mathcal{M}_U}^{-1})(\epsilon_T \times \epsilon_U)]^a{}_b * {F_{S\mu\nu}}^b - S_1{F_{S\mu\nu}}^a , \tag{2.11}$$

where the axion field a is defined by

$$\epsilon^{\mu\nu\rho\sigma}\partial_\sigma a = \sqrt{-g}e^{-\eta}g^{\mu\sigma}g^{\nu\lambda}g^{\rho\tau}H_{\sigma\lambda\tau} , \tag{2.12}$$

and where $S = S_1 + iS_2 = a + ie^{-\eta}$.

Thus T-duality transforms Kaluza-Klein electric charges $({F_S}^3, {F_S}^4)$ into winding electric charges $({F_S}^1, {F_S}^2)$ (and Kaluza-Klein magnetic charges into winding magnetic charges), U-duality transforms the Kaluza-Klein and winding electric charge of one circle $({F_S}^3, {F_S}^2)$ into those of the other $({F_S}^4, {F_S}^1)$ (and similarly for the magnetic charges) but S-duality transforms Kaluza-Klein electric charge $({F_S}^3, {F_S}^4)$ into winding magnetic charge $(\tilde{F_S}^3, \tilde{F_S}^4)$ (and winding electric charge into Kaluza-Klein magnetic charge). In summary we have $SL(2,Z)_T \times SL(2,Z)_U$ and $T \leftrightarrow U$ off-shell but $SL(2,Z)_S \times SL(2,Z)_T \times SL(2,Z)_U$ and an S–T–U interchange on-shell.

One may also consider the Type IIA action I_{TUS} and the Type IIB action I_{UST} obtained by cyclic permutation of the fields S, T, U. Finally, one may consider an action [11] where the S, T and U fields enter democratically with a prepotential

$$F = STU \tag{2.13}$$

which off-shell has the full STU interchange but none of the $SL(2,Z)$. All four versions are on-shell equivalent.

Following [9], it is now straightforward to write down an S–T–U symmetric Bogomolnyi mass formula. Let us define electric and magnetic charge vectors α_S^a and β_S^a associated with the field strengths ${F_S}^a$ and $\tilde{F_S}^a$ in the standard way. The electric and magnetic charges Q_S^a and P_S^a are given by

$$F_{S\,0r}^{\,a} \sim \frac{Q_S^a}{r^2} \qquad * F_{S\,0r}^{\,a} \sim \frac{P_S^a}{r^2}\,, \tag{2.14}$$

giving rise to the charge vectors

$$\begin{pmatrix} \alpha_S^a \\ \beta_S^a \end{pmatrix} = \begin{pmatrix} S_2^{(0)} \mathcal{M}_T^{-1} \times \mathcal{M}_U^{-1} & S_1^{(0)} \epsilon_T \times \epsilon_U \\ 0 & -\epsilon_T \times \epsilon_U \end{pmatrix}^{ab} \begin{pmatrix} Q_S^b \\ P_S^b \end{pmatrix}. \tag{2.15}$$

For our purpose it is useful to define a $2 \times 2 \times 2$ array a_{ABD} via

$$\begin{pmatrix} a_{000} \\ a_{001} \\ a_{010} \\ a_{011} \\ a_{100} \\ a_{101} \\ a_{110} \\ a_{111} \end{pmatrix} = \begin{pmatrix} -\beta_S^1 \\ -\beta_S^2 \\ -\beta_S^3 \\ -\beta_S^4 \\ \alpha_S^1 \\ \alpha_S^2 \\ \alpha_S^3 \\ \alpha_S^4 \end{pmatrix}, \tag{2.16}$$

transforming as

$$a^{ABD} \to {\omega_S}^A{}_{A'} {\omega_T}^B{}_{B'} {\omega_U}^D{}_{D'} a^{A'B'D'}\,. \tag{2.17}$$

Then the mass formula is

$$m^2 = \frac{1}{16} a^T ({\mathcal{M}_S}^{-1} {\mathcal{M}_T}^{-1} {\mathcal{M}_U}^{-1} - {\mathcal{M}_S}^{-1} \epsilon_T \epsilon_U - \epsilon_S {\mathcal{M}_T}^{-1} \epsilon_U - \epsilon_S \epsilon_T {\mathcal{M}_U}^{-1}) a\,. \tag{2.18}$$

This is consistent with the general $N = 2$ Bogomolnyi formula [34]. Although all theories have the same mass spectrum, there is clearly a difference of interpretation with electrically charged elementary states in one picture being solitonic monopole or dyon states in the other.

This $2 \times 2 \times 2$ array a_{ABD} is an example a "hypermatrix", a term coined by Cayley in 1845 [10] where he also introduced a "hyperdeterminant".

3 Cayley's hyperdeterminant

In 1845 Cayley [10] generalized the determinant of a 2×2 matrix a_{AB} to the *hyperdeterminant* of a $2 \times 2 \times 2$ *hypermatrix* a_{ABD}

$$\begin{aligned}
\text{Det } a = & -\frac{1}{2}\epsilon^{A_1A_2}\epsilon^{B_1B_2}\epsilon^{D_1D_4}\epsilon^{A_3A_4}\epsilon^{B_3B_4}\epsilon^{D_2D_3}a_{A_1B_1D_1}a_{A_2B_2D_2}a_{A_3B_3D_3}a_{A_4B_4D_4} \\
= & \; a_{000}^2a_{111}^2 + a_{001}^2a_{110}^2 + a_{010}^2a_{101}^2 + a_{100}^2a_{011}^2 \\
& -2(a_{000}a_{001}a_{110}a_{111} + a_{000}a_{010}a_{101}a_{111} \\
& +a_{000}a_{100}a_{011}a_{111} + a_{001}a_{010}a_{101}a_{110} \\
& +a_{001}a_{100}a_{011}a_{110} + a_{010}a_{100}a_{011}a_{101}) \\
& +4(a_{000}a_{011}a_{101}a_{110} + a_{001}a_{010}a_{100}a_{111}) \qquad (3.1) \\
= & \; a_0^2a_7^2 + a_1^2a_6^2 + a_2^2a_5^2 + a_3^2a_4^2 \\
& -2(a_0a_1a_6a_7 + a_0a_2a_5a_7 + a_0a_4a_3a_7 + a_1a_2a_5a_6 + a_1a_3a_4a_6 + a_2a_3a_4a_5) \\
& +4(a_0a_3a_5a_6 + a_1a_2a_4a_7) \qquad (3.2)
\end{aligned}$$

where we have made the binary conversion $0, 1, 2, 3, 4, 5, 6, 7$ for $000, 001, 010, 011, 100, 101, 110, 111$.

The hyperdeterminant vanishes iff the following system of equations in six unknowns p^A, q^B, r^D has a nontrivial solution, not allowing any of the pairs to be both zero:

$$\begin{aligned}
a_{ABD}p^Aq^B &= 0 \\
a_{ABD}p^Ar^D &= 0 \\
a_{ABD}q^Br^D &= 0 \qquad (3.3)
\end{aligned}$$

For our purposes, the important properties of the hyperdeterminant are that it is a quartic invariant under $[SL(2)]^3$ and under a triality that interchanges A, B and D. These properties are valid whether the a_{ABD} are complex, real or integer.

One way to understand this triality is to think of having three different metrics (Alice, Bob and Daisy)

$$\begin{aligned}
(\gamma_A)_{A_1A_2} &= \epsilon^{B_1B_2}\epsilon^{D_1D_2}a_{A_1B_1D_1}a_{A_2B_2D_2} \\
(\gamma_B)_{B_1B_2} &= \epsilon^{D_1D_2}\epsilon^{A_1A_2}a_{A_1B_1D_1}a_{A_2B_2D_2} \\
(\gamma_D)_{D_1D_2} &= \epsilon^{A_1A_2}\epsilon^{B_1B_2}a_{A_1B_1D_1}a_{A_2B_2D_2} \qquad (3.4)
\end{aligned}$$

Explicitly,

$$\gamma = \begin{pmatrix} 2(a_0a_6 - a_2a_4) & a_0a_7 - a_2a_5 + a_1a_6 - a_3a_4 \\ a_0a_7 - a_2a_5 + a_1a_6 - a_3a_4 & 2(a_1a_7 - a_3a_5) \end{pmatrix} \qquad (3.5)$$

$$\beta = \begin{pmatrix} 2(a_0a_3 - a_1a_2) & a_0a_7 - a_1a_6 + a_4a_3 - a_5a_2 \\ a_0a_7 - a_1a_6 + a_4a_3 - a_5a_2 & 2(a_4a_7 - a_5a_6) \end{pmatrix} \tag{3.6}$$

$$\alpha = \begin{pmatrix} 2(a_0a_5 - a_4a_1) & a_0a_7 - a_4a_3 + a_2a_5 - a_6a_1 \\ a_0a_7 - a_4a_3 + a_2a_5 - a_6a_1 & 2(a_2a_7 - a_6a_3) \end{pmatrix} \tag{3.7}$$

All are equivalent, however, since

$$det\ \alpha = det\ \beta = det\ \gamma = -\text{Det}\ a \tag{3.8}$$

If we make the identifications

$$\begin{aligned} a_0 &= \tfrac{1}{\sqrt{2}}(-P^0 + P^2) \\ a_1 &= \tfrac{1}{\sqrt{2}}(-Q^0 + Q^2) \\ a_2 &= \tfrac{1}{\sqrt{2}}(\ P^1 - P^3) \\ a_3 &= \tfrac{1}{\sqrt{2}}(\ Q^1 - Q^3) \\ a_4 &= \tfrac{1}{\sqrt{2}}(-P^1 - P^3) \\ a_5 &= \tfrac{1}{\sqrt{2}}(-Q^1 - Q^3) \\ a_6 &= \tfrac{1}{\sqrt{2}}(-P^0 - P^2) \\ a_7 &= \tfrac{1}{\sqrt{2}}(-Q^0 - Q^2) \end{aligned} \tag{3.9}$$

then we find the $O(2,2)$ scalar products

$$2(a_0a_6 - a_2a_4) = (P^0)^2 + (P^1)^2 - (P^2)^2 - (P^3)^2 = P^2$$

$$2(a_1a_7 - a_3a_5) = (Q_0)^2 + (Q_1)^2 - (Q_2)^2 - (Q_3)^2 = Q^2$$

$$a_0a_7 - a_2a_5 + a_1a_6 - a_3a_4 = (P^0Q_0) + (P^1Q_1) + (P^2Q_2) + (P^3Q_3) = P.Q$$

so

$$\gamma = \begin{pmatrix} P^2 & P.Q \\ P.Q & Q^2 \end{pmatrix} \tag{3.10}$$

and

$$-Det\ a = P^2Q^2 - (P.Q)^2$$

4 Black hole entropy

The STU model admits extremal black hole solutions satisfying the Bogomolnyi mass formula. As usual, their entropy is given by one quarter the area of the event horizon. However, to calculate this area requires evaluating the mass not with the asymptotic values of the moduli, but with their frozen values on the horizon which are fixed in terms of the charges [27]. This ensures that the entropy is moduli-independent, as it should be. The relevant calculation was carried out in [11] for the model with the STU prepotential. The electric and magnetic charges of that paper are denoted (p^0, q_0), (p^1, q_1), (p^2, q_2), (p^3, q_3). In these variables, the entropy is given by

$$S = \pi \left(W(p^\Lambda, q_\Lambda)\right)^{1/2} \tag{4.1}$$

where

$$W(p^\Lambda, q_\Lambda) = -(p \cdot q)^2 + 4((p^1q_1)(p^2q_2) + (p^1q_1)(p^3q_3) + (p^3q_3)(p^2q_2)) - 4p^0q_1q_2q_3 + 4q_0p^1p^2p^3 \,. \tag{4.2}$$

The function $W(p^\Lambda, q_\Lambda)$ is symmetric under transformations: $p^1 \leftrightarrow p^2 \leftrightarrow p^3$ and $q_1 \leftrightarrow q_2 \leftrightarrow q_3$. For the solution to be BPS we have to require $W > 0$.

If we make the identifications [1]

$$\begin{bmatrix} p^0 \\ p^1 \\ p^2 \\ p^3 \\ q_0 \\ q_1 \\ q_2 \\ q_3 \end{bmatrix} = \begin{bmatrix} -a_0 \\ -a_1 \\ -a_2 \\ a_4 \\ -a_7 \\ a_6 \\ a_5 \\ -a_3 \end{bmatrix} \tag{4.3}$$

we recognize from (3) that

$$W = -\text{Det } a \tag{4.4}$$

and hence the black hole entropy is given by

$$S = \pi\sqrt{-\text{Det } a} \tag{4.5}$$

Some examples of supersymmetric black hole solutions [29] are provided by the electric Kaluza-Klein black hole with $\alpha = (1,0,0,0)$ and $\beta = (0,0,0,0)$; the electric winding black hole with $\alpha = (0,0,0,-1)$ and $\beta = (0,0,0,0)$; the magnetic Kaluza-Klein black hole with $\alpha = (0,0,0,0)$ and $\beta = (0,-1,0,0)$; the magnetic winding black hole with $\alpha = (0,0,0,0)$ and $\beta = (0,0,-1,0)$. These are characterized by a scalar-Maxwell coupling parameter $a = \sqrt{3}$. By combining these 1-particle states, we may build up 2-, 3- and 4-particle bound states at threshold [29, 9]. For example $\alpha = (1,0,0,-1)$ and $\beta = (0,0,0,0)$ with $a = 1$; $\alpha = (1,0,0,-1)$ and $\beta = (0,-1,0,0)$ with $a = 1/\sqrt{3}$; $\alpha = (1,0,0,-1)$ and $\beta = (0,-1,-1,0)$ with $a = 0$. The 1-, 2- and 3-particle states all yield vanishing contributions to Det a. A non-zero value is obtained for the 4-particle example, however, which is just the Reissner-Nordstrom black hole.

5 The N=8 generalization

The black holes described by Cayley's hyperdeterminant are those of $N = 2$ supergravity coupled to three vector multiplets, where the symmetry is $[SL(2,Z)]^3$. One might therefore ask whether the black hole/information theory correspondence could be generalized. There are three generalizations we might consider:

1) $N = 2$ supergravity coupled to l vector multiplets where the symmetry is $SL(2, Z) \times SO(l-1, 2, Z)$ and the black holes carry charges belonging to the $(2, l+1)$ representation ($l+1$ electric plus $l+1$ magnetic).

2) $N = 4$ supergravity coupled to m vector multiplets where the symmetry is $SL(2, Z) \times SO(6, 6+m, Z)$ where the black holes carry charges belonging to the $(2, 12+m)$ representation ($m+12$ electric plus $m+12$ magnetic).

3) $N = 8$ supergravity where the symmetry is the non-compact exceptional group $E_{7(7)}(Z)$ and the black holes carry charges belonging to the fundamental 56-dimensional representation (28 electric plus 28 magnetic).

In all three case there exit quartic invariants akin to Cayley's hyperdeterminant whose square root yields the corresponding black hole entropy. If there is to be a quantum information theoretic interpretation, however, it cannot just be random entanglement of more qubits, because the general n qubit entanglement is described by the group $[SL(2, C)]^n$, which, even after replacing Z by C, differs from the above symmetries (except when $n = 3$, which correspond to case (1) above with $l = 3$, the case we already know.).

We note, however, that

$$E_{7(7)}(Z) \supset [SL(2, Z)]^7 \tag{5.1}$$

and

$$E_7(C) \supset [SL(2, C)]^7, \tag{5.2}$$

We shall now show that the corresponding system in quantum information theory is that of seven qubits (Alice, Bob, Charlie, Daisy, Emma, Fred and George). However, the larger symmetry requires that they undergo at most tripartite entanglement of a very specific kind. The entanglement measure will be given by the quartic Cartan $E_7(C)$ invariant [16, 17, 18, 19].

6 Decomposition of $E_{7(7)}$

Consider the decomposition of the fundamental 56-dimensional representation of $E_{7(7)}$ under its maximal subgroup

$$\begin{aligned} E_{7(7)} &\supset SL(2)_A \times SO(6,6) \\ 56 &\rightarrow (2,12) + (1,32) \end{aligned} \tag{6.1}$$

Further decomposing $SO(6,6)$,

$$\begin{aligned} SL(2)_A \times SO(6,6) &\supset SL(2)_A \times SL(2)_B \times SL(2)_D \times SO(4,4) \\ (2,12) + (1,32) &\rightarrow (2,2,2,1) \\ &+ (2,1,1,8_v) + (1,2,1,8_s) + (1,1,2,8_c) \end{aligned} \tag{6.2}$$

Further decomposing $SO(4,4)$,

$$\begin{aligned} SL(2)_A \times SL(2)_B \times SL(2)_D \times SO(4,4) &\supset SL(2)_A \times SL(2)_B \times SL(2)_D \\ &\times SO(2,2) \times SO(2,2) \end{aligned}$$

$$(2,2,2,1)+(2,1,1,8_v)+(1,2,1,8_s)+(1,1,2,8_c) \to$$
$$(2,2,2,1,1)+(2,1,1,4,1)+(2,1,1,1,4)$$
$$+(1,2,1,2,2)+(1,2,1,2,2)+(1,1,2,2,2)+(1,1,2,2,2) \tag{6.3}$$

Finally, further decomposing each $SO(2,2)$

$$SL(2)_A \times SL(2)_B \times SL(2)_D \times SO(2,2) \times SO(2,2) \supset$$

$$SL(2)_A \times SL(2)_B \times SL(2)_D \times SL(2)_C \times SL(2)_G \times SL(2)_F \times SL(2)_E$$
$$(2,2,2,1,1)+(2,1,1,4,1)+(2,1,1,1,4)$$
$$+(1,2,1,2,2)+(1,2,1,2,2)+(1,1,2,2,2)+(1,1,2,2,2) \to$$
$$(2,2,2,1,1,1,1)+(2,1,1,2,2,1,1)+(2,1,1,1,1,2,2)+$$
$$(1,2,1,2,1,1,2)+(1,2,1,1,2,2,1)+(1,1,2,2,1,2,1)+(1,1,2,1,2,1,2)$$

In summary,

$$E_{7(7)} \supset SL(2)_A \times SL(2)_B \times SL(2)_C \times SL(2)_D \times SL(2)_E \times SL(2)_F \times SL(2)_G \tag{6.4}$$

and the 56 decomposes as

$$56 \to$$
$$(2,2,1,2,1,1,1)$$
$$+(1,2,2,1,2,1,1)$$
$$+(1,1,2,2,1,2,1)$$
$$+(1,1,1,2,2,1,2)$$
$$+(2,1,1,1,2,2,1)$$
$$+(1,2,1,1,1,2,2)$$
$$+(2,1,2,1,1,1,2) \tag{6.5}$$

An analogous decomposition holds for

$$E_7(C) \supset [SL(2,C)]^7. \tag{6.6}$$

7 Tripartite entanglement of 7 qubits

We have seen that in the case of three qubits, the tripartite entanglement is described by $[SL(2,C)]^3$ and that the entanglement measure is given by Cayley's hyperdeterminant. Now we consider seven qubits (Alice, Bob, Charlie, Daisy, Emma, Fred and George) but where Alice has tripartite entanglement not only with Bob/Daisy but also with Emma/Fred and also with George/Charlie, and similarly for the other six individuals. So, in fact, each person has tripartite entanglement with each of the remaining three couples:

$$
\begin{aligned}
|\Psi\rangle = & \\
& a_{ABD}|ABD\rangle \\
& +b_{BCE}|BCE\rangle \\
& +c_{CDF}|CDF\rangle \\
& +d_{DEG}|DEG\rangle \\
& +e_{EFA}|EFA\rangle \\
& +f_{FGB}|FGB\rangle \\
& +g_{GAC}|GAC\rangle
\end{aligned} \tag{7.1}
$$

Note that:
1) Any pair of states has an individual in common
2) Each individual is excluded from four out of the seven states
3) Two given individuals are excluded from two out of the seven states
4) Three given individuals are never excluded

The entanglement may be represented by a heptagon with vertices A,B,C,D,E,F,G and seven triangles ABD, BCE, CDF, DEG, EFA, FGB, and GAC. See Figure 1. Alternatively, we can use the Fano plane. See Figure 2. The Fano plane corresponds to the multiplication table of the octonions as may be seen from the description of the state $|\Psi\rangle$ given in Table 1.

Each of the seven states transforms as a $(2,2,2)$ under three of the $SL(2)$'s and are singlets under the remaining four. Note that from (6.2) we see that the A-B-C triality of section 3 is linked with the $8_v - 8_s - 8_c$ triality of the $SO(4,4)$. For example, interchanging A and B leaves $|\Psi\rangle$ invariant provided we also interchange C and F. Individually, therefore, the tripartite entanglement of each of the seven states is given by Cayley's hyperdeterminant. Taken together however, we see from (6.5) that they transform as a complex 56 of $E_7(C)$. Their tripartite entanglement must be is given by an expression that is quartic in the coefficients a, b, c, d, e, f, g and invariant under $E_7(C)$. The unique possibility is the Cartan invariant I_4, and so the 3-tangle is given by

$$\tau_3(ABCDEFG) = 4|I_4| \tag{7.2}$$

If the wave-function (7.1) is normalized, then $0 \leq \tau_3(ABCDEFG) \leq 1$.

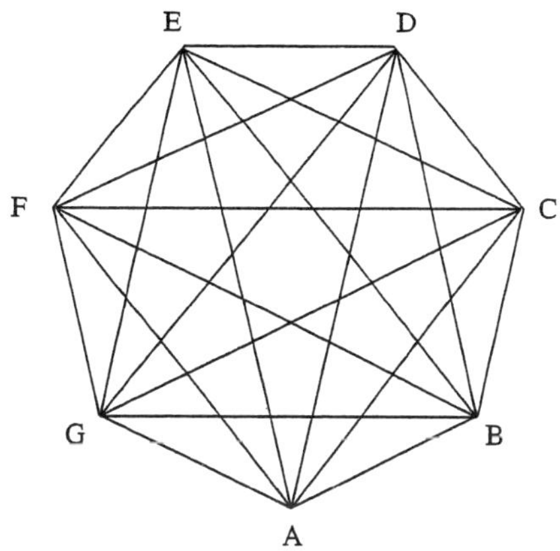

Figure 1: The E_7 entanglement diagram. Each of the seven vertices A,B,C,D,E,F,G represents a qubit and each of the seven triangles ABD, BCE, CDF, DEG, EFA, FGB, GAC describes a tripartite entanglement.

8 Cartan's $E_{7(7)}$ invariant

The Cremmer-Julia [17] form of the Cartan $E_{7(7)}$ invariant may be written as

$$I_4 \;=\; \mathrm{Tr}(Z\bar{Z})^2 - \tfrac{1}{4}(\mathrm{Tr}\, Z\bar{Z})^2 + 4(\mathrm{Pf}\ Z + \mathrm{Pf}\ \bar{Z})\ , \tag{8.1}$$

and the Cartan form [16] may be written as

$$I_4 \;=\; -\mathrm{Tr}(\,x\,y)^2 + \tfrac{1}{4}(\mathrm{Tr}\, x\,y)^2 - 4(\mathrm{Pf}\ x + \mathrm{Pf}\ y)\ . \tag{8.2}$$

Here

$$Z_{AB} = -\frac{1}{4\sqrt{2}}(x^{ab} + iy_{ab})(\Gamma^{ab})_{AB} \tag{8.3}$$

and

$$x^{ab} + iy_{ab} = -\frac{\sqrt{2}}{4} Z_{AB}(\Gamma^{AB})_{ab} \tag{8.4}$$

The matrices of the $SO(8)$ algebra are $(\Gamma^{ab})_{AB}$ where $(a\,b)$ are the 8 vector indices and (A, B) are the 8 spinor indices. The $(\Gamma^{ab})_{AB}$ matrices can be considered also as $(\Gamma^{AB})_{ab}$ matrices due to equivalence of the vector and spinor representations of the $SO(8)$ Lie algebra. The exact relation between the Cartan invariant in (8.2) and Cremmer-Julia invariant [17] in (8.1) was established in [20, 21]. The quartic invariant I_4 of $E_{7(7)}$ is also related to the octonionic Jordan algebra J_3^O [19].

In the stringy black hole context, Z_{AB} is the central charge matrix and (x, y) are the quantized charges of the black hole (28 electric and 28 magnetic). The relation between the entropy of stringy black holes and the Cartan-Cremmer-Julia $E_{7(7)}$ invariant was established

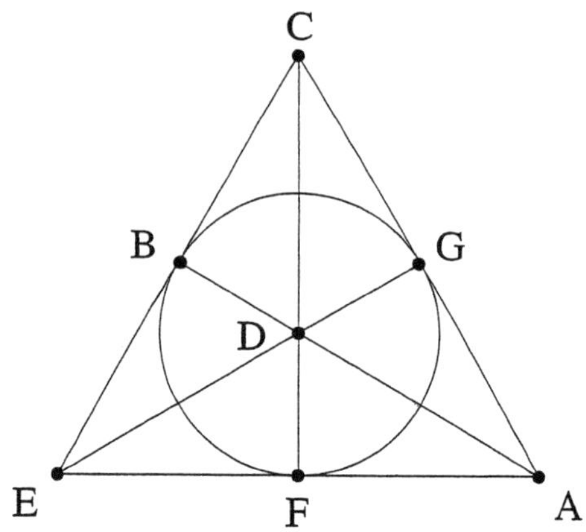

Figure 2: The Fano plane has seven points, representing the seven qubits, and seven lines (the circle counts as a line) with three points on every line, representing the tripartite entanglement, and three lines through every point.

in [18]. The central charge matrix Z_{AB} can be brought to the canonical basis for the skew-symmetric matrix using an $SU(8)$ transformation:

$$Z_{ab} = \begin{pmatrix} z_1 & 0 & 0 & 0 \\ 0 & z_2 & 0 & 0 \\ 0 & 0 & z_3 & 0 \\ 0 & 0 & 0 & z_4 \end{pmatrix} \otimes \begin{pmatrix} 0 & 1 \\ -1 & 0 \end{pmatrix} \tag{8.5}$$

where $z_i = \rho_i e^{i\varphi_i}$ are complex. In this way the number of entries is reduced from 56 to 8. In a systematic treatment in [22], the meaning of these parameters was clarified. From 4 complex values of $z_i = \rho_i e^{i\varphi_i}$ one can remove 3 phases by an $SU(8)$ rotation, but the overall phase cannot be removed; it is related to an extra parameter in the class of black hole solutions [23, 24]. In this basis, the quartic invariant takes the form [18]

$$\begin{aligned} I_4 &= \sum_i |z_i|^4 - 2\sum_{i<j} |z_i|^2|z_j|^2 + 4(z_1z_2z_3z_4 + \bar{z}_1\bar{z}_2\bar{z}_3\bar{z}_4) \\ &= (\rho_1+\rho_2+\rho_3+\rho_4)(\rho_1+\rho_2-\rho_3-\rho_4)(\rho_1-\rho_2+\rho_3-\rho_4)(\rho_1-\rho_2-\rho_3+\rho_4) \\ &\quad +8\rho_1\rho_2\rho_3\rho_4(\cos\varphi - 1) \end{aligned} \tag{8.6}$$

Therefore a 5-parameter solution is called a generating solution for other black holes in N=8 supergravity/M-theory. The expression for their entropy is always given by

$$S = \pi\sqrt{|I_4|} \tag{8.7}$$

for some subset of 5 of the 8 parameters mentioned above. Recently a new class of solutions was discovered, describing black rings. The maximal number of parameters for the known

	A	B	C	D	E	F	G
A		D	G	-B	F	-E	-C
B	-D		E	A	-C	G	-F
C	-G	-E		F	B	-D	A
D	B	-A	-F		G	C	-E
E	-F	C	-B	-G		A	D
F	E	-G	D	-C	-A		B
G	C	F	-A	E	-D	-B	

Table 1: The entanglement of the state $|\Psi\rangle$ coincides with the multiplication table of the octonions.

solutions is 7. The entropy of black ring solutions found so far was identified in [25, 26] with the expression (8.7) for a subset of 7 out of 8 parameters mentioned above.

Kallosh and Linde have shown that I_4 depending on 4 complex eigenvalues can be represented as Cayley's hyperdeterminant of a hypermatrix a_{ABD}. To see this, we that in x, y basis only the $SO(8)$ symmetry is manifest, which means that every term in (8.2) is invariant only under $SO(8)$ symmetry. However, it was proved in [16] and [17] that the sum of all terms in (8.2) is invariant under the full $SU(8)$ symmetry, which acts as follows

$$\delta(x^{ab} \pm iy_{ab}) = (2\Lambda^{[a}{}_{[c}\delta^{b]}{}_{d]} \pm i\Sigma_{abcd})(x^{cd} \mp iy_{cd}) \,. \tag{8.8}$$

The total number of parameters is 63, where 28 are from the manifest $SO(8)$ and 35 from the antisymmetric self-dual $\Sigma_{abcd} = {}^*\Sigma^{abcd}$. Thus one can use the $SU(8)$ transformation of the complex matrix $x^{ab} + iy_{ab}$ and bring it to the canonical form with some complex eigenvalues $\lambda_I, I = 1, 2, 3, 4$. The value of the quartic invariant (8.2) will not change.

$$(x^{ab} + iy_{ab})_{\text{can}} = \begin{pmatrix} 0 & \lambda_1 & 0 & 0 & 0 & 0 & 0 & 0 \\ -\lambda_1 & 0 & 0 & 0 & 0 & 0 & 0 & 0 \\ 0 & 0 & 0 & \lambda_2 & 0 & 0 & 0 & 0 \\ 0 & 0 & -\lambda_2 & 0 & 0 & 0 & 0 & 0 \\ 0 & 0 & 0 & 0 & 0 & \lambda_3 & 0 & 0 \\ 0 & 0 & 0 & 0 & -\lambda_3 & 0 & 0 & 0 \\ 0 & 0 & 0 & 0 & 0 & 0 & 0 & \lambda_4 \\ 0 & 0 & 0 & 0 & 0 & 0 & -\lambda_4 & 0 \end{pmatrix} \tag{8.9}$$

The relation between the complex coefficients λ_I, the parameters x^{ij} and y_{kl}, the matrix a_{ABD} and the black hole charges p^i and q_k [1] is given by the following dictionary:

$$\lambda_1 \;=\; x^{12} + iy_{12} \;=\; a_{111} + ia_{000} \;= -q_0 - ip^0$$

$$\begin{array}{rclclcl} \lambda_2 &=& x^{34}+iy_{34} &=& a_{001}+ia_{110} &=& -p^1+iq_1 \\ \lambda_3 &=& x^{56}+iy_{56} &=& a_{010}+ia_{101} &=& -p^2+iq_2 \\ \lambda_4 &=& x^{78}+iy_{78} &=& a_{100}+ia_{011} &=& p^3-iq_3 \end{array} \tag{8.10}$$

If we now write the quartic $E_{7(7)}$ Cartan invariant in the canonical basis (x^{ij}, y_{ij}), $i, j = 1, ..., 8$:

$$\begin{aligned} I_4 \quad &= -(x^{12}y_{12}+x^{34}y_{34}+x^{56}y_{56}+x^{78}y_{78})^2 - 4(x^{12}x^{34}x^{56}x^{78}+y_{12}y_{34}y_{56}y_{78}) \\ &+4(x^{12}x^{34}y_{12}y_{34}+x^{12}x^{56}y_{12}y_{56}+x^{34}x^{56}y_{34}y_{56}+x^{12}x^{78}y_{12}y_{78}+x^{34}x^{78}y_{34}y_{78} \\ &+x^{56}x^{78}y_{56}y_{78}) \;. \end{aligned} \tag{8.11}$$

then it may now be compared to Cayley's hyperdeterminant (3.1). We find

$$I_4 = -\text{Det } a \tag{8.12}$$

The above discussion of $E_{7(7)}$ also applies, mutatis mutandis, to $E_7(C)$.

To understand better the entanglement we note that, as a result of (6.5), Cartan's invariant contains not one Cayley hyperdeterminant but seven! It may be written as the sum of seven terms each of which is invariant under $[SL(2)]^3$ plus cross terms. To see this, denote a 2 in one of the seven entries in (6.5) by A, B, C, D, E, F, G. So we may rewrite (6.5) as

$$56 = (ABD)+(BCE)+(CDF)+(DEG)+(EFA)+(FGB)+(GAC) \tag{8.13}$$

or symbolically

$$56 = a+b+c+d+e+f+g \tag{8.14}$$

Then I_4 is the singlet in $56 \times 56 \times 56 \times 56$:

$$\begin{aligned} J_4 \sim a^4+b^4+c^4+d^4+e^4+f^4+g^4+ \\ 2[a^2b^2+b^2c^2+c^2d^2+d^2e^2+e^2f^2+f^2g^2+g^2a^2+ \\ a^2c^2+b^2d^2+c^2e^2+d^2f^2+e^2g^2+f^2a^2+g^2b^2+ \\ a^2d^2+b^2e^2+c^2f^2+d^2g^2+e^2a^2+f^2b^2+g^2c^2] \\ +8[bcdf+cdeg+defa+efgb+fgac+gabd+abce] \end{aligned} \tag{8.15}$$

where products like

$$\begin{aligned} a^4 &= (ABD)(ABD)(ABD)(ABD) \\ &= \epsilon^{A_1A_2}\epsilon^{B_1B_2}\epsilon^{D_1D_4}\epsilon^{A_3A_4}\epsilon^{B_3B_4}\epsilon^{D_2D_3}a_{A_1B_1D_1}a_{A_2B_2D_2}a_{A_3B_3D_3}a_{A_4B_4D_4} \end{aligned} \tag{8.16}$$

exclude four individuals (here Charlie, Emma, Fred and George), products like

$$\begin{aligned} a^2f^2 &= (ABD)(ABD)(FGB)(FGB) \\ &= \epsilon^{A_1A_2}\epsilon^{B_1B_2}\epsilon^{D_1D_4}\epsilon^{F_3F_4}\epsilon^{G_3G_4}\epsilon^{D_2B_3}a_{A_1B_1D_1}a_{A_2B_2D_2}f_{F_3G_3B_3}f_{F_4G_4B_4} \end{aligned} \tag{8.17}$$

exclude two individuals (here Charlie and Emma), and products like

$$\begin{aligned} abce &= (ABD)(BCE)(CDF)(EFA) \\ &= \epsilon^{A_1B_2}\epsilon^{B_1C_2}\epsilon^{D_1A_4}\epsilon^{C_3E_4}\epsilon^{D_3F_4}\epsilon^{E_2F_3}a_{A_1B_1D_1}b_{B_2C_2E_2}c_{C_3D_3F_3}e_{E_4F_4A_4} \end{aligned} \tag{8.18}$$

exclude one individual (here George).

9 The black hole analogy

In the STU stringy black hole context [1, 9, 11, 2] the a_{ABC} are integers (corresponding to quantized charges) and hence the symmetry group is $[SL(2,Z)]^3$ rather than $[SL(2,C)]^3$. However, as discussed by Levay [3], there is a branch of quantum information theory which concerns itself with real qubits, called *rebits*, for which the a_{ABC} are real. (One difference remains, however: one may normalize the wave function, whereas for black holes there is no such restriction on the charges a_{ABC}.) It turns out that there are three reality classes which can be characterized by the hyperdeterminant

$$\begin{aligned} &1)\ \mathrm{Det}\ a < 0 \\ &2)\ \mathrm{Det}\ a = 0 \\ &3)\ \mathrm{Det}\ a > 0 \end{aligned} \tag{9.1}$$

Case (1) corresponds to the non-separable or GHZ class [28], for example,

$$|\Psi\rangle = \frac{1}{2}(-|000\rangle + |011\rangle + |101\rangle + |110\rangle) \tag{9.2}$$

Case (2) corresponds to the separable (A-B-C, A-BC, B-CA, C-AB) and W classes, for example

$$|\Psi\rangle = \frac{1}{\sqrt{3}}(|100\rangle + |010\rangle + |001\rangle) \tag{9.3}$$

In the string/supergravity interpretation [1], cases (1) and (2) were shown to correspond to BPS black holes, for which half of the supersymmetry is preserved. Case (1) has non-zero horizon area and entropy ("large" black holes), and case (2) to vanishing horizon area and entropy ("small" black holes), at least at the semi-classical level. However, small black holes may acquire a non-zero entropy through higher order quantum effects. This entropy also has a quantum information interpretation involving bipartite entanglement of the three qubits [2].

Case (3) is also GHZ, for example the above GHZ state (9.2) with a sign flip

$$|\Psi\rangle = \frac{1}{2}(|000\rangle + |011\rangle + |101\rangle + |110\rangle) \tag{9.4}$$

In the string/supergravity interpretation, case (3) corresponds to non-BPS black holes [2]. With four non-zero charges (q_0, p^1, p^2, p^3) in (8.10), for example, an extreme but non-BPS black hole [29] may be obtained by flipping the sign [30] of one of the charges. The canonical GHZ state

$$|\Psi\rangle = \frac{1}{\sqrt{2}}|111\rangle + \frac{1}{\sqrt{2}}|000\rangle \tag{9.5}$$

also belongs to case (3).

In the $N = 8$ theory, "large" and "small" black holes are classified by the sign of I_4:

$$\begin{aligned} &1)\ I_4 > 0 \\ &2)\ I_4 = 0 \end{aligned}$$

$$3)\ I_4 < 0 \tag{9.6}$$

Once again, non-zero I_4 corresponds to large black holes, which are BPS for $I_4 > 0$ and non-BPS for $I_4 < 0$, and vanishing I_4 to small black holes. However, in contrast to $N = 2$, case (1) requires that only 1/8 of the supersymmetry is preserved, while we may have 1/8, 1/4 or 1/2 for case (2).

It is worth noting that the charge orbits corresponding to non-zero I_4 are associated with the following cosets:

$$\frac{E_{7(7)}}{E_{6(2)}} \tag{9.7}$$

and

$$\frac{E_{7(7)}}{E_{6(6)}} \tag{9.8}$$

The large black hole solutions can be found [31] by solving the $N = 8$ classical attractor equations [27] when at the attractor value the Z_{AB} matrix, in normal form, becomes

$$Z_{AB} = \begin{pmatrix} Z\epsilon & 0 & 0 & 0 \\ 0 & 0 & 0 & 0 \\ 0 & 0 & 0 & 0 \\ 0 & 0 & 0 & 0 \end{pmatrix} \tag{9.9}$$

for positive I_4 and

$$Z_{AB} = e^{i\pi/4}|Z| \begin{pmatrix} \epsilon & 0 & 0 & 0 \\ 0 & \epsilon & 0 & 0 \\ 0 & 0 & \epsilon & 0 \\ 0 & 0 & 0 & \epsilon \end{pmatrix} \tag{9.10}$$

for negative I_4. These values exhibit the maximal compact symmetries $SU(6) \times SU(2)$ and $USp(8)$ for the positive and negative I_4, respectively.

If the phase in (8.6) vanishes (which is the case if the configuration preserves at least 1/4 supersymmetry [22]), I_4 becomes

$$I_4 = \lambda_1 \lambda_2 \lambda_3 \lambda_4 \ , \tag{9.11}$$

where we have defined λ_i by

$$\begin{aligned} \lambda_1 &= \rho_1 + \rho_2 + \rho_3 + \rho_4 \\ \lambda_2 &= \rho_1 + \rho_2 - \rho_3 - \rho_4 \\ \lambda_3 &= \rho_1 - \rho_2 + \rho_3 - \rho_4 \\ \lambda_4 &= \rho_1 - \rho_2 - \rho_3 + \rho_4 \end{aligned} \tag{9.12}$$

and we order the λ_i so that $\lambda_1 \geq \lambda_2 \geq \lambda_3 \geq |\lambda_4|$. The charge orbits for the small black holes depend on the number of unbroken supersymmetries or the number of vanishing eigenvalues. The orbit is [19, 22, 32]

$$\frac{E_{7(7)}}{H_{1,2,3}} \tag{9.13}$$

where

$$H_1 = F_{4(4)} \ltimes T_{26} \qquad \lambda_1, \lambda_2, \lambda_3 \neq 0, \quad \lambda_4 = 0 \quad (1/8\ BPS)$$

$$H_2 = SO(5,6) \ltimes (T_{32} \times T_1) \qquad \lambda_1, \lambda_2 \neq 0, \;\; \lambda_3, \lambda_4 = 0 \quad (1/4\ BPS)$$
$$H_3 = E_{6(6)} \ltimes T_{27} \qquad \lambda_1 \neq 0, \;\; \lambda_2, \lambda_3, \lambda_4 = 0 \quad (1/2\ BPS) \tag{9.14}$$

For $N = 8$, as for $N = 2$, the large black holes correspond to the two classes of GHZ-type (entangled) states and small black holes to the separable or W class.

10 Subsectors

Having understood the analogy between $N = 8$ black holes and the tripartite entanglement of 7 qubits using $E_{7(7)}$, we may now find the analogy in the $N = 4$ case using $SL(2) \times SO(6,6)$ and the $N = 2$ case using $SL(2) \times SO(2,2)$.

For $N = 4$, as may be seen from (6.2), we still have an $[SL(2)]^7$ subgroup but now there are only 24 states

$$|\Psi\rangle = a_{ABD}|ABD\rangle + e_{EFA}|EFA\rangle + g_{GAC}|GAC\rangle \tag{10.1}$$

So only Alice talks to all the others. This is described by just those three lines passing through A in the Fano plane. Then the equations analagous to (8.13) and (8.14) are

$$(2,12) = (ABD) + (EFA) + (GAC) = a + e + g \tag{10.2}$$

and the corresponding quartic invariant, I_4, reduces to the singlet in $(2,12) \times (2,12) \times (2,12) \times (2,12)$.

$$I_4 \sim a^4 + e^4 + g^4 + 2[e^2 g^2 + g^2 a^2 + a^2 e^2] \tag{10.3}$$

If we identify the 24 numbers $(a_{ABD}, e_{EFA}, g_{GAC})$ with (P^μ, Q_ν) with $\mu, \nu = 0, \ldots 11$, this becomes [9, 23, 24]

$$I_4 = P^2 Q^2 - (P.Q)^2 \tag{10.4}$$

which is manifestly invarinat under $SL(2) \times SO(6,6)$.

For $N = 2$, as may be seen from (6.2), we only an $[SL(2)]^3$ subgroup and there are only 8 states

$$|\Psi\rangle = a_{ABD}|ABD\rangle \tag{10.5}$$

This is described by just the ABD line in the Fano plane. This is simply the usual tripartite entanglement, for which

$$(2,2,2) = (ABD) = a \tag{10.6}$$

and the corresponding quartic invariant

$$I_4 \sim a^4 \tag{10.7}$$

is just Cayley's hyperdeterminant

$$I_4 = -\text{Det}a \tag{10.8}$$

11 Conclusions

We note that the 56-dimensional Hilbert space given in (6.5) and (7.1) is not a subspace of the usual 2^7-dimensional seven-qubit Hilbert space given by $(2,2,2,2,2,2,2)$, but rather a direct sum of seven 2^3-dimensional three-qubit Hilbert spaces $(2,2,2)$. This is however, a subspace of the 3^7-dimensional seven-qutrit Hilbert space given by $(3,3,3,3,3,3,3)$. Under

$$[SL(3)]^7 \rightarrow [SL(2)]^7 \tag{11.1}$$

we have the decomposition

$$\begin{aligned}
&(3,3,3,3,3,3,3) \rightarrow \\
&1 \text{ term like } (2,2,2,2,2,2,2) \\
&7 \text{ terms like } (2,2,2,2,2,2,1) \\
&21 \text{ terms like } (2,2,2,2,2,1,1) \\
&35 \text{ terms like } (2,2,2,2,1,1,1) \\
&35 \text{ terms like } (2,2,2,1,1,1,1) \\
&21 \text{ terms like } (2,2,1,1,1,1,1) \\
&7 \text{ terms like } (2,1,1,1,1,1,1) \\
&1 \text{ term like } (1,1,1,1,1,1,1)
\end{aligned} \tag{11.2}$$

which contains

$$\begin{aligned}
&(2,2,1,2,1,1,1) \\
&+(1,2,2,1,2,1,1) \\
&+(1,1,1,2,2,1,2) \\
&+(2,1,1,1,2,2,1) \\
&+(1,2,1,1,1,2,2) \\
&+(2,1,2,1,1,1,2)
\end{aligned} \tag{11.3}$$

So the Fano plane entanglement we have described fits within conventional quantum information theory.

The Fano plane also finds application in switching networks that can connect any phone to any other phone. It is the 3-switching network for 7 numbers. However there also exists a 4-switching network for 13 numbers, a 5-switching network for 21 numbers, and generally an $(n+1)$-switching network for (n^2+n+1) numbers corresponding to the projective planes of order n [33]. It would be worthwhile pursuing the corresponding quantum bit entanglements.

Exceptional groups, such as $E_{7(7)}$, have featured in supergravity, string theory, M-theory and other speculative attempts at unification of the fundamental forces. However, it is unusual to find an exceptional group appearing in the context of qubit entanglement. It would be interesting to see whether it can be subject to experimental test.

12 Acknowledgement

We are grateful to Murat Gunaydin for pointing out the connection between our entanglement diagram, Figure 1, and the multiplication table of the octonions. MJD has enjoyed useful conversations with Leron Borsten. Help with the diagrams from Laura Andrianopoli is also gratefully acknowledged.

This work was supported in part by the National Science Foundation under grant number PHY-0245337 and PHY-0555605. Any opinions, findings and conclusions or recommendations expressed in this material are those of the authors and do not necessarily reflect the views of the National Science Foundation. The work of S.F. has been supported in part by the European Community Human Potential Program under contract MRTN-CT-2004-005104 Constituents, fundamental forces and symmetries of the universe, in association with INFN Frascati National Laboratories and by the D.O.E grant DE-FG03-91ER40662, Task C. The work of MJD is supported in part by PPARC under rolling grant PPA/G/O/2002/00474, PP/D50744X/1.

Weare grateful to Professor A. Zichichi for hospitality in Erice.

References

[1] M. J. Duff, "String triality, black hole entropy and Cayley's hyperdeterminant," arXiv:hep-th/0601134.

[2] R. Kallosh and A. Linde, "Strings, black holes, and quantum information," arXiv:hep-th/0602061.

[3] P. Levay, "Stringy black holes and the geometry of entanglement," Phys. Rev. D **74**, 024030 (2006) [arXiv:hep-th/0603136].

[4] M. J. Duff and S. Ferrara, "E_7 and the tripartite entanglement of seven qubits," arXiv:quant-ph/0609227.

[5] P. Levay, "Strings, black holes, the tripartite entanglement of seven qubits and the Fano plane," arXiv:hep-th/0610314.

[6] M. J. Duff and R. R. Khuri, "Four-Dimensional String / String Duality," Nucl. Phys. B **411**, 473 (1994) [arXiv:hep-th/9305142].

[7] C. M. Hull and P. K. Townsend, "Unity of superstring dualities," Nucl. Phys. B **438**, 109 (1995) [arXiv:hep-th/9410167].

[8] M. J. Duff, "Strong / weak coupling duality from the dual string," Nucl. Phys. B **442**, 47 (1995) [arXiv:hep-th/9501030].

[9] M. J. Duff, J. T. Liu and J. Rahmfeld, "Four-dimensional string-string-string triality," Nucl. Phys. B **459**, 125 (1996) [arXiv:hep-th/9508094].

[10] A. Cayley, "On the theory of linear transformations," Camb. Math. J. 4 193-209,1845.

[11] K. Behrndt, R. Kallosh, J. Rahmfeld, M. Shmakova and W. K. Wong, "STU black holes and string triality," Phys. Rev. D **54**, 6293 (1996) [arXiv:hep-th/9608059].

[12] V. Coffman, J. Kundu and W. K. Wootters, "Distributed Entanglement," Phys. Rev. A **61**, 052306 (2000) [arXiv:quant-ph/9907047].

[13] A. Miyake and M. Wadati, "Multipartite entanglement and hyperdeterminants," [arXiv:quant-ph/0212146].

[14] M. J. Duff, "Hidden symmetries of the Nambu-Goto action," arXiv:hep-th/0602160.

[15] F. Klein, "Zur Theorie der Liniencomplexe des ersten und zweiten Grades", Math. Ann. 2, 198-226, 1870.

[16] E. Cartan, "Oeuvres completes", (Editions du Centre National de la Recherche Scientifique, Paris, 1984).

[17] E. Cremmer and B. Julia, "The SO(8) Supergravity," Nucl. Phys. B **159**, 141 (1979).

[18] R. Kallosh and B. Kol, "E(7) Symmetric Area of the Black Hole Horizon," Phys. Rev. D **53**, 5344 (1996) [arXiv:hep-th/9602014].

[19] S. Ferrara and M. Gunaydin, "Orbits of exceptional groups, duality and BPS states in string theory," Int. J. Mod. Phys. A **13**, 2075 (1998) [arXiv:hep-th/9708025].

[20] V. Balasubramanian, F. Larsen and R. G. Leigh, "Branes at angles and black holes," Phys. Rev. D **57**, 3509 (1998) [arXiv:hep-th/9704143];

[21] M. Gunaydin, K. Koepsell and H. Nicolai, "Conformal and quasiconformal realizations of exceptional Lie groups," Commun. Math. Phys. **221**, 57 (2001) [arXiv:hep-th/0008063].

[22] S. Ferrara and J. M. Maldacena, "Branes, central charges and U-duality invariant BPS conditions," Class. Quant. Grav. **15**, 749 (1998) [arXiv:hep-th/9706097].

[23] M. Cvetic and D. Youm, "All the Static Spherically Symmetric Black Holes of Heterotic String on a Six Torus," Nucl. Phys. B **472**, 249 (1996) [arXiv:hep-th/9512127].

[24] M. Cvetic and A. A. Tseytlin, "Solitonic strings and BPS saturated dyonic black holes," Phys. Rev. D **53**, 5619 (1996) [Erratum-ibid. D **55**, 3907 (1997)] [arXiv:hep-th/9512031].

[25] I. Bena and P. Kraus, "Microscopic description of black rings in AdS/CFT," JHEP **0412**, 070 (2004) [arXiv:hep-th/0408186]

[26] I. Bena, P. Kraus and N. P. Warner, "Black rings in Taub-NUT," Phys. Rev. D **72**, 084019 (2005) [arXiv:hep-th/0504142].

[27] S. Ferrara, R. Kallosh and A. Strominger, "N=2 extremal black holes," Phys. Rev. D **52**, 5412 (1995) [arXiv:hep-th/9508072].

[28] D. M. Greenberger, M. Horne and A. Zeilinger, in *Bell's Theorem, Quantum Theory and Conceptions of the Universe*, ed. M. Kafatos (Kluwer, Dordrecht, 1989)

[29] M. J. Duff and J. Rahmfeld, "Massive string states as extreme black holes," Phys. Lett. B **345**, 441 (1995) [arXiv:hep-th/9406105].

[30] M. J. Duff and J. Rahmfeld, "Bound States of Black Holes and Other P-branes," Nucl. Phys. B **481**, 332 (1996) [arXiv:hep-th/9605085].

[31] S. Ferrara and R. Kallosh, "On N = 8 attractors," Phys. Rev. D **73**, 125005 (2006) [arXiv:hep-th/0603247].

[32] H. Lu, C. N. Pope and K. S. Stelle, "Multiplet structures of BPS solitons," Class. Quant. Grav. **15**, 537 (1998) [arXiv:hep-th/9708109].

[33] E. Pegg Jr, "The Fano Plane", MAA online,
http://www.maa.org/editorial/mathgames/mathgames_05_30_06.html

[34] A. Ceresole, R. D'Auria, S. Ferrara and A. Van Proeyen, "On electromagnetic duality in locally supersymmetric N=2 Yang-Mills theory," arXiv:hep-th/9412200.

CHAIRMAN: M. DUFF

Scientific Secretaries: J.N. Ihry, I. Rubinskiy

DISCUSSION I

- *Lopez-Mateos:*

Could you use the understanding of the symmetries of the polytope defined by Bell inequalities in n-qubit systems to increase the understanding of the symmetries of the hyper-determinant?

- *Duff:*

In theoretical physics it often helps to do the same problem in many different ways. Here we have three different physical systems, I have spoken about two of them, quantum information and the Nambu-Goto string. Tomorrow I will talk about black holes. The same mathematics, the same manipulations work in all three cases. When that happens, the ideal situation is that there is some physical connection. Or it could just mean that the same techniques that work over here, also work over there. Even in this case we may still learn new things about one particular system by looking at it from a different point of view.

In the case of black holes, the particular system that I will be looking at has SL(2,R) x SL(2,R) x SL(2,R) symmetry. In that case I can relate it to three qubit quantum entanglement. But that is just a special case of a more general situation that you have in string theory. For example, the kind of systems that I am looking at can be generalized in one of two ways. In one way, the group is SL(2,R) x SO(N,N), where N can be anything. And in another way of generalizing it, it becomes an exceptional group E7(7). So an interesting question, do these other groups which have a meaning in black hole physics and in superstring theory, also have a meaning in n-qubit quantum information theory. I am currently looking into this with Sergio Ferrara and we're optimistic that that may be the case. So it could be that there are things to be learned in quantum information theory that you wouldn't have thought of without string theory. But it's still very preliminary.

- *Aschieri:*

I am intrigued by the target space-time being O(2,2). Is it possible to understand this O(2,2), for example, not as a space-time but come from space-time and the introduction of an electro-magnetic field. That can be rephrased as giving a negative signature?

- *Duff:*

Why would that negative signature be connected to the electro-magnetic field?

-*Aschieri:*

I have in mind the work of Gibbons where Dirac-Born-Infeld action is shown, for static configurations, to have SO(2,4) symmetry and the two time components are due to the E and H fields. So the electric potential can be thought of as a coordinate in that case.

- *Duff:*

At the moment I don't see how. But if someone else can think of a way I would like to hear it.

- *Sonnenschein:*

I think that he was referring to changing the Nambu-Goto into a D-B-I action. Where instead of taking the determinant of the metric, it's the determinant of the metric and $F_{\mu,\upsilon}$. And I think that he's right, it can be considered as an additional coordinate in the electro-magnetic degrees of freedom.

- *Duff:*

Well that is interesting, but it's not a development that I have looked at. So, I can't answer your question. I know that the D-B-I is obtained by adding the F term, but I don't see how that helps.

- *Didenko:*

Can the hyper-determinant be expressed via ordinary determinants, constructed from these two a's?

- *Duff:*

Yes. Let's remind ourselves, first of all, what the ordinary determinant is. To construct the ordinary determinant you need two epsilons and two 2x2 matrices. To construct the hyper-determinant we have six epsilons and four of these matrices. But if you think of peeling off two of the epsilons, you can write that hyper-determinant as a determinant of something. And the something is, in fact, the worldsheet metric in the Nambu-Goto context. It is something constructed out of two a's. Because, if you take the determinant of something made with two a's you get something made out of four a's.

Here you have a 2x2 matrix, symmetric as it turns out, made out of two a's and two epsilons. If I take its determinant, I will get six epsilons and four a's. And those are precisely the six epsilons and four a's that go to make the hyper-determinant. The interesting thing is that I can do that in three different ways (Alice, Bob and Charlie). Instead of singling out the object with the doubly primed indices and taking its determinant, I could single out the same object with the singly primed indices or the same object with the unprimed indices.

- *Didenko:*

It is crucial, at this point that the indices run only two, from 1 to 2?

- *Duff:*

Yes, only two, from 0 to 1 or from 1 to 2 which ever you prefer. So the determinant of the alpha is equal to the determinant of beta is equal to the determinant of gamma and they are all equal to the hyper-determinant of a, up to a sign.

- *Golovnev:*

I would like to make a comment on the discussion of whether we can say that physics is a mathematical science or not. We study nature, therefore we need experiments. But, mathematics is not less important, because in order to compare something with experiments we first need to have this something. So we must think first about the theory, about its` mathematics, absence of contradictions in it, after that you can do all these comparisons. It's the only way. Moreover, if we have no theory, we can't make experiments, because we have to interpret them, we can have just plenty of observations without any knowledge if we don't have theory. Then if we are speaking about the logics, mathematics, about the laws of thinking, we should not forget that all of them are inside of our world too, inside the logics of nature, they so-to-say play their part in the physics of our world. Suppose we consider deformation quantization issues, this general works of Kontsevitch and others, can we verify them by experiments? No. Because it's not some definite prediction, it's not even some definite physical theory, it's just the whole category, the way of thinking about the world. Nevertheless, works of such kind contribute much to our understanding of quantum world. So, my conclusion is that we can definitely call physics a mathematical science.

- *Duff:*

Thank you for that very interesting comment. I pretty much agree with what you have just said. Let me add something. When I said that physics is a mathematical science, I did not intend to imply that it is not an empirical science, as I don't see them as something mutually exclusive (although some people might interpret it that way). I certainly agree with you that ideally first we need to make sure that our ideas are logically consistent, before we compare them to experiment, although we do not always achieve that ideal.

But this raises a broader issue, as Prof. Susskind mentioned in one of his lectures, whether theories like the string theory are likely to yield empirical test, and if so how soon. My own view, which might not be shared by all string theorists, is that we haven't departed from the traditional view that the ultimate test of a physical theory is whether or not it agrees with experiment. Most of the string theorists that I know are quite happy with that. How long will you agree to wait before your theory is testable? I see the whole String Program as a long-term program, and we will not get an instant gratification. To the colleagues that ask why we haven't predicted something easy to verify, I say - "Just wait".

I might add here that there are great theoretical ideas that took many years before they were verified empirically. For example, Bose-Einstein condensation took about 70 years from being predicted to being confirmed by experiment. Yang and Mills predicted non-abelian gauge bosons in 1954, but we did not see the W till 1983. What are we looking for today experimentally? Gravitational waves (predicted 1916), the cosmological constant (1917), extra dimensions (1926), the Higgs boson (1964), supersymmetry (1972). If you are going to throw away a theory for the reason that it does not give you instant empirical confirmation, you'll be throwing away lot's of the ideas the people are currently looking for, and I think that will be a mistake.

- *Colangelo:*

I'd like to give a comment. The two Dutch physisists [*George Uhlenbeck and Sam Goudsmit*] published their idea [*of spinning electron*] before they got to talk to Pauli, and after they talked to Pauli, it was already too late to withdraw the paper. So, they got the Nobel Prize for that work.

- *Duff*

That's right. My supervisor Abdus Salam used to tell a similar story, how Pauli asked him "not to waste his time on violating parity" when Salam put forward his two-component neutrino hypothesis.

- *Korthals-Altes:*

I have a minor correction. Bose-Einstein condensation was discovered with liquid helium in 1935 or something like that. (*editor: 1938*)

- *Krotov:*

In the last years there was some activity on interpretation of Hawking entropy in terms of entanglement of the states, which are inside and outside of the horizon. Is there any relation between what you are talking about and this idea?

- *Duff:*

Excellent question. The answer is "I don't know". This morning I talked about two states I called Alice (A) and Bob (B) and their entanglement. The concept I did not mention is the entanglement entropy . The entanglement emtropy is defined to be -Tr(rhoAlogrhoA), where rhoA is the density matrix formed by doing a partial trace over the subsystem A. It's also equal to -Tr(rhoBlogrhoB). The measure of the entanglement given by the determinent of ` a' is the 2-tangle which is sometimes referred to in the literature as the `concurrence'. It's not hard to show that det a equals det rhoA squared which also equals det rhoB squared . So there is a relation between the two-tangle and entropy of entanglement. Now, the quantity I am going to find, namely the black hole entropy is not quite this quantity. It's the 3-tangle, and the 3-tangle does not have an interpretation in terms of entanglement entropy. However, according to Kallosh and Linde, there are quantum corrections to the classical one quarter of the area of the event horizon of the famous Hawking-Bekenstein event horizon formula and, interestingly enough, these are given by the 2-tangles.

I was originally hoping that the similarity between the mathematics of black hole entropy and the mathematics of qubit entanglement would be more then just mathematics, and that there'd be some physical connection. I have to admit, that at the moment I don't see what it is. Nevertheless, there are reasons to hope for such a connection. The black holes we are talking about have a near-horizon geometry of 2D anti-de Sitter space times the 2-sphere. The work on AdS/CFT suggests a dual description in terms of a conformal field theory (CFT) and in the case of AdS2 it's just a 1D quantum mechanical system (though it's true to say that full details have not yet been worked out.) So you might think that there was some relation between the entanglement entropy of that quantum mechanical system and the entropy of the black hole whose near horizon geometry is AdS2 times S2. For that reason, it's not beyond the bounds of possibility that the kinds of mathematical connections I'm talking about might indeed have some physical basis. But it is not clear why 3 qubits should be singled out in this quantum mechanical theory.

- *Susskind:*

Isn't it true that the largest entanglement entropy for 3-qubits is a logarithm of 8? Maximum dimension of the Hilbert space is 8 and the log of the number of states is 8. It seems rather unreasonable to try to connect that directly with the entropy of a black hole.

- *Duff:*

Yes, first of all, as I said, the entropy of a black hole is given by a 3-tangle not the entanglement entropy. Note also that we are using unnormalized state vectors. So the entropy is not limited in the way you described.

CHAIRMAN: M.J. DUFF

Scientific Secretaries: A. Golovnev, J.N. Ihry

DISCUSSION II

- *Csanad:*

Is there a way to make some kind of quantum optical measurement to find something out about black holes with this analogy?

- *Duff:*

That is a very good question, but I don't know an answer. I'd like to think that this analogy between quantum information theory and black holes could be extended so that observations in the quantum realm would have some bearing on black holes. This is only a few months old so there have not been any attempts made in this area.

- *Lopez-Mateos:*

In your talk you said that small black holes have zero entropy classically, but that they may aquire some entropy through quantum corrections. I am not familiar with how this happens. Can you elaborate on how a system can gain entropy throught quantum corrections.

- *Duff:*

This isn't a field I have worked in myself, but my understanding is that it is to do with higher order derivative corrections to the effective action. The usual result that the black hole entropy is one quarter the area of the event horizon is the result that you would get with just the Einstein-Hilbert action, which contains two derivatives. In string theory, we know that the higher order corrections (alpha' corrections - string loop corrections) which would have r^4 terms in it, would have higher power of the curvature. It is these r^4 terms that modify the canonical result and give you corrections to the Berkenstein-Hawking rule. However, it is still true that the entropy is a U duality invariant. It may not be the area, but it is still invariant under the symmetry that the string theory had to begin with. In my case, it is still SL(2,Z)^3 invariant.

- *Durnford:*

You mentioned yesterday, that you could think about the quantum entanglement problem in terms of twisters, can you also think of the Nambu-Goto string with the same formulism? And, could that help generalize the theories away from the two time dimensions?

- *Duff:*

That's an interesting question and I haven't given it any thought,... till today. So, I invite you to have a look at it.

- *Machado:*

The expression you gave for the entropy of a small black hole in terms of the concurrence, are there any restrictions or conditions that the non-zero electric or magnetic charges must satisfy? Or is that an exact expression?

- *Duff:*

The relevant formulae (taken from the paper of Kalash and Linde) the concurrence was given in terms of the bi-partite entanglement between two of the qubits. So, any configurations of charges which gives you a nonzero answer for one of these concurrences, will give you a nonzero entropy. But, it doesn't need to be tri-partite. In fact, there are black holes which are big in the sense of having a non-vanishing tri-partite entanglement, and also have quantum corrections. But you also have small black holes which have no tri-partite entanglements at the classical level and require quantum corrections through the bi-partite corrections, so-to-speak.

- *Didenko:*

As far I have understood, you obtained this expression analyzing the square of the horizon and then noted it is SL(2)^3 invariant. Is there any group theoretical notion of that fact? Because one might expect it is a Casimir operator in the representation of this moduli space.

- *Duff:*

It is not miraculous, it's guaranteed. If you start with a Lagrangian that has the SL(2)^3 symmetry and you find a black hole solution, then its entropy has to have that symmetry.

- *Didenko:*

But this entropy is not guaranteed on shell. The Lagrangian is invariant under the symmetry.

- *Duff:*

Techniquely, what I should have said is that the equations of motion are invariant under SL(2)^3. However, the same remark still applies; given that the field equations, of SL(2)^3 symmetry, then the solutions of the black holes will have entropy that respect SL(2)^3 symmetry. It's true in that original paper by Burns et al., they just brute force calculated it and got the complicated expressions in terms of the charges. I don't think they recognized that it was SL(2)^3 invariant, but in hindsight one can see that it had to be. That is what we find when it is the Cayley hyper-determinant. Incidentally, the Cayley hyper-determinant is the only quartic invariant that has SL(2)^3 symmetry. Again, in hindsight, we know that the entropy has to be given by the Cayley hyper-determinant.

- *Krotov*

In your talk you mentioned the possibility to extract the black hole entropy from counting the string states. Could you explain, a little bit, how this can be done.

- *Duff*

Well, I don't think I did talk about that, but it is an interesting topic. Everything I've done in these lectures is the macroscopic black hole entropy which just comes from applying Hawking's ¼ the area of the event horizon rule. But in string theory using D-branes we can also give a microscopic derivation of the entropy. And it has been done for the cases that I've been talking about because you can put that the black holes that I described are configuration of D-branes where D-branes are warped around the cycles of extradimensions. And at least in the case of extremal black holes you reproduce the Bekenstein-Hawking result using this microscopic counting. That's very interesting but in a sort tangential to the theme of our lectures.

- *Aschieri*

Can you comment on the fermionic sector? What happens with this SL(2,Z)-cube symmetry on the fermionic sector of the theory? Is it trivial or do something interesting occur?

- *Duff*

Well, the Lagrangian I wrote down with N=2 supergravity coupled to 3 vector multiplets. So, there is a fermionic sector I did not bother to write down. And the entire, I wouldn't say Lagrangian, entire field equations including fermions have the SL(2)-cube invariance. Now, that did not have any direct bearing on the black hole solution because in the final black hole solution we have to put the fermions to zero but the symmetry is still there when you include fermions. Now, in the context of fermions one could ask a question about the Nambu Goto string because the bosonic string in the signature(2,2) is the bosonic sector of critical N=2 superstring. And that's the interesting theory because its field equations tell you that the Riemann tensor is self dual. So, what I have not done and what I think could be interesting object would be to extend analyses to N=2 superstring. And it have not been done yet.

- *Dudchenko*

You found hidden symmetries of the Nambu-Goto action. And is it possible to find new hidden symmetries or it seems to be a "game over" for the symmetries of these action?

- *Duff*

I already told you that. If I found more I would have told you about.

- *Dudchenko*

And rather stupid question: Why STU-model is called STU? Is it something common with Mandelstam stu-variables?

- *Duff:*

It's nothing to do with Mandelstam variables as far as I know. Now, STU is because 3 vector fields go by the names S, T and U. There is a reason for that. The S-field is one associated with the so called S-duality which is the electromagnetic duality. And the T and U vector fields are associated with is associated with the target space duality which interchanges big dimensions with small dimensions. So we have three fields and they are called S, T and U.

- *Marrani:*

I have two more remarks and two questions for you. The first remark is about the relation between the sign of the Caley superdeterminant and the kind of BPS-preserving feature of the black holes. Well, in global there is no relation that you mention in the sense that non-BPS attractors will vanish central charge that have been recently found. And such attractors have negative Caley superdeterminant as they are BPS-extremal attractors. It means that actually there are two general families of the STU black holes that you mentioned. The one is related with the negative Caley superdeterminant and it's the one of BPS. And the non-BPS will have vanishing central charge. And it has been shown that these two kind of black holes can be simply related by flipping some of the black hole charges such that the sign of the Caley superdeterminant remains the same. While the non-BPS STU black holes that you have mentioned this morning are the ones with non-vanishing central charges that were known since a few years.

- *Duff:*

You say there are non-BPS black holes with both signs of superdeterminant.

- *Marrani:*

Yes, and the sign is related to the vanishing of the central charge of the N=2 supersymmetry algebra. This has been very recently found by solving locally the operator equation for STU black holes in N=2 or 4 supergravity.

The second remark is that it seems to me there is a very intensity in geometric classification of black hole shift configuration that you were showing in that slide that related two systems with general three particle entanglement or not. In the sense that this can be easily classified by some orbits of the loop gravity group put is this case SU(1,1)-cube and this has been in general made recently for symmetric special kind of geometries. And it's very interesting because the large black holes, the ones with non-vanishing classical entropy, should be related to light-like or space-like orbits of SU(1,1)-cube depending on the sign of Caley superdeterminant. So, these large black holes as I mentioned are related to ones of BPS and non-BPS with non-BPS splitting into ones with vanishing central charge or non-vanishing. So, in some sense there should be a relation between general three particle entanglement and the non-degeneracy of the Hawking's SU(2) duality group. And this is correctly understood about this geometric interpretation.

And the simple question is if you think about the procedure of stabilizing the multicenter STU black holes that you have mentioned this morning by choosing entropy black hole check configuration is a kind of a fine tuning where you can always satisfy the antigravity bound such that there is an equilibrium between gravitational interaction and the Coulomb repulsion.

- *Duff:*

Thank you for your interesting comment but the good question is can you always find the multicenter solution that is stable.

- *Marrani:*

So, you always satisfy the antigravity bound in that case. You don't have to fine tune the black hole check configuration such that this antigravity bound is satisfied once and you know the multicenter solution.

- *Duff:*

In BPS you can always find a multicenter solution. And you can also prove a theorem, I can't do it, but I think it is true more or less.

- *Marrani:*

OK and the last question is you mentioned that recently Leray introduced a sort of real version of the interpretation of STU black holes. Do you think it can be related to a recent real formulation of special geometry done by Ferrara and Mazia. They use some Darboux special real coordinates for a local special Kahler geometry. So, may be it could be a geometric interpretation of what Leray says about the real version of SL(2).

- *Duff:*

Yes I know about that collaborators doing very interesting work on the attractor mechanism and I am sure the answer is yes. You can relate it.

- *Bourjaily:*

I have a very simple question about this slide right up now. In the small black hole case the quantum corrections is the entire entropy. So, it's natural to look at this. Nicely turns out that there is also SU(2)-cube case in it. I wonder if the quantum corrections to large black holes also keep SL(2) cube symmetry.

- *Duff:*

Yes. But for the small black holes it is the whole story. But according to Cardy-Verlinde for large black holes you simply add this to the Caley hyperdeterminant you already have.

The Status of Lattice QCD

Richard Kenway
School of Physics, The University of Edinburgh,
Edinburgh EH9 3JZ, Scotland

Abstract

It is now possible to formulate QCD on a space-time lattice with correct chiral and flavour symmetries, and to simulate the dynamical effects of the *u*, *d* and *s* quarks with masses close to their physical values. This short review for non-experts describes the status of the simulation results, and how they are helping us to understand the strong interaction, to test the Standard Model and to search for physics beyond it.

1. The Strong Interaction from First Principles

Quarks are confined. Although not yet rigorously proven from the theory, this empirical fact is strongly supported by QCD, and the quest for an appropriate theoretical tool was Ken Wilson's original motivation for Lattice QCD over 30 years ago. Specifically, his formulation of Yang-Mills theories on a regular hypercubic space-time lattice solved the problem of how to render gauge field theories finite without spoiling the local symmetry (at least at the lattice points). The corresponding lattice formulation of fermion fields proved challenging and was not satisfactorily resolved until recently. A crucial element is to ensure that the desired quantum field theory with the correct symmetries is recovered in the continuum limit, when the lattice spacing is taken to zero. The lattice technology has now reached a point where it can be used to define QCD, and a growing number of other quantum field theories, without relying on the assumption that some parameter is small (perturbation theory) and in such a way that it can be simulated on a computer.

Simulation begins from the path integral for QCD in four-dimensional Euclidean space. This analytic continuation of Minkowski space-time to imaginary time converts the path integral into what looks like a rather complicated statistical mechanics problem. The advantage is that, when the Euclidean space is replaced by a finite hypercubic lattice (i.e., with a finite volume), the resulting path integral becomes a finite-dimensional integral over the fields at each of the lattice sites that may be evaluated numerically using Monte Carlo methods. Michael Creutz recognised this and launched a huge computational effort, which, in turn, led to both the development of new algorithms and new computer architectures, in order to achieve the performance required to simulate QCD on realistic lattices, i.e., with large enough volumes to contain the hadrons of interest, small enough lattice spacings to be close to the continuum, and QCD parameters close to their physical values.

2. The Parameters of QCD

Ensuring proximity of the bare parameters in the QCD action to their physical values is itself a significant computational task. These values are the inputs to the simulations. Determining them is logically the first stage in being able to use lattice QCD to compute quantities that can be useful for phenomenology. The cause of this difficulty is the confinement of quarks, which hides the fundamental degrees of freedom (the quark and gluon fields) from direct experimental measurements. Experiments measure the properties of hadrons and it is from these that we have to infer the values of the QCD parameters: the quark masses, the quark-gluon coupling and the vacuum angle.

In principle, we would do this by guessing the parameter values, performing a simulation to compute a set of hadronic quantities, comparing these with their experimental values, and then adjusting the parameters and repeating until we get agreement with experiment. We could then conclude that the input parameters have their physical values. In practice, this process is complicated by the lack of computer power to run simulations actually at the physical values, so that some extrapolation from unphysical values is required.

The hadronic quantities used to fix the bare masses of the various quark flavours are typically the masses of mesons whose valence quarks have those flavours. If there are N_f flavours, we need to input N_f meson masses from experiment (usually, we assume isospin symmetry so that the u and d quark masses are equal and fixed by the pion mass; the s quark mass is then fixed by the kaon mass). One additional dimensionful quantity is needed to set the lattice scale. Conventionally, this is used to define the lattice spacing in physical units, a, by demanding that, for $N_f + 1$ meson masses, $M^{\text{lattice}}(m_1,\ldots, m_{Nf}) = aM^{\text{experiment}}$, where m_i denotes the mass of quark flavour i.

If we could perform simulations at the physical quark masses, this would be a direct iterative procedure. As it is, this is still too expensive, so chiral perturbation theory is used to model the dependence of meson masses on quark masses, so that simulation results can be extrapolated from the unphysically large u and d quark masses used in the simulations to their physical values. The lattice spacings available to today's computers correspond to a momentum cut-off comparable to the c quark mass, so that it is not possible simultaneously to simulate the c, b and t quarks without big discretisation effects. These flavours are only included as valence quarks (i.e., they don't appear in the sea).

The lattice spacing in physical units, a, tells us how close our simulation is to the continuum limit. If a is smaller than any scale relevant to the hadronic physics we are trying to compute, then that physics, expressed as dimensionless combinations of physical quantities, should become independent of a as it is made smaller. This is called 'scaling' and when a quantity 'scales' it is equal to its value in the continuum limit. The quark-gluon coupling, g, is the extra parameter in the QCD action (in addition to the quark masses) needed to vary the lattice spacing. Since non-abelian gauge theories are asymptotically free, the effective interaction strength vanishes at short distance. Turning this argument around, the lattice spacing vanishes with g. So, taking the continuum limit of a physical quantity corresponds to varying g until the quantity of interest becomes independent of it as g is made smaller.

3. Renormalisation

Having determined the bare parameters in the QCD action, we need to relate them to the values they would have in a continuum perturbative renormalisation scheme. This matching is needed because the analysis of experimental data uses matrix elements defined in continuum perturbation theory. This renormalisation process also defines the normalisations of the lattice fields, and of composite products of them, so that we can use the values of matrix elements of composite operators computed in lattice QCD in the corresponding continuum expressions.

Renormalisation introduces an unphysical momentum scale that, through a process called dimensional transmutation, eventually sets the scale for dimensionful quantities like hadron masses. The normalisation of the lattice fields must be done by introducing a momentum scale that fits on the lattice and so cannot exceed the momentum cut-off imposed by the inverse lattice spacing. Continuum renormalisation schemes rely on perturbation theory that is only good at relatively high momentum scales where the QCD coupling, g, is weak. One way to match the lattice normalisations to the continuum normalisations is to compute both in perturbation theory. However, the two results are needed at the same momentum scale and there may not be a momentum range in which both the lattice and continuum perturbation theory calculations are sufficiently under control. For this to be the case, requires simulations at small lattice spacings.

A more sophisticated approach carries out a sequence of lattice renormalisations and matches them to each other at momentum scales differing by a factor of two. Iterating this matching enough times, gives renormalisation factors at arbitrarily high momentum scales that can then be matched accurately to continuum schemes using perturbation theory.

4. The Continuum Limit

The continuum limit involves tuning g to a critical point of the theory, where physical lengths in units of the lattice spacing diverge. In QCD, fortunately, this occurs at $g = 0$, close to which perturbation theory is reliable and we are able to check that our lattice simulation results agree with the perturbative behaviour. Details of the physics on the scale of the lattice spacing should become 'washed out' as we approach a critical point, so that different lattice formulations may have the same continuum limit. This means we can add so-called 'irrelevant' interactions at the scale of the lattice spacing which affect the local physics, but which are suppressed by positive powers of the lattice spacing and vanish in the continuum limit. Lattice actions which differ only in this way, and hence have the same continuum limit, are said to belong to the same universality class. Generally, only the symmetries and dimensionality of a lattice action dictate which universality class it belongs to.

In practice, we exploit this freedom to construct lattice actions for QCD in which irrelevant interactions are added to try to cancel the leading-order corrections in powers of the lattice spacing, in the hope that scaling can be achieved at larger values of g, i.e., larger lattice spacings, where simulations are less costly. Since taking the continuum limit involves demonstrating control over the extrapolation of results from simulations at several lattice spacings, a lot of effort is focus on these 'improved' lattice actions.

In the case of QCD, it turns out that all the lattice interaction terms which break Lorentz symmetry are irrelevant and Lorentz invariance in recovered in the continuum limit 'for free'. Lorentz invariance is called an 'accidental symmetry' of lattice QCD. Wilson showed how to preserve local gauge invariance at non-zero lattice spacing. At strong coupling, i.e., large values of g, it is possible to prove that quarks are confined as a consequence of the gauge invariance. All of the evidence from simulations suggests that there is no phase transition to a deconfined phase as g is reduced towards zero, so that confinement remains a property of the continuum limit (at zero temperature). A rigorous proof of this is yet to be found.

Demonstrating that the correct chiral and flavour symmetries of QCD are also recovered in the continuum limit has proved much more elusive. Early lattice formulations of fermion fields broke one or other of these symmetries explicitly, complicating renormalisation due to operator mixings allowed by the reduced symmetry, and calling into question whether the lattice actions really belong to the universality class of continuum QCD. Now, a class of lattice actions (which obey the so-called Ginsparg-Wilson condition) have been formulated that preserve both symmetries, and the rigorous lattice formulation of QCD is essentially complete. It actually remains to prove the Osterwalder-Schrader conditions required for a Lorentz-invariant quantum field theory, a proof which is missing for most improved actions, but this is regarded purely as a technicality.

5. The Staggered Quark Controversy

Today, lattice QCD provides the most precise determination of the strong coupling at the scale of the Z mass, $\alpha_S = g^2/4\pi$ (see the Particle Data Group tables), and this precision is backed up by careful control of all the sources of systematic error, including consistency of the results using different lattice actions. The light quark masses, however, reveal that not all sources of uncertainty are understood. Results using the staggered approach to lattice quarks gives an s quark mass around 90MeV, whereas alternative Wilson formulations give values around 120MeV, both with estimated uncertainties of order 10%. Whether this is due to a problem with the staggered action, or an underestimate of one or other of the uncertainties, is not yet understood.

Staggered quarks are one way of dealing with the problem of how to formulate lattice fermion fields with given chiral and flavour symmetries. The controversy with them concerns whether they are an acceptable solution. Using a simple finite-difference approximation to the derivative in the Dirac action yields twice as many continuum fermion species as lattice species per dimension of space-time, i.e., 16 continuum species in four dimensions corresponding to a single lattice field. These species are degenerate in mass and obviously give the wrong continuum limit for QCD.

Early on, Wilson proposed adding an irrelevant term to the lattice fermion action which gives 15 of these species masses of order the lattice cut-off so that they effectively decouple, but which explicitly breaks chiral symmetry. In the continuum limit, chiral symmetry and the correct number of flavours should be restored, but the absence of chiral symmetry on the lattice allows additive renormalisation of the quark masses and the mixing of operators of different chirality. Both complicate the calculations of phenomenologically interesting matrix elements.

Recently, it was realised that by suitably generalising the definition of chiral symmetry and using a lattice Dirac operator which obeys the Ginsparg-Wilson relation, full chiral symmetry could be realised at non-zero lattice spacing. Several such Dirac operators have been found. They are not ultra-local, i.e., they couple the fields at all lattice sites, but they are local (interactions fall exponentially with distance) when the gauge fields are sufficiently smooth. All share the feature that they can be approximated by a five-dimensional ultra-local lattice theory, the most widely used called 'domain wall fermions'. Chiral symmetry breaking is exponentially suppressed by the size of the fifth dimension and so can be easily made small. However, there is a computational price for this – a factor roughly proportional to the size of the fifth dimension. This should be more than compensated for by having the full continuum symmetries to control mixing and renormalisation.

The staggered fermion approach exploits the fact that the finite-difference approximation to the Dirac action can be diagonalised to decouple the four spinor components of the lattice fermion field. If three of the four components are dropped, the resulting fermion doubling effect is reduced to four degenerate species in the continuum limit. This is still the wrong flavour structure for QCD and, to recognise this, the four species are called 'tastes', but there is a four-fold saving in computational cost from the reduction in the number of degrees of freedom. Each flavour corresponds to a different staggered lattice field and, hence, appears in the continuum with four tastes.

The controversial step is to adjust the numerical algorithm for simulating the staggered path integral to take the fourth root of the fermion determinant that arises from integrating over the fermion fields. In perturbation theory, this gives the correct number of tastes, i.e., one. However, it cannot be described by a local field theory and there is no guarantee that it is in the QCD universality class. Current wisdom is that the correct continuum limit is obtained, provided the zero-quark-mass limit is not taken first. The price is considerable technical complexity for all but the simplest physical quantities in simultaneously controlling the extrapolation to the physical u and d quark masses and the continuum limit.

Combined with improvement that helps decouple the unwanted tastes, staggered QCD has been hugely successful in computing some physical quantities with unprecedented precision in what its proponents claim is QCD with 2+1 light flavours (degenerate u and d quarks, with masses as low as one-tenth of the s quark, and the latter at its physical mass). Where comparisons can be made, these results agree with experiment. This is certainly strong circumstantial evidence that QCD correctly describes strong-interaction physics. However, lingering unease about the theoretical foundation and the technical complexity of the calculations mean that, should a discrepancy between staggered QCD and experiment emerge, we would not be able to conclude today whether this was due to new physics or something wrong with the calculation.

6. Computational Cost

The computational challenge is to bring all sources of error sufficiently under control that results for physical quantities are useful inputs to phenomenology. The size of error that can be tolerated depends on the quantity concerned – some require few percent accuracy, others would be useful if only known to within an order of magnitude.

The number of arithmetical operations involved depends on the choice of algorithm and the size of the ensemble of gauge field configurations required to get a good enough signal-to-noise ratio for the quantity concerned. The statistical uncertainty falls with the square root of the number of independent gauge field configurations, but the ensemble size required to achieve a given statistical error in a given quantity is highly dependent on the quantity itself.

Both the cost of generating the ensemble and computing the given quantity depend on the QCD parameter values of the simulation. The dependence on them can only be determined empirically. Since it is expensive to do this at physically interesting parameter values, we have to rely on extrapolating estimates of the cost from unphysical parameter ranges where the simulations are relatively cheap. This is the reason why lattice theorists tend to be vague about how much more computer power they need!

There has been very significant improvement in algorithm efficiency over the past two years. Together with the steady increase in computer performance, this has brought the chiral regime, where the *u* and *d* quarks are near their physical masses, within reach on lattices of roughly 3fm across, with lattice spacings of around 0.12fm. Roughly, the cost of today's best algorithm for the Monte Carlo generation of an ensemble of gauge field configurations with 2+1 light flavours grows like the fifth power of the linear lattice size, the sixth power of the inverse lattice spacing, and inversely with the mass of the *u* and *d* quarks (times the extent of the fifth dimension for Ginsparg-Wilson quarks).

The most powerful computers dedicated to QCD today are the purpose-built QCDOC ('QCD On a Chip') machines, developed jointly by Columbia and Edinburgh Universities with IBM, and their commercial successor, IBM's Blue Gene. QCDOC sustains one teraflop/s for QCD production codes on 4000 processors. An integrated sustained performance of 10 teraflop/s is available with the full set of QCDOC machines currently being exploited by UK and US researchers. With this scale of resource, ensembles of several hundred independent configurations on the above lattice sizes at three or four *u* and *d* quark mass values (to enable extrapolation to their physical values) can be generated in a year using domain-wall quarks with a fifth dimension of 16 lattice units. Staggered QCD ensembles can be generated correspondingly faster, allowing a greater range of lattice spacings to be explored. Thus, continuum limits are available for staggered results, but not yet for domain wall QCD.

7. Quenched QCD

The most computationally demanding part of the QCD Monte Carlo ensemble generation is the contribution of the fermion determinant, which results formally from integrating out the Grassmann quark fields. If there are N degenerate quark flavours, this determinant is raised to the power N. The quenched approximation to QCD sets $N = 0$ and amounts to ignoring the effects of virtual quark-antiquark pairs on the vacuum. It was used extensively in the early years of lattice QCD to cut the computational cost to a level that enabled meaningful lattice sizes and spacings to be simulated with the relatively limited computing resources available at the time. Some effect of the sea quarks is built in to renormalized quantities through the scale-setting procedure described earlier and this may explain why many of the results from quenched QCD agree so well with experiment, typically at the 10-20% level.

Hamber and Parisi's first determination of the nucleon mass in quenched QCD in 1982, which obtained 900(100) MeV, is an example of this and a lesson in how crucial it is to have control of the errors in lattice QCD. Eventually, calculations by the CP-PACS Collaboration completed in 1998 achieved sufficient precision to show that the quenched light hadron spectrum disagrees with experiment, proving, as expected, that quenched QCD is wrong. This example demonstrates how computer simulation can be used as a precision tool to falsify a theory, and it is the hope behind efforts to expose signals of new physics through precise determination of the CKM matrix elements, described below.

All serious simulations today include the effects of some number of dynamical quark flavours, albeit with u and d masses larger than in the real world. As a theoretical laboratory, the ability to vary the flavour content of QCD can be an advantage in attempting better to understand the physics involved. One archetypal feature of dynamical quarks is that the flux tube, which forms between a quark and an antiquark as they are separated, should break when the energy stored in it is enough to pop the lightest quark-antiquark pair out of the vacuum. This effect is now clearly seen in simulations.

8. The QCD Vacuum

The ensemble of gauge field configurations generated in a QCD simulation provides snapshots of the QCD vacuum. The Monte Carlo algorithm picks out configurations that dominate the path integral and so it is interesting to look within them for any coherent dynamical structures that might give clues to the confinement mechanism, e.g., monopoles and instantons. Gauge field configurations with non-zero topological charge are related to the chiral properties of quarks through the Atiyah-Singer index theorem. Our understanding is growing as simulations become more realistic, but no profound insights have emerged yet.

The symmetries of QCD allow a *CP*-odd term in the action which is proportional to the density of topological charge, with a coupling parameter called the 'vacuum angle', θ. Non-zero θ allows the neutron to acquire a non-zero electric dipole moment, something for which there is a tiny experimental upper bound. It appears that θ must be correspondingly tiny and, since we know of no physics that would set such a small scale, there is a conjecture that it is precisely zero. Why this happens is the 'strong *CP* problem'.

Lattice simulations with two flavours of domain wall quarks have computed the neutron dipole moment at leading order in the topological charge density, i.e., linear in θ, and find that, to satisfy the experimental bound, $|\theta| < 10^{-11}$. This indicates that our simulations are correctly capturing the physics, even if we don't yet understand it!

Before moving on to the use of lattice QCD to help search for physics beyond the Standard Model, let's summarise its status as a tool for studying QCD. The lattice now provides a rigorous non-perturbative definition of QCD, which correctly realises all of its internal symmetries and has the correct continuum limit. (It is not restricted to Yang-Mills theories and is increasingly also being used to study some chiral and supersymmetric quantum field theories.) There has been dramatic progress in algorithms which, together with the steady advance in computer performance, has expanded the range of parameter values that can be simulated directly. This gives us the means to reliably extrapolate our results, where necessary, to the physical parameter values using effective theories, such as chiral perturbation theory, and separately to control the various sources of systematic uncertainty. After more than thirty years, there is today a high degree of confidence that directly simulating QCD with light quarks is within reach, although the hope that this would give us insight into the origin of confinement remains unfulfilled.

9. The Parameters of the Standard Model

The parameters of the Standard Model encapsulate our lack of knowledge. To the extent that experimental measurements of them give consistent results, the Standard Model works as a description of physics at the most fundamental level accessible to us. Consequently, precision measurements of its parameters test the Standard Model and any inconsistency in the results would give a clue to new physics. Lattice QCD plays a vital role in this, because some parameters, such as the CKM matrix elements, relate directly to quarks, which are buried inside hadrons. Also, many experiments employ hadronic processes. Either way, there is a need for a first-principles and precise relationship between the quarks, in terms of which the parameters are defined, and the observable properties of hadrons on which experiments may be conducted.

Lattice QCD may be used to constrain physics beyond the Standard Model in a similar way, where low-energy experimental measurements on hadrons today can be related to models of new interactions, e.g., in Grand Unified Theories.

Although we do not yet have a full set of techniques available to compute all hadronic matrix elements, we can get a lot from the simplest correlation functions. The basic ingredients are composite operators built from the lattice quark and gluon fields, which have specific quantum numbers and, when acting on the vacuum, create states that have a non-zero overlap with a given hadron. Together with currents from the electroweak effective Hamiltonian, these enable us to construct and compute correlation functions of products of composite operators and currents at different space-time points.

A two-point correlation function, which is the expectation value of a product of two composite operators at different Euclidean times, decays exponentially at large time separations with a rate proportional to the lowest energy state that both operators overlap. Evaluated at zero momentum, this energy is the associated hadron mass. The amplitude of the exponential is related to the leptonic decay constant. From a suitable combination of three-point and two-point correlation functions, it is possible to extract the matrix element of a current between hadron states. This can be related to semileptonic decay form factors. These correlation functions enable us to compute decays in which there is zero or one hadron in the final state. General techniques for multi-hadron final states are yet to be developed.

10. Hadron Spectrum

As we have seen, the computation of some hadron masses is used to fix the quark masses and lattice spacing. Once this has been done, the simulations become predictive. The extent to which the hadron spectrum matches experiment is a test of QCD and it is just this test that quenched QCD eventually failed. Results for full QCD, typically with 2+1 light flavours, have not yet achieved comparable precision for enough states, particularly baryons, to constitute such a serious test, but there is no sign of a problem. In fact, the long-standing objective to compute the nucleon mass looks to be nearly achievable – domain wall QCD results for 2+1 flavours at fixed lattice spacing appear to be heading towards the experimental number as the u and d quark masses are decreased, with much smaller error bars than previously. This is encouraging a sustained new effort to investigate in detail the baryon spectrum, structure and decays.

In addition, QCD predicts many states which have not yet been identified experimentally and lattice simulations may aid the search for these. The most interesting unobserved hadrons are glueballs. In quenched QCD, Morningstar and Peardon have obtained what is widely accepted to be the complete continuum glueball spectrum up to about 5 GeV. However, this is little more than a theoretical 'tour de force', because glueballs are expected to mix strongly with singlet meson states in full QCD, leading to a quite different spectrum in reality. The lattice calculation of singlet states with dynamical quarks is very challenging, because of poor signal-to-noise ratios and the need to compute many more states than usual to control mixing.

One dramatic demonstration of the predictive power of lattice QCD was the prediction of the mass of the B_c meson by the HPQCD, Fermilab and UKQCD collaborations just months before it was observed experimentally. The lattice prediction agreed with the experimental mass and has smaller error bars.

11. Determination of the CKM Matrix

The strengths of the electroweak-mediated decays and mixings of quark flavours include some of the most poorly known parameters of the Standard Model. If there are only three generations of quarks and leptons, these parameters form a 3×3 unitary matrix, the CKM matrix, and so their values are inter-related. The relationships are usually expressed in terms of unitarity triangles, in which the vanishing of the scalar product of two different columns of the matrix is represented as three complex numbers making up a triangle in the complex plane. Whether or not the experimental values for the CKM parameters are consistent with such unitarity triangles, has long been the major objective of the B

physics programme. The area of a unitarity triangle is proportional to the size of *CP* violation resulting from the one independent complex phase in the CKM matrix. Thus, this experimental programme has been seeking clues for physics beyond the Standard Model and insight into the *CP* violation mechanism.

As it has turned out, all the evidence gathered so far points to the Standard Model working very well. Measurements of the CKM parameters from different experiments agree and the unitarity triangles close. This means that yet more precise measurements of the CKM parameters are needed. In many cases, the major source of uncertainty is not experiment but theory, and particularly the uncertainties coming from the lattice computation of QCD matrix elements. This is a major challenge for lattice QCD in the immediate future.

Given the recent progress towards correctly incorporating u, d and s sea quarks in simulations, with masses close to their experimental values, the main technical difficulty in computing matrix elements relevant to CKM studies is the incorporation of heavy quarks. Note the t quark is so massive that all of its effects may be treated in perturbation theory. So the issues concern c and b quarks, and are different for each of them.

With lattice cut-offs (due to non-zero lattice spacing) typically in the range 2-3GeV, and towards the lower end of this range when dynamical light flavours are included, the b quark is too massive to be simulated directly. There are two options. One is to use a heavy-quark or non-relativistic effective action for the b quark and include only the effects of valence b quarks. This requires some care, because such actions do not have a well-defined continuum limit. It is necessary to choose parameters in the action and improvement terms so that the results scale. The alternative is to simulate the b quark on a fine enough lattice, and consequently an unphysically small volume, then carry out a sequence of finite-size scaling steps to iterate the results to larger volumes where the box size does not significantly distort the light-quark physics.

The c quark mass is on the margin of what can be simulated directly with a relativistic action on today's lattice sizes. It is also not usually a good approximation to treat it non-relativistically. Given the precision experimental results coming from CLEO-c, charm physics presents a serious challenge to lattice QCD in terms of finer lattices and/or improved relativistic actions, and in terms of achieving higher precision results. It is likely to be an important testing ground over the next few years.

12. Leptonic Decays

The electroweak decay rate of a meson into a lepton and associated antineutrino is proportional to a (squared) CKM matrix element times the corresponding leptonic decay constant. The latter may be extracted from two-point lattice correlation functions. The π and K decay constants have been computed in the continuum limit of 2+1-flavour staggered QCD. The results are amongst the most precisely determined, with uncertainties of 2%, and are in complete agreement with experiment.

The D and D_s decay constants have recently been measured by CLEO-c with errors a little less than 10%. For the reasons mentioned above, the lattice calculations of them are not fully under control, with uncertainties of 10% or more. At this level, the results are in complete agreement, as is the SU(3) flavour-breaking ratio of the D_s to D matrix elements, where some systematic effects are likely to cancel, with a value of 1.24 ± 0.07 from lattice QCD and 1.26 ± 0.11 ± 0.03 from CLEO-c.

The Belle experiment obtained the first measurement of the B decay constant this year. This is interesting, because it is an important test of lattice calculations with heavy quarks and because the leptonic decay itself is sensitive to a charged Higgs. So comparing the measurement with the Standard Model prediction could show up new physics. Currently, the results agree, but the 20% experimental uncertainty and 10% lattice uncertainty are too big to conclude anything yet. These will be reduced in time, so this is a quantity that will continue to feature prominently in lattice heavy-quark phenomenology.

13. *CP* Violation in the Neutral Kaon System

The mixing of neutral kaon states with even and odd *CP* quantum numbers results in a long-lived physical kaon (mostly *CP* odd), which decays predominantly into three pions and occasionally into two, and a short-lived physical kaon (mostly *CP* even) which does the reverse. The *CP* odd eigenstate decays into two pions with a strength ε', violating *CP* conservation, while the admixture of the *CP* even eigenstate in the long-lived kaon is of size ε. The parameters ε' and ε are, respectively, the measures of 'direct' and 'indirect *CP* violation' in the kaon system. Apart from wanting to understand the mechanism for *CP* violation, which appears so rarely in Nature, yet is the key to the matter-antimatter asymmetry of the Universe, the value of ε is an important constraint on the unitarity triangle.

The QCD matrix element which describes the mixing of neutral kaons at leading order in the electroweak effective Hamiltonian is called B_K. Apart from known factors it is related to ε through a particular product of CKM matrix elements. Given the existing experimental constraints on these and assuming that the unitarity triangle closes, the value of B_K should be 0.68 ± 0.10. This matrix element has been well studied in quenched lattice QCD over many years and a consistent continuum limit has been obtained using several formulations. The result obtained is 0.81 ± 0.04 ± 0.08, where the last error is an estimate of the quenching effect. This is consistent with new results from 2+1-flavour QCD at fixed lattice spacing. The first 5% error is indicative of the uncertainties in lattice computations of such quantities today. For lattice QCD to have a real impact on CKM matrix tests of the Standard Model, these need to be further halved.

Direct *CP* violation in the neutral kaon system is one of the 'holy grails' of lattice QCD. It is quantified by the parameter ratio ε'/ε, which is now well-known experimentally, but it has proven to be the most difficult computation in lattice QCD that has been seriously undertaken to date. This is due to several factors. Firstly, the matrix elements contributing have similar magnitudes and opposite signs. So there is a delicate cancellation, which requires good numerical control to see a signal. Secondly, the relevant kaon decay has two pions in the final state, a situation where, initially, we lacked the theoretical tools.

Now, although we have a method to compute the amplitude directly, to work well numerically it requires much bigger lattices than are currently available. So chiral perturbation theory has been used to relate the two-pion amplitude to matrix elements whose final state is one pion or the vacuum.

These methods were employed a few years ago in quenched QCD by the CP-PACS and RBC Collaborations. They both obtained results that were of the same magnitude as the experimental value, but opposite sign! Although the former was very encouraging after years with no discernable signal, the latter was a puzzle. It is now understood that this is because, in quenched QCD, there is mixing with unphysical operators, which renders the calculation intrinsically ambiguous. Thus ε'/ε is a quantity that demands a full QCD simulation and this effort is now underway using domain wall QCD.

14. Neutral *B* Meson Mixing

The electroweak mixings of neutral *B* mesons, both B_d and, this year from CDF, B_s, have also become experimentally accessible. The 'box diagrams', which dominate the theoretical calculation, contain a loop involving virtual *t* quarks and *W* bosons, and could be sensitive to physics beyond the Standard Model. So this is another quantity where comparison between the Standard Model prediction and experiment could reveal a clue to new physics.

Usually, the mixing is expressed as the mass difference between the neutral *B* meson and its antiparticle. This is related to the *B* meson equivalent of B_K, called B_B, and the *B* leptonic decay constant, via a combination of CKM matrix elements. Both of these QCD matrix elements have been the subject of lattice computations over many years, although they have been complicated by the difficulties of correctly incorporating *b* quarks. In the ratio of the B_s to B_d mass differences many factors cancel, including some systematics affecting the lattice calculations, so that a theoretical uncertainty of around 5% is possible. This is beginning to provide a tightening of the constraint on the unitarity triangle, but, again, a reduction in the lattice uncertainty by a factor of two is needed for a big impact.

15. Semileptonic Decays

The semileptonic decays of *K*, *D* and *B* mesons provide a more elaborate example of the interaction between lattice QCD and experiment, in the quest for tighter constraints on CKM matrix elements. The reason is that these decays involve a momentum transfer, q^2, to the leptonic system (via the virtual *W*) and, consequently, the amplitudes have non-trivial q^2 dependence. The momenta that can be accurately represented on a finite lattice are bounded above (by the lattice spacing) and below (by the box size). So our ability to correctly simulate the decay process depends on the kinematics and the simulation parameters, adding an additional systematic effect. Where the full kinematic range is not spanned by the lattice momenta, some model dependence is introduced due to the need to fit the lattice data to model form factors.

Now that the chirally-symmetric lattice formulations are enabling a new attack on kaon physics, semileptonic *K* decays are being computed as a test of lattice QCD. These so-called *Kl*3 decays are well-determined experimentally, as is the associated CKM matrix element. Continuum limits for 2+1-flavour QCD are at least another year away, but success will give a big boost to lattice studies of kaon physics.

The maximum recoil of the kaon in $D \to K\ell\nu$ semileptonic decays is around 1 GeV and so the lattice data are able to span the entire kinematic range. The associated CKM matrix element is well measured and CLEO-c is producing high precision form factor measurements. So, D semileptonic decays are also becoming important test cases for lattice QCD, with the added challenge over K decays that the c quark must be simulated correctly.

While semileptonic decays of B mesons into D mesons can be analysed using heavy-quark effective theory, where the assumption that the c quark is heavy seems to work well enough as a leading-order approximation fixing the normalisation of the form factors, the semileptonic decays of B mesons into pions and ρ mesons have no such approximate symmetry. Consequently, lattice QCD provides the only theoretical tool for computing the form factors and, here, both the light and heavy quark physics must be simultaneously under control. Furthermore, the lattice kinematic range is restricted to near zero recoil, i.e., high q^2, currently above 16 GeV^2. This issue has been taken up by experimentalists, who now quote partial decay widths in this range in order to avoid significant model dependence when extracting the corresponding CKM matrix element. With continuum results for the form factors using 2+1 staggered flavours from the HPQCD and Fermilab Collaborations, the lattice results are now a crucial input to the unitarity triangle analyses.

16. The Impact of Lattice QCD on Flavour Physics

However, while making enormous progress in reducing our uncertainties in the values of the Standard Model parameters, the combined assault on the unitarity triangle constraints by experimentalists and lattice theorists, described above, has not yet delivered the hoped-for clue to physics beyond. Even, this year's first observation of oscillations in the B_s system and measurement of the corresponding mass difference, turned out to be fully consistent with the Standard Model. We have to conclude that flavour physics, including *CP* violation, is governed predominantly by the CKM picture. Sadly for lattice QCD, greater precision, which probably means at least halving the uncertainties in even the best determined quantities, will be necessary if it is going to lead to a discovery of new physics from precision tests of the flavour sector of the Standard Model.

17. Constraints on Physics Beyond the Standard Model

Experimental constraints on models of new physics are possible, in principle, from measurements of rare and forbidden decays in the Standard Model. Rare decays are those whose leading contributions in perturbation theory are loop diagrams, which may be sensitive to contributions from undiscovered virtual particles. This would be signalled by a difference between the Standard Model prediction and experiment. Forbidden decays are constrained by lifetime bounds. Both processes typically have amplitudes which, at leading order, are the product of QCD matrix elements and model-specific factors. Experimental measurement of the amplitude, combined with lattice QCD results for the dominant matrix elements, then can be used to constrain the model-dependent parts.

The example of a rare decay where both experimental and lattice results could be brought together is $B \to K^*\gamma$, which may receive loop contributions from supersymmetric particles. This has been rather neglected in recent lattice studies, but a significant improvement over early predictions should now be possible.

The lower bound from SuperKamiokande on the proton lifetime in the kaon channel, due to triplet Higgsino exchange which appears favoured by supersymmetric Grand Unified Theories, now exceeds 10^{33} years. The matrix elements of all the dimension-6 baryon-number violating operators allowed by the Standard Model symmetries, between proton and kaon states, have been computed in both quenched and 2-flavour lattice QCD. Combined with the lifetime bound, these results can be used to place a lower bound on the Higgsino mass. Already, this rules out the minimal SU(5) model and the bound is close to ruling out others. There is a strong dependence of the lattice results on the lattice spacing scale, so it is important to repeat these calculations for 2+1-flavour QCD with good control over the light-quark physics. The promise of lattice QCD playing an important role in Beyond Standard Model building is beginning to be realised.

18. Lattice QCD as a Phenomenological Tool

Over 30 years of theoretical effort has resulted in a formulation of lattice QCD which correctly realises all the continuum symmetries, and allows full control over all sources of systematic and statistical error in simulations. This has been accompanied by dramatic improvement in the efficiency of numerical algorithms and continual increase in computer performance through Moore's law. Although we are on the verge of exploiting all of these developments to carry out simulations of QCD with near-realistic u, d and s quark masses, without more breakthroughs, this programme will take several more years to carry out, even for the relatively straightforward set of quantities described in this article.

Accomplishing this much alone should have a big impact on CKM studies that are currently limited by the precision of lattice QCD. However, the combined effect of the availability of the full continuum symmetries to simplify renormalisation and operator mixing calculations (compared with the older lattice formulations where the reduced lattice symmetry introduces daunting complications) and near-realistic parameter values should motivate the computation of a much more ambitious range of matrix elements. We can expect a rapid expansion of the reach and relevance of lattice QCD, as the increasing experimental focus on discovering the physics beyond the Standard Model motivates the calculation of non-Standard Model processes.

Furthermore, our growing ability to simulate quantum field theories with massless fermions will open up new non-perturbative physics to computational study, e.g., in supersymmetric and chiral gauge theories. Just as soon as there is any hint of supersymmetry in the real world to justify the cost, lattice simulations of them will begin in earnest, and lattice QCD will become just one part of a much bigger enterprise.

CHAIRMAN: R.KENWAY

Scientific Secretary: D. Antonio

DISCUSSION I

- *Preghenella:*

How many CPU's does the six teraflops QCDOC computer have?

- *Kenway:*

About 25,000.

- *Susskind:*

How certain are you that the up quark mass is non-zero and how sure are you that the definition of the up quark mass that you use is the same as the one you would use in studying the CP problem?

- *Kenway:*

I showed some results in my lectures which certainly indicate that the up quark mass is non-zero, if you believe the error bars. Our ability to control the calculation of the up quark mass is dependent on controlling the extrapolation of our data down to very light quark masses. Today this is a large extrapolation and we don't really have compelling evidence that our data are in the chiral regime, where we can rely on chiral perturbation theory to guide the extrapolation.

As for the definition of the up quark mass, it is normally defined using the Axial Ward Identity. Is there another definition that you think we should be using?

- *Susskind:*

I don't know the answer, I don't know if anybody has ever done the analysis. The only actual fully experimental indication of a vanishing up quark mass, in the sense in which I mean, is that the CP angle is zero.

- *Kenway:*

Exactly. Lattice calculations of the neutron electric dipole moment show that the angle is very small, certainly consistent with zero. But that doesn't help us very much as we also find that the up quark mass is non-zero. So, at the moment, there are no clues from Lattice Quantum Chromodynamics (LQCD) as to what the answer to the strong CP problem is.

- *Csanad:*

How do you compute the quark masses from LQCD?

- *Kenway:*

The way to compute the quark masses is to start with a guess for the values of the bare masses in the Lagrangian. From the simulation you compute a set of hadron masses, such as the pion mass for the up and down quarks, or the kaon mass for the strange quark, then adjust the input parameter values until the output hadron masses agree with experiment. At that point, your lattice simulation has the correct bare input masses. But, what you want to know are the renormalised quark masses. Typically we use the renormalised Axial Ward Identity to compute what the simulation sees as the sum of the quark masses (expressed as a ratio of matrix elements) associated with the flavours which make up the axial vector current.

- *Csanad:*

If one can calculate the quark masses this way, why don't you use that when calculating the critical temperature or deconfinement?

- *Kenway:*

That's exactly what should be done. The calculations at non-zero temperature and baryon density are somewhat more difficult than the calculations I have described. They are inevitably behind in terms of running simulations close to the physical quark masses, because we have to get the results from the zero temperature simulations first.

- *Lockyer:*

What does lattice QCD have to say about the top quark?

- *Kenway:*

Just about nothing, for two reasons: a) there is no need to use non-perturbative methods to do calculations involving the top quark, because it is at an energy scale where perturbation theory (PT) works, and b) even if you tried to put the top quark on the lattice, the lattice spacing would have to be very small to ensure that the cutoff is significantly above the top mass. Consequently, the lattices would be very small and we would not be able to incorporate directly the effects of any of the other quarks. The problem with simulating all the flavours of quarks is the huge range of mass scales. You would need a very small lattice spacing and a large lattice to incorporate the whole range, which, at the moment, is beyond us.

- *Lockyer:*

So can I extrapolate that to all fundamental higher mass particles that are found? Is LQCD out of business?

- *Kenway:*

No, it depends what you're trying to do. The problem I described was one in which I wanted to simulate the top quark as well as the effects of the other quarks. In a world in which I only care about the top quark then I could use LQCD, but I would be using a difficult technique to solve a problem that would be much easier in perturbation theory.

If you mean a different heavy particle, from a different theory, then you would have to address all the problems associated with finding a critical point of the theory so that the continuum limit can be taken. Assuming you can do this, there is no reason why you cannot use lattice methods at higher energy scales.

- *Colangelo:*

You have been very positive about the rooting procedure and mentioned important developments during the last year – can you be more specific?

- *Kenway:*

The reason that it became such a hot topic is that, in the simulations that have been done over the last four or five years, the staggered approach was winning in terms of computational cost at light quark masses and a big industry developed. Most of the results that I quote in my lectures come from staggered simulations. However, there was a growing concern that using the 4th root of the determinant of the Dirac matrix to describe the sea quarks might not give a well defined quantum field theory. The problem is that you cannot write a local version of the rooting procedure on the lattice. However, in the last year Shamir has carried out a Renormalisation Group Blocking analysis of the rooted staggered theory and has shown that these non-locality and non-unitarity artifacts do vanish as you approach the continuum limit. So they are simply problems of the lattice theory itself. There is one caveat, however, that the quark masses must be kept away from zero as the continuum limit is approached. There are some cases where it is known that if you take the chiral limit first then you get the wrong answer. So, there appears to be no theoretical problems with the rooting procedure in QCD, but you have to make sure you really are controlling all the systematics as you approach the continuum limit. Also the chiral perturbation theory is very complicated for rooted staggered quarks (sometimes involving fits to as many as 40 free parameters), so taking the chiral limit is a technically demanding.

- *Susskind:*

Does the staggered approach, as Kogut and I first used it, to describe an SU(2) flavour world have these problems?

- *Kenway:*

No. The problem is the strange quark. Your approach works fine for a model world with SU(2) flavour symmetry (degenerate up and down quarks only). The strange quark requires a formulation with one single flavour and this requires rooting unless you use Wilson-like formalisms, which do not suffer in this way.

- *Castellani:*

You mentioned that other field theories can be investigated on a (finite) lattice. Could you comment on lattice formulations for gravity actions? Do you know of analogues of the Wilson action for gravity?

- *Kenway:*

There was a lot of activity 10 or 15 years ago in discrete models of gravity. At the time, the problem they hit was finding a critical point of the theory in order to take the continuum limit. Most attempts failed to get a sensible continuum limit of the model; the universes tended to scrunch up, becoming fractal-like, and didn't look anything like you'd hope.

- *Machado:*

One more successful lattice type approach is causal dynamical triangulations in which an appropriate continuum limit has been found.

- *'t Hooft:*

If you take the 4th root of the determinant of the Dirac matrix, what do you do with negative eigenvalues? Isn't that a fundamental ambiguity?

- *Kenway:*

Yes, it certainly is a problem for some formulations, such as Wilson fermions. For the staggered approach, there are no negative eigenvalues: the eigenvalue spectrum is bounded below by the quark mass. There is no such lower bound for Wilson quarks.

- *'t Hooft:*

I also have a comment. Conceptually it is easy to put fermions on a lattice such that all chiral and flavour symmetries are conserved: put the gauge fields on the lattice, find a unique continuation of the gauge field in the continuum in between the lattice sites and put the fermions in the continuum. I do not understand the difficulties lattice people seem to have with this. The apparent problem is that one might find it difficult to find a unique continuation of the gauge fields. But, if you think about that, it's not so difficult at all.

- *Kenway:*

There are moves in that general direction without going as far as you have suggested, for example in multigrid approaches, using a smaller lattice spacing for the gauge fields to ensure they are smooth. It may be that the lattice community has not responded to your question, but is edging towards it via a different route. The Ginsparg-Wilson approach, which describes quarks with nice properties, requires that they sit on smooth gauge backgrounds to ensure locality. On coarse lattices, smooth backgrounds may be obtained using Renormalisation Group blocking of the gauge fields. That goes some way to what you are suggesting: making a smooth gauge configuration, but with discrete (though not ultra local) fermions sitting on top of it.

- *Aschieri:*

You have discussed how internal symmetries can be implemented on a lattice. Can you say anything about putting electromagnetism on a lattice and checking electric-magnetic (EM) duality?

- *'t Hooft:*

I think that situation is very similar to the 2-dimensional Ising model where you can read off what duality is on the lattice. Duality in the 2-dimensional Ising model is the same as EM duality in 4 dimensions.

- *Baldini*:

How difficult is it to evaluate nucleon form factors on the lattice?

- *Kenway:*

There are no difficulties in principle and the calculations are similar to those I describe in my lectures. The renormalisation and operator mixing technology is all there for Wilson fermions The problem is that it's important to be in the light quark regime, because the dynamical quark simulations so far, with relatively heavy quarks, have not shown significant deviation from the quenched results, which disagree with experiment.

CHAIRMAN: R.KENWAY

Scientific Secretary: D. Antonio

DISCUSSION II

- Karliner:

Would it be possible to study monopole-induced proton decay on the lattice?

- Kenway:

It would be very difficult to simulate the monopole directly, but if you can express the effects of the monopole induced interactions as a low energy theory, then the amplitude for proton decay – at leading order – will be a combination of the dimension six operators whose matrix elements we are calculating in LQCD. This wouldn't be a first principles simulation of the process, but it would give you a model independent estimate of the QCD contributions to the process.

- Sonnenschein:

What is the state of the art for the determination of the Luescher term in lattice simulations? What is known about the Regge trajectories of glueballs and mesons, can the intercept be determined exactly?

- Kenway:

The string tension can be computed accurately out to quite large distances and I believe that it fits the Luescher form perfectly once you take into account discretisation effects at short distances.

- Sonnenschein:

Is there any principle whereby you can determine whether a bosonic string is right or a supersymmetric string is right? What is the number of transverse directions?

-Korthals-Altes:

M. Teper and H. Meyer have – during the last eight years – measured the Luescher term in 3 and 4 dimensions. There were no surprises. The precision of the data on the string tension is around 1%. The precision on the Luescher term is not quite this good, however.

- Csanad:

When will you be able to calculate the products of the collision of two nucleons on the lattice? Is it needed and, if so, when do you think that it will be possible?

- Kenway:

At present this is beyond our theoretical technology. There are several reasons for this. Consider just one nucleon: it is now becoming possible to put one nucleon on the lattice and get an accurate set of results for its mass spectrum and decays etc. (as I describe in the lectures). In order to do this, you need flavour and chiral symmetry to control the calculation, so efforts with the rooted staggered approach haven't been able to do it, because it is too complicated to define a nucleon in this approach. The chirally symmetric Wilson approach, which is fairly recent, can simulate a single nucleon, but simulations with this action have only recently been started. In this formulation, you can measure nucleon excited states, decays and structure (in the sense of parton distributions), but we need to be closer to the chiral limit than current simulations to get meaningful results. This will happen within the next two years.

If you want to put two hadrons in a box then there are serious theoretical issues to address. The simulations are in Euclidean space, so everything is real, and phase shifts, for example, cannot be extracted directly. Luescher realised that, if you put more than one hadron in a box, then the system will have a set of discrete energy levels and if you vary the size of the box, you can relate the way the spectrum changes to phase shifts and scattering lengths. It is technically very complicated and is starting to be done for relatively simple systems, e.g. Kaon decay to two pions.

A collision of two hadrons is an even more difficult task. If you want to do this at high momentum then you need to have a very fine lattice spacing. Personally, I don't think it will be possible for many years.

- *Karliner:*

There is very strong experimental evidence for at least one 4-quark meson – X(3872) – with mass exactly equal to the DD* threshold. There are also excited D_s states seen by B factories and by CLEO which are extremely narrow and do not fit into the standard quark-antiquark picture. What can we expect from the lattice regarding the understanding of these exotic states and possible predictions for additional exotics?

- *Kenway:*

The spectrum of these states has been computed on the lattice, following their experimental discovery. This is actually quite similar to the situation I described earlier where you have two hadrons in a box, i.e. would the 4-quark state fall apart into 2 hadrons, or is it bound? Actually the calculations that have been done varied the boundary conditions rather than the size of the box, but this has a similar effect on the spectrum. The calculations have to be considered as exploratory due to the fixed lattice spacing and large sea-quark masses etc, but they did not see any evidence for bound 4-quark states. The concern with the D_s mesons was that these were very narrow states that the quark model would have placed above the threshold for decay into a D and a Kaon, for example.

- *Karliner:*

It's more than that. They can decay into D_s and a pion, for example. A quark-antiquark state cannot decay into such products as it cannot have isospin, charm and strangeness; it's impossible. Either isospin is broken, or it is not a quark-antiquark state.

- *Kenway:*

According to the quark model some of the D_s states would be above threshold and narrow, which is puzzling. When you do the LQCD calculation, you find that the quark model is wrong and they are below threshold, which presumably explains why they are narrow.

- *Colangelo:*

In the calculation of the hadronic vacuum polarisation contribution to the muon (g-2), photons were not taken into account. Do you see any hope for doing this in the near future?

- *Kenway:*

No. Again this is a problem of scales. It is not going to be easy to have a first principles simulation of the QED effects on the same lattice as QCD. You have to work with QED perturbation theory to organise the calculation, and compute the hadronic matrix elements in LQCD. The result I showed is an early first attempt to calculate the hadronic contribution to the vacuum polarisation at leading order in QED. It actually neglected some contributions from disconnected diagrams that should have been included, so there is still some way to go before the calculation is under control. The result is interesting, because it appears that the theoretical uncertainty is comparable to the experimental error, in the estimate based on the hadron production from electron-positron annihilation. So there is some hope that LQCD will eventually enable us to reduce the hadronic uncertainties in the muon (g-2).

- *Colangelo:*

One problem is that the bare parameters when QED is included would be different from the bare parameters when you have QCD alone. I see this as a hard theoretical problem.

- *Kenway:*

Yes, in principle, you would need to retune the bare parameters but it is a small effect.

- *Susskind:*

I think that one of the problems when trying to deal with multi hadrons is that you are on a Euclidean lattice and therefore at finite temperature. It is therefore possible to ionise these things and the entropy, or statistical weight, may be dominated by things which are not bound. Does that suggest that if you really wanted to solve this problem that you might want to invest more time in Hamiltonian lattice gauge theory methods?

- *Kenway:*

The sorts of calculations I have described are aimed at producing results that are valid at zero temperature. Although in a finite box you are at non-zero temperature, if you take the box size to infinity, with all dimensions equal, you converge on to zero temperature.

Regarding Hamiltonian approaches, you may well be right that this provides a tool to address the multi hadron problem. The difficulty that these approaches ran into was taking the continuum limit: there is a roughening transition in the way, as you take the bare coupling to zero, and this is a serious impediment to pursuing such an approach.

- *Karliner:*

I have a comment about whether or not the roughening transition you describe is a problem in practice. In the pure glue sector, i.e. without fermions, glueball spectrum calculations were done in the Hamiltonian formulation with very accurate results, which agree with the best Euclidean Monte Carlo calculations of today. I was involved in some of these calculations. I don't think the roughening transition is so much of a problem in practice.

-.*Korthals-Altes:*

It is an essential singularity, therefore a very weak singularity.

- *Kenway:*

The results that were obtained from the strong coupling expansion stood the test of time well and for many years were in better agreement with experiment than the Monte Carlo simulation results, although this is no longer the case.

- *Neri:*

In your lectures you talked about the impact of LQCD on the Unitarity Triangle. Do you think, in the near future, that LQCD will have a major impact on the precise CP measurements or vice versa?

- *Kenway:*

There is a race on between LQCD trying to reduce the theoretical uncertainties and the experimentalists trying to reduce the experimental uncertainties. At the moment it is a synergistic relationship. For the sorts of quantities that I was describing this morning, the theoretical uncertainties are typically larger than the experimental errors, so the pressure is really on the lattice theorists to do better. The next generation of simulations will improve this situation. I don't envisage dramatic changes, however. The errors may be halved, but the experimentalists are still likely to be ahead. I expect that theoretical uncertainties will dominate for the next few years. Unless there are some indications that greater precision may expose a discrepancy in the Unitarity Triangle, it will be hard to motivate the effort and cost of doing yet more precise lattice calculations of those quantities.

The Landscape and its Physics Foundations How String Theory Generates the Landscape

Leonard Susskind

Department of Physics, Stanford University
Stanford, CA 94305-4060, USA

Abstract

If the observable universe really is a hologram, then of what sort? Is it rich enough to keep track of an eternally inflating multiverse? What physical and mathematical principles underlie it? Is the hologram a lower dimensional quantum field theory, and if so, how many dimensions are explicit, and how many "emerge?" Does the Holographic description provide clues for defining a probability measure on the Landscape?

The purpose of this lecture is first, to briefly review a proposal for a holographic cosmology by Freivogel, Sekino, Susskind, and Yeh (FSSY), and then to develop a physical interpretation in terms of a "Cosmic Census Taker:" an idea introduced in reference [1]. The mathematical structure–a hybrid of the Wheeler DeWitt formalism and holography–is a boundary "Liouville" field theory, whose UV/IR duality is closely related to the time evolution of the Census Taker's observations. That time evolution is represented by the renormalization-group flow of the Liouville theory.

Although quite general, the Census Taker idea was originally introduced in [1], for the purpose of counting bubbles that collide with the Census Taker's bubble. The "Persistence of Memory" phenomenon discovered by Garriga, Guth, and Vilenkin, has a natural RG interpretation, as does slow roll inflation. The RG flow and the related C-theorem are closely connected with generalized entropy bounds.

1 Introduction

Of all the "String Inspired" cosmological scenarios, only one seems to me to have an element of inevitability to it. The facts and principles that drive it are as follows:

- Observational evidence supports the existence of a period of slow-roll inflation during which the universe exponentially expanded by a factor no less than e^{50}. The universe grew to a size which is at least $1,000$ times larger (in volume) than the portion which is observable.
- A small residual vacuum energy of order $10^{-123}M_p^4$ remained at the end of inflation and now dominates the energy density of the universe. If this situation persists, then not only is the universe at least $1,000$ larger than what can be seen; it is $1,000$ larger than what can ***ever*** be seen [2].
- String Theory apparently gives rise to an immense Landscape of de Sitter vacua [3][5][4][6] with a very dense "discretuum" of vacuum energies. None of these vacua are absolutely stable: each can decay to vacua with smaller cosmological constant .
- Black Hole (or Observer) Complementarity, [7] [8] [9], and the Holographic Principle, [10] have been confirmed by string theory, at least in a certain wide class of backgrounds [11][12][13]. The implication is twofold. On the one hand, observer complementarity requires the identification of a causal patch; conventional quantum mechanics only makes sense within such a patch. The Holographic Principle requires that a region of space be described by boundary degrees of freedom whose number does not exceed the area, measured in Planck Units.
- Inflation, if it lasts long enough, has a tendency [14] to populate the Landscape with a great diversity of nucleated "pocket universes."

The first two items imply that all of observable cosmology consisted of a roll from one value of the vacuum energy (probably no bigger than $10^{-14}M_p^4$), to its final current value. How and why the universe began with such an unnatural energy density is not explained by any standard theory, but the Landscape suggests the following guess: At some point in the remote past the universe occupied a point on the Landscape with a much higher vacuum energy, perhaps of order one in Planck units. Rolling, unimpeded, to a vacuum energy of 10^{-14} without getting stuck in a local minimum is unlikely. (Think of rolling a bowling ball from the top of Mount Everest to sea level.) It is far more likely that the universe would get stuck in many minima, and have to tunnel [15] multiple times, before arriving at the very small vacuum energy required by conventional slow-roll inflation. We will not dwell on Anthropic issues in this paper, but I would point out that a long period of conventional inflation appears to be required for structure formation [16]. The argument is similar to the well-known Weinberg argument concerning the cosmological constant.

These considerations strongly suggest that the period of conventional slow-roll inflation was preceded by a tunneling event from a previous neighboring vacuum. In other words, the observed universe evolved by a sudden bubble nucleation from an ***"Ancestor"*** vacuum, once removed on the Landscape. It seems obvious that one of the next big questions for cosmology will be to find the theoretical and observational tools to confirm or refute the past existence of an Ancestor, and to find out as much as we can about it. If we are lucky and the amount of slow-roll inflation that followed bubble nucleation is as small as observational evidence allows, then we have a chance of seeing features of the Ancestor imprinted on the sky [16]. The two smoking guns would be:

- Negative spatial curvature: bubble nucleation leads to a negatively curved, infinite, FRW universe.
- Tensor modes in the CMB, but only in the lowest harmonics. Although the vacuum energy subsequent to tunneling (during conventional slow-roll inflation) was almost certainly too small to create observable tensor modes, the cosmological constant in the Ancestor was probably much larger. During the Ancestor epoch, large tensor fluctuations would be created by rapid inflation. A tail (diminishing rapidly with l) of those fluctuations could be visible if the number of slow-roll e-foldings is minimal.

If the observational evidence for an Ancestor is weak, so is the current theoretical framework. To many of us, eternal inflation, bubble nucleation, and a multiverse, seem all but inevitable, but it is also true that they have inspired what Bjorken[1] has called "the most extravagant extrapolation in the history of physics." Eternal inflation leads to an uncontrolled infinity of "pocket universes" which we have no good idea how to regulate–the inevitable has led to the preposterous. In my opinion, this situation reflects serious confusion, and perhaps even a crisis.

[1] James Bjorken, private communication.

Eternal inflation is not the only extravagance that we have had to tame in recent decades. I have in mind the fact that a naive but very compelling interpretation of black holes seemed, at one time, to imply that a black hole can absorb an infinite amount of information behind its horizon [17]. By feeding a black hole with coherent energy at the same rate that it evaporates, it would seem that an infinity of bits could be lost to the observable world.

I believe these two crises may be related. In both cases the infinities result from "cutting across horizons" and attempting to describe global space-like surfaces with ***independent*** degrees of freedom at each location. The cure is to focus attention on a single causal region, and to describe it by a Holographic set of degrees of freedom [7] [18].

In FSSY [19] the authors described one such holographic framework–call it holography in a hat–based on mathematical ideas that have become familiar from String Theory. At the same time, Shenker and collaborators [1] have developed an intuitive "gedanken observational" approach based on a fictitious observer called the "Census Taker." My purpose in this lecture is to explain the close connection between these ideas.

2 The Census Bureau

Let us begin with a precise definition of a causal patch. Start with a cosmological space-time and assume that a future causal boundary exists. For example, in flat Minkowski space the future causal boundary consists of $\mathcal{I}^+$ (future light like infinity) and a single point, time-like-infinity. For a (non-eternal) Schwarzschild black hole, the future causal boundary has an additional component: the singularity.

A causal patch is defined in terms of a point $\underline{\mathbf{a}}$ on the future causal boundary. I'll call that point the "Census Bureau[2]." The causal patch is, by definition, the causal past of the Census Bureau, bounded by its past light cone. For Minkowski space, one usually picks the Census Bureau to be time-like-infinity. In that case the causal patch is all of Minkowski space as seen in figure 1.

[2]This term originated during a discussion between myself and Steve Shenker in a Palo Alto Cafe. Neither of us will admit to having coined it first, but it wasn't me.

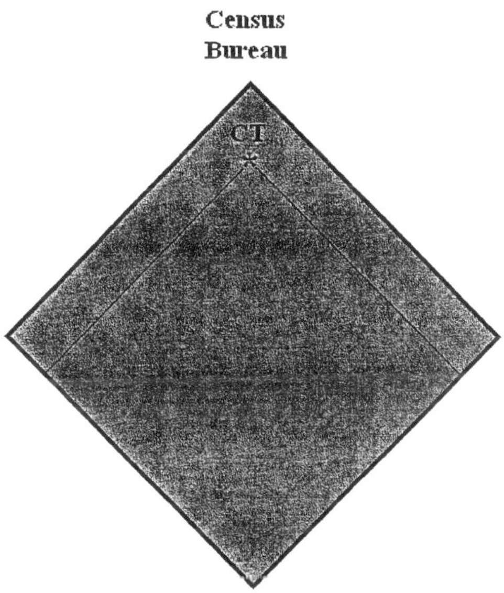

Figure 1: Conformal diagram for ordinary flat Minkowski space. The causal patch associated with the "Census Bureau" is the entire space-time. A Census Taker and his past light-cone are also shown.

In the case of the Schwarzschild geometry, **a** can again be chosen to be time-like-infinity, in which case the causal patch is everything outside the horizon of the black hole. There is no clear reason why one can't choose **a** to be on the singularity, but it would lead to obvious difficulties.

The term "Census Taker" was introduced [1]) to denote an observer, at a point inside a causal patch, who looks back into the past and collects data. He can count galaxies, other observers, hydrogen atoms, colliding bubble-universes, civilizations, or anything else within his own causal past. As time elapses the Census Taker sees more and more of the causal patch. Eventually all Census takers within the causal patch arrive at the Census Bureau where they can compare data.

De Sitter space has the well known causal structure as shown in figure 2. In this case all points at future infinity are equivalent: the Census Bureau can be located at any of them. However, String Theory and other considerations suggest that de Sitter minima are never stable. After a series of tunneling events they eventually end in terminal vacua with exactly zero or negative cosmological constant. The entire distant future of de Sitter space is replaced by a fractal of terminal bubbles.

Decay to negative cosmological constant always leads to a singular crunch. Barring governmental stupidity, this seems an unlikely place for a Census Bureau. The disadvantages (or advantages) of locating a government agency at a crunch are the same as at a black hole singularity.

Terminal vacua with zero cosmological constant seem more promising; the bubble then evolves to an open, negatively curved, FRW geometry, bounded by a "hat" [19]. The Census Bureau is at the tip of the hat.

In the case of the black hole, the degrees of freedom beyond the horizon, i.e., outside the causal patch, are redundant descriptions of degrees of freedom within the patch: they should not be double-counted. We assume that the same is true of the causal patch of a hat. In both cases the conventional rules of quantum mechanics are expected to apply ***only*** within the causal patch. Furthermore the rules should respect the Holographic Principle.

The reader may wonder about the relationship between hatted terminal geometries, and observational cosmology with a non-zero cosmological constant. There are two answers:

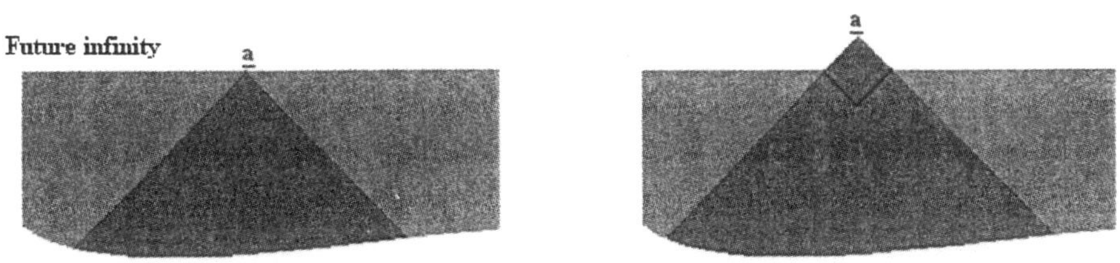

Figure 2: Conformal diagrams for eternal and metastable de Sitter space. The grey areas are causal patches associated with the points **a**. In the metastable case the causal patch is associated with the tip of a hat.

the first is that for many purposes, the current cosmological constant is so small that it can be set to zero. Later we will argue that the conformal field theory description of the approximate hat which results from non-zero cosmological constant is an ultraviolet cut-off version of the type of field theory that describes a hat.

The second answer was emphasized by Shenker, et al. [1] who argued that because our present de Sitter vacuum will eventually decay, a Census Taker can look back into our current vacuum from a point at or near the tip of a hat, and gather information. In principle the Census Taker can peek back, not only into the Ancestor vacuum (our vacuum in this case), but also into bubble collisions with other vacua of the Landscape. Much of this paper is about the gathering of information as the Census Taker's time progresses, and how it is encoded in the renormalization-group (RG) flow of a holographic field theory.

3 Open FRW and Euclidean ADS

The classical space-time in the interior of a Coleman De Luccia bubble, has the form of an open infinite FRW universe, Let $\mathcal{H}_3$ represent a hyperbolic geometry with constant negative curvature.

$$d\mathcal{H}_3^2 = dR^2 + \sinh^2 R \ \ d\Omega_2^2. \tag{3.1}$$

The metric of open FRW is

$$ds^2 = -dt^2 + a(t)^2 d\mathcal{H}_3^2, \tag{3.2}$$

or in terms of conformal time T (defined by $dT = dt/a(t)$)

$$ds^2 = a(T)^2(-dT^2 + d\mathcal{H}_3^2) \tag{3.3}$$

Note that in (3.1) the radial coordinate R is a hyperbolic angle and that the symmetry of the spatial sections is the non-compact group $O(3,1)$. This $O(3,1)$ symmetry plays a central role in what follows.

If the vacuum energy in the bubble is zero, i.e., no cosmological constant, then the future boundary of the FRW region is a hat. The scale factor $a(t)$ then has the early and late-time behaviors

$$a(T) \sim t \sim De^T. \tag{3.4}$$

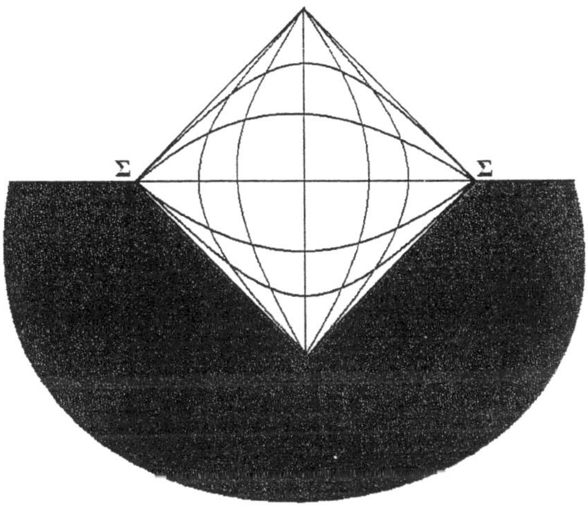

Figure 3: A Conformal diagram for the FRW universe created by bubble nucleation from an "Ancestor" metastable vacuum. The Ancestor vacuum is shown in green. The red and blue curves are surfaces of constant T and R. The two-sphere at spatial infinity is indicated by Σ.

For early time when $T \to -\infty$ the constant D is conveniently chosen to be the Hubble scale of the Ancestor, H^{-1}.

$$a(T) = H^{-1}e^{T} \qquad (\mathrm{T} \to -\infty) \tag{3.5}$$

At late time it is always larger. In the simplest thin-wall case D is given by the Ancestor Hubble-length at all times.

In figure 3, a conformal diagram of FRW is illustrated, with surfaces of constant T and R shown in red and blue. The green region represents the de Sitter Ancestor vacuum. Figure 4 shows the the Census Taker, as he approaches the tip of the hat, looking back along his past light cone.

Part of the inspiration for FSSY was the geometry of the spatial slices of constant T. Each slice, taken by itself, is a three dimensional, negatively curved, hyperbolic plane. It

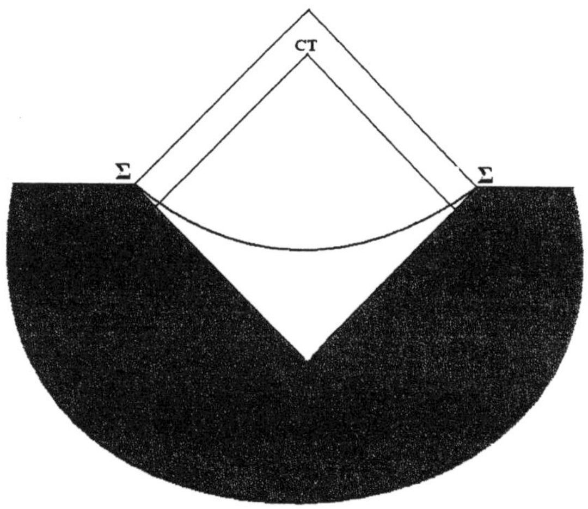

Figure 4: The Census Taker is indicated by the red dot. The blue lines represent his past light-cone.

Figure 5: Escher's drawing of the Hyperbolic Plane, which represents Euclidean anti de Sitter space or a spatial slice of open FRW. The green circle shows the intersection of the Census Taker's past light-cone, which moves toward the boundary with Census-Taker-time.

is very familiar to relativists and string theorists, being identical to 3-D Euclidean ***anti de Sitter space***. The best way that I know of for becoming familiar with the hyperbolic plane is to study Escher's drawing "Limit Circle IV." It is both a drawing of Euclidean ADS and also a fixed-time slice of open FRW. In figure 5, the green circle is the intersection of Census Takers past light cone with the time-slice. As the Census Taker advances in time, the green circle moves out, ever closer to the boundary.

A fact (to be explained later) which will play a leading role in what follows, concerns the Census Taker's angular resolution, i.e., his ability to discern small angular variation. If the time at which the CT looks back is called T_{CT}, then the smallest angle he can resolve is of order $\exp(-T_{CT})$. It is as if the CT were looking deeper and deeper into the ultraviolet structure of a quantum field theory on Σ.

The boundary of anti de Sitter space plays a key role in the ADS/CFT correspondence, where it represents the extreme ultraviolet degrees of freedom of the boundary theory. The corresponding boundary in the FRW geometry is labeled Σ and consists of the intersection of the hat $\mathcal{I}^+$, with the space-like future boundary of de Sitter space. From within the interior of the bubble, Σ represents ***space-like infinity.*** It is the obvious surface for a holographic description. As one might expect, the $O(3,1)$ symmetry which acts on the time-slices, also has the action of two dimensional conformal transformations on Σ. Whatever the Census Taker sees, it is very natural for him to classify his observations under the conformal group. Thus, the apparatus of (Euclidean) conformal field theory, such as operator dimensions, and correlation functions, should play a leading role in organizing his data.

In complicated situations, such as multiple bubble collisions, Σ requires a precise definition. The asymptotic light-cone $\mathcal{I}^+$ (which is, of course, the limit of the Census Takers past light cone), can be thought of as being formed from a collection of light-like generators. Each generator, at one end, runs into the tip of the hat, while the other end eventually enters the bulk space-time. The set of points where the generators enter the bulk define Σ.

4 The Holographic Wheeler DeWitt Equation

Supposedly, String Theory is a quantum theory of gravity, and indeed it has proved to be a remarkably powerful one, but only in certain special backgrounds. As effective as it is in describing scattering amplitudes in flat (supersymmetric) space-time, and black holes in anti de Sitter space, it is an inflexible tool which at present is close to useless for formulating a mathematical framework for cosmology. What is it that is so special about flat and ADS space that allows a rigorous formulation of quantum gravity, and why are cosmological backgrounds so difficult?

The problem is frequently blamed on ***time-dependence***. But time-dependent deformations of anti de Sitter space or Matrix Theory are easy to describe. Something else is the culprit. There is one important difference between the usual String Theory backgrounds and more interesting cosmological backgrounds. Asymptotically-flat and anti de Sitter backgrounds have a property that I will call ***asymptotic coldness***. Asymptotic coldness means that the boundary conditions require the energy density to go to zero at the asymptotic boundary of space[3]. Similarly, the fluctuations in geometry tend to zero.

[3]Note that asymptotic coldness refers only to conditions at spatial infinity. A violation of asymptotic coldness does not imply that the temperature remains finite as the time goes to infinity, although even

This condition is embodied in the statement that all physical disturbances are composed of normalizable modes. Asymptotic coldness is obviously important to defining an S-matrix in flat space-time, and plays an equally important role in defining the observables of anti de Sitter space.

But in cosmology, asymptotic coldness is never the case. Closed universes have no asymptotic boundary, and homogeneous infinite universes have matter, energy, and geometric variation out to spatial infinity; under the circumstances an S-matrix cannot be formulated. String theory at present is ill equipped to deal with ***asymptotically warm*** geometries. To put it another way, there is a conflict between a homogeneous cosmology, and the Holographic Principle which requires an isolated, cold, boundary.

The traditional approach to quantum cosmology–the Wheeler-DeWitt equation–is the opposite of string theory; it is very flexible from the point of view of background dependence–it doesn't require any definite boundary condition, it can be formulated for a closed universe, a flat or open FRW universe, de Sitter space, or for that matter, flat and anti de Sitter space spacetime–but it is not a consistent quantum theory of gravity. It is based on an obsolete approach–local quantum field theory–that fails to address the problems that String Theory and the Holographic Principle were designed to solve: the huge overcounting of degrees of freedom implicit in a local field theory.

FSSY suggested a way out of dilemma: synthesize the Wheeler-DeWitt philosophy with the Holographic Principle to construct a Holographic Wheeler-DeWitt theory. We will begin with a review of the basics of conventional WDW; For a more complete treatment, especially of infinite cosmologies, see [24].

The ten equations of General Relativity take the form

$$\frac{\delta}{\delta g_{\mu\nu}} I = 0 \tag{4.1}$$

where I is the Einstein action for gravity coupled to matter. The canonical formulation of General Relativity makes use of a time-space split [23]. The six space-space components are more or less conventional equations of motion, but the four equations involving the time index have the form of constraints. These four equations are written,

$$H^{\mu}(x) = 0. \tag{4.2}$$

---

this is a problem in geometries that contain de Sitter boundary conditions. Hats are somewhat better in that they become cold at late time.

They involve the space-space components of the metric g_{nm}, the matter fields Φ, and their conjugate momenta. The time component $H^0(x)$, is a local Hamiltonian which "pushes time forward" at the spatial point x. More generally, if integrated with a test function,

$$\int d^3x\ f(x)\ H^0(x) \tag{4.3}$$

it generates infinitesimal transformations of the form

$$t \to t\ +\ f(x). \tag{4.4}$$

Under certain conditions H^0 can be integrated over space in order to give a global Hamiltonian description. Since H^0 involves second space derivatives of g_{nm}, it is necessary to integrate by parts in order to bring the Hamiltonian to the conventional form containing only first derivatives. In that case the ADM equations can be written as

$$\int d^3x\ H = E. \tag{4.5}$$

The Hamiltonian density H has a conventional structure, quadratic in canonical momenta, and the energy E is given by a Gaussian surface integral over spatial infinity. The conditions which allow us to go from (4.2) to (4.5) are satisfied in asymptotically cold flat-space-time, as well as in anti de Sitter space; in both cases global Hamiltonian formulations exist. Indeed, in anti de Sitter space the Hamiltonian of the Holographic boundary description is identified with the ADM Energy, but, as we noted, cosmology, at least in its usual forms, is never asymptotically cold. The only recourse for a canonical description, is the local form of the equations (4.2).

When we pass from classical gravity to its quantum counterpart, the usual generalization of the canonical equations (4.2) become the Wheeler DeWitt equations,

$$H^\mu|\Psi\rangle = 0 \tag{4.6}$$

where the state vector $|\Psi\rangle$ is represented by a wave functional that depends only on the space components of the metric g_{mn}, and the matter fields Φ.

The first three equations

$$H^m|\Psi\rangle = 0 \quad (m = 1,\ 2,\ 3.) \tag{4.7}$$

have the interpretation that the wave function is invariant under spatial diffeomorphisms,

$$x^n \rightarrow x^n + f^n(x^m) \tag{4.8}$$

In other words $\Psi(g_{mn}, \ \Phi)$ is a function of spatial invariants. These equations are usually deemed to be the easy Wheeler-DeWitt equations.

The difficult equation is the time component

$$H^0|\Psi\rangle = 0. \tag{4.9}$$

It represents invariance under local, spatially varying, time translations. Not only is equation (4.9) difficult to solve; it is difficult to even formulate: the expression for H^0 is riddled with factor ordering ambiguities. Nevertheless, as long as the equations are not pushed into extreme quantum environments, they can be useful.

4.1 Wheeler DeWitt and the Emergence of Time

Asymptotically cold backgrounds come equipped with a global concept of time. But in the more interesting asymptotically warm case, time is an approximate, derived, concept [24, 25], which emerges from the solutions to the Wheeler-DeWitt equation. The perturbative method for solving (4.9) that was outlined in [24], can be adapted to the case of negative spatial curvature. We begin by decomposing the spatial metric into a constant curvature background, and fluctuations. Since we will focus on open FRW cosmology, the spatial curvature is negative, the space metric having the form,

$$ds^2 = a^2 \left(dR^2 + \sinh^2 R \ \ (d\theta^2 + \sin^2\theta d\phi^2)\right) + a^2 h_{mn} dx^m dx^n \tag{4.10}$$

In (4.10) a is the usual FRW scale factor and the $x's$ are $(R, \ \ \theta, \ \phi)$.

The first approximation, in which all fluctuations are ignored, is usually called the mini-superspace approximation, but it really should be seen as a first step in a semiclassical expansion. In lowest order, the Wheeler DeWitt wave function depends only on the scale factor a. To carry out the leading approximation in open FRW it is necessary to introduce an infrared regulator which can be done by bounding the value of R,

$$R < R_0 \ \ (R_0 >> 1). \tag{4.11}$$

Lets also define the total dimensionless coordinate-volume within the cutoff region, to be V_0.

$$V_0 = 4\pi \int dR \sinh^2 R \approx \frac{1}{2}\pi e^{2R_0}. \tag{4.12}$$

The first (mini-superspace) approximation is described by the action,

$$L = \frac{-aV_0\dot{a}^2 - V_0 a}{2} \tag{4.13}$$

Defining P to be the momentum conjugate to the scale factor a,

$$P = -aV_0\dot{a} \tag{4.14}$$

the Hamiltonian H^0 is given by [4],

$$H^0 = \frac{1}{2V_0}P\frac{1}{a}P + \frac{1}{2}V_0 a \tag{4.15}$$

Finally, using $P = -i\partial_a$, the first approximation to the Wheeler-DeWitt equation becomes,

$$-\partial_a \frac{1}{a}\partial_a \Psi - V_0^2 a\Psi = 0. \tag{4.16}$$

The equation has the two solutions,

$$\Psi = \exp(\pm iV_0 a^2), \tag{4.17}$$

corresponding to expanding and contraction universes; to see which is which we use (4.14). The expanding solution, labeled Ψ_0 is

$$\Psi_0 = \exp(-iV_0 a^2). \tag{4.18}$$

From now on we will only consider this branch.

There is something funny about (4.18). Multiplying V_0 by a^2 seems like an odd operation. $V_0 a^3$ is the proper volume, but what is $V_0 a^2$? The answer in flat space is that it is junk, but in hyperbolic space its just the proper area of the boundary at R_0. One sees from the metric (3.3) that the coordinate volume V_0, and the coordinate area A_0, of the boundary at R_0, are (asymptotically) equal to one another, to within a factor of 2.

$$A_0 = 2V_0. \tag{4.19}$$

---

[4]The factor ordering in the first term is ambiguous. I have chosen the simplest Hermitian factor ordering

Thus the expression in the exponent in (4.18) is $-\frac{1}{2}iA$, where A is the proper area of the boundary at R_0.

$$\Psi_0 = \exp(-2iA). \tag{4.20}$$

This is a suggestive indication of a Wheeler-DeWitt boundary-holography of open FRW.

To go beyond the mini-superspace approximation one writes the wave function as a product of Ψ_0, and a second factor $\psi(a, h, \Phi)$ that depends on the fluctuations.

$$\Psi(a, h, \Phi) = \Psi_0 \ \ \psi(a, h, \Phi) = \exp(-iV_0 a^2) \ \ \psi(a, h, \Phi). \tag{4.21}$$

By integrating the Wheeler-DeWitt equation over space, and substituting (4.21), an equation for ψ can be obtained.

$$i\partial_a \psi + \frac{1}{aV_0}\partial_a \frac{1}{a}\partial_a \psi = H_m \psi \tag{4.22}$$

In this equation H_m has the form of a conventional Hamiltonian (quadratic in the momenta) for both matter and metric fluctuations.

In the limit of large scale factor the term $\frac{1}{aV_0}\partial_a \frac{1}{a}\partial_a \psi$ becomes negligible and (4.22) takes the form of a Schrodinger equation.

$$i\partial_a \psi = H_m \psi \tag{4.23}$$

Evidently the role of a is not as a conventional observable, but a parameter representing the unfolding of cosmic time. One does not calculate its probability, but instead constrains it–perhaps with a delta function or a Lagrange multiplier. As Banks has emphasized [25], in this limit, and maybe ***only*** in this limit, the wave function ψ has a conventional interpretation as a probability amplitude.

4.2 Holographic WDW

All of this brings us to the central question of this lecture: what form does the correct holographic theory take in asymptotically warm cosmological backgrounds? The answer suggested in FSSY was a holographic version of the Wheeler-DeWitt theory, living on the space-like boundary Σ.

As we have described it, the Wheeler-DeWitt theory is a throwback to an older view of quantum gravity based on the existence of bulk, space-filling degrees of freedom. It has become clear that this is a tremendous overestimate of the capacity of space to contain quantum information.The correct (holographic) counting of degrees of freedom is in terms of the area of the boundary of space [10]. In the present case of open FRW, the special role of the boundary is played by the surface Σ at $R = \infty$.

Just as in the ADS/CFT correspondence [13], it is useful to define a regulated boundary, Σ_0, at $R = R_0$. In principle R_0 can depend on angular location on Ω_2. In fact later we will discuss invariance under gauge transformations of the form

$$R \rightarrow R + f(\Omega_2). \tag{4.24}$$

(the notation $f(\Omega_2)$. indicating that f is a function of location on Σ_0.)

The conjecture of FSSY is that the correct Holographic description of open FRW is a Wheeler-DeWitt equation, but one in which the degrees of freedom are at the boundary of space, i.e., on Σ, instead of being distributed throughout the bulk.

Thus we assume the existence of a set of boundary fields, that include a two dimensional spatial metric on Σ_0. The induced spatial geometry of the boundary can always be described in the conformal gauge in terms of a Liouville field $U(\Omega_2)$.

$$ds^2 = e^{2U(\Omega_2)} e^{2R_0(\Omega_2)} d\Omega_2^2 \tag{4.25}$$

U may be decomposed into a homogeneous term U_0, and a fluctuation; obviously the homogeneous term can be identified with the FRW scale factor by

$$e^{U_0} = a. \tag{4.26}$$

In section 8 we will give a more detailed definition of the Liouville degree of freedom.

In addition we postulate a collection of boundary "matter" fields. The boundary matter fields, y, are not the limits of the usual bulk fields Φ, but are analogous to the boundary gauge fields in the ADS/CFT correspondence. In this paper we will not speculate on the detailed form of these boundary matter fields.

4.3 The Wave Function

In addition to U and y, we assume a local Hamiltonian $H(x^i)$ that depends only on the boundary degrees of freedom (the notation x^i refers to coordinates of the boundary Σ), and a wave function $\Psi(U, y)$,

$$\Psi(U, y) = e^{-\frac{1}{2}S+iW}. \tag{4.27}$$

At every point of Σ, Ψ satisfies

$$H(x^i)\ \Psi(U, y). = 0 \tag{4.28}$$

In equation (4.27), $S(U, y)$ and $W(U, y)$ are real functionals of the boundary fields. For reasons that will become clear, we will call S the action. However, S should not in any way be confused with the four-dimensional Einstein action.

The local Hamiltonian $H(x^i)$, and the imaginary term W in the exponent, play important roles in determining the expectation values of canonical momenta, as well as the relation between scale factor and ordinary time. In this paper H and W will play secondary roles.

We make the following three assumptions about S and W:

- Both S and W are invariant under conformal transformations of Σ. This follows from the symmetry of the background geometry: open FRW .
- The leading (non-derivative) term in the regulated form of W is $-\frac{1}{2}A$ where A is the proper area of Σ_0,

  $$W = -\frac{1}{2}\int_{\Sigma} e^{2R_0}e^{2U} + \ldots \tag{4.29}$$

  This follows from (4.20).
- S and W have the form of ***local*** two dimensional Euclidean actions on Σ. In other words they are integrals, over Σ, of densities that involve U, y, and their derivatives with respect to x^i.

The first of these conditions is just a restatement of the symmetry of the Coleman De Luccia instanton. Later we will see that this symmetry is spontaneously broken by a number of effects, including the extremely interesting "Persistence of Memory" discovered in by Garriga, Guth and Vilenkin [20].

The second condition follows from the bulk analysis described earlier in equation (4.20). It allows us to make an educated guess about the dependence of the local Hamiltonian $H(x^i)$ on U. A simple form that reproduces 4.20 is

$$H(x) = \frac{1}{2}e^{-2U}\pi_U^2 - 2e^{2U} + \; ... \tag{4.30}$$

where π_U is the momentum conjugate to U. It is easily seen that the solution to the equation $H\Psi = 0$ has the form (4.20).

The highly nontrivial assumption is the third item–the locality of the action. As a rule quantum field theory wave functions are not local in this sense. That the action S is local is far from obvious. In our opinion it is the strongest (meaning the weakest) of our assumptions and the one most in need of confirmation. At present our best evidence for the locality is the discrete tower of correlators, including a transverse, traceless, dimension-two correlation function, described in the next section. In principle, much more information can be obtained from bulk multi-point functions, continued to Σ. For example, correlation functions of h_{ij} would allow us to study the operator product expansion of the energy-momentum tensor.

As I said, the assumption that S is local is a very strong one, but I mean it in a rather weak sense. One of the main points of this lecture is that there is a natural RG flow in cosmology (see section 6). By locality I mean only that S is in the basin of attraction of a local field theory. If it is true, locality would imply that the measure

$$\Psi^*\Psi = e^{-S} \tag{4.31}$$

has the form of a local two dimensional Euclidean field theory with action S, and that the Census Taker's observations could be organized not only by conformal invariance but by conformal field theory.

5 Data

The conjectured locality of the action S is based on data calculated by FSSY. The background geometry studied in [19] was the Minkowski continuation of a thin-wall Coleman

De Luccia instanton, describing transitions from the Ancestor vacuum to a hatted vacuum. For a number of reasons such a background cannot be a realistic description of cosmology. First of all, there is a form of spontaneous breaking of the $O(3,1)$ symmetry that Garriga, Guth, and Vilenkin call "The Persistence of Memory." In section 8 we will see that these type of effects are "dual" to effects expected in the theory of RG-flows.

More importantly, we do not live in a universe with zero cosmological constant . Observational cosmology has come close to ruling out vanishing cosmological constant , but also theoretical considerations rule it out; in the Landscape of String Theory the only vacua with exactly vanishing cosmological constant are supersymmetric. Nevertheless, hatted geometries are interesting in that they are the simplest versions of asymptotically warm geometries.

In FSSY, correlation functions were computed in the thin-wall, Euclidean, Coleman De Luccia instanton, and then continued to Minkowski signature. The more general situation, including the possibility of slow-roll inflation after tunneling, is presently under investigation with Ben Freivogel, Yasuhiro Sekino, and Chen Pin Yeh. Here we will mostly confine ourselves to the thin-wall case.

We begin by reviewing some facts about three-dimensional hyperbolic space and the solutions of its massless Laplace equation. An important distinction is between normalizable modes (NM) and non-normalizable modes (NNM); a scalar minimally coupled field χ is sufficient to illustrate the important points.

The norm in hyperbolic space is defined in the obvious way:

$$\langle \chi|\chi\rangle = \int dR d\Omega_2 \, \chi^2 \, \sinh^2 R \tag{5.1}$$

In flat space, fields that tend to a constant at infinity are on the edge on normalizability. With the help of the delta function, the concept of normalizability can be generalized to continuum-normalizability, and the constant "zero mode" is included in the spectrum of the wave operator, but in hyperbolic space the normalization integral (5.1) is exponentially divergent for constant χ. The condition for normalizability is that $\chi \to 0$ at least as fast as e^{-R}. The constant mode is therefore non-normalizable.

Normalizable and non-normalizable modes have very different roles in the conventional ADS/CFT correspondence. NM are dynamical excitations with finite energy and can be produced by events internal to the anti de Sitter space. By contrast NNM cannot be excited dynamically. Shifting the value of a NNM is equivalent to changing the boundary conditions from the bulk point of view, or changing the Lagrangian from the boundary perspective. But, as we will see, in the cosmological framework of FSSY, asymptotic warmness blurs this distinction.

5.1 Scalars

Correlation functions of massless (minimally coupled) scalars, χ, depend on time and on the ***dimensionless*** geodesic distance between points on $\mathcal{H}_3$. In the limit in which the points tend to the holographic boundary Σ at $R \to \infty$, the geodesic distance between points 1 and 2 is given by,

$$l = R_1 + R_2 + \log(1 - \cos\alpha) \tag{5.2}$$

where α is the angular distance on Ω_2 between 1 and 2. It follows on $O(3,1)$ symmetry grounds that the correlation function $\langle \chi(1)\chi(2)\rangle$ has the form,

$$\begin{aligned} \langle \chi(1)\chi(2)\rangle &= G(T_1,\, T_2,\, l_{1,2}) \\ &= G\{T_1,\, T_2,\, (R_1 + R_2 + \log(1 - \cos\alpha))\}. \end{aligned} \tag{5.3}$$

Before discussing the data on the Coleman De Luccia background, let us consider the form of correlation functions for scalar fields in anti de Sitter space. We work in units in which the radius of the anti de Sitter space is 1. By symmetry, the correlation function can only depend on l, the proper distance between points. The large-distance behavior of the two-point function has the form

$$\langle \chi(1)\chi(2)\rangle \sim \frac{e^{-(\Delta-1)l}}{\sinh l}. \tag{5.4}$$

In anti de Sitter space the dimension Δ is related to the mass of χ by

$$\Delta(\Delta - 2) = m^2. \tag{5.5}$$

We will be interested in the limit in which the two points 1 and 2 approach the boundary at $R \to \infty$. Using (5.2) gives

$$\langle \chi(1)\chi(2)\rangle \sim e^{-\Delta R_1} e^{-\Delta R_2} (1 - \cos\alpha)^{-\Delta}. \tag{5.6}$$

It is well known that the "infrared cutoff" R, in anti de Sitter space, is equivalent to an ultraviolet cutoff in the boundary Holographic description [13]. The exponential factors, $\exp(-\Delta R)$ in (5.6) correspond to cutoff dependent wave function renormalization factors and are normally stripped off when defining boundary correlators. The remaining factor, $(1 - \cos\alpha)^{-\Delta}$ is the conformally covariant correlation function of a boundary field of dimension Δ.

In FSSY it was claimed that in the Coleman De Luccia background, the correlation function contains two terms, one of which was associated with NM and the other with NNM. A third term was found, but ignored on the basis that it was negligible when continued to the boundary. In fact the third term has an interesting significance that we will come back to, but first we will review the terms studied in FSSY.

In [19] the correlation function was expressed as a sum of two contour integrals on the k plane–k being an eigenvalue of the Laplacian on $\mathcal{H}_3$. The integral involves a certain reflection coefficient $\mathcal{R}(k)$ for a Schrodinger equation, derived from the wave equation on the Coleman De Luccia instanton. The contour integral is

$$e^{-(T_1+T_2)} \oint_C \frac{dk}{2\imath} \mathcal{R}(k) e^{-ik(T_1+T_2)} \frac{\left(e^{-ikR} - e^{-ikR-2\pi k}\right)}{2\sinh R \sinh k\pi} \tag{5.7}$$

The contours of integration are shown in Figure 6. The integrand has poles at all imaginary values of k, with a double pole at $k = i$. In addition there may be other singularities in the lower half plane. FSSY studied only the terms coming from the upper contour labeled **a** in the figure. It contains two terms related to NM and NNM respectively.

The normalizable contribution, G_1, is an infinite sum, each term having the form (5.6) with T-dependent coefficients. For late times,

$$\begin{aligned} G_1 &= \sum_{\Delta=2}^{\infty} G_\Delta e^{(\Delta-2)(T_1+T_2)} \frac{e^{-(\Delta-1)l}}{\sinh l} \\ &\rightarrow \sum_{\Delta=2}^{\infty} G_\Delta e^{(\Delta-2)(T_1+T_2)} e^{-\Delta R_1} e^{-\Delta R_2} (1 - \cos\alpha)^{-\Delta} \end{aligned} \tag{5.8}$$

where Δ takes on integer values from 2 to ∞, and G_Δ are a series of constants which depend on the detailed CDL solution.

The connection with conformal field theory correlators is obvious; equation (5.8) is a sum of correlation functions for fields of definite dimension Δ, but with coefficients which

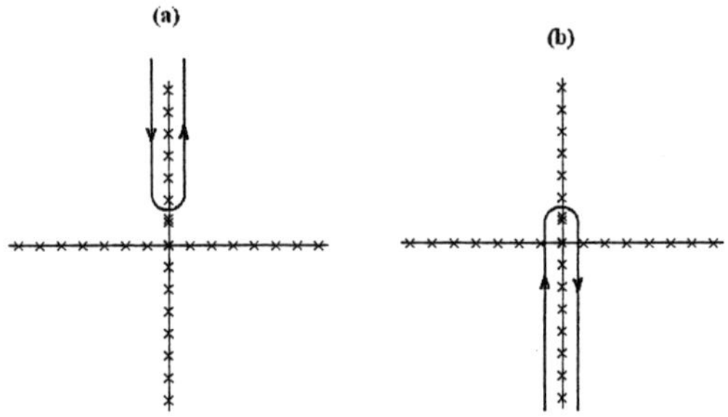

Figure 6: Contours of integration for the two contributions G_1, G_2.

depend on the time T. (It should be emphasized that the dimensions Δ in the present context are not related to bulk four dimensional masses by (5.5).) Note that the sum in (5.8) begins at $\Delta = 2$, implying that every term falls at least as fast as $\exp(-2R)$ with respect to either argument. Thus every term is normalizable.

Let us now extrapolate (5.8) to the surface Σ. Σ can be reached in two ways–the first being to go out along a constant T surface to $R = \infty$. Each term in the correlator has a definite R dependence which identifies its dimension.

Another way to get to Σ is to first pass to light-like infinity, $\mathcal{I}^+$, and then slide down the hat, along a light-like generator, until reaching Σ. For this purpose it is useful to define light-cone coordinates, $T^{\pm} = T \pm R$.

$$G_1 = e^{-(T_1^+ + T_2^+)} \sum_{\Delta} G_{\Delta} e^{(\Delta-1)(T_1^- + T_2^-)} (1 - \cos\alpha)^{-\Delta} \tag{5.9}$$

We note that apart from the overall factor $e^{-(T_1^+ + T_2^+)}$, G_1 depends only on T^-, and therefore tends to a finite limit on $\mathcal{I}^+$. If we strip that factor off, then the remaining expression consists of a sum over CFT correlators, each proportional to a fixed power of e^{T^-}. In the limit ($T^- \to -\infty$) in which we pass to Σ, each term of fixed dimension tends to zero as $e^{(\Delta-1)(T_1^- + T_2^-)}$ with the dimension-2 term dominating the others.

The second term in the scalar correlation function discussed by FSSY consists of a single term,

$$G_2 = \frac{e^l}{\sinh l}(T_1 + T_2 + l) \to \left\{T_1^+ + T_2^+ + \log(1 - \cos\alpha)\right\} \tag{5.10}$$

The contribution (5.10) does not have the form of a correlator of a conformal field of definite dimension. To understand its significance, consider a canonical massless scalar field in two dimensions. On a two sphere the correlation function is ultraviolet divergent and has the form

$$\log\left\{\kappa^2(1 - \cos\alpha)\right\} \tag{5.11}$$

where κ is the ultraviolet regulator momentum. If the regulator momentum varies with location on the sphere–for example in the case of a lattice regulator with a variable lattice spacing–formula (5.11) is replaced by

$$\log\left\{(1 - \cos\alpha)\right\} + \log\kappa_1 + \log\kappa_2 \tag{5.12}$$

Evidently if we identify the UV cutoff κ with T^+,

$$\log\kappa = T^+ \tag{5.13}$$

the expressions in (5.10) and (5.12) are identical. The relation (5.13) is one of the central themes of this paper, that as we will see, relates RG flow to the observations of the Census Taker.

That the UV cutoff of the 2D boundary theory depends on R is very familiar from the UV/IR connection [13] in anti de Sitter space. In that case the T coordinate is absent and the log of the cutoff momentum in the conformal field theory would just be R. The additional time dependent contribution in (5.13) will become clear later when we discuss the Liouville field.

The logarithmic ultraviolet divergence in the correlator is a signal that massless 2D scalars are ill defined; the well-defined quantities being derivatives of the field. When calculating correlators of derivatives, the cutoff dependence disappears. Thus for practical purposes, the only relevant term in (5.12) is $\log(1 - \cos\alpha)$.

The existence of a dimension-zero scalar field on Σ is a surprise. It is obviously associated with bulk field-modes which don't go zero for large R. Such modes are non-normalizable on the hyperbolic plane, and are usually not included among the dynamical variables in anti de Sitter space.

In String Theory the only massless scalars in the hatted vacua would be moduli, which are expected to be "fixed" in the Ancestor. For that reason FSSY considered the effect of adding a four-dimensional mass term, $\mu\chi^2$, in the Ancestor vacuum. The result on the boundary scalar was to shift its dimension from $\Delta = 0$ to $\Delta = \mu$ (for small μ less than the Ancestor Hubble constant the corresponding mode stays non-normalizable). However the correlation function was not similar to those in G_1, each term of which had a dependence on T^-. The dimension μ term depends only on T^+:

$$G_2 \to e^{-\mu T_1^+} e^{-\mu T_2^+} \log (1 - \cos\alpha)^{-\mu} \tag{5.14}$$

The two terms, (5.9) and (5.14) depend on different combinations of the coordinates, T^+ and T^-. It seems odd that there is one and only one term that depends solely on T^+ and all the rest depend on T^-. In fact the only reason is that FSSY ignored an entire tower of higher dimension terms, coming from the contour **b** that, like (5.14), depend only on T^+. From now on we will group all terms independent of T^- into the single expression G_2:

$$G_2 = \sum_{\Delta'} G_{\Delta'} e^{-(\Delta')(T_1^+ + T_2^+)}(1 - \cos\alpha)^{-\Delta'} \tag{5.15}$$

The Δ' include μ, the positive integers and whatever other poles appear for $ik < 1$. In the case $\mu = 0$, the leading term in G_2 is (5.10).

We will return to the two terms G_1 and G_2 in section 6.5.

5.2 Metric Fluctuations

To prove that there is a local field theory on Σ, the most important test is the existence of an energy-momentum tensor. In the ADS/CFT correspondence, the boundary energy-momentum tensor is intimately related to the bulk metric fluctuations. We assume a similar connection between bulk and boundary fields in the present context. In FSSY, metrical fluctuations were studied in a particular gauge which we will call the ***Spatially Transverse-Traceless*** (STT) gauge. The coordinates of region I can be divided into

FRW time, T, and space x^m where $m = 1,\ 2,\ 3$. The STT gauge for metric fluctuations is defined by

$$\begin{aligned} \nabla^m h_{mn} &= 0 \\ h^m_m &= 0 \end{aligned} \tag{5.16}$$

In the second of equations (5.16), the index is raised with the aid of the background metric (3.3). The main benefit of the STT gauge is that metric fluctuations satisfy minimally coupled, massless, scalar equations, and the correlation functions are similar to G_1 and G_2. However the index structure is rather involved. We define the correlator,

$$\begin{aligned} \langle\, h^\mu_\nu h^\sigma_\tau \rangle &= G \left\{ {}^{\mu\sigma}_{\nu\tau} \right\} \\ &= G_1 \left\{ {}^{\mu\sigma}_{\nu\tau} \right\} + G_2 \left\{ {}^{\mu\sigma}_{\nu\tau} \right\}. \end{aligned} \tag{5.17}$$

The complicated index structure of G was worked out in detail in FSSY. In this paper we quote only the results of interest–in particular those involving elements of $G \left\{ {}^{\mu\sigma}_{\nu\tau} \right\}$ in which all indecies lie in the two-sphere Ω_2. Thus we consider the correlation function $G_1 \left\{ {}^{ik}_{jl} \right\}$.

As in the scalar case, G_1 consists of an infinite sum of correlators, each corresponding to a field of dimension $\Delta = 2,\ 3,\ 4, \ldots$. The asymptotic T and R dependence of the terms is identical to the scalar case, and the first term has $\Delta = 2$. This is particulary interesting because it is the dimension of the energy-momentum tensor of a two-dimensional boundary conformal field theory. Once again this term is also time-independent.

After isolating the dimension-two term and stripping off the factors $\exp(-2R)$, the resulting correlator is called $G_1 \left\{ {}^{ik}_{jl} \right\} |_{\Delta=2}$. The calculations of FSSY revealed that this term is two-dimensionally traceless, and transverse.

$$\begin{aligned} G_1 \left\{ {}^{ik}_{il} \right\} |_{\Delta=2} = G_1 \left\{ {}^{ik}_{jk} \right\} |_{\Delta=2} &= 0 \\ \nabla_i G_1 \left\{ {}^{ik}_{jl} \right\} |_{\Delta=2} &= 0. \end{aligned} \tag{5.18}$$

Equation (5.18) is the clue that, when combined with the dimension-2 behavior of $G_1 \left\{ {}^{ik}_{il} \right\} |_{\Delta=2}$, hints at a local theory on Σ. It insures that it has the precise form of a two-point function for an energy-momentum tensor in a conformal field theory. The only ambiguity is the numerical coefficient connecting $G_1 \left\{ {}^{ik}_{jl} \right\} |_{\Delta=2}$ with $\langle\, T^i_j T^k_l \rangle$. We will return to this coefficient momentarily.

The existence of a transverse, traceless, dimension-two operator is a necessary condition for the boundary theory on Σ to be local: at the moment it is our main evidence. But there is certainly more that can be learned by computing multipoint functions. For example, from the three-point function $\langle hhh \rangle$ it should be possible verify the operator product expansion and the Virasoro algebra for the energy-momentum tensor.

Dimensional analysis allows us to estimate the missing coefficient connecting the metric fluctuations with T^i_j, and at the same time determine the central charge c. In [19] we found c to be of order the horizon entropy of the Ancestor vacuum. We repeat the argument here:

Assume that the (bulk) metric fluctuation h has canonical normalization, i.e., it has bulk mass dimension 1 and a canonical kinetic term. Either dimensional analysis or explicit calculation of the two point function $\langle hh \rangle$ shows that it is proportional to square of the Ancestor Hubble constant.

$$\langle hh \rangle \sim H^2. \tag{5.19}$$

Knowing that the three point function $\langle hhh \rangle$ must contain a factor of the gravitational coupling (Planck Length) l_p, it can also be estimated by dimensional analysis.

$$\langle hhh \rangle \sim l_p H^4. \tag{5.20}$$

Now assume that the 2D energy-momentum tensor is proportional to the boundary dimension-two part of h, i.e., the part that varies like e^{-2R}. Schematically,

$$T = qh \tag{5.21}$$

with q being a numerical constant. It follows that

$$\begin{aligned} \langle TT \rangle &\sim q^2 H^2 \\ \langle TTT \rangle &\sim q^3 l_p H^4. \end{aligned} \tag{5.22}$$

Lastly, we use the fact that the ratio of the two and three point functions is parametrically independent of l_p and H because it is controlled by the classical algebra of diffeomorphisms: $[T, T] = T$. Putting these elements together we find,

$$\langle TT \rangle \sim \frac{1}{l_p^2 H^2} \tag{5.23}$$

Since we already know that the correlation function has the correct form, including the short distance singularity, we can assume that the right hand side of (5.23) also gives the central charge. It can be written in the rather suggestive form:

$$c \sim Area/G \quad (G = l_p^2) \tag{5.24}$$

where $Area$ refers to the horizon of the Ancestor vacuum. In other words, the central charge of the hypothetical CFT is proportional to the ***horizon entropy of the Ancestor*** .

5.3 Dimension Zero Term

The term $G_2 \left\{ {ik \atop jl} \right\}$ begins with a term, which like its scalar counterpart, has a non-vanishing limit on Σ. It is expressed in terms of a standard 2D bi-tensor $t \left\{ {ik \atop jl} \right\}$ which is traceless and transverse in the two dimensional sense. If the correlation function were given just by $t \left\{ {ik \atop jl} \right\}$, it would be a pure gauge artifact. One can see this by considering the linearized expression for the 2D curvature-scalar C,

$$C = \nabla_i \nabla_j h^{ij} - 2\nabla^i \nabla_i \, Tr \; h. \tag{5.25}$$

The 2D curvature associated with a traceless transverse fluctuation vanishes, and since $t \left\{ {ik \atop jl} \right\}$ by itself is traceless-transverse with respect to both points, it would be pure gauge if it appeared by itself.

However, the actual correlation function $G_2 \left\{ {ik \atop jl} \right\}$ is given by

$$G_2 \left\{ {ik \atop jl} \right\} = t \left\{ {ik \atop jl} \right\} \{ R_1 + T_1 + R_2 + T_2 + \log(1 - \cos\alpha) \} \tag{5.26}$$

The linear terms in $R + T$, being proportional to $t \left\{ {ik \atop jl} \right\}$ are pure gauge, but the finite term

$$t \left\{ {ik \atop jl} \right\} \log(1 - \cos\alpha) \tag{5.27}$$

gives rise to a non-trivial 2D curvature-curvature correlation function of the form

$$\langle CC \rangle = (1 - \cos\alpha)^{-2}. \tag{5.28}$$

One difference between the metric fluctuation h, and the scalar field χ, is that we cannot add a mass term for h in the Ancestor vacuum to shift its dimension.

Finally, as in the scalar case, there is a tower of higher dimension terms in the tensor correlator, $G_2 \left\{ {ik \atop jl} \right\}$ that only depend on T^+.

The existence of a zero dimensional term in $G_2\left\{{ik \atop jl}\right\}$, which remains finite in the limit $R \to \infty$ indicates that fluctuations in the boundary geometry–fluctuations which are due to the asymptotic warmness–cannot be ignored. One might expect that in some way these fluctuations are connected with the field U that we encountered in the Holographic version of the Wheeler-DeWitt equation. In the next section we will elaborate on this connection.

That is the data about correlation functions on the boundary sphere Σ that form the basis for our conjecture that there exists a local holographic boundary description of the open FRW universe. There are a number of related puzzles that this data raises: First, how does time emerge from a Euclidean QFT? The bulk coordinate R can be identified with scale size just as in ADS/CFT but the origin of time requires a new mechanism.

The second puzzle concerns the number of degrees of freedom in the boundary theory. The fact that the central charge is the entropy of the Ancestor suggests that there are only enough degrees of freedom to describe the false vacuum and not the much large number needed for the open FRW universe at late time.

6 Liouville Theory

6.1 Breaking Free of the STT Gauge

The existence of a Liouville sector describing metrical fluctuations on Σ seems dictated by both the Holographic Wheeler-DeWitt theory and from the data of the previous section. It is clear that the Liouville field is somehow connected with the non-normalizable metric fluctuations whose correlations are contained in (5.26), although the connection is somewhat obscured by the choice of gauge in [19]. In the STT gauge the fluctuations h are traceless, but not transverse (in the 2D sense). From the viewpoint of 2D geometry they are not pure gauge as can be seen from the fact that the 2D curvature correlation does not vanish. One might be tempted to identify the Liouville mode with the zero-dimension piece of (5.26). To do so would of course require a coordinate transformation on Ω_2 in order to bring the fluctuation h_i^j to the "conformal" form $\tilde{h}\delta_i^j$.

This identification may be useful but it is not consistent with the Wheeler-DeWitt philosophy. The Liouville field U that appears in the Wheeler-DeWitt wave function is not tied to any specific spatial gauge. Indeed, the wave function is required to be invariant under gauge transformations,

$$x^\mu \to x^\mu + f^\mu(x) \tag{6.1}$$

under which the metric transforms:

$$g_{\mu\nu} \to g_{\mu\nu} + \nabla_\mu f_\nu + \nabla_\nu f_\mu. \tag{6.2}$$

Let's consider the effect of such transformations on the boundary limit of h_{ij}. The components of f along the directions in Σ induce 2D coordinate transformation under which h transforms conventionally. Invariance under these transformations merely mean that the action S must be a function of 2D invariants.

Invariance under the shifts f^R and f^T are more interesting. In particular the combination $f^+ = f^R + f^T$ generates non-trivial transformations of the boundary metric h_{ij}. An easy calculation shows that,

$$h_i^j \to h_i^j + f^+(\Omega_2)\delta_i^j. \tag{6.3}$$

In other words, shift transformations f^+, induce Weyl re-scalings of the boundary metric. This prompts us to modify the definition of the Liouville field from

$$U = T + \tilde{h} \tag{6.4}$$

to

$$U = T + \tilde{h} + f^+. \tag{6.5}$$

One might wonder about the meaning of an equation such as (6.5). The left side of of the equation is supposed to be a dynamical field on Σ, but the right side contains an arbitrary function f^+. The point is that in the Wheeler DeWitt formalism the wave function must be invariant under shifts, but in the original analysis of FSSY a specific gauge was chosen. Thus, in order to render the wave function gauge invariant, one must allow the shift f^+ to be an integration variable, giving it the status of a dynamical field.

A similar example is familiar from ordinary gauge theories. The analog of the Wheeler-DeWitt gauge-free formalism would be the unfixed theory in which one integrates over the time component of the vector potential. The analog of the STT gauge would be the Coulomb gauge. To go from one to the other we would perform the gauge transformation

$$A_0 \to A_0 + \partial_0\phi. \tag{6.6}$$

Integrating over the gauge function ϕ in the path integral would restore the gauge invariance that was given up by fixing Coulomb gauge.

Returning to the Liouville field, since both $\tilde{h}$ and f are linearized fluctuation variables, we see that the classical part of U is still the FRW conformal time.

One important point: because the effect of the shift f^+ is restricted to the trace of h, it does not influence the traceless-transverse (dimension-two) part of the metric fluctuation, and the original identification of the 2D energy-momentum tensor is unaffected.

Finally, invariance under the shift f^- is trivial in this order, at least for the thin wall geometry. The reason is that in the background geometry, the area does not vary along the T^- direction.

Given that the boundary theory is local, and includes a boundary metric, it is constrained by the rules of two-dimensional quantum gravity laid down long ago by Polyakov [28]. Let us review those rules for the case of a conformal "matter" field theory coupled to a Liouville field. Two-dimensional coordinate invariance implies that the central charge of the Liouville sector cancels the central charge of all other fields. We have argued in [19] (and in section 5) that the central charge of the matter sector is of order the horizon area of the Ancestor vacuum, measured in Planck Units. It is obvious from the 4-dimensional bulk viewpoint that the semiclassical analysis that we have relied on, only makes sense when the Hubble radius is much larger than the Planck scale. Thus we we take the central charge of matter to satisfy $c >> 1$. As a consequence, the central charge of the Liouville sector, c_L, must be large and negative. Unsurprisingly, the negative value of c is the origin of the emergence of time.

The formal development of Liouville theory begins by defining two metrics on Ω_2. The first is what I will call the reference metric $\hat{g}_{ij}$. Apart from an appropriate degree smoothness, and the assumption of Euclidean signature, the reference metric is arbitrary but fixed. In particular it is not integrated over in the path integral. Moreover, physical observables must be independent of $\hat{g}_{ij}$

The other metric is the "real" metric denoted by g_{ij}. The purpose of the reference metric is merely to implement a degree of gauge fixing. Thus one assumes that the real metric has the form,

$$g_{ij} = e^{2U} \hat{g}_{ij}. \tag{6.7}$$

The real metric–that is to say U–is a dynamical variable to be integrated over.

For positive c_L the Liouville Lagrangian is

$$L_L = \frac{Q^2\sqrt{\hat{g}}}{8\pi}\left\{\hat{\nabla}U\hat{\nabla}U + \hat{R}u\right\} \tag{6.8}$$

where $\hat{R}$, $\hat{\nabla}$, all refer to the sphere Ω_2, with metric $\hat{g}$. The constant Q is related to the central charge c_L by

$$Q^2 = \frac{c_L}{6} \tag{6.9}$$

The two dimensional cosmological constant has been set to zero for the moment, but it will return to play a surprising role. For future reference we note that the cosmological term, had we included it, would have had the form,

$$L_{cc} = \sqrt{\hat{g}}\lambda e^{2U}. \tag{6.10}$$

It is useful to define a field $\phi = 4QU$ in order to bring the kinetic term to canonical form. One finds,

$$L_L = \frac{\sqrt{\hat{g}}}{8\pi}\left\{\hat{\nabla}\phi\hat{\nabla}\phi + Q\hat{R}\phi\right\} \tag{6.11}$$

and, had we included a cosmological term, it would be

$$L_{cc} = \sqrt{\hat{g}}\lambda \exp\frac{\phi}{2Q}. \tag{6.12}$$

By comparison with the case of positive c_L, very little is rigorously understood about Liouville theory with negative central charge. In this paper we will make a huge leap of faith that may well come back to haunt us: we assume that the theory can be defined by analytic continuation from positive c_L. To that end we note that the only place that the central charge enters (6.11) and (6.15) is through the constants Q and γ, both of which become imaginary when c_L becomes negative. Let us define,

$$Q = i\mathcal{Q}. \tag{6.13}$$

Equations (6.11) and (6.15) become,

$$L_L = \frac{\sqrt{\hat{g}}}{8\pi}\left\{\hat{\nabla}\phi\hat{\nabla}\phi + i\mathcal{Q}\hat{R}\phi\right\}$$

$$= \frac{\sqrt{\hat{g}}}{8\pi}\left\{\hat{\nabla}\phi\hat{\nabla}\phi + 2iQ\phi\right\} \tag{6.14}$$

(where we have used $\hat{R} = 2$), and

$$L_{cc} = \sqrt{\hat{g}}\lambda \exp\frac{-i\phi}{2Q}. \tag{6.15}$$

Let us come now to the role of λ. First of all λ has nothing to do with the four-dimensional cosmological constant, either in the FRW patch or the Ancestor vacuum. Furthermore it is not a constant in the action of the boundary theory. Its proper role is as a ***Lagrange multiplier*** that serves to specify the time T, or more exactly, the global scale factor. The procedure is motivated by the Wheeler-DeWitt procedure of identifying the scale factor with time. In the present case of the thin-wall limit, we identify $\exp 2U$ with $\exp 2T$. Thus we insert a δ function in the path integral,

$$\delta\left(\int \sqrt{\hat{g}}(e^{2U} - e^{2T})\right) = \int dz \exp iz\left(\int \sqrt{\hat{g}}(e^{2U} - e^{2T})\right) \tag{6.16}$$

The path integral (which now includes an integration over the imaginary 2D cosmological constant z) involves the action

$$L_L + L_{cc} = \frac{\sqrt{\hat{g}}}{8\pi}\left\{\hat{\nabla}\phi\hat{\nabla}\phi + 2iQ\phi + 8\pi iz\exp\frac{-i\phi}{2Q} - 8\pi ize^{2T}\right\} \tag{6.17}$$

There is a saddle point when the potential

$$V = 2iQ\phi + 8\pi iz\exp\frac{-i\phi}{2Q} - 8\pi ize^{2T} \tag{6.18}$$

is stationary; this occurs at,

$$\begin{aligned} \exp\frac{-i\phi}{2Q} &= e^{2T} \\ z &= i\frac{Q^2}{8\pi}e^{-2T} \end{aligned} \tag{6.19}$$

or in terms of the original variables,

$$\begin{aligned} e^{2U} &= e^{2T} \\ \lambda &= \frac{Q^2}{8\pi}e^{-2T} \end{aligned} \tag{6.20}$$

Once λ has been determined by (6.20), the Liouville theory with that value of λ determines expectation values of the remaining variables as functions of the time. Thus, as we mentioned earlier, the cosmological constant is not a constant of the theory but rather a parameter that we scan in order to vary the cosmic time.

It should be noted that the existence of the saddle point (6.20) is peculiar to the case of negative c.

6.2 Liouville, Renormalization, and Correlation Functions

6.3 Preliminaries

There are two preliminary discussions that will help us understand the application of Liouville Theory to cosmic holography. The first is about the ADS/CFT connection between the bulk coordinate R, and renormalization-group-running of the boundary field theory. There are three important length scales in every quantum field theory. The first is the "low energy scales;" in the present case the low energy scale is the radius of the sphere which we will call L.

The second important length is the "bare" cutoff scale–where the underlying theory is prescribed. Call it a. The bare input is a collection of degrees of freedom, and an action coupling them. In a lattice gauge theory the degrees of freedom are site and link variables, and the couplings are nearest neighbor to insure locality[5] In a ferromagnet they are spins situated on the sites of a crystal lattice.

The previous two scales have obvious physical meaning but the third scale is arbitrary: a sliding scale called the renormalization or reference scale. We denote it by δ. The reference scale is assumed to be much smaller than L and much larger than a, but otherwise it is arbitrary. It helps to keep a concrete model in mind. Instead of a regular lattice, introduce a "dust" of points with average spacing a. It is not essential that a be uniform on the sphere. Thus the spacing of dust points is a function of position, $a(\Omega_2)$. The degrees of freedom on the dust-points, and their nearest-neighbor couplings, will be left implicit.

Next we introduce a second dust at larger spacing, δ. The δ-dust provides the reference scale. It is well known that for length scales greater than δ, the bare theory on the a-dust

[5]Nearest neighbor is common but not absolutely essential. However this subtlety is not important for us.

can be replaced by a renormalized theory defined on the δ-dust. The renormalized theory will typically be more complicated, containing second, third, and n^{th} neighbor couplings.

Generally, the dimensionless form of the renormalized theory will depend on δ in just such a way that physics at longer scales is exactly the same as it was in the original theory. The dimensionless parameters will flow as the reference scale is changed.

If there is an infrared fixed-point, and if the bare theory is in the basin of attraction of the fixed-point, then as δ becomes much larger than a, the dimensionless parameters will run to their fixed-point values. In that case the continuum limit ($a \to 0$) will be a conformal field theory with $SO(3,1)$ invariance.

Similar things hold in the theory of bulk anti de Sitter space, although in that case the discussion of the bare scale is less relevant–one might as well take the continuum limit $a \to 0$ from the start. In the boundary field theory the infrared scale is provided by the spherical boundary of ADS. From the bulk viewpoint the boundary is at infinite proper distance, at $R = \infty$. However, the time for a signal to reach the boundary and be reflected back to the bulk is finite. In that respect anti de Sitter space behaves like a finite cavity, requiring specific boundary conditions. To be definite, the bulk theory is infrared regulated by replacing Σ with a reference-boundary Σ_0, at finite R. Specifying the boundary conditions on Σ_0 is equivalent to specifying the field theory parameters at scale δ. In parallel with the field theory discussion, the cutoff R, can vary with angular position: $R = R_0(\Omega_2)$. We can now state the UV/IR connection by the simple identification,

$$\delta(\Omega_2) = e^{-R_0(\Omega_2)} \tag{6.21}$$

A useful slogan is that ***"Motion along the R direction is the same as renormalization-group flow."***

Now to the second preliminary–some observations about Lioville Theory. Again it is helpful to have a concrete model. Liouville theory is closely connected with the theory of dense, planar, "fishnet" diagrams [26] such as those which appear in large N gauge theories, and matrix models [27][29][30]. The fishnet plays the role of the bare lattice in the previous discussion, but now it's dynamical–we sum over all fishnet diagrams, assuming only that the spacing (on the sphere) is everwhere much smaller than the sphere size, L. As before, we call the angular spacing between neighboring points on the sphere, $a(\Omega)$.

Each fishnet defines a metric on the sphere. Let $d\alpha$ be a small angular interval (measured in radians). The fishnet-metric is defined by[6]

$$ds^2 = \frac{d\alpha^2}{a(\Omega)^2} \tag{6.22}$$

As before we introduce a reference scale δ. It can also be a fishnet, but now it is fixed, its vertices nailed down, not to be integrated over. We continue to assume that δ satisfies the inequalities, $a(\Omega) << \delta(\Omega) << L$, but otherwise it is arbitrary. The δ-metric is defined by

$$ds_\delta^2 = \frac{d\alpha^2}{\delta(\Omega)^2} \tag{6.23}$$

We can now define the Liouville field U. All it is is the ratio of the reference and fishnet scales:

$$e^U \equiv \delta/a \tag{6.24}$$

Using (6.24) together with $\delta = e^{-R}$, and $ds = \frac{d\alpha}{a}$, we see that U is also given by the relation,

$$ds = d\alpha \; e^{(R_0+U)}. \tag{6.25}$$

In (6.25) both R_0 and U are functions of location on Ω_2, but only U is dynamical, i.e., to be integrated over.

6.4 Liouville in the Hat

With that in mind, we return to cosmic holography, and consider the metric on the regulated spatial boundary of FRW, Σ_0. In the absence of fluctuations it is

$$ds^2 = e^{2R_0} e^{2T} d^2\Omega_2.$$

In general relativity it is natural to allow both R_0 and T to vary over the sphere, so that

$$ds^2 = e^{2R_0(\Omega_2)} e^{2T(\Omega_2)} d^2\Omega_2 \tag{6.26}$$

[6]Strictly speaking there is no need to introduce a discrete fishnet lattice at the scale δ. It is sufficient to just define a continuous function $\delta(\Omega_2)$, and from it define a reference metric by (6.22).

The parallel between (6.25 and (6.26) is obvious. Exactly as we might have expected from the Wheeler-DeWitt interpretation, the Liouville field, U, may be identified with time T, at least when both are large.

$$U \approx T \tag{6.27}$$

To summarize, Let's list a number of correspondences:

$$\begin{aligned} \delta &\leftrightarrow \mu^{-1} \leftrightarrow e^{R} \\ \lambda &\leftrightarrow e^{-T} \\ a &\leftrightarrow e^{-T^{+}} = e^{-(T+R)} \end{aligned} \tag{6.28}$$

One other point about Liouville Theory: the density of vertices of a fishnet is normally varied by changing the weight assigned to vertices. When the fishnet is a Feynman diagram the weight is a coupling constant g. It is well known that the coupling constant and Liouville cosmological constant are alternate descriptions of the same thing. Either can be used to vary the average vertex density–increasing it either by increasing g or decreasing λ. The very dense fishnets correspond to large U and therefore large FRW time, whereas very sparse diagrams dominate the early Planckian era.

6.5 Proactive and Reactive Objects in Quantum Field Theory

There are two kinds of objects in Wilsonian renormalization that correspond quite closely to the terms G_1 and G_2 that we have found in the section 5. I don't know if there is a term for the distinction, but I will call them ***"proactive*** and ***"reactive***. Proactive objects are not quantities that we directly measure; they are objects which go into the definition of the theory. The best example is the exact Wilsonian action, defined at a specific reference scale. The form of proactive quanities depends on that reference scale, and so does the value of their matrix elements; indeed their form, varies with δ in such a way as to keep the physics fixed at longer distances.

By contrast, reactive objects are observables whose value does not depend at all on the reference scale. They do depend on the "bare" cutoff scale a through wave function renormalization constants, which typically tend to zero as $a \to 0$. The wave function renormalization constants are usually stripped off when defining a quantum field but we will find it more illuminating to keep them.

The distinction between these two kinds of objects is subtle, and is perhaps best expressed in Polchinski's version of the exact Wilsonian renormalization group [32]. In that scheme, at every scale there is a renormalized description in terms of local defining fields $\phi(x)$, but the proactive action grows increasingly complicated as the reference scale is lowered.

Consider the exact effective action defined at reference scale δ. It is given by an infinite expansion of the form

$$L_W(\delta) = \sum_{\Delta=2}^{\infty} g_\Delta \mathcal{O}_\Delta \tag{6.29}$$

Where $\mathcal{O}_\Delta$ are a set of operators of dimension Δ, and g_Δ are dimensional coupling constants. The renormalization flow is expressed in terms of the dimensionless coupling constants,

$$\tilde{g}_\Delta = g_\Delta \delta^{(2-\Delta)}. \tag{6.30}$$

The $\tilde{g}$ satisfy RG equations,

$$\frac{d\tilde{g}}{d\log\delta} = -\beta(\tilde{g}), \tag{6.31}$$

and at a fixed point they are constant. Thus the dimensional constants g_Δ in the Lagrangian will grow with δ. Normalizing them at the bare scale a, in the fixed-point case we get,

$$g_\Delta = g_a \left\{\frac{\delta}{a}\right\}^{(\Delta-2)} \tag{6.32}$$

$$L_W(\delta) = \sum_{\Delta=2}^{\infty} \mathcal{O}_\Delta \left\{\frac{\delta}{a}\right\}^{(\Delta-2)}. \tag{6.33}$$

Now consider the two point function of the effective action, $\langle L_W(\delta) L_W(\delta)\rangle$, evaluated at distance scale $L >> \delta$

$$\langle L_W(\delta) L_W(\delta)\rangle = \sum_{\Delta=2}^{\infty} \langle \mathcal{O}_\Delta \mathcal{O}_\Delta \rangle \left(\frac{\delta}{a}\right)^{2(\Delta-2)} \tag{6.34}$$

Suppose the theory is defined on a sphere of radius L and we are interested in the correlator $\langle L_W(\delta) L_W(\delta)\rangle$ between points separated by angle α. The factor $\langle \mathcal{O}_\Delta \mathcal{O}_\Delta \rangle$ is the two-point function of a field of dimension Δ, in a theory on the sphere of size L, with an ultraviolet cutoff at the reference scale δ. Accordingly it has the form

$$\langle \mathcal{O}_\Delta \mathcal{O}_\Delta \rangle = \left(\frac{\delta}{L}\right)^{2\Delta} (1-\cos\alpha)^{-\Delta} \tag{6.35}$$

where the two factors of $\left(\frac{\delta}{L}\right)^{\Delta}$ are the ultraviolet-sensitive wave function renormalization constants. The final result is

$$\langle L_W(\delta)L_W(\delta)\rangle = \sum_{\Delta=2}^{\infty} C_\Delta \left(\frac{\delta}{a}\right)^{2(\Delta-2)} \left(\frac{\delta}{L}\right)^{2\Delta} (1-\cos\alpha)^{-\Delta} \tag{6.36}$$

Note the odd dependence of (6.36) on the arbitrary reference scale δ. That dependence is typical of proactive quantities.

Now consider a reactive quantity such as a fundamental field, a derivative of such a field, or a local product of fields and derivatives. Their matrix elements at distance scale L will be independent of the reference scale (although it will depend of the bare cutoff a) and be of order,

$$\langle\phi\phi\rangle \sim \left(\frac{a}{L}\right)^{2\Delta_\phi} \tag{6.37}$$

where Δ_ϕ is the operator dimension of ϕ. Thus we see two distinct behaviors for the scaling of correlation functions:

$$\left(\frac{\delta}{a}\right)^{2(\Delta-2)} \left(\frac{\delta}{L}\right)^{2\Delta} \quad \text{proactive} \tag{6.38}$$

and

$$\left(\frac{a}{L}\right)^{2\Delta_\phi} \quad \text{reactive} \tag{6.39}$$

The formulas are more complicated away from a fixed point but the principles are the same.

We note that the effective action is not the only proactive object. The energy-momentum tensor, and various currents computed from the effective action, will also be proactive. As we will see these two behaviors–proactive and reactive–exactly correspond to the dependence in (5.9) and (5.10).

Now we are finally ready to complete the discussion about the relation between the correlators of Section 5 and proactive/reactive operators. Begin by noting that in ADS/CFT, the minimally coupled massless (bulk) scalar is the dilaton, and its associated boundary field is the Lagrangian density. It may seem puzzling that in the present case, an entire infinite tower of operators seems to replace, what in ADS/CFT is a single operator. In the case of the metric fluctuations a similar tower replaces the energy-momentum tensor. The puzzle may be stated another way. The FRW geometry consists of an infinite number of Euclidean ADS time slices. At what time (or what 2D cosmological constant) should we evaluate the boundary limits of the metric fluctuations, in order to define the energy-momentum tensor? As we will see, a parallel ambiguity exists in Liouville theory.

Return now, to the three scales of Liouville Theory: the infrared scale L, the reference scale δ, and the fishnet scale a, with $L >> \delta >> a$. It is natural to assume that the basic theory is defined at the bare fishnet scale a by some collection of degrees of freedom at each lattice site, and also specific nearest-neighbor couplings–the latter insuring locality. Now imagine a Wilsonian integration of all degrees of freedom on scales between the fishnet scale and the referencc scale, including the fishnet structure itself. The result will be a proactive effective action of the type we described in equation (6.29). Moreover the correlation function of L_{eff} will have the form (6.36). But now, making the identifications (6.21) and

$$\frac{\delta}{a} = e^U = e^T, \tag{6.40}$$

we see that equation (6.38) for proactive scaling becomes (for each operator in the product)

$$e^{(\Delta-2)T} e^{-\Delta R} \tag{6.41}$$

This is in precise agreement with the coefficients in the expansion (5.8). Similarly the reactive scaling (6.39) is $e^{-\Delta T^+}$, in agreement with the properties of G_2.

It is not obvious to me exactly why the bulk fields should correspond to proactive and reactive boundary fields in the way that they do. I might point out the solutions to the wave equation in the bulk are generally sums of two types of modes,

$$\begin{aligned} \chi_- &\rightarrow g_-(\Omega_2)\ F_-(T^-)\ e^{-2T} \\ \chi_+ &\rightarrow g_+(\Omega_2)\ F_+(T^+). \end{aligned} \tag{6.42}$$

Evidently the two types of solutions couple to objects that are reactive and proactive under the RG flow.

What happens to the proactive objects if we approach Σ by sending $T^+ \rightarrow \infty$ and $T_- \rightarrow -\infty$? In this limit only the dimension-two term survives: exactly what we would expect if the matter action ran toward a fixed point. All of the same things hold true for the tensor fluctuations. Before the limit $T^- \rightarrow -\infty$, the energy-momentum tensor consists of an infinite number of higher dimension operators but in the limit, all tend to zero except for the dimension two term.

It should be observed that the higher dimension contributions to $G_1 \left\{ {ik \atop jl} \right\}$ are not transverse in the two-dimensional sense. This is to be expected: before the limit is taken, the Liouville field does not decouple from the matter field, and the matter energy-momentum is not separately conserved. But if the matter theory is at a fixed point, i.e., scale invariant, the Liouville and matter do decouple and the matter energy-momentum should be conserved. Thus, in the limit in which the dimension-two term dominates, it should be (and is) transverse-traceless.

The RG flow is usually thought of in terms of a single independent flow-parameter. In some versions it's the logarithm of the bare cutoff scale, and in other formulations it's the log of the renormalization scale. In the conventional ADS/CFT framework, R can play either role. One can imagine a bare cutoff at some large R_0 or one can push the bare cutoff to infinity and think of R as a running renormalization scale.

However, for our purposes, it is better to keep track of both scales. One can either think of a one-dimensional (logarithmic) axis–we can call it the "Wilson line"–extending from the infrared scale to the fishnet scale a, or a two dimensional R, T plane. In either case the effective action as a function of two independent variables. Figure 7 shows a sketch of the Wilson line and the two dimensional plane representing the two directions R, and T.

The two independent parameters can be chosen to be a and δ, or equivalently R and T. Yet another choice is to work in momentum space. The reference energy-scale is usually called μ.

$$\mu = e^R \tag{6.43}$$

And in the case of negative central charge, the two-dimensional cosmological constant λ can replace T.

In this light, it is extremely interesting that the distinction between proactive and reactive scaling, corresponds to motion along the two light-like directions T^- and T^+ as depicted in figure 8.

It is important to understand that the duality between FRW cosmology and Liouville 2D gravity does not only involve the continuum fixed point theory. As long as T is finite the theory has some memory of the bare theory. It's only in the limit $T \to \infty$ that the theory flows to the fixed point and loses memory of the bare details. We will come back

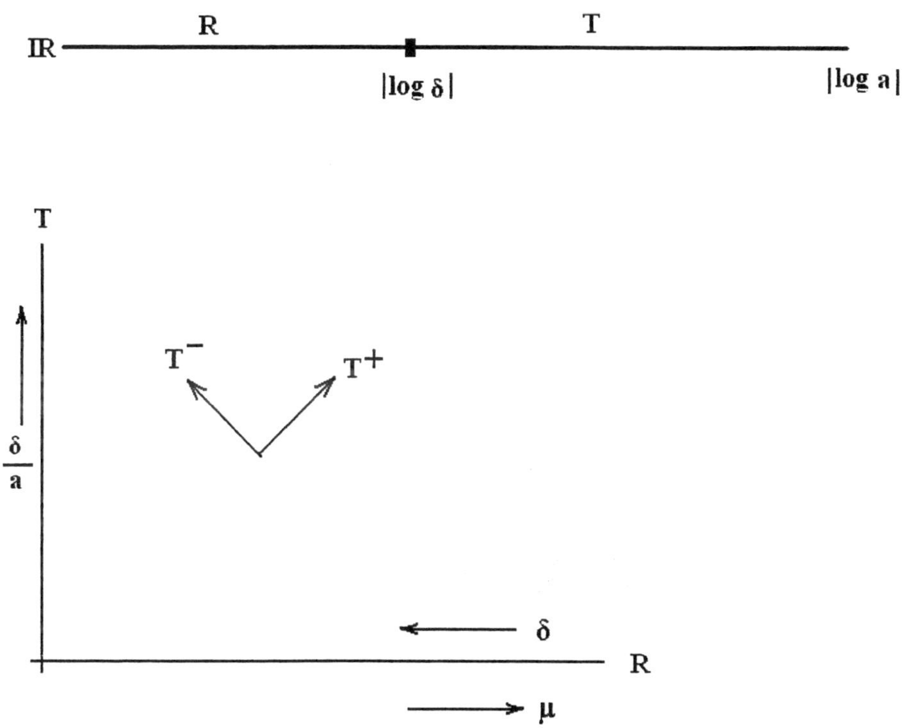

Figure 7: The Wilson line of scales, and the two dimensional R, T plane.

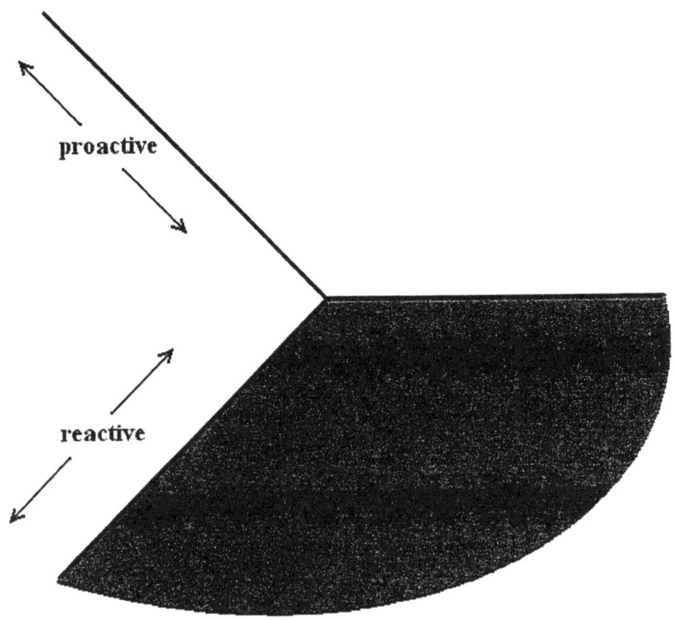

Figure 8: Proactive and reactive quantities scale with the two light-like directions T^- and T^+

to this point in section 8 when I discuss the Garriga, Guth, Vilenkin [20] "Persistence of Memory" phenomenon.

6.6 Boussovian Bounds and The c-Theorem

When a conformal matter theory is coupled to a Liouville field, the two sectors decouple, except for the constraint that the total central charge vanish,

$$c + c_L = 0, \tag{6.44}$$

where c and c_L refer to matter and Liouville respectively. Moreover the matter central charge is constant since a conformal theory is, by definition, at a fixed-point. From now on when I speak of the central charge I will be referring to the matter sector only.

There is one caveat to this rule that c is constant at a fixed point: it applies straightforwardly as long as the reference scale is much smaller than the infrared scale, i.e., the size of the sphere. However, the finite (coordinate) size of the boundary sphere provides an infrared cutoff that is similar to the confinement scale in a confining gauge theory. As the reference scale becomes comparable to the total sphere, the theory runs out of lower energy degrees of freedom and, with some definition, the c-function will go to zero.

The central charge is a measure of the number of degrees of freedom in an area-cell of size e^{-2T^+}. Naively, the holographic principle would say it is the area in such a cell in Planck units. However, as we have seen in the past, this is not always the case [21] [22].

Recall that motion up and down the T^- axis, at fixed T^+, corresponds to the usual RG flow of the Wilsonian action (keeping the bare scale fixed and letting the reference scale vary). Thus, at a fixed point, the area along a fixed T^+ line should be constant. In the thin wall limit the area is given by

$$A = e^{2T} \sinh^2 R. \tag{6.45}$$

As long as $R >> 1$ the area is indeed constant for fixed T^+. However as we approach $R \sim 1$ the area quickly tends to zero, consistent with the remarks above.

More generally, away from fixed points the Zamolodchikov c-theorem requires c to decrease with increasing reference-scale δ. This seems to suggest that the area must monotonically decrease as T^- increases, with T^+ held fixed, which, as we will see raises a paradox; the area is not monotonic beyond the thin-wall approximation.

The study of how area varies along light sheets–"Boussology"–has a long and celebrated history which I will assume you are familiar with [22]. Rather than deal with the equations that determine how area varies I will draw some Bousso diagrams and tell you the conclusions. First the thin-wall case: Figure 9 shows the FRW patch of a thin-wall Coleman De Luccia nucleation. In fact it is the forward light-cone of a point in flat Minkowski space. The red line is a light-sheet of constant T^+. The Bousso wedges indicate the light-like directions along which the area decreases. The entire geometry consists of a single region in which all wedges point toward the origin–the vertical side of the triangle–at $R = 0$. The area is maximum at "*enter*": it is almost constant from "*enter*" to a and then b, but by c it starts to significantly decrease. All of this is in accord with the c-theorem.

Next, in figure 10 some more interesting cosmological examples are shown. On the left is pure radiation dominated FRW. Unlike the thin-wall case it consists of two regions separated by the bold black line. In this case the area ***increases*** from "*enter*" to a, reaching a maximum at b and then decreasing very imperceptibly to c. At c it starts a quick decent to zero. This looks dangerous.

On the right side of figure 10 the blue region is a patch with a finite vacuum energy. It is intended to model an era of slow-roll inflation. For simplicity, at reheating (horizontal edge of the blue region) I have attached it to flat space-time. More realistically one might attach it to the radiation dominated case but the result would be the same.

In this slow-roll case the area starts an exponential increase at "*enter*" and again reaches a maximum at b. Beyond that the behavior is the same as for thin-wall. In fact this behavior of an increase of area, followed by a decrease is very generic. It would seem to violate the c-theorem.

The point, however, is that the identification of c with area is wrong. The number of degrees of freedom of a system is a direct measure of the maximum entropy of that system. The right measure of c, for example at the point a, is not the area of a, but rather the maximum entropy that can pass through the light-sheet from a to the origin at $R = 0$. According to Bousso, that is given by the maximum entropy on the interval between a and b, plus the maximum entropy between b and the origin. Calling the areas of the various points $A(a)$, $A(b)$, $A(c)$ the maximum entropy of the light-sheet bounded by a is given by,

$$S_{max}(a) = (A(b) - A(a)) + (A(b)). \tag{6.46}$$

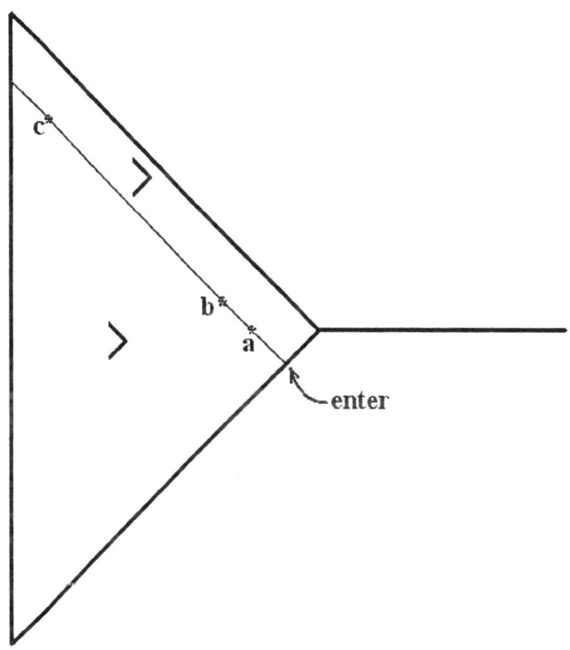

Figure 9: Bousso diagram for the FRW geometry resulting from a thin-wall CDL bubble. The entire FRW geometry consists of a single region in which the contracting light-sheets all point in the same directions. The light-like red line is a surface of constant T^+

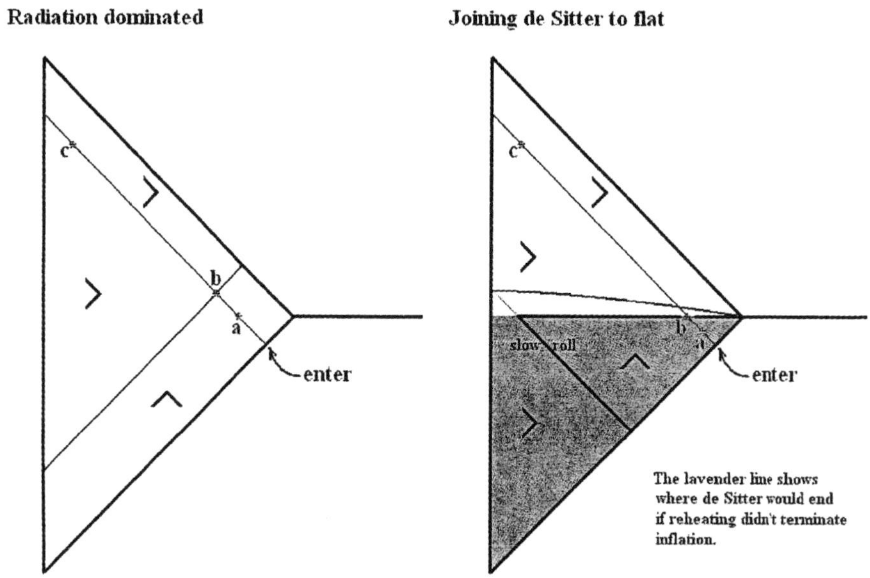

Figure 10: Bousso diagrams for radiation dominated and slow-roll universes. The slow-roll case has been simplified by attaching a de Sitter region to a patch of flat space-time. The blue area represents the inflating region. The light-like red line is a surface of constant T^+.

It is maximized at "*enter*" and monotonically decreases to to zero at $R = 0$. Thus, identifying $c = S_{max}$ restores the monotonicity of the central charge.

It's interesting to compare the naive behavior, $c = A$ with the correct formula $c = S_{max}$ in the slow-roll inflationary case. The naive formula would have c exponentially increasing from "*enter*" to b. The correct formula has it exponentially decreasing toward its (inflated) value at b.

7 Scaling and the Census Taker

7.1 Moments

Now we come to the heart of the matter: the intimate connection between the scaling behavior of two-dimensional quantum field theory and the observations of a Census Taker as he moves toward the Census Bureau. In order to better understand the connection between the cutoff-scale and T^+, let's return to the similar connection between cutoff and the coordinate R in anti de Sitter space. We normalize the ADS radius of curvature to be 1; with that normalization, the Planck area is given by $1/c$ where c is the central charge.

Consider the proper distance between points 1 and 2 given by (5.2). The relation $l = R_1 + R_2 + \log(1 - \cos\alpha)$ is approximate, valid when l and $R_{1,2}$ are all large. When $l \sim 1$ or equivalently, when

$$\alpha^2 \sim e^{-(R_1+R_2)} \tag{7.1}$$

equation (5.2) breaks down. For angles smaller than (7.1) the distance in anti de Sitter space behaves like

$$l \sim e^R \alpha. \tag{7.2}$$

Thus a typical correlation function will behave as a power of $(1 - \cos\alpha)$ down to angular distances of order (7.1) and then fall quickly to zero.

The angular cutoff in anti de Sitter space has a simple meaning. The solid angle corresponding to the cutoff is of order e^{-2R} while the area of the regulated boundary is e^{+2R}. Thus, metrically, the cutoff area is of order unity. This means that in Planck units, the cutoff area is the central charge c of the boundary conformal field theory.

Now consider the cutoff angle implied by (5.12) and (5.13). By an argument parallel to the one above, the cutoff angle becomes

$$\alpha^2 \sim e^{-(T_1^+ + T_2^+)} \tag{7.3}$$

Once again this corresponds to a proper area on Σ_0 (the regulated boundary) which is time-independent, and in Planck units, of order the central charge.

Consider the Census Taker looking back from some late time T_{CT}. For convenience we place the CT at $R = 0$. His backward light-cone is the surface

$$T + R = T_{CT}. \tag{7.4}$$

The CT can never quite see Σ. Instead he sees the regulated surfaces corresponding to a fixed proper cutoff (figure 4). The later the CT observes, the smaller the angular structure that he can resolve on the boundary. This is another example of the UV/IR connection, this time in a cosmological setting.

Let's consider a specific example of a possible observation. The massless scalar field χ of section 5 has an asymptotic limit on Σ that defines the dimension zero field $\chi(\Omega)$. Moments of χ can be defined by integrating it with spherical harmonics,

$$\chi_{lm} = \int \chi(\Omega) Y_{lm}(\Omega) d^2\Omega \tag{7.5}$$

It is worth recalling that in anti de Sitter space the corresponding moments would all vanish because the normalizable modes of χ all vanish exponentially as $R \to \infty$. The possibility of non-vanishing moments is due entirely to the asymptotic warmness of open FRW.

We can easily calculate [7] the mean square value of χ_{lm} (It is independent of m).

$$\langle \chi_l^2 \rangle = \int \log(1 - \cos\alpha) P_l(\cos\alpha) \sim \frac{1}{l(l+1)} \tag{7.6}$$

It is evident that at a fixed Census Taker time T_{CT}, the angular resolution is limited by (7.3). Correspondingly, the largest moment that the CT can resolve corresponds to

$$l_{max} = e^{T_{CT}} \tag{7.7}$$

Thus we arrive at the following picture: the Census Taker can look back toward Σ but at any given time his angular resolution is limited by (7.3) and (7.7). As time goes on more and more moments come into view. Once they are measured they are frozen and cannot

[7] I am indebted to Ben Freivogel for explaining equation (7.6) to me.

change. In other words the moments evolve from being unknown quantum variables, with a gaussian probability distribution, to classical boundary conditions that explicitly break rotation symmetry (and therefore conformal symmetry). One sees from (7.6) that the symmetry breaking is dominated by the low moments.

But I doubt that this phenomenon ever occurs in an undiluted form. Realistically speaking, we don't expect massless scalars in the non-supersymmetric Ancestor. In Section 5 we discussed the effect of a small mass term, in the ancestor vacuum, on the correlation functions of χ. The result of such a mass term is a shift of the leading dimension[8] from 0 to μ. This has an effect on the moments. The correlation function becomes

$$e^{-\mu T_1^+} e^{-\mu T_2^+} (1 - \cos\alpha)^{-\mu}. \tag{7.8}$$

and the moments take the form

$$\langle \chi_l^2 \rangle = e^{-2\mu T_{CT}} \int (1 - \cos\alpha)^{\mu} \, P_l(\cos\alpha) \tag{7.9}$$

The functional form of the l dependence changes a bit, favoring higher l, but more importantly, the observable effects decrease like $e^{-2\mu T_{CT}}$. Thus as T_{CT} advances, the asymmetry on the sky decreases exponentially with conformal time. Equivalently it decreases as a power of proper time along the CT's world-line.

7.2 Homogeneity Breakdown

Homogeneity in an infinite FRW universe is generally taken for granted, but before questioning homogeneity we should know exactly what it means. Consider some three-dimensional scalar quantity such as energy density, temperature, or the scalar field χ. Obviously the universe is not uniform on small scales, so in order to define homogeneity in a useful way we need to average χ over some suitable volume. Thus at each point X of space, we integrate χ over a sphere of radius r and then divide by the volume of the sphere. For a mathematically exact notion of homogeneity the size of the sphere must tend to infinity. The definition of the average of χ at the point X is

$$\overline{\chi(X)} = \lim_{r\to\infty} \frac{\int \chi d^3x}{V_r} \tag{7.10}$$

[8] In the de Sitter/CFT correspondence [34], the dimension of a massive scalar becomes complex when the mass exceeds the Hubble scale. In our case the dimension remains real for all μ. I thank Yasuhiro Sekino for this observation.

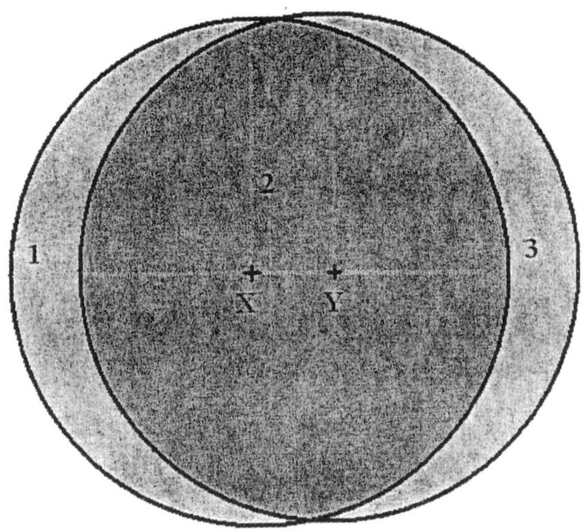

Figure 11: Two large spheres centered at X and Y.

Now pick a second point Y and construct $\overline{\chi(Y)}$. The difference $\overline{\chi(X)} - \overline{\chi(Y)}$ should go to zero as $r \longrightarrow \infty$ if space is homogeneous. But as the spheres grow larger than the distance between X and Y, they eventually almost completely overlap. In figure 11 we see that the difference between $\overline{\chi(X)}$ and $\overline{\chi(y)}$ is due to the two thin crescent-shaped regions, 1 and 3. It seems evident that the overwhelming bulk of the contributions to $\overline{\chi(X)}$, $\overline{\chi(Y)}$ come from the central region 3, which occupies almost the whole figure. The conclusion seems to be that the averages, if they exist at all, must be independent of position. Homogeneity while true, is a triviality.

This is correct in flat space, but surprisingly it can break down in hyperbolic space [9]. The reason is quite simple: despite appearances the volume of regions 1 and 3 grow just as rapidly as the volume of 2. The ratio of the volumes is of order

$$\frac{V_1}{V_2} = \frac{V_3}{V_2} \sim \frac{l}{R_{curvature}} \tag{7.11}$$

and remains finite as $r \longrightarrow \infty$.

To be more precise we observe that

$$\begin{aligned} \overline{\chi(X)} &= \frac{\int_1 \chi + \int_2 \chi}{V_1 + V_2} \\ \overline{\chi(Y)} &= \frac{\int_3 \chi + \int_2 \chi}{V_3 + V_2} \end{aligned} \tag{7.12}$$

and that the difference $\overline{\chi(X)} - \overline{\chi(Y)}$ is given by

$$\overline{\chi(X)} - \overline{\chi(Y)} = \frac{\int_1 \chi}{V_1 + V_2} - \frac{\int_1 \chi}{V_1 + V_2} \tag{7.13}$$

which, in the limit $r \longrightarrow \infty$ is easily seen to be proportional to the dipole-moment of the boundary theory,

$$\overline{\chi(X)} - \overline{\chi(Y)} = l \int \chi(\Omega) \cos\theta d^2\Omega = l\ \chi_{1,0}, \tag{7.14}$$

where l is the distance between X and Y.

Since, as we have already seen for the case $\mu = 0$, the mean square fluctuation in the moments does not go to zero with distance, it is also true that average value of $|\overline{\chi(X)} - \overline{\chi(Y)}|^2$ will be nonzero. In fact it grows with separation.

---

[9]L.S. is grateful to Larry Guth for explaining this phenomenon, and to Alan Guth for emphasizing its importance in cosmology.

However there is no reason to believe that a dimension zero scalar exists. Moduli, for example, are expected to be massive in the Ancestor, and this shifts the dimension of the corresponding boundary field. In the case in which the field χ has dimension μ, the effect (non-zero rms average of moments) persists in a somewhat diluted form. If a renormalized field is defined by stripping off the wave function normalization constants, $\exp(-\mu T^+)$, the squared moments still have finite expectation values and break the symmetry. However, from an observational point of view there does not seem to be any reason to remove these factors. Thus it seems that as the Census Taker time tends to infinity, the observable asymmetry will decrease like $\exp(-2\mu T_{CT})$.

8 Bubble Collisions and Other Matters

The Census Taker idea originated with attempts to provide a measure on the Landscape. By looking back toward Σ, the Census Taker can see into bubbles of other vacua–bubbles that in the past collided with his hatted vacuum. By counting the bubbles of each type on the sky, he can try to define a measure on the Landscape. Whether or not this can be done, it is important to our program to understand the representation of bubble collisions in the language of the boundary holographic field theory.

Long ago, Guth and Weinberg [35] recognized that a single isolated bubble is infinitely unlikely, and that a typical "pocket universe" will consist of a cluster of an unbounded number of colliding bubbles, although if the nucleation rate is small the collisions will in some sense be rare. To see why such bubble clusters form it is sufficient to recognize why a single bubble is infinitely improbable. In figure 12 the main point is illustrated by drawing a time-like trajectory that approaches Σ from within the Ancestor vacuum. The trajectory has infinite proper length, and assuming that there is a uniform nucleation rate, a second bubble will eventually swallow the trajectory and collide with the original bubble. Repeating this process will produce an infinite bubble cluster.

More recently Garriga, Guth, and Vilenkin [20] have argued that the multiple bubble collisions must spontaneously break the $SO(3,1)$ symmetry of a single bubble, and in the process render the (pocket) universe inhomogeneous and anisotropic. The breaking of symmetry in [20] was described, not as spontaneous breaking, but as explicit breaking due to initial conditions. However, spontaneous symmetry breaking is nothing but the memory of a temporary explicit symmetry breaking, if the memory does not fade with

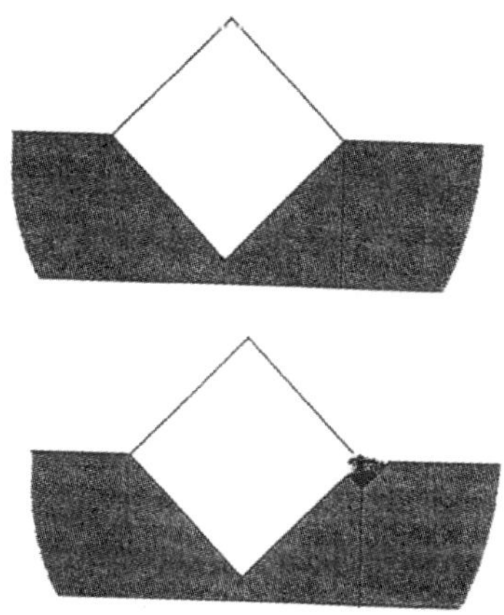

Figure 12: The top figure represents a single nucleated bubble. The red trajectory is a time-like curve of infinite length approaching Σ. Because there is a constant nucleation rate along the curve, it is inevitable that a second bubble will nucleate as in the lower figure. The two bubbles will collide.

time. For example, a small magnetic field in the very remote past will determine the direction of an infinite ferromagnet for all future time. Spontaneous symmetry breaking ***is*** "The Persistence of Memory."

The actual observability of bubble collisions depends on the amount of slow-roll inflation that took place after tunneling. Much more than 60 e-foldings would probably wipe out any signal, but our interest in this paper is conceptual. We will take the viewpoint that anything within the past light-cone of the Census Taker is in principle observable.

In the last section we saw that perturbative infrared effects are capable of breaking the $SO(3,1)$ symmetry, and it is an interesting question what the relation between these two mechanisms is. The production of a new bubble would seem to be a non-perturbative effect that adds to the perturbative symmetry breaking effects of the previous section. Whether it adds distinctly new effects that are absent in perturbation theory is not obvious and may depend on the specific nature of the collision. Let us classify the possibilities.

8.1 Collisions with Identical Vacua

The simplest situation is if the true-vacuum bubble collides with another identical bubble, the two bubbles coalescing to form a single bubble, as in the top of figure 13.

The surface Σ is defined by starting at the tip of the hat and tracking back along light-like trajectories until they end–in this case at a false vacuum labeled **F**. The collision is parameterized by the space-like separation between nucleation points. Particles produced at the collision of the bubbles just add to the particles that were produced by ordinary FRW evolution. The main effect of such a collision is to create a very distorted boundary geometry, if the nucleation points are far apart. When they are close the double nucleation blends in smoothly with the single bubble. These kind of collisions seem to be no different than the perturbative disturbances caused by the non-normalizable mode of the metric fluctuation. Garriga, Guth, and Vilenkin, compute that the typical observer will see multipole moments on the sky, but as we've seen, similar multipole moments can also occur perturbatively.

In the bottom half of figure 13 we see another type of collision in which the colliding bubbles correspond to two different true vacua: red (r) and blue (b). But in this case red

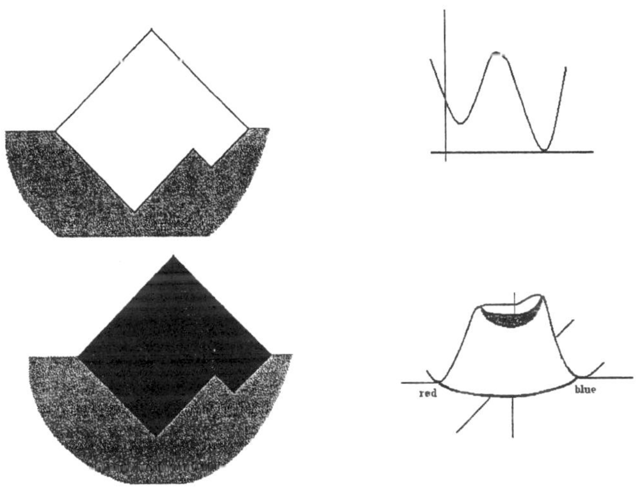

Figure 13: In the top figure two identical bubbles collide. This would be the only type of collision in a simple landscape with two discrete minima–one of positive energy and one of zero energy. In the lower figure a more complicated situation is depicted. In this case the false vacuum F can decay to two different true vacua, "red" and "blue," each with vanishing energy. The two true vacua are connected by a flat direction, but CDL instantons only lead to the red and blue points.

and blue are on the same moduli-space, so that they are connected by a flat direction[10]. Both vacua are included within the hat. In the bulk space-time they bleed into each other, so that as one traverses a space-like surface, blue gradually blends into purple and then red.

On the other hand the surface Σ is sharply divided into blue and red regions, as if by a one dimensional domain wall. This seems to be a new phenomenon that does not occur in perturbation theory about either vacuum.

As an example, consider a case in which a red vacuum-nucleation occurs first, and then much later a blue vacuum bubble nucleates. In that case the blue patch on the boundary will be very small and The Census Taker will see it occupying tiny angle on the sky. How does the boundary field theorist interpret it? The best description is probably as a small blue instanton in a red vacuum. In both the bulk and boundary theory this is an exponentially suppressed, non-perturbative effect.

However, in a conformal field theory the size of an instanton is a modulus that must be integrated over. As the instanton grows the blue region engulfs more and more of the boundary. Eventually the configuration evolves to a blue 2D vacuum, with a tiny red instanton. One can also think of the two configurations as the observations of two different Census Takers at a large separation from one another. Which one of them is at the center, is obviously ambiguous.

The same ambiguous separation into dominant vacuum, and small instanton, can be seen another way. The nucleation sites of the two bubbles are separated by a space-like interval. There is no invariant meaning to say that one occurs before the other. A element of the de Sitter symmetry group can interchange which bubble nucleates early and which nucleates later.

Nevertheless, a given Census Taker will see a definite pattern on the sky. One can always define the CT to be at the center of things, and integrate over the relative size of the blue and red regions. Or one can keep the size of the regions fixed–equal for example–and integrate over the location of the CT.

From both the boundary field theory, and the bubble nucleation viewpoints, the probability for any finite number of red-blue patches is zero. Small red instantons will be

[10]I assume that there is no symmetry along the flat direction, and that there are only two tunneling paths from the false vacuum, one to red, and one to blue.

sprinkled on every blue patch and vice versa, until the boundary becomes a fractal. The fractal dimensions are closely connected to operator dimensions in the boundary theory. Moreover, exactly the same pattern is expected from multiple bubble collisions.

But the Census taker has a finite angular resolution. He cannot see angular features smaller than $\delta\alpha \sim \exp(-T_{CT})$. Thus he will see a finite sprinkling of red and blue dust on the sky. At T_{CT} increases, the UV cutoff scale tends to zero and the CT sees a homogeneous "purple" fixed-point theory.

The red and blue patches are reminiscent of the Ising spin system (coupled to a Liouville field). As in that case, it makes sense to average over small patches and define a continuous "color field" ranging from intense blue to intense red. It is interesting to ask whether Σ would look isotropic, or whether there will be finite multipole moments of the renormalized color color field (as in the case of the χ field). The calculations of Garriga, Guth, and Vilenkin suggest that multipole moments would be seen. But unless for some reason there is a field of exactly zero dimension, the observational signal should fade with Census Taker time.

There are other types of collisions that seem to be fundamentally different from the previous. Let us consider a model landscape with three vacua–two false, B and W (Black and White); and one true vacuum T. Let the the vacuum energy of B be bigger than that of W, and also assume that the decays $B \to W$, $B \to T$, and $W \to T$ are all possible. Let us also start in the Black vacuum and consider a transition to the True vacuum. The result will be a hat bounded by Σ.

However, if a bubble of W forms, it may collide with the T bubble as in figure 14. The W bubble does not end in a hat but rather, on a space-like surface. By contrast, the true vacuum bubble does end in a hat. The surface Σ is defined as always, by following the light-like generators of the hat backward until they enter the bulk–either Black or White–as in figure 14.

In this case a portion of the boundary Σ butts up against B, while another portion abuts W. In some ways this situation is similar to the previous case where the boundary was separated into red and blue regions, but there is no analogue of the gradual bleeding of vacua in the bulk. In the previous case the Census Taker could smoothly pass from red to blue. But in the current example, the CT would have to crash through a domain wall in order to pass from T to W. Typically this happens extremely fast, long before the CT

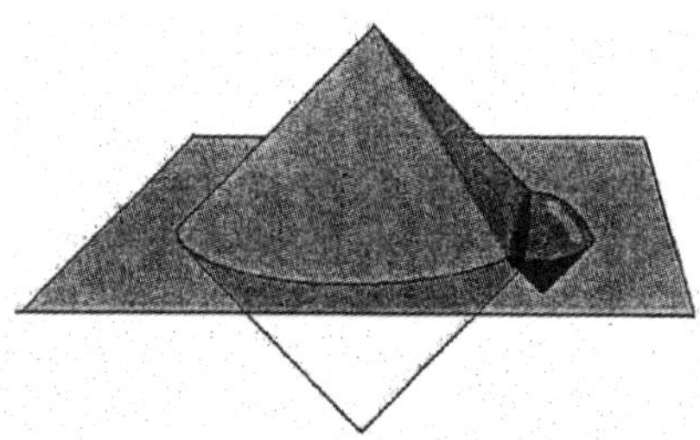

Figure 14: A bubble of True vacuum forms in the Black false vacuum and then collides with a bubble of White vacuum. The true vacuum is bounded by a hat but the White vacuum terminates in a space-like surface. Some generators of the hat intersect the Black vacuum and some intersect the White. Thus Σ, shown as the red curve, is composed of two regions.

could do any observation. In fact if we define Census Takers by the condition that they eventually reach the Census Bureau, then they simply never enter W.

From the field theory point of view this example leads to a paradox. Naively, it seems that once a W patch forms on Σ, a B region cannot form inside it. A constraint of this type on field configurations would obviously violate the rules of quantum field theory; topologically (on a sphere) there is no difference between a small W patch in a B background, and a small B patch in a W background. Thus field configurations must exist in which a W region has smaller Black spots inside it. There is no way consistent with locality and unitarity to forbid bits of B in regions of W.

Fortunately the same conclusion is reached from the bulk point of view. The rules of tunneling transitions require that if the transition $B \to W$ is possible, so must be the transition $W \to B$, although the probability for the latter would be smaller (by a large density of states ratio). Thus one must expect B to invade regions of W.

As the Census Taker time advances he will see smaller and smaller spots of each type. If one assumes that there are no operators of dimension zero, then the pattern should fade into a homogeneous average grey, although under the conditions I described it will be almost White.

The natural interpretation is that the boundary field theory has two phases of different free energy, the B free energy being larger than that of W. The dominant configuration would be the ones of lower free energy with occasional fluctuations to higher free energy.

8.2 The Persistence of Memory

The "Persistence of Memory" reported in [20] had nothing to do with whether or not the Census Taker's sees a fading signal: Garriga, Guth, and Vilenkin were not speaking about Census Taker time at all. They were referring to the fact that no matter how long after the start of eternal inflation a bubble nucleates, it will remember the symmetry breaking imposed by the initial conditions; not whether the signal fades with T_{CT}. Returning to figure 12, one might ask why no bubble formed along the red trajectory in the infinitely remote past. The authors of [20] argue that eternal inflation does not make sense without an initial condition specifying a past surface on which no bubbles had yet formed. That surface invariably breaks the O(3,1) symmetry and distinguishes a "preferred Census Taker" who is at rest in the frame of the initial surface. He alone sees an isotropic sky whereas all the other Census Takers see non-zero anisotropy. What's more the effect persists no matter how late the nucleation takes place.

As before, when the Census Taker's time advances, the asymmetry should become diluted if there are no dimension zero operators, but the existence of a preferred Census Taker at finite time makes this symmetry breaking seem different than what we have discussed up to now.

Let us consider how this phenomenon fits together with the RG flow discussed earlier[11]. Begin by considering the behavior for finite δ in the limit of small a. It is reasonable to suppose that in integrating out the many scales between a and δ, the theory would run to a fixed point. Now recall that this is the limit of very large T. If in fact the theory has run to a fixed point it will be conformally invariant. Thus we expect that the symmetry $O(3,1)$ will be unbroken at very late time.

On the other hand consider the situation of δ/a near 1. The reference and bare scales are very close and very few degrees of freedom have been integrated out. There is no reason why the effective action should be near a fixed point. The implication is that at very early time (recall, $\delta/a = e^T$) the physics on a fixed time slice will not be conformally invariant. Near the beginning of an RG flow the effective action is strongly dependent on the bare theory. The implication of a breakdown of conformal symmetry is that there is no symmetry between Census Takers at different locations in space. In such situations the center of the (deformed) anti de Sitter space is, indeed, special.

Shenker and I suggest that the GGV boundary condition at the onset of eternal inflation is the same thing as the initial condition on the RG flow. In other words, varying the GGV boundary condition is no different from varying the bare fishnet theory.

Is it possible to tune the bare action so that the theory starts out at the fixed point? If this were so, it would be an initial condition that allowed exact conformal invariance for all time. Of course it would involve an infinite amount of fine tuning and is probably not reasonable. But there may be reasons to doubt that it is possible altogether, even though in a conventional lattice theory it is possible.

The difficulty is that the bare and renormalized theories are fundamentally different. The bare theory is defined on a variable fishnet whose connectivity is part of the dynamical

[11]These observations are based on work with Steve Shenker.

degrees of freedom. The renormalized theory is defined on the fixed reference lattice. The average properties of the underlying dynamical fishnet are replaced by conventional fields on the reference lattice. Under these circumstances it is hard to imagine what it would mean to tune the bare theory to an exact fixed point.

The example of the previous subsection involving two false vacua, B and W, raises some interesting questions. First imagine starting with GGV boundary conditions such that, on some past space-like surface, the vacuum is pure Black and that a bubble of true vacuum nucleates in that environment. Naively the boundary is mostly black. That means that ***in the boundary theory*** the free energy of Black must be lower than that of White.

But we argued earlier that white instantons will eventually fill Σ with an almost white, very light grey color, exactly as if the initial GGV condition were White. That means that White must have the lower potential energy. What then is the meaning of the early dominance of B from the 2D field theory viewpoint?

The point is that it is possible for two rather different bare actions to be in the same broad basin of attraction and flow to the same fixed point. The case of Black GGV conditions corresponds to a bare starting point (in the space of couplings) where the potential of B is lower than W. During the course of the flow to the fixed point the potential changes so that at the fixed-point W has the lower energy.

On the other hand, White GGV initial conditions corresponds to starting the flow at a different bare point–perhaps closer to the fixed point—where the potential of W is lower.

This picture suggests a powerful principle. Start with the space of two-dimensional actions, which is broad enough to contain a very large Landscape of 2D theories. With enough fields and couplings the space could probably contain everything. As Wilson explained [31], the space divides itself into basins of attraction. Each initial state of the universe is described either as a GGV initial condition, or as a bare starting point for an RG flow. The endpoints of these flows correspond to the possible final states-the hats–that the Census Taker can end up in.

We have not exhausted all the kinds of collisions that can occur–in particular collisions with singular, negative cosmological constant vacua. A particularly thorny situation results if there is a BPS domain wall between the negative and zero CC bubbles, then as shown by Freivogel, Horowitz, and Shenker [36] the entire hat may disappear in a catastrophic crunch. A possible interpretation is that the catastrophe is due to the existence of a relevant operator which destabilizes the fixed point. These and other issues will be taken up in a paper with Shenker.

8.3 A Remark about Supersymmetry

Most likely, the only 4D vacua with exactly vanishing cosmological constant are supersymmetric. Does that mean that the boundary theory on Σ is also supersymmetric? The answer is no: correlators on Σ are largely determined by the properties of the non-supersymmetric Ancestor vacuum. For example, the gravitino will be massive in the Ancestor, and the methods of [19] would give different dimensions, Δ, for the graviton and gravitino fluctuations.

In fact Σ is contiguous with the de Sitter Ancestor and has every reason to strongly feel the supersymmetry breaking. It is the region close to the tip of the hat where the physics should be dominated by the properties of the supersymmetric terminal vacuum. If one looks at the expansion (5.9), it is clear that the tip of the hat is dominated by the asymptotically high dimensional terms. Thus we expect supersymmetry to manifests itself asymptotically, in the spectrum and operator products of high dimensional operators.

8.4 Flattened Hats and other Tragedies

In a broad sense this paper is about phenomenology: the Census Taker could be us. If we lived in an ideal thin-wall hat we would see, spread across the sky, correlation functions of a holographic quantum field theory. We could measure the dimensions of operators both by the time dependence of the received signals, and their angular dependence. Bubble collisions would appear as patches resembling instantons.

Unfortunately (or perhaps fortunately) we are insulated from these effects by two forms of inflation–the slow-roll inflation that took place shortly after bubble nucleation–and the current accelerated expansion of the universe. The latter means that we don't live in a true hatted geometry. Rather we live in a flattened hat, at least if we ignore the final decay to a terminal vacuum.

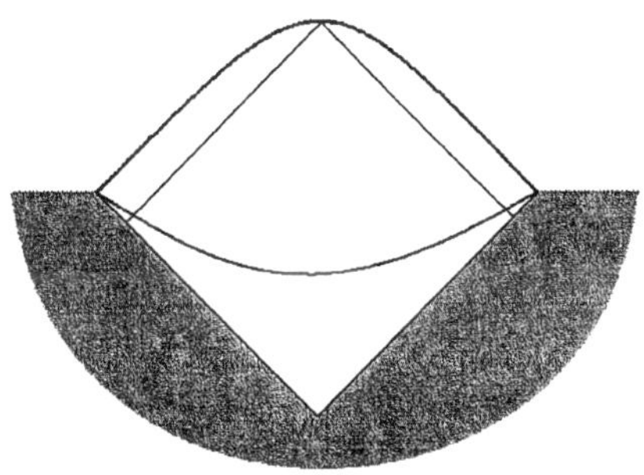

Figure 15: If a CDL bubble leads to a vacuum with a small positive cosmological constant, the hat is replaced by a rounded space-like surface. The result is that no Census Taker can look back to Σ.

The Penrose diagram in figure 15 shows an an Ancestor, with large vacuum energy, decaying to a vacuum with a very small cosmological constant. The important new feature is that the hat is replaced by a space-like future infinity. Consider the Census Taker's final observations as he arrives at the flattened hat. It is obvious from figure 18 that he cannot look back to Σ. His past light cone is at a finite value of T^+. Thus for each time-slice T, there is a maximum radial variable $R = R_0(T)$ within his ken, no matter how long he waits. In other words there is an unavoidable ultraviolet cutoff. It is completely evident that a final de Sitter bubble must be described by a theory with no continuum limit; in other words not only a non-local theory, but one with no ultraviolet completion.

This suggests that de Sitter Space may have an intrinsic imprecision. Indeed, as Seiberg has emphasized, the idea of a metastable vacuum is imprecise, even in condensed matter physics where they are common[12].

The more tragic fact is that all of the memory of a past bubble may, for observational purposes, be erased by the slow-roll inflation that took place shortly after the Coleman De Luccia tunneling event–unless it lasted for the minimum permitted number of e-foldings [16]. In principle the effects are imprinted on the sky, but in an exponentially diluted form.

8.5 Note About W

What I have described is only half the story: the half involving S, the real part of $\log \Psi$. The other half involves W, the phase of the wave function. Knowing S is enough to compute the expectation values of the fields y at a given value of time, or, strictly speaking, at a value of the scale factor. Scanning over scale factors can be done by varying the Liouville cosmological constant.

However, quantum mechanics cannot be complete without the phase of the wave function. In particular, the values of conjugate momenta requires knowledge of W. The same is true for products of fields at different times.

9 Some Conclusions

I've given some circumstantial evidence that there is a duality between cosmology on the Landscape, and two-dimensional conformal field theory, with a Liouville field. The

[12]I am grateful to Nathan Seiberg for discussions on this point.

data that supports the theory are the computations done in [19], but more importantly, a compelling physical picture accompanies the data. The most pertinent observation is that the sky is a two-sphere. Moreover, it is covered with interesting observable correlations; in principle we can look back through the surface of last scattering and observe these correlations at any past time. The Liouville field, or alternately, the Liouville cosmological constant (not to be confused with the four-dimensional cosmological constant) is the dual of that time, along the observers backward light-cone.

As we view the deep sky from increasingly late times, we can in principle see greater angular detail on Σ_0, i.e., spatial "almost-infinity". The increasing angular resolution defines a renormalization group flow that begins with some bare action, and ends at an infrared fixed point. The starting point of the flow is equivalent to the boundary condition, whose memory, Garriga, Guth, and Vilenkin have argued, is persistent. The phenomena of symmetry breaking–the picking out of a special Census Taker at the center of things–by the GGV's initial condition, is equivalent to the breaking of conformal symmetry, at the start of a RG flow. As in anti de Sitter space, breaking conformal symmetry makes the center of ADS (at $R = 0$) special.

Our considerations are based on the Holographic Principle but with a new twist. The asymptotic warmness of space requires a field in order to represent geometric fluctuations at infinity. By now we are used to one or more spatial directions emerging holographically, but this new Liouville degree of freedom creates a new emergence–of time.

Examples of cosmic phenomena that can be simply interpreted as two-dimensional field theory phenomena are the bubble collisions of Guth and Weinberg [35], which appear as instantons on the two-dimensional sky; the fading of the initial conditions with Census Taker time is connected with the spectrum of conformal dimensions–in particular the lack of dimension zero scalars; and ordinary slow-roll inflation corresponds to an exponential decrease in the Zamolodchikov c-function [33]. It will be interesting to try to interpret CMB fluctuations in this language but this has not been studied–at least that I know of.

10 Warning to All Census Takers

Memo from the Director

Keep in mind that when you look back toward Σ, what you see will be influenced not *only by conditions on the regulated boundary, but also by the gravitational field between you and Σ_0. Angular separations detected by you, must be corrected for nearby gravitational distortions such as lensing and gravitational waves. Please make all corrections before reporting your data.*

11 Acknowledgments

I am enormously grateful to my collaborators, Ben Freivogel, Yasuhiro Sekino, Chen-Pin Yeh, Steve Shenker, and Alex Maloney. This lecture represents an ongoing discussion with them. Any (positive) reference to it should be accompanied by references to [1] and [19].

Discussions with Larry Guth and Alan Guth were important to my understanding of the breaking of symmetries in Hyperbolic space.

I would also like to acknowledge helpful conversations with Raphael Bousso, another believer in the causal patch; Simeon Hellerman; and Matt Kleban for various insights; and finally, Nathan Seiberg for explaining the intrinsic imprecision of metastable states.

Any conceptual or computational errors in the paper are of course my own.

References

[1] A. Maloney, S. Shenker, L. Susskind, unpublished

[2] N. Kaloper, M. Kleban and L. Sorbo, "Observational implications of cosmological event horizons," Phys. Lett. B **600**, 7 (2004) [arXiv:astro-ph/0406099]; M. Kleban, "Inflation unloaded," [arXiv:hep-th/0412055].

[3] Raphael Bousso, Joseph Polchinski, Quantization of Four-form Fluxes and Dynamical Neutralization of the Cosmological Constant, hep-th/0004134, JHEP 0006 (2000) 006

[4] S. Kachru, R. Kallosh, A. Linde and S. P. Trivedi: De Sitter vacua in string theory. Phys. Rev. D 68, 046005 (2003), hep-th/0301240.

[5] L. Susskind, The Anthropic Landscape of String Theory [arXiv:hep-th/0302219]

[6] Michael R. Douglas, The statistics of string/M theory vacua, JHEP 0305 (2003) 046, hep-th/0303194

[7] L. Susskind, L. Thorlacius, J. Uglum, The Stretched Horizon and Black Hole Complementarity, Phys.Rev. D48 (1993) 3743-3761, arXiv:hep-th/9306069

[8] Y. Kiem, Herman L. Verlinde, Erik P. Verlinde Quantum horizons and complementarity. *Trieste 1994, High energy physics and cosmology* 586-606.

[9] Tom Banks, Willy Fischler M theory observables for cosmological space-times. hep-th/0102077

[10] G. 't Hooft, Dimensional Reduction in Quantum Gravity, Dimensional Reduction in Quantum Gravity, gr-qc/9310026

L. Susskind, The World as a Hologram, J.Math.Phys. 36 (1995) 6377-6396, hep-th/9409089

[11] Juan Maldacena, The Large N Limit Of Superconformal Field Theories And Supergravity, Int.J.Theor.Phys.38:1113-1133,1999, hep-th/9711200

[12] Edward Witten, Anti-de Sitter space and holography. Adv.Theor.Math.Phys.2:253-291,1998, hep-th/9802150

[13] Susskind, L., and E. Witten, 1998, The holographic bound in Anti-de Sitter space, hep-th/9805114.

[14] J.R. Gott, Nature 295, 304 (1982); A.H. Guth and E. Weinberg, Nucl. Phys. B212, 321 (1983); P.J. Steinhardt, in The Very Early Universe, ed. by G.W. Gibbons, S.W.Hawking and S.T.C. Siklos (Cambridge University Press, Cambridge, 1983); J.R. Gott and T.S. Statler, Phys. Lett. 136B, 157 (1984); M. Bucher, A.S. Linde, A D 1986 Eternal chaotic inflation, Mod. Phys. Lett. A 1, 81-5; Linde, A D 1986 Eternally existing selfreproducing chaotic inflationary universe, Phys.

[15] S. R. Coleman and F. De Luccia, "Gravitational Effects On And Of Vacuum Decay," Phys. Rev. D **21**, 3305 (1980).

[16] Ben Freivogel, Matthew Kleban, Maria Rodriguez Martinez, Leonard Susskind, Observational Consequences of a Landscape, JHEP 0603 (2006) 039 , arXiv:hep-th/0505232

[17] S. W. Hawking, Comm .Math. Phys. 43 (1975), 199, Phys. Rev. D14 (1976), 2460.

[18] Raphael Bousso, Ben Freivogel, I-Sheng Yang, Eternal Inflation: The Inside Story, Phys.Rev. D74 (2006) 103516 arXiv:hep-th/0606114

[19] Ben Freivogel, Yasuhiro Šekino, Leonard Susskind, Chen-Pin Yeh, "A Holographic Framework for Eternal Inflation," Phys.Rev. D74 (2006) hep-th/0606204

[20] Jaume Garriga, Alan H. Guth, Alexander Vilenkin, Eternal inflation, bubble collisions, and the persistence of memory, arXiv:hep-th/0612242

[21] Fischler, W., and L. Susskind, 1998, Holography and cosmology, hep-th/9806039.

[22] R. Bousso, The holographic principle, Rev. Mod. Phys. 74 (2002) 825 [arXiv:hep-th/0203101]

[23] R. Arnowitt, S. Deser and C. W. Misner, Gravitation: an introduction to current research, Louis Witten ed. (Wiley 1962), Chapter 7, pp 227-265, [arXiv:gr-qc/0405109].

[24] Tom Banks , W. Fischler , Leonard Susskind, Quantum Cosmology In (2+1)-Dimensions And (3+1)-Dimensions. Nucl.Phys.B262:159,1985.

[25] Tom Banks, T C P, Quantum Gravity, the Cosmological Constant and All That... Nucl.Phys.B249:332,1985.

[26] A. Kraemmer, H.B. Nielsen and L. Susskind Nucl. Phys. B28 (1971), p. 34.

[27] G 't Hooft, Nucl.Phys.B72:461,1974

[28] Alexander M. Polyakov, Quantum Geometry of Bosonic Strings. Phys.Lett.B103:207-210,1981.

[29] Michael R. Douglas, Stephen H. Shenker, Nucl.Phys.B335:635,1990.

[30] David J. Gross, Alexander A. Migdal, Phys.Rev.Lett.64:127,1990.

[31] K. G. Wilson, Rev.Mod.Phys.47:773,1975

[32] J. Polchinski, Nucl.Phys.B231:269-295,1984

[33] A.B Zamolodchikov, "Irreversibility' Of The Flux Of The Renormalization Group In A 2-D Field Theory, JETP Lett.43:730-732,1986.

[34] Andrew Strominger, The dS/CFT Correspondence, arXiv:hep-th/0106113, Journal-ref: JHEP 0110 (2001) 034

[35] Alan H. Guth , Erick J. Weinberg, Could the Universe Have Recovered from a Slow First Order Phase Transition? Published in Nucl.Phys.B212:321,1983.

[36] Ben Freivogel, Gary T. Horowitz, Stephen Shenker, Colliding with a Crunching Bubble, arXiv:hep-th/0703146

CHAIRMAN: L. SUSSKIND

Scientific Secretaries: D. Lopez Mateos, C. Durnford

DISCUSSION I

- *Bourjaily:*

The landscape contains theories, as you discussed, that are very qualitatively different at low energy. I mean, different numbers of families, different gauge structures, also different couplings…

- *Susskind:*

And maybe much worse than that.

- *Bourjaily:*

And the reason of course is that these are all properties that string theory doesn't, let's say, single out. So I have a criticism, and I'd like you to comment on this. It seems that there are at least two points that are being assumed that String Theory obviously needs to single out. And that is that because we are looking at the landscape of flux compactifications of Calabi-Yaus, we are looking at things where we have high-scale supersymmetry and also 3+1 because we are compactifying on 6-dimensional manifolds. In the landscape we're not considering G_2 compactifications to 2+1, or Spin(7) compactifications to 1+1 so, can I ask why we're not?

- *Susskind:*

I have not thus far used the A word, but I guess I have to now introduce it. The universe appears to be very big: much bigger than the 10 billion light-years that we see. We don't know how big. We don't know how uniform it is. Let's make the assumption, try it out, that the universe is exceedingly big and exceedingly inhomogeneous, meaning, with lots and lots of kinds of structures, lots and lots of kinds of laws of nature and so forth; in other words, that this landscape is populated. Then it becomes no mystery as to why we are in a part of the universe which we can exist in. There's no mystery: we are where we are because we couldn't be anywhere else. Most of the universe is probably extremely hostile.

I believe that any other dimension, other than 3+1 is probably hostile, probably very hostile. In 2+1 you simply can't stick more than a Planck's worth of energy before it rolls itself up. 4+1, 5+1 and so forth can't have atoms as far as we know. You also can't have stable solar systems. So, 3+1 may be the only region of this large landscape and also of this large multiverse or whatever you want to call it where we can exist.

Another example, incidentally, people never seem to ask this question: why don't we live in a fully supersymmetric vacuum? Wouldn't that be the simplest thing? To live in a world that was truly supersymmetric. Well, the answer is: you cannot live in a world that is truly supersymmetric. Let me tell you why. It's chemistry that would break down. Imagine an atom, any atom, with several valence electrons. What would happen to it? Well, if there was a massless photino, and there would be a massless photino if there were massless photons, then the outer electrons could decay to the massless photino. Where would it get the energy from? From dropping down into the interior of the atom, where it would no longer be Fermi-blocked. So, what would happen to all atoms? All atoms would just collapse to Bose condensates near the nucleus. You probably couldn't live in a supersymmetric world or a world of dimension different from 3+1. At least not our kind of life, life as we know it. That's the only reasons I can think of for not considering these options.

- *Bourjaily:*

OK, if it's going to be an anthropic argument, it's OK, and, presumably, then all the other models in the string theory landscape end up anthropically excluded; but getting pass that for a minute, if not having low-scale supersymmetry is an anthropic condition, then, why are we not considering non-Calabi-Yau 6-manifolds?

- *Susskind:*

Why are we not? Don't ask me. I don't know. I think we should be considering everything which is consistent. To some extent, of course, we are doing lamp-post physics. Lamp-post physics is to look in the region where there is enough light, because we have enough experience that we can make some claims on the basis of the basic kind of physics that we knew that came before it. Lamp-post physics is not necessarily a good guide towards what the real physics should be.

Now, the only other thing I would say is Michael talked a great deal about high-scale supersymmetry. I was involved in that. I made some of the arguments about why there should be high-scale supersymmetry. If you ask me what I really think, I would guess that there is low-energy supersymmetry. I notice Michael smiling there… And the reason is the unification of coupling constants… Now, I am so tired I forgot what your question was. Why did I bring that up? I don't remember. Please…

- *Bourjaily:*

I think I wanted to know whether you would call it lamp-post physics.

- *Susskind:*

Yes. If there is low-energy supersymmetry, and the scale of supersymmetry breaking is low enough, then it's quite likely that the right kind of vacuum should be close to the supersymmetric vacua, which would be Calabi-Yaus. So we might, luck-out. So I would say one should hope, if you're a physicist like myself, that we do find supersymmetry because it may mean we are near the lamp-post.

- *Aschieri:*

What's the difference between the approximately 10^{1000} vacua of string theory nowadays?

- *Susskind:*

Why it may not be such an impossible situation? First of all I pointed out to you that the biological landscape is something like 10 to the hundred million, and I just don't find that terribly intimidating. One of the reasons, incidentally, among many others, is that an awful lot of the base pairs in DNA are 'junk' DNA – they don't have to do with anything terribly important. Here's a kind of simplified picture of our universe that I think Michael would give:

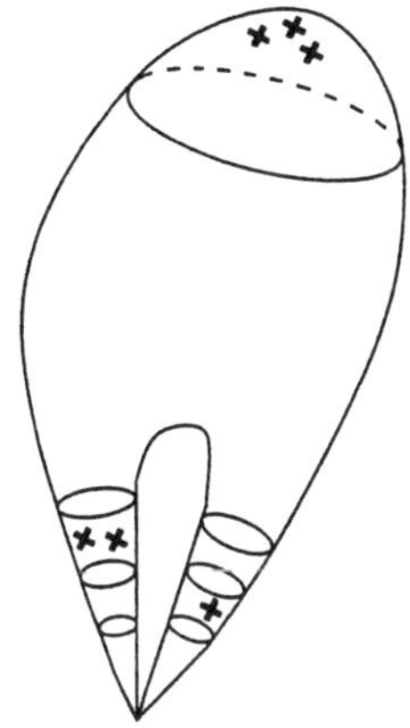

supposing other compactification looked something like this. Some kind of compact manifold that has some holes in it, and also contains some D-branes, the D-branes being where the Standard Model - and all its cousins, hidden sectors and other things – live. (Indicated by crosses at the TOP of the figure). But it could well be that most of the tens of thousands of vacua have to do with fluxes down here (indicating crosses at BOTTOM). They have to do a great deal, for example, with the cosmological constant/vacuum energy/dark energy, and not a lot more. I would then say that a lot of this is junk DNA from the point of view of Standard Model physics: the only thing it really has to do with is the vacuum energy, and at the end of the day, that's pretty much where we started. All of that complexity, or a large fraction of that complexity, perhaps 10^{1000}, might have to do with things that we're not all that interested in as particle physicists, and the interesting stuff may be up here (indicating crosses at TOP) and it may be of much, much smaller numerical size. It think that this is roughly the content of the statement that there is some statistical independence between, for example, the cosmological constant, and other properties that we might be interested in. Thus far it's turned out in the search for vacua that the stuff which is interesting to particle physics might be isolated from vacuum energy. So that's one answer to why 10^{1000} vacua might not be all that bad. All it would do is to tell you that there are enough vacua to allow you to get the cosmological constant right.

- *Aschieri:*

Previously we thought there were different string theories and nowadays we have learned to consider them as different sections of a bigger theory. Aside from the fruitfulness of this approach, can you provide examples of two different theories that are not related in this way? Is it always possible in principle to construct a bigger hyper-theory that encompasses (as special sections) the two different theories?

- *Susskind:*

I'd like to postpone that til the lecture where I talk about cosmology, but let me not completely postpone it. In the example that I gave which had two parallel railroad tracks that particles could move along, one being the phi-fourth oscillator and one being the phi-squared oscillator, the answer is: if you wait long enough in one single theory, tunnelling can happen back and forth between them. So if you wait long enough the harmonic oscillator will turn into the other thing, or the other thing will turn into the harmonic oscillator. There will be transitions. A world with a single quantum field theory or, lets say, a single kind of vacuum – lets suppose that it had a cosmological constant – all it could do is sit there and never change. It could never change from one thing to another, but if it has this possibility, or the property of being part of a bigger landscape, then transitions could happen. The transitions that I'm interested in are the bubble nucleations which could be the things which populate the landscape. So the possibility of transitions between different behaviors will characterize the difference between a landscape and bunch of different disconnected theories.

- *Colangelo:*

You mentioned that the size of the universe that we see may actually be a small fraction of the actual size...

- *Susskind:*

Well, that is certainly true. Experimentally, here's what we know: the universe is at least ten times larger than the horizon size, at least, at minimum.

- *Colangelo:*

Then, you hinted the fact that the universe is very different in other regions of space. Do you have any notion of a natural size of the uniform part of the universe, so that we can check that the fact that we see everything uniform is to be expected?

- *Susskind:*

Let me postpone that till I discuss cosmology. I am going to discuss cosmology, bubble nucleation and things like that, and what kind of universes they are likely to lead to, so I am going to postpone that till then. I do have an answer to the following question: assuming that there really does exist a dark energy (by dark energy I mean the cosmological constant), then there is a fundamental limitation to how far we can ever see, and that is about 12 or 13 billion light-years. Just the fact that there is a cosmological constant means that we are living in a world with a horizon and we will never ever be able to see beyond that horizon and we know that the world is pretty homogeneous up to that distance. So I am afraid that the answer is there is no possibility of seeing far out enough to see these other regions. The possibility that we do have is of looking into the past. If it is true that the world was born out of a bubble nucleation, out of a nucleation from a different previous vacuum – tunnelling through from a different point in the landscape – then, in principle there could be an imprint in our world of that tunnelling event. So, we have a possibility of looking into the past. Then I can tell you how long we have to look into the past: 10 to 12 billion years. The possibility of looking out in space and seeing other worlds, I think it's not going to happen.

- *'t Hooft:*

If it is true what you say, that all 10^{500} vacua are smoothly connected by domain walls and such, shouldn't all positive energy vacua decay into the negative energy ones, and shouldn't we this way have returned to the problem string theorists have been addressing for a long time: which vacuum is the "true" vacuum, the one with lowest energy?

- *Susskind:*

There are surely vacua with zero energy, those are the supersymmetric vacua, the ones that we really know about. They are the centre-of-the-lamp-post. They really have zero energy, and we can call them ground states. But we also know that there are states with lower energy than that – negative cosmological constant – and those are the ones that we call anti-de Sitter spaces. Certainly string theory has them, and when you tunnel into them it's not a happy ending. When you tunnel into them you go through a big crunch rather quickly. What was your question again?!

- *'t Hooft:*

Whether or not these different vacuum solutions can be connected continuously. It also sounds a little bit at odds with the idea that the vacuum is counted by the number of fluxes and the like: how can the number of fluxes vary continuously when going from one region of the universe to the other?

- *Susskind:*

Well, you can decay…

- *'t Hooft:*

Can a flux line decay in a Calabi-Yau?

- *Susskind:*

Oh yes. They decay by forming boundaries which are D-branes for example. Remember that the lowest energy vacuum is likely to be one with negative cosmological constant, so it's not going to be a stable state, there's going to be a big crunch. Tunnelling through to a state with negative cosmological constant does not lead to a stable anti-de Sitter space, it leads to a big crunch. The reason this doesn't cause problems for string theory is that we know that vacua with zero cosmological constant are absolutely stable.

This is a correct statement, although it sounds contradictory: there are vacua with energy below zero but the zero energy ones are absolutely stable. They cannot decay down. This is because of supersymmetry. I'll discuss this more tomorrow.

If A can decay into B, it does not necessarily imply that B can decay into A, but then I would say the A is connected to B. It's an equivalence thing. If one can decay into the other then I'll say they that they are equivalent and it is true that there are vacua which can decay into negative cosmological constants. They can also decay into zero cosmological constants, so they're in the same class. The zero cosmological constant vacua which are truly supersymmetric cannot decay, but many different things can decay into them, and many of those things can also decay into the negative cosmological constant, so they're connected by a big network, even though some of these vacua are endpoints which can no longer decay.

- *Preghenella:*

Is it possible to think about string landscapes just starting from the current known physics?

- *Susskind:*

Do you mean: if you were to take all of the know physics – the Standard Model, gravity, quantum mechanics, black holes, the quantum mechanics of black holes etc – take it alone and forget string theory, could you make an argument for such a large landscape to exist? The answer is, I cannot but it is one of the things that I keep emphasizing: Forget string theory. Try to take all of the things that we know, including gravity and quantum mechanics and try to argue that this landscape exists without even using the words string theory. It would be nice and interesting to discover that this landscape is just a consequence of theories which have quantum mechanics and gravity. I think people should think about that, think very hard about that.

- *Didenko:*

Could you clarify the conditions that lead to knowing that there are a finite number of Calabi-Yau manifolds?

- *Susskind*:

No, I cannot. I'm not even sure that that is true, though I guess you mean a finite number of different topologies. The number of different geometries is certainly continuously infinite because of the moduli. Ask a mathematician.

CHAIRMAN: L. SUSSKIND

Scientific Secretaries: D. Lopez Mateos, C. Durnford

DISCUSSION II

\- *Goldstein:*

I asked you a question during your talk about the size of the Calabi-Yau manifold and you responded that it has something to do with the flux lines that wrap around the funny shapes. Yet, you also talked about junk DNA and those flux perhaps being irrelevant for looking for lambda close to zero solutions. So, am I interpreting your metaphors too literally or did I miss something?

\- *Susskind*:

No, I think there may be some fluxes which have to do with features of the Calabi-Yau manifolds which do have to do with properties of the Standard Model and other particle physics things; and other fluxes which are wrapped around cycles, which are in some sense far enough away or out of the way enough of the physics of the Standard Model that they may not make much difference to it. Those are called metaphorically 'junk DNA', but yes you are right, everything counts a little bit. So some features of certain fluxes are unlikely, for example, to make significant changes in the ratio of the electron to the muon mass or other interesting things of that type, that when I say interesting I mean experimentally interesting. I call those 'junk DNA'. They may still make differences to other parts of physics such as the cosmological constant, or hidden matter sectors or things of that nature. So, it's a metaphorical use of the term.

\- *Bourjaily:*

I just had a comment relating to your observation about the fundamental limit that colliders could possibly probe in the Planck scale and that if you get above that in energy, we are probing lower energies. I wasn't sure whether you were being optimistic or pessimistic from the reductionist point of view. But I wanted to point out that I at least would find that as a rather expected result from the reductionist point of view. And that's that if it were the other case, that there weren't a fundamental limit, then it could be true that you would always need to build another collider and that you would never actually get to the end of knowledge. But in the case we're in, where the Planck scale is a fundamental limit of the collider physics, then there really is a point when all collider physics is exhausted and we've learned everything we possibly could from those types of experiments.

\- *Susskind:*

Except for very subtle things about black holes which are too subtle probably to ever measure. I do agree with that. I was sort of responding to a casual remark that Nino made that at every scale we may expect to see reductionism, that is smaller things. And of course, what he was talking about was ordinary particle physics scales; and I thought I would just mention that the logic of things inside things inside things will eventually break down. But, yes, I agree with you. As a reductionist I would like the idea that there is an end to the whole thing.

The thing which makes me more pessimistic about the idea of an ultimate reductionism is this fact that as you move over the landscape, from one place to another, there seems to be this morphing of what is fundamental in one place, being composite in another place. So, for example, in one part of the landscape one-dimensional strings are the things that are fundamental, where fundamental strings is a term referring to things with certain quantum numbers; and as you move across the landscape in other places you find that those fundamental strings are big and fat and start behaving like solitons and other strings that are very solitonic in nature shrink down and get small. And so it is very, very difficult. We see a place for a new kind of theory in which there is something very elusive about what's fundamental and what's derived. And it's really that what makes me a little suspicious about the reductionist plan of things.

- *Lockyer:*

This morning you said that the heterotic landscape was most like the real world. Could you expand on that?

- *Susskind:*

Old heterotic string theory and the newer version of it, which is called M-theory, or better yet Horava-Witten theory, or maybe even Mike Duff theory, that theory does have a certain similarity with what appears to be the real world in that it has gauge groups which look like the kind of gauge groups that physicists want. It has an E_6 which looks roughly like O(10) and has the standard model embedded in it, and it has a hidden group which is disconnected from this except for the gravitational interactions. For many years before string theory was discovered, supersymmetrists were looking for exactly that kind of structure: a grand unified theory coupled together with some sort of hidden sector which would break supersymmetry and transfer through to the low energy sector. These are the features that the heterotic string seems to have quite naturally, and it was for that reason I said it looked like ordinary nature. This doesn't mean very much, but what *is* true is that the landscape of this theory is one of the most poorly understood and difficult to analyze mathematically, and it could be very important to us.

I'd also like to make another comment about what Haim said. He accused me of thinking everything, including his own height, sort of *random*. I don't believe that. I think it's our problem now to figure out which are the things that are environmental in nature and could be different in different places, and which are the things which are more hard wired and for which there are good reasons. The neutrino masses could be one for which there are good reasons for why they are what they are, so I am not somebody who believes that *everything* is random.

- *Smith:*

Is it not a problem for the completeness of theories today that there is missing – dark – matter and energy which we know nothing about?

- *Susskind:*

The fact that dark energy is *there* is not a mystery. The mystery is why there is so little of it. Theoretical physics says that there should be an enormously large amount of dark energy: the puzzle is why it is so small. Don't forget that there really was a prediction that dark energy should be there. It was made by Steven Weinberg, and should be taken as one of the successful predictions of this field of physics. At the moment, that is the one prediction that this version of physics has solidly given us that seems correct.

Einstein initially said that the cosmological constant was zero, then he said it was not zero, but in both cases he had absolutely no idea why, and it seems he had no idea that it had anything to do with vacuum energy. Well, he may have understood that it was the energy of the vacuum is some sense, but he had no idea that it was in any way connected with quantum mechanics, so I think the Einstein history can be removed from the whole story here. As far as I know it was actually Veltman who was the first one to recognize that modern theory, in particular the Higgs theory, suggests that there ought to be a vacuum energy. Even earlier, David Finkelstein, calculated the energy stored in the electromagnetic field coming from energies down to ordinary atomic physics, and he calculated that the curvature of the universe should be something like the distance from the Earth to the Sun – that's if the energy in the universe is at least as big as from quantum electrodynamics.

The problem got really nasty when people started realizing that there is a quantum contribution to the energy of the vacuum and that this energy contribution has no reason to be small. That's where things really started, it didn't start with Einstein. It's not that he didn't know what he was doing, but he said: "Yes," "No," "Maybe," "Yes!" "It's not beautiful," and so forth, but really he did not have any good physical reason to believe it was there or not there.

- *Zichichi:*

The reason why Einstein introduced Lambda was to avoid what happened to Newton, namely the collapse of the universe. Einstein put in an arbitrary repulsion due to nothing but to empty space; this has nothing to do with the quantum energy of the vacuum. His reason was to avoid the catastrophe of Newton.

- *Susskind:*

That's correct. The question was why gravity didn't attract the universe and collapse it. Well, Einstein realized that he could introduce in his own equation a term called the cosmological constant, which from the Newtonian point of view would be a long range repulsion, and he said we should take this into the balance. He was actually wrong because it's an unstable equilibrium. There were many reasons he was wrong about it. The worst of the reasons is that the universe *isn't* static. He then looked at it and said, "Who needs this ugly son of a gun? Let's throw it away." What he didn't understand is that there are good quantum reasons why it ought to be there. The great mystery is not *why* there is dark energy, but why there is so *little* of it.

- *Parihar:*

I'd like to know whether CPT is conserved in string theory or whether there are any crazy compactifications which would lead to CPT non-conservation.

- *Susskind:*

I'm not expert in this at all, but as far as I know all of the honest string theory compactifications are CPT invariant, I can't think of any examples, or reasons to believe that string theory would violate CPT. My understanding is that string theory is fully expected to conserve CPT, one of the reasons is that Lorentz invariance is expected. Much of string theory is based on the assumption that you can turn Lorentzian space into Euclidean space by the usual Wick rotation, and once you Wick rotate Lorentz invariance, there you are: you've got CPT. So I don't know of any examples of string theory violating CPT – I know there were papers by some people, I tended to dismiss them and not read them very carefully.

- Zichichi:

This subject was discussed here a few years ago in lectures by Brian Greene and the conclusion was that it was not possible to prove a CPT theorem in string theory.

- Susskind:

We are unable to prove *anything* in string theory because we don't know what its equations are. Brian Greene wrote a book that was more famous than mine, but I am the Father of string theory(!), so as the Father of string theory I hereby declare seniority. I don't think Brian Greene knew what he was talking about when he said that, and I'm going to SPANK him when I get home.

- Krotov:

When Calabi-Yaus were presented, most of them had a conic-like singularity. Is this necessary for the construction?

- Susskind:

No, I don't think it's necessary for the construction, but one of the things that Denef and Douglas did on their studies of the statistics of the Calabi-Yau manifolds with fluxes is to show that these pointy-like structures are extremely generic, and that there's a great deal of probability under the parts of the landscape that contain these pointy-like structures. So, generically I draw them in. I don't think there is any reason why they have to be there. Whether they have to be there for experimental reasons, they are very, very useful for making hierarchies. I don't think they have anything to do with the ordinary gauge hierarchy, but perhaps the hierarchy between the Planck scale and the GUT scale could well be approached like that. So they are useful for explaining some hierarchies and they tend to be somewhat generic in the various constructions. Remember Michael drew a diagram in which there was a large lump of probability and he said something about it, I don't remember what it was, but I believe that large lump of probability was the piece that contained the spike-like structure. So there is a lot of probability in the spike-like structure. Is Denef here? Am I getting it right?

- Denef:

Yes.

- Susskind:

OK, good.

- Korthals-Altes:

This morning you explained why the supersymmetric ground state is stable against formations of negative-lambda bubbles. Now, there is something I missed. There is also entropy when you create a bubble, and entropy usually helps to have more bubbles, right?

- *Susskind:*

But the Euclidean description of instantons I think contains that entropy. I think that the Euclidean description of the instanton classes which allows the decay is really the free energy including the entropy. So thinking from a Euclidean point of view, not just an energy point of view in 3-dimensional space but as a 4-dimensional figure. And that was intended to be a 4-dimensional figure.

However, this is not a story I'm making up. This is a story string theorists began worrying some years ago and I am reporting my understanding of what they explained to me. These things were also discussed in the Coleman-De Lucia paper long ago.

- *Keren-Zur:*

Could you please clarify the idea in the landscape – motion with respect to what time axis?

- *Susskind:*

In general relativity time is a funny thing – it's local. Time in one place is measured differently to time in another place and there's no such thing as global time. Let me give you an example. We could be sitting here and suddenly discover a bubble of another type of vacuum, materializing and swallowing us up. We, with our clocks and our time would say a change of vacuum occurs.

- *Keren-Zur:*

Could one universe have different regions with different Calabi-Yau manifolds?

- *Susskind:*

Yes, a region could open up over here with a different Calabi-Yau manifold, or the same Calabi-Yau but different fluxes. It would take a tunneling process, but a lump could form over here with a different Calabi-Yau structure, and if it had lower energy it would start to expand.

- *Keren-Zur:*

Then what would happen to the excess of energy?

- *Susskind:*

It goes accelerating the bubble walls – the kinetic energy of the bubble walls.

CHAIRMAN: L.SUSSKIND

Scientific Secretaries: V. Didenko, A. Matveev

DISCUSSION III

-Durnford:

I'm trying to imagine what it's like to be alive in a bubble cosmology universe. As I travel along my time-like world line, what do I see?

-Susskind:

For most of the time on world line you have an environment with a cosmological constant which is so big to pull atoms apart. For most of it you will not be there. You may live in a green valley and wait, wait, wait long enough and then you'll make a transition to something else. And what can this something else be? Most likely it will be a thing with smaller cosmological constant. Our cosmological constant is very small and it suggested either it could be zero or negative. Either is it a disaster? Zero in string theory really does mean supersymmetry. Supersymmetry means your atoms go to hell. It means they all collapse to Bose condensates. Negative cosmological constant is really a disaster; it means a big crunch in the future.

-Preghenella:

You said we are living in a tiny bubble inside the Universe. Suppose a new tiny bubble appears right now in our world. Do we detect anything?

- Susskind:

Let the bubble has a slightly smaller cosmological constant and everything else more or less the same. And cosmological constant changes just a tiny bit, let say. I don't think we'll see very much because after all cosmological constant itself only has implications for physics on cosmic scales. It is repulsion. It's long range repulsion, a force which increases with distance. Only large distances are important. So what would happen to us in this room if for all of us tiny cosmological constant suddenly makes a shift and it's still positive? I don't think very much would happen to us as long as that change didn't involve anything which is important to us, such as the structure of our atoms.

- Krotov:

Suppose we have chosen a C-Y compactification. It induces some complicated potential. It is clear how one can tunnel between different vacua on the certain C-Y. How can one treat tunneling between different C-Y?

- Susskind:

With a lot of work that nobody has really worked out yet. In certain low energy situation we know how to work out the superpotentials. The superpotentials tell us what the potentials are. So in many low energy situations we know how to calculate the potential in detail. Example is KKLT construction in which the details of potentials are worked out. Once you know potential you can calculate the tunneling amplitude. Other cases which are less geometric and less under control it would be very much harder. But at least in some cases you know how to do the calculations.

- *Machado:*

How causally connected are the bubble-nucleated universes to their "parent" universes?
Is there any information within a bubble universe which would allow one to determine from which "parent" universe it originated?

- *Susskind:*

Other bubble universes are simply outside of our horizon and we can not directly see them. However, what in principle we could directly see is the fact that in our past we might have existed in another kind of environment then we are existing now. We might have a reason from a bubble nucleation in a background which is different than our own. That is something in principle we could see. We can not see bubbles which are space like relative to us, because of the existence of horizons and space like relative to us are outside of horizon. So we have a chance of looking back into the past and discovering that was a complex past doing which we made transition from other kinds of environments. This is a principle statement.

- *Aschieri:*

Can you construct analogy between ensemble of all possible C-Y with fluxes and the task of finding the one describing our world and the possible microstate of a gas and is it the task of finding the ground energy?

- *Susskind:*

I'm not sure what you mean by finding the ground state, first of all. We are not particularly looking for the ground states or anything. We are looking for the states that we can live in. And it's the end of the story. But let's suppose we won't looking for the ground state and let me ignore the question of a negative cosmological constant, which means we are looking for the states with the smallest positive cosmological constant. So, using the statistical mechanics methods it may be shown that time scale is exponentially large. But I'm not an expert in this particular case.

- *Csanad:*

Could a bubble exist which is small enough not to explode forever and we can detect?Or could the one we live in collapse, so that we can cross its boundary?

- *Susskind:*

I don't know any example where the bubble would start to expand and become large enough for anything interesting to exist in it, in which subsequently it would start to collapse. Quite the opposite tends to happen. Once the bubble becomes big enough than the potential energy that you save by increasing the size of the potential, by increasing the volume with lower energy. The potential energy that you save goes into an accelerating wall outward. And so the typical situation and the only situation I know is that if the bubble is too small so that surface energy dominate volume energy then you will collapse. But presumably that the situation in which bubble is so small that nothing interesting happens in it that is called quantum fluctuations. Once the bubble happens by accident to grow big enough that volume becomes more important than surface area then the bubble starts to grow and as far as I know every situation you can think of the saving of potential energy by increasing the volume goes into accelerating outward the bubble boundary. So I just don't know any case where bubble will recollapse on you.

- *Tsallis:*

The landscape idea is very beautiful while you don't speak about the whole Universe. You say somehow that there are a lot of "Universes" to which we will never have access. Is it then Science, to talk about that?

- *Susskind:*

We make up the best theories to explain what we see as best we can and nobody knows we can and can't do. And we just have that hope that smarter people, smarter young people at this meeting at this school, twenty five years from now, thirty five year from now, forty years from now will figure out waste to test them. I don't have anything else to say about it.

CHAIRMAN: L. SUSSKIND

Scientific Secretaries: D. Krotov, E. Skvortsov

DISCUSSION IV

-Durnford:

Quick cosmology question: I think that inflation was introduced to solve the horizon and flatness problem and I remember that this needs 60 e-folding of inflation to work. You say we need 60 e-foldings to achieve the observed CMB anisotropy. Are these calculations related?

-Susskind:

Phenomenologically in order the Universe to be that flat as we observe it you need approximately 60 e-foldings that's a convention to a certain extent. Depending on the initial conditions it could be 65 e-foldings, 63 e-foldings that's not so important for the moment. In order to create structure you also need a certain number of e-foldings, because for not their number of e-foldings you will not flat the Universe out enough. The invariant quantity is the difference between the number of e-foldings you actually observe and the number of e-foldings you need to create structure. So that's difference is about 59.5 and 62 which is the real issue. One can use different conventions and the 62 goes to 71. The difference between the number of e-foldings needed for life verses the number of e-foldings needed for observation is 2.5.

I need to make a one remark: about additional phenomenological signals of what one might be able to see from a bubble nucleation.

In the inflaton potential which I have drawn there reheating is done on the right-hand side somewhere while on the left-hand side there is possibly a connection with some piece of a landscape which might tunnel out from some other minimum and then roll slow down while inflation takes place. There is a one to one correspondence between how long are you roll and the size-scale of the structures, which are produced by inflation. The size you can usually describe in terms of angle or inverse of angle (angular-momentum). The largest scale structures are produced first while the smaller structures are produced later. There is an interesting question: what would be the implication if instead of the inflaton field being perfectly flat and scale invariant you will stand on the top of the hill and roll down. In this case the velocity would be higher. In the fluctuation spectrum velocity appears in the denominator, so you might expect that the answer for a slow roll inflation might have some deviation from the scale invariance. And the deviation from scale invariance could be smaller fluctuations, suppressed fluctuations at the very low L and in fact that of course is seen in data. That is the famous plot of the fluctuation spectrum. The fluctuation spectrum which we expect form scale invariant theory would be perfectly flat but in fact the experimental data have a funny deeping at very very low L which is the quadrupole and octupole. Which is due to the fact that the Universe is produced only ones. Another probe for the very low L would be gravitational waves, these are tensor modes. What is interesting is that tensor modes ought to be enhanced at very low L. And the reason is if you tunnel from a vacuum with larger cosmological constant, the larger the cosmological constant the higher tensor modes fluctuations. So even during inflation the cosmological constant during inflation is quite low it is still possible you can see tensor modes coming from the events which are occurred just after tunneling. There is a chance that you can see decreasing signals from the tensor modes. They will allow to get some information from the early days of the Universe.

-Machado:

I was going to ask you to expand on your statement this morning that we could determine in principle: whether or not our Universe comes from a bubble nucleation by looking on its past light cone, but in way the question has already been answered.

-Keren-Zur:

Could you explain the source of the surface tension and the physics at the surface of a bubble?

-Susskind:

It depends on the particular properties of the bubble and the two vaccua which are connected. One particular example is when two vaccua can have different fluxes, in that case it would be what we call a D-brane. D-branes are boundaries between different values of fluxes.

-Lopez Mateos:

Could you again explain why bubble nucleation implies negative curvature? If a small negative curvature is found, is anything else that would explain it apart from internal inflation?

-Susskind:

If we discover a positive curvature that is a falsification of these ideas. Bubble nucleation leads to an open Universe. If a closed Universe is discovered this particular model of bubble nucleation could not be the answer to how the Universe is populated.

-Aschieri:

Consider the SM, inside this theory we have the prediction of Higgs with the mass in a given window. This is a well accepted prediction, on the other hand predictions based on anthropic principle are more controversial. Why? Is the fact that it is not inside just one theory? Where do you put the line between prediction tautology like homosapiens are more adopted to the world than Neanderthal because we are alive and they are not?

-Susskind:

The existence of homosapiens implies two things to me, that the basic equations which define how the planet should look like which allow human beings to exist since it is a lot of fine tuned sets of conditions. The "planet equations" have enough solutions to guarantee that it will be appeared in the proper range of parameters. But it is not enough for equations to have many solutions. There must also be a lot of planets. So it is not a tautology to say we are here because we are here. The fact we are here tells us something

about the logic of nature: we have enough solutions to stay in the anthropic range and a big enough Universe to contain a vast variety.
The next question is whether you can use the anthropic principle to make predictions, at the moment I do not think so. But it can help you to distinguish the questions which are foolish from questions which are not foolish. The foolish question is to look for an eigen value equation for the temperature of the Earth which is between freezing and boiling. It would be a damned approach to issue of the temperature of the Earth, it could be also foolish to search for the fundamental mathematical principle defining the cosmological constant being ten to the minus one hundred and twenty in Plank units, because it can vary from place to place.

-Didenko:

What are the physical grounds of using scalar field in inflation scenario.

-Susskind:

If idea of the landscape is right it is a many dimensional space and it means that our world can be deformed in a large number of ways. It suggests that there are a lot of ways to deform the compactification. This means there are a lot of scalar fields. Generic implications of a large landscape is that at sufficiently large energies we would discover not a one scalar field but possibly seven scalar fields, one hundred, or two thousand scalar fields of course all of them rather massive.

-Svyatkovsky:

The number of e-foldings mentioned in your lecture is from what moment to what moment?

-Susskind:

The Universe is very curvature dominated it could be seen from FRW equation. There are several terms in it and one goes like one over "a" squared, when "a" is small in the very beginning it is the biggest term, later on some other terms become important, actually the vacuum energy term becomes as important as the curvature term. And accidentally this takes a relatively long time at least ten to the seven Plank times. During this time the Universe is curvature dominated and Universe is expanded in the way which is not look like inflation. It is from the point where the term one over "a" squared becomes nearly equal to the inflationary vacuum energy.

-Hisamatsu:

You have mentioned about the possibility of gravitational wave experiment but I could find out the "scale" from the experimental point of view concerning current sensitivity how far are we now?

-Susskind:

These are extremely long cosmically large wave length gravitational waves which are not the ones we are measuring in LIGO, the ones we can only measure by polarization of CMB.

-Karliner:

You described a branching process in which a Universe spawns "the next generation" of Universes. Now suppose that you have a universe which is almost livable, but quite. What is the probability that the next generation will be more livable? In other words, does it work like evolution?

-Susskind:

I see nothing in this kind of bubble nucleation that would tell you, for example, that the neighboring vacua which might be separated by the domain wall looks anything like the vacuum on the other side of the domain wall. So, the next Universe may have properties which are distinctly different, such that life is completely impossible and out of question. In this sense this is not a kind of evolution in the ordinary sense of evolution.

C. TSALLIS

"Complexity and Nonextensive Statistical Mechanics – Theory, Experiments, Observations and Computer Simulations"

The Content of the Lectures can be found in:

- *"Asymptotically scale-invariant occupancy of phase space makes the entropy Sq extensive"*, C. Tsallis, M. Gell-Mann and Y. Sato, Proc. Natl. Acad. Sc. USA 102, 15377-15382 (2005);

- *"A generalization of the central limit theorem consistent with nonextensive statistical mechanics"*, S. Umarov, C. Tsallis and S. Steinberg, Milan J. Math. (2008), in press [cond-mat/0603593v4];

- *http://tsallis.cat.cbpf.br/biblio.htm*

CHAIRMAN: C. TSALLIS

Scientific Secretaries: A.Naumenko, M.Serbyn

DISCUSSION I

- *Susskind:*

Suppose you have a big system which is described by q-entropy. Are the fluctuations of the small part of this system also governed by the q-entropy or they are governed by the conventional entropy?

- *Tsallis:*

I think they are not convolved by the normal entropy because the domain of applicability of q-entropy is when you have big correlations. So either your system is big or small the correlations will always be there.

- *Didenko :*

What kind of interactions in thermodynamical system will lead to statistic with $q \neq 1$?

- *Tsallis :*

The general scenario is whenever you have long correlations in either space or in time.

- *Didenko:*

Can you give an example of the system in which you can directly derive q-statistics from the type of the interactions as the generalization to the standard Gibbs formula?

- *Tsallis:*

There are a few systems with everything checked within this scheme. Those systems are of the type of simple maps. There we can calculate everything and we see appearing q-exponentials, and other formulas I was mentioning. Another class of such systems is those with a long ranged interactions, e.g. self-gravitating systems. For those we only have a set of hints that may suggest (they do suggest for some people) that there is some relationships with the theory I developed in lecture. But for other people they are not sufficiently convincing. The ultimate goal would be if for a system with long-range interaction we could have an analytical expression for one of those q's that I was referring to as a function of alpha and d, alpha characterizing the length of interaction decay distance, d – dimension of the space in which you're working. However we do not have it today for any system with long-range interaction.

- *Csanad*

You showed a lot of expressions covariant of the value of q. Can q be still determined for the given system?

- *Tsallis*

I showed in the morning lecture how q is determined for the maps I referred to in my answer to the Didenko's question. q is given by microscopic dynamics, and if you have this dynamics you can calculate it. However, in the case when the system is too complex we do not have that information. Also it appears to that non-q-describable systems exist.

- *Keren-Zur*

Besides being mathematically compatible with Boltzmann distribution, is there a physical logic behind this q-distribution? For Boltzmann distribution you count the states of the system and you reach the exponential.

- *Tsallis*

Whenever the number of configurations of a system increase exponentially with N you want to use Boltzmann statistical mechanics. But when you have a system whose visited configurations grow not exponentially with N but as N^{ρ}, where ρ is some positive number, then you don't want to use Boltzmann statistical mechanics. So the question can be reformulated as follows: what is the kind of correlations that makes the system not to visit everything it is allowed to, but only a small part of it.

- *Susskind*

Entropy is adiabatic invariant. It doesn't change when you slowly change a control parameter. If the control parameter happens to be a magnitude or a coefficient of a long-range force, so that you vary a system adiabatically between having no long-range force, and having a long-range force. If system have to jump from q=1 statistic to q="whatever" statistics, your entropy would be no longer adiabatic invariant. Please, comment on this.

- *Tsallis*

When you have a short-range interaction you can speak about a weak interaction, so this is the limit in which Boltzmann was thinking. Your coupling constant is very small, and because the range of the interactions is very short, you don't have anything particularly complex. When you have long range interactions, the situations with coupling constant equal to zero and slightly different from zero are completely different. Because for coupling constant even slightly different from zero the system is not integrable.

- *Susskind*

I'm just pointing out that if your control parameter was itself long range force, you have to give up your idea that entropy is an adiabatic invariant.

- *Tsallis*

What probably happens in this kind of situation is a crossover: along some amount of time you can work with one entropy and then if you go to the limit, if changes to another entropy. You have non-permutation of two limits: long time limit and N-size limits are not commutable. This kind of crossover could fit the situation that you describe.

- *Aschieri*

You spoke about q-statistics. Do those systems also have q-symmetries?

- *Tsallis*

The infinite family of q-parameters which exist in every non-extensive system has 2 symmetries which appear systematically, and the reason for this is not known. These are $q \to 2-q$ and $q \to 1/q$.

CHAIRMAN: C. TSALLIS

Scientific Secretaries: C. Cattuto, F. Sylos Labini

DISCUSSION II

- *Keren-Zur:*

If you take a q-gaussian and fourier-transform it using a different q, do you still get another kind of q-gaussian?

- *Tsallis:*

You don't. You might remember in the definition of the q-Fourier transform you have the function f(x), then you have the q-product, and the q-exponential. You might ask what happens if I generalize in a such way to have a q-product and a q' exponential, and I apply this to a q'' Gaussian. I'm just generalizing your question. Shall I get outside a q''' gaussian? You don't, we have checked this. However you get something incredibly close to a q''' Gaussian. If you were doing fittings, you would be easily fooled.

- *Beck:*

Why do q-triplet and duality tranformation work for solar wind?

- *Tsallis:*

I said it this morning, that there is a strange convergence of algebras. If we take the generalized central limit theorem we obtain an algebra. Transforming a value of q in an infinite set, a countable infinite number of qs. If you do a renormalization group procedure on some simple type of Hamiltonian, you get another algebra, which happens to be the same as the previous one. If then you just construct the the algebra with additive duality and the multiplicative duality, you get another algebra which happens to be again a sub-algebra of the previous two that I was refererring. Finally if you take the analytical results of Robledo and collaborators concerning a variety of one-dimensional maps, edge of chaos, pitchfork bifurcations, tangent bifurcations, again you get the same expressions. Why does this happen? Why, somehow, they all point to a single algebra, which is the one that appears in the generalized limit theorem. Why this happens is a kind of mystery to myself. With regard to the q-triplet that was observed by NASA, we found a connection essentially because it is so natural when you work in the kind of subject that most people work here, it is so natural when you have a symmetry to try and construct an algebra related to those symmetries. I have a conjecture about this additive duality and multiplicative duality and it was so natural for Gell-Mann to try to mix them to make algebras, and he did that for quarks and many of you do that all the time. So he did that again, and we found an expression that the central elements of those algebras are zero, plus one and minus one, reproduced the numbers of the q-triplets observed by NASA. I would like to understand better why.

I would like to put a question to the audience. Yesterday evening, we were discussing whether there are three or four universal constants in Physics. There is some discrepancy there. Some people think there are three, which would be the velocity of light, Planck's constant and the gravitational constant. And some other people like myself and Christian Beck, who think there are four, the fourth being Boltzmann,'s constant. Are there in Nature three or four fundamental constants and what is the definition of an universal constant is one of the points that was raised.

- *Duff:*

The answer is that you have as few or as many dimensionful constants as you like, depending on your choice of units. In fact the International Bureau of Standard has a set of units in which there are seven fundamental units. This is an entirely arbitrary human convention. Physics doesn't care how many dimensionful constants you have, it only cares about dimensionless constants. The whole purpose of the dimensionful parameter is to convert from one set of units to the other.

- *'t Hooft:*

Lorentz invariance allows you to convert seconds into meters. If you don't use that symmetry, they're independent dimensionful constants. I think it's similar in the case of Boltzmann's constant. There, again, the question is whether we use the logarithm of two, or say bits and bytes separately, or whether we use that physical phenomena, the physical interpretation of entropy. If you decide not to use it you have three dimensionful constants. The same holds for quantum phenomena that relate energy to the inverse of time or momentum to the inverse of a length – that's Planck's constant. If you don't know quantum mechanics there are separate dimensionful units. As soon as you realized that there's a fundamental physical phenomena relating them into some sort of universal way, then you might decide to convert one into the other by definition, and that removes the velocity of life as a freely adjustable parameter. Something similar works for Boltzmann's constant.

- *Tsallis:*

The question that was put one day by Gell-Mann to me was: if I choose the Planck's constant equal to one, the velocity of light equal to one, and G equal to one, what would be the value of Boltzmann's constant? I thought a couple of days about that, and then I told him it's impossible to find an answer to that question. If you choose Planck's constant, c and G to be equal to one – which you have a right to do – you cannot express Boltzmann's constant as a number, because it is an independent physical universal constant. No way. You cannot.

- *Susskind:*

I think you're wrong. The definition of temperature that I might use, for example for Centigrade temperature is the difference between frozen ice and steam – that could be calculated from first principles in some other units, for example in Planck's units. Once I have calculated what the centigrade unit was in Planck's units, then the value of Boltzmann's constant would be determined.

- *'t Hooft*

This question is a mix between a linguistic question and a power struggle between different fields of physics. When you say that Planck's constant is equal to one, you have elevated quantum mechanics as a superiod theory relating things. Instead of putting Planck's constant equal to one, you might choose the proton mass equal to one, or I could put the distance of the Earth from the Moon equal to one. But then you say, somehow the planetary size that determines the distance between the Earth and the Moon is less fundamental than quantum mechanics. Therefore we put hbar equal to one, and so it's a power struggle with statistical mechanics. But then also it's a linguistic question, how to describe the laws of Nature. We use units for convenience, but we could use any procedure we like. Nowadays we're so much busy with gravity that we like to put the gravitational constant equal to one. This is not a very wise choice because this constant is poorly known while other constants are much better known.

- *Tsallis:*

How many constants are you allowed, independently, to put equal to one in contemporary physics? The fundamental question that is behind this is whether the concept of entropy is something else or not. So it's very different to say that the basic entropy, the quantum of entropy is logarithm of two, or Boltzmann's constant times the logarithm of two. That makes a big difference. The whole work of Landauer is around that difference: does it create a new branch of Physics or not?

- *Shebulin:*

Isn't Boltzmann's constant just a measure of classical energy scale to quantum (atomic) energy scales? It relates the amount of heat required to raise a millimeter of mercury to, say, the average kinetic energy of a room temperature molecule. It relates the large to the small.

- *Tsallis:*

It relates to something more than that. What do you mean by that amount of mercury? Do you mean the amount of atoms of mercury, the quarks inside those molecules? Which level of the description do you mean? They're human conventions. You will say I'm not interested in quarks, I'm only interested in atoms. Then this level of information is something that you decide, and that gives you some entropy, which is Boltzmann's constant times to logarithm of something. In the same physical object - a droplet of mercury – there are zillions of quarks and you might be interested in describing the configuration of those quarks. The entropy changes. How does it change? The logarithm of that something, not the constant that is in front of it.

CHAIRMAN: C. TSALLIS

Scientific Secretaries: C. Cattuto, F. Sylos Labini

DISCUSSION III

-Csanad:

Under what circumstances can you make a statement on the independence or dependence of the elementary probabilities of a given system, based on its q-value ?

-Tsallis:

The first situation is when you know the microscopic dynamics of the system, which of course is quite rare - for example we don't know the microscopic dynamics of earthquakes. When you know that, you can check - in the conditions that you're interested in – whether the maximal Lyapunov exponent is positive, zero, or negative. If it is positive, then you want to use Boltzmann-Gibbs statistical mechanics, because you're going to have strong chaos, from chaos you're going to have $q = 1$. If it is negative, you're going to have simple orbits, simple attractors and you don't want to use statistical mechanics at all. If you have zero, then you have a strong candidate for non-extensive statistical mechanics. For that case, if the maximal Lyapunov exponent is zero, you want to check the sensitivity to the initial conditions, and it's probably going to be asymptotically a power of time. The exponent is $1 / (1\text{-}q)$, and then you know q. So this is the basic procedure. When you don't know the microscopic dynamics, perhaps you know the mesoscopic dynamics. Perhaps you know the Langevin equation of the system, or the Fokker-Planck equation, or a master equation for the system. If it so, then you can still handle the problem. If you don't even have that, then you just use the functional forms that appear within that theory, and you see whether it is one of the cases. In pretty much the same sense that if you are an astronomer and you want to describe the orbit of Mars, you start by using the functional form of an ellipse, because this is what Newtonian mechanics provides. But then, for the parameters of that particular orbit, you cannot calculate them, not because Newtonian mechanics is incomplete, but because nobody has given to us the initial conditions of the planetary system. Since we don't have enough information, we use the functional form of Keplerian orbits and we fit the rest.

- Parihar:

Could you comment about anomalous entropy systems, for example black holes where the entropy goes like E^2 and hence it's impossible to use the canonical ensemble at high energies. In such cases is microcanonical ensemble the only choice?

- Tsallis:

If you want to speak about black holes, I don't have a reliable answer. If you want to talk about negative specific heat, I can repeat some of the things that I said today about long-range interactions, when I presented the molecular dynamical simulation. There, for that system, when it is microcanonical (that is when it is isolated), you use Newton equations (because it is classical) and there is indeed a region where the specific heat is negative, and I showed it today. Of course at first sight a negative specific heat is surprising, and I remind the audience that the Boltzmann-Gibbs statistical dynamics specific heat is always positive, in canonical and grand canonical ensemble, but it has no definite sign in the microcanonical one. In that region where you have negative specific heat, Boltzmann-Gibbs statistical mechanics does not apply when N, the number of elements, is very large. There is long-standing quasi-stationary state which is by no means Boltzmannian. The ordering of the limits t to infinity and N to infinity do not commute in such a case.

- *Sylos-Labini:*

Regarding fundamental constants: in a gravitational simulation I need the value of the gravitational constant, but in order to characterize an equilibrium state, I may introduce Boltzmann's constant, but I don't need it before, so in this sense it seems that Boltmann's constant is less fundamental than the gravitational constant.

- *Tsallis:*

Well, it depends on your choice. If you're going to work out the problem at the mechanical level only, then you don't need to speak about entropy, temperature, certainly not about Boltzmann's constant. If you want to treat something in quantum mechanics, you only need Planck's constant, the velocity of light, and you solve it. If you want to change your description into a statistical mechanical description, then the game is different. It's up to you. If you change the description into something macroscopic, then you need Boltzmann's constant.

- *Sylos-Labini:*

Which kind of two-point correlations do you expect if the distribution is non-Gaussian, if it's a q-distribution? Is there a classification that we can do?

- *Tsallis:*

These kinds of correlations are precisely at the very heart of the entire idea of non-extensive statistical mechanics, so this is a crucial question. Let me try to answer this in general. If you have here the phase space of the system and the interactions in this system are short-range, then you start in a point and the system, quite quickly, goes more or less everywhere in phase space. This is the justification for equal probabilities in Boltzmann-Gibbs theory. If you start from another initial condition, the same thing occurs, so basically the system forgets its initial condition and this is why you can do statistical mechanics. What happens with those systems that have long-range interactions, which is presumably the case of the gravitational problem? You have a phase space, you start from a point, but the system doesn't go everywhere. Things go like in the case of Continental airlines, located mainly in Houston, because it has started in Houston. But then you start it in Paris and you will have Air France, a structure which is quite similar to the other one but doesn't go everywhere because of the long-range correlation. So you will stay in a region of phase space that has basically zero Lebesgue measure. And then you have to make some special kind of configurational average that might give you some soft function that could be a q-exponential. Each of those is a mess, it's a network, but you have to make an average over all those possible networks and perhaps the outcome is a soft function, which could be close to an exponential, which could be a q-exponential. This is the basic idea of how it is expected to work.

- *Preghenella:*

You showed that the cosmic-ray flux plot could be fitted by using a q-Gaussian. Maybe it means nothing physically, but suppose it does. What could it mean?

- *Tsallis:*

Well, first of all, the system is clearly not in thermal equilibrium. Second, it suggests that it has some scale invariance. Don't ask me what is the scale invariance that exists for cosmic rays, because I have no idea. But the fact that q-statistics gives you a function which fits very well the whole thing over so many decades, that gives a strong hint that you should expect some sort of scale invariance of the process. This is not a full answer, but this is a guide, a path. You should look for some mechanism that exhibits some sort of scale invariance.

- *Aschieri:*

Concerning q-statistics, a place where q is not 1, is the deformation of quantum integrable systems, like X-Y spin chains. Shouldn't be that a natural place to find q-statistics? I understand that q-groups and q-statistics have different origins, but the same q-functions, like the q-exponential, arise in both fields. Could therefore those integrable systems give examples of q-statistics?

- *Tsallis:*

The physical logic is quite different for both, but it happens that for q-quantum groups, if you calculate how you compose the number of particles, that happens to have the same type of compositional law as the entropy I was talking about. So they are two different areas of physics, where the same mathematical structure appears.

- *Didenko:*

Is it possible to formulate a kinetic Boltzmann equation within the q-statistical approach?

- *Tsallis:*

I've not worked personally on that, but some people have. If you write the most basic Boltzmann equation, in general you don't know how to solve it. Boltzmann made the famous molecular chaos hypotesis. So you have a strange probability there, which due to this hypothesis factorizes into two probabilities. Doing that he was able to solve the integral and he was able to find stationary states, and this is Boltzmann statistical mechanics. Somebody has generalized the molecular chaos hypothesis by using, instead of factorizing probabilities, some functional form (which reproduces molecular chaos hypothesis for q=1). And with that he was again able to find the stationary solution, and that solution is a q-exponential instead of being Boltzmann's factor.

- *Loehr:*

Regarding the description of jet broadening in e+e- reactions, can two parameters describe pQCD + hadronization? Do you think that these parameters have any physical interpretation, or is it rather that the variety of functions you have with these q-Gaussians is big enough that any reasonable function can be fitted by that?

- *Tsallis:*

First of all it's not big. It's just one more parameter, in fact. I think the question you are addressing is the following: why should we use Boltzmann-Gibbs statistical mechanic to describe the gas, if I can use for example Langevin equation, a stochastic equation that also gives you the correct answer. Or why use any of them if I can use, for a classical system, Newton's equations? We expect that if you use Newton's law here, you are going to find Boltzmann factors. So what you are telling me is that you have some handle on the microscopic phenomena that are behind that particular process. This does not make useless to adopt a Langevin approach with stochastic differential equations or Boltzmann factors, or just classical thermodynamics. Those are different levels of description of the system – it depends on how much you want to describe. Now for the specific problem of your question, clearly if q is different from one, it has to reflect special correlations. Not being particularly familiar with those specific processes, I have no idea about what these correlations are.

- *Goldstein:*

The example of hadronic jet structure fits the q-Gaussian description nicely. Is there something deeper about this, fitted values versus the energy of jets?

- *Tsallis:*

You can consider the description a kind of phenomenological one, where q replaces non-trivial correlations that I'm pretty sure are inside the mechanisms our colleague was describing: non-Markovian processes, who knows what.

- *Zichichi:*

In the production of particles in colliding protons we have shown, using the effective energy, that there are no long-range correlations. I'm speaking about production of jets in hadronic interactions and e+e- collisions. There are no long-range correlations. You should not link what you do to what we are doing. We're not learning more from what you are doing than we do learn – once you find that there the results agree excellently well with the predictions based on the radiating emission of gluons, production of quarks-antiquarks systems, etc. you don't need to carry out your analysis. It happens that your analysis fits the data, but this has nothing to do with what we are doing.

- *Zichichi:*

Are you referring back to the observation that when you adjusted the energy appropriately all those jets cross-sections follow the same line?

- *Tsallis:*

Yes.

- *Zichichi:*

The effective energy makes all processes identical, no matter what the initial state is.

- *Goldstein:*

But here we are looking, I think, at the multiplicity of pions.

- *Zichichi:*

The multiplicity of pions is a consequence of this. If you don't use the effective energy, it depends on the initial state and this is simply not true. It depends on the initial state if you don't analyze the final state in terms of effective energy. QCD cannot explain this. G. 't Hooft is the only one who tried to work out the effective energy from the planar diagrams. This is something to be understood.

If you consider a phenomenon that you think you understand, which you prove is a related to a process that you understand theoretically, then Tsallis' analysis is not going to teach us anything more. It happens that in this series of phenomena his analysis agrees with the data.

- *Tsallis:*

Is it easy to calculate, with those mechanisms you describe, a specific heat? Because a specific heat within statistical mechanics is very easy to compute: you calculate the partition function, you take its logarithm – in this case the q-logarithm – you make a double derivative and that's it. Let me ask you: what advantage is there, in having Boltzmann's statistical mechanics if you can say that the molecule here obeys to a Lennard-Jones potential. To know PV=NkT is extremely useful, but clearly it does not give you any specific information about the molecules of the air, Oxygen, Nitrogen or whatever. It shows to you some property which has some degree of universality, not the precise mechanism.

- *Zichichi:*

I'm speaking about fundamental processes. Billions of molecules represent an enormous number of elementary objects – which is difficult to describe because they're very many. The way in which we have understood Nature is by studying the simplest processes as well as possible. So when you introduce billions of molecules, I respect what you're doing, I like it very much all of your conclusions, but don't tell me that from this you work out quantum electrodynamics.

- *Tsallis:*

I never said that. The q that works here is roughly 11/9. The q that works for cosmic rays was the same. Is it not interesting? They appear to have something in common.

- *Zichichi:*

And so what? What do you understand out of this?

- *Tsallis:*

I understand that there is scale-invariance in the specific mechanism.

- *Kaur:*

Since the dynamical correlations are seen as a sign of non-extensive entropy, and as shown by you in the case of hadronic jets the temperature becomes a constant as a function of energy, does it indicate the onset of fluctuations pointing to the quark-gluon plasma phase?

Are your results valid for a system – say a system of magnetic dipoles – that attains a negative temperature?

- *Tsallis:*

I cannot answer the question about quark-gluon plasma. What I can answer is that it seems reasonable to look for mechanisms that have scale invariance. It's not a full answer, it's a hint.

About negative temperatures, that's not a big deal. When it happens, it's related to the fact that you populate the excited states instead of the low-energy states. With respect to positive or negative temperature, everything I said remains the same. Nothing fundamental changes.

- *Susskind:*

If I understand it correctly, you said that q-statistics applies to situations which are not really ergodic, where you get stuck in a certain place in phase space. And then, afterwards, you average over various places. Roughly speaking, might you say that you do Boltzmann statistics in the neighborhood of given points in phase space, but then you average them all up?

- *Tsallis:*

Everything you said until this last statement is absolutely agreeable. About the last statement, perhaps you'd like to formulate it in a slightly different way. For some systems it certainly is true, but not in the sense you mentioned, of working in the neighborhood of a given point in phase space.

- *Susskind:*

This reminds me of the difference between quenched and annealed systems. In quenched systems, you do some statistics, Boltzmann statistics, and then you go to some nearby region of parameter space and do another Boltzmann statistics, and in then you average them all together and define something which is not really Boltzmann statistics at all, it's quenched systems' statistics. Can you say that q-statistics has something to do with quenched systems?

- *Tsallis:*

In some cases I think it does. This is the idea of superstatistics. Basically it's a kind of Boltzmann statistics where beta fluctuates itself, either temporally or also spatially. We have in the room one of the world specialists on that approach.

- *Beck:*

There are many ways in which q-statistics can arise. Long-range interactions are one way, but the other processes are precisely what you describe. And there are many other ways, I think it's a very general things and there are many sources of q-statistics.

COMPLEXITY AT THE FUNDAMENTAL LEVEL: CONSEQUENCES FOR LHC

Antonino Zichichi

CERN, Geneva, Switzerland
Enrico Fermi Centre, Rome, Italy
INFN and University of Bologna, Italy

Abstract

This lecture – delivered at the 2006 Erice School – is closely linked to the one delivered in 2004, where the existence of Complexity at the fundamental level has been fully discussed. Here the consequences of Complexity for LHC are treated and the QGCW Project is presented. The references have been updated since the final edition of the lecture has been printed on March 2008. If complexity exists at the fundamental level the expectations must fall outside all possible predictions. The QGCW should be the source for totally unexpected phenomena.

CERN – Geneva – March 2008

CONTENTS

COMPLEXITY AT THE FUNDAMENTAL LEVEL: CONSEQUENCES FOR LHC

1 SEVEN DEFINITIONS OF COMPLEXITY

People speak of 'Complexity' as a source of new insights in physics, biology, geology, cosmology, social sciences and in all intellectual activities which look at the world through the lens of a standard analysis in terms of either Simplicity or Complexity. **But 'Complexity'** is ill-defined, as shown by the existence of at least seven definitions of Complexity.

DEFINITION NUMBER 1

Complexity is a property of systems that are somewhere in between a completely random and a completely regular state, often described by a highly non linear set of equations but sometimes not describable by equations at all.

DEFINITION NUMBER 2

Bad ones:

1) Chaos.
2) The need for lengthy calculations.
3) The need for many distinct variables.

Better ones:

4) Unexpected difficulty when attempting to describe something in a precisely formulated theory.
5) What is left over after all systematic approaches failed.

But it could also be that: Complexity is an excuse for sloppy thinking.

DEFINITION NUMBER 3

The Complexity of a theory (problem) is the minimum amount of computer time and storage required to simulate (solve) it to a specified level of precision.

<u>Definition Number 4</u>

If we admit that biological or linguistic evolution, or financial dynamics are complex phenomena, then their typical dynamics is somehow between strong chaos (i.e. positive Lyapunov exponents) and simple orbits (i.e. negative Lyapunov exponents). In other words, Complexity (or at least some form of it) is deeply related to the edge of chaos (i.e. vanishing maximal Lyapunov exponent). Since the edge of chaos appears to be related paradigmatically to an entropy index 'q' different from unity, there must be some deep connection between Complexity and generalized entropies such as 'S_q'.

<u>Definition Number 5</u>

From the mathematical point of view:

- A problem can be polynomial, which means that it is not to hard to predict surprises.
- A problem can be NP or NP-complete, which represent different degrees of difficulty in predicting surprises.
  - Surprises means: UEEC event.
  - That degree of difficulty can be associated with the level of Complexity.

<u>Definition Number 6</u>

A system is 'complex' when it is no longer useful to describe it in terms of its fundamental constituents.

<u>Definition Number 7</u>

The simplest definition of Complexity: '*Complexity is the opposite of Simplicity*'. This is why we have studied the platonic Standard Model (Addendum 1) and its extension to the platonic Superworld (Addendum 2).

These seven definitions of Complexity must be compared with the whole of our knowledge (see later) in order to focus our attention on the key features needed to study our real world.

2 COMPLEXITY EXISTS AT ALL SCALES

The Logic of Nature allows the existence of a large variety of structures with their regularities and laws which appear to be independent from the basic constituents and fundamental laws of Nature which govern their interactions.

But, without these laws it would be impossible to have the real world which is in front of us and of which we are part of. A series of complex systems is shown in figure 1.

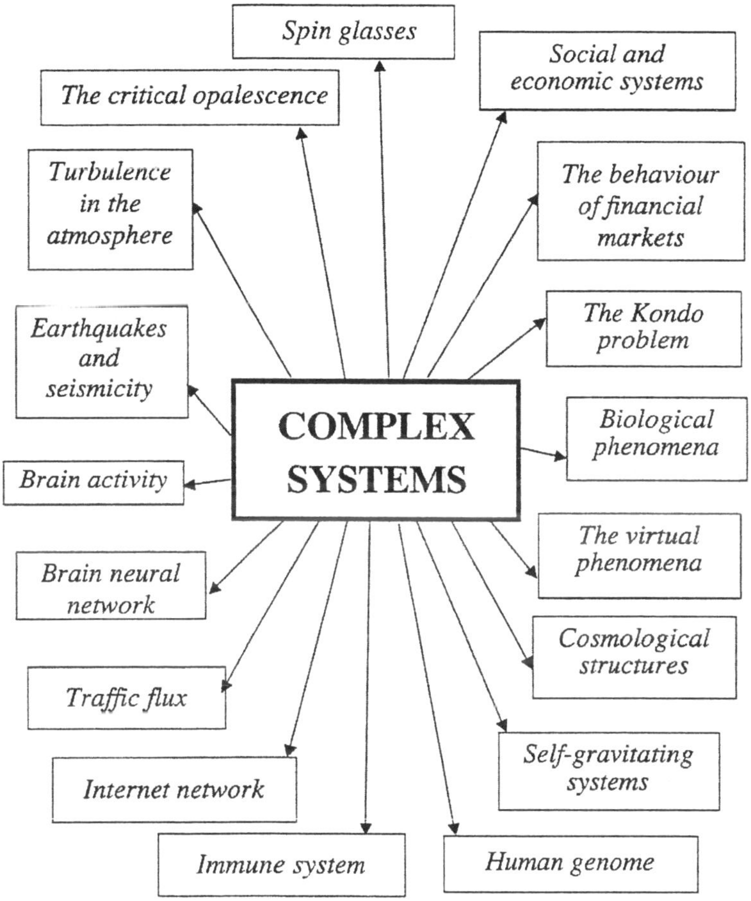

Figure 1

As you can see, we go from traffic flux, to the internet network, to earthquakes and seismicity, to social and economic systems, to the behaviour of financial markets, to the study of cosmological structures, and so on.

The experimental evidences for the **existence** of **Complexity are two**:

1) The Anderson-Feynman-Beethoven-type phenomena (**AFB**) i.e. phenomena whose laws and regularities ignore the existence of the Fundamental Laws of Nature from which they originate;
2) The Sarajevo-type effects, i.e. Unexpected Events of quasi irrelevant magnitude which produce Enormous Consequences (**UEEC**).

The only certainty about Complexity is the existence of these experimentally observable effects. The **AFB** will be discussed in chapter 3 and the **UEEC** in chapter 4. These effects exist at all scales, and therefore Complexity exists at all scales, as illustrated in figure 2.

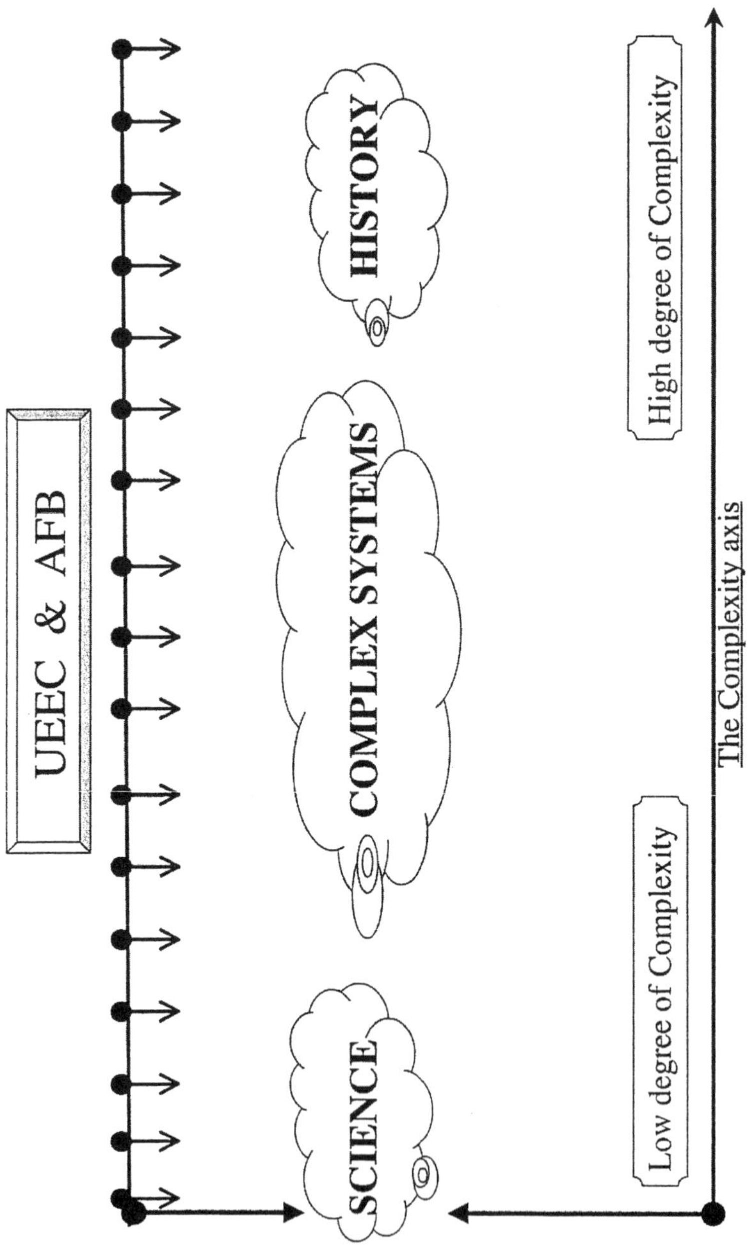

Figure 2

3 AFB PHENOMENA FROM BEETHOVEN TO THE SUPERWORLD

Beethoven and the laws of acoustics.

Beethoven could compose superb masterpieces of music without any knowledge of the laws governing acoustic phenomena.

But these masterpieces could not exist if the laws of acoustics were not there.

The living cell and QED.

To study the mechanisms governing a living cell, we do not need to know the laws of electromagnetic phenomena whose advanced formulation is QED.

All mechanisms needed for life are, to a great extent, examples of electromagnetic processes. If QED was not there, Life could not exist.

Nuclear physics and QCD.

Proton and neutron interactions appear as if a fundamental force of nature is at work: the nuclear force, with its rules and its regularities.

These interactions ignore that protons and neutrons are made with quarks and gluons.

Nuclear physics does not appear to care about the existence of Quantum ChromoDynamics (QCD), the fundamental force acting between quarks and gluons at the heart of the subnuclear world.

Nuclear physics ignores QCD but all phenomena occurring in nuclear physics have their roots in the interactions of quarks and gluons.

In other words, protons and neutrons behave like Beethoven: they interact and build up nuclear physics without 'knowing' the laws governing QCD.

The most recent example of Anderson-Feynman-Beethoven-type phenomenon: **the World could *apparently* not care less about the existence of the Superworld.**

4 UEEC EVENTS, FROM GALILEI UP TO SM&B

In figure 3 there is a sequence of **UEEC** events from Galilei to Fermi-Dirac and the 'strange particles'.

In figures 4, 5, 6 from Fermi-Dirac to the construction of the Standard Model and in figure 7 a synthesis of the UEEC events in what we now call the Standard Model and Beyond (**SM&B**). These figures (4, 5, 6) cover the first fifty years of Subnuclear Physics, whose detailed description can be found in my book whose front page is reproduced below.

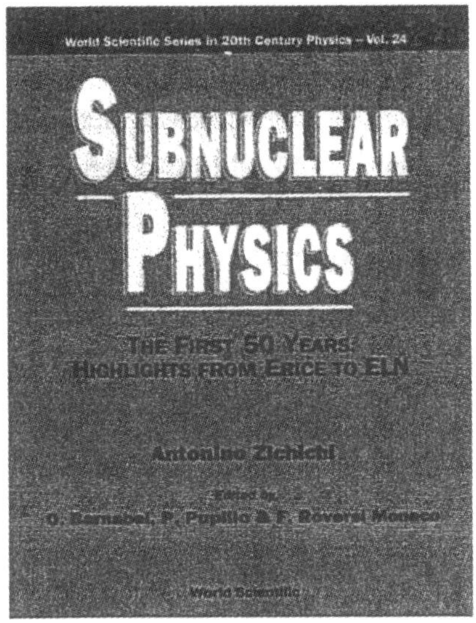

'UEEC'

TOTALLY UNEXPECTED DISCOVERIES

FROM GALILEI TO FERMI-DIRAC AND THE 'STRANGE' PARTICLES

I	Galileo Galilei discovery of $F = mg$.
II	Newton discovery of $F = G\dfrac{m_1 \cdot m_2}{R_{12}^2}$
III	Maxwell discovers the unification of electricity, magnetism and optical phenomena, which allows him to conclude that light is a vibration of the EM field.
IV	Planck discovery of $h \neq 0$.
V	Lorentz discovers that space and time cannot be both real.
VI	Einstein discovers the existence of time-like and space-like worlds. Only in the time-like world, simultaneity does not change, with changing observer.
VII	Rutherford discovers the nucleus.
VIII	Hess discovers the cosmic rays.
IX	Dirac discovers his equation, which opens new horizons, including the existence of the antiworld.
X	Fermi discovers the weak forces.
XI	Fermi and Dirac discover the Fermi–Dirac statistics.
XII	The 'strange particles' are discovered in the Blackett Lab.

Figure 3

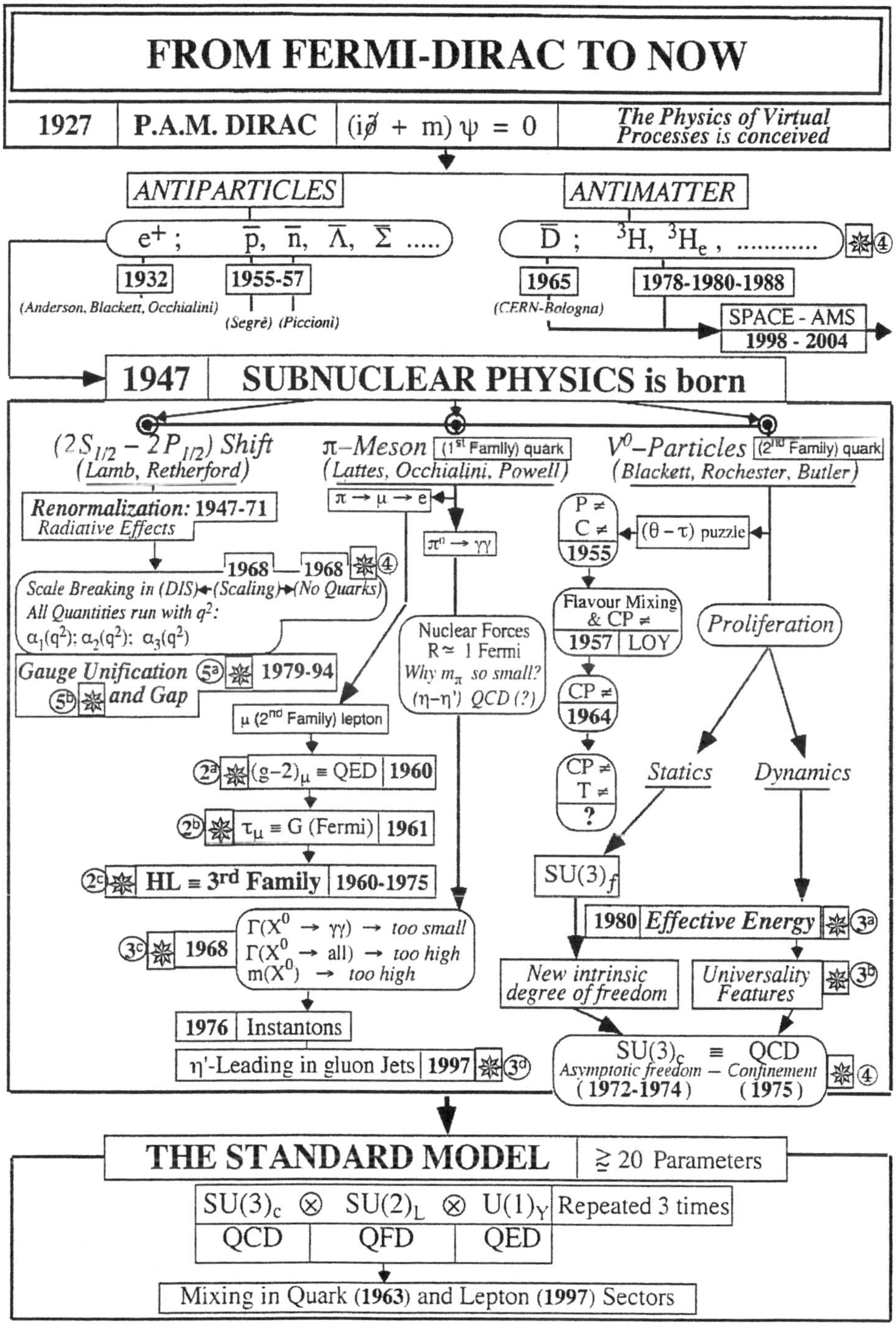
FROM FERMI-DIRAC TO NOW
1927
P.A.M. DIRAC
(i∂̸ + m) ψ = 0
The Physics of Virtual Processes is conceived
ANTIPARTICLES
ANTIMATTER
e+ ; p̄, n̄, Λ̄, Σ̄ .....
D̄ ; ³H, ³He , ...
1932
1955-57
(Anderson, Blackett, Occhialini)
(Segrè) (Piccioni)
1965
1978-1980-1988
(CERN-Bologna)
SPACE - AMS
1998 - 2004
1947
SUBNUCLEAR PHYSICS is born
(2S1/2 – 2P1/2) Shift
(Lamb, Retherford)
π–Meson
(1st Family) quark
(Lattes, Occhialini, Powell)
V0–Particles
(2nd Family) quark
(Blackett, Rochester, Butler)
Renormalization: 1947-71
Radiative Effects
π → μ → e
π0 → γγ
P ≠
C ≠
1955
(θ – τ) puzzle
1968
1968
Scale Breaking in (DIS)←(Scaling)→(No Quarks)
All Quantities run with q2:
α1(q2): α2(q2): α3(q2)
Flavour Mixing
& CP ≠
1957
LOY
Proliferation
Nuclear Forces
R ≃ 1 Fermi
Why mπ so small?
(η–η') QCD (?)
Gauge Unification
1979-94
and Gap
CP ≠
1964
μ (2nd Family) lepton
(g–2)μ ≡ QED
1960
CP ≠
T ≠
?
Statics
Dynamics
τμ ≡ G (Fermi)
1961
HL ≡ 3rd Family
1960-1975
SU(3)f
Γ(X0 → γγ) → too small
Γ(X0 → all) → too high
m(X0) → too high
1968
1980
Effective Energy
New intrinsic degree of freedom
Universality Features
1976
Instantons
η'-Leading in gluon Jets
1997
SU(3)c ≡ QCD
Asymptotic freedom – Confinement
(1972-1974)
(1975)
THE STANDARD MODEL
≳ 20 Parameters
SU(3)c ⊗ SU(2)L ⊗ U(1)Y
Repeated 3 times
QCD
QFD
QED
Mixing in Quark (1963) and Lepton (1997) Sectors

Figure 4

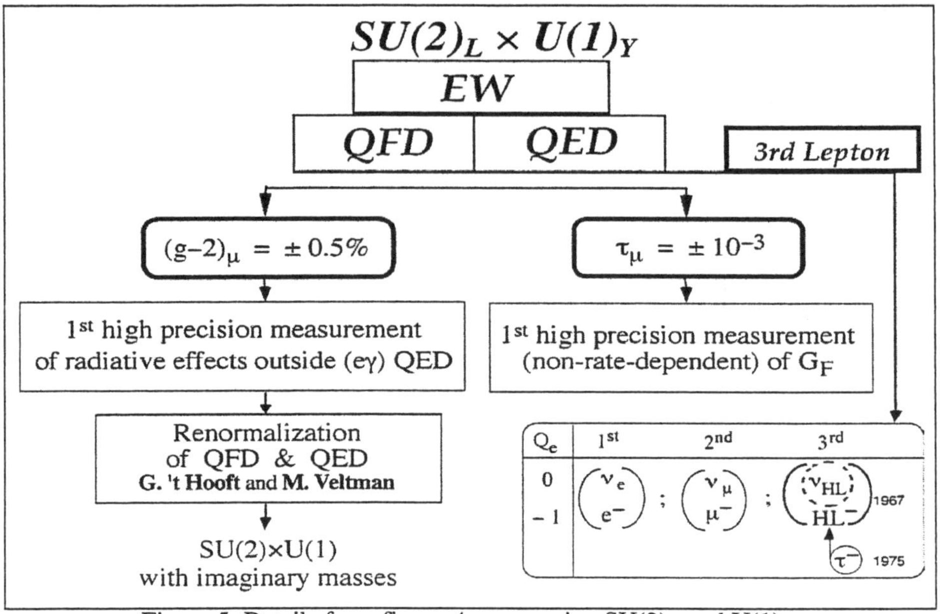

Figure 5: Details from figure 4, concerning $SU(2)_L$ and $U(1)_Y$.

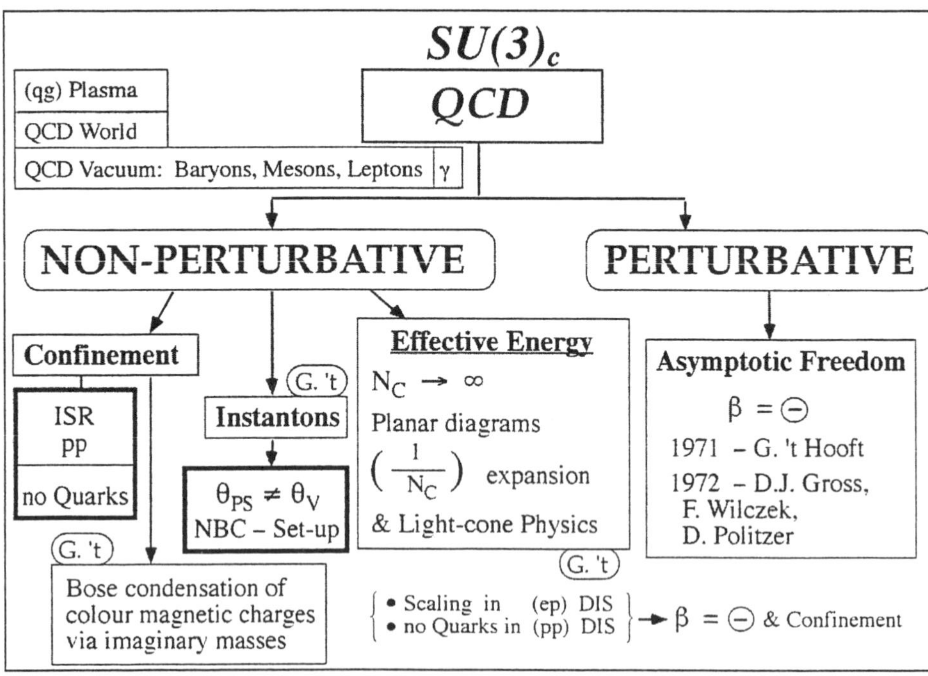

Figure 6: Details from figure 4, concerning $SU(3)_c$.

SM&B

THE STANDARD MODEL AND BEYOND

① RGEs (α_i ($i \equiv 1, 2, 3$); m_j ($j \equiv q, l, G, H$)) : $f(k^2)$.

- GUT ($\alpha_{GUT} \cong 1/24$) & GAP ($10^{16} - 10^{18}$) GeV.
- SUSY (to stabilize $m_F/m_P \cong 10^{-17}$).
- RQST (to quantize Gravity).

② Gauge Principle (hidden and expanded dimensions).

— How a Fundamental Force is generated: SU(3); SU(2); U(1) and Gravity.

③ The Physics of Imaginary Masses: SSB.

— The Imaginary Mass in SU(2)×U(1) produces masses ($m_{W^\pm}$; m_{Z^0} ; m_q; m_l), including $m_\gamma = 0$.

— The Imaginary Mass in SU(5)⇒SU(3)×SU(2)×U(1) or in any higher (not containing U(1)) Symmetry Group ⇒ SU(3)×SU(2)×U(1) produces Monopoles.

— The Imaginary Mass in $SU(3)_c$ generates Confinement.

④ Flavour Mixings & CP ≠ , T ≠ .

— No need for it but it is there.

⑤ Anomalies & Instantons.

— Basic Features of all Non-Abelian Forces.

Note:

q	≡	quark and squark;	m_F	≡	Fermi mass scale;
l	≡	lepton and slepton;	m_P	≡	Planck mass scale;
G	≡	Gauge boson and Gaugino;	k	≡	quadrimomentum;
H	≡	Higgs and Shiggs;	C	≡	Charge Conjugation;
RGEs	≡	Renormalization Group Equations;	P	≡	Parity;
GUT	≡	Grand Unified Theory;	T	≡	Time Reversal;
SUSY	≡	Supersymmetry;	≠	≡	Breakdown of Symmetry Operators.
RQST	≡	Relativistic Quantum String Theory;			
SSB	≡	Spontaneous Symmetry Breaking.			

The five basic steps in our understanding of nature. ① The renormalization group equations (RGEs) imply that the gauge couplings (α_i) and the masses (m_j) all run with k^2. It is this running which allows GUT, suggests SUSY and produces the need for a non point-like description (RQST) of physics processes, thus opening the way to quantize gravity. ② All forces originate in the same way: the gauge principle. ③ Imaginary masses play a central role in describing nature. ④ The mass-eigenstates are mixed when the Fermi forces come in. ⑤ The Abelian force QED has lost its role of being the guide for all fundamental forces. The non-Abelian gauge forces dominate and have features which are not present in QED.

Figure 7

Let me dedicate some attention to discuss UEEC events in nuclear physics.

Nuclear Physics and the UEEC events.

It is considered standard wisdom the fact that nuclear physics is based on perfectly sound theoretical predictions. People forget the impressive series of UEEC events discovered in what I have decided to call the 'Yukawa gold mine'.

Let me quote just three of them:

1 The first experimental evidence for a cosmic ray particle believed to be the Yukawa meson was a lepton: the muon.

2 The decay-chain: $\pi \rightarrow \mu \rightarrow e$ was found to break the symmetry laws of Parity and Charge Conjugation.

3 The intrinsic structure of the Yukawa particle was found to be governed by a new fundamental force of Nature, Quantum ChromoDynamics: QCD.

As you know 2007 was the centenary of the birth of Hideki Yukawa, the father of theoretical nuclear physics. In 1935 the existence of a particle, with mass intermediate (this is the origin of 'mesotron' now 'meson') between the light electron, m_e, and the heavy nucleon (proton or neutron), m_N, was proposed by Yukawa [1].

This intermediate mass value was deduced by Yukawa from the range of the nuclear forces. Contrary to the general wisdom of the time, Yukawa was convinced that the particles known (electrons, protons, neutrons and photons), could not explain how protons and neutrons are bound into the extremely small dimensions of a nucleus.

In order to make this 'prediction', Yukawa needed the Heisenberg uncertainty principle: a totally unexpected theoretical discovery.

The origin of it was the totally unexpected discovery of the dual nature of the electron (wave and particle) and of the photon (wave and particle).

Heisenberg himself tried to explain the binding forces between the proton and the neutron, via the exchange of electrons, in order not to postulate the existence of a new particle. The very light electron, m_e, could not stay in the very small dimension of the nucleus.

The author of the uncertainty principle and father, with Dirac and Pauli, of Quantum Mechanics, did not realise this contradiction. The need for a new particle was the reason. What no-one was able to predict is the 'gold-mine' hidden in the production, the decay and the intrinsic structure of this 'particle'.

This 'gold-mine' is still being explored nowadays and its present frontier is the Quark-Gluon-Coloured-World (QGCW) [2].

I have recently described [3] the unexpected conceptual developments coming from the study of the production, the decay and the intrinsic structure of the Yukawa particle.

Let me just quote the most relevant UEEC events: chirality–invariance, spontaneous symmetry breaking, symmetry breaking of fundamental invariance laws (P, C, T), anomalies, and 'anomaly-free condition', existence of a third family of fundamental fermions, gauge principle for non-Abelian forces, instantons and existence of a pseudoscalar particle made of the quanta of a new fundamental force of Nature acting between the constituents of the Yukawa particle.

A few cases (seven) where I have been directly involved are summarised in figure 8. Each **UEEC** event is coupled with a **despite**, in order to emphasize the reason why the event is unexpected.

UEEC EVENTS
IN THE CONSTRUCTION OF THE
SM&B ≡ MY PERSONAL EXPERIENCE

① ***The 3rd lepton***, **HL** (now called τ) with its own neutrino, ν_{HL} (now called ν_τ),
despite the abundance of neutrinos: ν_e and ν_μ.

② ***Antimatter***
despite S-matrix and C, P, CP, T breakings.

③ ***Nucleon** Time-like EM structure*
despite S-matrix.

④ ***No quarks** in violent (pp) collisions*
despite scaling.

⑤ ***Meson mixings***
$\theta_V \neq \theta_{PS}$: $(51^\circ) \neq (10^\circ) \neq 0$ **despite** $SU(3)_{uds}$.

⑥ ***Effective energy:*** the QCD-light
despite QCD.

⑦ ***The running** of α_1 α_2 α_3 versus **energy** at* **a point** E_{GU}.
(1979) (1991) **despite** straight line convergence.
EGM

Figure 8

The **SM&B** is the greatest synthesis of all times in the study of the fundamental phenomena governing the Universe in all its structures. The basic achievements of the **SM&B** have been obtained via UEEC events; moreover the **SM&B** could not care less about the existence of Platonic Simplicity.

An example is shown in figure 9 where the straight line (small dots) would be the Platonic simple solution towards the Unification of all Fundamental Forces.

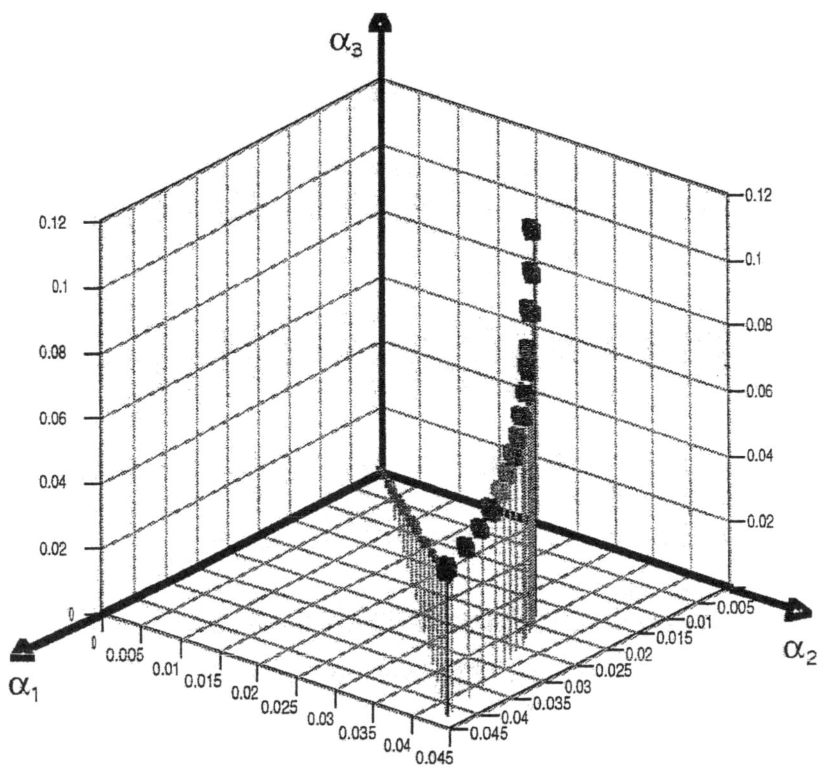

The points have a sequence of 100 GeV in energy. The last point where the 'ideal' platonic straight line intercepts the theoretical prediction is at the energy of the Grand Unification. This corresponds to $E_{GU} = 10^{16.2}$ *GeV. Other detailed information on the theoretical inputs: the number of fermionic families,* N_F, *is 3; the number of Higgs particles,* N_H, *is 2. The input values of the gauge couplings at the* Z^0*-mass is* $\alpha_3 (M_Z) = 0.118 \pm 0.008$*; the other input is the ratio of weak and electromagnetic couplings also measured at the* Z^0*-mass value:* $\sin^2 \theta_W (M_Z) = 0.2334 \pm 0.0008$.

Figure 9

Nevertheless the effective unification is expected to be along the sequence of points (the big ones) computed using the Renormalization Group Equations (**RGEs**) [4].

Platonic Simplicity for the unification of all forces is again reported in Addendum 1 and Platonic Supersymmetry for the existence of the Superworld is discussed in Addendum 2. We will see how many times the platonic Simplicity is violated when we construct the Superworld.

In Addendum 3 there is a synthesis of UEECC events needed in the process of construction of the SM&B [5]. The numerous violations of Simplicity are the proof that Complexity exists at the fundamental level of scientific knowledge where we have proved that **AFB** phenomena and **UEEC** events are present.

The conclusion is that Complexity exists at the elementary level. In fact, starting from Platonic Simplicity, the **SM&B** needs a series of 'ad hoc' inputs [5].

5 THE TWO ASYMPTOTIC LIMITS: HISTORY AND SCIENCE

The real world seems to be characterized by two basic features, which are one on the opposite side of the other: *Simplicity* and *Complexity*.

It is generally accepted that *Simplicity* is the outcome of *Reductionism*, while *Complexity* is the result of *Holism*.

The most celebrated example of *Simplicity* is *Science* while the most celebrated example of *Complexity* is *History*.

Talking about asymptotic limits, the general trend is to consider ***History*** as the asymptotic limit of *Holism* and of *Complexity*; ***Science*** as the asymptotic limit of *Reductionism* and of *Simplicity*, as illustrated in figure 10.

The Logic of Nature allows the existence of ***Science*** (the asymptotic limit of Simplicity) and of ***History*** (the asymptotic limit of Complexity), which **share a property, common to both of them**.

It is interesting to define ***Science*** and ***History*** in terms of this property, **probably the only one**, which they share; i.e. **Evolution**.

- Science is the Evolution of our Basic Understanding of the laws governing the world in its Structure ≡ **EBUS**.
- History is the Evolution of the World in its Real Life ≡ **EWRL**.

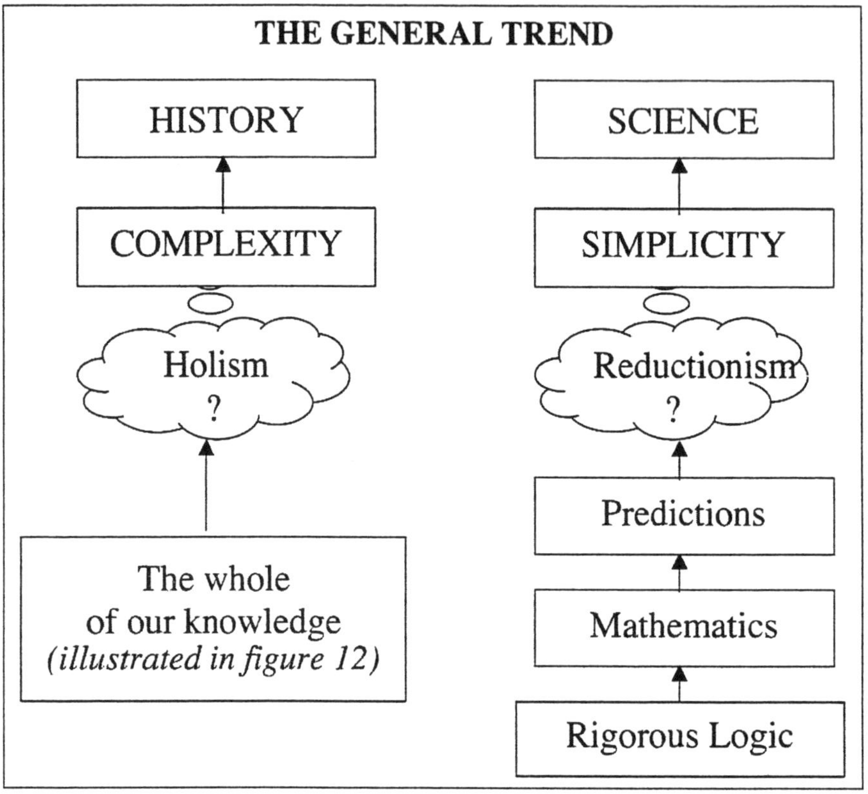

Figure 10

In Table 1 we compare these two supposedly asymptotic limits — History and Science — on the basis of 'What if?'; a condition elaborated by the specialists in what is now known as 'virtual history' [6].

On the basis of 'What if?' these specialists conclude [6] that the world would not be as it is, if one, or few, or any number of 'What if?' had not been as History tells us. This is not the case for Science. The world would have exactly the same laws and regularities, whether Galileo Galilei or somebody else had discovered $F = mg$ ($F \equiv$ force; $m \equiv$ mass; $g \equiv$ acceleration due to gravity), and so on for all the other scientific discoveries.

It is in the consequences of 'What if?' that the two asymptotic limits of Simplicity and Complexity seem to diverge, despite the fact that the sequence of 'What if?' in Science belongs to the 'totally unexpected events' (UEEC) exactly like the others listed in the column of History.

TABLE 1

'WHAT IF?'

	In History ≡ EWRL		In Science ≡ EBUS
I	What if Julius Caesar had been assassinated many years before?	*I*	What if Galileo Galilei had not discovered that $F = mg$?
II	What if Napoleon had not been born?	*II*	What if Newton had not discovered that $F = G\frac{m_1 \cdot m_2}{R_{12}^2}$?
III	What if America had been discovered few centuries later?	*III*	What if Maxwell had not discovered the unification of electricity, magnetism and optical phenomena, which allowed him to conclude that light is a vibration of the EM field?
IV	What if Louis XVI had been able to win against the 'Storming of the Bastille'?	*IV*	What if Planck had not discovered that $h \neq 0$?
V	What if the 1908 Tunguska Comet had fallen somewhere in Europe instead of Tunguska in Siberia?	*V*	What if Lorentz had not discovered that space and time cannot be both real?
VI	What if the killer of the Austrian Archduke Francisco Ferdinand had been arrested the day before the Sarajevo event?	*VI*	What if Einstein had not discovered the existence of time-like and space-like real worlds? Only in the time-like world, simultaneity does not change, with changing observer.
VII	What if Lenin had been killed during his travelling through Germany?	*VII*	What if Rutherford had not discovered the nucleus?
VIII	What if Hitler had not been appointed Chancellor by the President of the Republic of Weimar Paul von Hindenburg?	*VIII*	What if Hess had not discovered the cosmic rays?
IX	What if the first nuclear weapon had been built either by Japan before Pearl Arbour (1941) or by Hitler in 1942 or by Stalin in 1943?	*IX*	What if Dirac had not discovered his equation, which opens new horizons, including the existence of the antiworld?
X	What if Nazi Germany had defeated the Soviet Union?	*X*	What if Fermi had not discovered the weak forces?
XI	What if Karol Wojtyla had not been elected Pope, thus becoming John Paul II?	*XI*	What if Fermi and Dirac had not discovered the Fermi–Dirac statistics?
XII	What if the USSR had not collapsed?	*XII*	What if the 'strange particles' had not been discovered in the Blackett Lab?

6 THE BASIC POINTS ON COMPLEXITY AND PREDICTIONS

What are the experimental evidences for **Complexity** to **exist**, and for **Predictions** to **exist**?

In the previous chapters 3 and 4 we have discussed the experimental basis for the existence of Complexity, i.e., AFB and UEEC events.

We will now discuss the experimental evidence for the existence of predictions and the sequence which correlates UEEC and predictions.

Predictions.

The experimental evidences for the **existence** of **Predictions are** the very many results of scientific reproducible experiments.

Quantum Electro-Dynamics, QED, is the best example. The anomalous magnetic moments, in symbols (g–2), of the electron (e) and of the muon (μ):

$$(g-2)_{e,\,\mu}$$

are theoretically computed at an extraordinary level of precision (few parts in ten billion parts for the electron) and are experimentally verified to be correct. Could the

$$(g-2)_{e,\,\mu}$$

be theoretically predicted before the discovery of the Maxwell equations and the existence of Quantum Electro-Dynamics (QED)? The answer is obviously no.

The sequence which correlates UEEC events and Predictions.

Predictions at the **fundamental level of scientific knowledge** depend on **UEEC events.**

For example: it is the discovery of the laws governing electric, magnetic and optical phenomena (all totally unpredicted) which produced the mathematical structure called QED.

The mathematical structure was not discovered before the innumerable series of **UEEC** events was found in electricity, magnetism and optics. This series of UEEC events allowed Maxwell to express 200 years of experimental discoveries in a set of 4 equations.

The mathematical formalism comes **after** a totally unexpected discovery: an **UEEC event** which no one was able **to predict**.

In the whole of our knowledge rigorous predictions exist only in Science. These predictions are based on the mathematical description of a single UEEC event or a series of UEEC events. This description can either be the result of new mathematics (example the Dirac δ-function) or the use of existing mathematical formalism (example: the Einstein use Ricci tensor calculus).

The UEEC event at the origin of the Dirac equation is the fact that the electron was not a 'scalar' particle but a spin ½ object.

The UEEC event at the origin of Einstein mathematical formulation of the gravitational forces are the discoveries of

$$\text{Galilei } (F = mg)$$

and of

$$\text{Newton } \left(F = G\frac{m_1 \cdot m_2}{R_{12}^2}\right).$$

These are just two examples of the fact that the greatest steps in the **progress of Science** come from totally unpredicted discoveries.

This is the reason why we need to perform experiments, as Galileo Galilei realized 400 years ago.

Even when we have a mathematical formalism coming from a series of UEEC events, if this formalism opens a new frontier, as it is the case for the Superworld, the experimental proof is needed to verify the validity of the new theoretical frontier.

Today we have a reasonable mathematical formalism to describe the **Superworld**, but in order to know if the Superworld exists we need the experimentally reproducible proof for its existence.

7 THE LESSON NEEDED FOR THE FUTURE

We have proved that AFB and UEEC – which are at the origin of Complexity, with its consequences permeating all our existence, from molecular biology to life in all its innumerable forms up to our own, including History – do exist at the fundamental level [7–10] and [5].

It turns out that Complexity in the real world exists, no matter the mass-energy and space-time scales considered.

Therefore the only possible prediction is that:

- **Totally Unexpected Effects** should **show up**.

- **Effects**, which are impossible to be predicted on the basis of **present knowledge**.

We should be prepared with **powerful experimental instruments, technologically at the frontier of our knowledge**, to discover Totally Unexpected Events in all laboratories, the world over (including CERN in Europe, Gran Sasso in Italy, and other facilities in Japan, USA, China and Russia). All the pieces of the Yukawa gold mine [3] could not have been discovered if the experimental technology was not at the frontier of our knowledge.

Example: the cloud-chambers (Anderson, Neddermeyer), the photographic emulsions (Lattes, Occhialini, Powell), the high power magnetic fields (Conversi, Pancini, Piccioni) and the powerful particle accelerators and associated detectors for the discovery – the world over – of the **SM&B** as synthetically reported in chapter 4.

This means that we must be prepared with the most advanced technology for the discovery of totally unexpected events like the ones found in the Yukawa gold mine.

The mathematical descriptions, and therefore the predictions – for new phenomena to be discovered in the field opened by the given UEEC event – come after the UEEC event, never before.

<u>**Recall**</u>:

- The **discoveries in Electricity**, **Magnetism** and **Optics** (UEEC).
- **Radioactivity** (UEEC).
- The **Cosmic Rays** (UEEC).
- The **Weak Forces** (UEEC).
- The **Nuclear Physics** (UEEC).
- The **Strange Particles** (UEEC).
- The **3 Columns** (UEEC).
- The **origin of the Fundamental Forces** (UEEC).

The present status of Science is reported in figure 11.

It could be that Science will be mathematically proved to be ‘NP-complete’. This is the big question for the immediate future [11].

It is therefore instructive to see how Science fits in the whole of our knowledge as reported in figure 12.

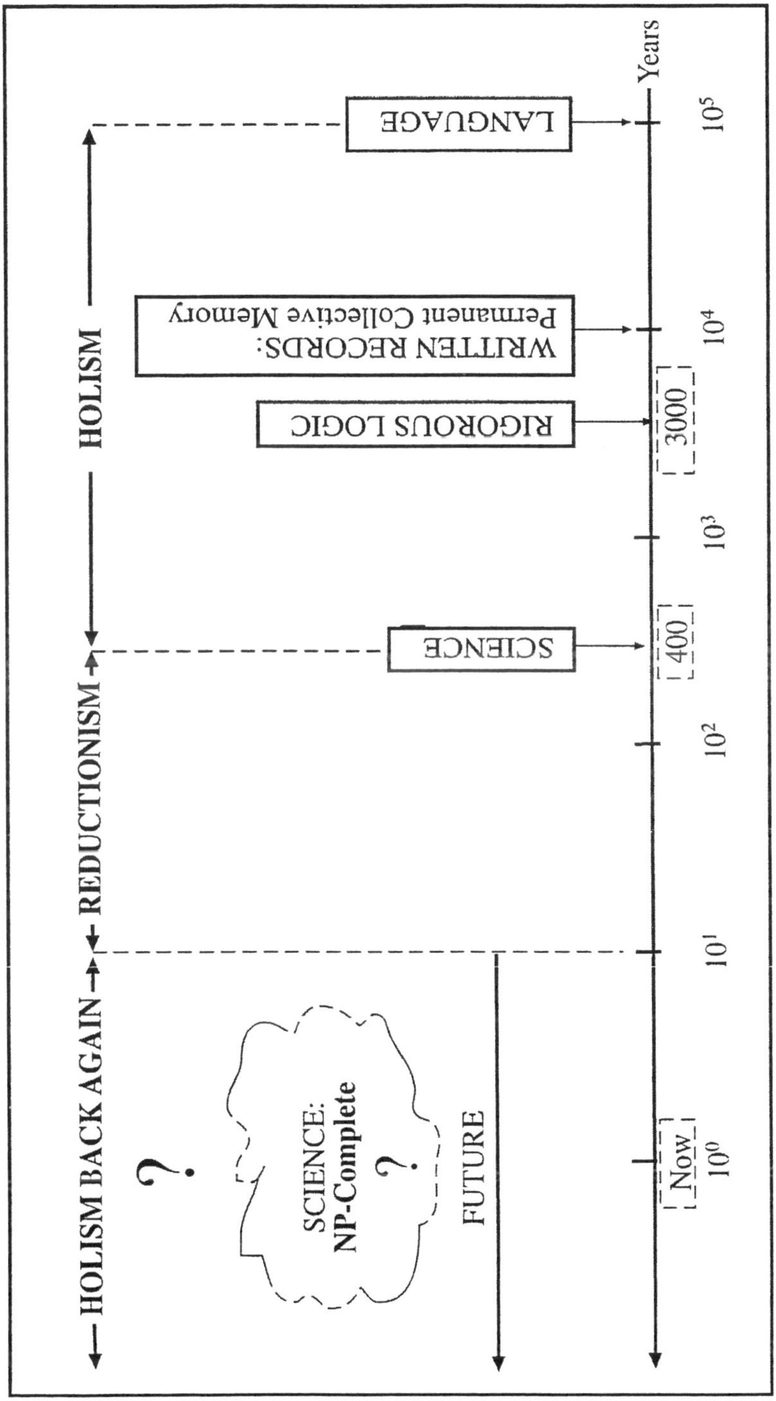

HOLISM BACK AGAIN
REDUCTIONISM
HOLISM
LANGUAGE
WRITTEN RECORDS:
Permanent Collective Memory
RIGOROUS LOGIC
SCIENCE
?
SCIENCE:
NP-Complete
?
FUTURE
Years
10^5
10^4
3000
10^3
400
10^2
10^1
Now
10^0

Figure 11

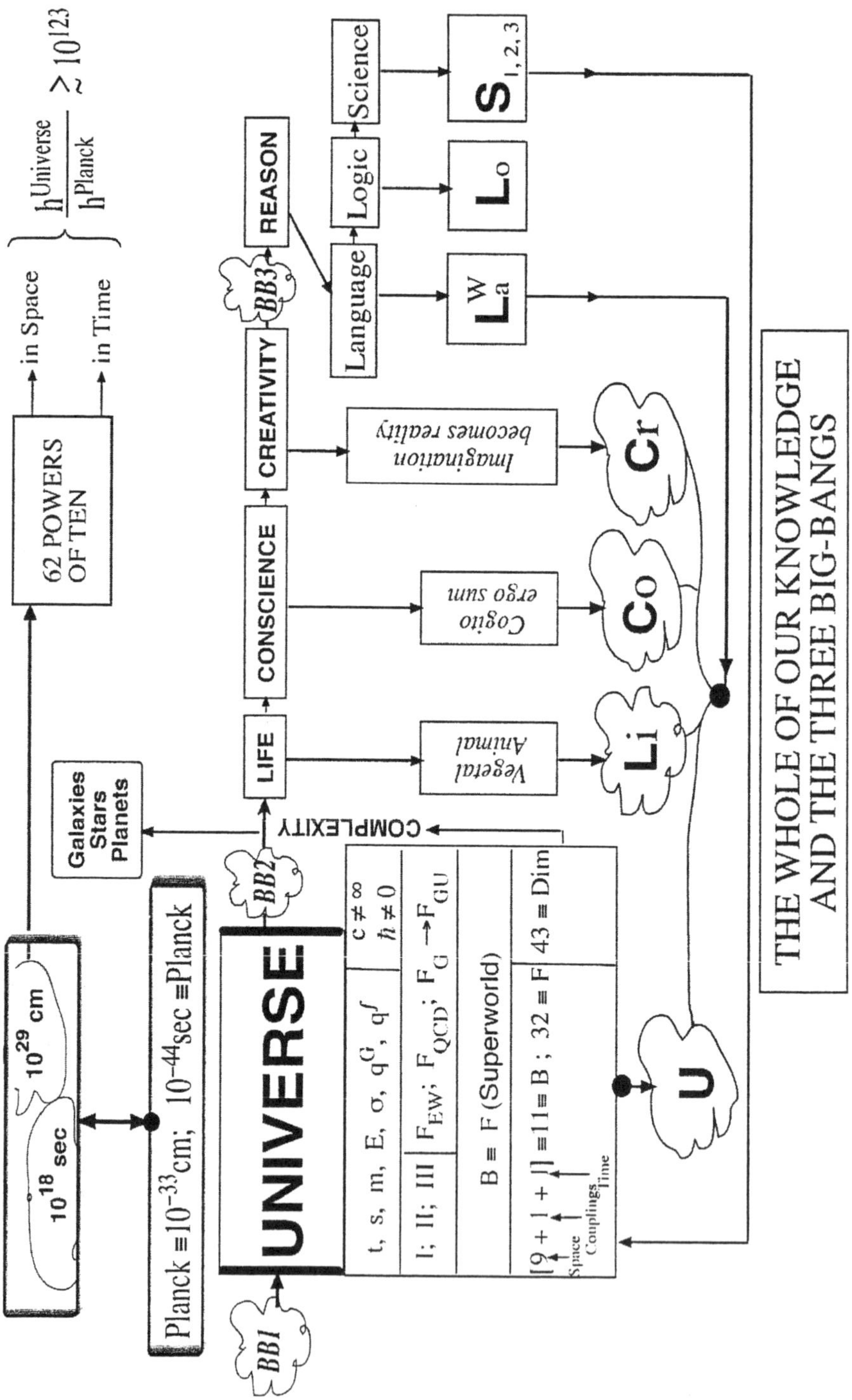
THE WHOLE OF OUR KNOWLEDGE
AND THE THREE BIG-BANGS
UNIVERSE
BB1
BB2
BB3
Planck ≡ 10^−33cm; 10^−44sec ≡ Planck
10^18 sec
10^29 cm
62 POWERS OF TEN
in Space
in Time
Galaxies Stars Planets
COMPLEXITY
LIFE
CONSCIENCE
CREATIVITY
REASON
Vegetal Animal
Cogito ergo sum
Imagination becomes reality
Language
Logic
Science
Li
Co
Cr
U
B ≡ F (Superworld)
Space
Couplings
Time

Figure 12

Let me point out that Science is the consequence of us being the only form of leaving matter endowed with Reason, from where the sequence of Language–Logic–Science has been originated [12]. The time-sequence of Language–Logic–Science is shown in figure 13.

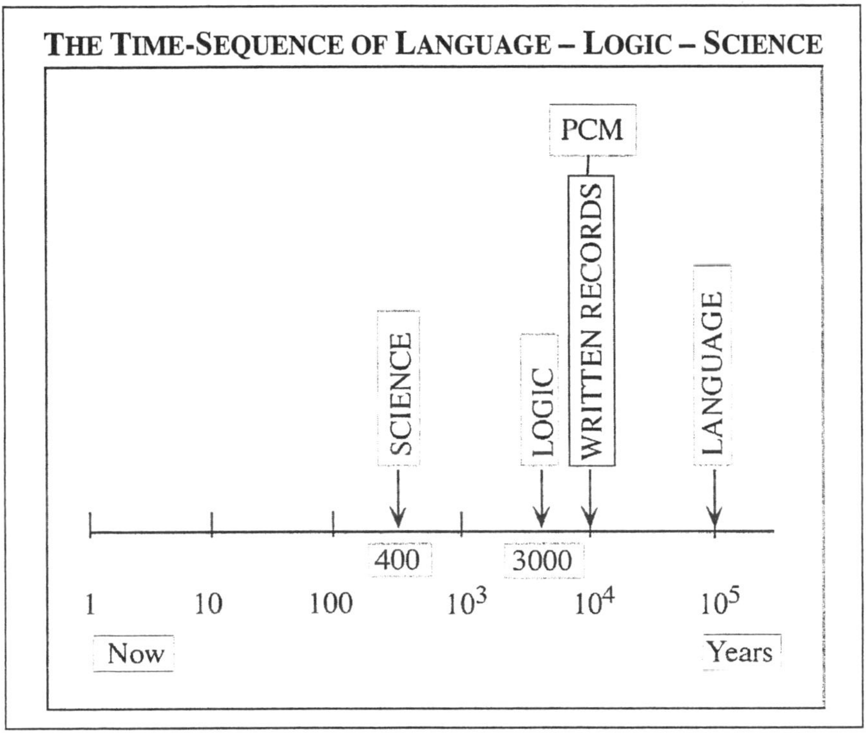

Figure 13

The experimental evidence is that UEEC events dominate our life as the evolution of the world in its real life (EWRL ≡ History) and the evolution of our basic understanding of the laws governing the world (EBUS ≡ Science) do show (see Table 1).

We confront the present status of physics with Ten Challenges in Addendum 4.

The next UEEC event must be outside these Ten Challenges.

We should be aware of the fact that it would be great if, **for the first time in the 400 years of Galilean Science, the sequence of UEEC** events could enjoy a **formidable stop**.

The final question is: why the greatest achievements of Science have always been originated by totally unexpected events?

8 FROM PLANCK TO COMPLEXITY

Four centuries of Galilean research work based on Reductionism, i.e. on the identification of the simplest elements in the study of Nature, has allowed us to get the greatest achievement of Science, i.e. the so called **Standard Model** and its extension (**SM&B**), illustrated before in figure 7.

This extension predicts GUT (the Grand Unification Theory), the existence of the Superworld and the resolution of the quantum-gravity problem via the powerful theoretical structure of RQST (Relativistic Quantum String Theory).

All these developments started thirty years ago when a great scientific novelty came; all experimental discoveries obtained with our powerful accelerators were to be considered only matters of extremely low energy.

The scale of energy on which to direct the attention to understand the Logic that rules the world, from the tiniest structures to the galactic ones, had to be shifted at a much higher level: to the mass-energy named after Planck, E_{Planck}, something like seventeen powers of ten above the Fermi scale, E_{Fermi}, that already seemed to be an extremely high level of energy.

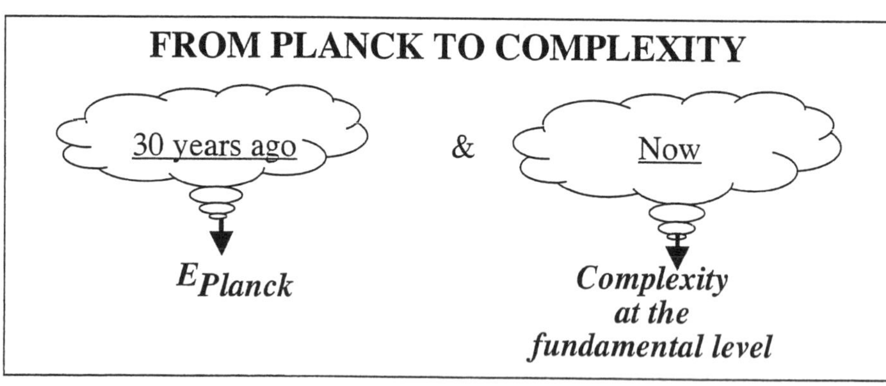

Figure 14

Now, after thirty years, it comes about the novelty of our time, illustrated in figure 14: Complexity exists at the fundamental level [5]. In fact, **AFB** and **UEEC** events exist at all scales, as reported all along this lecture and summarized in figure 2.

This result is corroborated from the mathematical structure (the only one) to be in a position of describing all that happens at the Planck scale: the Relativistic Quantum String Theory (**RQST**).

This mathematical structure produces innumerable minima of energy, named **Landscape**.

The theoretical discovery of the **Landscape** (Leonard Susskind) [13], has been followed by another formidable discovery in mathematical physics: the most rigorous model of RQST (Raphael Bousso and Joseph Polchinski) is **NP-complete** (Michael R. Douglas and Frederik Denef) [14].

This discovery corroborates all that we have put in evidence during the last five years [7–10]: **Complexity exists at the fundamental level** [5].

We do not know what will be the final outcome of String Theory.

What we know is that: '*The world appears to be complex at every scale.* ***Therefore*** *we must expect a continued series of surprises that we cannot easily predict'*.

But, with the advent of the LHC it will be possible to study the properties of the Quark-Gluon-Coloured-World (QGCW), which is a world totally different from all we have been dealing with since the origin of Science.

9 CONSEQUENCES FOR LHC: THE QGCW PROJECT

9.1 INTRODUCTION

With the advent of the LHC supercollider at CERN we propose to study the properties of the '**new world'** which is produced in a collision between heavy nuclei ($_{208}Pb^{82+}$) at the maximum energy so far available in our planet, i.e. 1150 TeV (1.15×10^{15} eV). The 'new world' is the **Quark-Gluon-Coloured-World** (QGCW).

We avoid in purpose to call it 'quark-gluon plasma' since, in the extremely high energy collision between heavy ions, many QCD open-colour-states should be produced.

The number of these QCD open-colour-states is by far higher that the number of baryons and mesons so far known, since these baryons and mesons have to obey the condition of being QCD-colourless.

In chapter 9.2 we discuss four basic new problems addressed by the QGCW project. In chapter 9.3 how to detect the properties of the new world (QGCW), in chapter 9.4 we discuss the two most difficult technical problems to solve in order to study the QGCW; in chapter 9.5 the synchronization issues, including four relevant details. The conclusions in chapter 10.

9.2 THE PHYSICS OF THE QUARK-GLUON-COLOURED WORLD (QGCW): THE NEW WORLD

The basic purpose of the project is to study the Quark-Gluon-Coloured World (QGCW) which is totally different from our world made of QCD vacuum with colourless baryons and mesons. We want to search for specific effects due to the fact that the colourless condition is avoided.

1st problem – In the QGCW there are all states allowed by the $SU(3)_c$ colour group. The number of possible states is by far more numerous than the number of colourless baryons and mesons which have so far been built in all Labs, since the colourless condition is not needed. **Question**: What are the consequences on the properties of the QGCW?

2nd problem – Light quarks versus heavy quarks. Are the coloured quark masses **the same** as the values we derive from the fact that baryons and mesons need to be in a colourless state? It could be that all six quark flavours are associated with nearly 'massless' states like those of the 1st family (u, d). In other words the reason why the 'top' quark appears to be so heavy ($\simeq 10^2$ GeV) could be due to the fact that it must satisfy some, so far unknown, condition related to the fact that the final state must be QCD-'colourless'.

We know that confinement produces masses of the order of a GeV. Therefore, according to our present understanding, the QCD 'colourless' condition could not explain the heavy quark mass, but since the origin of the quark masses is still not known, it cannot be excluded that in a QCD coloured world, the six quarks are all nearly massless.

If this was the case, the masses we measure are heavier than the effective coloured quark masses. In this case all possible states generated by 'heavy' quarks would be produced in the **QGCW** at much less temperature than the one needed in our world made with baryons and mesons, i.e. QCD colourless states.

Here again we should try to see if with masses totally different from those expected, on the basis of what we know about colourless baryons and mesons, new **effects** could be detected due to the existence of all six flavours at relatively low temperature in the QGCW world.

3rd problem – To search for effects on the thermodynamic properties of the QGCW. Are these properties going to be along the 'extensivity' and / or 'non-extensivity' conditions?

4th problem – Derive the equivalent Stefan-Boltzmann Radiation Law for the QGCW. The relation between **energy density** at emission **U**, and **Temperature** of the source **T**, is

$$U = cT^4$$

in classical Thermodynamics.

In the QGCW the correspondence should be

U $\equiv$ $p_\perp$ (transverse momentum)

T $\equiv$ average energy in the CM system.

In the QGCW the production of 'heavy' flavours should be studied versus $\langle p_\perp \rangle$ and versus $\langle E \rangle$. The expectation is

$$\langle p_\perp \rangle \equiv C \cdot \langle E \rangle^4$$

and any deviation would be extremely important.

The study of the properties of the QGCW should produce the correct mathematical structure able to correctly describe the QGCW; the same mathematical formalism should allow to go from QGCW to the Physics of Baryons and Mesons (PBM) and from here to a restricted component of PBM, namely Nuclear Physics, where all properties of the nuclei should find a correct description.

9.3 HOW TO STUDY THE NEW WORLD: QGCW

An example is illustrated in the figure 15 below where beams of known particles (p, n, π, K, μ, e, γ, ν) bombard the QGCW and a special set of detectors measures the properties of the outcoming particles.

Totally unexpected effects should show up if Nature follows the Logic of Complexity at the Fundamental level.

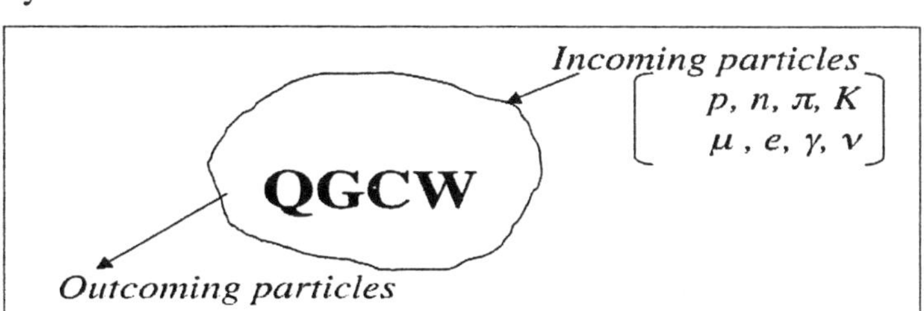

Figure 15

9.4 THE NEW TECHNOLOGY NEEDED TO STUDY THE QGCW

The technology needed covers two fields: one is the accelerator technology, the other is the detectors technology.

The first step in the accelerator technology refers to the availability of a proton beam able to bombard the QGCW produced in the lead-lead collisions.

The LHC physics program foresees lead-lead collisions with a design luminosity of 10^{27} cm^{-2} s^{-1}. For this to be achieved an upgrade of the ion injector chain comprising Linac3, LEIR, PS and SPS machine is needed. Each LHC ring will be filled in 10 min by almost 600 bunches, each of 7×10^{7} lead ions. Central to the scheme is the Low Energy Ion Ring (LEIR), which transforms long pulses from Linac3 into high brilliance bunches by multi-turn injection, electron cooling and accumulation.

The total collision energy between heavy ions, $_{208}Pb^{82+}$ (fully stripped), is 1150 TeV. Table 1 shows the basic parameters of the lead-ion injectors.

Table 1: NOMINAL PARAMETERS OF THE LEAD ION INJECTORS

	Linac3	LEIR	PS	SPS
Output energy	4.2 MeV/n	72.2 MeV/n	5.9 GeV/n	177 GeV/n
^{203}Pb charge state[1]	27+/54+	54+	54+/82+	82+
Output Bρ [Tm][1]	2.28/1.14	4.80	86.7/57.1	1500
# Batches to fill next machine	4–5	1	13,12,8	12
Bunches/ring		2 (1/8 PS)	4	52,48,32
Ions per pulse[2]	1.15×10^{9}	9×10^{3}	4.8×10^{3}	$<4.7\times10^{9}$
Ions/LHC bunch	1.15×10^{9}	2.25×10^{2}	1.2×10^{2}	9×10^{7}
Bunch spacing [ns]		352	100	100
ε^{*}_{rms} [μm] $= (\beta\gamma)_{rel}\,\sigma^2/\beta_{twitss}$	0.25	0.7	1.0	1.2
ε_1 [eVs/u/bunch]		0.05	0.05	0.24
4 σ bunch length		200 ns	4 ns	2 ns
Repetition time [s]	0.2–0.4	3.6	3.6	~50

[1]Values before/after stripping.
[2]50 eμA × 200 μs Pb^{54+} Linac3 output after stripping.

Once the lead-lead collision is available, the problem is to synchronize the 'proton' beam with the QGCW produced. This problem is at present under study.

The detector technology is also under intense R&D since the synchronization needed is at a very high level of precision; the status of the problems is reported in the following chapter 9.5.

9.5 SYNCHRONIZATION ISSUES RELATED TO THE QGCW PROJECT

The present limit of precision timing is given by the stability of master clocks and of optical fiber signal transmission lines.

Present timing signal stabilities is of order **nanoseconds.** Synchronization to the level down to **femto-seconds (10^{-15} sec)** between distributed equipment, accelerators RF systems, and between accelerator beam bunches will in future be required.

Synchronization at CERN LHC

The CERN network of accelerators requires the transmission, over long distances, of precise timing pulses. These timing pulses are derived from a master clock and therefore the ultimate time accuracy of the pulses is determined by the quality of the transmission of the clock frequency itself. A similar problem arises when two machines need to be synchronized RF-wise. In all cases the transmission of a CW wave in the frequency range of several hundred MHz over distances up to several kilometers is the key to a proper synchronization.

Optical fibers are used as transmission media and a 4 km optical fiber link was installed already in 1978 between the CERN SPS and its injector for the purpose of synchronizing the two machines. In preparation for the LHC, its timing distribution and its synchronization challenges, a special R&D project (RD12) had been set-up to study solutions for machine and experiments. A common solution to the timing, trigger and control (TTC) system requirement for machine and experiments leads to an important economies of scale and permits a rationalization of the requirements for development, operation, and support efforts.

The common systems allows to control the detector synchronization and delivers the fast signals and messages that are phased with the LHC clock, orbit or bunch structure. These include the bunch-crossing clock, level-1 trigger decisions, bunch and event numbers, as well as test signals and broadcast commands.

In the framework of the LHC common project RD12 the development and test of a multi-function optoelectronic TTC system was launched. The TTC system needed to meet the requirement of central signal broadcasting and local distribution at the different sub-detectors of the experiments.

A laser transmitter, modulator, encoder, VME-bus interface and machine interface have been developed as well as a subminiature radiation-hard optical fiber connector, active device mount and photo-detector / preamplifier.

A radiation-hard timing receiver ASIC is being designed which will generate the full range of decoded signals for electronics controllers from a single input and a PMC receiver module is being developed to facilitate initial applications.

The system incorporates programmable coarse and desktop facilities to compensate for different particle flight times and detector, electronics, propagation and test generator delays.

It can also transmit asynchronous slow controls and data such as individually addressed channel enables and calibration parameters to several thousand destinations. More details are published by the RD12 collaboration in their final report.

Synchronization challenges of the QCGW project

For the QCGW project it will be mandatory to improve the LHC timing system. One needs to provide a much shorter and predictable constant time delay distribution of RF signals than presently available at LHC. One should aim at carrying time information with resolution and stability to order of femtoseconds over distances of kilometers.

The R&D for the QCGW project needs to concentrate on research of **stability properties of optical fibers** for signal transport, on the development of an **improved microwave oscillator** (master clock), and on beam instrumentation.

LHC heavy ion particle bunches at high energies have rise-times of a few nanoseconds, the rise-time of proton bunches could be as low as 100 picoseconds, for photons and lasers one is in the femtosecond range. Therefore, referring to the rising or front edge of a bunch requires timing precision covering nanoseconds, picoseconds or even femtoseconds.

The key timing component in a synchrotron is the RF generator of the accelerating RF cavity. Charged particles moving through the cavity will see the longitudinal electrical field in the cavity gap. As a result, they are captured in the 'RF bucket' as bunches and thus are localized inside the synchrotron vacuum pipe.

By definition a 'bunch phase' is the momentary longitudinal position of the leading front or, in electrical terms, the rising edge of the moving particle bunch.

So, a bunch phase refers to the local position where it is momentary seen. The bunch phase can also be described in angular terms when referring it to a geometric reference point of the ring accelerator. To capture a bunch in a cavity implies that there is a relation between the bunch phase to the cavity RF phase. The presence of a charged bunch of particles is detected by a capacitive pick up in the vacuum pipe. These 'Beam Position Monitors' allow to measure the bunch phase in real-time and thus enables the cavity synchronization by controlling the phases of the RF-cavities. It is essential for proper acceleration to maintain phase differences of less than a few degrees between all cavities in the accelerator complex.

A precise reference signal common to all RF cavity stations is needed to operate with the same signal references during acceleration ramps. Otherwise, all RF-cavities will have arbitrary phases. At each group cavity system one has to select one reference signal for phase locking the cavity phases.

This is not only valid for a single accelerator synchronization. It is a rather direct extrapolation if one has to control a whole complex build out of synchrotrons and storage rings: Each of the named 'phase references' in a single accelerator system can be referenced to a wider spanned 'campus reference' in the same manner.

A system that transmits standard time- and frequency signals at the same time using signal-multiplexing techniques, could distribute more common time and frequency reference signals as well. These signals will all have the same predictable delay and link all local generated RF signals and additional time- and trigger-signals (TTC) reference planes of the whole campus site together. Thus, a common coherent campus synchronous network is built. This has been the basic idea for the LHC synchronization scheme.

The communications industry has created an RF signal transport technique called 'optical RF link' [15]. These devices are intended for linking broadband RF signals between mobile telephone base stations.

These optical links utilize direct modulated laser diode sources, which feed optical single mode fibers of several kilometers length and fast photodiodes at the receiving end of the fiber. The bandwidth of the analog link transmitter/receiver pair is beyond 2GHz in base-band mode. The RF signals are transmitted in base-band, they cover the frequency band of 900-1900 MHz as 'block' without up/down converting in frequency. The capabilities of these links match with present accelerator signal transport tasks: (i) analog transmission of multiple RF signals, (ii) constant delay properties of the transport fibers, the thermal length deviation coefficient of 'from the stock fibers' is about 7 ppm/K.

For a 1000m run of optical fiber (5ns/mgroup delay) a deviation of 35ps/K can be derived. For the target value of 40 K temperature variation (over the whole length) the total delay change is 1.4 ns.

For the QCGW project this value will have to be improved.

Transporting femto-second accelerator timing signals over long distances requires a new approach. One has to overcome effects of time jitter, typically induced by microphony, electromagnetic noise and temperature dependent variation of fiber cable length.

A possible solution could be based on the idea of a reference signal feedback system to stabilize the signal transmission. We think of using an extremely stable laser as an optical oscillator.

The laser provides reference signals (master clock) and in combination with a piezo-electric fiber stretcher one stabilizes the signal frequency transmitted through a long signal transport fiber cable in feedback mode. This idea is under study for X-FEL pulses and will be worked out in more detail for the QCGW project [15].

ANNEX 1: DETAILS ON TIMING DISTRIBUTION AT THE LHC

Extract from **Timing Distribution at the LHC** by Bruce.Taylor@cern.ch

The timing signals for each ring of LHC will be encoded and transmitted over optical links from the RF system to the PCR, where beam-synchronous messages will be added. High power laser transmitters will then broadcast the signals over single-mode optical fibers to the four LHC experiments, to the test beam areas, and to the beam instrumentation located around the LHC ring and on the SPS transfer lines. At the experiment areas, trigger information and local synchronous commands and data will be added. The regenerated signals will then be broadcast over multimode passive optical networks to several thousand destinations.

At the LHC, 'fast' timing signals must be distributed to all experiments and to the beam instrumentation of the machine. These signals are derived from the LHC RF generators and will be synchronous with the circulating beams, so that their frequencies will vary a little during acceleration. At 7 TeV, the bunch clock frequency will be about 40.07897 MHz while the orbit frequency will be 11.2455 kHz.

They are distinct from the 'slow' LHC timing signals, having a granularity of 1 ms, which will signal machine events and distribute UTC time for data tagging and post mortem applications.

At the LHC, distributing correctly synchronized signals to several thousand electronics channels presents some interesting challenges. The R&D work has been done in the framework of the RD12 TTC Project [16], which comprised members from all the LHC experiments, the Microelectronics and Beam Instrumentation Groups and two industrial partners.

The unified approach to TTC distribution developed by RD12 provides for the broadcasting of the fast timing signals through all the transmission stages from the RF generators of the LHC machine to the outputs of the timing receiver ASICs at the experiment and beam instrumentation destinations. That general path will be followed in this review of the system.

ANNEX 2: DETAILS ON LHC BUNCH STRUCTURE

Commencing with the timing of the LHC machine, it should first be noted that there has been an important change to the bunch structure described in the Yellow Book [17]. Initially it was proposed to accelerate trains of 84 bunches in the PS, 81 of which would be injected into the SPS, 3 being lost in the PS ejector. The difficulty with this configuration is that it would be dirty in the PS and SPS machines and there would be longitudinal stability problems in the PS with the 84-bunch trains.

Various solutions to these problems have been proposed and the one that has been retained as the current baseline foresees the acceleration of PS trains of 72 bunches which will be entirely injected into the SPS. In order to maintain an acceptable filling factor in the LHC with the 72-bunch PS trains, the SPS batches, which will be injected into the LHC, will comprise groups of 3, 3 and then 4 PS trains. So, whereas formerly we had quasi 12-fold symmetry in the LHC (the last PS train in the last SPS batch being suppressed to allow for the rise-time of the LHC extraction kicker), we now have quasi 4-fold symmetry with corresponding implications for the TTC synchronization algorithms [18].

As a result of this change the number of bunch crossings per orbit will be reduced from 2835 to 2808. Note that this applies only to ATLAS and CMS – since these experiments are diametrically opposite each other it is possible to phase the beams to make the LHC extractor gaps coincide at both of them. That is not possible at ALICE and LHCb, which will result in the loss of a further 188 bunch crossings per orbit at these experiments. It should also be noted that it is now expected that there may be quite a substantial initial running period with 75 ns instead of 25 ns bunch spacing. The expected rms collision length remains about 180 ps.

In the case of the LHC bunch structure for heavy ions, there may be several re-synchronizations during each orbit and there could be gaps, which are a non-integer number of bunch intervals in length.

But neither of these factors should be a cause for alarm, for in this case the bunch spacing concerned will be 100 ns or 125 ns. The TTC system will continue to distribute a 40.079 MHz clock during this mode of operation and the bunch crossings will remain in phase with this clock.

ANNEX 3: DETAILS ON CLOCK ARTEFACTS

On the other hand, at times there may be some artefacts in the distributed clocks. There could be a 1 ms hole in the SPS RF/5 clock occurring once before each transfer from the PS, because the LHC machine will be the master of the timing and the SPS has to be synchronized such that the SPS batches are injected into the correct part of the LHC orbit. That will be done by calculating back to the PS, so the PS and SPS have to be re-synchronized before each injection and during this time there may be an interruption in the clock.

The situation with colliding beams at the LHC will be more comfortable. In that case there could be a 1 ms hole in the bunch clock, which will occur once, and once only, before the very first injection from the SPS into the LHC. The reason for this is that a general RF system reset will be made prior to each LHC run in order to ensure that all the dividers have the correct phase and there may be an interruption to the clock while this is applied.

During these clock holes, the TTC system will continue to distribute a 40.079 MHz clock to the experiments. But developers should be aware that there may be a momentary phase perturbation when the system re-synchronizes with the real clocks when they are restored after the interruptions.

ANNEX 4: DETAILS ON DISTRIBUTION ARCHITECTURE

At present the RF timing generators are located in the BA3 Faraday Cage adjacent to the Prevessin Control Room (PCR).

Four clocks are available: the constant frequency 40.079 MHz LHC bunch clock, a pseudo LHC orbit signal obtained by dividing the clock by 3564, the real SPS orbit signal and the ramping SPS 40 MHz clock obtained by dividing the SPS RF by 5.

The PCR transmitters can each broadcast only one orbit and one clock signal simultaneously. The selected pair are encoded and used to modulate a high power laser, the output from which is split by a 1:32 optical tree coupler and broadcast via optical fibers to different destinations around the CERN sites.

At present these destinations include the test beam areas in the North and West halls and labs where beam instrumentation and TTC development work are being done. Finally they will include the LHC experiment areas. At the experiment areas the signals will be received by a TTC machine interface in which they are decoded.

10 CONCLUSIONS

We must be prepared with the most advanced technology for the discovery of totally unexpected events like the ones found in the Yukawa gold mine.

The occasion of the Yukawa Centenary (2007) has been of great value in order to draw attention to the impressive series of conceptual developments linked with his meson: chirality–invariance, spontaneous symmetry breaking, symmetry breaking of fundamental invariance laws, anomalies, and 'anomaly-free condition', existence of a third family of fundamental fermions, gauge principle for non-Abelian forces, instantons and existence of a pseudoscalar particle made of the quanta of a new fundamental force of Nature acting between the constituents of the Yukawa particle.

All the pieces of the Yukawa gold mine **could not have been discovered** if the experimental technology was not at the frontier of our knowledge, as already reported in chapter 7: the cloud-chambers (Anderson, Neddermeyer), the photographic emulsions (Lattes, Occhialini, Powell), the high power magnetic fields (Conversi, Pancini, Piccioni) and the powerful particle accelerators and associated detectors for the discovery – the world over – of the intrinsic structure of the Yukawa particle (quarks and gluons) which has brought us to QCD and now to the QGCW Project.

ADDENDUM 1
THE PLATONIC GRAND UNIFICATION

We report here again figure 9 from page 282 since this is the best examples of Platonic Grand Unification. The points have a sequence of 100 GeV in energy. The last point where the 'ideal' platonic straight line intercepts the theoretical prediction is at the energy of the Grand Unification. This corresponds to $E_{GU} = 10^{16.2}$ GeV. Other detailed information on the theoretical inputs: the number of fermionic families, N_F , is 3; the number of Higgs particles, N_H , is 2. The input values of the gauge couplings at the Z^0-mass is α_3 $(M_Z) = 0.118 \pm 0.008$; the other input is the ratio of weak and electromagnetic couplings also measured at the Z^0-mass value: $\sin^2 \theta_W (M_Z) = 0.2334 \pm 0.0008$.

The Platonic Grand Unification should be along the straight line (in blu) but Nature seems to follow the red points.

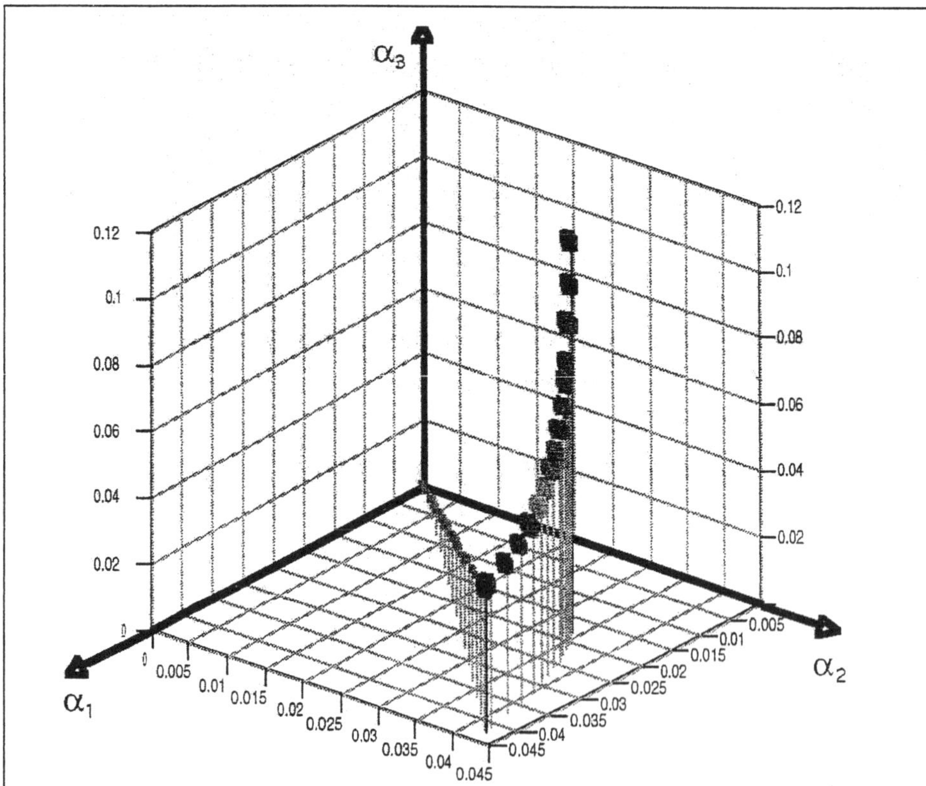

Figure 9 (from page 282)

ADDENDUM 2
THE PLATONIC SUPERSYMMETRY

THE PLATONIC CONCEPT OF SUPERSYMMETRY

The **Gauge Principle** should generate a
Gauge Force ⟶ **Gauge Bosons**
If *NATURE* was platonically *SUPERSYMMETRIC*
Supersymmetry Transformation should generate **Gauginos**

1st **DEVIATION FROM PLATONIC SIMPLICITY**
OUR FERMIONS ARE NOT THE GAUGINOS

2nd **DEVIATION FROM PLATONIC SIMPLICITY**
THE FUNDAMENTAL FERMIONS ARE OF TWO DIFFERENT CLASSES: LEPTONS AND QUARKS

3rd **DEVIATION FROM PLATONIC SIMPLICITY**
THERE IS NOT ONLY ONE BUT THREE FAMILIES OF FUNDAMENTAL FERMIONS

4th **DEVIATION FROM PLATONIC SIMPLICITY**
THE FUNDAMENTAL FERMIONS BECAME MIXED WHEN THE WEAK FORCES ARE SWITCHED ON: MIXINGS EXIST

5th **DEVIATION FROM PLATONIC SIMPLICITY**
THERE ARE DIFFERENT MIXINGS

Figure 16

ADDENDUM 3
EXAMPLES OF UEEC EVENTS IN THE CONSTRUCTION OF THE STANDARD MODEL AND BEYOND

There are very many UEEC events in the construction of the Standard Model and Beyond (SM&B). Let me select a group of problems where I have been directly involved: 1) the mass ≠ matter problem; 2) the mesonic mixings; 3) the Gribov QCD-light; 4) the Gap between the GUT energy and the string unification energy.

ADDENDUM 3–1
THE MASS≠MATTER PROBLEM

The fact that mass and matter had to be two different physical quantities, i.e. the mass ≠ matter problem, started with the Einstein discovery $E = mc^2$. The symbol ‘m’ originally was considered to represent ‘matter’ and thus the Einstein discovery become the problem of explaining the stability of matter. The meaning of ‘m’ had to be different from ‘matter’. This is how the distinction between ‘matter’ and ‘mass’ come in the forefront of fundamental physics. Einstein proposed to solve the problem mass ≠ matter with the existence of just one ‘charge’, the electromagnetic one. But, seven decades and the sequence of UEEC events reported in figure 17 were needed to understand the stability of matter. The ‘charge’ needed for the mass ≠ matter problem is not the ‘gauge’ charge, like the electromagnetic one, but a set of ‘flavour’ charges; and these are 12, as reported in figure 18 where it is also specified that there are three classes of ‘masses’.

ADDENDUM 3–2
THE PSEUDOSCALAR AND VECTOR MESONIC MIXINGS

The problem started when experimental physics was dominated by bubble chambers and the ‘mixing’ was determined using mass–formulae: i.e. a tautology. I designed and built a non-bubble-chamber detector, NBC; it consisted of an original neutron missing mass spectrometer coupled with a powerful electromagnetic detector which allowed to clearly identify all final states of the decaying mesons into (e^+e^-) or $(\gamma\gamma)$ pairs. The mass of the meson (be it pseudoscalar or vector) was measured by the neutron missing mass spectrometer. The two ‘mixing angles’, the pseudoscalar θ_{PS} and the vector θ_V, where directly measured (without using the masses) to be, not as expected by $SU(3)_{uds}$, i.e. $\theta_{PS} = \theta_V = 0$, but, $\theta_{PS} \neq 0$, $\theta_V \neq 0$ and totally different $\theta_{PS} \neq \theta_V$. Many years were needed and Gerard 't Hooft instantons to explain why $\theta_{PS} \simeq 10°$ and $\theta_V \simeq 51°$.

ADDENDUM 3–3
THE GRIBOV QCD LIGHT

When the physics of strong interactions finally became the physics of quarks and gluons, QCD had a problem, defined by Gribov as being its 'hidden QCD side': i.e., the large number of different final states produced by different pairs of interacting particles, such as (πp, pp, $\bar{p}p$, Kp, e^+e^-, νp, μp, ep, etc.). I did not limit myself to suggesting that a totally different approach was needed to put all these final states on the same basis. I found what this basis could be and this is how the 'Effective Energy' became the correct quantity to be measured in each interaction. The 'Effective Energy' was not predicted by QCD. To perform this study, it was necessary to analyze tens of thousands of (pp) interactions at the ISR. This was done despite all the difficulties to be overcome. And this is how what Vladimir Gribov defined the 'QCD light' was discovered (figures 20 and 21). Gribov pointed out what follows. Newton discovered that QED light is the sum of different colours. In QCD we have quarks and gluons interacting and producing Jets made of many pions, as for example in the (pp) reaction

$$pp \rightarrow \pi + X$$

whose spectrum is shown in figure 20. The horizontal axis is for the fractional energy of the pion (also called Feynman x), while the vertical axis is for the number of pions having fractional energy x_F. The spectrum in figure 20 is the sum ($\sum$) of all spectra shown if figure 21 where each one corresponds to a single value of the 'effective energy' (defined in terms of $2E_{had}$).

ADDENDUM 3–4
THE GAP BETWEEN E_{GUT} AND E_{SU}

The exact use of the RGEs for the running of the three gauge couplings (α_1 α_2 α_3) brings to the conclusion that the three gauge couplings (α_1 α_2 α_3) converge at E_{GUT} which is two powers of ten below the String Unification Energy E_{SU}. This is shown in figure 22 and the details in figure 23.

The lines are the result of calculations executed with a supercomputer using a system of three weakly coupled differential non-linear equations:

$$\mu \frac{d\alpha_i}{d\mu} = \frac{b_i}{2\pi} \alpha_i^2 + \sum_j \frac{bij}{8\pi^2} \alpha_j \alpha_i^2 .$$

describing the evolution of all phenomena including the superworld, from the maximum level of energy (Planck scale) to our world at the minimum of energy.

THE INCREDIBLE SERIES OF UEEC EVENTS
NEEDED TO EXPLAIN THE STABILITY OF MATTER
SEVEN DECADES: FROM THE ANTIELECTRON TO ANTIMATTER
AND THE UNIFICATION OF ALL GAUGE FORCES

• **The validity of C invariance from 1927 to 1957.**
After the discovery by Thomson in 1897 of the first example of an elementary particle, the Electron, it took the genius of Dirac to theoretically discover the Antielectron thirty years after Thomson.

1927 → Dirac equation [19]; the existence of the antielectron is, soon after, theoretically predicted. Only a few years were needed, after Dirac's theoretical discovery, to experimentally confirm (Anderson, Blackett and Occhialini [20]) the existence of the Dirac antielectron.

1930-1957 → **<u>Discovery of the C operator</u>** [(charge conjugation) H. Weyl and P.A.M. Dirac [21]]; discovery of the P Symmetry Operator [E.P. Wigner, G.C. Wick and A.S. Wightman [22, 23]]; discovery of the T operator (time reversal) [E.P. Wigner, J. Schwinger and J.S. Bell [24, 25, 26, 27]]; discovery of the CPT Symmetry Operator from RQFT (1955-57) [28].

1927-1957 → Validity of C invariance: e^+ [20]; $\bar{p}$ [29]; $\bar{n}$ [30]; $K_2^0 \rightarrow 3\pi$ [31] but see LOY [32].

• **The new era starts: C ≠ ; P ≠ ; CP ≠ (*).**

1956 → Lee & Yang P ≠; C ≠ [33].

1957 → Before the experimental discovery of P ≠ & C ≠, Lee, Oehme, Yang (LOY) [32] point out that the existence of the second neutral K-meson, $K_2^0 \rightarrow 3\pi$, is proof neither of C invariance nor of CP invariance. Flavour antiflavour mixing does not imply CP invariance.

1957 → C.S. Wu et al. P ≠; C ≠ [34]; CP ok [35].

1964 → $K_2^0 \rightarrow 2\pi \equiv K_L$: CP ≠ [36].

1947-1967 → QED divergences & Landau poles.

1950-1970 → The crisis of RQFT & the triumph of S-matrix theory (i.e. the negation of RQFT).

1965 → Nuclear antimatter is (experimentally) discovered [37]. See also [38].

1968 → The discovery [39] at SLAC of Scaling (free quarks inside a nucleon at very high q^2) but in violent (pp) collisions no free quarks at the ISR are experimentally found [40]. Theorists consider Scaling as being evidence for RQFT not to be able to describe the Physics of Strong Interactions. The only exception is G. 't Hooft who discovers in 1971 that the β-function has negative sign for non-Abelian theories [41].

1971-1973 → β = – ; 't Hooft; Politzer; Gross & Wilczek. The discovery of **non-Abelian** gauge theories. Asymptotic freedom in the interaction between quarks and gluons [41].

1974 → All gauge couplings α_1 α_2 α_3 run with q^2 but they do not converge towards a unique point.

1979 → A.P. & A.Z. point out that the new degree of freedom due to SUSY allows the three couplings α_1 α_2 α_3, **<u>to converge towards a unique point</u>** [42].

1980 → QCD has a 'hidden' side: the multitude of final states for each pair of interacting particles: (e^+e^- ; $p\bar{p}$; πp; Kp; νp; pp; etc.)
The introduction of the Effective Energy allows to discover the Universality properties [43] in the multihadronic final states.

1992 → All gauge couplings converge towards a unique point at the gauge unification energy: $E_{GU} \cong 10^{16}$ GeV with $\alpha_{GU} \cong 1/24$ [44, 45].

1994 → The Gap [4] between E_{GU} & the String Unification Energy: $E_{SU} \cong E_{Planck}$.

1995 → **<u>CPT loses its foundations at the Planck scale (T.D. Lee)</u>** [46].

1995-1999 → **<u>No CPT theorem from M-theory (B. Greene)</u>** [47].

1995-2000 → A.Z. points out the need for new experiments to establish if matter-antimatter symmetry or asymmetry are at work.

(*) The symbol ≠ stands for 'Symmetry Breakdown'.

Figure 17

Mass ≠ Matter

$|m_i\rangle \equiv$ Mass $\equiv$ Antimass $\equiv |\overline{m}_i\rangle$

i ≡ **1** (Intrinsic); i ≡ **2** (Confinement); i ≡ **3** (Binding)

$C\,|m_i\rangle = |m_i\rangle$ ✻✻✻

i ≡ 1, 2, 3

$|m_i Q_j\rangle \equiv$ Matter $\neq$ Antimatter $\equiv |m_i \overline{Q}_j\rangle$

$Q_j \equiv$ Flavour Charges

$$\mathbf{j} \equiv (\mathbf{u}\ \mathbf{d}\ \mathbf{c}\ \mathbf{s}\ \mathbf{t}\ \mathbf{b}) \equiv (1, 2, 3, 4, 5, 6)$$

$$(\nu_e\ e^-\ \nu_\mu\ \mu^-\ \nu_{HL}\ HL^-) \equiv (7, 8, 9, 10, 11, 12)$$

$$\nu_{HL} \downarrow \nu_\tau \qquad HL^- \downarrow \tau^-$$

$C\,|m_i Q_j\rangle = |m_i \overline{Q}_j\rangle$ ✻✻✻

i ≡ 1, 2, 3 ; J ≡ 1, 2, 3, 4, 5, 6, 7, 8, 9, 10, 11, 12.

Figure 18

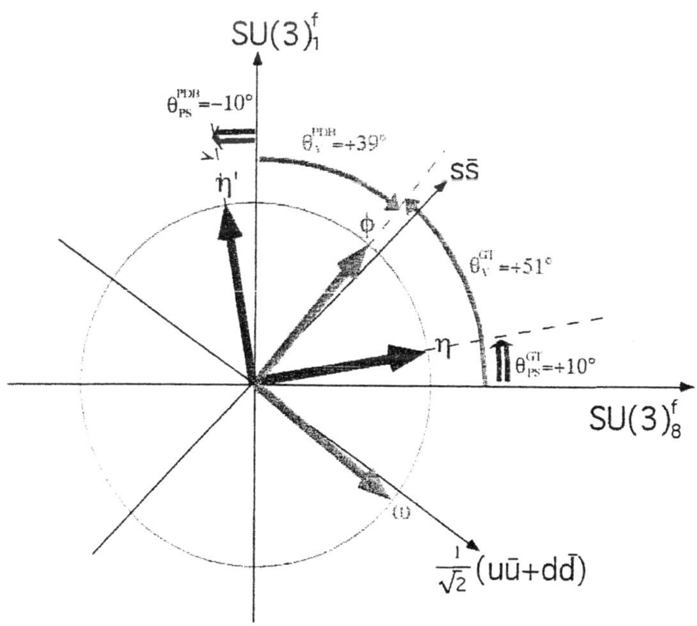

PDB ≡ Particle Data Book

GT ≡ Gerardus 't Hooft

$\theta_{PS} \neq \theta_V$

Direct: Not Using Mass Formulae

Figure 19

p-p $\rightarrow \pi^{+} + X$
Nominal Energy of the (pp) collision $= \sqrt{s} = 24$ *GeV*

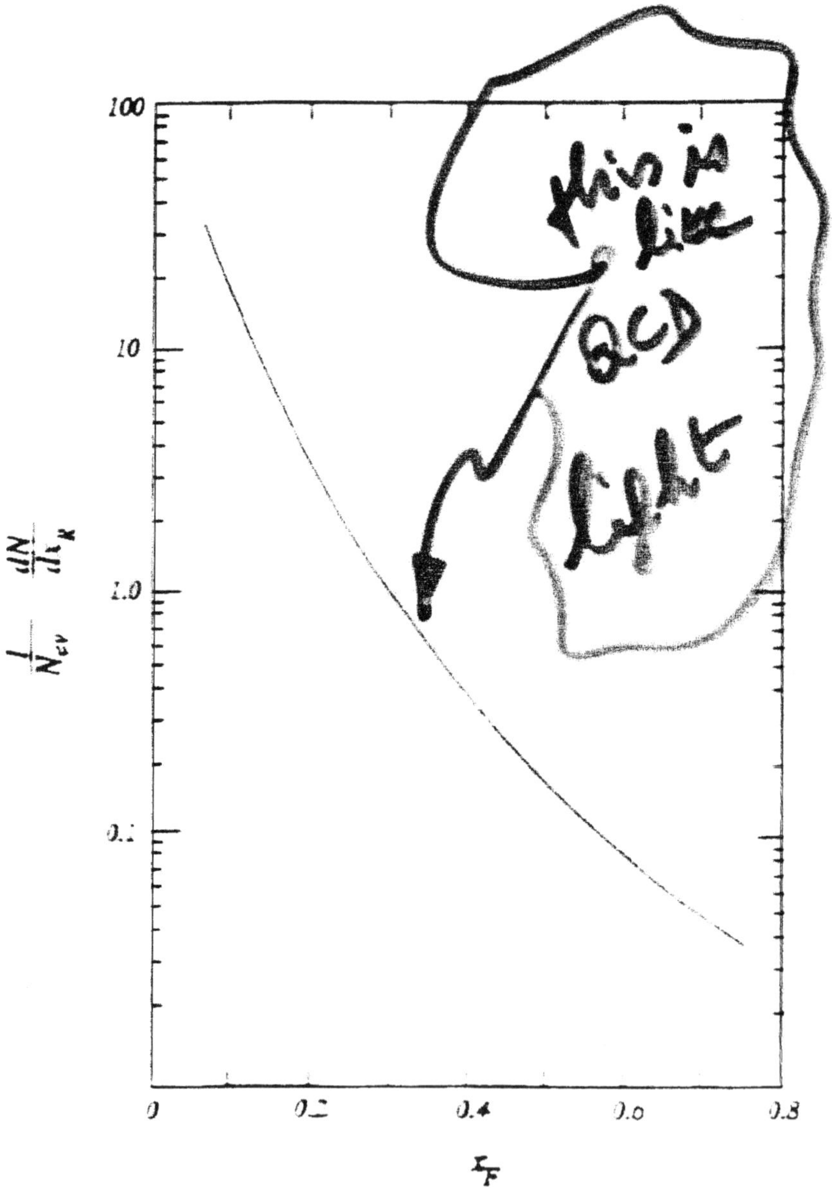

Figure 20

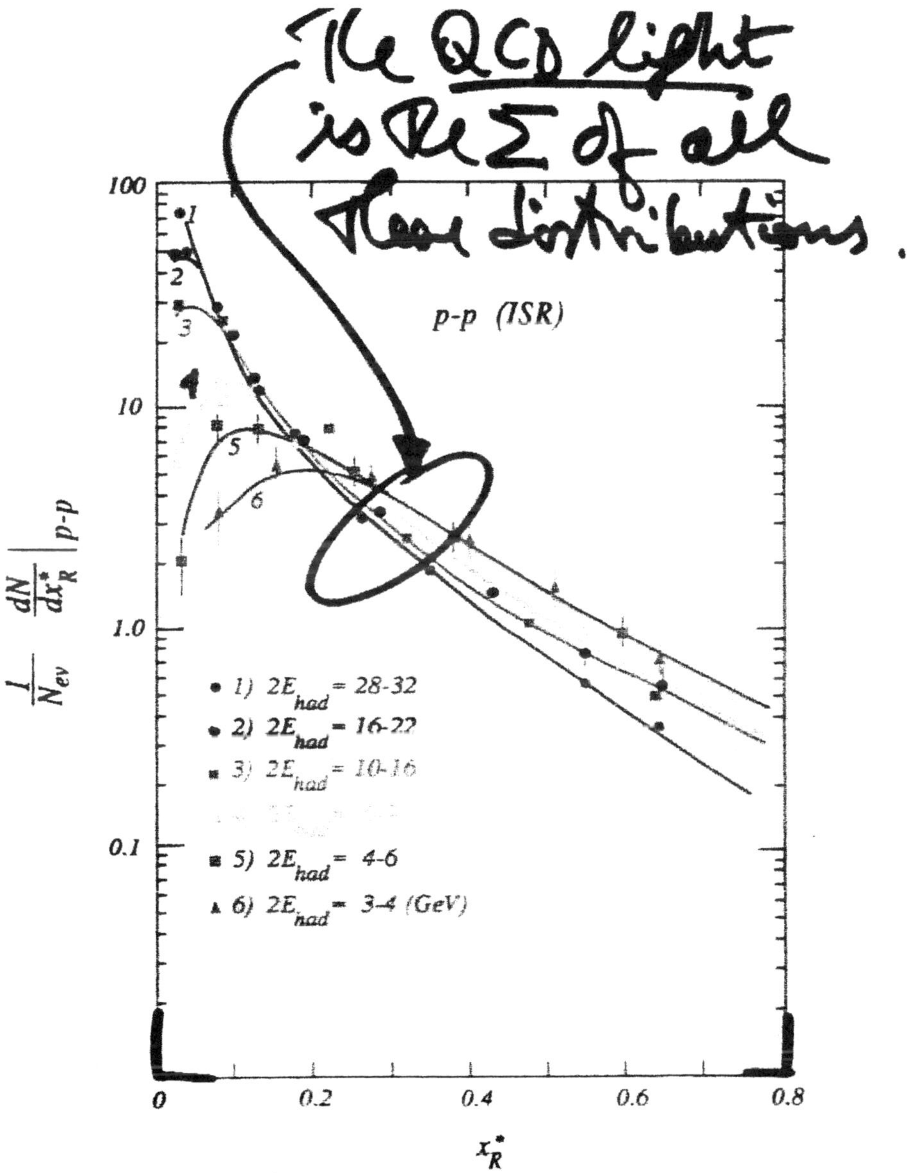
The QCD light is the Σ of all these distributions.
p-p (ISR)
100
10
1.0
0.1
0
0.2
0.4
0.6
0.8
x_R^*
$\frac{1}{N_{ev}} \frac{dN}{dx_R^*}\Big|_{p-p}$
1) $2E_{had}$ = 28-32
2) $2E_{had}$ = 16-22
3) $2E_{had}$ = 10-16
5) $2E_{had}$ = 4-6
6) $2E_{had}$ = 3-4 (GeV)

Figure 21

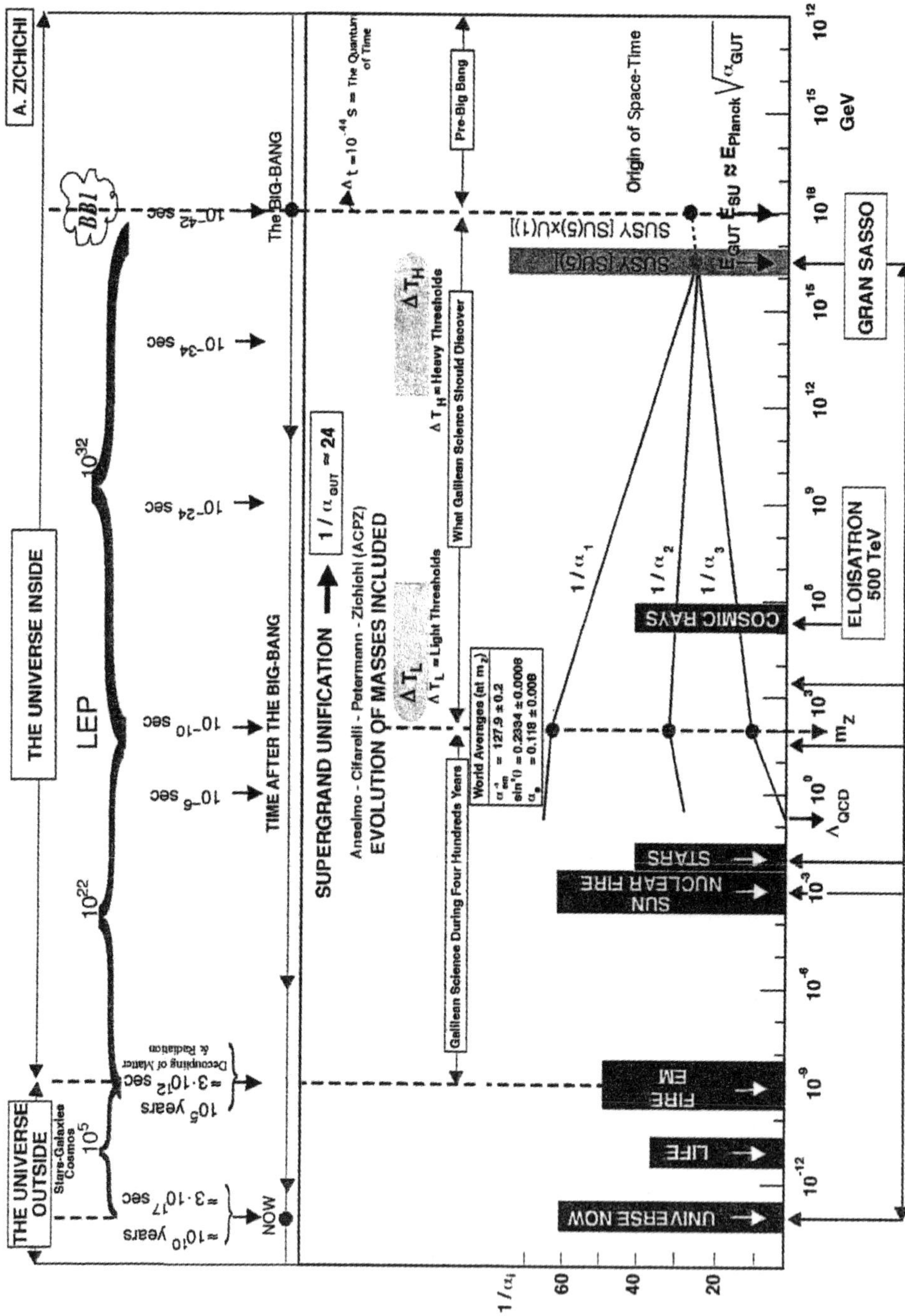
A. ZICHICHI
THE UNIVERSE INSIDE
THE UNIVERSE OUTSIDE
Stars-Galaxies Cosmos
LEP
TIME AFTER THE BIG-BANG
The BIG-BANG
NOW
SUPERGRAND UNIFICATION
1 / α GUT ≈ 24
Anselmo - Cifarelli - Petermann - Zichichi (ACPZ)
EVOLUTION OF MASSES INCLUDED
Pre-Big Bang
What Galilean Science Should Discover
Galilean Science During Four Hundreds Years
ΔT_H = Heavy Thresholds
ΔT_L = Light Thresholds
Origin of Space-Time
SUSY [SU(5)]
SUSY [SU(5)xU(1)]
World Averages (at m_Z)
UNIVERSE NOW
LIFE
FIRE EM
SUN NUCLEAR FIRE
STARS
COSMIC RAYS
ELOISATRON 500 TeV
GRAN SASSO
GeV
m_Z
Λ_QCD
1 / α_1
1 / α_2
1 / α_3

Figure 22

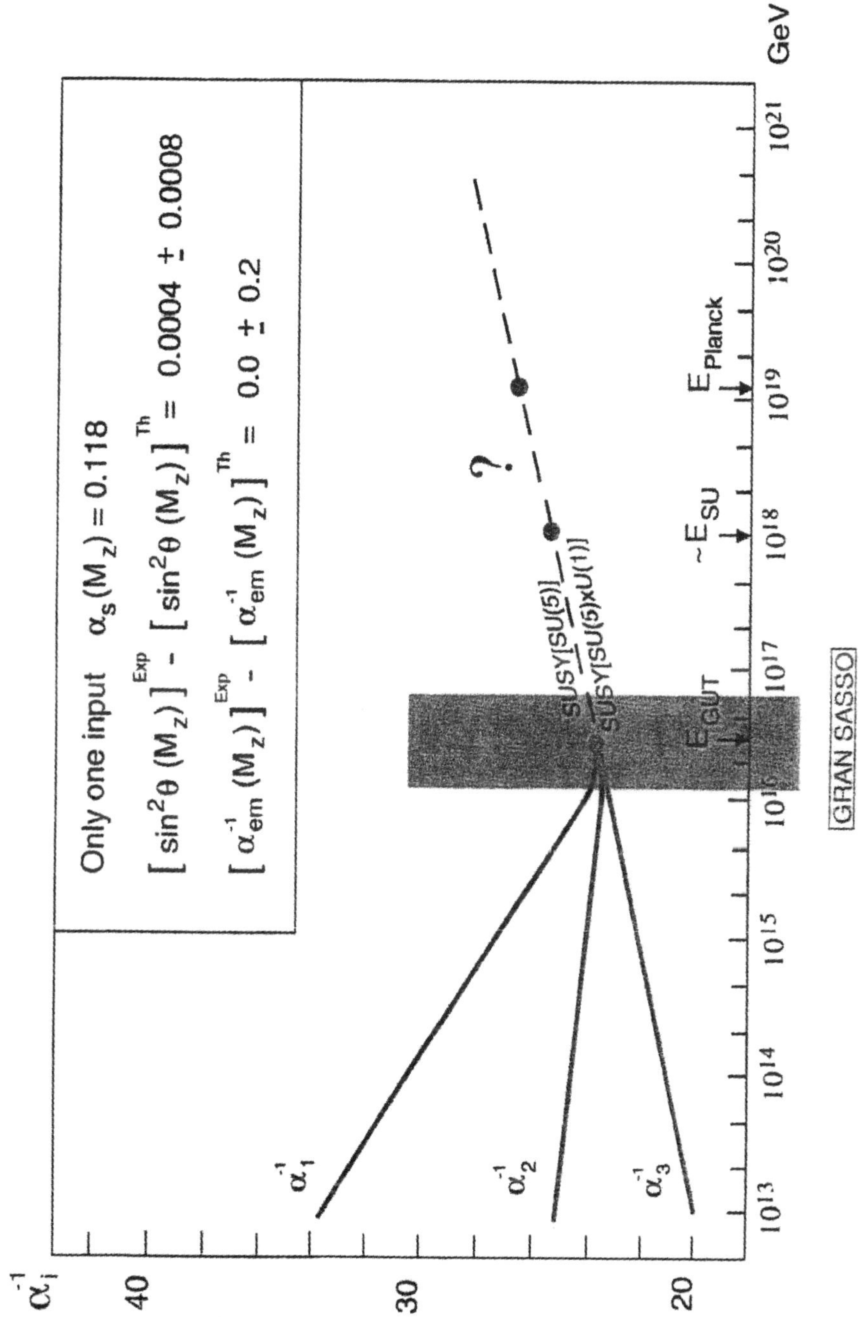
Only one input $\alpha_s(M_Z) = 0.118$
$[\sin^2\theta(M_Z)]^{Exp} - [\sin^2\theta(M_Z)]^{Th} = 0.0004 \pm 0.0008$
$[\alpha_{em}^{-1}(M_Z)]^{Exp} - [\alpha_{em}^{-1}(M_Z)]^{Th} = 0.0 \pm 0.2$
?
SUSY[SU(5)]
SUSY[SU(5)xU(1)]
α_1^{-1}
α_2^{-1}
α_3^{-1}
E_{GUT}
$\sim E_{SU}$
E_{Planck}
GRAN SASSO
α_i^{-1}
40
30
20
10^{13}
10^{14}
10^{15}
10^{16}
10^{17}
10^{18}
10^{19}
10^{20}
10^{21}
GeV

Figure 23

ADDENDUM 4
THE TEN CHALLENGES OF SUBNUCLEAR PHYSICS

Here is the list.

1 Non-perturbative QCD.
2 Anomalies and Instantons.
3 The Physics of NSSB (non-Spontaneous Symmetry Breaking: CP ≠, T ≠, CPT ≠ (*) Matter-Antimatter Symmetry).
4 The Physics of Imaginary Masses: SSB (part of this is the Higgs particle/particles).
5 The Physics of 43 dimensions (part of this is Supersymmetry).
6 Flavour mixing in the quark sector.
7 Flavour mixing in the leptonic sector.
8 The problem of the missing mass in the Universe.
9 The problem of the Hierarchy.
10 The Physics at the Planck scale and the number of expanded dimensions. Here the most interesting consequence would be that for a given expanded dimension it could be that the Planck scale goes down to the range of the Fermi scale, as illustrated in figure 24.

(*) The symbol ≠ means that a Symmetry law is non spontaneously broken as it happens with C, P, CP and T). [C (charge conjugation, i.e. interchange of charges with anti-charges); P (parity, i.e. interchange of left and right); T (inversion of the arrow of Time)]. The products CP and CPT mean the simultaneous Symmetry laws for all operations CP and CPT, respectively. The existence of Matter-Antimatter Asymmetry would be a proof of CPT ≠ .

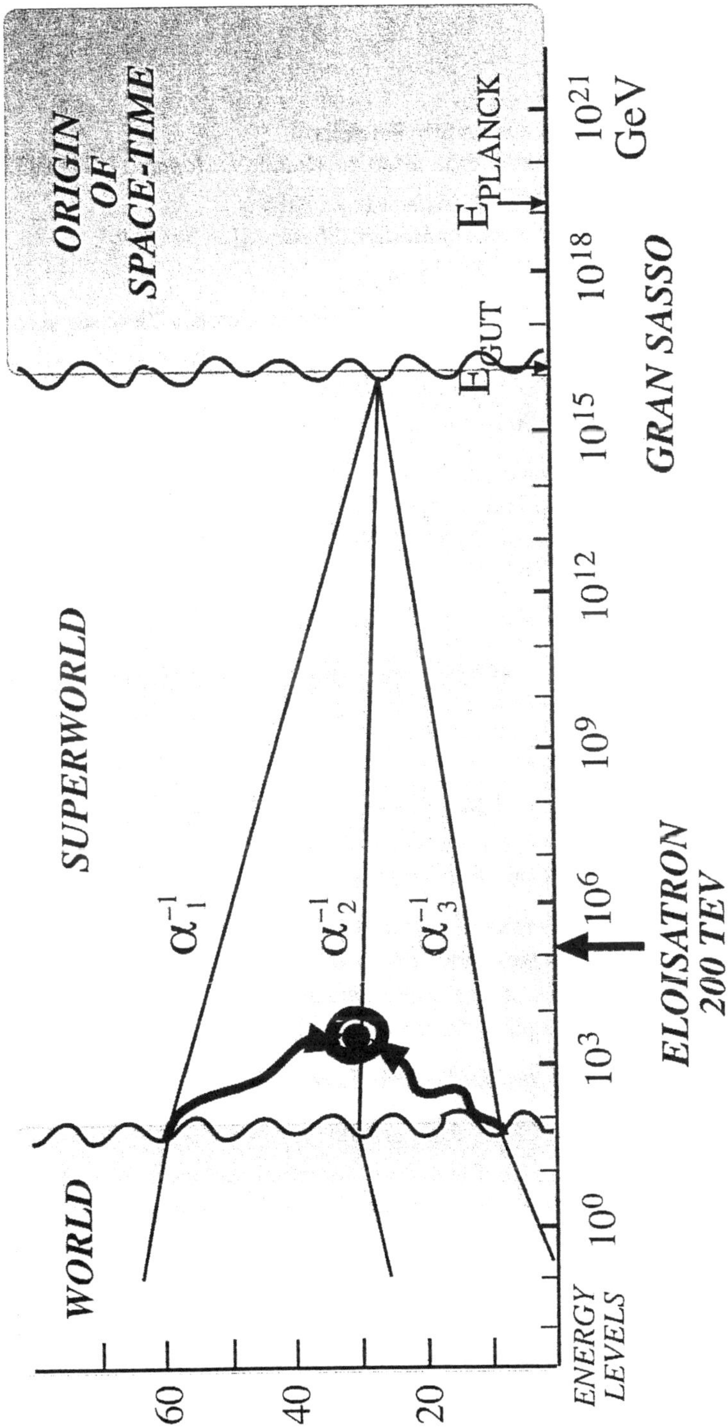

ORIGIN
OF
SPACE-TIME
SUPERWORLD
WORLD
α_1^{-1}
α_2^{-1}
α_3^{-1}
E_{GUT}
E_{PLANCK}
10^0
10^3
10^6
10^9
10^{12}
10^{15}
10^{18}
10^{21}
GeV
GRAN SASSO
ELOISATRON
200 TEV
ENERGY
LEVELS
60
40
20

Figure 24

REFERENCES

[1] *Interaction of Elementary Particles*
H. Yukawa, Part I, Proc. *Physico-Math. Soc. Japan* 17, 48 (1935);

Models and Methods in the Meson Theory
H. Yukawa, *Reviews of Modern Physics* 21, 474 (1949).

[2] *The QGCW Project*
A. Zichichi et al., CERN-LAA Preprint, October 2006; see also

Logical Reasoning in Experimental Physics: Past and Future
A. Zichichi, in *Gerardus 't Hooft Liber Amicorum to celebrate his 60th anniversary* (2006).

[3] *From the Yukawa Particle to the QGCW*
A. Zichichi, in Proceedings of the *'Symposium for the Centennial Celebration of Hideki Yukawa'*, International Nuclear Physics Conference, Tokyo, Japan, June 3-8 (2007).

[4] *A Study of the Various Approaches to* M_{GUT} *and* α_{GUT}
F. Anselmo, L. Cifarelli and A. Zichichi, *Nuovo Cimento* 105A, 1335 (1992).

[5] *Complexity exists at the Fundamental Level*
A. Zichichi in Proceedings of the 2004–Erice Subnuclear Physics School *'How and Where to Go Beyond the Standard Model'*, World Scientific (2007) pages 251-333.

[6] *Virtual History: Alternatives and Counterfactuals*
Niall Ferguson ed., Basic Books, New York (1999).

The References [7–10] refer to the various occasions where I have presented papers on highly specialized topics and discussed the connection of these topics with Complexity. Each title on the upper part refers to the connection with Complexity while the specialized topic is reported in the detailed references.

[7] *Complexity at the Fundamental Level*
A. Zichichi
presented at:

- International Conference on 'Quantum [un]speakables' in Commemoration of John S. Bell, International Erwin Schrödinger Institut (ESI), Universität Wien (Austria), November 2000, *'John Bell and the Ten Challenges of Subnuclear Physics'*.

- 40th Course of the International School of Subnuclear Physics, Erice (Italy), September 2002, *'Language Logic and Science'*.

- 31st, 32nd and 33th Course of the International School of Solid State Physics, Erice (Italy), July 2004, *'Complexity at the Elementary Level'*.

- 42nd International School of Subnuclear Physics, Erice (Italy), August - September 2004, '*Complexity at the Elementary Level*'.
- Trinity College, Dublin (Ireland), February 2005, '*Complexity at the Elementary Level*'.
- Department of Physics, University of Padova (Italy), March 2005, '*Complexity at the Elementary Level*'.
- 43th Course of the International School of Subnuclear Physics, Erice (Italy), September 2005, '*Complexity at the Elementary Level*'.
- Italian Physics Society (SIF) XCI Annual National Congress, University of Catania (Italy), September 2005, '*Complexity at the Elementary Level*'.
- Desy, Hamburg, November 2005, '*Complexity at the Fundamental Level*'.
- 44th Course of the International School of Subnuclear Physics, Erice (Italy), September 2006, '*Complexity at the Fundamental Level*'. These proceedings.

[8] *The Logic of Nature and Complexity*
A. Zichichi
presented at:

- Pontificia Academia Scientiarum, The Vatican, Rome (Italy), November 2002, '*Scientific Culture and the Ten Statements of John Paul II*'; '*Elements of Rigour in the Theory of Evolution*'.
- The joint Session of:
6th Course of the International School of Biological Magnetic Resonance; Erice (Italy), July 2003, '*Language Logic and Science*'.
- 2nd Workshop on Science and Religion of the Advanced School of History of Physics; Erice (Italy), July 2003, '*Language Logic and Science*'.
- 10th Workshop of the International School of Liquid Crystals; Erice (Italy), July 2003, '*Language Logic and Science*'.
- International School on Complexity, 1st Workshop on Minimal Life, Erice (Italy), December 2004, '*Evolution and Complexity at the Elementary Level*'.

[9] *Complexity and New Physics*
A. Zichichi
presented at:

- INFN-Alice Meeting, University of Catania (Italy), January 2005, '*Complexity at the Elementary Level*'.
- INFN Eloisatron Project 'The 1st Physics ALICE Week', Erice (Italy), December 2005, '*Complexity and New Physics with ALICE*'.
- 50th Anniversary of INFN Bologna - ALICE Week, Bologna (Italy), June 2006, '*Complexity at the Fundamental Level*'.

[10] *Complexity and Planetary Emergencies*
A. Zichichi
presented at:

- 27th Sessions of the International Seminars on Planetary Emergencies, Erice (Italy), August 2002, '*Language, Logic and Science*'.
- 28th Sessions of the International Seminars on Planetary Emergencies, Erice (Italy), August 2003, '*Language Logic and Science, Evolution and Planetary Emergencies*'.
- 36th Sessions of the International Seminars on Planetary Emergencies, Erice (Italy), August 2006, '*Complexity and Planetary Emergencies*'.

[11] *The Logic of Nature, Complexity and New Physics: From Quark-Gluon Plasma to Superstrings, Quantum Gravity and Beyond*
Proceedings of the 2006–Erice Subnuclear Physics School, World Scientific. These proceedings.

[12] *Language, Logic and Science*
A. Zichichi, proceedings of the 26th Session of the International Seminar on Nuclear War and Planetary Emergencies, Erice 18-26 August 2002 (World Scientific, 2003).

[13] *The Landscape and its Physics Foundations*
How String Theory Generates the Landscape
L. Susskind
Proceedings of the 2006–Erice Subnuclear Physics School, World Scientific. These proceedings.

[14] *Complexity and Landscape in String Theory*
M.R. Douglas and F. Denef
Proceedings of the 2006–Erice Subnuclear Physics School, World Scientific. These proceedings.

[15] Harald Klingbeil
GSI, private communication.

[16] RD12 collaboration: *Timing, Trigger and Control Systems (TTC) for LHC Detectors, spokesman*
Bruce G. Taylor 1992 – 2000 CERN/ Preprint 1999-12-01
Further references RD12
DRDC/90-62/P15, DRDC/91-22/P15 Add.1, DRDC/92-10/SR, DRDC/93-22/SR, DRDC/94-16/SR, LHCC/95-26/SR, LHCC/97-29/LEB/SR, LHCC/98-38/LEB/SR LHCC 2000-002/LEB/SR.

[17] *Yellow Book*
LHC Conceptual Design CERN/AC/95-05 (LHC), (October 1995).

[18] *Trigger, Timing and Control (TTC) Systems for the LHC*
LHC parameters and commissioning plan
Paul Collier/ CERN AB http://ttc.web.cern.ch/TTC/intro.html.

[19] P.A.M. Dirac, '*The Quantum Theory of the Electron*', *Proc. Roy. Soc.* (London) A117, 610 (1928); '*The Quantum Theory of the Electron, Part II*', *Proc. Roy. Soc.* (London) A118, 351 (1928).

[20] *The Positive Electron*
C.D. Anderson, *Phys. Rev.* 43, 491 (1933);

Some Photographs of the Tracks of Penetrating Radiation
P.M.S. Blackett and G.P.S. Occhialini, *Proc. Roy. Soc.* A139, 699 (1933).

[21] H. Weyl, '*Gruppentheorie und Quantenmechanik*', 2nd ed., 234 (1931).

[22] E.P. Wigner, '*Unitary Representations of the Inhomogeneous Lorentz Group*', *Ann. Math.*, 40, 149 (1939).

[23] G.C. Wick, E.P. Wigner, and A.S. Wightman, '*Intrinsic Parity of Elementary Particles*', *Phys. Rev.* 88, 101 (1952).

[24] E.P. Wigner, '*Über die Operation der Zeitumkehr in der Quanten-mechanik*', *Gött. Nach.* 546-559 (1931). Here for the first time an anti-unitary symmetry appears.

[25] E.P. Wigner, *Ann. Math.* 40, 149 (1939).

[26] J. Schwinger, *Phys. Rev.* 82, 914 (1951).

[27] J.S. Bell, '*Time Reversal in Field Theory*', *Proc. Roy. Soc.* (London) A231, 479-495 (1955).

[28] To the best of my knowledge, the CPT theorem was first proved by W. Pauli in his article '*Exclusion Principle, Lorentz Group and Reflection of Space-Time and Charge*', in 'Niels Bohr and the Development of Physics' [Pergamon Press, London, page 30 (1955)], which in turn is an extension of the work of J. Schwinger [*Phys. Rev.* 82, 914 (1951); '*The Theory of Quantized Fields. II.*', *Phys. Rev.* 91, 713 (1953); '*The Theory of Quantized Fields. III.*', *Phys. Rev.* 91, 728 (1953); '*The Theory of Quantized Fields. VI.*', *Phys. Rev.* 94, 1362 (1954)] and G. Lüders, '*On the Equivalence of Invariance under Time Reversal and under Particle-Anti-particle Conjugation for Relativistic Field Theories*' [*Dansk. Mat. Fys. Medd.* 28, 5 (1954)], which referred to an unpublished remark by B. Zumino. The final contribution to the CPT theorem was given by R. Jost, in '*Eine Bemerkung zum CPT Theorem*' [*Helv. Phys. Acta* 30, 409 (1957)], who showed that a weaker condition, called 'weak local commutativity' was sufficient for the validity of the CPT theorem.

[29] *Observation of Antiprotons*
O. Chamberlain, E. Segrè, C. Wiegand, and T. Ypsilantis, *Physical Review* 100, 947 (1955).

[30] *Anti-Neutrons Produced from Anti-Protons in Charge Exchange Collisions*
B. Cork, G.R. Lambertson, O. Piccioni, W.A. Wenzel, *Physical Review* 104, 1193 (1957).

[31] *Observation of Long-Lived Neutral V Particles*
K. Lande, E.T. Booth, J. Impeduglia, L.M. Lederman, and W. Chinowski, *Physical Review* 103, 1901 (1956).

[32] *Remarks on Possible Noninvariance under Time Reversal and Charge Conjugation*
T.D. Lee, R. Oehme, and C.N. Yang, *Physical Review* 106, 340 (1957).

[33] *Question of Parity Conservation in Weak Interactions*
T.D. Lee and C.N. Yang, *Phys. Rev.* 104, 254 (1956).

[34] *Experimental Test of Parity Conservation in Beta Decay*
C.S. Wu, E. Ambler, R.W. Hayward, D.D. Hoppes, *Phys. Rev.* 105, 1413 (1957);

Observation of the Failure of Conservation of Parity and Charge Conjugation in Meson Decays: The Magnetic Moment of the Free Muon
R. Garwin, L. Lederman, and M. Weinrich, *Phys. Rev.* 105, 1415 (1957);

Nuclear Emulsion Evidence for Parity Non-Conservation in the Decay Chain $\pi^+\mu^+e^+$
J.I. Friedman and V.L. Telegdi, *Phys. Rev.* 105, 1681 (1957).

[35] *On the Conservation Laws for Weak Interactions*
L.D. Landau, *Zh. Éksp. Teor. Fiz.* 32, 405 (1957).

[36] *Evidence for the* 2π *Decay of the* K_2^0 *Meson*
J. Christenson, J.W. Cronin, V.L. Fitch, and R. Turlay, *Physical Review Letters* 113, 138 (1964).

[37] *Experimental Observation of Antideuteron Production*
T. Massam, Th. Muller, B. Righini, M. Schneegans, and A. Zichichi, *Nuovo Cimento* 39, 10 (1965).

[38] *The Discovery of Nuclear Antimatter*
L. Maiani and R.A. Ricci (eds), Conference Proceedings 53, Italian Physical Society, Bologna, Italy (1995); see also A. Zichichi in '*Subnuclear Physics - The first fifty years*', O. Barnabei, P. Pupillo and F. Roversi Monaco (eds), a joint publication by University and Academy of Sciences of Bologna, Italy (1998); World Scientific Series in 20th Century Physics, Vol. 24 (2000).

[39] The first report on 'scaling' was presented by J.I. Friedman at the 14th International Conference on *High Energy Physics* in Vienna, 28 August-5 September 1968. The report was presented as paper n. 563 but not published in the Conference Proceedings. It was published as a SLAC preprint. The SLAC data on scaling were included in the Panofsky general report to the Conference where he says «... the apparent success of the parametrization of the cross-sections in the variable ν / q^2 in addition to the large cross-section itself is at least indicative that point-like interactions are becoming involved». '*Low* q^2 *Electrodynamics, Elastic and Inelastic Electron (and Muon) Scattering*', W.K.H. Panofsky in Proceedings of 14th International Conference

on *High Energy Physics* in Vienna 1968, J. Prentki and J. Steinberger (eds), page 23, published by CERN (1968). The following physicists participated in the inelastic electron scattering experiments: W.B. Atwood, E. Bloom, A. Bodek, M. Breidenbach, G. Buschhorn, R. Cottrell, D. Coward, H. DeStaebler, R. Ditzler, J. Drees, J. Elias, G. Hartmann, C. Jordan, M. Mestayer, G. Miller, L. Mo, H. Piel, J. Poucher, C. Prescott, M. Riordan, L. Rochester, D. Sherden, M. Sogard, S. Stein, D. Trines, and R. Verdier. For additional acknowledgements see J.I. Friedman, H.W. Kendall and R.E. Taylor, '*Deep Inelastic Scattering: Acknowledgements*', *Les Prix Nobel 1990*, (Almqvist and Wiksell, Stockholm/Uppsala 1991), also *Rev. Mod. Phys.* 63, 629 (1991). For a detailed reconstruction of the events see J.I. Friedman '*Deep Inelastic Scattering Evidence for the Reality of Quarks*' in '*History of Original Ideas and Basic Discoveries in Particle Physics*', H.B. Newman and T. Ypsilantis (eds), Plenum Press, New York and London, 725 (1994).

[40] *Quark Search at the ISR*
T. Massam and A. Zichichi, *CERN preprint*, June 1968;

Search for Fractionally Charged Particles Produced in Proton-Proton Collisions at the Highest ISR Energy
M. Basile, G. Cara Romeo, L. Cifarelli, P. Giusti, T. Massam, F. Palmonari, G. Valenti and A. Zichichi, *Nuovo Cimento* 40A, 41 (1977); and

A Search for quarks in the CERN SPS Neutrino Beam
M. Basile, G. Cara Romeo, L. Cifarelli, A. Contin, G. D'Alì, P. Giusti, T. Massam, F. Palmonari, G. Sartorelli, G. Valenti and A. Zichichi, *Nuovo Cimento* 45A, 281 (1978).

[41] A. Zichichi in '*Subnuclear Physics - The first fifty years*', O. Barnabei, P. Pupillo and F. Roversi Monaco (eds), a joint publication by University and Academy of Sciences of Bologna, Italy (1998); World Scientific Series in 20th Century Physics, Vol. 24 (2000).

[42] *New Developments in Elementary Particle Physics*
A. Zichichi, *Rivista del Nuovo Cimento* 2, n. 14, 1 (1979). The statement on page 2 of this paper, «*Unification of all forces needs first a Supersymmetry. This can be broken later, thus generating the sequence of the various forces of nature as we observe them*», was based on a work by A. Petermann and A. Zichichi in which the renormalization group running of the couplings using supersymmetry was studied with the result that the convergence of the three couplings improved. This work was not published, but perhaps known to a few. The statement quoted is the first instance in which it was pointed out that supersymmetry might play an important role in the convergence of the gauge couplings. In fact, the convergence of three straight lines ($\alpha_1^{-1}\alpha_2^{-1}\alpha_3^{-1}$) with a change in slope is guaranteed by the Euclidean geometry, as long as the point where the slope changes is tuned appropriately. What is incorrect about the convergence of the couplings is that, with the initial conditions given by the LEP results, the change in slope needs to be at $M_{SUSY} \sim 1$ TeV as claimed by some authors not aware in 1991 of what was known in 1979 to A. Petermann and A. Zichichi.

[43] V.N. Gribov, G. 't Hooft, G. Veneziano and V.F. Weisskopf '*The Creation of Quantum ChromoDynamics and the Effective Energy*', L.N. Lipatov (ed), a joint publication by the University and the Academy of Sciences of Bologna, Italy (1998); World Scientific Series in 20th Century Physics, Vol. 25 (2000).

[44] *The Effective Experimental Constraints on M_{SUSY} and M_{GUT}*
F. Anselmo, L. Cifarelli, A. Petermann and A. Zichichi, *Nuovo Cimento* 104A, 1817 (1991).

[45] *The Simultaneous Evolution of Masses and Couplings: Consequences on Supersymmetry Spectra and Thresholds*
F. Anselmo, L. Cifarelli, A. Petermann and A. Zichichi, *Nuovo Cimento* 105A, 1179 (1992).

[46] *Are Matter and Antimatter Symmetric?*
T.D. Lee, in Proceedings of the '*Symposium to celebrate the 30th anniversary of the Discovery of Nuclear Antimatter*', L. Maiani and R.A. Ricci (eds), Conference Proceedings 53, page 1, Italian Physical Society, Bologna, Italy (1995).

[47] *String Theory: the Basic Ideas*
B. Greene, Erice Lectures - Discussion 1999 in '*Basics and Highlights in Fundamental Physics*', A. Zichichi (ed), World Scientific (2001).

CHAIRMAN: A.ZICHICHI

Scientific Secretaries: C.Sbarra, S.Bufalino

DISCUSSION

- *Parihar*

What is NP? What is NP complete? Do you believe that physics at present is NP complete?

- *Zichichi*

What literally NP means is "Non Polynomial". NP complete means that you have no chance to solve the problem. I don't think that physics is NP complete. As I said in my lecture, F.Denef and M.R.Douglas have proved that the most rigorous model of RQST (R.Bousso and J.Polchinski) is NP complete.

-*Lopez Mateos*

You spoke of Platonic simplicity. Could you give a more precise definition of the word 'Platonic' in this context?

-*Zichichi*

The answer is very simple. Platonic simplicity is what Plato has formulated when he was thinking of fundamental concepts: things must be as simple as possible.
Our Grand Unified Theory is extremely complicated (running coupling constants, differential non linear equations coupled, ...). Platonic simplicity is the simplest way to formulate a model. It is not what we define as Grand Unified Theory. For example the SU(5) has a big number of particles. Instead, the Platonic unified theory is based on one and only one fundamental particle. So a basic fundamental particle exists and it could be either a boson or a fermion because of the Supersimmetry and this is the Platonic Supersimmetry. I constructed the Grand Unified Platonic Model in order to show that to obtain the Standard Model we need many deviations from Platonic simplicity. The Platonic Supersimmetry implies that the only allowed fermions are the gauginos, but in the real world we have quarks and leptons and they are not gauginos. In the real world you have many unexpected deviations from Platonic simplicity: lepton families, different mixings, etc...

-*Lopez Mateos*

But then, your platonically simple models of Nature are not compatible with life?

-*Zichichi*

This is absolutely correct. Complexity in fact exists at the fundamental level and not purely in fields that carry the reputation of being complex. The first approach to understand Nature was the Holistic one. From Galileo Galilei to now we have discovered the laws of Nature not through the Olistic approach but via the Galilean Reductionism.

-Tsallis

Why did you say that there are three, not four, Universal Constants?

-Zichichi

Because I think that the fundamental constants are h, c, and G: the Planck action, the speed of light and the Newton constant. I did not mention the Boltzmann constant that represents the minimal amount of Entropy (like the Planck constant, that is the minimal amount of action).

-Goldstein

T. Kuhn says that the anomalies should lead to revolutions in physics. What are the anomalies in the Standard Model? Cosmology tells us that 96% of matter-energy is missing, that is an anomaly. What do we need to go beyond the Standard Model?

-Zichichi

Cosmology, until now, has discovered nothing in the sense that it is an interpretation of the observations. For example the Dark Matter is an invention of the theoreticians but the discoveries are what we have found experimentally. The real scientific revolutions are the discoveries of completely new and totally unpredicted things. In this sense I disagree with Popper and Kuhn that, instead, thought of revolutions as falsifications of existing theories.We are the children of Galileo Galilei: reductionism, reproducibility and the experimental proofs are essential conditions for Science. This is the reason why Dark Matter and Dark Energy are to be first experimentally proved to exist.

For example even if Cosmology is fascinating, what we know about the Cosmo is due to the work of nuclear and subnuclear physicists. For example, the Neutron Stars are there because Chadwick in 1932 discovered the neutron. Now cosmologists speak of Quark stars. These new objects come from the discovery of the quarks in our laboratories.

Thinking at the Superworld, it is so far a set of mathematical formulae. The Superworld needs experimental proofs.

PRESENT AND FUTURE OF THE GRAN SASSO UNDERGROUND LABORATORY

Eugenio COCCIA
INFN Gran Sasso National Laboratory and U. of Rome Tor Vergata, Italy
E-mail: eugenio.coccia@lngs.infn.it

ABSTRACT

Today the future of underground science appears challenging and rich. This is true in particular for the INFN Gran Sasso Laboratory whose main characteristics, present scientific activity and future perspectives I present here.

1. Underground physics

Very high energy phenomena, such as p-decay and $0\nu\beta\beta$, happen spontaneously, but at extremely low rates. The study of neutrino properties from natural and artificial sources and the detection of dark matter candidates requires capability of detecting extremely weak effects. Underground laboratories provide the necessary low radioactive background environment to investigate these processes. So these laboratories appear complementary to those with accelerators in the basic research of the elementary constituents of matter, of their interactions and symmetries. It is remarkable that the only clear evidence today for physics beyond the standard model (SM) comes from underground experiments.

In general one can consider two classes of motivations to push beyond the SM. There are "particle physics" reasons: the SM does not truly unify the elementary interactions, it leaves the problem of fermion masses and mixings unsolved and it exhibits the gauge hierarchy problem in the scalar sector. The second class of reasons finds its origin in "astroparticle and underground physics" issues: the problems of the solar and atmospheric neutrinos deficits, Majorana or Dirac neutrinos, baryogenesis, dark matter. These issues can be faced by contemporary experimental physics and constitute a formidable motivation for the underground laboratories activity.

In underground physics the struggle to advance the high energy frontier and to go beyond the standard model is the struggle for background control and reduction. Environment is the principal source of background. The environmental backgrounds of the laboratory depend on the depth and on the nature of the surrounding rocks and, as a consequence, may differ in the different facilities.

The importance of the different background sources clearly depends on the experiment. For example the energy directly deposited by the muons in the detector is more relevant for $\beta\beta$ than for dark matter searches, while the neutron backgrounds and cosmogenesis are important for both; in any case a depth of 1000 m or a little more is enough. Even atmospheric neutrinos, interesting by themselves as they are,

are a background; the principal one already at shallow depth, for proton decay experiments, which, as a consequence, need not to be too deep.

Frontier experiments do not require to the laboratory only a low background environment, but also technological support, easy and safe access and general as well as specific support structures.

I review here the present activity and future perspectives of the INFN Laboratori Nazionali del Gran Sasso. More information in the WEB-site http://www.lngs.infn.it.

2. The Gran Sasso National Laboratory

Figure 1: The underground facilities and the Gran Sasso massif geology

The INFN Laboratori Nazionali del Gran Sasso (LNGS) is the largest underground laboratory in the world. It is located besides a freeway tunnel under the Apennines, in central Italy near the town of L'Aquila, 120 km from Rome.

The mission of the Laboratory is to host experiments that require a low background environment in the field of particle and astroparticle physics and nuclear astrophysics and other disciplines that can profit of its characteristics and of its infrastructures. The geographical location (inside the National Park of Gran Sasso - Monti della Laga) and the special operating conditions (underground, near a highway tunnel and in close proximity to a large water basin) demand that special attention is paid to the safety and environmental aspects of the activities.

The proposal to build a large, high technology, underground laboratory was advanced in 1979 by A. Zichichi, then the President of INFN, and approved by the Italian Parliament in 1982. Since the original project the orientation of the halls was towards CERN, in order to host detectors to study neutrino physics on a beam produced at that laboratory.
Civil engineering works, under the responsibility of ANAS, the Italian Road Department, started in autumn 1982 and were completed by 1987.

The horizontal access allows easy transportation and installation of large pieces of apparatus. The underground facilities consist of three experimental halls and a set of connecting tunnels and service areas, for a total surface area of 18000 m^2. The three halls are approximately 100x18x18 m^3. An almost angle-independent 1400m rock overburden provides a μ flux attenuation of 10^{-6}. The neutron flux is $= 3.7 \pm 0.3 \times 10^{-2} m^{-2} s^{-1}$.

The infrastructures of the laboratory are completed by a number of buildings on the surface, near the western entrance of the tunnel, hosting offices, laboratories, shops, library, canteen, etc. General services include the headquarters, the secretariats and user support offices, the administrative offices, the prevention, safety, security and environmental protection services, the general, electrical and safety plants services and, in direct support to the experiments, low activity techniques (including very low background facilities with Ge detectors underground), cryogenics, mechanics, chemistry, electronics, computing and networking. Last but not least, the laboratory has an important activity in the outreach and in the diffusion of the scientific culture and in the training of young people.

The permanent staff is composed by 70 persons. Presently, the user are about 800 in number, from 25 different countries.

After almost 20 years of operational life, several first generation experiments have been completed and decommissioned, like MACRO and Gallex-GNO. A programme for the second phase has been implemented.

Main research topics of the present scientific programme are: neutrino physics with neutrinos naturally produced in the Sun and in Supernova explosion and neutrino oscillations with a beam from CERN (CNGS program), search for neutrino mass in neutrinoless double beta decays, dark matter search, studies of nuclear reactions of astrophysical interest.

Let us summarise the recent activity of the main research lines.

Solar neutrino physics is one of the traditional research sectors of the laboratory. After the glorious life of GALLEX and GNO, the focus is now on Borexino, which is dedicated mainly to the measurement of the Be line component of the solar neutrino spectrum. This experiment is now filling the detector sphere with liquid scintillator and will start the data taking in May 2007.

The solar models are based on data and extrapolations; in particular the thermonuclear cross sections of the involved reactions are not measured in the relevant energy range but rather extrapolated from higher energies. The direct measurements are made very difficult by the very low values of the cross sections. Using the new 400 kV accelerator, LUNA is continuing his activity for the measurements of the cross section of nuclear reaction of astrophysical interest.

Stars substantially more massive than the Sun end their life through spectacular stellar explosions known as core collapse supernovae. Supernovae are among the most violent phenomena in the universe. The huge amount of the binding energy of the collapsed core is emitted in the form of neutrinos and anti-neutrinos of all flavours. The observation of neutrinos released in the various stages of the explosion will provide direct information about the dynamics at the centre of the Supernova and therefore of the explosion mechanism. Neutrinos are in fact a unique tool to study the central region of the star, and hence of nuclear matter in the extreme state of temperature and density encountered in the Supernova core. Neutrinos originating from Supernavae explosions are continuously monitored, with 100% live time, by the LVD detector, constituted by 1000 tons of liquid scilntillator.

LVD is part of the SNEWS (Supernovae Early Warning System), which involves an international collaboration of experimenters representing several of the current supernova neutrino detectors in the world. In the future. The goal of SNEWS is to provide the astronomical community with a prompt alert of the occurrence of a Galactic core collapse event and enable optical observation of the early phases of the Supernova.

Elementary particles are different from their antiparticles because their charges - not only the electric one, but all of them - are opposite. The standard model assumes that neutrinos have only one charge, the lepton number. But, if this charge is not conserved, neutrinos and antineutrinos can be two states of the same particle. In this case well-specified nuclides would decay through the neutrino-less double beta channel. The Laboratory hosts today experiments searching for these very rare decays, employing different and complementary techniques, and is preparing new important activities. The Heidelberg-Moscow experiment which with a sensitive mass of 11 kg of enriched 76Ge was the most sensitive experiment in the world and accumulated data for a 75 kg y exposure. CUORICINO, which employs TeO2 bolometers for a total of 42 kg and started taking data data in 2003. It is in operation and is the most sensitive running experiment today. The Laboratory approved two new experiments: CUORE, which brings the CUORICINO technique to a mass of more than 400 kg of TeO2 bolometers, and GERDA planning to employ 500 kg of enriched 76Ge. These activities represent the state of the art of the field.

From astronomical observations, we know that most of the matter in the Universe is not made of nuclei and electrons as normal matter. It is called dark matter, because it does not emit light, and its nature is unknown. Probably, its constituents, elementary particles that interact only very weakly with the rest (they are called WIMPs,) are not yet discovered; they are around us, invisible, waiting to be discovered. The search for WIMPs is very difficult and requires a very low background environment and the development of advanced background reduction techniques. The search is going on in many experiments worldwide. At Gran Sasso several experiments, using different techniques, are active and new experiments have been approved and started operating in 2004.

DAMA/LIBRA employs NaI crystals to detect the WIMPs by means of the flash of light produced in the detector by an Iodine nucleus recoiling after having been hit by a WIMP, a very rare phenomenon. To distinguish these events from the background, DAMA searches for an annual modulation of the rate, a behaviour that has several aspects that are peculiar of the searched effect and not of the main backgrounds. With about 100 kg sensitive mass DAMA was the only experiment world wide sensitive to the annual modulation signature. After the conclusion of the experiment, results were published confirming a signal of annual modulation. The larger experiment LIBRA, with 250 kg sensitive mass, is continuing regularly to take data.

CRESST searches for WIMPs with a cryogenic technique, looking for a very tiny temperature increase in the detector, due to the energy deposited by nuclei hit by the WIMPs. After the final design, the CRESST2 CaWO2 detector has been installed.

Two new dark matter experiments, WARP and XENON have started their operation during 2006. They are based on the simultaneous detection of the ionization and scintillation signals in, respectively, liquid argon and liquid xenon. WARP performed preliminary measurements underground with a small (2.3 l) prototype and XENON is taking data with a 10 l apparatus. Both experiments appear very promising for the future of the field.

One of the major committments of the Gran Sasso laboratory in the next years will be the search of tau neutrino appearance on the artificial neutrino beam being built at CERN in Geneva, the CERN Neutrinos to Gran Sasso (CNGS) project. A beam of protons, with an energy of 400 GeV, produced in the SPS accelerator of CERN, collides with a target of graphite to produce a secondary beam consisting partially of pions () and kaons (K), charged particles with short lifetimes. After having been focused and directed towards the Gran Sasso, the + and K+ decay, giving origin to neutrinos that maintain practically the same direction of the particles that generated them. The neutrino beam contains almost exclusively muon neutrinos with an average energy of 17.4 GeV and a contamination on the level of a few percent of anti-muon neutrinos, electron and anti-electron neutrinos. The beam is directed through the Earth crust to Gran Sasso at 732 km distance and reaches the Gran Sasso Laboratory with a flux of thousands of billions per day.

The OPERA experiment is designed for the direct observation of tau neutrinos resulting from oscillations of the muon neutrino of the beam. This search requires both micrometer scale resolution, obtained with modern emulsion techniques and large sensitive mass (1800 t) obtained with Pb sheets interleaved by emulsion layers. The experimental apparatus is made up of two supermodules, each formed by a target followed by a muon spectrometer. The target consists of 62 walls containing over 200,000 bricks, where the neutrino interactions occur. The walls are interleaved with layers of Target Trackers, detectors that electronically track the event, permitting the determination in real time of the position. The measurement of the position of the neutrino interaction allows the identification of the brick in which the interaction occured. That brick can be extracted using the BMS and, after developing the photographic emulsions, can be analyzed using automated microscopes. While the events concerning muon neutrinos are relatively frequent (about 50 per day), only a few tau neutrino events are foreseen each year. The production and installation of the 200 000 bricks containing the Pb sheets and the emulsion layers is now proceeding regularly.

The first neutrino beam arrived in August 18, 2006. The first interactions with the Gran Sasso rocks, and the consequent production of muons, were immediately observed, as expected, by the LVD and OPERA detectors (see figures).

ICARUS is a general-purpose detector, with a broad physics programme, based on the novel concept of the liquid Argon time-projection chamber. After the development of the "definitive project" and of the risk analysis, completed in 2003, the 600 ton module was transported from Pavia to the underground Laboratory in december 2004 and is now in the phase of installation and completion of the cryogenic infrastructure. The experiment should be ready for the beginning of 2008.

The main activity of the theory group, staff and visitor scientists, has been focused on astroparticle physics, including solar and Supernova neutrinos, massive neutrinos, ultra high energy cosmic rays, topological defects and relativistic astrophysics. Important activity took place also in particle phenomenology and computer simulations of Lattice Field Theories. Finally it must me mentioned that a number of experiments in different disciplines, mainly geology and biology are profiting of the peculiar conditions of the laboratory.

3. Coordination in Europe

The directors of the Western Europe underground laboratories (Gran Sasso in Italy, Modane in France, Canfranc in Spain and Boulby in UK) initiated actions to co-ordinate activities and to optimise the use of the available resources, taking into account the different characteristics of the infrastructures.

A first action had been done by LNGS, which has been recognized by the European Union (EU) as an European Large Scale Facility. A contract between EU and INFN, started in December 2002, funded access to LNGS for (new) European users. Gradually EU is taking more responsibilities in basic research, a sector that had been previously left to the member states, with the declared aim of the Commission to

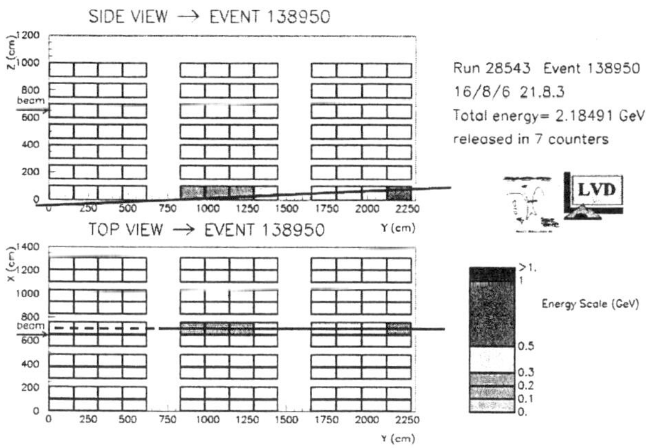

Figure 2: An event of the LVD experiment

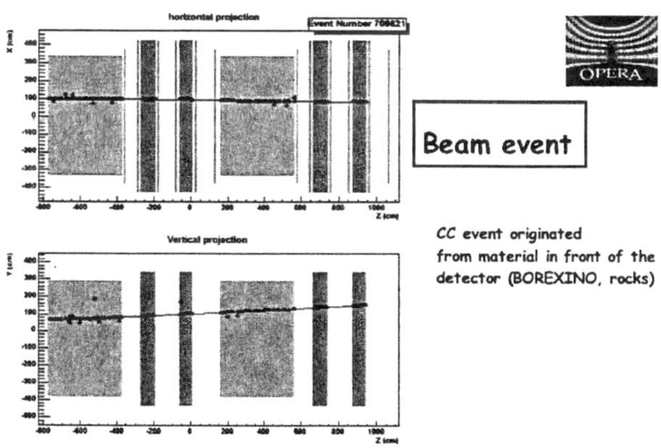

Figure 3: An event of the Opera experiment

build a "common European research area". This new policy has been fully embedded in the 6th "framework programme". In this framework, the ILIAS project was submitted to EU by ApPEC, the European inter-agency committee for astroparticle physics. The EU approved the project in the parts of underground science, gravitational waves and theory.
ILIAS contains several elements: the co-ordination of a number of activities of the laboratories (such as environmental background measurement and control, safety procedures, outreach activities, etc.), the funding for access of new users and common R&D in $\beta\beta$ and dark matter search. Detectors will evolve toward much larger masses, experimental setups and analysis tools will become much more complex, the experimental collaborations, used to work as rather small groups, most evolve toward larger organisations. ILIAS is a good opportunity to foster the necessary convergence and its program will likely continue in the seventh framework program.

4. Conclusions

Experiments in underground laboratories have opened a new window in neutrino physics, beyond the standard model.

The activity of the Gran Sasso Laboratory gives an idea of how exciting are the perspectives for underground physics in the next decade. The challenges are the following:

- Determine neutrinos absolute masses and nature (Majorana vs. Dirac), searching for $0\nu\beta\beta$ decay. Ton to multi-ton scale detectors and drastic background reduction are needed.

- Dark matter search. Requires ton-scale detectors with reduced background and good stability.

- Solar Be and pp neutrino fluxes should be measured with flavour sensitivity and high (few percent) precision. Weak Interactions may be still richer than what we know today (remember lessons of J. Bahcall and of neutrino oscillations).

- Measuring or limiting $|\theta_{13}|^2$. Requires a very intense ν_μ beam at low, but not too low, energy (of the order of 1 GeV) and a very massive Cherenkov detector or liquid argon TPC. May be at shallow depth. This step is necessary before seriously launching a project for a neutrino factory.

- p-decay. Needs Megaton scale Cerenkov detector or tens of kiloton liquid argon TPC. May be at shallow depth.

There is an increasing interest in realizing a large particle detector based on liquid Argon technology to afford exciting searches as proton decay and neutrino physics with natural and artificial sources. Such a project can drive the realization of a new shallow depth laboratory in the Gran Sasso area and exploit again a dedicated CERN to Gran Sasso neutrino beam.

CHAIRMAN: E. COCCIA

Scientific Secretaries: I. Gnesi, P. Jindal

DISCUSSION

- *Hisamatsu:*

I couldn't find any other neutrino beam monitor detector other than the muon monitor in your movie. Isn't there a near-detector at OPERA to measure the neutrino flux before oscillations?

- *Coccia:*

There is no near detector. At CERN there is only a muon monitoring before the neutrinos leave CERN territory, and this is sufficient for the aim of the OPERA experiment. Talking of the beam, the drawing - the manuscript signed by A. Zichichi, Italy and CERN, and the Direction of Gran Sasso - I showed this morning not only indicated the possibility of performing this experiment, but also guided Nino to orient the three experimental halls towards Geneva, so that now, for instance, OPERA and LVD as well as other future experiments offer all their lengths in the direction of CERN, they are all oriented towards CERN. However, to end my answer, there is no short-near detector at CERN.

- *Hisamatsu:*

There was an accident at HORN in K2K experiment which terminated the experiment. What kind of improvements are there at HORN in CNGS?

- *Coccia:*

Concerning the accident you mentioned, which I didn't remember, all the works in CERN (Geneva) for construction of the beam and at LNGS for realization of the experiments are subject to European laws, and all the safety issues are respected. As you know, an underground laboratory has very special safety problems as compared to other laboratories, and we suffered from an incident (it was not a safety accident involving people), when a spillage of liquid scintillator from the BOREXINO experiment occurred in 2002. The laboratory got into a lot of trouble. The spillage caused 50 liters of pseudocumene going into a river. I think not even a single fish died, but the "green people" were angry, the hostility of these local people caused a lot of trouble, because there are politicians who are ready to immediately strumentalize this hostility and the laboratory suffered for seals and sequestration of an entire hall. So, even if we didn't suffer any accident with persons, we know that all safety measures must be strictly respected in an underground laboratory. Otherwise the importance of any such accident will be seriously over estimated. So I think that we cannot afford to have another such accident, even a small one, so we are very careful to fulfill all necessary safety measures.

- *Lopez Mateos*:

For OPERA there is a veto layer shown in your slides in the way of the beam line. What do you have to veto in that direction?

- *Coccia:*

I think that probably there will be a veto, if there is a big shower or something like that. Then, they know there is not a single track going inside the beam. I don't know if there is someone from OPERA here, but I think they said something like that.

- *Lopez Mateos:*

In the LVD what is your energy resolution, and how good must it be to measure the tau neutrino appearance (to detect it)?

- *Coccia:*

In the transparency concerning the 'first CNGS event', you can see the energy scale, while the color here refers to the energy deposition. I think that is a very simple explanation.

- *Sartorelli:*

We didn't say that with LVD we detected the tau neutrino appearance, we are doing the monitor of the beam, and we are looking for muons produced in the rock in front of the detector or inside the detector. So we are talking about the detection of muons.

- *Lopez:*

So it's purely monitoring.

- *Sartorelli:*

Yes, we are just monitoring for the moment.

- *Zichichi:*

Not only monitoring, but measuring the disappearance, which is not trivial.

- *Sartorelli:*

However, just for the moment, we have to make analysis for sure.

- *Neri:*

The DAMA experiment claimed evidence of dark matter detecting signals of WIMP particles. The results seem to be contradicted by the CDMS experiment. Can you comment on the results, and if there can be a different interpretation of the results?

- *Coccia:*

The claim as you said is evidence of annual modulation of the signals detected by the NaI detector versus time from many years of data with DAMA, and they are continuing with larger mass with the name DAMA/LIBRA. There is a strong evidence for annual modulation. To go back and determine the candidate, and which is its coupling, which is the spin interaction with the NaI, this is a very different subject. To my knowledge when people from CDMS or from Edelweiss say they exclude DAMA, they refer to a particular model. So this exclusion is not for all possible couplings and spin dependent or independent interactions. Secondly, they are restricting the field, but I think since these are different detectors with different materials, and DAMA is the only one with NaI there is no competition for this material. There is a possibility that this can be a real candidate. However, of course the next measurements with many other detectors, even detectors that can discriminate between neutral massive particles with respect to others can help a lot.

- *Neri:*

So are you saying that the only possibility that the results of the two experiments can agree is based on a different spin of the WIMP?

- *Coccia:*

Yes, it is based on the fact that the possible candidate has some spin dependent coupling which then differs in one detector with respect to the other.

- *Csanad:*

How broad is the neutrino beam in the CNGS experiment, and how hard is it to tune it not to miss the target, does it shift year after year and is it easy to tune it?

- *Coccia:*

Before sending the beam for the first measurement, there, of course, has been a commissioning phase, during which small adjustment of the beam direction was performed, which was pointing towards Gran Sasso. Adjustment is possible because there is beam monitoring at the end of the line at CERN, which tells us the direction of the muons, so they can allow adjustment during this commissioning phase.

Regarding the beam spread, it is several hundred meters, when it arrives in Gran Sasso, and the maximum intensity is achieved for a width, say, of more than one hundred meters, and I think it is very well pointed. In fact, both OPERA and LVD, which have a distance of more than one hundred meters between them, report the number of interactions equal to the one expected. The beam is very well directed toward Gran Sasso and the spread is the nominal one that is a few hundred meters.

- *Csanad:*

But from this picture of the cells that shows the energy deposited by the neutrino beam ...

- *Coccia:*

But this is a single muon. The beam is not a line. When the beam arrives in Gran Sasso, it has front, so this is one of the muons. Each one of them can be traced back to CERN.

- *Csanad:*

So this is a single event?

- *Coccia:*

Yes, one event.

- *Brandt:*

How often can you check the emulsions in OPERA?

- *Coccia:*

Mainly continuously. When a candidate event is detected, the robot takes that particular brick, precisely this one. This brick is then developed in Gran Sasso. OPERA has an automatic imaging technique for investigating all the tracks; the track lengths and so on. This is done continuously, so there are millions of emulsions ready for five years of the experiment.

- *Brandt:*

What is the cost per event?

- *Coccia:*

The emulsions cost about 20 million euros. This is the contribution of the Japanese part of the collaboration (the Nagoya University, mainly). The emulsions are made especially by Fuji; their sensitivity is about 4000 ASA.

- *Aschieri:*

Can you tell us how the Gran Sasso laboratory was chosen for receiving the neutrinos from CERN? There are other underground laboratories in Europe and elsewhere. What were the variables relevant for making such a decision?

- *Coccia:*

To have the nu-mu to nu-tau probability of conversion at the energy attainable by the SPS you must choose a distance of about 700 km. The beam energy was finally tuned to have a maximum probability for this transition. For the energy of the beam extracted from SPS, Gran Sasso was at the right distance.

- *Preghenella:*

Is there any particular method to analyze the OPERA emulsions?

- *Coccia:*

Well, the same method was used before for the CHORUS experiment at CERN. There are electronic microscopes which are used for reconstruction of the tracks and also automatic imaging techniques. I must say that this automatic imaging technique has been especially developed for this kind of experiments, and it is now being used by physicians for cancer diagnosis. It allows having numerous pictures of the affected region. About 100000 pictures can be combined and computer-analyzed together instead of having to analyze each picture one by one.

- *Kaur:*

What information will be obtained from the emulsion detector which shall evidently be analyzed off-line? Can this information be connected with the on-line information coming from the neutrino or other interactions?

- *Coccia:*

Well, the most important, fundamental information for the OPERA experiment is the signature of the tau particles in the emulsions. If a tau neutrino appears in the beam, then it must generate a tau particle, which in turn decays showing a kink in the track of very small length, thus determining this kind of event. An emulsion detector like OPERA has a granularity of the order of a micrometer and can detect such signature.

- *Kaur:*

So the emulsion is exposed only once and then developed?

- *Coccia:*

Yes, and immediately substituted with another brick. We already have millions of emulsions ready.

- *Bourjaily:*

You showed the plot of events versus the azimuthal angle which has a large "easily removable" background of cosmic muons. Can OPERA contribute to the atmospheric neutrino data using these events?

- *Coccia:*

OPERA is not competitive with other experiments like SUPERKAMIOKANDE which have determined the deficit of atmospheric neutrinos. However, surely they are receiving this kind of neutrinos and they could also confirm that. But the goal of OPERA is just to detect the tau particle.

- *Sens:*

Could you put into words the arguments that led to the vital decision to go for emulsions in the equipment? People who work at large machines and equipment are generally very hesitant to use anything like emulsion near the machine because of the non recoverability.

- *Coccia:*

The community which is behind OPERA was in part the same as at CERN under the West-area neutrino facility in the experiment CHORUS. In CHORUS and NOMAD they made experiments with a short base-line, without having the result of the neutrino tau appearance. Now we know OPERA is the right experiment to do. CHORUS mastered the emulsion technique and could approach micrometer granularity, so it was natural for OPERA to continue to apply this emulsion technique.

- *Lockyer:*

What is the advantage of liquid argon over a water Cherenkov detector?

-Coccia:

Well, the liquid argon technique detector is competitive with the water Cherenkov detector, but with much less volume, so for a water Cherenkov detector you need a megaton of water, while in the case of argon it is about 50000 tons. If you want a megaton Cherenkov detector, you need an incredibly large laboratory. There is a proposal to create such a detector under the Frejus, which means to perform a very big excavation. Similar kind of results is obtained with a much smaller detector, when liquid argon is used. Basically, there is this difference in terms of cost of the project, number of personnel and the apparatus required.

- *Preghenella:*

How much time does it take to replace one OPERA brick?

- *Coccia:*

This is just a question of few minutes. There is a robot that takes the particular brick for developing and replaces it with a new one.

- *Loehr:*

What is the time structure of the beam and the time resolution of OPERA?

- *Coccia:*

The time structure of the beam as shown before is a 10.5 microsecond bunch with a cycle of 2, separated by 50 milliseconds. The OPERA time resolution is of the order of picoseconds.

- *Loehr:*

Can every event be associated with a unique bunch of the beam?

- *Coccia:*

Yes.

From BABAR to the Future

M.A. GIORGI

INFN Sezione di Pisa and Università di Pisa Dipartimento di Fisica "Enrico Fermi"
Largo B.Pontecorvo, 3
I 56127 Pisa, Italy
E-mail: giorgi@pi.infn.it

Summary. — The B Factories PEP-II and KEKB operating since 1999 have produced a rich harvest of beautiful and somehow intriguing experimental results in the flavor sector. After the discovery of the indirect (2001). and the direct (2004) *CP* violation [1-4] in the B sector, the present experimental situation is in a phase of precision measurements of sides and angles of the Unitarity Triangle, potentially able to produce more precise tests of CKM model. As well known, the *CP* violation accounted by CKM in the standard model fails in the explanation of amount of deficit of antimatter with respect to the matter in the universe. If we still assume *CP* as the origin of such a deficit, one should expect contribution coming from phases related to **New Physics**. Lepton Flavor Violation (**LFV**) in τ decay and in general rare τ and B decays would allow some evidence of **New Physics**. The status of B Factories measurements is reported here, together with ideas of a future facility operating at a luminosity $\geq 10^{36} cm^2 s^{-1}$, much higher than the present B Factories PEP-II and KEKB.

1. – Introduction

In the present paper does not appear any description of the two experiments BABAR and BELLE, neither their subsystems. The two detectors (Figure 1) are installed in the interaction region of two $e^{\pm}$ asymmetric B Factories PEP-II and KEKB operating at $\Upsilon(4S)$.

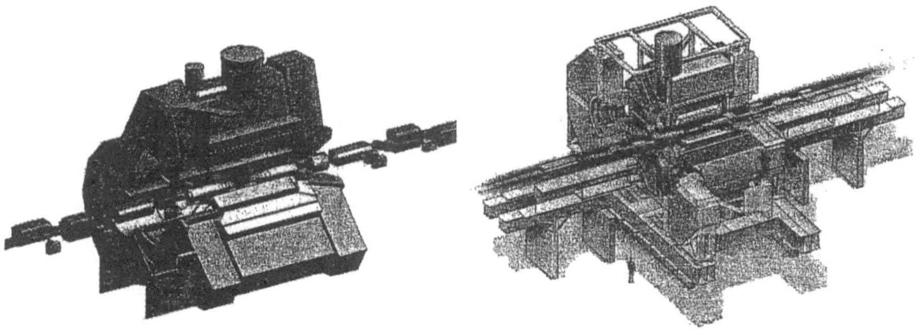

Fig. 1. – BABAR and BELLE detectors

The experimental methods to extract signal are not described in any detail, neither the basic of *CP* violation, namely the mixing formalism, the three kind of *CP* violation, the CKM model inside the SM, for which we refer to other general description contained for instance in [5], [6] or in the proceedings of this school [7], [8].

1·1. *Luminosity achievements by PEP-II and KEKB* . – The two existing asymmetric B Factories PEP-II [9] and KEKB [10] have started their operations in 1999, have exceeded by far their design peak luminosity . PEP-II is running with a peak luminosity $\mathcal{L} \geq 9.2 \times 10^{33}\,\mathrm{cm}^{-2}\,\mathrm{s}^{-1}$ a factor of three higher than the design and KEKB a factor 1.4 higher than its design ($\mathcal{L} = 10^{34}\,\mathrm{cm}^{-2}\,\mathrm{s}^{-1}$. These huge increase in luminosity has been obtained mainly by increasing the circulating currents of electrons and positrons, but also by means of the continuous beam injection that both laboratories have assumed as normal operation mode.

The remarkable increase in the integrated luminosity of both machines came soon after the start of operations in 1999; very soon both machines initiated to work in factory mode with peak luminosity $\mathcal{L} > 10^{33}\,\mathrm{cm}^{-2}\,\mathrm{s}^{-1}$. The integrated luminosity of both B Factories doubled every 2 years as shown in Figure 2. At present the total integrated luminosity. recorded by the two experiments $323\,\mathrm{fb}^{-1}$ of BABAR and the $551\,\mathrm{fb}^{-1}$ of BELLE) corresponding to approximately one billion of produced $B\overline{B}$, $c\bar{c}$, $\tau^+\tau^-$ pairs.

2. – 2006 status of B Factories measurements

2·1. *CKM parameters* [11], [12]: *update of constraints from new measurements of sides and angles.* – In the last two years the measurement of $sin2\beta$ from time dependent analysis of $b \to c\bar{c}\, s$ decays has become a precision measurement program for both e^+e^- asymmetric B Factory. An update of the results in $sin2\beta$ [13], [14] are reported in the

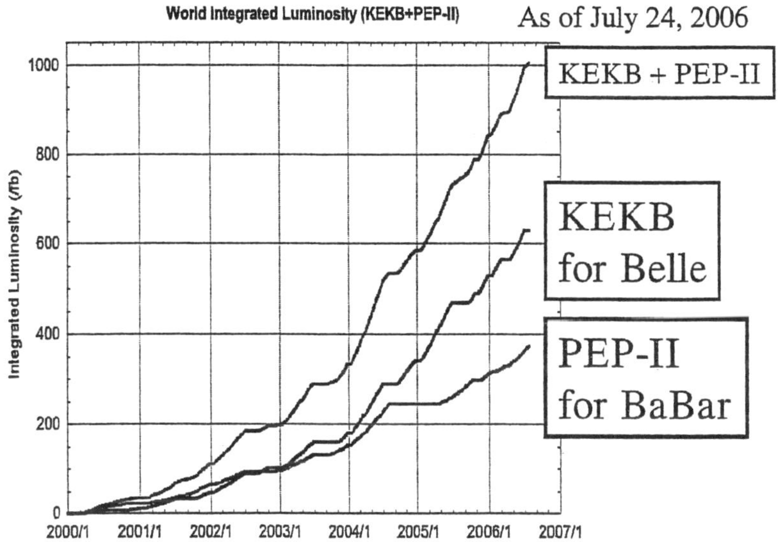

Fig. 2. – Integrated luminosities delivered by PEP-II and by KEKB.In red the top curve shows the total luminosity ($PEP-II$ + KEKB)the blue curve is the KEKB luminosity and the brown lower curve shows PEP-II luminosity, the low integrated luminosity of PEP-II is due to the accident involving an electrician, that implied a stop for one year of the accelerator operation.

HFAG [15] table shown in Figure 3, the updated averaged value is:

$$\sin 2\beta = 0.67 \pm 0.028.$$

In addition the measurements of cos2β of BABAR [16], [17] and BELLE [18], [19] from different decay channels (see Figure 4) the last are coming from $B^0 \to D^0 h^0$ and have been presented tho the 2006 ICHEP Conference in Moscow.

These new measurements give for BELLE:

$$\cos 2\beta > 0 \text{ at } 98.3 \text{ \% CL}$$

and for BABAR:

$$\cos 2\beta > 0 \text{ at } 87.0 \text{ \% CL}$$

then having solved the ambiguity, the angle β is determined with a quite high precision, better than 5% as shown in Figure 3:

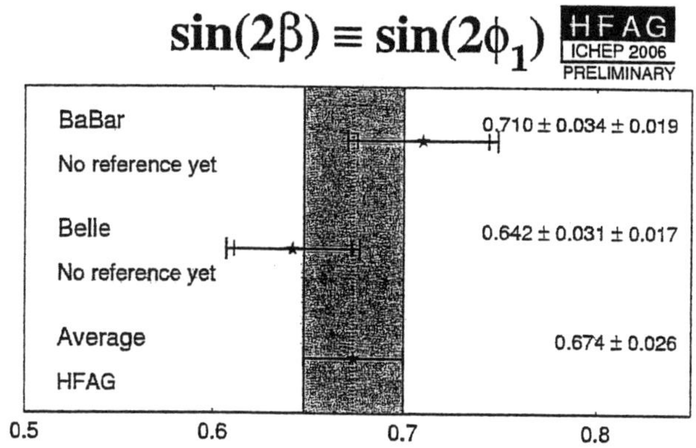

Fig. 3. – Status of $sin2\beta$ measurements as presented at ICHEP06

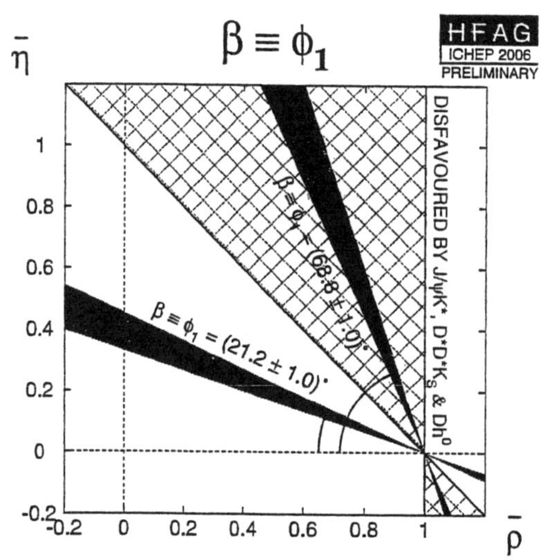

Fig. 4. – The $cos2\beta$ measurements exclude one of the two solution allowing the measurement of angle β.

$$\beta = (21.2 \pm 1.0)^{\circ}.$$

The extraction of β together with the measurement of the other angles ad sides allows to set better constraints to the Unitarity Triangle in the $(\overline{\rho}, \overline{\eta})$ plane, but the extraction of the other two angles γ and α is more difficult. As well known the measurement of $\sin 2\alpha$ is not as clean as the measurement of $\sin 2\beta$, since in general in all the charmless decays of B^0 as : $B^0 \rightarrow \pi^+\pi^-$, $B^0 \rightarrow \rho\pi$, $B^0 \rightarrow \rho\rho$ or $B^0 \rightarrow a_1\pi$ there is more than one amplitude contributing to the decay. Amplitudes with a loop "*penguin*" are also contributing to the above processes in addition to the amplitudes at *tree* level, the extraction of α angle from time dependent analysis is now more difficult than the measurement of β, in actual fact a strong phase δ is present, so that the asymmetry in time dependent analysis gives a $\sin 2\alpha_{eff}$, where $\alpha_{eff} = \alpha + \delta$. A set of measurements, the use of other experimental information and a quite elaborate isospin analysis of the final states can allow the extraction of α (all this matter and the extraction methods have been widely discussed in the literature, for an extensive reference see for instance [5], [6],in the proceedings of this school [7], [8]) or in the original papers [20-23].
In the past few years both experiments BABAR and BELLE had performed isospin analysis in several charmless modes as : $B^0 \rightarrow \pi^+\pi^-$, $B^0 \rightarrow \rho^+\pi^-$, $B^0 \rightarrow \rho\rho$ or $B^0 \rightarrow a_1\pi$.
The Figure 5 from the **CKM fitter** group [24] shows the result of the isospin analysis for the $B^0 \rightarrow \pi\pi$ decay modes, [25, 26] clearly it doesn't constrain the value of the angle and in fact the use of the other modes (essentially $B^0 \rightarrow \rho\rho$) is necessary to allow the extraction of α . The fit of all the decay channels adding the information from both experiments [27, 28] the other decay channels gives:

$$\alpha = [93^{+11}_{-9}]^{\circ}$$

In the figure is also reported with its uncertainty the value of α as obtained from the indirect general fit of all the CKM parameters excluding the measured value of α.

The use of this measurement has been included in the general fit of the Unitarity Triangle parameters and as is evident it starts to contribute in constraining the apex of the triangle in the $(\overline{\rho}\ \overline{\eta})$ plane as can be seen in Figure 6

The extraction of γ angle is more complicated, at $\Upsilon(4S)$ it is possible from direct *CP* violation asymmetries as pointed out by several authors. γ in fact is the phase between $(b \rightarrow u) \propto V_{ub}$ and $(b \rightarrow c) \propto V_{cb}$, it can be extracted with different methods [29-32], comes out in the interference between two amplitudes contributing to the same final state in the decay of $B^{\pm}$. One of the most promising method is the so called Dalitz Method [32]. In Figure 7 are shown the Feynman graphs of two amplitude of the decays $B^- \rightarrow D^0K^-$and $B^- \rightarrow \overline{D}^0K^-$ showing the same final state $D^0/\overline{D}^0 \rightarrow [K_s\pi^+\pi^-]$. It is clear that the size of the two amplitudes is quite different , the two diagrams contributing to the same final state are of the same order of magnitude in terms of CKM parameters [33], but one of them as can be seen seen in the figure is actually color suppressed.

The ratio $r_B = \frac{|A(B^- \rightarrow \overline{D}^0 K^-)|}{|A(B^- \rightarrow D^0 K^-)|} \sim 0.1$
between the two amplitudes enters in the fit of data with resonances and continuum on the Dalitz plot (scatter plot).

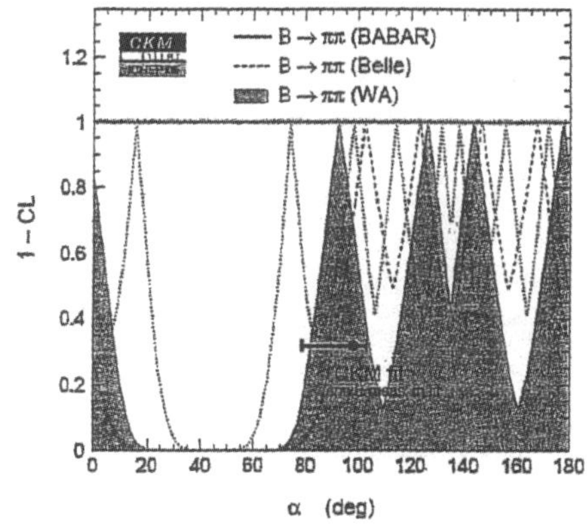

inputs

$$\left.\begin{array}{l}\mathcal{B}(\pi^+\pi^0) = (5.75 \pm 0.42)\\ \mathcal{B}(\pi^+\pi^-) = (5.20 \pm 0.25)\\ \mathcal{B}(\pi^0\pi^0) = (1.30 \pm 0.21)\end{array}\right\} \times 10^{-6}$$

$$\mathcal{A}(\pi^0\pi^0) = +0.35 \pm 0.33$$

$$\mathcal{S}(\pi^+\pi^-) = -0.59 \pm 0.09$$

$$\mathcal{A}(\pi^+\pi^-) = +0.39 \pm 0.07$$

Fig. 5. – Full analysis using all the informations available for the extraction of α from BELLE and BABAR

The "Dalitz " distributions to fit from data are expressed in the following way

$$B^+ : |f(m_+^2, m_-^2) + r_+ e^{i(\delta+\gamma)} f(m_-^2, m_+^2)|^2$$

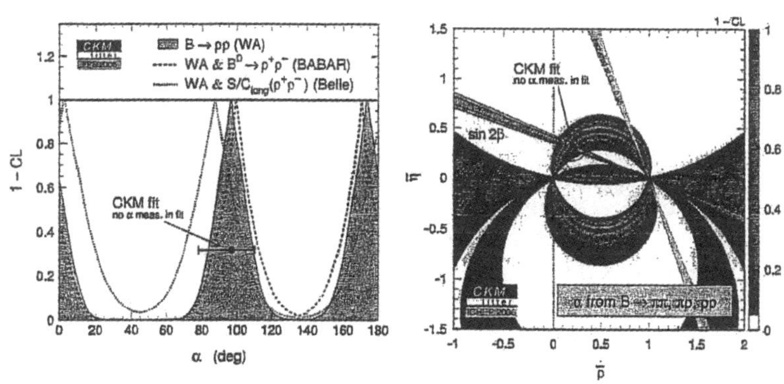

a) b)

Fig. 6. – The $\overline{\rho}\,\overline{\eta}$ plane with the information of the α angle as measured by the two asymmetric B Factory

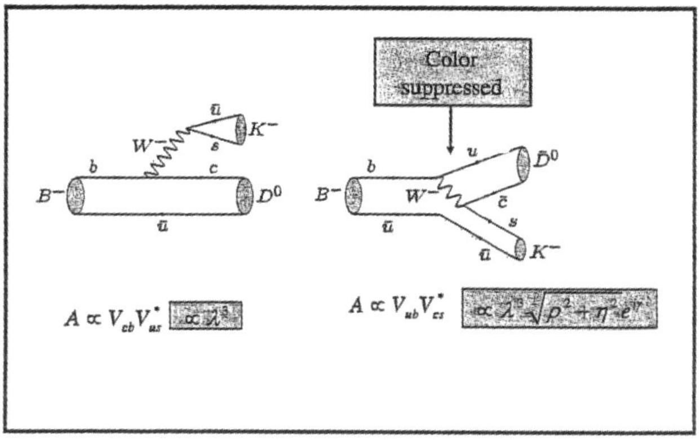

Fig. 7. – Feynman graphs for the decay amplitudes of the decay process that can be used with a so called Dalitz analysis for the extraction of angle γ. The asymmetry used in this analysis indicates a direct CP violation in the decay of charged B's

$$B^- : |f(m_-^2, m_+^2) + r_- e^{i(\delta-\gamma)} f(m_+^2, m_-^2)|^2$$

where $f(m_-^2, m_+^2)$ is the D decay amplitude as a function of the Dalitz plot variables $m_\pm^2 = m^2(K_S^0 \pi^\pm)$. CP is assumed to be conserved in the D decay in the formulae above. The size of the asymmetry is given by the ratio

$$r_- = \frac{|B^- \to \overline{D}^0 K^-|}{|B^- \to D^0 K^-|} \qquad r_+ = \frac{|B^+ \to D^0 K^+|}{|B^+ \to \overline{D}^0 K^+|}.$$

We expect that of course $r_+ = r_- \equiv r$ and $r \sim 0.1$ according to simple CKM matrix elements evaluation. We keep them as separate free parameters to allow a better statistical behavior.
The physics parameters r, δ and γ are quite strongly correlated and then they don't allow for a reasonable estimate of the errors. BABAR has introduced two new variables

$$x_\pm = \Re(r_\pm e^{i(\delta_B \pm \gamma)}) \text{ and } y_\pm = \Im(r_\pm e^{i(\delta_B \pm \gamma)}),$$

which are almost uncorrelated .

Such a fit allows in principle the extraction of r_B, the angle γ and the strong phase δ. They are not independent variables, r_B is in fact correlated with the phases and very small values of it would require an infinite statistics to extract γ, the dependance of the error on γ from the fit and the value of r_B is well illustrated in Figure 8. In the side a) of the same figure we can see the two regions delimited by the 1σ contour around the best fitted ($x^\pm, y^\pm$) interested by the data in the decay of charged B mesons. The distance between the two circles is representing amount of direct CP violation .

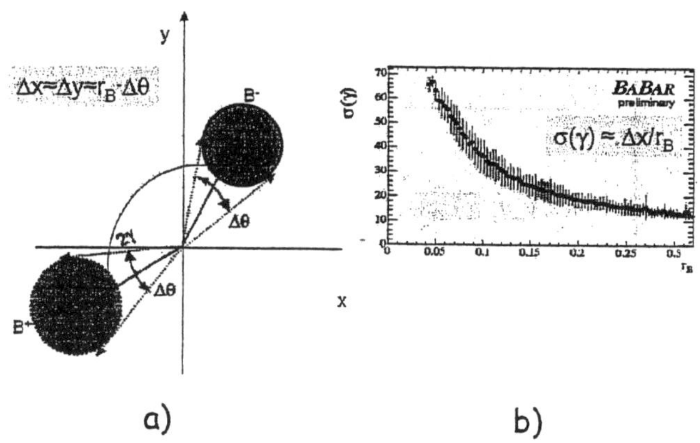

Fig. 8. – In a) is shown the expected solution in (x,y) plane ,1 σ circles around the best values when $\Delta x = \Delta y$ the *CP* asymmetry is graphically represented by the distance between the centers of the two regions. The b) part of the Figure shows The error on γ from the fit as a function of the r_B parameter, at small values of r_B the uncertainty on the value of the angle becomes larger and larger preventing the determination of γ.

The measurement of γ with the Dalitz method has shown so far to be the most promising at least with the present statistics at the asymmetric B Factory, it can be seen in Figure 9 the results as reported by **HFAG** group for the Dalitz analysis on $B^{\pm}$ decay to D^0 or $\bar{D}^{*0}$.These results have been recently presented at ICHEP06 by BABAR [34]:

$$\gamma = (92 \pm 41_{\text{Stat.}} \pm 11_{\text{Syst.}} \pm 12_{\text{Res.Model}})^{\circ}$$
$$r_B < 0.142\delta_B(D^0K^*) = (118 \pm 63 \pm 19 \pm 36)^{\circ}$$

and by BELLE [35] :

$$\gamma = (53^{+15}_{-8}{}_{\text{Stat.}} \pm 3_{\text{Syst.}} \pm 9_{\text{Res.Model}})^{\circ}$$
$$0.01 < r_B^* < 0.203\delta_B(D^0K^*) = (298 \pm 59 \pm 18 \pm 10)^{\circ}$$

The average for the measurements on D^* channel gives $\gamma = (78 \pm 30)^{\circ}$
the error does not contain the model used to fit distributions on the Dalitz plane.

The measurements of $|V_{ub}|$ is correlated to the measurement of γ as can be seen in Figure 10

The analysis on semileptonic decay channels $b \to cl\nu$ and $b \to ul\nu$ for the measurement of V_{cb} [36, 37], [38, 39] and V_{ub} [40, 41], [42, 43]. Both experiments have followed two

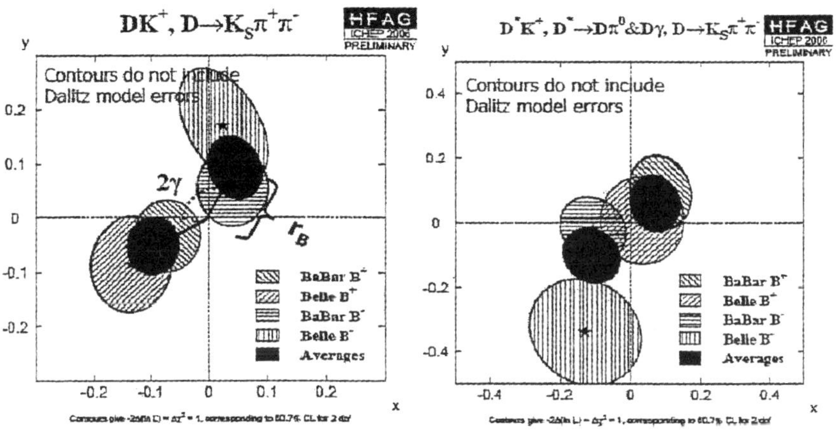

Fig. 9. – The **HFAG** average of the $(x, y)^{\pm}$ for the $B^{\pm} \to D^0/\overline{D}^0 K^{\pm}$ a) and $B^{\pm} \to D^{*0}/\overline{D}^{*0} K^{\pm}$

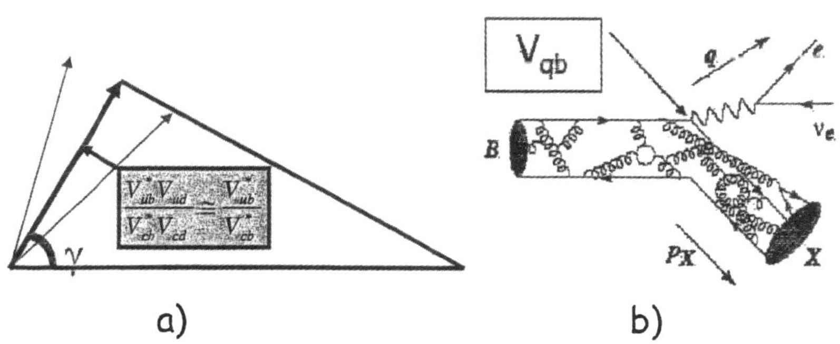

Fig. 10. – One of the sides of angle γ is V_{ub}. Its extraction is not easy and comes out of the measurement of the branching fraction of semileptonic decay $b \to ul\nu$

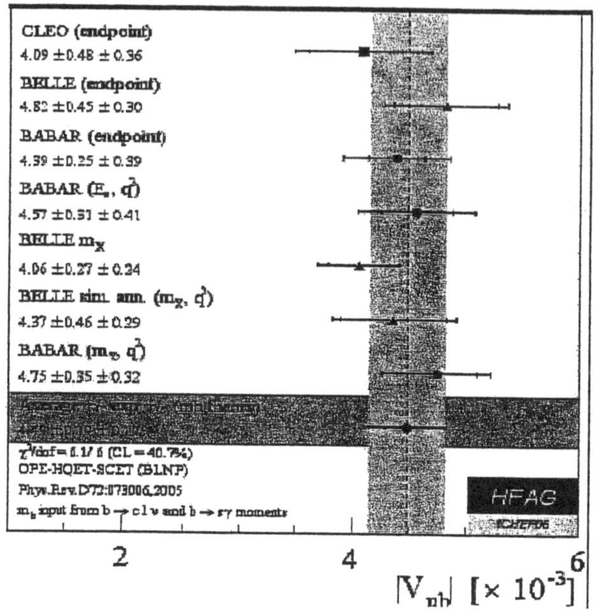

Fig. 11. – V_{ub} results where the contribution from B Factories is very important and mainly coming from inclusive analysis

different methods : exclusive and inclusive semileptonic decays . The inclusive analysis of course is more challenging from the experimental point of view, but certainly cleaner and much less affected by theoretical uncertainties. In Figure 11 the world results on $V_{ub}$are reported. As shown in Figure 10 the angle γ and the side(whose size is $\sim \frac{|V_{ub}|}{|V_{ub}|}$) are physically correlated, their independent measurement can strictly constrain the Unitarity Triangle in the $\overline{\rho}\ \overline{\eta}$ plane. The Figure 12 reporting the results of fitting analysis the **UT fitter** group [44]. The part a) shows how the semileptonic B decay allow already a precision measurement of the side as indicated in the circle , while the blue beams defining the experimental limits in γ don't contribute with the present statistics to constrain the Unitarity Triangle. Figure 12 b) is showing the status of the art on CKM as resulting from the measurements of BABAR and BELLE at the $e^{\pm}$ asymmetric B Factory.

2·2. *Measurements of* $\sin 2\beta$ *via decays described by penguin diagrams* . – There are however some decay channels as $b \rightarrow (s\bar{s}s)$ and $b \rightarrow (sd\bar{d})$ that can only/also proceed through diagrams with the so called "*penguin*". Some of these channels are:

- $B^0 \rightarrow \phi K^0$ $B^0 \rightarrow (KK)_{CP}K^0$
- $B^0 \rightarrow \eta\prime K^0$ $B^0 \rightarrow f_0 K^0$

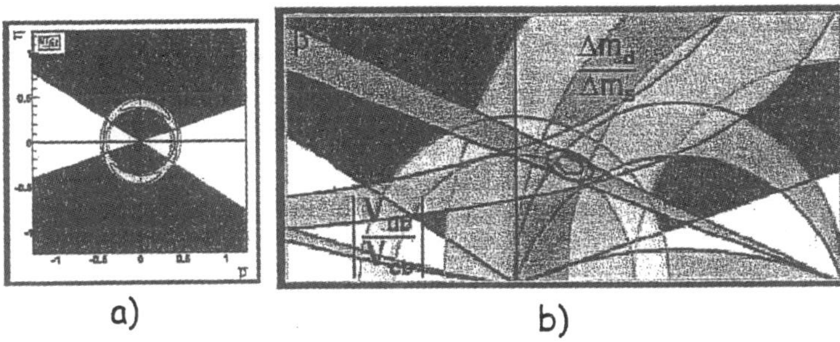

Fig. 12. – $\frac{|V_{ub}|}{|V_{ub}|}$ and γ measurements are reported in the part a) of the figure. The present status of Unitarity Triangle parameters shown in the part b) of the figure. No evidence with the present statistics of deviations from the Standard Model

- $B^0 \to \pi^0 K^0$ $B^0 \to \rho^0 K^0$
- $B^0 \to \omega K^0$

The above channels

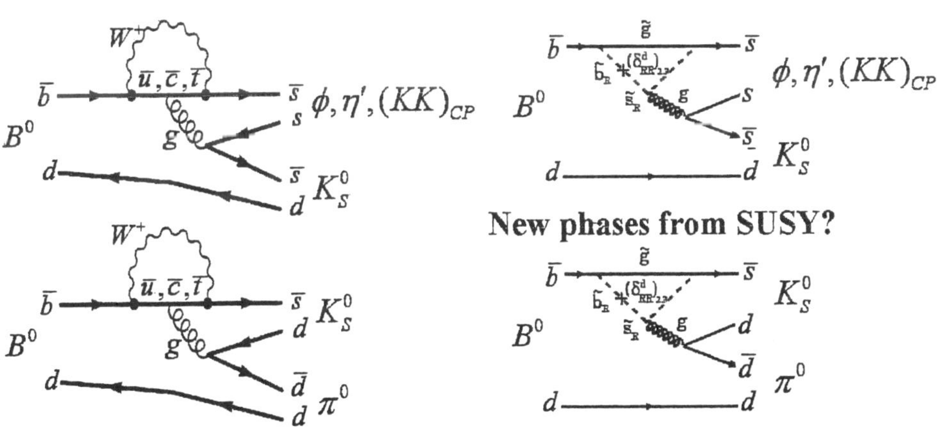

Fig. 13. – Possible **New Physics** particles contribution in the loops to $b \to (s\bar{s}s)$ and $b \to (sd\bar{d})$ diagrams.

take contributions from the same CKM elements of charmonium channels $b \to (c\bar{c}s)$, as can be seen in the left hand part of Figure 13. The interference ($B^0 Mixing - Decay - K^0 Mixing$) should give the same phase factor $e^{-2i\beta}$ of charmonium decay, assuming that the only relevant contribution to the amplitude is coming from the heaviest fermion (top quark) in the loop

(1) $$\lambda = \frac{V_{tb}^* V_{td}}{V_{tb} V_{td}^*} \frac{V_{tb}^* V_{ts}}{V_{tb} V_{ts}^*} \frac{V_{cs}^* V_{cd}}{V_{cs} V_{cd}^*} = e^{-2i\beta}$$

as can be easily checked comparing eq. (1) with the corresponding one for the charmonium modes

(2) $$\lambda = \frac{V_{tb}^* V_{td}}{V_{tb} V_{td}^*} \frac{V_{cs}^* V_{cb}}{V_{cs} V_{cb}^*} \frac{V_{cd}^* V_{cs}}{V_{cd} V_{cs}^*} = e^{-2i\beta}$$

and using the Wolfenstein parametrization [33] we can see that apart from V_{td} the other terms in the expressions are real matrix elements and disappear so that the only term that survives is $\frac{V_{td}}{V_{td}^*}$.

If new heavy quanta contribute to the loops, as an example in the right hand side of Figure 13 are presented possible SUSY contributions, new phases can contribute to the asymmetry and the S coefficient of the time dependent analysis could be substantially different from $\sin 2\beta$.

However some of these channels are affected by theoretical uncertainties at the level of a few per cent, on some others instead the uncertainty level is higher up to tens per cent. Let us call $\sin 2\beta^{eff}$ the result obtained in the time dependent analysis of channel that are *penguin* dominated In Figure 14 is shown the summary table prepared by **HFAG**, it contains the values of $\sin 2\beta^{eff}$ from the *CP* asymmetries in decays dominated by loops and the world average value of $\sin 2\beta$ as measured in processes (b $\to$ charmonium), dominated by *tree* diagrams.

For example about the cleanest channels, $B^0 \to \phi K^0$ and $B^0 \to (KK)_{CP} K^0$, we must consider a S.M. uncertainty due to the contribution of a similar graph with the insertion of an up quark in the loop. This uncertainty is of the order of $\lambda^2 \approx 5\%$. For the other decay channels the uncertainty could be even greater than 5% up to 10% or even 20%. These channels can receive contributions also from color suppressed *tree* diagrams ($b \to (u\bar{u}s)$). However as shown in the right hand side b) of Figure 14 recent QCD estimate on the uncertainties on the corrections to the Standard Model value are almost all shifted on one side. Assuming that than a discrepancy should appear since so far most of central values of $\sin 2\beta_{eff}$ are smaller (on the left) of the world average value of $\sin 2\beta$. It is clear that we are at the beginning of a very interesting season when we could start probing the Standard Model in the flavor sector with a statistics much higher than what would be obtainable with the present asymmetric B Factory. The statistics needed to try to make this comparison really possible is to reduce the statistical error below the theoretical uncertainties would be of the order of 40 ab^{-1} or higher.

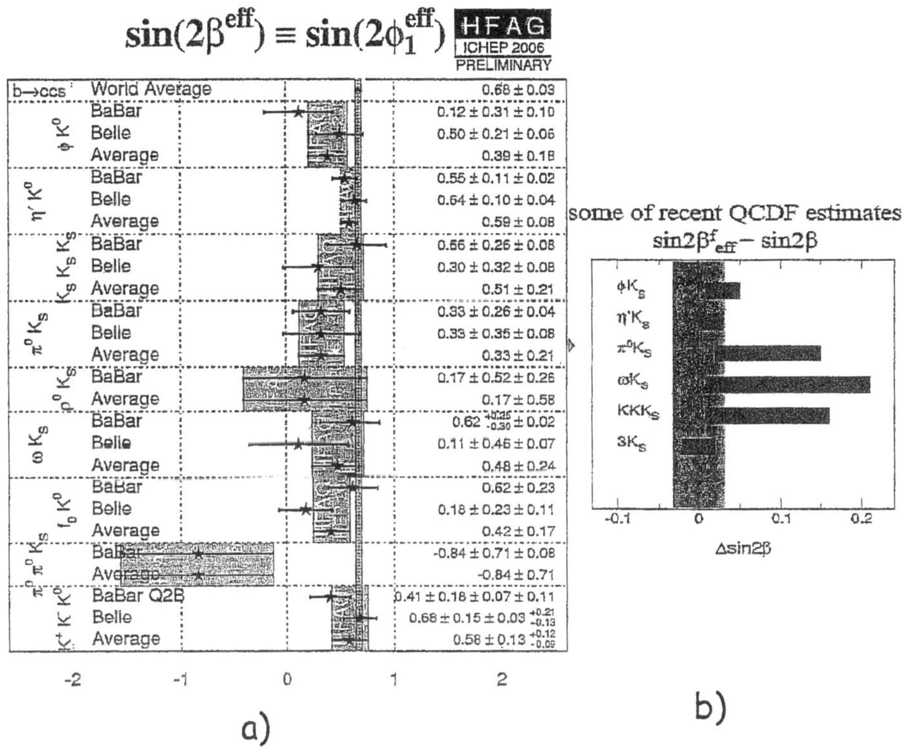

Fig. 14. – Present status of comparison between $\sin 2\beta$ measured in charmonium and in channels dominated by *penguin* diagrams

2·3. *Precision measurement of very rare decays.* – The search for **New Physics** with $e^{\pm}$ B Factories proceeds via the the measurements of CKM parameters as discussed in the previous subsection, looking at slight discrepancies with respect with Standard Model predictions, but proceeds also via the precise measurements in very rare decays, otherwise excluded or very highly suppressed by the Standard Model. The tables IV and III, contain the estimate of the present theoretical uncertainties on some of these channels. Some measurements based with the present statistics of BABAR and BELLE set already according to SUSY models as MSSM important exclusion regions in the plane ($\tan\beta$, Mass of the SUSY insertion). In the Figure 15 is synthetically presented the status of the measurement of $BF(B^+ \to \tau^+\nu_\tau)$ [45],

$$\text{BELLE}: BF(B^+ \to \tau^+\nu_\tau) = (1.79^{+0.56}_{-0.49}\text{Stat.}^{+0.39}_{-0.46}\text{Syst.}) \times 10^{-4}$$

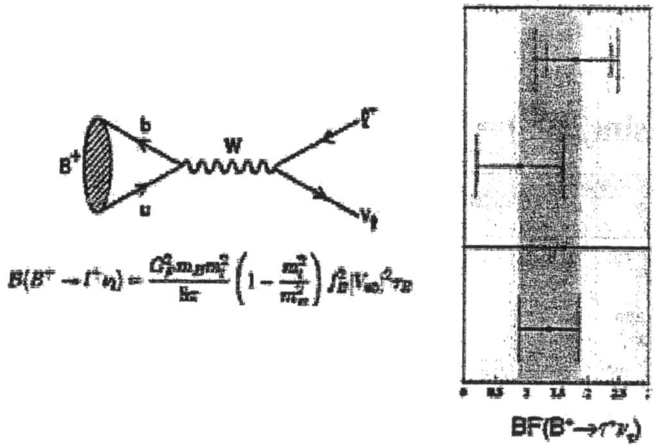

Fig. 15. – On the left of the figure the Standard Model diagram is shown, once $|V_{ub}|$ is determined, the Branching Fraction of this channel allows the extraction of f_b , the right end side of the figure presents the status of the experimental results of $\mathbf{BF}(B^+ \to \tau\nu_\tau)$.

BABAR: $BF(B^+ \to \tau^+\nu_\tau) = (0.88^{+0.68}_{-0.67}\mathrm{Stat.} \pm 0.11\mathrm{Syst.}) \times 10^{-4}$ $BF < 1.8090\%CL$

with an average value of $(1.36 \pm 0.48) \times 10^{-4}$.

On the other hand if we consider a SUSY model with 2 Higgs doublet model then the Standard Model prediction is altered by a factor: $r_H = (1 - \frac{m_B^2}{m_H^2} tan^2\beta)^2$.

In Figure 16 the limits on the parameter plane $(M_{H^\pm}, \tan\beta)$ from the asymmetric B Factoryexperiments are shown, together with the region excluded by LEP.

The search for **LFV** in τ decay is one of the most intriguing studies in the two experiments BABAR and BELLE. From neutrino oscillation experiments it is clear that the violation of the leptonic flavor in the decay of charge leptons is allowed ia intermediate states with neutrinos, with a Branching Fraction tremendously suppressed such as this effect could not be visible at all in accelerator experiments. A possible observation of decay modes violating the leptonic flavor would be a clear indication of **New Physics.** One dedicated experiment MEG is going to start soon at PSI (Zurich) looking for the decay of muons. The B Factories are in fact also tau factories since at the $\Upsilon(4S)$ the cross section to produce tau pairs is of the same order for producing B pair. So far there is not any clue of **LFV** and the limits are still far from challenging the most popular SUSY models for **New Physics.** BABAR and BELLE have recently updated their results [46, 47] , here we report only the limits obtained by the two experiments in $\tau \to \mu/e\gamma$, as shown in the table I)

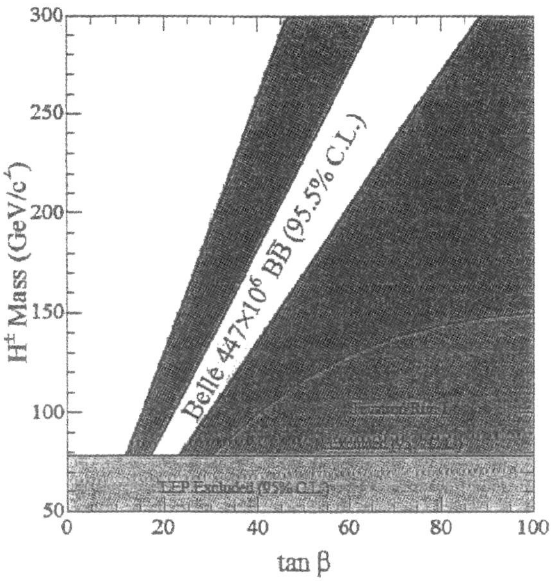

Fig. 16. – The exclusion regions as determined by LEP and by the measurement of $\mathbf{BF}(B^+ \to \tau^+\nu_\tau)$ in the $e^\pm$ B Factories

Parameter	BABAR	BELLE
$\mathbf{BF}(\tau \to \mu\gamma)$	$< 6.7 \times 10^{-8}$	$< 4.1 \times 10^{-8}$
$\mathbf{BF}$ $(\tau \to e\gamma)$	$< 11.0 \times 10^{-8}$	$< 12.0 \times 10^{-8}$

TABLE I. – ***LFV*** *limits (90 % CL) obtained by* BABAR *and* BELLE *with the present statistics. At this level the limit are still scaling linear with the inverse of luminosity.*

The Figure 17, from the BELLE colaboration , shows the excluded region in the plane $(M_{SUSY}, \tan\beta))$.

2·4. *Spectroscopy* . – The good luminosity performance of the asymmetric B Factories has allowed to discover new states, this adventure was started in 2003 by BABAR with the famous discovery of $D_{sJ}(2019)$ soon after confirmed by BELLE and by **Cleo**. The careful study of decay products of B mesons or decays of charmonium has produced the discovery of many new particles in BABAR and in BELLE. In addition, more recently, a very powerful tool for spectroscopy, the ISR (initial state radiation) method, has been extensively employed to find new states. It consists in looking at the invariant mass of

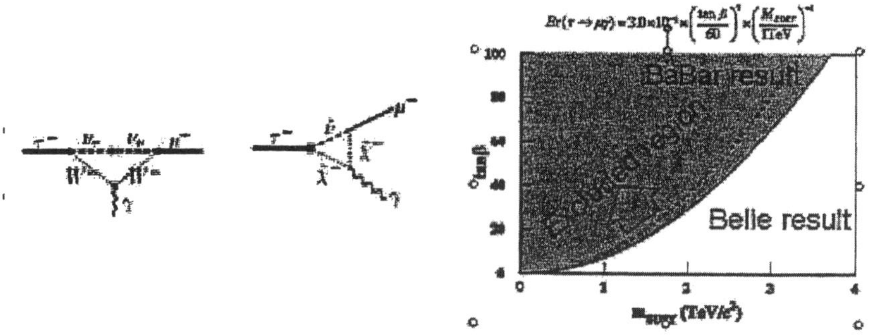

Fig. 17. – Lepton flavor Violation measurements from the experiments at the B Factories starts to constrain the parameters of SUSY $b \to (s\bar{s}s)$ and $b \to (sd\bar{d})$ diagrams.

the final state once an energetic gamma radiated in the initial state is removed from the event $e^+e^- \to \gamma M_X$. The Figure 18 shows two new states found by BABAR, out of plenty recently discovered in both the experiments . Y(4260) [48] (has been seen in the decay in $J/\psi\ \pi^+\pi^-$) with :

$$m = (4259 \pm 8^{+2}_{-6})\,\mathrm{MeV}/c^2 \text{ and } \Gamma = (88^{+6}_{-4})\,\mathrm{MeV}/c^2$$

also confirmed by Cleo [49] in ($J/\psi\ \pi^0\ \pi^0$ decay) and the Y(4350) [50] ($D\bar{D}$ decay), both have been found in ISR analysis , their mass is above $D\bar{D}$ threshold and they look like heavy charmonium states not fully explained by theory, intriguing and suggesting interesting solutions as 4 quarks bound states see for instance [51].

3. – Super B Factory

One of the most interesting program with a very high luminosity B Factory is related to the very rare decays of b quarks and τ leptons that appears as the unique tool in Flavor Physics to find experimental evidence of **New Physics** beyond the Standard Model .

At the moment the only existing project of a sort of "Super B Factory" has been developed at KEK. The goal of this project is to increase continuously the luminosity of their machine up to $4.10^{35}\,\mathrm{cm}^{-2}\,\mathrm{s}^{-1}$, capable of integrating $40\,\mathrm{ab}^{-1}$ by the end of the next decade.

In tables II, III, IV are reported the expected precision for *CP* violation in time dependent analysis in *penguin* dominated channels, for **FCNC** decays of B's and for extremely rare decays. All of them are potentially sensitive to **New Physics**.

This last table contains the **LFV** channel $\tau \to \mu\gamma$, that is one of cleanest channel to look for **New Physics**, its sensitivity is depending on the correct evaluation of the ultimate systematic error, whose evaluation at the moment still quite uncertain.

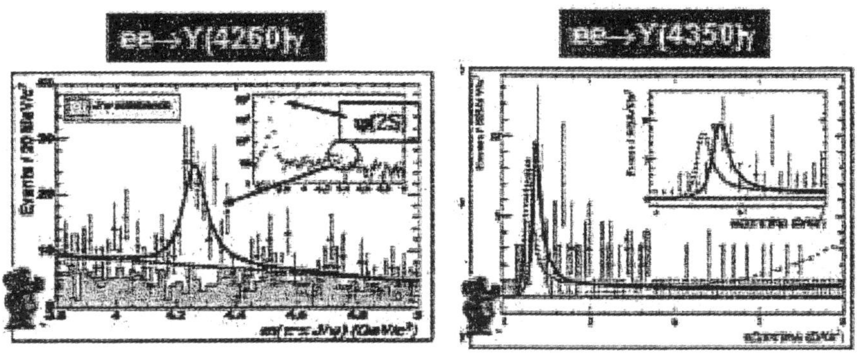

Fig. 18. – Two of the new states discovered by BABAR above the D D threshold in ISR events

It is clear that a data set collected with an integrated luminosity of $\sim 50\,\mathrm{ab}^{-1}$ is perhaps compatible with the discovery of new physics, however to leave some headroom and being a bit conservative a data set bigger than that (obtainable with more than $75\,\mathrm{ab}^{-1}$ or even $100\,\mathrm{ab}^{-1}$, accumulated in 5 years, as in the Snowmass prescription) should allow to control the uncertainties at the level needed. This ambitious goal can

TABLE II. – *Measurement precision for CP asymmetries in rare decays sensitive to New Physics.*

CPV **in Rare Decays**		**e^+e^- Precision**		
Measurement	**Goal**	**3/ab**	**10/ab**	**50/ab**
$S(B^0 \to \phi K^0_S)$	$\approx 5\%$	16%	8.7%	3.9%
$S(B^0 \to \eta' K^0_S)$	$\approx 5\%$	5.7%	3%	1%
$S(B^0 \to K^0_S \pi^0)$		8.2%	5%	4%
$S(B^0 \to K^0_S \pi^0 \gamma)$	SM: $\approx 2\%$	11%	6%	4%
$A_{CP}\ (b \to s\gamma)$	SM: $\approx 0.5\%$	1.0%	0.5%	0.5%
$A_{CP}\ (B \to K^*\gamma)$	SM: $\approx 0.5\%$	0.6%	0.3%	0.3%

TABLE III. – *Measurement precision for FCNC processes $s\ell^+\ell^-$, $K\ell^+\ell^-$, $K^*\ell^+\ell^-$ decays.*

$s\ell^+\ell^-$, $K^{(*)}\ell^+\ell^-$ Decays	e^+e^- Precision		
Measurement	**3/ab**	**10/ab**	**50/ab**
$\mathcal{B}(B \to K\mu^+\mu^-)/\mathcal{B}(B \to Ke^+e^-)$	$\sim 8\%$	$\sim 4\%$	$\sim 2\%$
$A_{CP}(B \to K^*\ell^+\ell^-)$ (all) (high mass)	$\sim 6\%$ $\sim 12\%$	$\sim 3\%$ $\sim 6\%$	$\sim 1.5\%$
$A_{FB}(B \to K^*\ell^+\ell^-)$: $\hat{s}_0$: A_{CP}	$\sim 20\%$	$\sim 9\%$	$\sim 9\%$
$A_{FB}(B \to s\ell^+\ell^-)$: $\hat{s}_0$: C_9, C_{10}	$\sim 27\%$ $36 - 55\%$	$\sim 15\%$ $20 - 30\%$	$\sim 7\%$ $9 - 13\%$

be achieved with a machine with a peak luminosity above 10^{36} cm^{-2} s^{-1} able to delivery 75 ab^{-1} in 5 years. Hitting such a target would be very difficult with a less performing

TABLE IV. – *Measurement precision for rare decays sensitive to New Physics.*

Rare Decays		e^+e^- Precision		
Measurement	**Goal**	**3/ab**	**10/ab**	**50/ab**
$\lvert V_{td}\rvert/\lvert V_{ts}\rvert \sim \sqrt{\frac{\mathcal{B}(b\to d\gamma)}{\mathcal{B}(b\to s\gamma)}}$		19%	12%	5%
$\mathcal{B}(B \to D^*\tau\nu)$	$\mathcal{B} = 8 \times 10^{-3}$	10%	5.6%	2.5%
$\mathcal{B}(B \to s\nu\bar{\nu})$ $(K^{-,0}, K^{*-,0})$	1 exclusive: $\sim 4 \times 10^{-6}$	$\sim 1\sigma$ (per mode)	$> 2\sigma$ (per mode)	$> 4\sigma$ (per mode)
$\mathcal{B}(B_d \to$ invisible)		$< 2 \times 10^{-6}$	$< 1 \times 10^{-6}$	$< 4 \times 10^{-7}$
$\mathcal{B}(B_d \to \mu\mu)$	$\sim 8 \times 10^{-11}$	$< 3 \times 10^{-8}$	$< 1.6 \times 10^{-8}$	$< 7 \times 10^{-9}$
$\mathcal{B}(B_d \to \tau\tau)$	$\sim 1 \times 10^{-8}$	$< 10^{-3}$	$\mathcal{O}(10^{-4})$	
$\mathcal{B}(\tau \to \mu\gamma)$			10^{-9}	$10^{-9} - 10^{-10}$

machine.

3.1. *Tools for High Luminosity* . – The classical method towards high luminosity is based on the increase of the currents of the circulating beams .

In an asymmetric B Factorythe increase of the circulating currents in high and low energy rings produces the increase of the luminosity, but not for free . High current and high luminosity in a traditional circular machine means high current through the interaction region and therefore the increase of the background in the detector.

The evaluation of the effects on detector of the increased luminosity in a conventional e^+e^- asymmetric B Factory have been studied by BABAR and BELLE and extensively discussed in 2 joint workshops at Hawaii held in 2004 and 2005 respectively.

The conclusions is that with a peak luminosity higher than $10^{35}\,\mathrm{cm}^{-2}\,\mathrm{s}^{-1}$ would be hard to stand in such an environment dominated by a background 10 to 100 times worst than what BABAR and BELLE have now.

To make an example the occupancy in the vertex detector would reach a level that would exceed the capability of the present detectors and the read-out technology.

The design of an electromagnetic calorimeter more robust than the present CsI crystal calorimeters of BABAR and BELLE would be also problematic although some solutions, that need an aggressive R&D, could be explored. A different way to the very high luminosity of a e^+e^- machine has been presented by P.Raimondi at the 2005 Workshop of Hawaii. In the original model high current e^+ and e^- were circulating inside two damping rings where beams are cooled and very low emittance bunches are extracted with a fast extractor, focused to a very small transverse dimensions (tens of nm) and collided inside the apparatus. Such an option had as an unpleasant consequence the need for high wall electric power ($\geq$ 100 MW), that would make problematic the operation cost of the machine. In Figure 19 a more convenient schematic design of the machine is presented. The machine is now circular and the rings are in fact acting as dumping rings.

The working principle is the same as before, it is based on small beam size, low emittance and a final focus as the one considered in the International Linear Collider (ILC) project. The high luminosity is obtained by squeezing the beams at the collision according to the formula:

$$\mathcal{L} = \frac{N_+ N_- n_b f_c}{4\pi\sigma_x\sigma_y} H_d \; ; \qquad \sigma = \sqrt{\beta\epsilon}$$

After the collision the bunches are recirculated and cooled down in the rings. The wall power needed is now down to 23 MW. The machine studies are still at the preliminary stage, but a team of accelerator physicists from different countries is involved inside an effort promoted by INFN for a Super Flavor Factory "SuperB" [52].

3.2. *Some advantages of "SuperB"*. – In the model proposed the current circulating inside the detector is at the level of that of PEP-II, then even running with luminosity

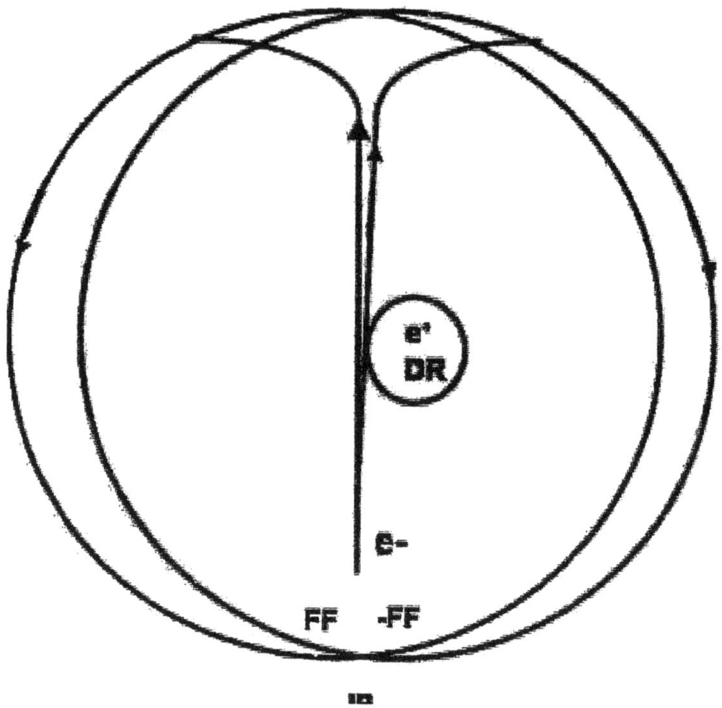

Fig. 19. – Conceptual design of the SuperB based on low emittance, large Piwinski angle and the final focus concept of the International Linear Collider .

higher than $10^{36}\,\mathrm{cm}^2\,\mathrm{s}^{-1}$ its background is expected of the same order as at present in BABAR. The Figure 20 contains the information about the beam-beam interaction, the size of beams and the power needed that not exceed what needed in PEP-II.

This important characteristic allows an easy design of the experimental apparatus, compatible with present technology, and perhaps allowing the reuse of large part of components of BABAR.

In addition due to the small transverse dimension of bunches the beam pipe radius can be reduced easily from the present 3 cm (BABAR) to 1.5 cm or less. Such a reduction could allow a better hermeticity of the detector. The geometrical acceptance can vary from the present 85% to 95% and the resolution on the vertexing will improve (at the first order it will scale linearly with the distance of the first layer of the Vertex Detector from the interaction point. The tracking and vertexing efficiencies will drastically improve with a positive impact on physics.

A very important peculiarity of such a concept of SuperB is the synergy with the ILC.

Energy (GeV)	4	7
C (m)	2762	2762
B_w (T)	1.4	1.05
L_{bend}(m)	2.1	10.8
N. bends	96	96
B_{bend} (T)	0.439	0.144
Uo (MeV/turn)	2.3	4.1
Wiggler sections	4	4
σ_z (mm)	7.0	7.0
τ_s (ms)	17	17
ε_x (nm)	0.79	0.71
Emittance ratio	0.25%	0.25%
ΔE	1.0×10^{-3}	1.1×10^{-3}
Momentum compaction	1.85×10^{-4}	3.90×10^{-4}
ν_s	0.012	0.026
Vrf (MV)	5.5	19
N_{part} ($\times10^{10}$)	3.31	1.89
I_{beam} (A)	2.5	1.44
Tousheck lifetime (min)	95	1100
P_{beam}(MW)	5.7	5.9
Frf (MHz)	476	
$N_{bunches}$	4167	
Gap	5%	
P_{wall} (MW) (50% eff) 2 rings	23.2	

β_x	20mm
σ_x	4μm
σ_{xp}	200μrad
β_y	200μm
σ_y	20nm
σ_{yp}	100μrad
σ_z	7mm
$2^*\theta$	30mrad

$L > 10^{36}\ cm^{-2}\ s^{-1}$

Fig. 20. – The SuperB parameters show a low circulating current , the small beam dimensions and the very low emittance

Another advantage of this exotic approach is the extreme flexibility of this machine where the energy of the beams can be varied allowing operations at the center of mass energies of 10.5 and 4 GeV, corresponding to $\Upsilon(4S)$ and to the open charm threshold.

It could in fact operate very well as a Super charm Factory or as a Super B Factory.

REFERENCES

[1] B. Aubert *et al.* [BABAR Collaboration], Phys. Rev. Lett. **87**, 091801 (2001)
[2] K. Abe *et al.* [Belle Collaboration], Phys. Rev. Lett. **87**, 091802 (2001) [arXiv:hep-ex/0107061].
[3] B. Aubert *et al.* [BaBar Collaboration], Phys. Rev. Lett. **93**, 131801 (2004) [arXiv:hep-ex/0407057].
[4] K. Abe *et al.* [Belle Collaboration], Phys. Rev. Lett. **93**, 021601 (2004) [arXiv:hep-ex/0401029].
[5] I.I.Bigi,A.I.Sanda *CP* violation, University Press Cambridge (2000)
[6]
[6] M.Giorgi,I,Mannelli,A.I.Sanda , *CP* violation from Quarks to Leptons,Proceedings of the International School "Enrico Fermi" -Course CLXIII- Ios Press Amsterdam (2006).
[7] M.A.Giorgi "Highlights from BABAR". Subnuclear Series Vol.41: "From Quarks to Black Holes" ed.A.Zichichi. World Scientific (June 2005) ISBN 978-981-270-179-4
[8] M.A.Giorgi "Highlight update from Babar" Subnuclear Series Vol.43: "Towards New Milestones in our Quest to go beyond the Standard Model" ed.A.Zichichi. World Scientific (November 2007) ISBN 978-981-277-911-3
[9] "PEP-II: An Asymmetric B Factory. Conceptual Design Report. June 1993," SLAC-R-418.
[10] "KEKB B-Factory Design Report", KEK Report 95-7, 1995.
[11] N. Cabibbo, *Phys. ReV. Lett.* **10**, 531 (1963)
[12] M. Kobayashi and T. Maskawa, *Prog. Theor. Phys.* **49**, 652 (1973)
[13] B. Aubert *et al.*,arXiv:hep-ex/0607107
[14] K.F.Chen *et al.*,Phys. Rev. Lett. **98**, 031802 (2007)
[15] **HFAG**,arXiv:hep-ex/07043575 or see(http://www.slac.stanford.edu/xorg/hfag/docs)
[16] B. Aubert *et al.*,Phys. Rev. D **71**, 032005 (2005)
[17] B. Aubert *et al.*,arXiv:hep-ex/0607105
[18] R.Itoh*et al.*,Phys. Rev. Lett. **95**, 091601 (2005)
[19] P.Krokovny*et al.*,Phys. Rev. Lett. **97**, 081801 (2006)
[20] M. Gronau and D. London,Phys. Rev. Lett. **65**, 3381 (1990)
[21] Y. Grossman and H. R. Quinn,Phys. Rev. D **58**, 017504 (1998)
[22] A. E. Snyder and H. R. Quinn,Phys. Rev. D **48**, 2139 (1993)
[23] M. Gronau and J. Zupan, Phys. Rev. D **70**, 074031 (2004)
[24] CKM fitter see: http://www.slac.stanford.edu/xorg/ckmfitter or htpp://ckmfitter.in2p3.fr
[25] B. Aubert *et al.*,arXiv:hep-ex/0607106
[26] H. Ishino *et al.*,arXiv:hep-ex/0608035
[27] B. Aubert *et al.*,arXiv:hep-ex/0607098
[28] K. Abe *etal.*,arXiv:hep-ex/0702009
[29] M. Gronau and D. Wyler,Phys. Lett. B **265**, 172 (1991)
[30] M. Gronau and D. London.,Phys. Lett. B **253**, 483 (1991)
[31] D. Atwood, I. Dunietz and A. Soni,Phys. Rev. Lett. **78**, 3257 (1997)
[32] A. Giri, Y. Grossman, A. Soffer and J. Zupan,Phys. Rev. D **68**, 054018 (2003)
[33] L. Wolfenstein,Phys. Rev. Lett. **51**, 1945 (1983)
[34] B. Aubert *et al.*,arXiv:hep-ex/0607104
[35] A. Poluetkov *et al.*,Phys. Rev. D **73**, 112009 (2006)
[36] B. Aubert *et al.*,arXiv:hep-ex/0607076
[37] A. Poluetkov *et al.*,Phys. Lett. B **526**, 247 (2002)
[38] B. Aubert *et al.*,Phys. Rev. D **74**, 091105 (2006)
[39] P. Urquijo *et al.*,Phys. Rev. D **75**, 032001 (2007)
[40] B. Aubert *et al.*,Phys. Rev. Lett. **92**, 071802 (2004)

[41] I. Bizjak *et al.*,Phys. Rev. Lett. **95**, 241801 (2005)
[42] B. Aubert *et al.*(update of the 2004 paper),arXiv:hep-ex/057017
[43] H. Kakuno *et al.*,Phys. Rev. Lett. **92**, 071802 (2004)
[44] **UTfit** see: http://www.utfit.org/
[45] B. Aubert *et al.*,Phys. Rev. Lett. **95**,041804(2005), B. Aubert *et al.*,arXiv:hep-ex/0611019, K.Ikado *et al.*,Phys. Rev. Lett. **97**, 251806 (2006)
[46] B. Aubert *et al.*,Phys. Rev. Lett. **95**,041802(2005)
[47] K. Abe *et al.*,arXiv:hep-ex/0609049
[48] B. Aubert *et al.*,Phys. Rev. Lett. **95**, 142001 (2005)
[49] Phys. Rev. D **74**, 091104 (2006), Phys. Rev. Lett. **96**, 162003 (2006)
[50] B. Aubert *et al.*,arXiv:hep-ex/0610057
[51] L. Maiani *et al.*,Phys. Rev. D **74**,091104(2006)
[52] J. Albert *et al.*,arXiv:physics/0512235v1 [physics.ins-det]

CHAIRMAN: M. GIORGI

Scientific Secretaries: N. Neri, S. Pacetti

DISCUSSION

– *Brandt:*

How is SuperB Factory complementary to LHC and to LHCb ?

– *Giorgi:*

Regarding LHC, if the Super World exists, then LHC will have the possibility to discover new s-particles and to measure their mass. The SuperB project will be complementary in the sense that it will have the possibility to measure the couplings of those s-particles and phases of the off diagonal terms of an equivalent s-mass matrix, which is a fundamental information to understand the theory.

Regarding LHCb, Jim Libby from LHCb presented a comparison of the sensitivities for several benchmark analyses between LHCb and SuperB after one year running.

That comparison shows how the SuperB sensitivities are manifestly better in the B_d sector. The comparison assumes that the LHC machine is working at the design luminosity and that they will have a properly behaving LHCb detector. Instead it considers the SuperB sensitivities with 15 ab^{-1}, which is the luminosity achievable with such a kind of machine, that I have presented this morning, in less than one year.

The goal of SuperB, in fact, is to go towards 100 ab^{-1}, i.e. 110 billions $B\overline{B}$ pairs, however LHCb will take an equivalent amount of B mesons in 2010. The assumption that SuperB experiment will be ready in 2020 is based on the Japanese estimate for a possible machine at KEK, named superKEK.
There are several very rare B decay modes in addition to all those with neutrinos, that are precluded to LHCb. Most of them are sensible to New Physics effects.
Well of course there is some physics, which is only achievable with the hadron machines. In fact we know very well the success of the Tevatron at Fermilab in measuring Δm_s, which was a very important ~~to~~ constraint to the Unitarity Triangle.

– *Hisamatsu:*

In case of $\tau \to \mu\gamma$ mode, severe background is a big problem. You have presented the prospect of this mode. Will you give more information on topics or strategy at SuperB machine for rejecting the background, other than increasing Luminosity ?

– *Giorgi:*

So far Belle has improved their results and they presented at ICHEP, in Moscow, the new value, which is better than the previous value presented by BaBar. The previous value for the Branching Ratio limit was $<6.7\times10^{-8}$ and now it is $<4.2\times10^{-8}$. In actual fact, so far, the scaling of this limit was almost linear with Luminosity, but if you remember my projection (see slide), I have stated that this limit is going down as $1/\sqrt{L}$ and not linearly, because there is a background that you cannot distinguish. Then you can improve the BR limit by a factor of 10 instead of 100. Now the measured value was made with 0.6 ab^{-1} and my projection for SuperB is $3.4x10^{-9}$, now it is $4.2x10^{-8}$. I set a task force inside the International Study Group to look for all the handles in τ physics and what we could do. It is not an easy task, because there are irreducible backgrounds, as the one for example coming from a channel where you are loosing neutrinos (i.e. $\tau^- \rightarrow \mu^- \overline{\nu}_\mu \nu_\tau \gamma$). If the neutrinos have low energy, that background cannot easily be kinematically rejected. A possible improvement could be to increase the acceptance of the apparatus. In addition we are also thinking to other improvements i.e. the reduction of the boost (from: 3.1 GeV vs. 9.0 GeV down to: 4 GeV vs. 7 GeV). Such a reduction gives a better angular resolution in the laboratory, but, on the other hand, we cannot cancel the boost because it is crucial in order to make time-dependent analyses. We are also planning to improve the vertexing at least by a factor of three and then we are allowed to reduce the boost and we will have more spherical events.

– *Pacetti:*

About the possible future SuperB machine, do you know if there are competitors in this business ?

– *Giorgi:*

At the moment there are two projects, one is a project carried on by an international community, with people working in Canada, France, Germany, Italy, Russia, United Kingdom and United States and the other is a Japanese project, which is based on a upgrade of the present machine KEKB, which was very successful. Such a project is feasible, if it will be approved and founded, after the modification of some components of the detector, they plan to reach 15 ab^{-1} of integrated luminosity in 2020. In this project the main issue is the increasing of the circulating current, this implies a wall power between 85 and 100 MW. The instant luminosity, in this scheme, was $5x10^{35}$ cm^{-2} s^{-1}. Now they are updating their project, incidentally they also intend to consider the crab waist that we are studying, and with the present revisited project they can achieve a higher luminosity up to $8x10^{35}$ cm^{-2} s^{-1}. I believe that we could go higher than 10^{36} cm^{-2} s^{-1} to be able to have some room for new physics. In fact when you are marginal or you are almost to the limit, it is very difficult experimentally to open new doors, when you have some wide room you have the possibility for the discovery. My group is considering a more ambitious project, i.e. an instant luminosity higher than 10^{36} cm^{-2} s^{-1}, and possibly going to 10^{37} cm^{-2} s^{-1}, by keeping the wall power at a reasonable level, not exceeding the electricity bill of SLAC. In addition we consider the possibility of using the present component of PEP-II of the SLAC machine.

– *Pacetti:*

About the ISR measurements at the B-factories, BaBar has already measured the proton time-like magnetic form factor, what about the possibility of performing a similar measurement on the neutron form factor ?

– *Giorgi:*

First of all for people who are interested to this kind of physics, spectroscopy but also measurement of form factors, the good new is that with such a high luminosity the rate of interesting events that you expect is terribly high. It would be higher by order magnitudes with respect to devoted machines. In fact also upgrading some of the present machines as for instance in Novosibirsk and Frascati, which work at the energy for these kinds of measurements you hardly can achieve luminosities higher than 10^{33} cm^{-2} s^{-1} to be compared with the 10^{37} cm^{-2} s^{-1}. Even if you loose luminosity, because of the initial radiation, you get a statistics at least between one and two order of magnitude higher. Of course we need to re-design the apparatus, the present idea is to use as many components as possible of BaBar or Belle. It is clear that in the future there will be only one SuperB Factory. Infact we are ready now to welcome, before the end of this month, the participation of people from KEK to our studying group to join in the effort, to consider the possibilities and to arrive to some project. Of course when you have a project site and a time schedule, then the realization will depend on the funding agencies. Concerning the site, it should be in Europe or in Japan, the SLAC hypothesis seems more difficult, in fact the laboratory has taken the decision to move to a completely different topic: the photon physics. Concerning Italy, the campus of Tor Vergata University (Roma2), near to the National Laboratories of Frascati, should be a possibility.

– *Neri:*

Do you consider an important issue the fact that the future SuperB collider could operate at different energies not only at the Υ(4s)?

– *Giorgi:*

I have already mentioned the flexibility of this machine, which indeed should be called a Super-Flavour Factory, with the possibility of special runs for charm and τ. In addition an important option is to have one polarized beam, which, producing τ–τ in the final state, allows to measuring T-violation in the lepton sector.

– *Sens:*

$B^0 \to K\pi$ CP-violation was observed and this process has three interfering "penguins", in $B^+ \to K\pi$, instead, CP-violation is not observed, in this process the interference vanishes. Is there a casual relation between these facts ?

– *Giorgi:*

The situation is more puzzling than that. BaBar discovered in 2004, for the first time, direct CP-violation in the B-sector, comparing the decays of neutral and charged B's. In the decay of B_d we have a huge direct CP-violation, while in case of B_u we do not observe such effect. The interpretation is not simply :- "There is no interference between penguins and trees". This can be due to isospin combinations in the final state, data are confirming this interpretation. Since there is also strong interaction, isospin plays a crucial role.

– *Antonio:*

Some of the numbers that are calculated in LQCD can also be extracted from experimental data by having them as free parameters in the UT fit. Do you think that will be a reduced need for lattice calculations in these areas ?

– *Giorgi:*

No, I do not think so. That because strong interaction is crucially affecting most of the data, indeed what we measure are not quark decays but particle decays, i.e. we have particles in the final states.

Evidence for a Quark-Gluon Plasma at RHIC

John W. Harris
Physics Department, Yale University,
P.O. Box 208124, 272 Whitney Avenue, New Haven CT, U.S.A. 06520-8124
E-mail: John.Harris@Yale.edu

Revised (revised date)

Collisions of heavy ions at the Relativistic Heavy Ion Collider (RHIC) form a hot system at energy densities greater than 5 GeV/fm^3, where normal hadrons cannot exist and a quark-gluon plasma is expected to be formed. The system cools rapidly to a temperature T $\sim$ 175 MeV at the quark-hadron phase boundary predicted by lattice QCD where hadrons coalesce from quarks. A large amount of collective flow at the quark level provides evidence for strong pressure gradients in the initial partonic stage of the collision when the system is dense and highly interacting prior to coalescence into hadrons. This flow and the subsequent distributions of hadrons are described by ideal hydrodynamics with zero viscosity, suggesting a nearly perfect fluid is formed. Such low viscosities approach a universal lower bound postulated from string theory. The suppression of both light (u, d, s) and heavy (c, b) hadrons at large transverse momenta and the quenching of di-jets provide evidence for extremely large energy loss of partons as they propagate through a dense, strongly-coupled medium.

1. Introduction

A few micro-seconds after the Big Bang all matter in the Universe existed in the form of quarks, leptons and the gauge bosons that carry the fundamental forces of Nature. Then as the Universe cooled, a quark-hadron phase transition occurred and the nuclear particles formed from quarks and gluons. Quantum Chromodynamics (QCD) on a lattice reveals such a quark-hadron phase transition at a temperature of 1.75×10^{12} Kelvin (175 MeV).[1] Above this temperature lattice QCD calculations predict that hadrons "melt" into a form of hot QCD matter consisting of quarks and gluons, known as the Quark-Gluon Plasma (QGP). Fig. 1 shows a schematic phase diagram of matter as a function of temperature and baryon density. Several different phases of QCD matter are indicated. A first order phase transition is expected along a curve up to the critical point.[1,2] The QGP phase is expected at higher temperatures. The early Universe cooled down from higher temperatures close to the vertical axis. The region of high baryon density appears rich with structure and is presently an area of intensive theoretical investigation.[3] Shown are hypothesized regions of color superconductivity and color flavor locked (CFL) structures. These phase transitions may have implications for nuclear physics, astrophysics, cosmology and particle physics.

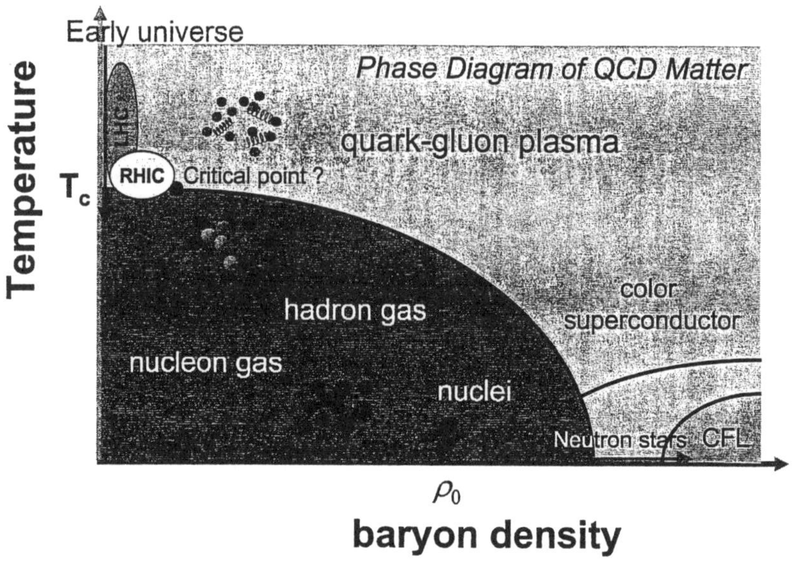

Fig. 1. Schematic phase diagram of QCD matter as a function of temperature T and baryon density. ρ_0 is normal nuclear matter density and $T_c \sim 175$ MeV.

The equation of state of hot QCD matter and its properties depend critically on the number of flavors and on the quark masses that are used in the lattice calculations.[1,2] Calculations of the energy density on the lattice as a function of temperature are displayed in Fig. 2. A relatively sharp deconfinement transition occurs at a temperature $T_c \sim 175$ MeV in 2-flavor QCD and at about 20 MeV lower temperature for 3-flavors (both in the chiral limit). There has also been recent success in implementation of techniques to calculate on the lattice at small but finite baryon density.[4]

The Relativistic Heavy Ion Collider (RHIC) at Brookhaven National Laboratory collides nuclei at ultra-relativistic energies to create hot, dense QCD matter, to study its properties and better understand the quark-hadron phase transition. The RHIC facility, which commenced operation for physics in the year 2000, accelerates and collides ions from protons to the heaviest nuclei over a range of energies, up to 250 GeV for protons and 100 A-GeV for Au nuclei. In the six production runs for physics, RHIC has collided Au + Au at center-of-mass energies $\sqrt{s_{NN}} = 19.6$, 62, 130, and 200 GeV, Cu + Cu at $\sqrt{s_{NN}} = 62$ and 200 GeV, d + Au at $\sqrt{s_{NN}} =$ 200 GeV, and p + p at $\sqrt{s} = 200$ GeV. RHIC and its four experiments (BRAHMS, PHENIX, PHOBOS, and STAR) are described comprehensively in Ref. [5]. The Large Hadron Collider (LHC) heavy ion program will start in a few years and is expected to explore farther into the high energy density regime of QCD.

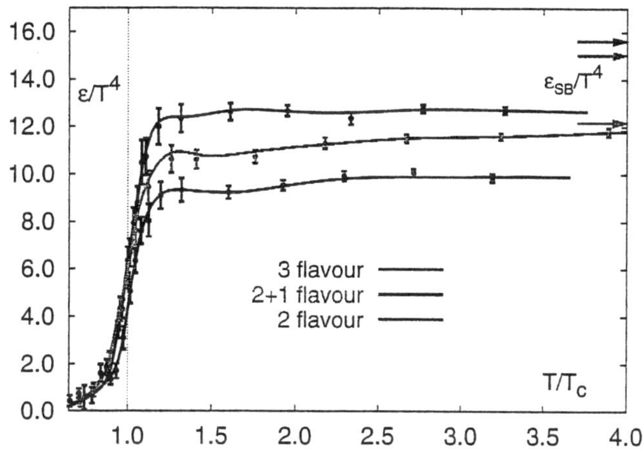

Fig. 2. Lattice QCD results for the energy density ϵ/T^4 as a function of the temperature T relative to the critical temperature T_c. The flavor dependence is shown by the three curves, from top to bottom depicting results for 3 light quarks (u,d,s), 2 light quarks (u,d) plus 1 heavier (s) quark, and 2 light quarks (u,d). The Stephan-Boltzmann values of ϵ_{SB}/T^4 are depicted by the arrows on the right for these three cases.

2. Extreme Energy and Particle Densities Created at RHIC

An initial objective of RHIC was to determine the energy density in the initial colliding system and to establish whether it surpasses the critical energy density from lattice QCD that is necessary for creating the QGP phase transition. Measurements of the transverse energy per unit pseudorapidity $dE_T/d\eta$,[6,7] and the total particle multiplicity density and mean transverse momentum per particle[8] were used to estimate the energy density assuming a Bjorken longitudinal expansion scenario.[9] The energy density can be estimated by $\epsilon_{Bj} = \frac{1}{\tau_o \pi R^2} \times \frac{dE_T}{dy}$, with dE_T/dy the transverse energy per unit rapidity, R the transverse radius of the system, and τ_o the formation time. Assuming a maximum value for the formation time $\tau_o = 1$ fm/c, a conservative estimate of the minimum energy density for the 5% (2%) most central Au + Au collisions at $\sqrt{s_{NN}} = 130$ GeV is 4.3 GeV/fm^3 [8] (4.6 GeV/fm^3)[6] and 4.9 GeV/fm^3 [7] for the 5% most central Au + Au collisions at $\sqrt{s_{NN}} = 200$ GeV. Note that the energy density derived from the RHIC experiments using the Bjorken formlation is a lower limit and may be much larger ($\sim$ ten times), since τ_o at RHIC is expected to be significantly less than 1 fm/c. This lower limit on the energy density at RHIC is approximately twenty-five times normal nuclear matter density ($\epsilon_{n.m.} = 0.17$ GeV/fm^3) and seven times the critical energy density ($\epsilon_c =$ 0.6 GeV/fm^3) predicted by lattice QCD for formation of the QGP.

Hadrons with transverse momentum less than 2 GeV/c are produced abundantly in collisions at RHIC. The measured charged hadron multiplicity density at midrapidity is $dn_{ch}/d\eta \mid_{\eta=0} = 700 \pm 27(syst)$ [10] for the 3% most central collisions of

Au + Au at $\sqrt{s_{NN}} = 200$ GeV. This corresponds to a hadron multiplicity density $dn_{total}/d\eta\ |_{\eta=0} = 1050$ and a total hadron multiplicity in the most central events of ~ 7000.[10,11] In terms of the number of created quarks, consider the case where all observed hadrons in the final state are mesons, each containing a quark and anti-quark. These 7,000 hadrons correspond to 14,000 quarks and anti-quarks. A lower limit for the number of created quarks and anti-quarks can be obtained by subtracting off the valence quarks that originally enter the collision in the incident nuclei. The number of original valence quarks in a head-on collision of two Au nuclei is 2 (Au nuclei) x 197 nucleons/Au x 3 quarks/nucleon $\sim$ 1200. Thus, more than 90% of the more than 14,000 quarks and anti-quarks in the final state are produced in the collision.[a]

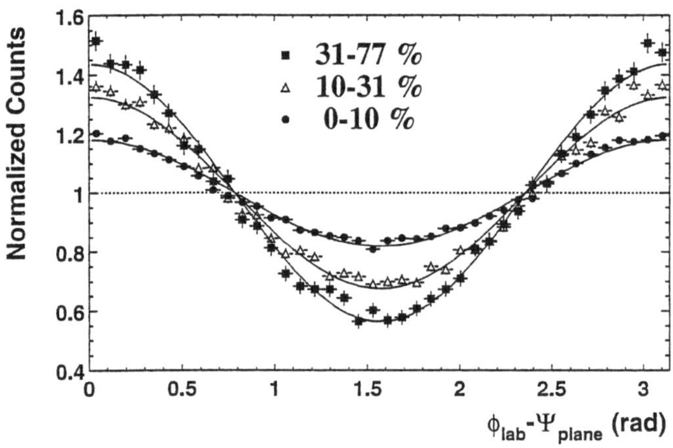

Fig. 3. Azimuthal correlations of charged hadrons as a function of the azimuthal angle relative to the reaction plane for three different centrality ranges (as denoted in the legend by percent of geometric cross section) in 130 GeV/n Au + Au collisions.

3. Observation of Strong Elliptic Flow at RHIC

The observation of an extremely large elliptic flow at RHIC has led to exciting consequences for understanding the dynamical evolution of these collisions. Unlike the case for collisions of elementary particles, nuclei colliding with non-zero impact parameter have an inherent spatial asymmetry associated with the asymmetric region of overlap. The larger the impact parameter, the larger the asymmetry perpendicular to the reaction plane.[b] Displayed in Fig. 3 are the azimuthal angular

[a]Note that the large number of gluons in the system has not been considered in this estimate. Taking these into consideration increases the number of produced quarks and anti-quarks.
[b]The reaction plane is the plane containing the incident beam and impact parameter vectors.

distributions for collisions over three different impact parameter ranges at RHIC.[12] Hadrons are emitted preferentially in the reaction plane (at 0, π) providing evidence for large pressure gradients early in the collision process that generate the elliptic flow. The pressure gradients and elliptic flow in-plane increase with increase of the initial spatial asymmetry out-of-plane. These results indicate that the initial spatial asymmetry is transformed efficiently into the observed momentum-space anisotropy during the brief traversal time of the incident nuclei (< 1 fm/c at these energies). Thus, the system must be dense and highly interacting to accomplish this transformation efficiently in such short time.

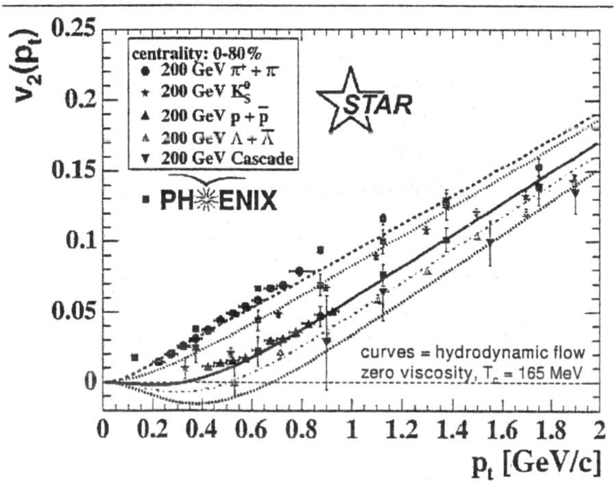

Fig. 4. Elliptic flow v_2 for $\sqrt{s_{NN}} = 200$ GeV Au + Au minimum bias collisions as a function of p_T for $\pi^{\pm}$, K^0_s, $p + \bar{p}$, $\Lambda + \bar{\Lambda}$, and $\Xi + \bar{\Xi}$. Curves are predictions of ideal hydrodynamics.

To study this azimuthal anisotropy in quantitative detail the second Fourier harmonic component of the azimuthal distribution of particles in momentum space is constructed with respect to the reaction plane, $v_2 = \langle \cos(2\phi) \rangle$ where $\phi = \text{atan}\,(p_y/p_x)$. The v_2 is called the elliptic flow. Displayed in Fig. 4 is v_2 for $\pi^{\pm}$, K^0_s, $p + \bar{p}$, $\Lambda + \bar{\Lambda}$, and $\Xi + \bar{\Xi}$ as a function of p_T in $\sqrt{s_{NN}} = 200$ GeV Au + Au minimum bias collisions.[13,14,15] Also shown are predictions from hydrodynamics. The elliptic flow is well described at these low transverse momenta by hydrodynamical models incorporating a softening of the equation of state due to quark and gluon degrees of freedom[16,17] and zero viscosity.[18] Such low viscosities were not expected nor ever before observed for hadronic or nuclear systems.

The v_2 measured for $\pi^{\pm}$, K^0_s, $p + \bar{p}$, $\Lambda + \bar{\Lambda}$, and $\Xi + \bar{\Xi}$ at higher p_T are displayed in Fig. 5a for $\sqrt{s_{NN}} = 200$ GeV Au + Au minimum bias collisions.[14,15,19] The mesons and baryons cross over near $p_T = 2$ GeV/c, above which the baryons exhibit higher values of v_2 than the mesons. This is more easily seen in Fig. 5b where the v_2 is plotted as a function of the transverse kinetic energy (KE_T) of the identified hadrons, where the data converge to two distinct groups, one for baryons and the other for mesons. Larger values of v_2 for baryons than mesons at higher p_T may result from the creation of hadrons in soft processes and boosted to higher p_T by collective flow. Coalescence of quarks to form composite hadrons may also contribute to this effect.

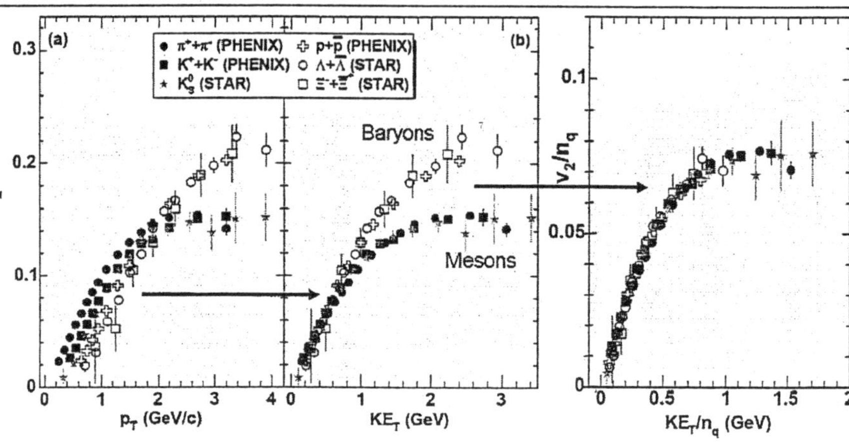

Fig. 5. a) Elliptic flow v_2 as a function of p_T measured for $\pi^\pm$, K^0_s, $p + \bar{p}$, $\Lambda + \bar{\Lambda}$, and $\Xi + \bar{\Xi}$. b) Elliptic flow v_2 as a function of transverse kinetic energy KE_T. c) v_2 per quark (v_2/n_q) as a function of KE_T per quark (KE_T/n_q). All data are for $\sqrt{s_{NN}} = 200$ GeV Au + Au minimum bias collisions.

Displayed in Fig. 5c is v_2 per quark (v_2/n_q) measured by STAR[14,19] and PHENIX[15] for $\pi^\pm$, K^0_s, $p + \bar{p}$, $\Lambda + \bar{\Lambda}$, and $\Xi + \bar{\Xi}$ as a function of KE_T per quark (KE_T/n_q) in $\sqrt{s_{NN}} = 200$ GeV Au + Au minimum bias collisions. When v_2 is plotted per quark (v_2/n_q) for baryons and mesons, the values of v_2/n_q scale with KE_T/n over the entire range of KE_T. This is consistent with a quark coalescence picture for hadrons[20] and is evidence for early collective flow at the quark level. This observation coupled with the extremely low to non-existent viscosity has led to descriptions of the system in terms of a nearly perfect liquid (non-viscous) of quarks and gluons.[21]

Given the subject matter of this School, I am compelled to mention the recent application of string theoretical concepts to describe the observation of an extremely low viscosity at RHIC. There is a general conjecture[22] of the equivalence between a string theory in an n-dimensional space and a quantum field theory on the n-1 dimensional conformal boundary of this space. This equivalence between our 4-dimensional (4D) Universe and its 5-dimensional (5D) space-time volume has been termed the Holographic Principle by t'Hooft[23] and Susskind.[24] [c] Also, a black hole

[c]see G. t'Hooft and L. Susskind lectures at this School

in a particular 5D volume was shown to correspond to a hot system of quarks and gluons on the 4D space-time boundary.[25] Therefore, the so-called AdS/CFT (anti-De-Sitter space/conformal field theory) correspondence is a conjectured equivalence between a string theory with gravity in a space-time volume, and a quantum field theory without gravity on the conformal boundary of this volume. This AdS/CFT correspondence can be used to determine the viscosity in a 4-dimensional world, assuming a strongly coupled N = 4 super-symmetric Yang-Mills (SUSY YM) conformal field theory, from a completely different approach with a tractable calculation and solution in a 5-dimensional space (AdS_5). The result is that there is a universal lower bound on the shear viscosity (η) to entropy (s) ratio of $\eta/s = 1/4\pi$.[26] This lower bound is similar in value to what is observed at RHIC, when attempts are made to quantify this ratio,[18] and considerably lower than other known fluids. The extent to which AdS/CFT can explain hot QCD, particularly with the hypothesis of N=4 SUSY YM, is debatable and will continue to be investigated.

4. Is the System Thermalized?

If the system can be described in terms of equilibrium thermodynamics, the ratios of the various types of particles must be reproduced with a consistent set of thermodynamic variables. Statistical and thermodynamic models reproduce the measured ratios using as variables the chemical freeze-out temperature (T) and the baryochemical potential (μ_B). These models employ hadronic degrees of freedom in a grand canonical ensemble. See Ref. [27] for a comprehensive review. Displayed in Fig. 6 are the particle ratios measured at RHIC along with results of a statistical-thermal model fit.[28,29] The particle ratios for $\sqrt{s_{NN}} = 130$ GeV Au + Au can be fit with the parameters T = 176 MeV and $\mu_B = 41$ MeV. For $\sqrt{s_{NN}} = 200$ GeV Au + Au, T = 177 MeV and $\mu_B = 29$ MeV are required. The statistical-thermal model fits reproduce the data extremely well. When the same approach is applied to lower energy data[30,31,32] chemical freeze-out follows a curve in (μ_B, T) space as depicted in Fig. 7. When drawn on a (μ_B, T) plane, the values of (μ_B, T) from statistical model fits to the experimental data approach at the higher energies the deconfinement phase transition boundary predicted by lattice QCD.[33]

5. Suppression of Large Transverse Momentum Particles

Hard scattering can be used to probe the medium through which the hard-scattered partons propagate. The radiation energy loss of a parton traversing a dense medium is predicted to be significant and is sensitive to the gluon density of the medium.[34,35,36] In order to investigate parton energy loss in the medium, the RHIC experiments have measured hadron spectra and azimuthal correlations of hadrons with large transverse momentum.

In order to compare results from relativistic heavy ion collisions to elementary p + p interactions, a nuclear modification factor R_{AA} is defined as $R_{AA}(p_T) =$

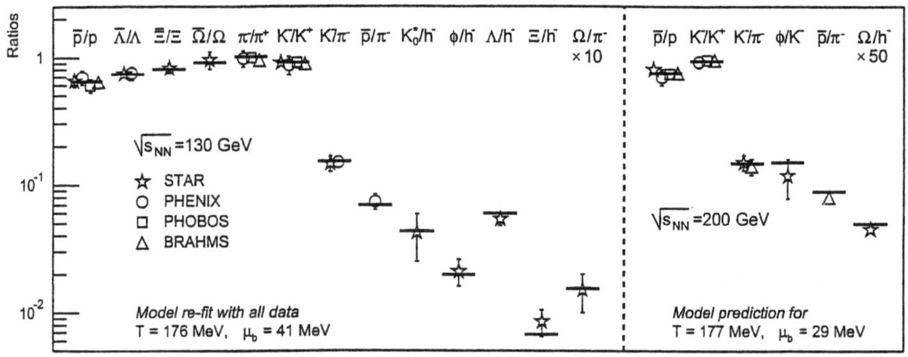

Fig. 6. Particle ratios measured in RHIC experiments denoted by symbols in the legend for central collisions of Au + Au at $\sqrt{s_{NN}}$ = 130 GeV (left panel) and 200 GeV (right panel). Results from a statistical-thermal model fit to the entire set of data are shown as a horizontal line for each ratio. Parameters (μ_B, T) for the best fit at each energy are shown at the bottom of each panel.

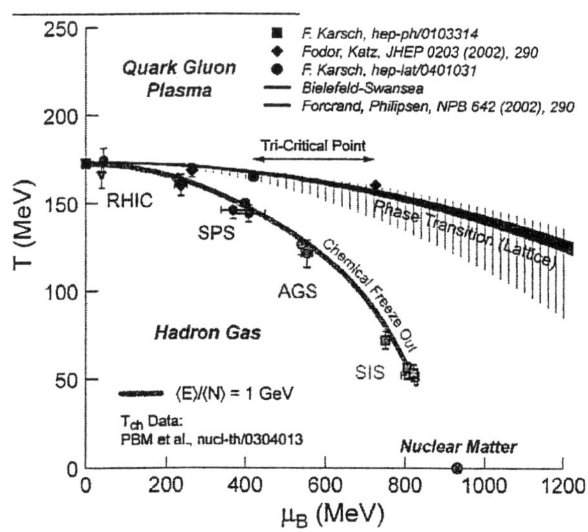

Fig. 7. Nuclear phase diagram extracted from a thermal model analysis of experimental results at various energies and heavy ion facilities (SIS at GSI, AGS at BNL, SPS at CERN, and RHIC at BNL). The lower curve corresponds to chemical freeze-out at a constant baryon density. The shaded curve corresponds to the deconfinement phase boundary predicted by lattice QCD (with lattice references given in the legend).

$\frac{dN^2/dp_T dy\ (AA)}{\langle N_{binary}\rangle dN^2/dp_T dy\ (NN)}$. $\langle N_{binary}\rangle$ is the number of binary collisions in a geometrical model in order to scale from elementary nucleon-nucleon (NN) collisions to nucleus-nucleus (AA) collisions. When R_{AA} = 1, AA collisions can be described as an incoherent superposition of NN collisions, as predicted by perturbative QCD (pQCD). This corresponds to scaling with the number of binary collisions (binary scaling).

5.1. *Light Hadrons*

Suppression of the hadron spectra at large transverse momenta $p_T > 2$ GeV/c has been observed in the nuclear modification factors measured in central Au + Au collisions at RHIC.[37,38,39] Displayed in Fig. 8 is the nuclear modification factor as a function of transverse momentum for central Au + Au collision data and for d + Au collisions from PHENIX (left panel)[40] and STAR (right panel).[41] The ratios are taken relative to measured p + p collision data scaled by the number of binary collisions. These data exhibit a clear suppression by a factor 4-5 in the central Au + Au case. The central Au + Au ratio remains rather flat up to the statistical limits of the data at 10 GeV/c transverse momentum, where no sign of reaching the pQCD limit of binary scaling is observed. The peripheral collision data (not shown) exhibit no nuclear modification, i.e. $R_{AA} = 1$, within errors.

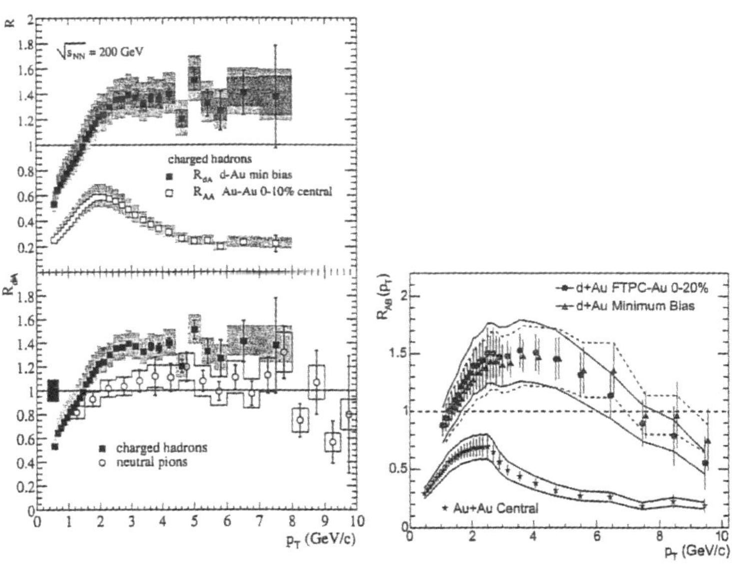

Fig. 8. Nuclear modification factor for $\sqrt{s_{NN}} = 200$ GeV minimum bias d + Au and for central Au + Au from PHENIX (left) for charged hadrons and neutral pions; and minimum bias d + Au, central d + Au and central Au + Au from STAR (right) for charged hadrons.

The Au + Au relative to p + p yields can be reproduced by pQCD calculations incorporating parton energy loss in dense matter[42] or by a model incorporating initial-state gluon saturation.[43] To distinguish these two theoretical approaches, final state energy loss versus initial state gluon saturation, measurements were made in d + Au where no parton energy loss in a dense medium (final state suppression) is expected. The ratio $R_{dAu}(p_T) > 1$ in d + Au as seen in Fig. 8 indicates a Cronin enhancement (initial state multiple scattering) for $1 < p_T < 7$ GeV/c with no suppression. The d + Au charged hadron data when compared to scaled p + p data[40,41,44,45] rule out initial-state gluon saturation leaving only final state effects as a cause of the suppression in the Au + Au data at mid-rapidity at RHIC. Furthermore, PHENIX has found[46] that direct photons are not suppressed, as expected since they do not interact strongly. This is seen in Fig. 9 where the measured R_{AA} of γ, π^0 and η are displayed as a function of p_T for central Au + Au collisions. The $R_{AA}(\gamma) \sim 1$, while the $R_{AA}(\pi^0)$ and $R_{AA}(\eta)$ are suppressed by ~ 4-5 like the charged hadrons. Also shown in Fig. 9 is a calculation of parton energy loss with a gluon density of $dn_{gluon}/dy = 1100$.[42] In order to describe the suppression of light hadrons at RHIC a gluon density of $dn_{gluon}/dy \sim 1000$ is generally required.[47] This is equivalent to an energy loss per unit length that is approximately 15 times that of normal nuclear matter. Initial calculations of jet quenching have been performed using the AdS/CFT correspondence.[48] The results give $\hat{q} = 4.5, 10.6, 20.7$ GeV^2/fm for a range of temperatures of the medium of T = 300, 400, 500 MeV, respectively. This is remarkably close to what is expected at RHIC (see discussion below for RHIC $\hat{q}$ values).

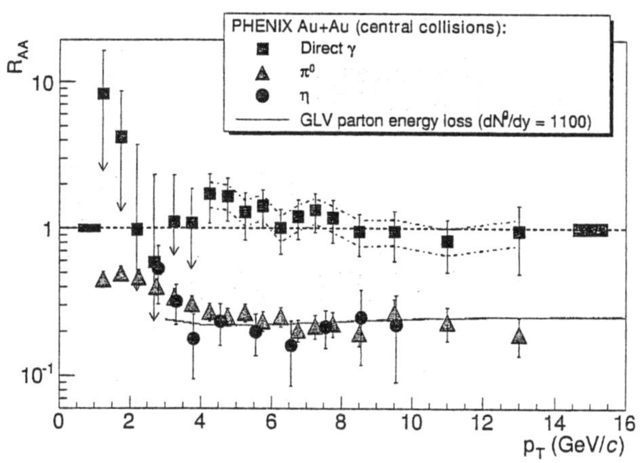

Fig. 9. Nuclear modification factor for $\sqrt{s_{NN}} = 200$ GeV central Au + Au from PHENIX for γ, π^0 and η as a function of p_T from PHENIX. The curve is a pQCD calculation with parton energy loss.

Displayed in Figs. 10 and 11 are the ratios R_{CP} of identified particles (specified in the legends) produced in central collisions relative to those produced in peripheral collisions scaled by the number of binary collisions in each data set. The data in Fig. 10 are strange particle data and charged hadrons from STAR.[49] The R_{CP} ratios for mesons reach a maximum at p_T between 1 - 2 GeV/c with ratios R_{CP} considerably less than 1 and continue to be suppressed ($R_{CP} < 1$) up to $p_T \sim 7$ GeV/c. The R_{CP} ratios for baryons peak near unity and are suppressed relative to binary-scaling only above around 3 GeV/c. The PHENIX data on π^0 and $p+\bar{p}$ are shown in Fig. 11.[50] The π^0 are suppressed by a factor of $\sim$ 4 - 5, while the protons do not deviate from binary scaling up to $p_T \sim 4$ GeV/c.

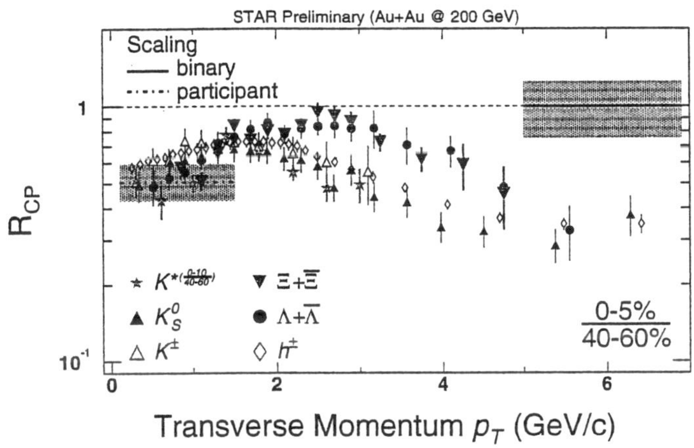

Fig. 10. R_{CP} ratios for $\sqrt{s_{NN}} = 200$ GeV Au + Au collision data measured in STAR.

5.2. *Charm and Beauty Hadrons*

Heavy (charm and beauty) quarks are expected to lose less energy than light quarks while propagating through a dense colored medium. The large heavy quark mass reduces the available phase space for gluon radiation. This has been called the "dead cone effect".[51] PHENIX and STAR have measured the spectra of non-photonic electrons in $\sqrt{s_{NN}} = 200$ GeV p + p and Au + Au collisions over a range of impact parameters. After subtraction of electrons from photon conversions and light hadron decays, the resulting non-photonic electron spectra are predominantly from semi-leptonic decays of heavy quarks (D- and B-meson decays). The R_{AA} for non-photonic electrons measured in central collisions are displayed in Figs. 12 and 13 for PHENIX[52] and STAR,[53] respectively, as a function of p_T. The non-photonic electrons from decays of D- and B- mesons are clearly suppressed to about the same degree as the light hadron spectra.

The curves in Figs. 12 and 13 are predictions of various models. In Fig. 12 the curves 1 a, b , and c are from Ref. [55] and include decays of D-mesons only, whereas

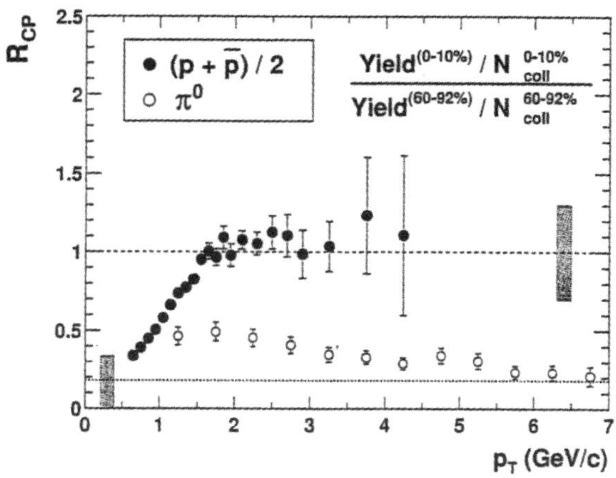

Fig. 11. R_{CP} ratios for $\sqrt{s_{NN}}$ = 200 GeV Au + Au collision data measured in PHENIX.

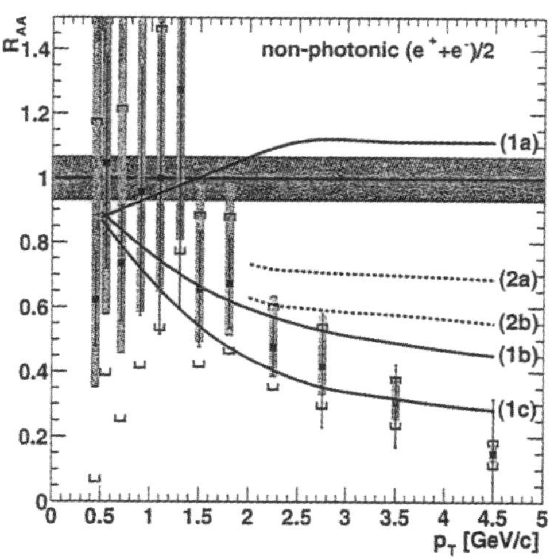

Fig. 12. R_{AA} for non-photonic electrons as a function of transverse momentum (p_T) in $\sqrt{s_{NN}}$ = 200 GeV central Au + Au compared to p + p in PHENIX for 0-10% centrality. See text for description of model curves.

curves 2a and b include B-decays also. Curves 1 (a, b, and c) are for various values of the transverse momentum transfer squared per unit length (0 - no medium, 4, 14 GeV^2/fm, respectively) of the parton traversing the medium and is proportional to the density of scatterers. Curves 2 (a and b) are for different values of the gluon

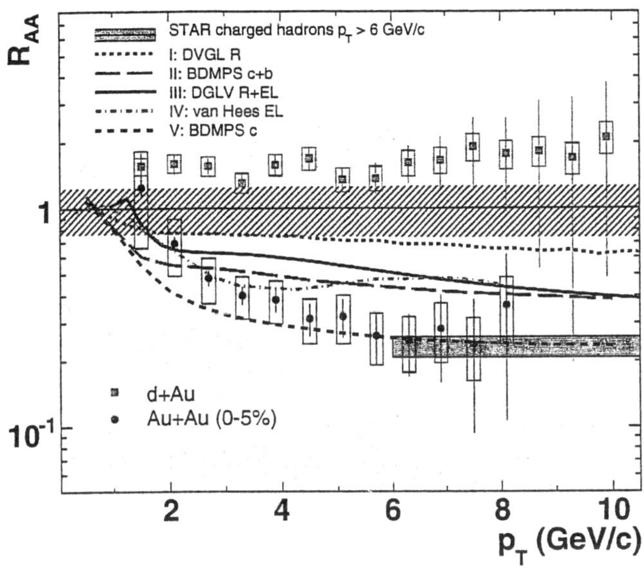

Fig. 13. R_{AA} for non-photonic electrons as a function of transverse momentum (p_T) in $\sqrt{s_{NN}}$ = 200 GeV d + Au and central (0-5%) Au + Au compared to p + p in STAR. See text for description of model curves.

density of the medium (1000 and 3500). Clearly only curve 1c approximates what is seen in the data, but this curve does not include B-meson decays that are expected to become important above about 3 GeV/c.

In Fig. 13 the curve labeled "DVGL-R" corresponds to a calculation utilizing energy loss from induced gluon radiation with a gluon density of $dn_{gluon}/dy = 1000$ that is consistent with the observed light quark suppression.[54] The curve labeled "BDMPS c+b" uses BDMPS radiative energy loss for charm and beauty from multiple soft collisions with a transport coefficient $\hat{q} = 14$ GeV^2/fm, also consistent with light quark suppression although only weakly constrained by it.[55] Both curves do not produce sufficient energy loss and suppression to reproduce the data. This is suggestive of the need for collisional energy loss. The curve labeled "DGLV-R+EL" includes both gluon radiation and elastic scattering energy loss, although this too is still insufficient to account for the measured heavy quark suppression and energy loss.[56] The curve labeled "van Hees EL" utilizes c and b quark resonances in a strongly coupled medium, but this result also does not fit the data.[57] The curve labelled "BDMPS c" is the same as "BDMPS c+b" but with charm contributions only. This curve agrees best with the data.[54] Overall, the lack of influence of B-meson decays on the spectra when comparing to theory is perplexing and calls into question our understanding of the propagation of heavy quarks in the medium. Further experimental results separating charm and beauty are necessary to resolve this dilemma, future theoretical breakthroughs not-withstanding.

6. Effects of Jet Quenching on Azimuthal Correlations

Two-particle correlations of high p_T hadrons can be used to extract the hard scattering component from the soft background at RHIC. The fragmentation of partons produces jets, a phase space clustering of particles in a small region of relative azimuth ($\Delta\phi$) and pseudorapidity ($\Delta\eta$). As seen in Fig. 14 an enhancement is observed at $\Delta\phi = 0$ in azimuthal correlations near midrapidity upon correlating charged hadrons having $4 < p_T^{trig} < 6$ GeV/c with all other charged hadrons having 2 GeV/c $< p_T < p_T^{trig}$ in the same event. This has been seen in collisions of Au + Au at $\sqrt{s_{NN}} = 130$ and 200 GeV ranging from minimum bias to central collisions, and in p + p and d + Au at $\sqrt{s} = 200$ GeV.[12,58] The peak at small relative azimuthal angle is indicative of short range correlations, i.e. jets. Data from p + p, d + Au, Cu + Cu and peripheral Au + Au exhibit a distinct back-to-back correlation, a characteristic of hard-parton scattering, with peaks at $\Delta\phi = 0$ and π, indicating di-jets. This is shown in Fig. 14 for p + p and d + Au reactions. However, for the most central Au + Au collisions the backward ($\Delta\phi = \pi$) correlation disappears. These results provide further evidence that in central Au + Au collisions a strong interaction results in quenching of high p_T partons.

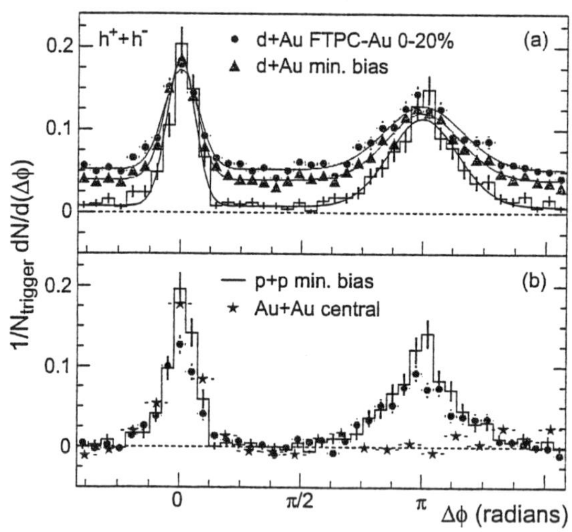

Fig. 14. a) Charged hadron correlation functions (see text for details) for p + p interactions, and minimum bias and central collisions of d + Au at $\sqrt{s_{NN}} = 200$ GeV. b) Comparison of correlations for p + p, central d + Au, and central Au + Au collisions with background subtracted.

When the p_T of the charged particles correlated with the trigger particle is lowered, the away-side peak re-appears, but broadened and possibly split.[59,60] This has led to descriptions of the broadening of this away-side particle distribution in terms of Mach-shock waves,[61] conical flow,[62] Cherenkov-like gluon radiation [63] and color wakes.[64] Further such experimental and theoretical investigation into the behavior of particles on the away-side of a high momentum trigger particle or jet may lead to a better understanding of the medium, its properties and its response to large energy deposition.

7. J/ψ

An early prediction of a deconfinement signature was the suppression of J/ψ production in a deconfined medium.[65] J/ψ suppression has been observed in central nucleus-nucleus collisions at the SPS [66] and initial measurements at RHIC.[67]

A suppression of J/ψ production has been observed in $\sqrt{s_{NN}} = 17.3$ GeV Pb + Pb collisions at the SPS [68] and attributed to a deconfinement phase transition as predicted by Ref. [65]. Alternative models based on interactions with hadronic co-movers have been proposed, but have problems reproducing the suppression as a function of centrality. Expectations at RHIC vary between the different models. If the suppression at the SPS is due to deconfinement, then the J/ψ should also be suppressed at RHIC energies. On the other hand, coalescence of $c\bar{c}$ pairs [69] and statistical-thermal models [70,71] predict significant J/ψ production at RHIC. First measurements of J/ψ have been reported by PHENIX at RHIC in $\sqrt{s_{NN}} = 200$ GeV p + p and Au + Au collisions.[67] These results disfavor significant enhancement due to $c\bar{c}$ coalescence[69] and thermal $c\bar{c}$ production[71], while still not yet being able to distinguish between models for suppression relative to binary scaling. The high statistics measurements from PHENIX for $J/\psi \to e^+e^-$ at mid-rapidity and $J/\psi \to \mu^+\mu^-$ at forward rapidities will provide data to adequately address the J/ψ production and evolution mechanisms.

8. Summary and Conclusions

The completely unexpected observation of strong elliptic flow in Au + Au collisions on the first day at RHIC, the subsequent detailed measurements of the variation of the elliptic flow for different particles, and the scaling of the flow with the numbers of quarks in the various mesons and baryons have led to a description of the medium created in these collisions as a nearly perfect liquid of quarks and gluons. The observed ratios of the many different hadrons produced at RHIC represent equilibrium abundances predicted from thermal models. The final state hadrons are formed and thermally distributed at the universal hadronization temperature (175 MeV) with a collective flow derived from their quark predecessors. The suppression of hadrons at large transverse momentum in Au + Au collisions, absent in d + Au, provides evidence for a high density medium, consistent with the initial determination of large energy densities created at RHIC. The observed quenching of the away-side jet (and its absence in central collisions) in Au + Au collisions requires a strongly-coupled colored medium, rather than the perturbative one expected from the initial lattice calculations. This nearly perfect liquid of quarks and gluons has behavior similar to that of other strongly-coupled systems such as degenerate fermi gases of atoms at low temperatures.[72] Furthermore, the RHIC liquid of quarks and gluons has a viscosity (divided by the entropy density) that approaches a quantum lower bound for shear viscosity derived using strongly-coupled N = 4 super-symmetric Yang-Mills theory.[26]

However, much still needs to be understood. The new results on suppression of heavy quark hadrons defy predictions. Addressing the question of deconfinement by measuring the extent to which charmonium is suppressed is just getting underway. A complete quarkonium program (measuring J/ψ, ψ' and Upsilon states) at the LHC and upgraded RHIC luminosities will be instrumental in determining the initial temperatures and degree of deconfinement via comparisons to more sophisticated future lattice calculations. So far, experimental signatures of chiral symmetry restoration have been elusive at RHIC. This will be investigated further through measurements of low mass di-leptons to search for medium modification of hadronic resonances. New insight may also be gained from a better understanding of hadronization and fragmentation in vacuum and in medium.

The strongly-coupled quark-gluon liquid is an exciting and unexpected phenomenon that must be understood. The RHIC results require a more sophisticated, dynamical approach in order to extract the nuclear matter equation of state of the liquid, its transport properties (e.g. shear viscosity, heat conductivity, speed of sound), and determine its constituents. Similar theoretical techniques are being used in attempts to understand results (e.g. strongly-interacting atomic gases, the strongly-coupled quark-gluon liquid, and black holes) from vastly different fields. On the experimental side, new measurements at RHIC and the LHC will be undertaken.

References

1. F. Karsch, Nuc. Phys. A698 (2002) 199c.
2. K. Kanaya, Nuc. Phys. A715 (2003) 233c.
3. see for example M. Alford J. Phys. G: Nucl. Part. Phys. 30 (2004) S441.
4. for review of results at finite temperature and density see Z. Fodor, Nuc. Phys. A715 (2003) 319c.
5. "The Relativistic Heavy Ion Collider and its Detectors", eds. M. Harrison, T. Ludlam, and S. Ozaki, Nuc. Inst. Meth. A499 (2003) p. 235-824.
6. K. Adcox et al. (PHENIX) Phys. Rev. Lett. 87 (2001) 052301.
7. J. Adams et al. (STAR), nucl-ex/0407003.
8. C. Adler et al. (STAR), Phys. Rev. Lett. 87 (2001) 112303.
9. J.D. Bjorken, Phys. Rev. D27 (1983) 140.
10. B. Back et al (PHOBOS), Phys. Rev. C65 (2002) 061901R.
11. B. Back et al (PHOBOS), Phys. Rev. Lett. 87 (2001) 102303.
12. C. Adler et al. (STAR) Phys. Rev. Lett. 90 (2003) 032301.
13. C. Adler et al (STAR), Phys. Rev. Lett. 87 (2001) 182301 and Phys. Rev. Lett. 89 (2002) 132301.
14. J. Adams et al (STAR), Phys. Rev. Lett. 95 (2005) 122301.

15. S.S. Adler et al (PHENIX), Phys. Rev. Lett. 91, 182301 (2003).
16. P.F. Kolb et al., Phys. Lett. B500 (2001) 232; P.F. Kolb et al., Nucl. Phys. A696 (2001) 197.
17. P. Huovinen et al., Phys. Lett. B 503 (2001) 58.
18. D. A. Teaney, J. Phys. G30 (2004) S1247.
19. J. Adams et al (STAR), Phys. Rev. C 72 (2005) 014904.
20. C. Nonaka, R. Fries, and S. Bass, nucl-th/0308051.
21. T. Hirano and M. Gyulassy, nucl-th/0506049v2.
22. J.M. Maldacena, Adv. Theor. Math. Phys. 2 (1998) 231.
23. G. t'Hooft, gr-qc/9310026.
24. L. Susskind, J. Math. Phys. 36 (1995) 6377.
25. E. Witten, Adv. Theor. Math. Phys. 2 (1998) 253.
26. P.K. Kovtun, D.T. Son, A.O. Starinets, Phys. Rev. Lett. 94 (2005) 111601.
27. P. Braun-Munzinger, K. Redlich, and J. Stachel, preprint nucl-th/0304013.
28. P. Braun-Munzinger, D. Magestro, K. Redlich, and J. Stachel, Phys. Lett. B518 (2001) 41 and D. Magestro (private communication).
29. for another statistical approach see F. Becattini, J. Phys. G28 (2002) 1553.
30. P. Braun-Munzinger, I. Heppe, and J. Stachel, Phys. Lett. B465 (1999) 15.
31. see also F. Becattini et al., Phys. Rev. C64 (2001) 024901.
32. F. Beccatini, M. Gazdzicki, and J. Sollfrank, Eur. J. Phys. C5 (1998) 143.
33. Z. Fodor and S.D. Katz, preprint hep-lat/0106002 (2001).
34. M. Gyulassy and M. Plümer, Phys. Lett. B243 (1990) 432.
35. X. N. Wang and M. Gyulassy, Phys. Rev. Lett. 68 (1992) 1480.
36. R. Baier, D. Schiff, and B. G. Zakharov, Annu. Rev. Part. Sci. 50 (2000) 37.
37. K. Adcox et al (PHENIX), Phys. Rev. Lett. 88 (2002) 022301.
38. C. Adler et al (STAR), Phys. Rev. Lett. 89 (2002) 202301.
39. B.B. Back et al. (PHOBOS), nucl-exp/0302015v1.
40. K. Adcox et al (PHENIX), Phys. Rev. Lett. 91 (2003) 072303.
41. J. Adams et al (STAR), Phys. Rev. Lett. 91 (2003) 072304.
42. I. Vitev and M. Gyulassy, Phys. Rev. Lett. 89 (2002) 252301; and I. Vitev, J. Phys. G 30 (2004) S791.
43. D. Kharzeev, E. Levin and L. McLerran, Phys. Lett. 561 (2003) 93.
44. B.B. Back et al. (PHOBOS), Phys. Rev. Lett. 91 (2003) 072302.
45. I. Arsene et al. (BRAHMS), Phys. Rev. Lett. 91 (2003) 072305.
46. S.S. Adler et al. (PHENIX), Phys. Rev. Lett. 96 (2006) 202301.
47. for a review of jet energy loss see M. Gyulassy et al., nucl-th/0302077v1.
48. H. Liu, K. Rajagopal and U. A. Wiedemann, Phys. Rev. Lett. 97 (2006) 182301.
49. M.A.C. Lamont et al (STAR), nucl-ex/0403059; J. Adams et al (STAR), nucl-ex/0306007.
50. S.S. Adler et al (PHENIX), nucl-ex/0307022.
51. Y.L. Dokshitzer and D.E. Kharzeev, Phys. Lett. B519 (2001) 199.
52. S.S. Adler et al. (PHENIX), Phys.Rev.Lett. 96 (2006) 032301).
53. B.I. Abelev et al. (STAR), Phys. Rev. Lett. 98 (2007) 192301.
54. M. Djordjevic et al., Phys. Lett. B632 (2006) 81.
55. N. Armesto et al., Phys. Lett. B 637 (2006) 362.
56. S. Wicks et al., Nucl. Phys. A784 (2007) 426.
57. H. van Hees, V. Greco, and R. Rapp, Phys. Rev. C 73 (2006) 034913.
58. C. Adler et al. (STAR) Phys. Rev. Lett. 89 (2002) 202301.
59. N. Grau et al. (PHENIX) nucl-ex/0612019.
60. F. Wang et al. (STAR) nucl-ex/0610011.

61. H. Stoecker, Nucl. Phys A750 (2005) 121.
62. J. Casalderrey-Solana, E.V. Shuryak and D. Teaney, Nucl. Phys. A774 (2006) 577.
63. V. Koch, A. Majumbder and X.N.Wang, Phys. Rev. Lett 96 (2006) 172302.
64. J. Ruppert and B. Muller, Phys. Lett. B618 (2005) 123.
65. T. Matsui and H. Satz, Phys. Lett. B 178 (1986) 416.
66. M.C. Abreu et al (NA50), Phys. Lett. B 477 (2000) 28; ibid. Phys. Lett. B 521 (2001) 195.
67. S.S. Adler et al (PHENIX), Phys. Rev. C69 (2004) 014901.
68. L. Ramello et al (NA50), Nucl. Phys. A715 (2003) 243c.
69. R.L. Thews, M. Schroedter, and J. Rafelski, Phys. Rev. C 63 (2001) 054905.
70. L. Grandchamp and R. Rapp, Nucl. Phys. A709 (2002) 415.
71. A. Andronic et al, nucl-th/0303036.
72. R.M. OHara et al., Science 298 (2002) 2179.

CHAIRMAN: J.W. HARRIS

Scientific Secretaries: O. Catu, B. Keren-Zur

DISCUSSION I

-Csanad:

Numeric Hydro codes fail to describe the source of the observables when they describe v2, but analytic hydro solutions describe several observables simultaneously.

-Harris:

I agree with you, and I thank you for your comment. Basically the hydro models need to be developed to where they include viscosity in the full three dimensions.

-Csanad:

Quark fluid instead of QGP might be a better notation since we have a fluid rather than a plasma.

-Harris:

I agree, in terms of plasma. We might argue whether a fluid or liquid is the right word. But in terms of plasma, it's known historically as that, we're getting away from that description. Soup!

-Lopez Mateos:

How do you reconcile the experimental fact that the QGP is strongly interacting and the lattice QCD theoretical fact that alpha_s goes to 0 at small distances as the temperature of the fluid increases beyond the critical temperature?

-Harris:

I said that it is strongly coupled. But you'll note that this is four times the critical temperature and let me give you a little bit more perspective which I didn't say that this is way beyond where we are at RHIC. If you take the hydrodynamics codes and you try to reproduce various aspects of the collisions, you find that we may be at 2 times the significant temperature at the very initial stage and then it cools down after that. It spends time through, you know, these lower temperatures ... and then it cools down...In the end we see the results of the quark-hadron phase transition and so, here it starts to develop a stronger alpha_s. You don't expect that we are really down here (small alpha_s at high temperature) at all, and that's why I added this (line on the figure where chiral symmetry may be restored) ... which is a whole other question about chiral symmetry. We haven't been able to see it, and one of the early questions asked earlier today - how would one look for chiral symmetery, someone said look for chiral partners. If the chiral vector gets disoriented then one would expect very different ratios of neutral to charged pions or something like that. At RHIC we looked for those things, that was predicted for 10 years or so. We have not seen them yet. It may be that we are not quite up that temperature, so this is something we might look forward to at the LHC, with heavy ions, where one would expect at least to get a factor of 2 higher temperatures.

-Preghenella:

You said RHIC matter is a quark-gluon fluid with deconfined quarks by using flow arguments. Did you measure J/psi supression or funny behaviour?

-Harris:

This is a good question... I read here two questions: Have we looked at J/psi suppression? Is there Debye-screening and is there deconfinement which one associates with Debye-screening of the color charge? The Phenix experiment which was designed for leptons and photons, has just last year measured J/psi with good statistics, and in fact I'm going to talk about that with their results; supression like that is seen at the SPS with low energies. For those who are unaware of this, I'm jumping to the next lecture. Let me give a short overview, and I'll show you the data next lecture. The point is like this. If you have a gas of quarks and gluons, and you make a ccbar pair, a J/psi for example, and then the ccbar pair gets Debye screened because they are in the middle of a medium with all other color charges, and you make less J/psi than you would if you had a hadronic or e+e- collision. They get, what we call, screened. Phenix, which I'll show the data for next time has measured that. There's an issue though, in that there are a lot of different factors, that it is not as simple as I've just said. Because there are nuclear effects that actually attenuate the ccbar making J/psi and before I say anything more, there's also possible ...thermally producing a ccbar, which would be more applicable I would say at the high energy of the LHC, but it might be already coming in at RHIC. It has been measured for the first time at RHIC last year, I'll show you the data next time.

-Noferini:

In the v_2 plot there is a discrepancy between data and hydro-predictions. I expect that hydro cannot be correct in that regime because hard scattering dominates.

-Harris:

We tried to understand it, and yes, we only expect that the hydro fits in lower p_t. Remember I said it looks like it fits, this, let's call it, mass structuring, as a function of p_t of the v_2 out to 1 GeV/c or so. Now you asked the question - why do we look here, where it is dramatically different as we go up in transverse momentum? Well, first of all we expect that hydro fails here. What I try to do is divide the conversation into what happens to the bulk (today), and then the day after tomorrow I'm going to say-ok, we're going to use these high transverse momentum particles up here above 2 GeV/c, which are less than one percent and they are usually products of hard scattering (via fragmentation): we are going to use these hard scattered partons, and their eventual hadron distributions, to actually probe THIS stuff. That's the next lecture. So, as the question asked, why do we plot this, it's a part of the calculation, we don't expect it to work out above 1 or 1.2 GeV/c.

-Noferini:

Anyway, there is a strong effect of flow. What is the explanation?

-Harris:

That's an even more interesting question. If you are considering only this part up here (above 2 GeV/c), so what happens is that there is an anisotropy, not as strong as hydro, but you don't expect hydro to be working up here. So for practical purposes, maybe we shouldn't plot this, but I'll show you tomorrow that if you have 2 hard scattered partons, it's 2 body kinematics, and then they fragment and we try to study it as a mutual distribution, because if those partons didn't interact in the bulk of this stuff, then they would come out back to back. And in fact we find that they don't, (I'm giving away the whole Jet-quenching) the away side parton gets quenched. In fact it disappears if you look for high momentum products at the away side. But if you now reduce and go down to the lower transverse momenta on the away side, then you actually see it come back. But it's spread out over more particles on the way side. So we are understanding that, but this actually has some effect on the v_2 distributions out here in this region (above 2 GeV/c). And this gets more complicated and into the details. When you make particles out here (2 7 GeV/c range), it's still not clear whether they are not just fragments of hard scattered partons. But it appears that up to 7 or 8 GeV/c there is a recombination of a hard scattered parton let's say a quark moving fast through the medium, and it picks up a slow quark which is sitting in the medium. And there are these various influences that can give you some azimuthal anisotropy out here.

-Csanad:

How can we interpret the fact that hydro scaling stops right there (2 GeV/c) where suppression starts – i.e. they compete with each other?

-Harris:

I think it's fortuitous in the sense that hydro works at small p_t and hard scattering and fragmentation works at high p_t and there is this intermediate range which has various things like I just said – recombination etc. In fact we're still waiting for a reasonable dynamical model to predict that.

-Csanad:

But this suppression starts already at lower p_t…

-Harris:

Yes. So you're saying that suppression – I'm still giving away the next lecture – that the strong coupling affects the partons that are going through dramatically lowering there momenta and even if they were out here (high p_t), and they fragment and make hadrons down here (low p_t).

-Arcelli:

You mention that RHIC experiments would like also to explore the QCD phase diagram in the region of the tri-critical point. I would like to know some more details on the experimental strategy to do that and which are the observables which would indicate the presence of a critical point.

-Harris:

To get out here – high in baryon chemical potential – in fact you have to lower the beam energy because at small baryon density at RHIC and LHC you're swamped by particle production, pair production. So in order to get to this phase (at the tri-critical point) you need some finite chemical potential or finite baryon number and you go down in energy, probably near the former SPS energies of 10 GeV center of mass energy. And that's the point of the accelerator running basically at injection energy for both ions and then the experiments will go and measure things that are predicted near a tri-critical point - for example fluctuations will change dramatically – because this (on the right in the diagram) is a first order phase transition, this (on the left) is a cross-over supposedly – perhaps various ratios of K to π will also change dramatically – I think the fluctuation issue is the real one to look for, you get fluctuations in all the observables and see if it changes in the same experimental setup as you go down in energy (to higher baryon chemical potentials).

-Susskind:

I'm wondering if anything is known about the electrical properties of the quark gluon plasma, in particular its conductivity?

-Harris:

There are a lot of theoretical approaches, but experimentally not, electromagnetic I would say is pretty weak compared to color and so...

-Susskind:

There is a black hole type prediction for electrical conductivity which is one of the oldest black hole predictions and I'm wondering if you have any idea whether that can be tested.

-Harris:

That's a good question, I should think about it. I haven't thought about the electrical properties. In general, I should say, almost as a disclaimer, we need, over the next years to measure the properties of whatever it is – and that includes the diffusion coefficients, the color conductivity, all the transport coefficients, and so electrical conductivity is a new one for me to think about.

-Abramowicz:

What is the relation between centrality and transverse momenta of particles, are those two sifferent parameters in your considerations?

-Harris:

We try to keep them independent, but I have to say that if you trigger on a high p_t hadron you tend to favor events that are more central, we've studied our trigger biases and that in fact is the case. What we do in terms of transverse momentum is we do have a special set of triggers that triggers on high transverse momentum and then we really have to see what bias that gives us for the regular distribution of triggers. There is a bias towards higher centrality.

-Abramowicz:

So the azimuthal flow you showed is also dependent on the centrality? So these are central events you showed?

-Harris:

I showed three centrality bins and it (the magnitude of azimuthal flow) changes with centrality and that centrality is just the geometrical trigger that's based on Glauber geometry.

CHAIRMAN: J.W. HARRIS

Scientific Secretaries: O. Catu, B. Keren-Zur

DISCUSSION II

-Parihar:

What are the total cross-sections in the Au-Au collisions?

-Harris:

The collider people measured it, because all the experiments have the same zero degree counters which measure the total cross-section. It is somewhere around 7 barns for Au + Au.

-Lopez-Mateos:

In the plot where you show the QCD phase-space diagram with the points probed by AGS, SPS, etc. there are also what seem to be predictions of where the tri-critical point is. Do we know whether there is a tri-critical point at all?

-Harris:

Those are not predictions for the tri-critical point, those are just predictions for where the phase transition would be as you go away from the mu (baryo-chemical potential) = 0 axis. This is the mu = 0 axis, this is Karsch's lattice paper of '04 and this is Fodor Katz in '02 and there's no calculation for the location of the tri-critical point. It could be anywhere in between and the hope is that if RHIC is going to get there it has to be closer this way (towards smaller mu) than that way (at larger mu).

-Pacetti:

Do you see vector meson broadening? Could you make some comments about this effect?

-Harris:

That's a good question that has to do with chiral symmetry restoration that hasn't been mentioned here. There was a question a couple of days ago about how would one see it, and one possibility (although it's still being debated a lot) is that the resonances like the omega, rho, phi, will possibly have shifted masses or be broadened. There were measurements at the SPS where they saw low mass electron pairs, from decays of low mass vector mesons, and it was very broad. There was a flurry of papers from Gerry Brown (and all these guys) about broadening and chiral symmetry restoration, not totally settled though. There has not yet been such an experiment at RHIC. The PHENIX experiment plans to put in a hadron-blind detector to look at e+ e- from low mass vector meson decays, and in fact will investigate whether such masses are broadened or shifted.

-Noferini:

About same-side charge correlation in two particle distributions of jets, if I understood correctly you said that a possible explanation is the "charge ordering" in the jet fragmentation. Is it possible that also the products of the decay of the resonances affect this measurement?

-Harris:

This correlation is not necessarily due to charge ordering. It's just the jet. Forget about the charge. It's there whether it is ++/--/+-. It's stronger for +- correlations by a factor of ~ 2.5 than for ++ or --. So there is charge ordering. But it's the whole jet, regardless of the charge, that makes this same-side peak after you substract off the long range correlations. Your question, if I understood correctly, is whether neutral resonances could cause that. If you're thinking of the +- correlations, the neutral resonances could do it, but if you take ++ or -- and you still see that peak, then you have trouble explaining that with decays of resonances (which would require the resonances to be doubly-charged for two-body decays). Now, perhaps you're asking: maybe the enhancement for +- is due to resonance decays. My understanding from the LEP data is that it's not. It is consistent with string breaking.

-Csanad:

Charm flow was measured and it is important.

-Harris:

I want to clarify – not charmonium flow, but open charm flow. The flow was measured by PHENIX. STAR did a measurement. We thought we saw it, but we didn't publish it because it wasn't statistically significant, and we're still working on it.

-Csanad:

The critical point is being measured. We need to go down in energy for that.

-Harris:

RHIC ran as low in energy as at the SPS energy, at 10 GeV, for one day. We looked at it. This sort of real search, if there's a tri-critical point close enough down in mu (baryo-chemical potential), would be at low energy and we haven't really run so much there yet. It is anticipated that there will be a dedicated RHIC run for this search at some point.

-Csanad:

Jets: A Mach-cone like structure is observed by PHENIX, and may help to measure speed of sound.

-Harris:

I did show it (the PHENIX result), but probably too quickly… I said that if you go from proton-proton collisions to very central gold on gold and you require the associated particles away from the trigger particle to have greater than 2 GeV/c, then you do not see the back-to-back correlation. But if you lower the associated particle momentum down to

200 MeV/c, then you see a structure that has a broad shape, and there are all kinds of discussions about color-wakes, Cherenkov radiation and Mach cones. The next slide is in fact the PHENIX data that you referred to for different centralities, where the most peripheral is in blue here, where the away-side looks like a broadened peak, and as they go more central they pick up a dip in what was the middle of the away-side jet. This was one of the things that started people thinking about some Mach cone structure on the away-side. In STAR we can't really say that we have a dip there, as you can see. When I talked about AdS5 /CFT, I said that they're also trying to address this sort of thing now, i.e. the response of the medium.

-Lopez-Mateos:

If you look at single events showing jet quenching, is the back to back correlation still present? And if not, does that say anything about the multiple scattering inside the medium?

-Harris:

For single events – when you make a cut – we look at events over a range of impact parameters, we can put them into categories of impact parameter - we first look at any event that has a particle with p_t greater than 4 GeV/c in this particular correlation and then we look in that event and correlate any other particles with that and those particles must have – in this analysis – p_t greater than 2GeV. And then you see jety-like event because all that other stuff is soft. And then we take events in the smallest impact parameter and put them all together and make the correlation – but it's event by event analysis. Now you ask: can we learn anything about the multiple scattering in the medium of the parton that is going on the other side, was that it?

-Lopez-Mateos:

Right, I was wondering if whether in a single event instead of exactly pi the parton would come out at say 3 radians?

-Harris:

It's a little complicated to do single events because you have statistical fluctuations and we only measure charged particles in this type of analysis while there are also neutrals. If you have EM and hadronic calorimetry together then you might be able to do that but still the fluctuations of small numbers of particles will always get you there. But to address the question I thought I heard – when you look on the away-side, I think that's going to be - I don't want to call it an industry - but something that the theorists and the experimentalists will look at very closely over the next few years and make comparisons. Because on the away-side you should be able to learn about the medium and its response. You know, we don't know what makes the broadening on the away-side. Whether it is just gluon bremsstrahlung, I don't think so, or scattering in the medium maybe for the heavy quarks or something else.... but the theorists are talking about Cherenkov radiation, Mach cones etc., those things can tell you about the medium itself, the velocity of sound in the medium, the diffusion coefficient, etc. So I think that's the approach we'll be taking over the next 5 years, both at RHIC and ALICE, study the response of the medium to high momentum parton probes.

-Neri:

What were the experimental techniques used to distinguish between signals from b quarks and from c quarks at RHIC? And what is the rate and efficiency?

-Harris:

We didn't distinguish between b and c quarks. So what we did is we looked at non-photonic electrons and there we cannot distinguish b from c. We subtract all the known sources for electrons and after so doing we're left with electrons that come from both B and D decays – we don't distinguish – it's really incumbent upon us experimentally to do that over the next 5 years but that will require both STAR and PHENIX to have better vertex detectors – there we would then go and distinguish the short lifetimes ($c\tau$) of 100-300 microns and look at the hadronic decay modes via displaced secondary vertices to distinguish D and B, or charm and beauty.

-Bourjaily:

How can ALICE complement RHIC? In particular about the measurement of the multiple scattering, you mentioned that you could also measure neutral particles with a hadronic calorimeter – is that something that ALICE is going to be able to do?

-Harris:

So the question is how will ALICE complement all of this – I could give a lecture on that. Point to note: at least the US contingent hopes to be able to provide an electromagnetic calorimeter for a reasonably large solid angle to measure jets. We – the US – are coming in somewhat late and we're just getting approval for funding for such a device and we're hoping to have it installed each of the next years by sometime in 2009. The point is that one can already see from this figure that the cross-section as a function of p_t is for hadrons themselves and for jets is considerably larger at LHC than at RHIC because of the higher energy, of course. Likewise, the production of b's at LHC is about a factor of 100 higher than RHIC – remember the center of mass energy is about 25 times higher at LHC for heavy ions than at RHIC. So we're talking at LHC of 5.5 TeV/nucleon for Pb-Pb as compared to 200 GeV/nucleon for Au-Au at RHIC. And then the charm cross-section is about a factor of 10 higher than at RHIC because it's already starting to come up at RHIC. And so the really interesting thing is to trigger on a photon and study the away-side jet, as when you have a quark-gluon scattering going to a gamma and jet in the final state. The photon will give you the energy of the hard scattering because it doesn't interact in the medium and then you can really start to understand what's happening on the away-side – the photon will give you the direction precisely and the energy and then you look at the jets on the other side, you chose maybe b-tagged jets to really get the quark component since most of it at the LHC in fact is gluon scattering.

-'t Hooft:

I have just a comment: at the end of your lecture you made some connection with string theory and black holes and you were thinking about what all these theories would have to say about the quark-gluon plasma But I'm interested in the converse, which is: the quark-gluon plasma results could be useful for other physics, from Black holes to string theory. It may be in the limit that you have ions that collide at infinite energy, then the Lorentz contraction is complete and the interaction takes place at one undefined moment in time, which means that you're now talking about only a 2-dimentional theory, only the transverse coordinates, since the z coordinate and the time coordinate essentially disappeared in the limit and that is very interesting from the point of view of Black holes, where at the horizon where entering particles enter at infinite velocities, outgoing ones enter with also infinite blue-shift there, they will be red-shifted in the very end, so that kind of physics is important at the Black hole horizon. So what I'm wondering about is whether you or anybody else knows about a theoretical investigation about this limiting case where you have just a 2-dimentional theory, where operators just live on a surface, and refer to ingoing as well as outgoing states, which is an aspect of Black hole physics I'm interested in. This should have been studied by people who are interested in quark-gluon plasma, because they have nearly the same situation there.

-Harris:

This is a great question from our Chair and let me say two things; one is that when you get to those infinite energies you're at very small x and you just have this – whatever the most saturated gluon states are – colliding, so that's one thing to keep in mind. What's really going on, whether it's this color glass condensate of gluons or not, I don't know, but you're really doing that. And the other thing is – to be able to use this string theory, the 't Hooft coupling or g^2m – at RHIC is already 10, which is fairly large. I think it only gets larger, so you can do this sort of thing, but I don't know of anyone working on this, this is the first I've heard of it, you're saying it.

-Sonnenschein:

Just a short comment: one shouldn't expect to get information about ordinary Black holes, because they are talking about 5-dimensional AdS Black holes, which are in nature quite different than the shifted Black holes, but this idea of holography is clearly a very essential ingredient in relating the string and the gauge side.

-Pinfold:

It just occurred to me that you get very high energy heavy ion collisions in cosmic rays, so maybe those people have thought about these topics, rather than experimental heavy ion people who work at much lower energy. The ultra-high energy cosmic ray people doing studies there would consider much higher energy particles.

INTERNATIONAL LINEAR COLLIDER

Nigel S. Lockyer
TRIUMF/UBC, Vancouver, BC, Canada

Abstract

The International Linear Collider (ILC) is a proposed electron-positron collider that will explore the Terascale. The ILC can be characterized as a precision machine that will study the discoveries made at the Large Hadron Collider (LHC), at CERN. These two giant machines are highly complementary. The concept of Electro-weak symmetry breaking and the Higgs Boson are the Holy Grail of particle physics. It is expected that the Higgs Boson will be discovered at the LHC. The ILC will be a superb instrument to measure precisely the properties of the Higgs Boson and thus provide a discovery window into the physics beyond the Standard Model. Even greater discoveries are anticipated at the LHC, such as Super-symmetry and evidence for extra dimensions of space. The basic configuration of the accelerator is presented. The ILC accelerator R&D program developing around the world is discussed.

Physics Motivation for the ILC

Particle physics is preparing to answer a series of powerfully incisive questions. If successful, the information will help physicists illuminate our understanding of the world around us. Key to our present understanding of the universe, both the large and short distance scales, has been the success of the Standard Model of particle physics. This picture, although it describes much of what we observe and measure in the laboratory, is obviously incomplete in a number of important ways. The physicists' key questions illustrate the incompleteness in our understanding. What is the origin of mass for the intermediate vector bosons and the elementary particles? What is the dark matter? Are there extra dimensions of space? Are all the forces of nature unified?

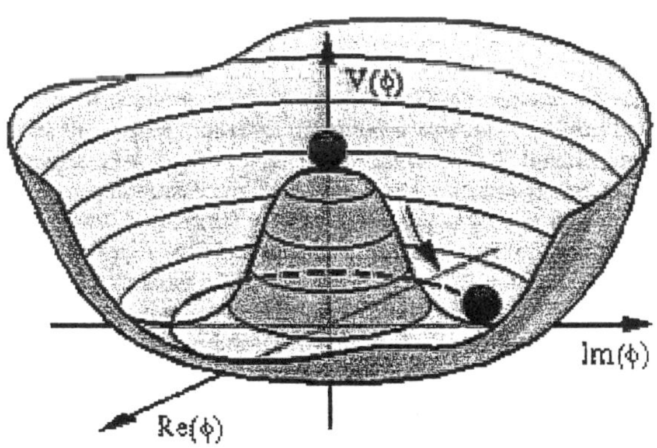

Figure 1. The famous "Mexican Hat" Higgs potential.

The International Linear Collider (ILC) will be about 30 kilometres long and consists of two long linear accelerators, aimed at each other, which collide electrons and positrons at 500 GeV in the centre of mass. The initial state created by the annihilation of the positron and electron is well understood and thus allows physicists to study with greater certainty the resulting particles and their behaviour than we can with accelerators such as the Large Hadron Collider (LHC), where the uncertainty in the quark and gluon momenta inside the colliding protons limits the precision of the experiments. Of course the two accelerators are complementary because the protons allow discoveries of particles that are produced by the strong interaction while the ILC allows discoveries associated with particles produced with the electromagnetic interaction. The Higgs potential illustrating Electro-weak symmetry breaking is shown in Figure 1.

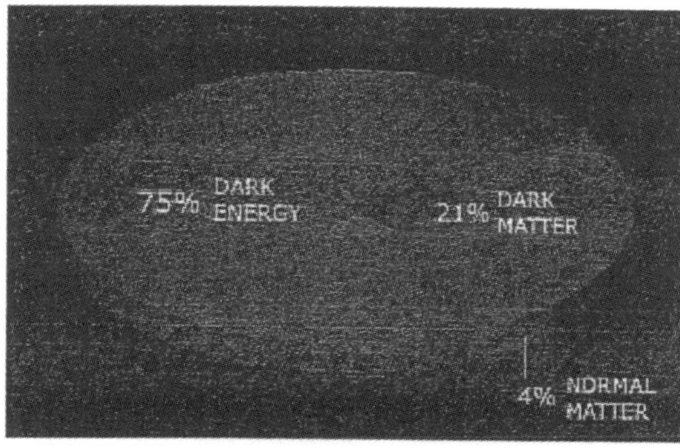

Figure 2. Energy fractions in the Universe.

The Holy Grail for particle physics, the first order of business at the LHC will be to discover the Higgs Boson. The Higgs field prefers to acquire a non-zero value to minimize its energy. During the earliest moments of the universe, around 10^{-12} after the Big Bang, the Higgs field suddenly filled space and caused the electroweak force to separate into two forces, the weak and electromagnetic force. At this point, the W and Z bosons acquire mass from their interactions with the Higgs field. Once the Higgs is discovered, the most important properties need to be measured. The coupling of the Higgs to the quarks can be determined at the ILC with great precision. In addition, the Standard Model predicts the values of these couplings will be proportional to the mass of the particles into which the Higgs will decay. It is crucial to know whether the Higgs is "standard" or is it different in some way. Models beyond the Standard Model will cause deviations from what is expected. For example, the Higgs may couple to the quarks with a different strength than that predicted by the Standard Model. The ILC is the ideal accelerator to measure the Higgs couplings to the quarks and therefore probe these interesting questions about possible new physics.

As can be seen in Figure 2, Dark Matter comprises about 21% and Dark Energy is nearly 75% of the energy in the universe. All the physics of the Standard Model has been focused on the 4% we see with our eyes aided by telescopes, which we refer to as normal matter. Particle physics may be able to help solve these mysteries. The calculation of the Higgs mass using Quantum Field Theory (QFT) results in a number which is much too high in value. The Large Electron Positron Collider (LEP) at CERN and the data from Fermilab's Tevatron strongly hint that the mass of the Higgs Boson is less than 192 GeV, much lower than the QFT calculation.. Some theoretical physicists believe that there must be new physics such that it "stabilizes" the mass of the Higgs Boson in the QFT calculations. One prospect is Super-symmetry

(SUSY), which should also be discovered at the LHC. Precision SUSY measurements will be done at the ILC. The LHC will measure the mass of the SUSY particles to about 10%, while the ILC will be able to measure their mass to 1% if they can be produced. An important point here is that the LHC, with its higher energy, will be able to produce particles that the ILC may not be able to measure precisely because they cannot be produced at the lower energy reach of the ILC. There is a "high precision" versus "energy reach" benefit ratio that needs to be optimized before the ILC is built. Once the energy scale of the new physics is understood at the LHC, the energy reach of the ILC will be clearer. The importance of mass precision cannot be overstated in this case. If SUSY is shown to exist in nature at the LHC, and the SUSY parameters realized also provide a dark matter candidate, then the dark matter candidate that will be produced at the LHC can only be interpreted as a new particle. There is not enough resolution to determine if the particle found is dark matter, however, the determination that this "new particle" is indeed the dark matter that corresponds to what is observed to influence the formation of galaxies, requires a combination of theoretical calculations on the dark matter relic abundance as well as high precision in measuring the mass. The ILC will be able to verify the "new particles" are dark matter. Perhaps the discovery of the century!

The unification of the four fundamental sources at high energy is an interesting idea, the kernel of which was pursued by Einstein for much of his career. Shown in Figure 3 is a graphic representation of the breaking apart of the forces in nature as the universe cools. The ILC will study with high precision, electroweak symmetry breaking, which occurs at the time period of about 1 pico-second after the Big Bang.

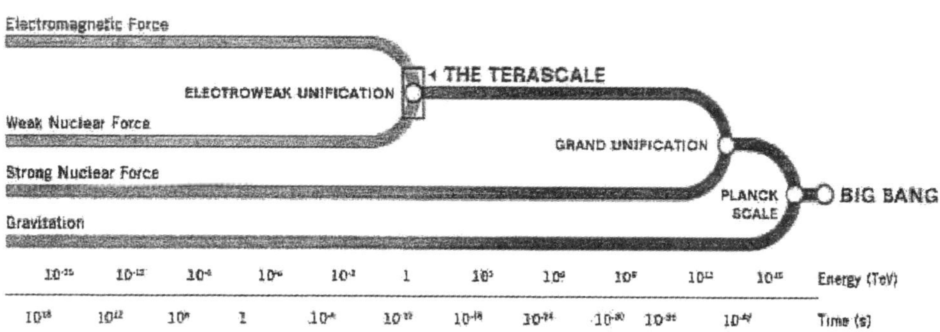

Figure 3 The evolution of the four forces as the universe cools.

ILC Machine Parameters

The main requirements for an e^+e^- collider that will be capable of addressing the physics goals was discussed by the world-wide particle physics community and endorsed by the ILC Steering Committee (ILCSC) [1]. The main parameters for the baseline ILC design are

- E_{cm}= 500 GeV adjustable from 200 – 500 GeV
- Luminosity → $\int L dt$ = 500 fb^{-1} in 4 years
- Energy stability and precision below 0.1%
- Electron polarization of at least 80%. (machine design allows for ~50% positron polarization.)
- The machine design should be upgradeable to 1 TeV

The ILC parameters are given in Table 1. The final operating point is almost certainly going to be within the space defined by these parameters, but will not necessarily correspond to anyone of them. The initial design luminosity of the collider is $2x10^{34}$ $cm^{-2}s^{-1}$.

	Min	Nom	Max	unit
Bunch Charge	1	2	2	X1e10
No of Bunches	1330	2820	5640	
Bunch interval	154	308	461	ns
Bunch length	150	300	500	μm
Vertical emit.	0.03	0.04	0.08	mm.mrad
IP β_x 500 GeV	10	21	21	μm
β_y	0.2	0.4	0.4	μm

Table 1. Baseline Parameters

Baseline Configuration

Since the ILC technology selection in August 2004, the Global Design Effort (GDE) has been working to define the baseline machine and needed R&D for the construction of the ILC. The main design issues were outlined at the first meeting held at KEK in November 2004 followed by baseline and alternate design and technology selection for each major subcomponent of the accelerator during the Snowmass meeting in August 2005:

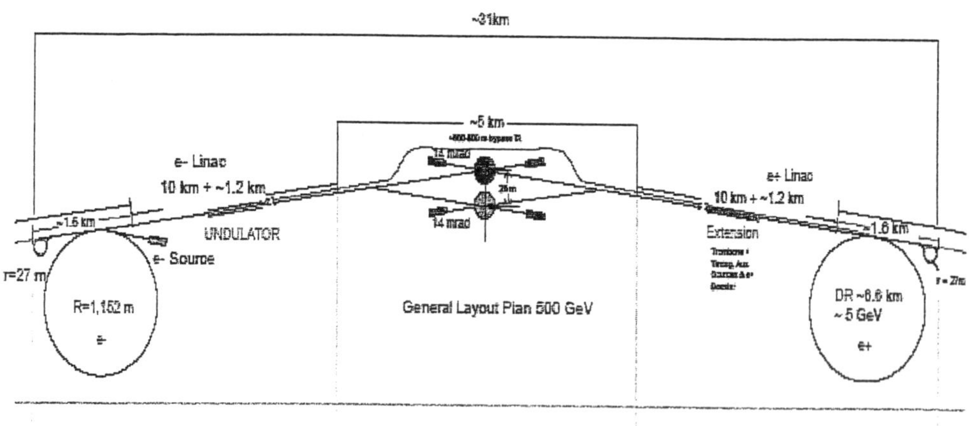

Figure 4. The initial ILC Baseline Configuration

The collaboration approved the ILC accelerator baseline configuration during its meeting at Frascati, Italy in December 2005. Since then considerable modifications to the baseline configuration have been made to optimize the design and reduce the cost of the collider. The current baseline configuration design is shown in Figure 4. The Baseline Configuration Document (BCD) [2] defines the machine parameters for a 0.5 TeV e^+e^- collider that could be upgraded to 1 TeV. The ILC-GDE has been working on developing the cost methodology and the cost of the baseline ILC. The ILC-GDE is developing a Reference Design Report (RDR) with cost estimate to be published in early 2007.

The electron source for the ILC will use conventional technology, a DC titanium-sapphire drive laser with 2 ns pulse. The long electron microbunches will be bunched with a two sub-harmonic buncher before injection into a room temperature linac and diagnostic section that will accelerates the beam to 100 MeV for injection into a standard ILC Main Linac section. The superconducting linac will accelerate the beam to 5 GeV for injection into a Damping Ring. Figure 5 shows the schematic layout of the Electron Source for the ILC.

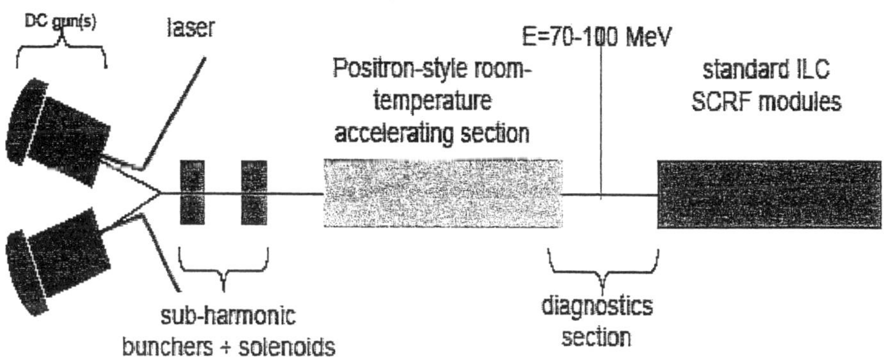

Figure 5. Schematic of the electron source

A helical undulator-based polarized positron source with a keep-alive system is the baseline design for the ILC, Figure 6. The 100-meter long undulator will be placed at the 150 GeV point of the main electron linac. This will allow for constant charge operation across the foreseen center-of-mass energy operation of the collider. The electron beam passing through the undulator generates circularly polarized photons. These photons interact with a 0.5 radiation length Ti target. The produced particles are captured and accelerated by a superconducting linac to 250 MeV. The positron is separated from the captured electrons by a magnetic chicane and further accelerated to 5 GeV using the standard Main Linac cryomodules. The positron beam is transported to the injection point of the Main Linac. The positron source also has an auxiliary e^- source for a "keep-alive" positron source at 10% of nominal intensity.

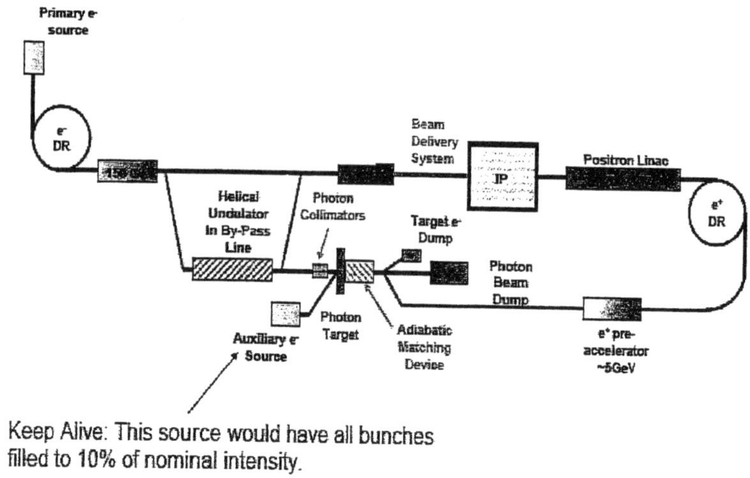

Figure 6. Schematic of the positron source

The ILC Damping Ring will have one ring for the each of the electron and positron beams at 5 GeV. The rings are ~6.6 km in circumference. The circumference is essentially driven by the kicker technology and electron-cloud effects. In the baseline configuration they are located at the two ends of Main Linac, but the exact location of the Damping Ring could be site dependent. The Damping Ring lattice is designed with TME cells and is wiggler dominated. The superconducting 650 MHz RF cavities have been selected for ease of injection and extraction. The wigglers and RF cavities are optimally located with beam dynamic considerations. Extensive accelerator physics studies have been carried out to study the dynamic aperture, vertical emittance tuning, space-charge, electron cloud and fast-ion effects, and preservation of the beam polarization. The calculated normalized emittance of the beam is 8x0.02 μm (H x V) with a bunch charge of about 2e10 electrons. The fill pattern in the accelerator has been optimized to reduce the size of the Damping Ring but requires development of very fast injection and extraction kickers. At present extensive R&D is being carried out on understanding Damping Ring beam stability and development of wigglers and kickers.

The cooled beam from the Damping Ring is transferred to Main Linac using a collection of beam lines called Ring to Main Linac (RTML). The RTML will be used to match the beam parameters between the Damping Ring and the Main Linac, compress the bunch length, compensate for the beam jitter by a feed forward system, collimate beam halo and transformation of beam polarization.

The ILC Main Linac will accelerate the cooled beam from 5 GeV to 250 GeV. It is designed to keep the beam emittance growth to minimum; the emittance budget for the whole linac is less than 100%. The linac is constructed with a basic building block called an ILC RF Unit. Figure 7 shows a schematic view of the ILC RF Unit.

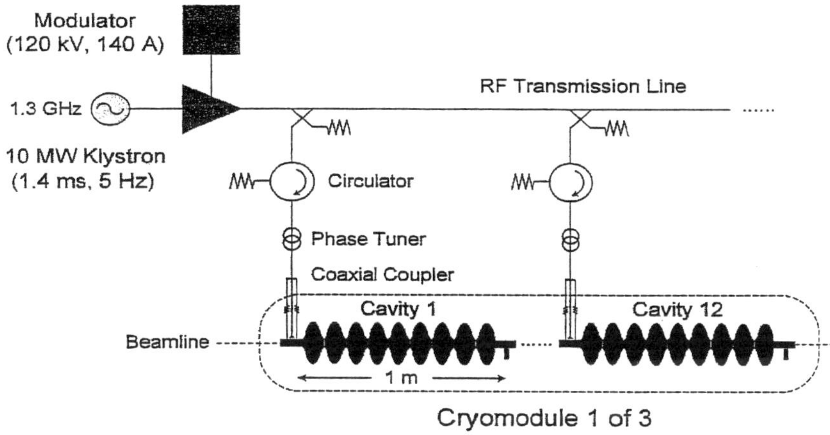

Figure 7. ILC Main Linac RF Unit

The ILC accelerating structures are superconducting cavities of the TESLA shape. The design accelerating gradient of these cavities are 35 MV/m at a Q = 1e10. A cryomodule is made up of a string of 8 cavities and an RF unit has 3 cryomodules. The middle cryomodule in an RF Unit will have a quadrupole, correction dipole and beam position monitor package. The RF power is injected into 24 cavities using a RF distribution system, a pair of 10 MWatt klystrons and a modulator. The 10 MWatt multi-beam klystrons being developed by Thales, CPI and Toshiba are the baseline choice. An alternate design using a sheet beam klystron technology is being developed by SLAC. The Fermilab designed modulator has been in operation at DESY for a decade and is being fabricated by industry. Recently Fermilab and SLAC have made modifications to this design to include more reliable pulse-transformer and the latest generation of high power transistors for the switches. SLAC is also developing a Marx modulator essentially aimed at cost reduction and modular operation. A high power input coupler (0.5 MW) based on the twin cylindrical window architecture couples the room temperature RF to the superconductive cavity. The distribution and control of RF power from one klystron to 24 cavities is done using a LLRF system. The central frequency of the cavity changes due to the variation in length during cooldown from room temperature to 2 deg K, from the RF (Lorentz Force Detuning), and the beam. The central frequency is controlled by use of a "slow" blade tuner connected to stepping motor and "fast" Piezo crystal tuners. The 250 GeV Linac of ILC will have about 330 such RF units operating at an average gradient of 31.5 MV/m at a Q=1e10.

The beam delivery system (BDS) receives the beams from the two linacs (250 GeV in Phase-1), collimates them, focuses them, and collides them at the interaction points and safely disposes of the exhaust beam at the beam dump.

The ILC conventional facilities and siting studies are being carried out with sample sites from each region. These sample sites have been selected so that a study of the construction of the ILC tunnels, conventional facilities and layout of accelerator elements can be completed in order to understand how distinctive site features affect the design, cost and performance. The sample sites will not necessarily be potential candidates for the final site. The US sample site is located near Fermilab, the European sample sites are located close to CERN and DESY and the Asian site is located in Japan. The details of the accelerator layout are still being developed by the three regions. Figure 8 shows the general layout of the twin tunnel design of the Main Linac.

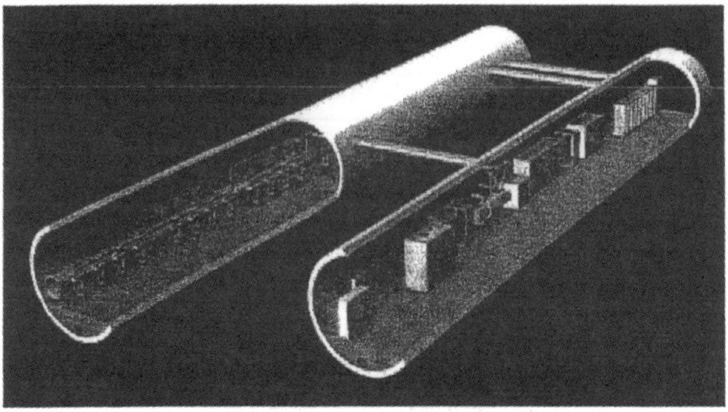

Figure 8. Conceptual design of main linac dual tunnel.

ILC R&D

Extensive R&D is being carried out by the ILC collaboration to address several crucial questions related with the baseline design of the ILC. The ILC R&D Board has compiled a list of major R&D topics [3] and prioritized them. In order to address the most crucial R&D topics the ILC-GDE has formed several task forces for Main Linac and Damping Ring R&D. The detailed goals and plans for these task forces are still being developed but the current focus is on following topics

- Develop cavity processing parameters for reproducible cavity gradients of 35 MV/m; improve the yield of 9-cell cavities for gradients of 35 MV/m in vertical tests. Carry out parallel/coupled R&D on cavity material, fabrication, and processing to identify paths to success.
- Assemble and test one or more cryomodules with average gradient > 31.5 MV/m.
- Build and test one or more ILC RF units with a beam operating with ILC parameters, high gradient, and full pulse repetition rate.
- Develop plans for an ILC Main Linac system test consisting of several RF units.
- Develop a plan for a coordinated R&D for the Damping Ring with specific goals and milestones.

The reliable demonstration of the ILC Main Linac accelerating gradient of 35 MV/m remains the highest priority for the ILC. Considerable progress is being made towards achieving this goal in all the three regions of the world with activities centered at DESY, KEK, JLAB, Cornell, and Fermilab. Figure 9 shows the compilation of the recent high gradient cavity performance results from DESY. These cavities have gone through standard surface processing and vertical testing. DESY has proceeded to use eight of these cavities to build a cryomodule to be tested soon in TESLA Test Facility (TTF/FLASH). It is expected the cryomdoule will reach an average gradient of 31.5 MV/m.

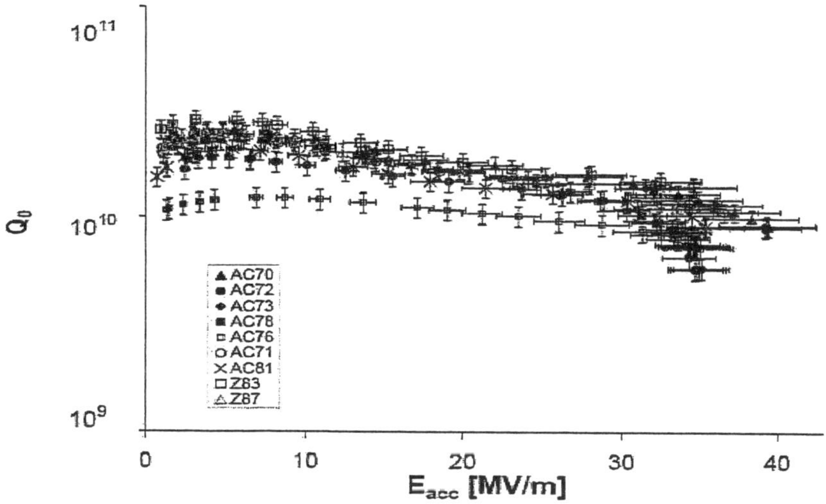

Figure 9. Low Power test results of DESY cavities.

The basic technology for achieving the accelerating gradient exists; the R&D needs to focus on reproducibility and increasing the yield. Alternate cavity shapes, i.e. Low Loss and Reentrant shapes, have achieved gradients of about 50 MV/m in single-cell test. Material R&D with large grain and single crystal niobium is showing good surface processing results.

Summary

The ILC is the internationally agreed upon next major accelerator. The main particle physics study groups from the three regions of the world, Asia, the Americas, and Europe have identified the ILC as the next major project in particle physics. The machine construction, however, will have to wait for the physics discoveries expected at the LHC. Knowledge of the physics discoveries at the LHC will allow machine designers to optimize the ILC parameters. Discovery of the Higgs Boson at LHC will trigger debate as to whether one can start construction of the ILC. The ILC as a "Higgs Factory" is very appealing. The Higgs would be the first fundamental scalar discovered and would be a very interesting object to study. Observation of new physics, either SUSY or something even more exotic, at low energy, will be the undisputed trigger to push hard for the commencement of ILC construction. Observation of new physics above one TeV will spark debate about a higher energy design for the ILC. Either way, the future is going to be interesting.

References

[1] ICFA Subcommittee Report, "Parameters for the Linear Collider", September 2003, http://www.fnal.gov/directorate/icfa/LC_parameters.pdf

[2] ILC Baseline Configuration Document, July 2006 http://www.linearcollider.org/wiki/lib/exe/fetch.php?cache=cache&media=bcd%3Abcd.13jul2006.pdf

[3] ILC R&D Topic List
http://www.linearcollider.org/wiki/lib/exe/fetch.php?cache=cache&media=rdb%3Ardb_external%3Ard_master20060407.xls

[4] Cavity and Cryomodule R&D Priorities and Plans at Fermilab, http://ilc-dms.fnal.gov/Workgroups/Presentations/ILC%20Presentation%20Shekhar%20Mishra/Cavity%20and%20Cryomodule%20RnD%20Priorities%20and%20Plans%20at%20Fermilab_.doc

Acknowledgements

I would like to thank Professor Antonino Zichichi for his very kind hospitality and a stimulating environment for discussion with students and distinguished scientists. I would also like to thank my colleague Shekhar Mishra from Fermilab for help preparing this paper.

CHAIRMAN: N. LOCKYER

Scientific Secretaries: G. Brandt, O. Smith

DISCUSSION I

-Neri:

Regarding the possibility to perform time dependent measurements using CP eigenstates like $B_s \rightarrow J/\psi,\phi$. What is the sensitivity you expect using the whole data sample and what will be the impact on supersymmetric models?

-Lockyer:

The decay of B into $J/\psi,\phi$ is one of the more interesting modes for us to pursue. We published a paper in 2000 (that predated the B factories) which was a 2σ measurement of $J/\psi,K_s$. The analysis of these two decay modes is very similar (phi of course is spin one), and this is just a matter of us accumulating enough statistics in order to make that measurement.
The expectation is not to observe an asymmetry to $J/\psi,\phi$, it has zero CP. Any observed asymmetry will give us indication of new sources of CP violation. This measurement is definitely high on our list, and now that the 'same side' tagging is working well, there is a good chance of doing it in the next couple of years.

-Lopez-Mateos:

What is the meaning of the disagreement between the top mass measurements in different channels and what will happen once the statistics get better?

-Lockyer:

For the record I first want to say that we are not in disagreement with ourselves or with D0, the error bars are sufficiently large. What you would like to see is a difference in the cross section between, for example, the dileptons and the lepton+jets. That would signify that something interesting is probably going on.

I would guess it is going to take a long time for us to get to the point where we will observe a significant difference. There is a slight trend towards higher cross sections in the dileptons and one cannot ignore it. We will continue to work on it. The hard thing is understanding the background. The processes in the background we do not understand are not modeled in our data, and then we have to rely on Monte Carlo which is not sufficiently accurate. I do not see us making any statements regarding this issue in the foreseeable future with that approach. If we were going to see something it would have happened right away and it would have to have been quite significant. I think this applies for the top mass measurements as well. They are all roughly within the same significance and so it is hard to separate them.

-Brandt:

I guess the reach in Q^2 at LHC and TeVatron is higher than at HERA. The quark radius at HERA was measured to be < 1fm. Can, and if yes how, it be measured in $p\bar{p}/pp$ and what is the expected limit at LHC / Tevatron?

-Lockyer:

The TeVatron is probing $Q^2 \sim 10^6 GeV^2$, I would guess HERA is around $Q^2 \sim 10^4 GeV^2$ and the LHC will be around $Q^2 \sim 10^8 GeV^2$. So yes the reach should be higher and this could be interpreted in terms of the quark radius but I'm not sure the measurement of the quark radius was done.

-Lopez-Mateos:

In the B_s oscillation measurements, why don't we see a perfect sinusoidal?

-Lockyer:

We have not exactly fit the oscillation; we used a method established by LEP, an amplitude method, so that we can compare with them when we set the limits. What you would like to see is amplitude vs. time. This is what we will do in the near future. Afterwards you will be able to judge for yourself if it is sinusoidal or not. I expect it to be completely sinusoidal. There is no reason for it not to be. It is an obvious thing to do.

-Sens:

I noticed in your talk that you concentrated on decays of B to D. Why did you not measure the decay $B \rightarrow K\pi$ as well? *(written addendum: For ~40 years CP-violation has been limited to K^0 decays. Although the c and b quarks have been discovered in 1974 resp. 1977 no further evidence for CP violation has thus far been found, not in $B^0 \rightarrow D$ nor in B^0_d nor in B^0_s nor in $B^{+-} \rightarrow K\pi$. Is this a correct assessment of the situation now, Sept. 2006?)*

-Lockyer:

CDF has measured the decay to $B_d^0 \rightarrow K^+\pi^-$. It is consistent with zero, the errors are too large. We hope to repeat this measurement for the B_s. The modes have been found, but we have not gone through the process of measuring CP. We are looking for CP in direct modes such as the $B_s \rightarrow K\pi$, and the asymmetry is expected to be large, maybe ~30% but with large theoretical uncertainty. The trigger I have described was really designed to pick up those modes. Two body modes are strengths of CDF in some sense; we should be able to do those.

-Sens:

That is why I was surprised to see that you have done the B→D mode and not the simple two-decay modes.

-Lockyer:

The effect of B_s mixing was expected to be large in the decay to $D_s\pi$ and that is in fact what we found. It is consistent with SM predictions. Whereas for the other modes, especially the charged modes, the theoretical predictions are, I think, poor, and so less reliable. People are looking into those decays, but they have not received the same effort. Now that we had all the machinery in place to make B→D measurements (which was a huge effort) it will be possible to do other ones as well.

-Pinfold:

In Run 1 data the *top* dilepton channel showed an excess of high missing energy events. Is there any evidence for such an excess in the latest data?

-Lockyer:

In Run 1 in the discovery period of CDF we had about 11 dilepton events and there was an anomalously high number of $e\mu$ events in that sample, and those events have an unusually high missing E_T as well. That was one of the first things we looked at in Run 2. The first sample we examined had around 20 events, all of them had a distribution that fitted Standard Model predictions. We never got those high missing E_T events we were hoping to see in that region. We now have around 55 events and it still looks normal.

There is another thing that looks unusual, though: In the tri-lepton search, there are 8-9 analyses ($eeX, \mu\mu X...$). The 'like sign' sample that looked for $\mu^+\mu^+, \mu^-\mu^-$ and a third lepton can go to very low lepton momentum for the third lepton. We examine the P_T of the leading lepton of those events, 4 in number, and they do not fit well to the SM predictions. People did notice it as being anomalous but we can not make any claims on the basis of four events. In general people like to believe the *top* quark is going to be different. We have around 100 top events, in the lepton+jets channel, (in the so called *top* factories) and most of the papers we publish show the kinematic distributions, so people can see for themselves if they agree with the SM, but we need a larger sample in order to push this approach i.e. looking for kinematic deviations from the Standard Model, for the discovery of new physics.

-Lopez-Mateos:

About the *top* mass measurement: the shape in the di-lepton channel was a nice Gaussian. What shape is the curve fitted to the all-jets final state distribution for the measurement of the *top* mass?

-Lockyer:

Shown in the plot is a set of templates. They are shapes for various *top* masses expected from the Monte Carlo, including full detector simulation etc. You fit your data to the template that fits best. This is called the template method.

-Lopez-Mateos:

So the deviation from a Gaussian is mostly detector effects?

-Lockyer:

Sure! The *top* mass peak itself is extremely narrow and is a Breit-Wigner, but we see a Gaussian from detector effects.

-Goldstein:

A follow-up question to that: Is the normalization of that template arbitrary?

-Lockyer:

The template is just a shape, it is normalized to number of events,---we simply fit for the fraction of signal and background. *A comment to the Higgs search at TeVatron:* CDF must first observe *WZ* production where $W \to l\nu$ and $Z \to bb$. This just like *WH*, with the same decay modes. We are close to seeing *WZ* decay into leptons, and I believe D0 has seen a signal.

CHAIRMAN: N. LOCKYER

Scientific Secretaries: M. Garbini, R. Preghenella

DISCUSSION II

- Zichichi:

Before asking for vital questions I would like to call your attention on the fact that in the following years, LHC time, we will see if the complexity at the fundamental level is really there because if complexity exists, as I claimed it should exist at the fundamental level, the UEEC (Unexpected Event with Enormous Consequences) event should be discovered at LHC. Unexpected Event with Enormous Consequences which means nothing of what we know is going to happen, is going to be discovered at the LHC. If this is true then what Professor Lockyer has presented to us, in this very interesting lecture, is the analytic continuation of what we know, which we do need to take into account. But CLIC which is the same project on e^+-e^- collider is related to the UEEC, something spectacularly new totally unexpected, so the following years will be extremely exiting to see if this is going to happen or not.

- Susskind:

Let me tell you about the concern that keeps coming up among some of us at Stanford in particular, I don't know if it is a question but it is a really concern, the original motivation for Supersymmetry which incidentally is due to Olsen was the Gauge hierarchy, it's difficult to stabilize the Higgs boson, etc. and etc., and that Supersymmetry is an ingredient which keeps the Higgs boson light. Suppose that there is something wrong with that logic and for various reasons at Stanford we keep thinking that there is something wrong with that logic, then the original argument that Supersymmetry should be at the TeV level, goes down a drain. But there is another argument about Supersymmetry and it's the unification of coupling constants, and it's hard to get away from, and so you can ask how high can the Supersymmetry breaking scale be and still unify coupling constants. The answer tends to be something like about 10 TeV, so the question I'm asking is if the Supersymmetry is breaking scale somehow manages to be, somewhere in the neighbourhood 8, 10, 6 TeV, low enough that unification can still take place but outside the direct range of LHC how is ILC a useful and interesting instrument to probe this range, in particular with high precision study of the Higgs?

- Lockyer:

I would have to think about that. My first reaction to this is it will depend on what we are probing. I said we can probe new physics of the type of Z' up to 5 TeV so there are certain things you can look at. I think this is just the type of question you have to go after and really see what you can do. I said earlier this is not the energy I would do if I could do it. I think you would like something higher but if you talk to accelerator physicists, which I'm not, the first thing they say is if you do ask to build the 3 TeV machine now they wouldn't do it because you got to go in steps, so there are certain limits to how far you can extrapolate, so 1 TeV machine is believable by everybody and maybe even 1.5 TeV. But I don't think anybody in the room says "we can build the 3 TeV machine with the technology we can make it work because you have to extrapolate too far from where you are".

- *Susskind:*

The question is when a TeV machine is used to study the high precision studies of Higgs would it be useful in studying a Supersymmetry breaking scale at 5 TeV?

- *Lockyer:*

If the Higgs scalar mixes with the SUSY sector, that is, does the lowest lying Higgs in that higher mass SUSY group mix with the low mass Higgs I'm studying at the ILC then the answer is it should, and so the new result is I have to look and see how much mixing there is and so on. I think the mixing of the scalar Higgs, which of course Professor Susskind says must exist, with new scalars, which exist in almost all extensions of the standard model, is a powerful tool to probe at the ILC.

- *Susskind:*

It seems to me that this should be studied.

-*Lockyer:*

I agree completely.

- *Zichichi:*

In answer to Professor Susskind's remark I like to recall that the 20 TeV range for Supersymmetry is the result of high precision study of the renormalization group equations ignoring the mass running. We have done 1991 the most precise study of the renormalization group equations using the running of the masses and we have shifted the limit by three orders of magnitude so the 20 TeV is totally out of question, is wrong as a prediction coming from the unification. So this is in support of professor Lockyer studies but if the range was 20 TeV the answer is no.

- *Susskind:*

What about 5 TeV?

- *Zichichi:*

No we have shown that the Fermi scale should be degenerate with the Supersymmetry scale. It is perfectly consistent, in fact it is favoured, of course assuming something at this scale must happen.

- *Wenninger:*

I just want to comment what you said and explain it was obviously the basic idea of Bjorn Wiik proposing the TESLA technology developed at DESY which was taken over after the LEP program to DESY, was to have a complementary machine at the same time as LHC as we always were lucky in particle physics to have electron machine and proton machine more or less simultaneously running to really do the precise measurements to see the new things. Now there is, I understand, this agreement to develop now for the 20-35 MeV/m to eventually higher gradients, but the theoretical limit we know all is 60 MeV/m ,so you can only go up to 1 TeV. And the decision was based on the fact that at the moment of the decision this was the technology compatible with the energy range you could cover with accelerator techniques however you mentioned in a little phrase that there is still ongoing one technique, which is CLIC, the effort going to this is marginal compared to what

DESY has invested in the cold technology and therefore once LHC shows what to do one has to reconsider and then maybe put the effort on the other higher energy but this is really what is the strategy, I guess, so I want to tell.

- *Lockyer:*

I want to comment to that, I think that is the position that CERN has taken certainly for a long time and I get the impression, this is my personal opinion, that the schedule which is written down by the ILC, the GDE people, is very aggressive because of the reason you just stated, there is an hidden competition that is not totally on the table between ILC and CLIC and some people think that CLIC is a dream for the reasons you said, is not having so much effort into and therefore the unknowns are so much larger versus the ILC which is a better bet in terms of technology. I guess I'm in the camp that I want just build the ILC, I wouldn't wait. I think CLIC has its own technology problems to solve, its own possible future. It is obviously an option far beyond the ILC. I like to see something at 5 TeV as the future machine. Once you build the ILC tunnel, which is 1/3 of the cost of the whole thing, so now once you have that tunnel you can upgrade, you can upgrade sections we can do all kind of things, so I view that this is a very scalable machine once you build it.

- *Wenninger:*

I agree with you, I only want to say for the young talents there are still options which they I have to consider when LHC shows things which we didn't expect.

- *Lockyer:*

Sure, absolutely.

- *Lopez Mateos:*

I just want to add to the comment made by professor Susskind I think there is a pretty full agreement in the community that even if the SUSY scale is at 5 TeV and it is not directly found at LHC the SUSY Higgs sector, which is much richer than the Standard Model sector, which might not be accessible at LHC just because of a matter of resolution would be definitively accessible at an ILC at 500 GeV, at least the two lighters Higgs: the CP even and CP odd neutral Higgses.

- *Goldstein:*

I have a naïve theorist question, I used to think that to make higher energy electron accelerators you just make them longer and longer, so for my naïve theoretical mind and probably for a lot of other non accelerator people here, the first question is why doesn't it work but the second question is what about a linear electron accelerator hitting the Fermilab proton ring?

- *Lockyer:*

There was a proposal at Fermilab to do an e-p machine long ago and it was not the choice as you know, rather the Tevatron was built. I suspect people think about colliding electrons with the LHC machine and it will be an obvious thing to do if you want to do that but I don't see there is too much enthusiasm and motivation for that at the moment. The answer to your first question is yes when you have a machine you can build as long as you want it is just a money problem. You are working hard to keep the gradient going up it's not only driven by length but there is certainly a factor in the parameterization of the costs for length, but also power, and you could make it longer and get higher energy if you want to, but it's not the cost effective way to do it.

- Pinfold:

You compared I think with the LHC and we are talking about time scale of 2015. And it seems to me that there are, well, plans on the way now for the super LHC of factor 10 increase in luminosity. Have you any particular comment on the Higgs coupling, Higgs self coupling I think, this sound like it could be done at super LHC?

-Lockyer:

I haven't been told about how you do that at the LHC, so LHC I think has to demonstrate that it can be done My reaction to that is you have to see how things work out before you worry about upgrades. I think if the environment proves to be very clean then you are making a statistic argument I'm making an argument I don't know if you can even make the measurement due to the environment, backgarounds, systematics etc. It depends where the Higgs mass is: if it's low mass Higgs, like is favoured, then LHC is going to have an hard time just finding it. I'm not too worried about ILC measurements falling off the list just because the upgrades at the LHC can get more luminosity. It's not going to happen and what I'm really talking about here is measuring couplings to all the various quarks to a precision that's well beyond what I imagine at a hadron machine and even as an enthusiast for hadron colliders I will not say it's possible. The ILC is an order of magnitude improvement in terms of the ability of measuring masses and so on than LHC.

- Loehr:

I want to make a remark to the statement that you have just to elongate the linear collider and you get higher energies; it seems to be true at the first look at it but in reality it is impracticable for some point. The number of components get more and more and you have to ask the question what is in the end the availability of the machine because of interruptions due to failures and for the present version of the ILC it is assumed that it is really at the limit that the components have a reliability of the order of 10^{-4} nevertheless it is already for ILC a big question whether the machine can be available for physics for let's say more than 50% of the time and this is was one of the decisive reasons to build the second tunnel because you have the access to repair the things while the machine is running if you would not have this second tunnel you would have shut down the machine for a period so you could not run at all.

- Lockyer:

I agree with that. Thank you.

- Bourjaly:

I would like to have a complementary linear collider with 500 GeV to go on with the LHC and then have the same time scale. But for a practical and some aspects critical point of view I was really surprised how fast the TDR is expected to come out: by early 2009 I will be really surprised with new and not for the LHC to be able to make decisive statements about what energy scale and what it's to probe and how well we are looking by the LHC because maybe favourable Higgs mass region, we can actually and we have to measure a lot of things and then there are a lot of

motivation question to change maybe going to 600 GeV, you maybe not even do a 500 GeV machine.
I was surprised that there is an expected TDR before we know what's going on at LHC.

-Lockyer:

No comment.

- Brandt:

I have a question for Professor Zichichi: I saw a design for an accelerator with a centre of mass energy of 200 TeV. Can you tell what happened to this project?

- Zichichi:

You are talking about the Eloisatron Project. It was taken with great interest by the wise international community of physicists. We are not wise because if our community was wise we should have already 15 years ago at least agreed on a unique machine of 300 km ring and put more magnets more radiofrequencies with time. We have followed the opposite channel: you want 16 TeV build the machine, do you want now 50 TeV another machine and so on but this is impossible. So our community is not wise: the best choice as it is documented here would have been not now but 15 years go to agree on a unique tunnel 300 km long. There was the money ready on the other hand when people speak about where to go with these new colliders? The answer is very simple: if you want the collider in country "A" then the government of country "A" should be prepared to pay. The illusion that you can have the machine in country "A" with money of country "B", "C" and "D" it is just an illusion, it's a waste of time. There will be never an agreement because our projects compete with going to the Moon, going to Mars, it's finished the era of this little gadget that professor Lockyer showed we are competing with serious enemies which are much more influential at the decision making level than we are and we must be careful in what we do for example this is the reason why this year we are talking about complexity and all the enormous consequences of this new trend which can not be ignored. Imagine somebody else will come up and say what do we have been talking about ourselves, people already say that reductionism is out of phase, it's just old, forget about as if holism was a great discovery. If we keep silent there will be no Subnuclear Physics in the next years so we are confronting with very important problems in terms of being a community that should be allowed to go on. If you look at the past this is most logical consequence but apparently this has not been understood. I will organize here in the near future some wise people conference like we did about 30 years ago and this has been decisive for the success of European projects. At least one government should be supporting what we are doing, one is enough because then the others follow. At present we have zero governments but Italy. So this is the choice: when they speak about ILC, who is paying? Nobody does. Where is the money? It is nowhere. What does it mean? People meet among them, they talk so what? Nothing happens. We must be realistic from the intellectual point of view we can not sleep so we must listen to people. Suppose somebody would say to a journalist: you know Douglas and Denef had discovered that string theory, no matter what model you propose is NP complete, all mathematicians would go up and say stop giving money to the idiots doing Subnuclear Physics they will never understand the logic of nature. This is what is going on, so we have to open our field to these new frontiers and we must realize that we need the analytic continuation of what we know otherwise we're just in an other limit of being lost. But we do need also to consider how to defend ourselves and the best defense I can think of is to prove that complexity exists at the fundamental level, therefore the prediction is that LHC should discover something fantastic and see what happens that we must be prepared not to be only limited to the analytic continuation of what we know and I consider CLIC a jump into the UEEC event.

- *Wenninger:*

I just want to comment to the question yes if you are still here on Tuesday I will talk a bit about the idea to go to 10^{35} and what is the time scale involved and what are the problems involved and certainly LHC will go as far as it can to reach as far physics as possible and also I think the time scale are here said moves a bit to really the moment that one knows what LHC provides and 2012 is maybe more realistic than 2009 but R&D has to continue on CLIC, on TESLA, on detectors and this is really what we are talking about to prepare the future but the decision is not for ....

Diffraction at HERA on the Quark and Gluon Scale

Bernd Löhr
Deutsches Elektronen Synchrotron DESY

1 Kinematics of Deep Inelastic Scattering and Diffractive Scattering

In deep-inelastic electron-proton scattering (DIS) at the HERA collider [1], we are dealing with the process which is sketched in figure 1. The word electron is further used generically for electrons and positrons. The incoming electron emits a virtual photon which interacts with one of the quarks in the proton. The struck quark receives a transverse momentum and separates from the remnant of the proton. The proton remnant and the struck quark fragment into hadrons. Between the proton remnant and the struck quark a colour string develops which eventually breaks up and forms additional hadrons. At a fixed center of mass energy, the kinematics of inclusive DIS is described by the following variables.

$s = (k+p)^2$: center of mass energy squared of the e-p system,

$Q^2 = -q^2 = -(k-k')^2$: negative momentum transfer squared,

$W^2 = M_h^2 = (p+q)^2$: cms energy of the virtual photon and the proton,
mass of the hadronic system in the final state,

$x = \dfrac{Q^2}{2p \cdot q}$: the fraction of the proton carried by the struck quark,

$y = \dfrac{p \cdot q}{p \cdot k}$: fraction of the electron momentum transferred to the proton in its rest system.

These variables are not all independent. They are connected by the relation $Q^2 = x \cdot y \cdot s$. Commonly used are the two variables Q^2 and x to characterize the reaction.

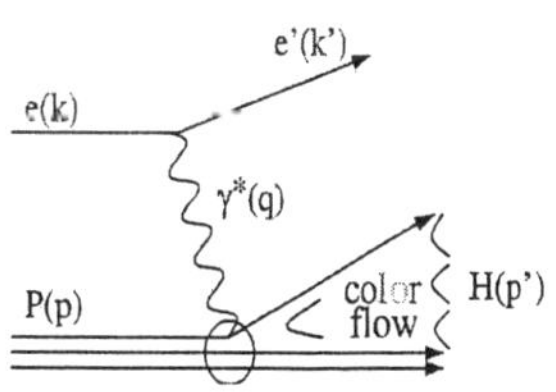

Figure 1: Diagram for inclusive deep-inelastic e-p scattering

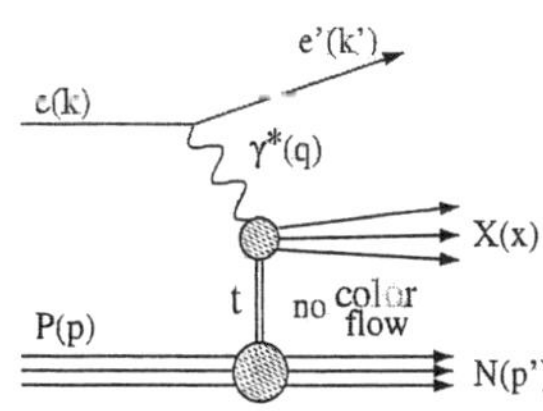

Figure 2: Diagram for diffractive e-p scattering

In diffractive scattering, the interaction of the virtual photon occurs with a colourless object which is emitted from the proton as shown in figure 2. This colourless object is called a pomeron. The proton remains intact or dissociates into a low mass hadronic system N. The virtual photon and the pomeron form a hadronic system X. Because the systems X and p/N are not connected by a coloured string the hadronic system X is well separated from the proton.

This leads to a gap in (pseudo)rapidity, $\eta = -ln\,tan(\Theta/2)$, between the proton and the system X. Here Θ is the angle between the proton direction, called forward, and the first detected particle from the system X. Additional variables are needed to describe diffractive scattering. These are the following ones:

M_X : mass of the diffractively produced hadronic system X,

$t = (p-p')^2$: four momentum transfer squared at the proton vertex,

$x_{I\!P} = \frac{(p-p')\cdot q}{p \cdot q} = \frac{M_X^2+Q^2}{W^2+Q^2}$: fraction of the proton momentum carried by the exchanged pomeron,

$\beta = \frac{Q^2}{2(p-p')\cdot q} = \frac{Q^2}{M_X^2+Q^2} = \frac{x}{x_{I\!P}}$: momentum fraction of the exchanged pomeron that is involved in the scattering process.

2 Regge Phenomenology versus Perturbative QCD

Hadronic reactions in peripheral processes have been studied extensively and can be described in the framework of Regge phenomenology. In Regge theory [2], the exchanged object between the target proton and the incoming projectile is a 'Regge trajectory'. A Regge trajectory describes the exchange of a system of generalized particles with continuous spin but otherwise the same quantum numbers. Trajectories are parametrized to be linear

$$\alpha_{I\!R}(t) = \alpha_{I\!R}(0) + \alpha' \cdot t\,.$$

At high energies, diffractive and total cross-sections are expressed in terms of the pomeron trajectory: (In DIS W takes the role of s in hadron-hadron collisions.)

$$\frac{d\sigma}{dt} \propto e^{b(W)\cdot t}\left(\frac{W}{W_0}\right)^{4(\alpha_{I\!P}(t)-1)} \qquad \sigma_{tot} \propto \left(\frac{W}{W_0}\right)^{2(\alpha_{I\!P}(0)-1)} \qquad \text{with} \quad b(W) = b_0 + 4\alpha' \cdot ln\left(\frac{W}{W_0}\right)\cdot t\,.$$

All the trajectories on which known particle lie have an intercept $\alpha_{I\!R}(0)$ smaller than one. This means that their contribution to the total cross-section falls with energy. The experimental fact that, at high energies, the total cross sections rise with energy required to invent a hypothetical new trajectory with an intercept greater than one, the pomeron trajectory [3], which carries the quantum number of the vacuum. At high enough energies, this pomeron contribution will dominate because the contributions from all other Regge trajectories will die out with energy. Diffractive reactions and in particular elastic scattering occur by the exchange of the colourless pomeron with the quantum nunmbers of the vacuum. From the analyses of many peripheral hadronic processes the pomeron trajectory was found to be [4]:

$$\alpha_{I\!P}(t) = 1.08 + 0.25 \cdot t\,.$$

The predictions from Regge theory for soft diffraction are a power-law behavior of the total diffractive γ^*-p cross-section as $\sigma(W) \propto W^\delta$ with $\delta \approx 0.22$ and an exponential drop of the differential cross section as a function of t, because t is negative, whith an increasing slope, $b(W)$, as W increases. This last fact is called shrinkage. The slope parameter b is proportional to r^2 where r is a measure for the extension of the interacting objects. With increasing W also r increases which means that the scattering objects get bigger.

Diffractive deep-inelastic scattering at high photon virtualities Q^2 (hard diffraction) can be described in the framework of perturbative QCD as shown in figures 3 and 4. The virtual photon splits into a quark-antiquark pair and in the simplest approach two gluons which form a colour singlet are emitted from the proton and couple to the quark-antiquark pair. In higher orders of pQCD, the exchanged gluon system is commonly treated as a BFKL-type ladder [5]. Various pQCD inspired models exist for hard diffractive scattering. All these models predict little or no shrinkage.

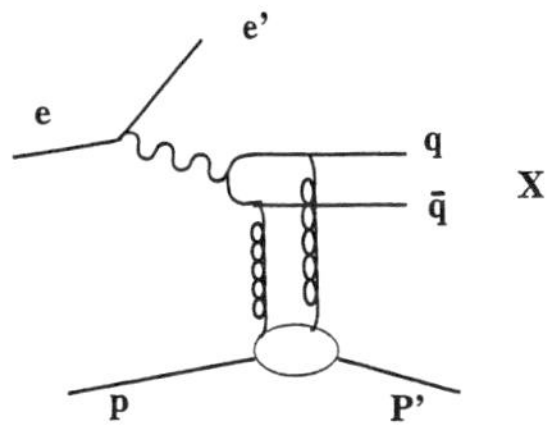

Figure 3: Diffractive e-p scattering in pQCD by two gluon exchange.

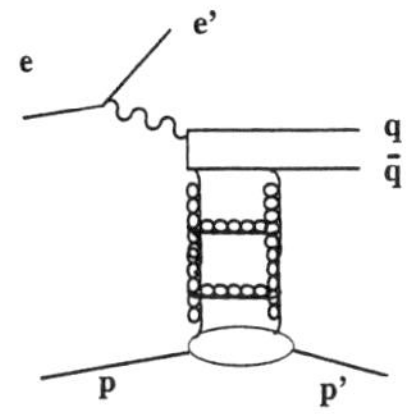

Figure 4: Diffractive e-p scattering in pQCD by BFKL-type ladder exchange.

3 Exclusive Vector-Meson Production

Exclusive vector-meson production is a diffractive process and has been studied extensively. Theoretically, the vector-meson dominance model (VMD) [6] plus Regge theory provide a framework in which exclusive vector-meson production is understood as a quasi-elastic scattering where the incoming vector meson is off mass shell. The situation is graphically shown in figure 5.

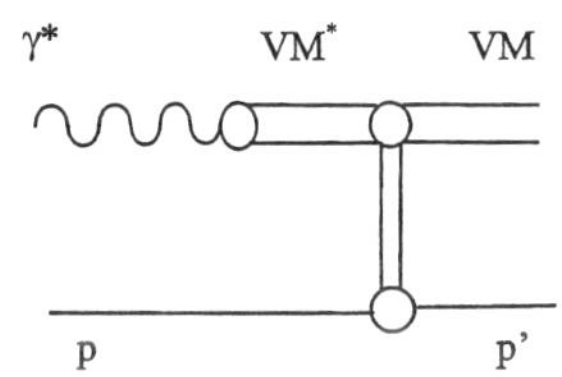

Figure 5: Exclusive vector-meson production in the VMD-Regge framework.

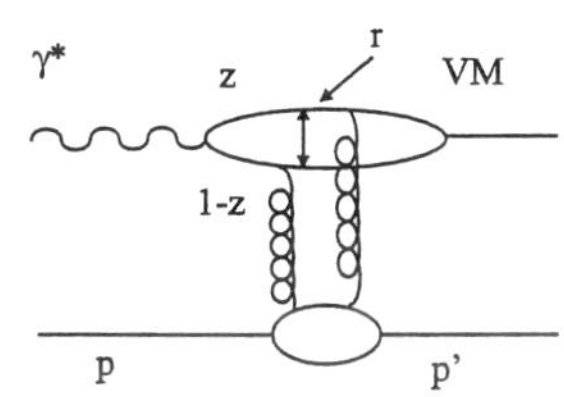

Figure 6: Exclusice vector-meson production as a pQCD process.

In pQCQ, exclusive vector-meson production proceeds according to figure 6. The virtual photon splits into a quark-antiquark pair with relative momentum fractions z and $1-z$. In the lowest order, two gluons are exchanged between the proton and the quark-antiquark pair which recombines afterwards to form the vector meson. The transverse separation, r, of the quark and antiquark is given in pQCD by :

$$r^2 = [z(1-z)Q^2 + m_q^2]^{-1} .$$

Here m_q is the mass of the quark(antiquark) of which the vector meson is composed. Perturbative QCD is expected to be applicable when the transverse dimension, r, of the quark-antiquark sytem gets small. This happens when either Q^2 or m_q get big. Perturbative QCD models predict a rise of the cross section like $\sigma(W) \propto W^{\delta}$ with $\delta \approx 0.8$ which is faster than expected from Regge theory. The slope of the t distribution is predicted to be $b \approx 4$ GeVand $\alpha' \approx 0$. This means no or little shrinkage. The conditions under which exclusive vector-meson production is a hard diffractive process that can be described by pQCD models will be investigated in the rest of this section.

3.1 Can the Vector-Meson Mass be a Hard Scale ?

Data from photoproduction, that means $Q^2 = 0$, permit to test the behaviour of exclusive vector-meson production as a function of the meson mass. Figure 7 shows the cross sections for production of ρ, ω, ϕ, J/Ψ, $\Psi(2S)$, and Υ as functions of W [7] [8] [9] [10] [11] [12]. The lines through the data points are only to guide the eye and are not fit results.

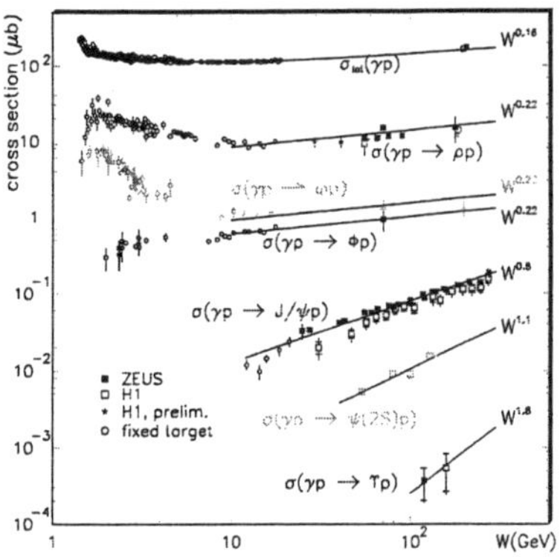

Figure 7: Photoproduction cross-section for ρ, ω, ϕ, J/Ψ, $\Psi(2S)$, and Υ.

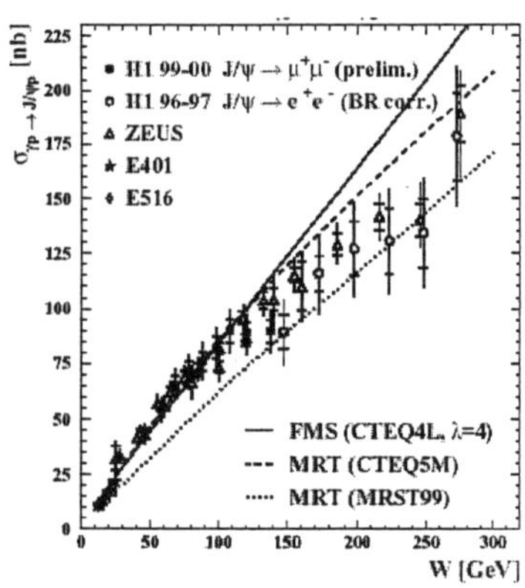

Figure 8: Cross section of J/Ψ photoproduction as a function of W compared to pQCD models.

The W dependence of the light vector-mesons (ρ, ω, ϕ) can be described by a slope $\delta \approx 0.22$ in agreement with Regge phenomenology. For higher vector-meson masses, e.g. for the J/Ψ, the rise with W gets steeper. This indicates the onset of hard diffraction. Therefore exclusive production of J/Ψ meson should be described by pQCD model calculations already from $Q^2 = 0$ on. In figure 8, the photoproduction cross-section for J/Ψ mesons as a function of W is compared to pQCD model calculations [13]. These calculations are able to describe the data qualitatively.

The t dependence of the cross section is given by the parameter $b(W) = b_0 + 4\alpha' \cdot ln(W/W_0) \cdot t$. Figure 9 shows measurements of slope parameter b for photoproduction of ρ mesons. From these measurements one extracts $\alpha' = 0.3 \pm 0.4\,\text{GeV}^{-2}$. Within the large error this is compatible with the value of 0.25 expected for soft processes. For the photoproduction of J/Ψ-mesons the same is shown in figure 10. Now one finds $\alpha' = (0.164 \pm 0.028 \pm 0.030)\,\text{GeV}^{-2}$ which is much smaller than 0.25. This again indicates that photoproduction of J/Ψ mesons is not a soft process and the vector-meson mass can provide a hard scale which makes pQCD applicable.

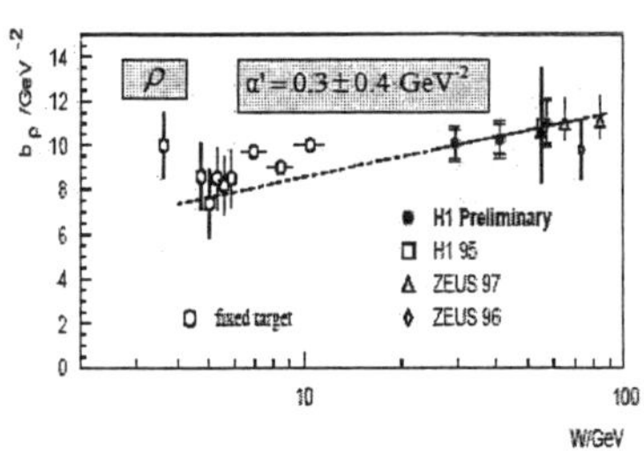

Figure 9: Slopes of the t depence for photoproduction of ρ mesons.

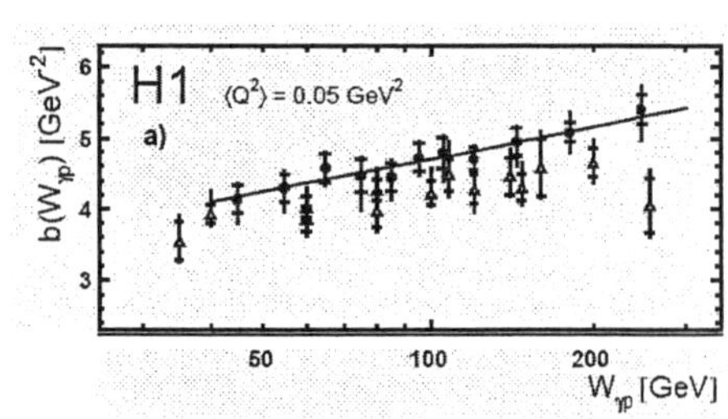

Figure 10: Slopes of the t depence for photoproduction of J/Ψ mesons.

3.2 Can Q^2 provide a Hard Scale ?

The cross section for exclusive production of ρ mesons at $Q^2 = 0$ behaves like $\sigma(W) \propto (W/W_0)^\delta$ with $\delta = 0.22$ as shown in section 3.1. Figure 11 shows this cross section as a function of W for higher values of Q^2. The exponent

δ increases with Q^2. This indicates the transition from a soft process to hard one. In figure 12, the cross section for J/Ψ productions is shown for different values of Q^2. Already at $Q^2 = 0$ the exponent δ is bigger than 0.22 and hardly changes with Q^2. The different W dependences of the ρ and J/Ψ cross sections as functions of Q^2 are even more clearly visible in figure 13 and figure 14. Figure 13 shows the slope of the W dependence as a function of Q^2 for DIS ρ-production [14]. Starting from a value compatible with a soft pomeron trajectory it rises with Q^2 to considerably higher values. For DIS J/Ψ-production [15], the W slope is independent of Q^2 at a value not compatible with a soft pomeron trajectory. This is shown in figure 14. The δ value for J/Ψ production is approximately equal to the value for ρ production at high Q^2.

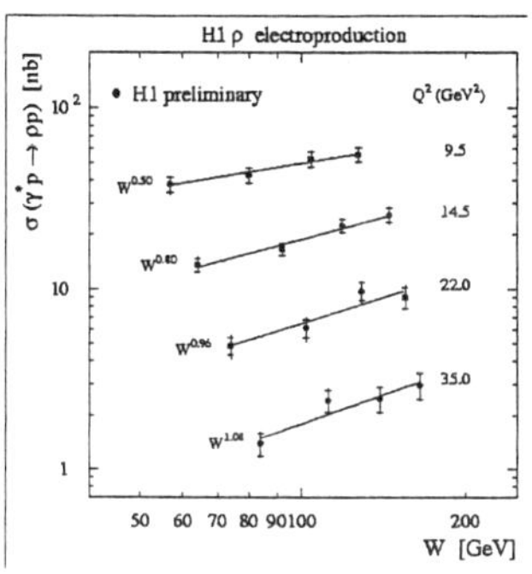

Figure 11: The W dependence of DIS ρ-production for different values of Q^2.

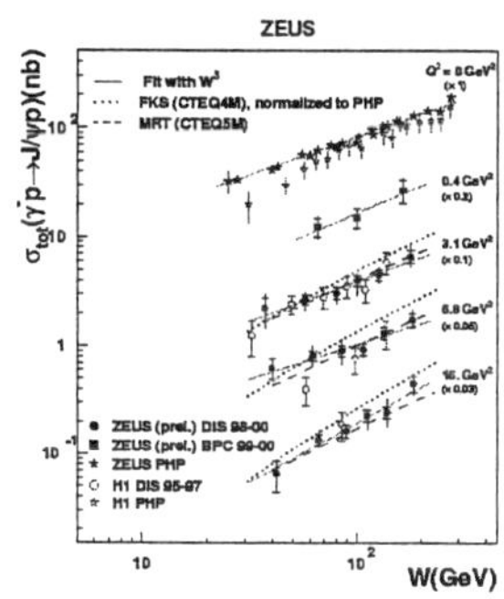

Figure 12: The slope δ of the W dependence of DIS J/Ψ-production.

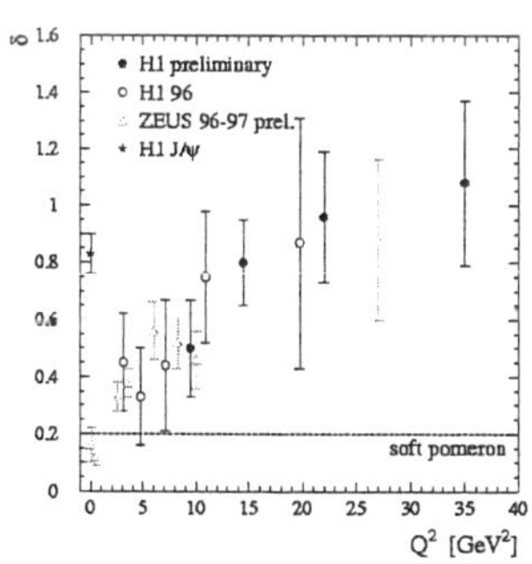

Figure 13: The slope δ of the W dependence of DIS ρ-production.

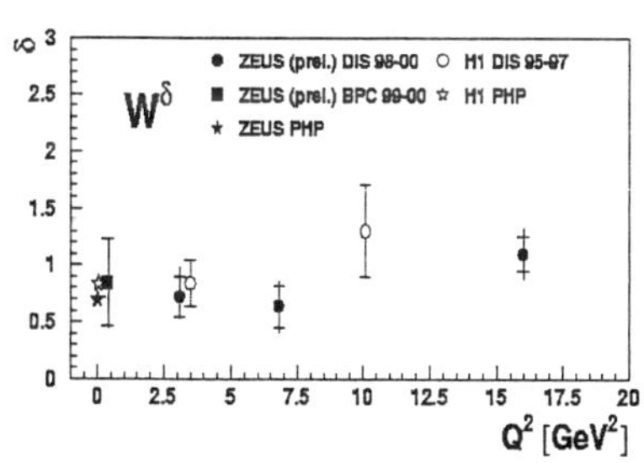

Figure 14: The slope δ of the W dependence of DIS J/Ψ-production.

Another parameter which is supposed to change in a transition from the Regge regime to pQCD is the slope b of the t distribution. The figures 15 and 16 show this slope as a function of Q^2 for ρ and J/Ψ production. For ρ production, the t slope decreases with increasing Q^2 to a value of about 4 as expected from pQCD models. For J/Ψ production this slope is constant with Q^2 at the level of about 4 which reflects the fact that J/Ψ production is a hard process from $Q^2 = 0$ on. One concludes that the initially soft ρ-production becomes a hard process with increasing Q^2.

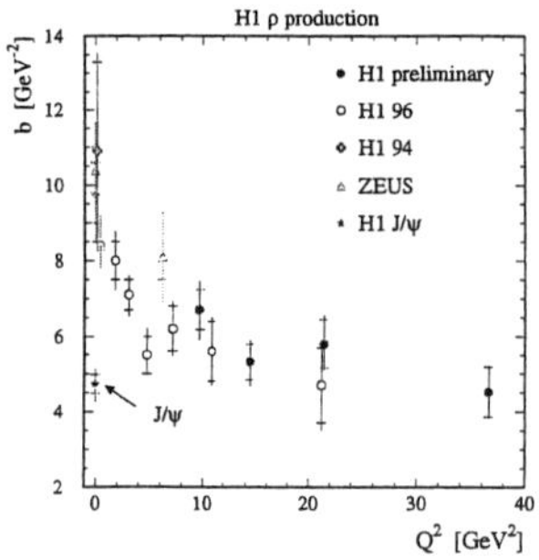

Figure 15: The t slope parameter b of DIS ρ-production as a function of Q^2.

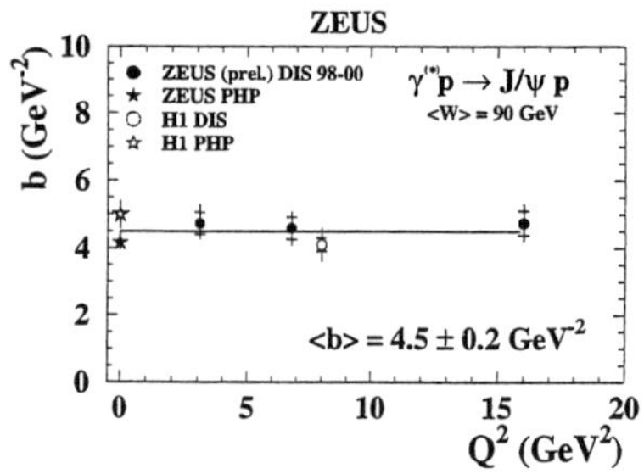

Figure 16: The t-slope parameter b of DIS J/Ψ-production as a function of Q^2.

Finally, figure 17 shows the cross section for J/Ψ production as a function of Q^2. The data are well described by pQCD models [16] [17] even from $Q^2 = 0$ on. These models use parton-density functions derived from inclusive deep-inelastic scattering. An empirical fit of the form $(Q^2 + M_{J/\Psi})^{-n}$ with $n = 2.72 \pm 0.10$ also describes the data at higher values of Q^2.

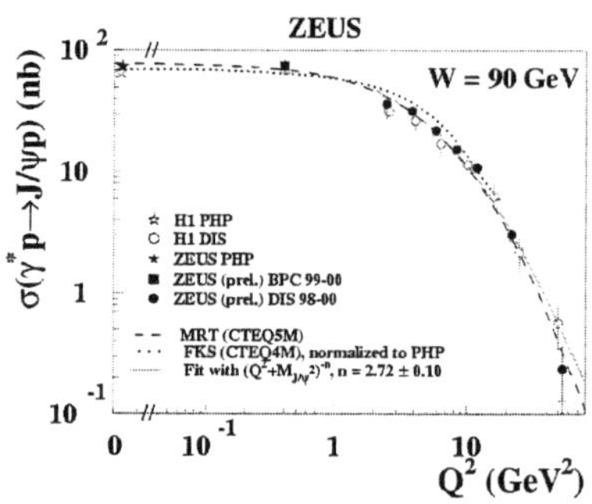

Figure 17: The cross section of J/Ψ production as a function of Q^2. The data are compared to pQCD models.

3.3 Can $|\, t \,|$ provide a Hard Scale ?

In a similar way as the square of the momentum transfer from the electron to the vector meson, Q^2, leads eventually to a hard scale which justifies the application of pQCD one would expect that also the square of the momentum transfer from the proton to the vector meson, $|\, t \,|$, can serve as a hard scale. To study this, photoproduction of vector mesons at high $|\, t \,|$ has been investigated. Experimentally, this leads to a small complication. All studies presented so far have been performed with data integrated over $|\, t \,|$. Since the $|\, t \,|$ dependence is exponentially falling mainly very small $|\, t \,|$ values dominate these data. At higher $|\, t \,|$ values, it becomes more and more likely that the proton will not stay intact but will dissociate into a hadronic system as shown in figure 18. At a low mass of the hadronic system Y, e.g. Y being a nuclear resonance, the particles emerging from this sytem will leave the detector under very small angles through the beam pipe without being detected. These events cannot be distinguished experimentally from events in which the proton stayed intact. At higher masses of Y, some of the particles emerge with high enough transverse momenta to be seen in the detector. These events can be recognized and excluded from the dataset. Thus at higher $|\, t \,|$ values, one deals with mixture of proton dissociatives and nondissociative processes. At high $|\, t \,|$ values, the proton-dissociative events finally dominate. In order to draw conclusions from such events for the process of diffraction one has to assume vertex factorization, i.e. that the ratio

$\sigma_{\gamma p \to \rho Y}/\sigma_{\gamma p \to \rho p}$ only depends on M_Y, W, and t and not on Q^2. Data on this ratio are shown in figure 19. Within the experimental uncertainties, vertex factorization holds.

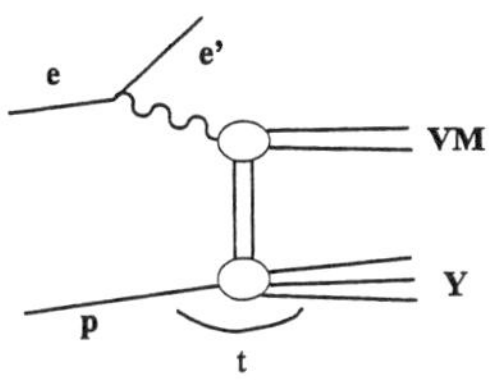

Figure 18: Schematic diagramm of vector-meson production with proton dissociation.

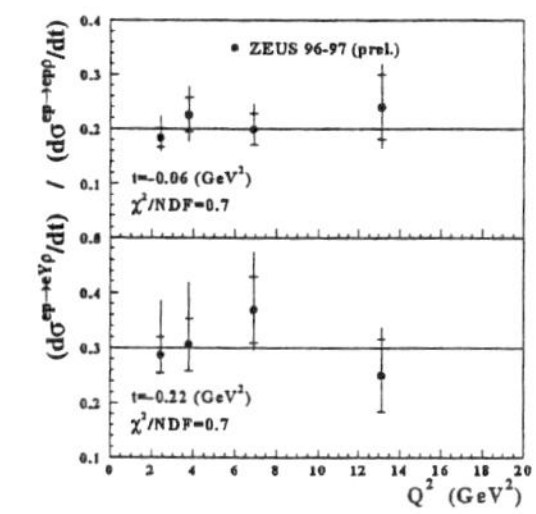

Figure 19: The ratio of proton-dissociative to proton nondissociative cross-sections.

In figure 20, t distributions are shown for proton-dissociative photoproduction of ρ, ϕ, and J/Ψ mesons from ZEUS. The data are well described by by fits of the form :

$$\frac{d\sigma_{\gamma p \to VY}}{dt} \propto | t |^{-n} \ .$$

The fit results for the exponents are given in the figures. The data are also compared to pQCD models of Bartels et al. [18] and Ivanov et al. [19]. Data from H1 on proton-dissociative photoproduction of J/Ψ mesons are displayed in figure 21. Here, the data are compared to pQCD fits according to BFKL and DGLAP. It follows from the H1 and ZEUS data that photoproduction of vector mesons at high $| t |$ are well described by pQCD calculations. Also large $| t |$ provides a hard scale.

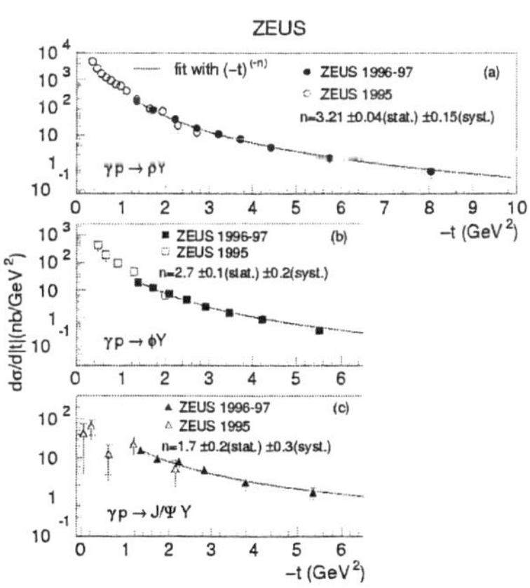

Figure 20: Differential cross-section as a function of t for proton-dissociative photoproduction of vector mesons from the ZEUS experiment.

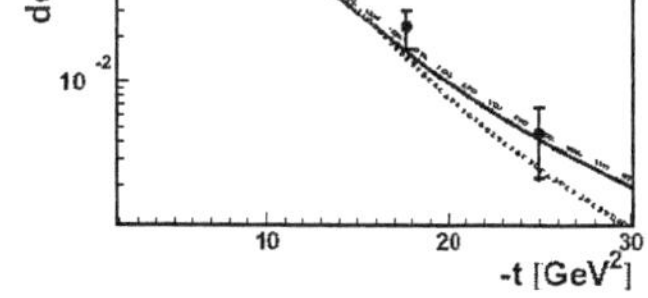

Figure 21: Differential cross-section as a function of t for proton-dissociative photoproduction of J/Ψ mesons from the H1 experiment.

3.4 The Pomeron Trajectory

Staying within the framework of Regge theory, the best way to determine the pomeron trajectory is to extract $\alpha_{I\!P(t)}$ for different t bins from the W dependence of the data according to :

$$\frac{d\sigma}{dt} \propto e^{b(W)\cdot t} \left(\frac{W}{W_0}\right)^{4(\alpha_{I\!P}(t)-1)} .$$

Figure 22 shows the so determined pomeron trajectories for photoproduction of ρ, ϕ, and J/Ψ mesons as well as the one from DIS ρ-production and the soft-pomeron trajectory. The following trajectories are derived from the data :

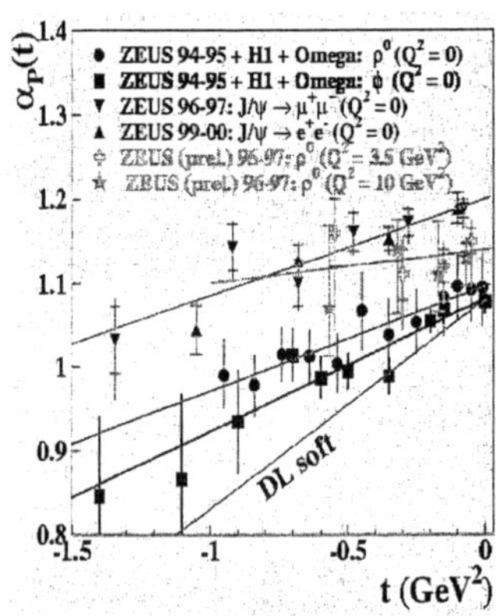

Figure 22: Pomeron trajectories from vector-meson production and soft hadronic processes.

ρ photoproduction	:	$\alpha_{I\!P}(t) = 1.10 + 0.13 \cdot t$,
ϕ photoproduction	:	$\alpha_{I\!P}(t) = 1.08 + 0.16 \cdot t$,
DIS ρ production	:	$\alpha_{I\!P}(t) = 1.14 + 0.04 \cdot t$,
J/Ψ photoproduction	:	$\alpha_{I\!P}(t) = 1.20 + 0.12 \cdot t$,
soft pomeron	:	$\alpha_{I\!P}(t) = 1.08 + 0.25 \cdot t$.

The $\alpha_{I\!P}(0)$ values of ρ and ϕ photoproduction are compatible with the soft-pomeron trajectory. The $\alpha_{I\!P}(0)$ values of DIS ρ- and J/Ψ-production are definitely higher. The t slopes of the pomeron trajectories as a function of t are shown in figure 23. The slope parameter $\alpha'_{I\!P}$ depends on t. This means that the trajectories are not linear. However, one must keep in mind that at high $|\, t \,|$ the proton-dissociative processes dominate.

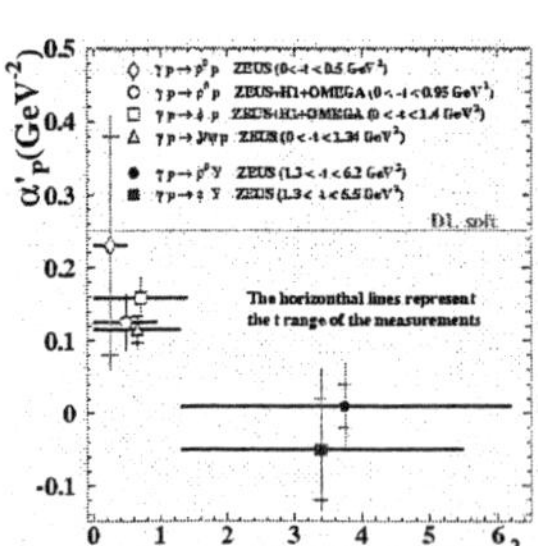

Figure 23: Slopes of the pomeron trajectories as a function of $-t$.

4 Deeply-Virtual Compton Scattering

Deeply-virtual Compton scattering [20] is the QCD process described in figure 24. A virtual photon scatters off the proton and appears as a real photon in the final state. This is a process very close to true elastic scattering.

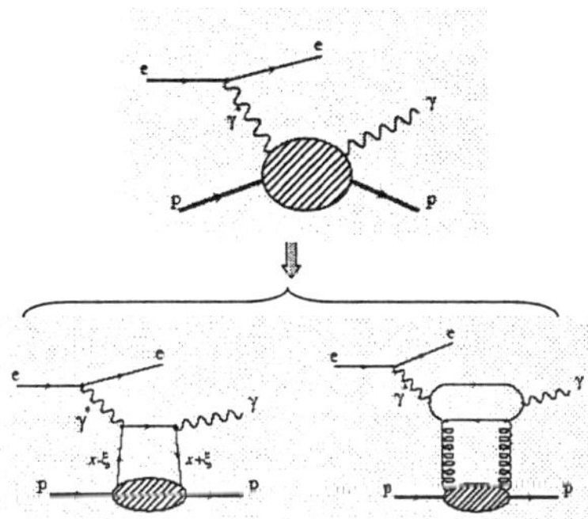

Figure 24: The graph for DVCS and its interpretation in perturbative QCD.

Bethe-Heitler reactions, which are elastic e-p scattering with radiation of photons from the incoming or outgoing leptons, lead to the same final state. This is shown in figure 25.

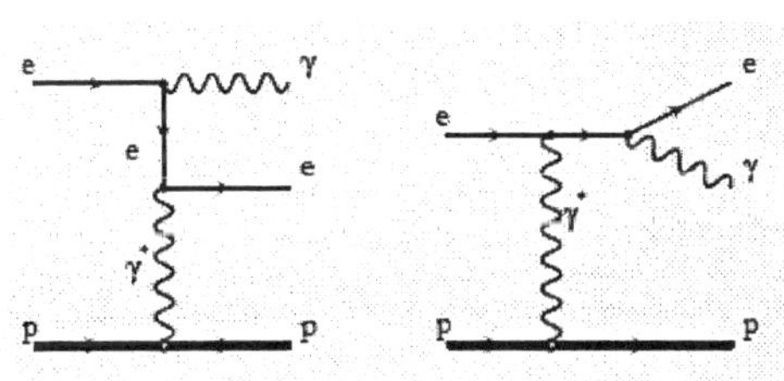

Figure 25: Graphs for Bethe-Heitler processes.

The measurement of the DVCS cross-section is an experimental challenge because the Bethe-Heitler cross-section is much larger than the DVCS one. In figure 26, the two possible final-state topologies are shown. In the case where the lepton appears under a small scattering angle both Bethe-Heitler and DVCS processes contribute. The topology in which the photon appears under the smaller scattering angle is only populated by the Bethe-Heitler process. In this way the Bethe-Heitler rate can be measured, extrapolated and subtracted to receive the pure DVCS rate. The interference of the QED and QCD processes gives access to the real part of the QCD amplitude, ReA_{QCD}. In lowest order of pQCD, two diagrams contribute to the DVCS process. They are shown in figure 27.

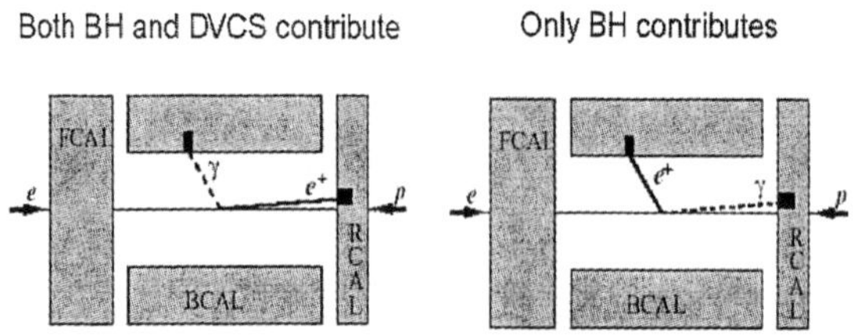

Figure 26: Experimental topologies to which Bethe-Heitler and DVCS processes contribute.

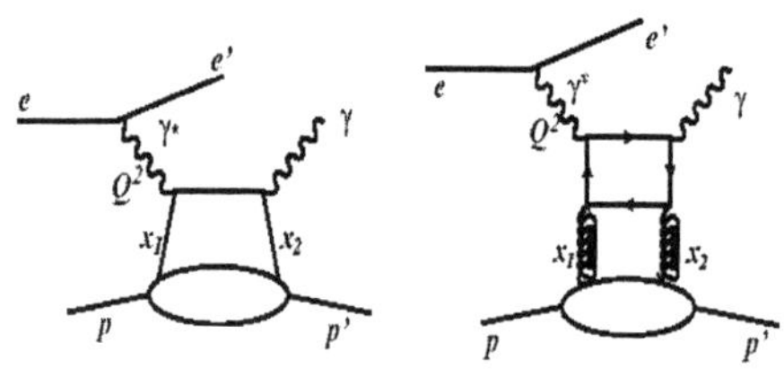

Figure 27: QCD diagrams which contribute to deeply-virtual Compton scattering.

A novel feature appears here. The two partons, emitted and reabsorbed by the proton, are at different values of Bjorken x. This leads to generalized parton distributions (GPD)[21]. The data for DVCS [22] are shown in figures 28 and 29. They are compared to pQCD models based on generalized parton distributions. These models can describe the W and Q^2 dependence well. The W dependence is stronger than expected from a soft-pomeron exchange. The available experimental data are not yet sufficient to measure angular distributions from which the interference term could be extracted.

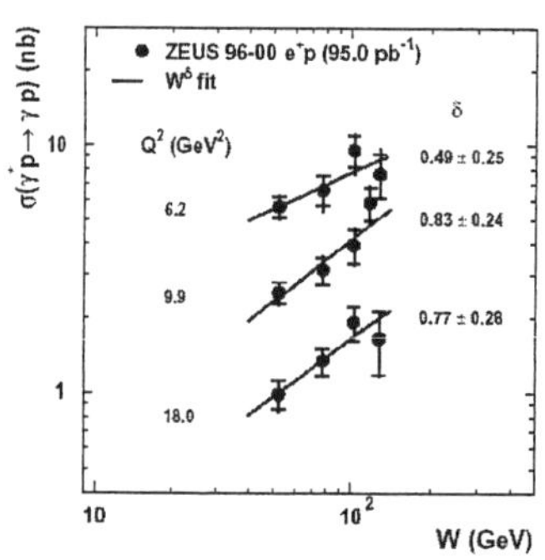

Figure 28: The DVCS cross-section as a function of W for various values of Q^2.

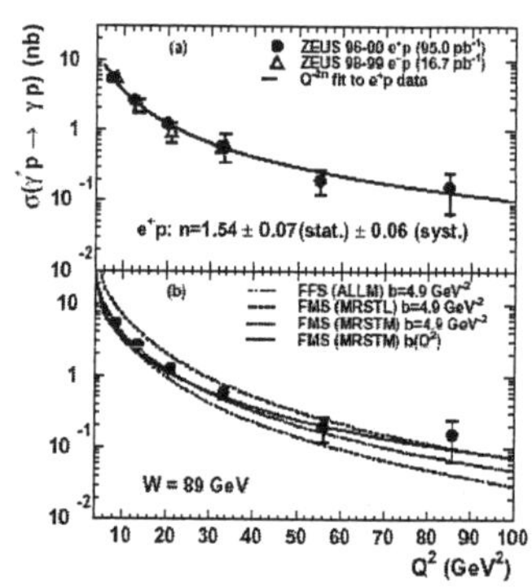

Figure 29: The DVCS cross-section as a function of Q^2. The data are compared to pQCD models.

5 Inclusive Diffraction

The process of inclusive diffraction is illustrated in figure 30 in which the pomeron has a partonic structure [23]. The virual photon scatters off one quark/antiquark in the pomeron. The scattered quark/antiquark together with the remainder of the pomeron form the diffractively produced system X. The proton may dissociate in a low mass hadronic system Y of which most particles will escape detection in the very forward direction.

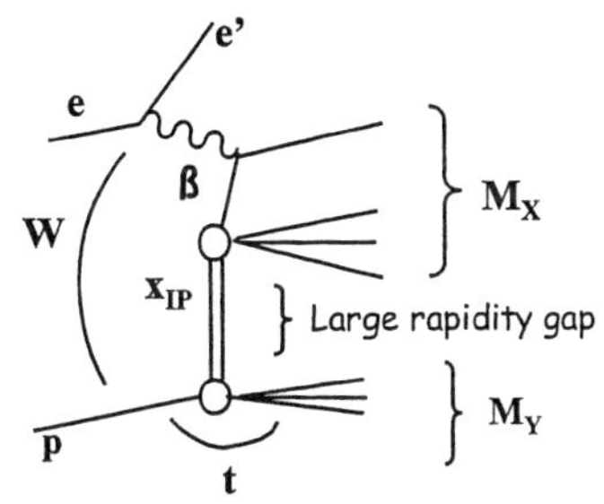

Figure 30: Diagram for inclusive diffractive-scattering when the pomeron has a partonic structure.

The two variables $x_{I\!P}$ and β acquire the following meaning in the partonic-pomeron picture:

$$x_{I\!P} = \frac{M_X^2 + Q^2}{W^2 + Q^2}$$ momentum fraction of the proton carried by the pomeron,

$$\beta = \frac{x}{x_{I\!P}} = \frac{Q^2}{M_X^2 + Q^2}$$ momentum fraction of the pomeron carried by the struck quark .

In a diffractive reaction vacuum quantum numbers are exchanged by a system which is generically called a pomeron. Experimentally three ways exist to select inclusive diffractive events.

1.) The η_{max} method :

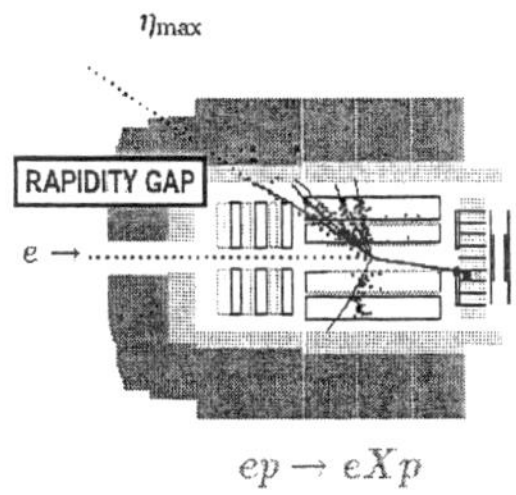

Figure 31: A large rapidity-gap event in the H1 detector.

This method makes explicit use of the rapidity gap in a diffractive event as shown in figure 31. A measure of the rapidity gap is the maximum pseudorapidity seen in the detector from a track or an energy deposit in the calorimeter.

$$\eta_{max} = -ln\ tan(\Theta_{min}/2)$$

Here Θ_{min} is the angle of the most forward detected track or energy deposit. Requesting an η_{max} value smaller than a certain value selects diffractive events.

2.) Detection of the outgoing proton:

A diffractive event is selected by identifying and measuring the scattered proton from the diffractive reaction. As an example for the experimental method figure 32 shows the leading-proton spectrometer (LPS) of the ZEUS experiment.

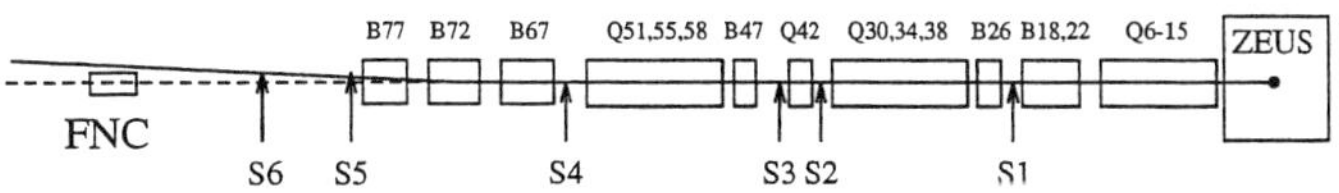

Figure 32: Schematic picture of the leading-proton spectrometer (LPS) of the ZEUS experiment.

Six stations of silicon-strip detectors (S1-S6) are placed very close to the proton beamline of the HERA storage ring. Their distance from the interaction point of the most downstream station (S6) is about 96m. Together with the magnets of the proton beamline the stations form a spectrometer which measures the momenta of diffractively scattered protons. Figure 33 shows a measured spectrum of the fraction $x_L = 1-x_{I\!P}$ of the proton-beam momentum carried by the detected scattered proton.

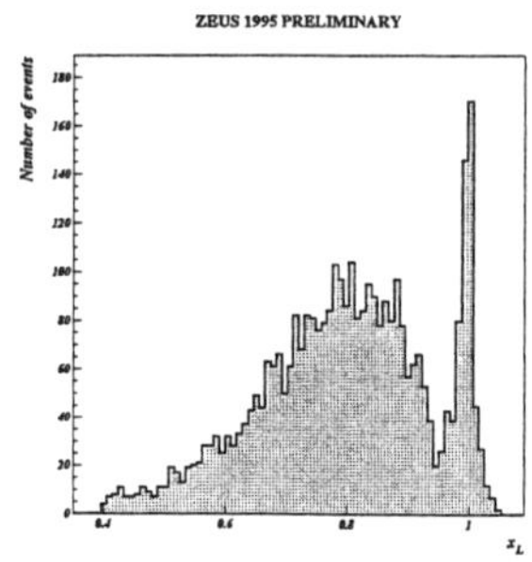

Figure 33: Schematic picture of the leading-proton spectrometer (LPS) of the ZEUS experiment.

Diffractive events in which the scattered proton stays intact show up as a peak around $x_L = 1$. The events from lower x_L values are partly from diffractive proton-dissociative events or from nondiffractive events. The detection of the diffractively scattered proton is the only method to measure the t distribution in diffractive events:

$$t = -\frac{p_T^2}{x_L} - \frac{(1-x_L)^2}{x_L} M_p^2 .$$

3.) The M_X method [25]:

This method exploits the different shapes of the $ln\ M_X^2$ distributions for nondiffractive and diffractive events.

(i) Nondiffractive events :

The rapidity of a produced particle is defined as $y = \frac{1}{2} ln \frac{E+p_z}{E-p_z}$ where E is the particle's energy and p_z its longitudinal momentum. For a given center of mass energy W of the reaction, maximum and minimum values of the rapidity are related to W by $W^2 = c_0 \exp{(y_{max} - y_{min})}$. The real detector has a limited acceptance and particles from the proton remnant in the very forward direction leave the detector unmeasured. The acceptance starts only at y_{limit}^{det}. The mass of the detected hadronic system is then given by $M_X^2 = c_0 \exp{(y_{limit}^{det} - y_{min})}$. For uncorrelated particle emission in nondiffractive events the rapidity distribution is flat [24]:

$$\frac{dN_{part}}{dy} = \lambda = const.$$

This is sketched in figure 34. The Poisson probability to find no particles in a region Δy close to the proton direction is given by :

$$P(0) = e^{-\lambda \Delta y} .$$

This leads to a $ln\ M_X^2$ distribution of the form :

$$\frac{dN_{nondiff}}{d\ lnM_X^2} = c \cdot e^{b \cdot ln\ M_X^2} .$$

The slope b must be determined from data. An experimental $ln\ M_X^2$ distribution is shown in figure 35. The slope b is indicated as a line rising with $ln\ M_X^2$. Ideally this rise continues to the highest $ln\ M_X^2$ value but in reality detector effects make the distribution turn over and go to zero.

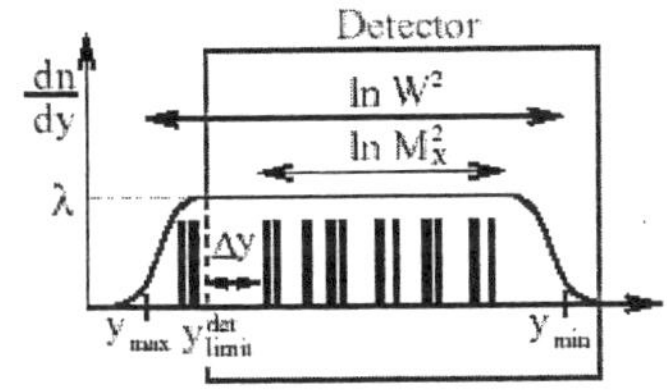

Figure 34: Rapidity distribution for nondiffractive events. Indicated is the full length of the rapidity plateau, the acceptance limits of a detector and a possible rapidity gap due to statistical fluctuations.

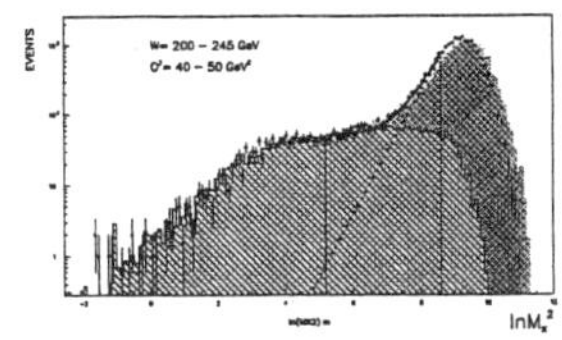

Figure 35: Measured $ln\ M_X^2$ distribution. The hatched area shows the MC-simulated contribution from diffractive events. The dotted line shows the exponential fall-off of the nondiffractive contribution.

(ii) Diffractive events :

It has been found experimentally that at high energies for diffractive events with not too low M_X the following relation is valid :

$$\frac{dN_{diff}}{d\ lnM_X^2} \propto \frac{1}{(M_X^2)^n} \qquad \text{with} \qquad n \approx 1\ .$$

This leads to a flat $ln\ M_X^2$ distribution :

$$\frac{dN_{diff}}{d\ lnM_X^2} = \text{const.} = D.$$

(iii) Sum of diffractive and nondiffractive events:

Inclusive DIS events consist of diffractive plus nondiffractive events. Their $ln\ M_X^2$ distribution is therefore

$$\frac{dN_{diff}}{d\ lnM_X^2} = D + c \cdot e^{b \cdot ln\ M_X^2} \qquad \text{for} \qquad lnM_X^2 \leq lnW^2 - \eta_0\ .$$

An MC simulation of the diffractive contribution is shown in figure 35. For the extraction of the number of diffractive events, the slope b of the distribution of nondiffractive events has been fitted to the data, the number of nondiffractive events has been evaluated and then has been subtracted from the total number of measured events.

5.1 Comparison of the Methods to select Diffractive Events

The goal is to measure the cross section for the inclusive diffractive-reaction

$$\gamma^* + p \rightarrow X + p$$

where X is the diffractively produced hadronic system which results from the exchange of a 'pomeron' with the quantum nunbers of the vacuum. None of the three methods gives directly this result. They all contain background contributions and must be corrected.

Detection of the scattered proton:

In this method, the selected event sample is practically free of proton-dissociative events. However, the sample may contain contributions from reggeon-exchange processes which are nondiffractive.

The large rapidity-gap method: Events selected with the large rapidity gap method contain in general contributions from proton-dissociative events as well as from reggeon exchanges. Contributions from reggeon exchanges become relevant usually only for $x_{I\!P} > 0.01$.

The M_X method: In the M_X method nondiffractive events are statistically substracted. This holds also for reggeon-exchange processes. Proton-dissociative processes are in general not excluded. Their contribution can be partially determined and subtracted. This is possible only for masses of the proton-dissociative system greater than a certain value which is given by the detector acceptance in the proton direction. This applies as well to the large rapidity-gap method.

5.2 Diffractive Cross-Section and Diffractive Structure-Functions

The differential cross-section for diffractive processes can be expressed in terms of a diffractive structure-function :

$$\frac{d^4\sigma}{dQ^2 dt dx_{I\!P} d\beta} = \frac{2\pi\alpha_{em}}{\beta Q^2}[1-(1-y)^2] \cdot F_2^{D(4)}(Q^2, t, x_{I\!P}, \beta) .$$

Here the contribution from the longitudinal structure-function has been neglected. The H1 Collaboration use instead the reduced diffractive cross-section :

$$\sigma_r^D = F_2^D - \frac{y^2}{1+(1-y)^2} F_L^D .$$

The longitudinal contribution becomes sizable only at very high y-values. If the variable t is not measured but integrated over the the cross section is:

$$\frac{d^3\sigma}{dQ^2 dx_{I\!P} d\beta} = \frac{2\pi\alpha_{em}^2}{\beta Q^4}[1-(1-y)^2] \cdot F_2^{D(3)}(Q^2, x_{I\!P}, \beta) .$$

In diffractive deep-inelastic scattering, QCD factorization of the following form has been proven [26]:

$$\sigma^{diff} \propto f_q^{diff}(Q^2, t, x_{I\!P}, \beta) \cdot \hat{\sigma}_{pQCD} .$$

Here $f_q^{diff}(Q^2, t, x_{I\!P}, \beta)$ are universal diffractive parton-densities and $\hat{\sigma}_{pQCD}$ is the perturbatively calculabe cross section for the hard parton-parton scattering. Another factorization is commonly used in the picture where the pomeron has a partonic structure [23]. This is illustrated in figure 36.

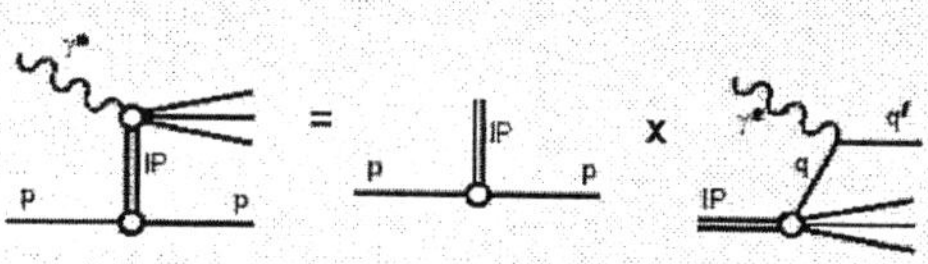

Figure 36: Schematic picture of Regge factorization.

The diffractive cross-section is expressed by a pomeron structure-function $F_2^{I\!P}$ and a pomeron-flux factor which comes from Regge theory :

$$\sigma_{diff} \propto f_{I\!P/p}(t, x_{I\!P}) \cdot F_2^{I\!P}(Q^2, \beta) \qquad \text{with} \qquad f_{I\!P/p}(t, x_{I\!P}) = \frac{e^{B\cdot t}}{x_{I\!P}^{2\alpha(t)-1}} .$$

This Regge factorization is an assumption. No proof exists for it.

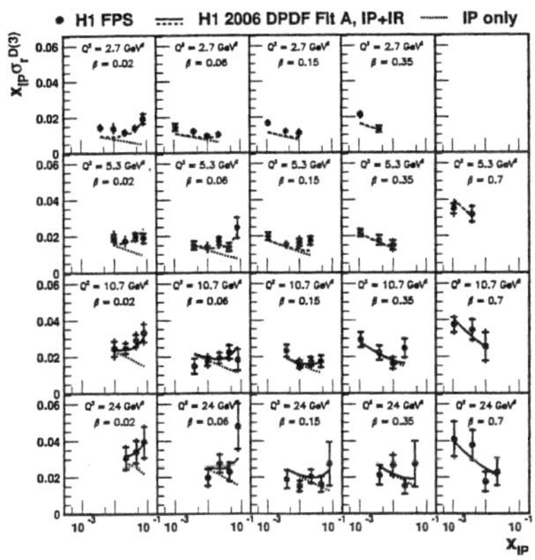

Figure 37: H1 FPS-results for $x_{I\!P}\sigma_r^{D(3)}$. See text for the fits.

5.3 Results on Inclusive Diffraction

5.3.1 Results from the proton-detection method

The experiments H1 and ZEUS both have detector systems to measure protons scattered under very small angles. They are called forward-proton spectrometer (FPS) for H1 and leading-proton spectrometer (LPS) for ZEUS. Figure 37 displays $x_{I\!P}\sigma_r^{D(3)}$ as a function of $x_{I\!P}$ for various Q^2 and β bins as measured by the H1 collaboration [27]. Shown is also a DGLAP fit (see later) for $F_2^{D(3)}$ and a reggeon contribution. The full lines are the fit to the sum of both contributions, the dashed lines are the diffractive contributions only.

The ZEUS experiment has published results for $x_{I\!P}F_2^{D(3)}$ from the LPS mesurements [28]. They are shown in figure 38. The fit is from a BEKW dipole-model which is explained later.

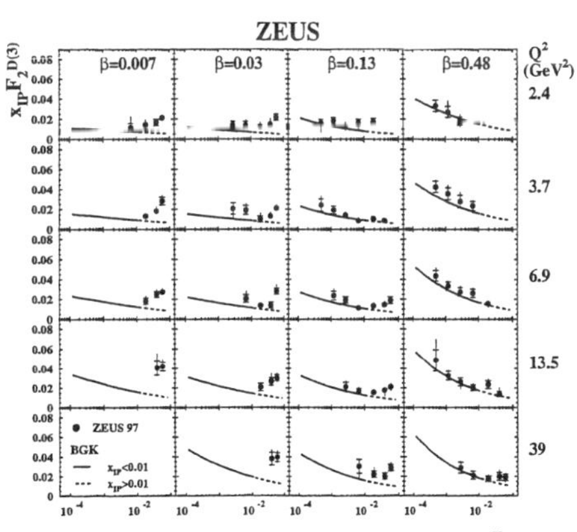

Figure 38: ZEUS LPS-results for $x_{I\!P}F_2^{D(3)}$. See text for the fit.

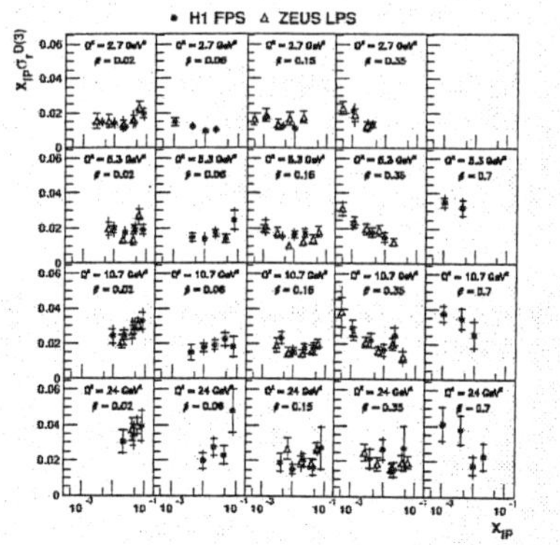

Figure 39: Comparison of the H1 FPS- and the ZEUS LPS-results for $x_{IP}\sigma_r^{D(3)}$

The comparision between the H1 and ZEUS results are shown in figure 39. In the kinematical region of these measurements the contribution from the longitudinal structure-function can be neglected. Therefore $x_{IP}F_2^{D(3)}$ is equivalent to $x_{IP}\sigma_r^{D(3)}$.

5.3.2 Results from the large rapidity-gap method (LRG)

Results for inclusive diffraction determined with the rapidity-gap method are available from the H1 experiment. Figure 40 shows the measured $x_{IP}\sigma_r^{D(3)}$ as a function of x_{IP} for Q^2 values from 3.5-1600 GeV^2 at different β-bins.

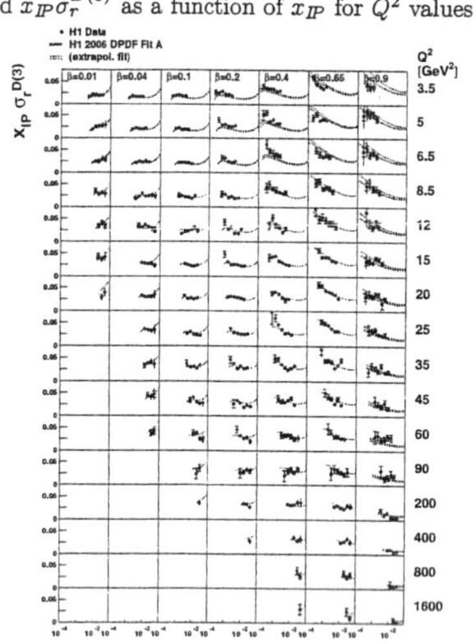

Figure 40: The reduced diffractive cross-section $x_{IP}\sigma_r^{D(3)}$ from the H1 experiment.

In the kinematical region where results are available also from the H1-FPS and the ZEUS-LPS method the data are compared in figres 41 and 42.

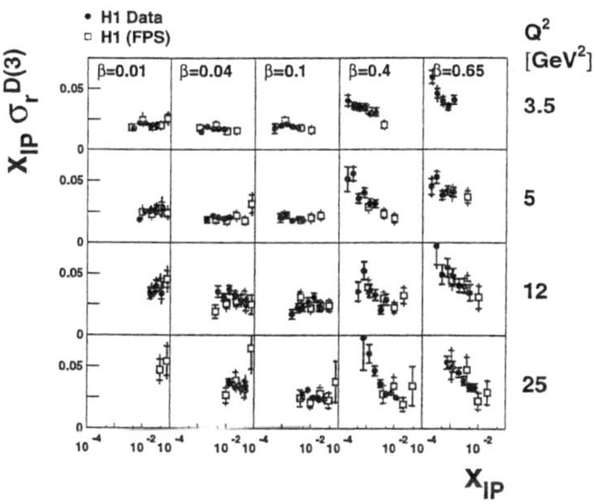

Figure 41: $x_{I\!P} \sigma_r^{D(3)}$ results from the H1-LRG method with results from the H1-FPS method.

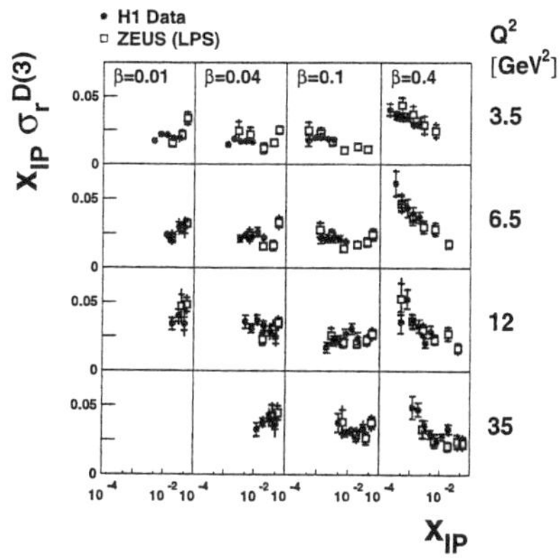

Figure 42: $x_{I\!P} \sigma_r^{D(3)}$ results from the H1-LRG method with results from the ZEUS-LPS method.

5.3.3 H1 fits of the diffractive structure-functions

The data selected by the LRG method contain contributions from reggeon exchanges which are nondiffractive. Therefore the $x_{I\!P} \sigma_r^{D(3)}$ results are fitted to a sum of a pomeron and of a reggeon contribution.

$$F_2^{D(4)}(x_{I\!P}, t, \beta, Q^2) = f_{I\!P}(x_{I\!P}, t) \cdot F_2^{I\!P}(\beta, Q^2) + n_{I\!R} f_{I\!R}(x_{I\!P}, t) \cdot F_2^{I\!R}(\beta, Q^2)$$

For the reggeon contribution the pion structure-function and the flux from the pion trajectory are used. The pomeron flux is taken from Regge theory as explained in section 5.1. The pomeron structure-function is expressed as a sum over the parton distribution-functions for $u, d, s, \overline{u}, \overline{d}.\overline{s}$ and gluon.

$$F_2^{I\!P}(\beta, Q^2) = \sum_i f_i^D(\beta, Q^2)$$

The parton distribution-functions $f_i^D(\beta, Q^2)$ are considered universal. They are fitted by a DGLAP evolution fit with a parametrization at Q_0^2 like:

$$z f_i(z, Q^2) = A_i z^{B_I} (1-z)^{C_i} \cdot e^{\frac{0.01}{1-z}}$$

Here z is the longitudinal momentum fraction of the parton entering the hard subprocess. For a lowest order quark-parton model process $z = \beta$, for higher orders $0 < \beta < z$. H1 performed two fits called A and B [29]. For the gluon parton distribution-function the parameter B has been set to zero in fit A and the parameter C in fit B. Figure 43 shows the fit results from thes two fits for the singlet structure-function $\Sigma(z) = u(z) + d(z) + s(z) + \overline{u}(z) + \overline{d}(z) + \overline{s}(z)$ and the gluon structure-function.

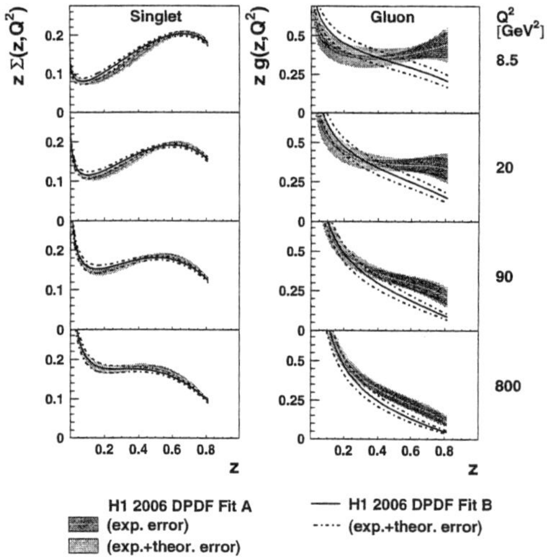

Figure 43: Results from the H1 DGLAP fits of the pomeron structure-function for the singlet structure-function and the gluon structure-function

5.3.4 Results from the ZEUS M_X-method

The Zeus collaboration published results from the M_X-method on inclusive diffraction for a range $2.7\,\mathrm{GeV}^2 < Q^2 < 55\,\mathrm{GeV}^2$ [30] referred to as FPCI . New preliminary data, referred to as FPCII [31], are avalible for the range $25\,\mathrm{GeV}^2 < Q^2 < 320\,\mathrm{GeV}^2$. The diffractive structure-function $x_{I\!P}F_2^{D(3)}$ from both datasets is shown as a function of $x_{I\!P}$ in figure 45. The data are fitted by a modified version of the BEKW model [32]. This is a dipole model which has built in saturation. It contains contributions from the transverse and longitudinal $q\overline{q}$ components and from a $q\overline{q}gluon$ component as shown in figure 44.

$$x_{I\!P}F_2^{D(3)}(\beta, x_{I\!P}, Q^2) \quad = \quad c_T \cdot F_{q\overline{q}}^T + c_L \cdot F_{q\overline{q}}^L + c_g \cdot F_{q\overline{q}g}^T,$$

where

$$\begin{aligned}
F_{q\overline{q}}^T &= (\frac{x_0}{x_{I\!P}})^{n_T(Q^2)} \cdot \beta(1-\beta), \\
F_{q\overline{q}}^L &= (\frac{x_0}{x_{I\!P}})^{n_L(Q^2)} \cdot \frac{Q_0^2}{Q^2+Q_0^2} \cdot [\ln(\frac{7}{4} + \frac{Q^2}{4\beta Q_0^2})]^2 \cdot \beta^3(1-2\beta)^2, \\
F_{q\overline{q}g}^T &= (\frac{x_0}{x_{I\!P}})^{n_g(Q^2)} \cdot \ln(1+\frac{Q^2}{Q_0^2}) \cdot (1-\beta)^{\gamma}.
\end{aligned}$$

The powers $n_T(Q^2), n_L(Q^2), n_g(Q^2)$ were assumed by BEKW to be of the form $n(Q^2) = n_0 + n_1 \cdot \ln[1+\ln(\frac{Q^2}{Q_0^2})]$. The ZEUS data support taking all powers to be equal and of the form $n(Q^2) = n_1 \ln(1+\frac{Q^2}{Q_0^2})$, i.e. $n_0 = 0$, resulting in a modified model BEKW(mod). Taking x_0=0.01 and Q_0^2=0.4 GeV^2 gives a good description of the data with the 5 parameters $c_T, c_L, c_g, n_1, \gamma$ with a χ^2/n_D of 0.71 for more than 400 data points. The full line in figure 45 is the BEKW(mod) fit, the dashed line the transverse $q\overline{q}$ contribution, the dotted line the longitudinal $q\overline{q}$ contribution and the dashe-dotted line the $q\overline{q}gluon$ contribution.

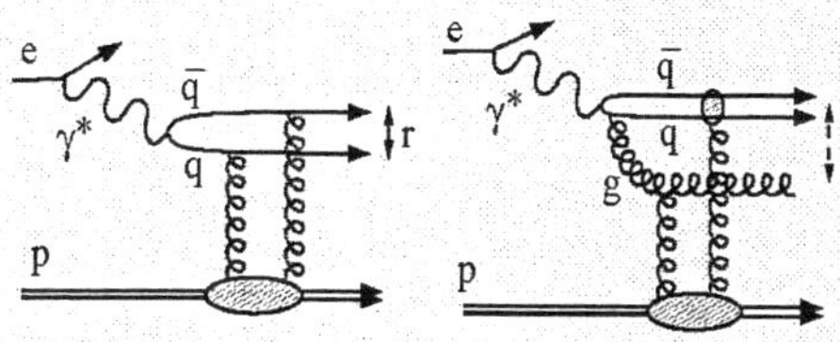

Figure 44: Picture of the BEKW dipole-model contributions.

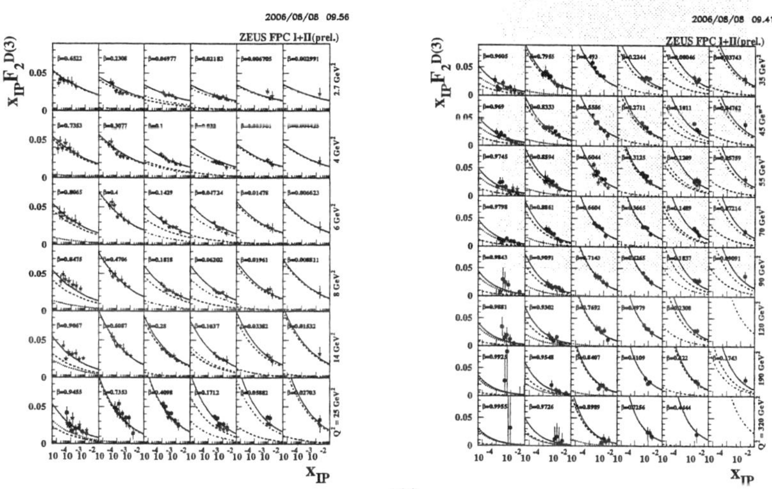

Figure 45: Diffractive structure-function $X_{I\!P}F_2^{D(3)}$ from the ZEUS M_x-method. Shown is also the results of a BEKW(mod) fit which is described in the text.

In figure 46 the $x_{I\!P}F_2^{D(3)}$ is shown for a fixed $x_{I\!P} = 0.01$ and Q^2 values from 25 GeV^2 to 190 GeV^2 as a function of β. There is a broad maximum at about $\beta = 0.5$ which results from the transverse $q\overline{q}$ term. The rise towards small β originates from the $q\overline{q}gluon$ contribution. At very high β the contribution from the longitudinal $q\overline{q}$ term is visible. The BEKW model contains a built in DGLAP-like evolution which also leads to scaling violations. This can bee seen from figure 47 where $x_{I\!P}F_2^{D(3)}$ is plotted as a function of Q^2 for various $x_{I\!P}$ and β bins. The data as well as the BEKW(mod) fit show clear scaling violations.

5.3.5 Ratio of diffractive DIS to the total DIS cross-section

The ZEUS collaboration calculate a diffractive cross section σ^{diff} by integrating $d\sigma/dM_x$ from $M_x = 0.28\,\text{GeV}$ to $M_x = 35.0\,\text{GeV}$ at $W = 220\,\text{GeV}$. This W value has been chosen because it gives the best coverage of the data. It is a good approximation of the total diffractive cross section at that W value. From the same data the total DIS

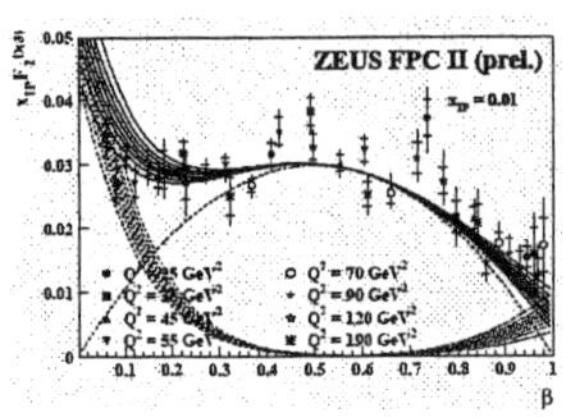

Figure 46: The structure function $x_{IP}F_2^{D(3)}$ versus β at x_{IP}=0.01 for 25 GeV^2 < Q^2 < 190 GeV^2. The curves are results from the BEKW(mod) fit (see text)

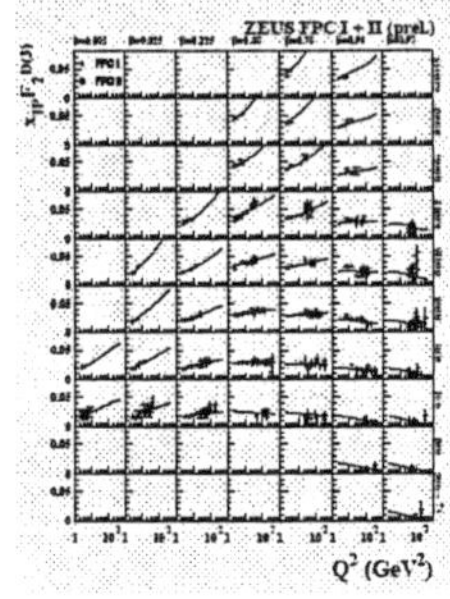

Figure 47: The structure function $x_{IP}F_2^{D(3)}$ versus Q^2 for different x_{IP} and β bins. The curves are the results of a BEKW(mod) fit.

cross-section σ^{tot} has been evaluated. The ratio $r = \sigma^{diff}(0.28\,\text{GeV} < M_x < 35\,\text{GeV})/\sigma^{tot}$ is plotted against Q^2 in figure 48. The ratio can well be fitted by $r = a - b \cdot ln(1 + Q^2)$ with a=0.22 and b=0.034. This ratio varies only logarithmically with Q^2. If diffractive reactions would be higher twist processes one expects a power-law behaviour as a function of Q^2. The logarithmic dependence is an indication that inclusive diffraction in DIS is a leading-twist process.

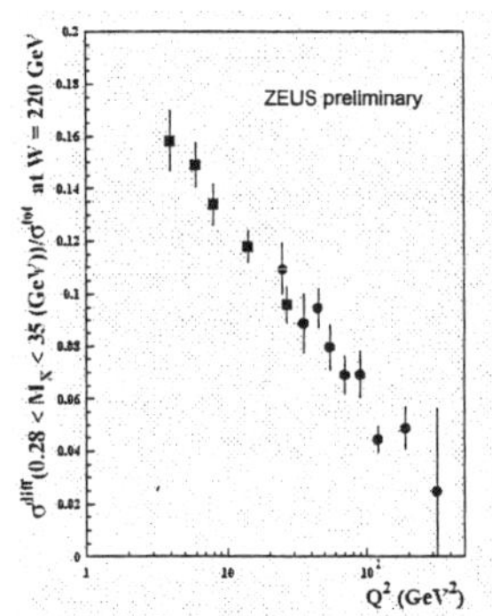

Figure 48: Ratio of the inclusive diffractive cross-section $\sigma^{diff}(0.28\,\text{GeV} < M_x < 35\,\text{GeV})$ at W of 220 GeV to the total DIS cross-section.

5.3.6 Are diffractive parton-distributions (dpdf) universal ?

The diffractive parton distribution-functions determined by the H1 DGLAP-fit to the reduced cross-section of inclusive diffraction are supposed to be universal in the sense that they can be used in perturbative QCD calculations to predict other diffractive processes. Figure 49 shows results on diffractive dijet-production from H1 compared to NLO predictions using the H1 dpdfs from fit A and fit B.

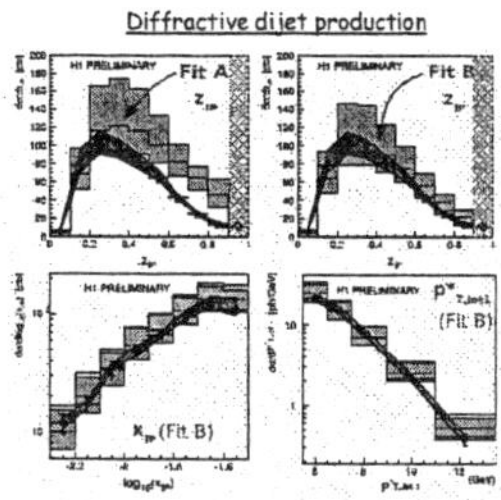

Figure 49: H1 results on diffractive dijet production. Data are shown with error bands and NLO predictions calculated using diffractive structure-functions from the H1 fits A and B and their theoretical uncertainties.

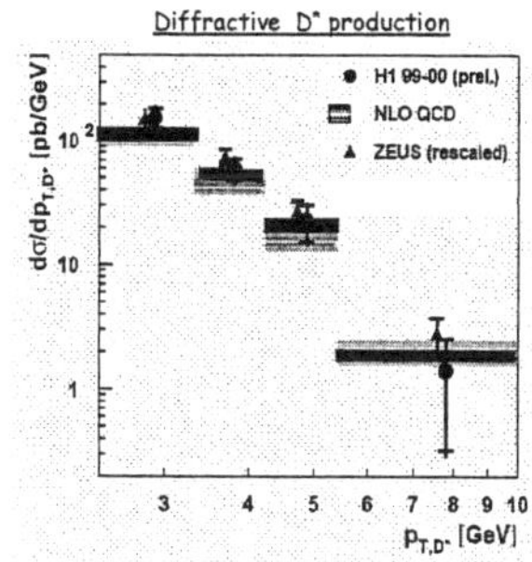

Figure 50: Results from H1 and ZEUS on diffractive D^* production. The ZEUS data have been rescaled to match the kinematical range of the H1 data. Shown are also results of a NLO calculation using diffractive structure-functions from a fit to the H1 2002 LRG results.

The predictions are the dotted lines and the corresponding error bands. The fit B gives a reasonable description of the H1 diffractive dijet-production. In figure 50 data on diffractive D^*-production from H1 and ZEUS are shown together with NLO predictions using an older H1 dpdf fit from the 2002 data [33]. The ZEUS results have been rescaled to the same kinematical region as for the H1 data. Also here the NLO calculation gives a satisfactory description of the data. This is a validation of the QCD factorization theorem and indicates that the dpdfs are universal in diffractive DIS.

5.3.7 HERA dpdfs and prediction for data from the Tevatron

At the Tevatron, the CDF experiment measured diffractive dijet-production in $p\bar{p}$ collisions [34]. The dominant contribution to this process is shown schematically in figure 51. A gluon from the proton(antiproton) collides with a gluon from the pomeron in a hard process and produces two jets.

The measured diffractive structure-function F_{JJ}^D is shown in figure 51 as a function of β. It has the form

$$F_{JJ}^D(\beta) = \beta[g(\beta) + \frac{4}{9}q(\beta)].$$

The measurements are compared to pQCD predictions which use the diffractive parton-distributions as measured by H1. It is obvious that these predictions are higher by a factor 5-7 than the data. The QCD factorization-theorem is not valid. In fact this theorem has been proven only for deep inelastic lepton-proton scattering. There is no proof for hadron-hadron scattering. The reason for this failure is the high probability for final state interactions in the production process which leads to a rapidity-gap survival probability smaller than one.

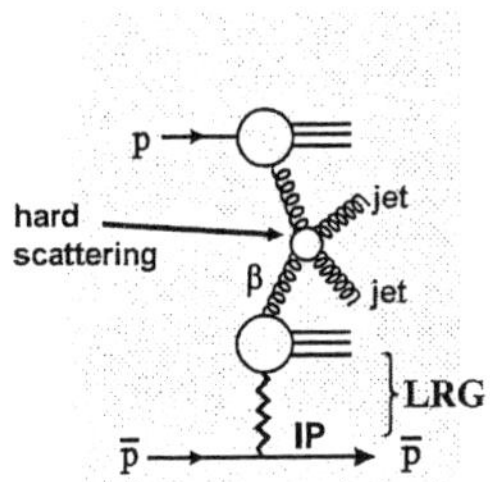

Figure 51: The process of dijet production in $p\bar{p}$ collisions

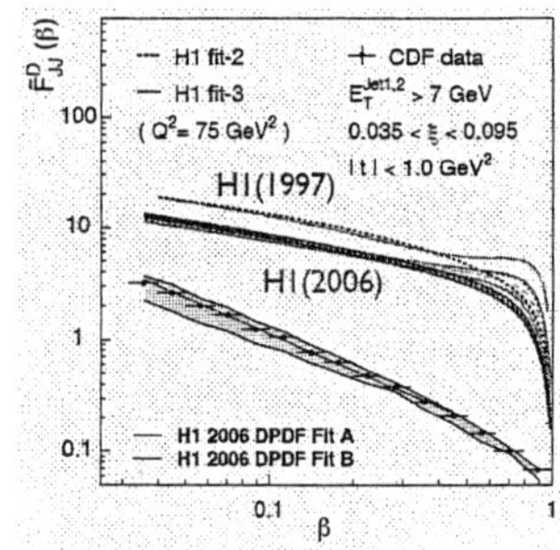

Figure 52: The structure function $F_{JJ}^{D}(\beta)$ from the CDF experiment and predictions from the H1 diffractive structure-functions.

6 Summary

The exclusive production of vector mesons in DIS is a good testing ground for the transition from soft diffractive processes to hard ones. Light vector-meson production at low Q^2 values is a soft process which can be described by the exchange of a soft pomeron with the trajectory as determined from soft hadronic processes. The vector-meson mass, the virtuality Q^2, and the four-momentum transfer squared t at the proton vertex each can provide a hard scale which makes perturbative QCD calculations of the process applicable. Deeply-virtual Compton scattering (DVCS) is a hard process and can be described by pQCD in terms of generalized parton-distributions. For inclusive diffraction three methods exist to select diffractive data samples and determine the diffractive structure-function $F_2^{D(3)}(Q^2, x_{I\!P}, \beta)$. Only the method which detects directly the diffractively scattered proton can determine $F_2^{D(4)}(Q^2, t, x_{I\!P}, \beta)$. Results from the H1 and the ZEUS experiments are in reasonable agreement taking into account that all methods correct only partly and in a different way for background contributions from proton dissociation and reggeon exchanges. The H1 collaboration determined diffractive parton distribution-functions (dpdf) from their data. Using these dpdfs in pQCD calculations to predict diffractive dijet-production and diffractive D^*-production in DIS leads to predictions which are in reasonable agreement with data for these processes measured by H1 and ZEUS. This confirms the theoretically proven QCD factorization theorem for DIS. The ZEUS collaboration use a modified BEKW model to fit their data on $F_2^{D(3)}(Q^2, x_{I\!P}, \beta)$ from the M_x method. The BEKW(mod) dipole-model gives a good description of the data and enables to separate the contributions from transverse and longitudinal $q\bar{q}$ and from $q\bar{q}gluon$ contributions. The diffractive structure-function $F_2^{D(3)}(Q^2, x_{I\!P}, \beta)$ shows considerable scaling violations which indicates a large gluon contribution in the colourless exchange at small β values. Inclusive diffraction is a leading-twist process. Predictions for diffractive dijet-production in $p\bar{p}$ processes using the dpdfs from diffractive DIS fail by a large factor as shown by measuremnts of the CDF experiment. The failure results from final state interactions leading to a rapidity-gap survival probability smaller than one.

References

[1] A Proposal for a Large Electron-Proton Colliding Beam Facility at DESY, DESY HERA 81-10 (1981).

[2] T.Regge, Nuov.Cim.14 (1959) 951; Nuov.Cim **18** (1960) 947 ;
P.Collins, An Introduction to Regge Theory and High Energy Physics, Cambridge University Press, Cambridge (1977).

[3] G.Chew,S.Frautschi, Phys.Rev.Lett. **7** (1961) 394;
N.Gribov, Sov.Phys.JETP **14** (1962) 478 ; ibid. 1395.

[4] A.Donnachi and P.V.Landshoff, Nucl.Phys. **B244** (1984) 322 ;
A.Donnachi and P.V.Landshoff, Phys.Lett. **B296** (1992) 227.

[5] L.N.Lipatov, Sov.J.Nucl.Phys. **23** (1976) 338 ;
E.A.Kuraev, L.N.Lipatov and V.S.Fadin, Sov.Phys.JETP **45** (1977) ;
Y.Y.Balitsky and L.N.Lipatov, Sov.J.Nucl.Phys. **28** (1978) 822.

[6] J.J.Sakurai, "Currents and Mesons", University of Chicago Press, Chicago (1969);
H.Fraas and D.Schildknecht, Nucl.Phys. **B14** (1969) 543.

[7] C. Berger et al., Phys. Lett. **39B** (1972) 659;
SWT Collab., Y. Eisenberg et al., Phys. Rev. **D5** (1972) 15;
J. Park et al., Nucl. Phys. **B36** (1972) 404;
SBT Collab., J.Ballam et al., Phys. Rev. **D5** (1972) 545;
SBT Collab, J.Ballam et al., Phys. Rev. **D7** (1973) 3150;
G.E. Gladding et al., Phys. Rev. **D8** (1973) 3721;
W. Struczinski et al., Nucl. Phys. **B108** (1976) 45;
R. M. Egloff et al, Phys. Rev. Lett. **43** (1979) 657.
OMEGA Photon Collab., D. Aston et al., Nucl. Phys. **B209** (1982) 56;
H1 Collab., S. Aid et al., Nucl.Phys. **B463** (1996) 3;
ZEUS Collab., J.Breitweg et al., Eur.Phys.J. **C2** (1998) 247.

[8] H.R. Crouch et al., Phys. Rev. **155** (1967) 1468;
ABBHHM Collab., R. Erbe et al., Phys. Rev. **175** (1968) 1669;
M. Davier et al., Phys. Rev. **D1** (1970) 790;
SWT Collab., Y. Eisenberg et al., Phys. Rev. **D5** (1972) 15;
SBT Collab., J. Ballam et al., Phys. Rev. **D7** (1973) 3150;
W. Struczinski et al., Nucl. Phys. **B108** (1976) 45;
R.M. Egloff et al., Phys. Rev. Lett. **43** (1979) 1545 and **44** (1980) 690 (Erratum);
A.M. Breakstone et al., Phys. Rev. Lett. **47** (1981) 1782;
D. Aston et al., Nucl. Phys. **B209** (1982) 56;
LAMP2 Group, D.P. Barber et al., Z. Phys. **C26** (1984) 343;
OMEGA Collab., M. Atkinson et al., Nucl. Phys. **B231** (1984) 15;
J. Busenitz et al., Phys. Rev. **D40** (1989) 1.

[9] J. Busenitz et al., Phys. Rev. **D40** (1989) 1;
R. Erbe et al., Phys. Rev. **175** (1968) 1669;
C. Berger et al., Phys. Lett. **B39** (1972) 659;
J. Ballam et al., Phys. Rev. **D7** (1973) 3150;
H.J. Besch et al., Nucl. Phys. **B70** (1974) 257;
H.J. Behrend et al., Phys. Lett. **B56** (1975) 408;
D.P. Barber et al., Phys. Lett. **B79** (1978) 150;
D. Aston et al., Nucl. Phys. **B172** (1980) 1;
M. Atkinson et al., Z. Phys. **C27** (1985) 233;
ZEUS Collab., M. Derrick et al., Phys.Lett. **B73** (1996) 259.

[10] B. Knapp et al., Phys. Rev. Lett. **34** (1975) 1040;
U. Camerini et al., Phys. Rev. Lett. **35** (1975) 483;
B. Gittelman et al., Phys. Rev. Lett. **35** (1975) 1616;
T. Nash et al., Phys. Rev. Lett. **36** (1976) 1233;
BFP Collab., A.R. Clark et al., Phys. Rev. Lett. **43** (1979) 187;
EMC Collab., J.J. Aubert et al., Nucl. Phys. **B213** (1983) 1;
NA14 Collab., R. Barate et al., Z. Phys. **C33** (1987) 505;
E687 Collab., P.L. Frabetti et al., Phys. Lett. **B316** (1993) 197;
NMC Collab., M. Arneodo et al., Phys. Lett **B332** (1994) 195;
E401 Collab., M. Binkley et. al., Phys Rev. Lett. **48** (1982) 73;
E516 Collab., B.H. Denby et al., Phys. Rev. Lett. **52** (1984) 795;
ZEUS Collab., J.Breitweg et al., Z.Phys. **C75** (1997) 215;

ZEUS Collab., S.Chekanov et al., Eur.Phys.J. **C24** (2002) 345;
H1 Collab., C.Adloff et al., Phys.Lett. **B483** (2000) 23;
H1 Collab., C.Adloff et al., Contribution to DIS 2003, St.Petersburg, Russia.

[11] H1 Collab., C.Adloff et al., Phys.Lett. **B541** (2002) 251;
ZEUS Collab., Contribution to International Europhysics Conference on High Energy Physics, July 2001, Budapest, Hungary.

[12] H1 Collab., C.Adloff et al., Phys.Lett. **B483** (2000) 23;
ZEUS Collab., J.Breitweg et al., Phys.Lett. **B437** (1998) 432.

[13] L.Frankfurt, M.McDermott and M.Strikman, Journal of High Energy Physiscs 103 (2001) 45;
CTEQ Collab., H.L.Lai et al., Eur.Phys.J. **C12** (2000) 375;
A.D.Martin et al., Eur.Phys.J. **C4** (1998) 463.

[14] H1 Collab., C.Adloff et al., Eur.Phys.J. **C13** (2000) 371;
H1 Collab., S.Aid et al., Nucl.Phys. **B468** (1996) 3;
Zeus Collab. Contributed Paper 594 to European Physical Society Conference 2001, Budapest, Hungary;
ZEUS Collab., J.Breitweg et al., Eur.Phys.J. **C6** (1999) 603.

[15] ZEUS Collab., Contributed Paper 813 to ICHEP 2002, Amsterdam, The Netherlands;
ZEUS Collab., J.Breitweg et al., Eur.Phys.J. **C6** (1999) 603;
H1 Collab., S.Aid et al., Nucl.Phys. **B468** (1996) 3.

[16] A.D.Martin, M.G.Ryskin and T.Teubner, Phys.Rev. **D62** (2000) 14022;
CTEQ Collab., H.L.Lai et al., Eur.Phys.J. **C12** (2000) 375.

[17] L.Frankfurt, W.Koepf and M.Strikman, Phys.Rev. **D57** (1998) 512.

[18] J.Bartels et al., Phys.Lett. **B375** (1996) 301.

[19] D.Yu.Ivanov, Phys.Lett. **B478** (2000) 101;
D.Yu.Ivanov, Phys.Rev. **D53** (1996) 3564.

[20] E.Wieczorek, V.A.Matveev and D.Robaschik, Theor.Mat.Phys. **19** (1974) 315;
J.Bartels and M.Loewe, Z.Phys. **C12** (1982) 263;
X.Ji, Phys.Rev. **D58** (1997) 7114 ;
A.V.Radyushkin, Phys.Rev. **D58** (1998) 114008.

[21] L.L.Frankfurt, A.Freund and M.Strikman, Phys.Rev. **D58** (1998) 114001,
Erratum ibid. **D59** (1999) 119901;
X.Ji, Phys.Lett. **B78** (1997) 610;
K.Golec-Biernat and A.Martin, Phys.Rev. **D59** (1999) 14029.

[22] ZEUS Collab. S.Chekanov et al., DESY-03-59, hep-ex/0305028, submitted to Phys.Lett.B.

[23] G.Ingelman, P.Schlein, Phys.Lett. **B152** (1985) 256.

[24] R.P.Feynman, "Photon-Hadron-Interactions", Benjamin, N.Y. (1972), lectures 50-54.

[25] Zeus Collab.,M.Derrick et al., Z.Phys. **C70** (1996) 391;
Zeus Collb., J.Breitweg et al., Eur.Phys.J. **C6** (1999) 43.

[26] J.Collins, Phys.Rev. **D57** (1998) 3051 and erratum-ibid. **D62** (2000) 19902.

[27] H1 Collab., A.Aktas et al., Eur.Phys.J. **C48** (2006) 749

[28] ZEUS Collab. S.Chekanov et al., Eur.Phys.J. **C38** (2004) 43.

[29] H1 Collab., A.Aktas et al., Eur.Phys.J. **C48** (2006) 715

[30] ZEUS Collab. S.Chekanov et al., Nucl.Phys. **B713** (2005) 3.

[31] B.Löhr, Proceedings of Science, Diffraction06, International Workshop on Diffraction in High Energy Physics, Adamantes, Milos Island, Greece, September 5-10, 2006

[32] J.Bartels, J.Ellis, H.Kowalski, M.Wuesthoff, Eur.Phys.J. **C7** (1999) 443.

[33] H1 Collab., Contributed Paper 980 to ICHEP2002, Amsterdam, July 2002

[34] CDF Collab., T.Affolder et al., Phys.Rev.Lett. **84** (2000) 5043

CHAIRMAN: B. LÖHR

Scientific Secretaries: C. Andrei, G. Brandt

DISCUSSION

- *Smith*:

What are the expectations of seeing diffractive events in the LHC?

- *Löhr*:

There has been a whole industry started to look for diffractive events at LHC. Actually, diffractive events at LHC gained attention after a rough calculation that was ready before the machine, because of the massive energy flow close to the beam direction. There is a group preparing detectors at LHC, up to 420 meters down in the tunnel. This is a collaboration between already various partners. At the moment the UK group has the lead in that. It is not first day physics to be done, there is an upgrade needed for the machine but the machine people, as well as experimentalists are working on that.

- *Lopez-Mateos*:

What is the difference between Reggeon and Pomeron?

- *Löhr*:

Reggeons (or reggeon trajectory in spin-mass-space) are realistic objects in the sense that they are objects on a trajectory on which one observes physical particles. Instead of exchanging a single physical particle, Regge theory leads to a formalism in which the whole row of particles is being exchanged. As I have told you all these reggeon trajectories have an intercept $\alpha(0)$ smaller than one. As the energy dependence of the total cross section in hadron-hadron-interaction is proportional to $s^{\alpha(0)-1}$ all of these exponents are negative. That means that the cross section falls with the energy. But we know experimentally, that the total cross section does not fall, it rises logarithmically.
This led to the introduction of a purely hypothetical trajectory which was then called the Pomeron trajectory with an intercept of slightly above one. As I always say the Pomeron (IP) is the parameterization of our ignorance.

- *Csanad*:

Is there any kind of Bose-Einstein correlation (HBT) measurements made at H1 or ZEUS similar to the ones that were done at L3 or UA1?

- *Löhr*:

No, not for diffractive events. There may be analysis going on, of which I don't know, but there are no results published yet. (Addition: There are publications on Bose-Einstein correlations for DIS events by the H1 Collaboration in 1997 and by the ZEUS Collaboration in 2003.)

- *Csanad*:

And there is also no plan yet to ...

- *Löhr*:

At least I don't know. I can make this statement only for ZEUS. I don't know of any.

- *Pacetti*:

Could you, in principle, see a final state like *epX* where *X* represents two mesons instead of a unique vector meson?

- *Löhr*:

Probably only at very special conditions, because you start out with something which has the quantum numbers of the photon and something which has the quantum numbers of the vacuum and now you want me to make two vector mesons out of it. Maybe you can arrange the quantum numbers in such a way that this is possible. I don't know out of my head.

- *Pacetti*:

But, for instance, if you have $\Phi\eta$ or $\Phi\eta'$ in the final state it seems like this final state can be mediated by vector mesons. In principle you can have this kind of a state.

- *Löhr*:

What I can say clearly: of course there are events where you have vector mesons plus anything else, including vector mesons prime. This of course exists. Whether what you suggest is possible, I would have to think of all kinds of combinations of numbers to see if this is possible or not.

- *Pinfold*:

The "Brother" of the Pomeron is the Odderon. HERA should be a great place for the Odderon. Could you briefly summarize the status of the search?

- *Löhr*:

Yes, it is relatively difficult to look for the Odderon, but nevertheless the H1 collaboration has done it, with negative results. However this is not statistically very strong. So the answer is: we haven't found anything, but to make an absolute statement that this Odderon does not exist, I mean you don't expect it really to exist, I think we cannot make yet.

- *Brandt:*

There was a special detector for that installed in H1. It was decided to replace it by another piece of equipment.

- *Goldstein*:

(Comment: Pomeron was introduced before the total cross section was known to rise.). Is it true that for very small x inclusive vector meson production at high Q^2 and high $|t|$, $d\sigma/dt$ behaves like QCD rather than Regge?

- *Löhr*:

First of all let me make a small correction: I didn't show anything about inclusive vector mesons. I have shown inclusive diffraction. It was the system X that was produced and we haven't looked at it to see what it is. For that I have shown results. And what I have shown you were fits which have been done this way, so the three dimensional structure function just derived from the number of measured events as a function of x_{IP}, β and Q^2 is parameterized for two components: one coming from the Pomeron and one contribution from the Reggeon. For these structure functions of the Reggeon, the pion structure function was assumed. The flux for the Reggeon was parameterized as I have shown before, on the previous slide. The flux of the Pomeron was also specified before and then this formula was fitted to the data and what you get out is, of course, the $F_2^{D(3)}$. Now, what is plotted in this figure (slide 22) is the whole fit: the Pomeron and the Reggeon contributions. And therefore you see that this is going up here at higher x_{IP} and this is the Reggeon contribution. Here (slide 22) they plot all their data and you can see that the Reggeon contribution is there, but the part where you don't expect Reggeon contribution, which means at $x_{IP} < 10^{-2}$, there you have Pomeron contribution only. And these are actually the QCD fits to this Pomeron part. That means that they have been parameterized at $Q^2 = 8$ GeV^2. This parameterization of the parton density functions has been evolved according to DGLAP, and then you arrive at this parton density function and from this parton density function you arrive at this kind of singlet structure function and gluon structure function. This is what this plot shows. So far you can say that the Pomeron part, or rather call it the diffractive part, let's forget about this Pomeron, can be fitted by QCD evolved parton densities.

- *Goldstein*:

So these parton densities are what we used to call "the Reggeon residues"?

- *Löhr*:

Yes.

- *Sens*:

Without wanting to make any priority claims I would like to mention for the record that at the CERN ISR in 1972-78 at $25 < \sqrt{s} < 50$ GeV extensive studies were made of elastic, inelastic and diffractive events with a.o. the result that the "flat" rapidity distributions were in fact rising in the center of the rapidity plot. The data were obtained with a 30m long, 5 magnet spectrometer at small angles, of the CERN-Holland-Lancaster-Manchester (CHLM) collaboration partly in coincidence with the British-Scandinavian spectrometer at large angles.

- *Löhr*:

Thank you for this remark but I understood my presentation here, clearly, as the development at DESY. There are many places in the world where diffractive reactions are measured, not only at DESY. I have shown just one slide for the Tevatron, but there is a whole industry going on at the Tevatron experiments measuring diffractive reactions. The purpose of this presentation was not to give a review of everything that we know today on diffraction. That would have probably taken days.

- *Kaur*:

What is the level of inefficiency in separating diffractive and non-diffractive events in ZEUS data?

- *Löhr*:

In the ZEUS data we have actually separated them at a statistical level as you have seen (Slide 17, right plot). And now, I have to say a little bit more. Of course we have non-diffractive data which goes underneath here, which we cut out. We have verified experimentally rather clearly that the non-diffractive data follows this line which is derived from the fit. This includes usual non-diffractive data, and this includes also Reggeon contributions. They also lie beyond here. What we are doing then, we subtract what is below here in bins of M_x, to determine the cross section $d\sigma/dM_x$. And we take only bins in the analysis were the diffractive part is at least 50%, which means that we subtract less then 50%. In the next bin, because this is an exponential distribution, it is already very much less.
Now, what you are probably asking is: what is the statistical error in subtracting these things. This is typically at the order of a few percent, up to 10% at the highest M_x where we take data, and it goes very quickly down, so this is not a concern. This is not the highest systematic error in the analysis

- *Lopez-Mateos*:

Why does perturbative QCD predict no shrinkage?

- *Löhr*:

I have said that QCD inspires models. The QCD inspiration is that, of course, we would like to understand the Pomeron in terms of physical quantities that we understand. And this, of course, means understanding it through the exchange of gluons and quarks. But models based on this understanding predict little or no shrinkage. They do not make a definite statement, they all say that α' is less then 0.25.

- *Lopez-Mateos*:

Is that a sign of new physics?

- *Löhr*:

No. The shrinkage is just a phenomenon which is seen in soft peripheral processes and it is no longer there in hard peripheral processes. The interesting thing that I haven't been able to work out is that with these diffractive processes we see the transitions from soft interactions (not described by perturbative QCD) to hard interactions that can be described by perturbative QCD. So there is of course a transition from here to there. I mean: make this diagram as complicated as you want and you are in the non-perturbative region. Now the interesting thing is that starting from models that describe this initial diagram you may find a way, and that is the hope, to come deeper down into the non-perturbative region and try to understand more about techniques to calculate things in the non-perturbative region.

- *Lopez-Mateos*:

Is there any effective field theory that might predict it?

- *Löhr*:

To my knowledge not, but my knowledge could be very limited in this case.

- *Machado*:

Has baryon number flow in large rapidity gaps been studied in HERA?

- *Löhr*:

No, not that I know of. These measurements, at least at ZEUS but also at H1, are now probably very difficult since we have upgraded our detectors and the rapidity gap in the forward direction becomes more and more complicated and the special detector that was installed in ZEUS in the forward direction is gone because there are now superconducting magnets of the machine. So, unfortunately, we can still probably do something but it is not clear yet to what extent.

- *Sens*:

I would propose to seize the occasion of the presence of several practitioners of diffractive studies to summarize the data on the total and diffractive reactions of *pp* and *ep* collisions as a function of $\sqrt{s}$.

- *Löhr*:

That is asking a little bit too much, but what I can show you here is the total *ep* cross-section as a function of $W_{p\gamma}$ and that is plotted for Q^2 equals zero, that means photoproduction $\gamma p \rightarrow X$.

- *Sens*:

So you concluded that there is a slight rise there?

- *Löhr*:

Yes, there is a slight rise in that cross section.

- *Sens*:

Now, if you take the diffractive part of that, which in the *pp* case is about 10% of the total, there has never been an explanation for the rise of that total cross section in *pp*. How does that stand here? Do you have a number for the diffractive part of *ep* cross section?

- *Löhr*:

Yes, as a function of Q^2. It is this plot here (Slide 29), which has been made at a particular $W_{\gamma p}$. The ratio is actually constant with $W_{\gamma p}$. This is at a particular $W_{\gamma p}$ and against Q^2. You see, more or less, the total diffractive cross-section because we have integrated over all masses from 0.2 to 35 GeV^2 . This falls logarithmically, from which we conclude that inclusive diffraction is of leading twist. And here you see what the fraction is: at low Q^2 it goes up to about 20% and then is falls down. So, when you integrate over that Q^2, you get about 20-25%.

2006 International School of Subnuclear Physics in Erice

Lecture notes by Horst Wenninger held on 5 September 2006

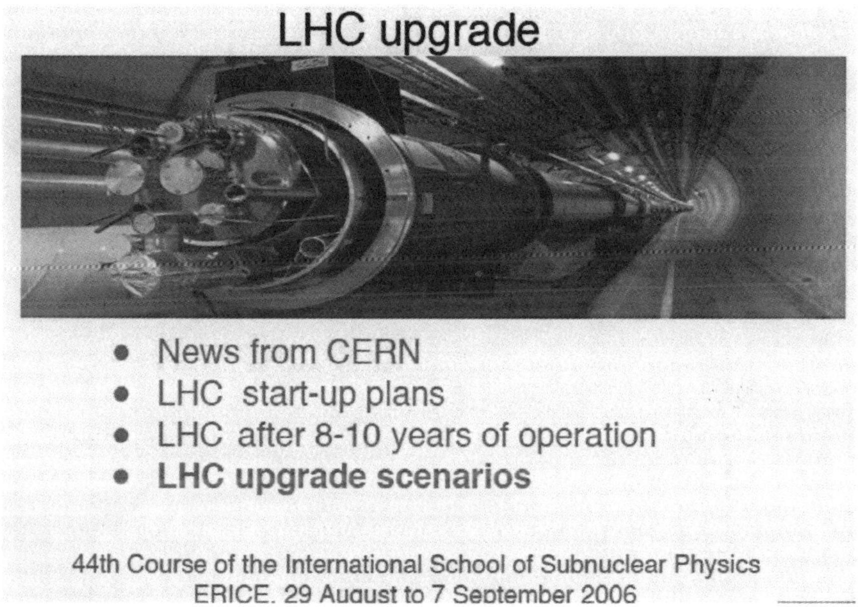

My lecture on the LHC upgrade scenarios includes a report about the latest news from the CERN Management to the CERN Council confirming the intension to start LHC commissioning in the fall of the year 2007. The superconducting magnets for LHC have all been delivered by end 2006 and after installation the machine should be closed by August with first beam in the ring towards the end of 2007.

The LHC Construction and Installation schedule can be followed on-line on the CERN LHC WEB page; also presented is the list of machine parameters and projected maximum design luminosity for LHC of 10^{34} cm^{-2} s^{-1}. This luminosity is indeed very large compared to present collider performances (1000 times LEP and 50 times Tevatron II) and will only be reached after several years of running.

Together with the LHC machine construction progress, impressive news can be reported on all detector construction and installation works. (example shown below: the production of the Time-of-Flight system of the ALICE detector at the University of Bologna). A series of LHC reports summarize the key questions of particle physics, addressed by LHC experiments, illustrate the signatures used to identify interesting events from large backgrounds, and outline major cross-sections and production rates of physics processes, that can be studies with the four major LHC detectors.

Turning to future LHC upgrade plans, it is worthwhile to underline the LHC challenges resulting from the high energy and luminosity of its proton beams. Young-Kee Kim from The University of Chicago has compared LHC with the present-day most powerful accelerator, the Tevatron at Fermilab.

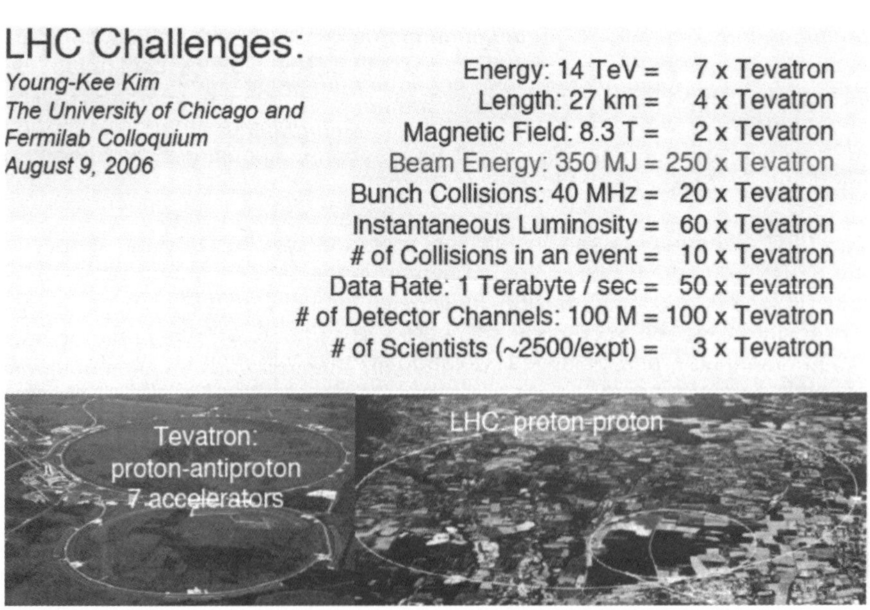

The challenge lies in the amount of stored beam energy of LHC, which is 250 times that of the Tevatron. Careful LHC beam commissioning is obviously mandatory to understand proper beam collimation and dumping and to avoid damage or quenching of the superconducting accelerator ring.

An expert group, set-up by the CERN Director-General, has studied various LHC upgrade scenarios.

LHC luminosity upgrade: why and when?

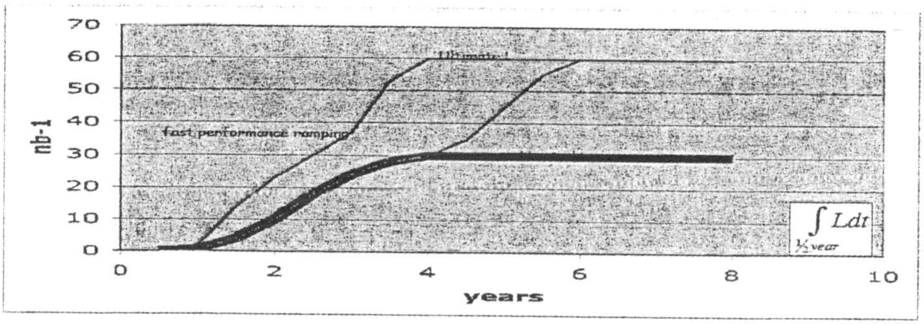

How fast performance is expected to increase:

4 y up to nominal *L*
4 y up to nominal *L* & 2 y up to ultimate *L*
4 y up to ultimate

Their recommendations are in order of priority: (i) acute need for consolidation of the LHC injector chain; following four years of running-in and operations, a new linac should be installed for the PS. This needs to be prepared as of today by intensive R&D and should improve operations stability; (ii) after eight years of running most urgent physics questions should have found answers by LHC results

LHC Luminosity/Sensitivity Evolution?

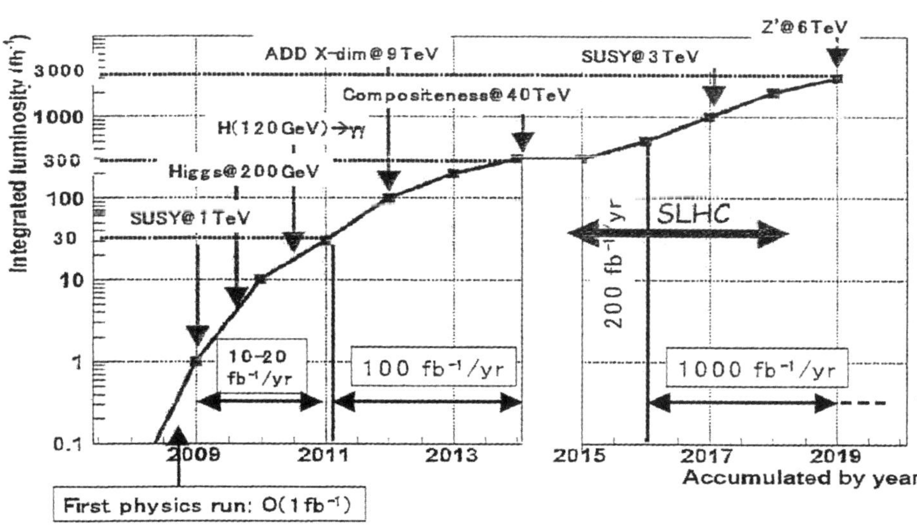

By that time both the machine and detectors in the interaction region will have reached a level of radiation damage that requires replacements of components.

So, around 2015 will be a good time to further upgrade LHC either in beam energy and/or in luminosity. Various scenarios are outlined in the following self-explaining slides (courtesy by Scandale).

scenarios for the luminosity upgrade

Phase 0:

steps to reach ultimate performance without hardware changes:

1) collide beams only in IP1 and IP5 with alternating H-V crossing
2) increase Nb up to the beam-beam limit L = 2.3 x 10^{34} cm-2 s-1
3) increase the dipole field from 8.33 to 9 T Emax = 7.54 TeV

The ultimate dipole field of 9 T corresponds to a beam current limited by cryogenics and/or by beam dump/machine protection considerations.

Bunch scheme

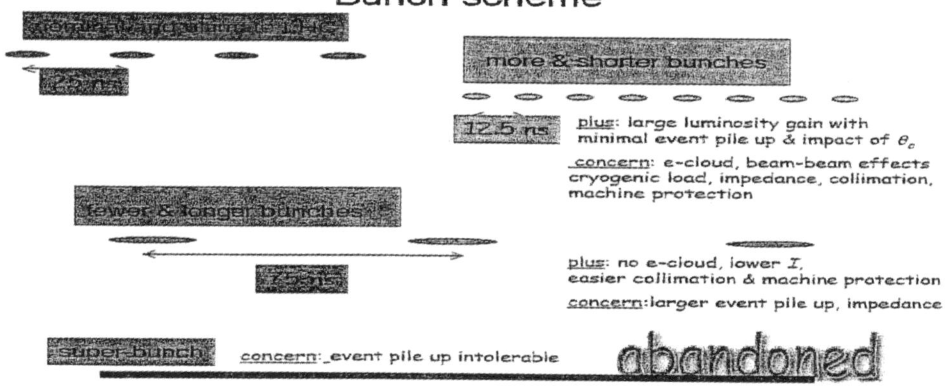

LHC performance and parameters

Parameter [units]	Nominal	Ultimate	Short bunch	Long bunch
No. of bunches n_b	2808	2808	5616	936
p^+ × bunch N_b [10^{11}]	1.15	1.7	1.7	6.0
Bunch spacing Δt_{sep} [ns]	25	25	12.5	75
Beam current [A]	0.58	0.86	→ 1.72	1.0
E_{beam} [MJ]	366	541	→ 1085	631
Beta at IP β^* [m]	0.55	0.50	0.25	0.25
Xing angle θ_c [μrad]	285	315	445	430
Bunch length [cm]	7.55	7.55	3.78	14.4
Piwinski ratio $\theta_c \sigma_s/(2\sigma^*)$	0.64	0.75	0.75	2.8
L lifetime τ_L [h]	15	10	6.5	4.5
L_{peak} [10^{34}cm^{-2}s^{-1}]	1.0	2.3	9.2	8.9
$T_{turnaround}$ [h]	10	10	5	5
Events per Xing	19.2	44.2	88	→ 510
$\int_{one\ year} L\, dt$ [fb^{-1}]	66.2	131	560	410

ε_n = 3.75 mm in all the options

scenarios for the luminosity upgrade

Phase 1:

steps to reach maximum performance with IR and RF changes:

1) modify the insertion quadrupoles and/or layout $\beta^* = 0.25$ m
2) increase crossing angle by $\sqrt{2}$ (445 μrad)
3) increase Nb up to ultimate luminosity $L = 3.3 \times 10^{34}$ cm^{-2}s^{-1}
4) halve sz with high harmonic RF system $L = 4.6 \times 10^{34}$ cm^{-2}s^{-1}
   - new RF system in LHC providing an accelerating voltage of 43MV at 1.2GHz
   - a power of about 11MW/beam - large estimated cost
   - a longitudinal beam emittance reduced to 1.78 eVs
   - horizontal Intra-Beam Scattering (IBS) growth time will decrease by about $\sqrt{2}$
5) double no. of bunches nb (increasing crossing angle) $L = 9.2 \times 10^{34}$cm^{-2}s^{-1}
6) (exceeding ultimate beam intensity)
   - upgrade LHC cryogenics, collimation and beam dump systems
   - upgrade the electronics of beam position monitors
   - possibly upgrade the SPS RF system and other equipments in injector chain

LHC machine and detector upgrades are due in some 8 to 10 years after LHC start-up and need funds and R&D preparation as of today.

Concluding remarks

A vigorous R & D programme is required on

- optics, beam control, machine protection, collimation
- high gradient high aperture SC quadrupoles
  - → Nb_3Sn SC wire and cable
  - → radiation-hard design
- RF & crab-cavities
- SC fast ramping magnets
  - → Nb-Ti SC wire and cable
  - → High speed energy removal & radiation-hard design
- for energy upgrade
  - → Nb_3Sn SC wire and cable high field (> 15 T)
  - → Low cost SC & magnets
- detector upgrade should be planned to handle higher L and larger radiation level

Refences

Lyn Evans, LHC Project Leader private communication

EPAC 2006 report presented by Walter Scandale

S. Gilardoni, E. Métral, O. Ullaland, A. De Roeck, K. Jakobs, G. Ross, A. Pic
CERN Summerstudent lectures 2006; LHC experiments web page fur public

Theoretical Aspects of the Behaviour of Beams in Accelerators and Storage Rings: International School of Particle Accelerators of the 'Ettore Majorana' Centre for Scientific Culture, 10–22 November 1976, Erice, Italy, M.H. Blewett (ed.), CERN report 77-13 (1977)
http://preprints.cern.ch/cgibin/setlink?base=cernrep&categ=Yellow_Report&id=77-13

CERN Accelerator School http://cas.web.cern.ch/cas/

CHAIRMAN: H. WENNINGER

Scientific Secretaries: I. Gnesi, P. Jindal

DISCUSSION

- *Csanad:*

When there are already two upgrade programs at RHIC, is it realistic to also upgrade LHC and to build ILC?

- *Wenninger:*

I think the community will not be so unwise to invest in LHC and then not exploit it to the maximum of its potential. I think that 10 years after the running there is a possibility to get another factor of 10 in luminosity, and 20 to 30% in mass range. From 2015 to 2020 you will do the upgrade, because you will not have another facility ready at that time to replace LHC.

- *Csanad:*

Basically, because the tunnel is already there and it's...

- *Wenninger:*

I think it is clear that we have probably seen that they closed ISR, although there was still potential to do good physics, in order to get LEP, and we closed LEP, although there was still potential, in order to get LHC, so what you ask yourself is 'Will the same thing happen to LHC in 10 years; to have a new project will we sacrifice LHC?' Is this your question?

- *Csanad:*

So basically you will continue with RHIC and CERN, or CERN and BNL, so that they both get upgraded every10 years.

- *Wenninger:*

I think we have to be prepared for the upgrade, it means we have to do all the necessary R&D from now to 2015, when it's a good moment to do the upgrade. I personally don't believe the accelerator project of a machine twice as expensive as LHC will come in that time scale. To just use the money of LHC operations to built the next machine is not feasible nor reasonable.

- *Neri*:

Do you think that the "Crab waist" technique developed in the context of the Super B Factory project by P.Raimondi will be helpful for the LHC luminosity upgrade?

- *Wenninger:*

Absolutely! Because, as you go up in beam current and luminosity, and you have to work on the long-range beam-beam effects, you have to find a way how to keep the crossing angles while optimizing the overlap of bunches. Actually we had crab cavities in the Omega beam at CERN 20 years ago to separate Kaons. It is the same idea behind this, and we are happy if it is now tested in the case of LHC. It is really a question, how you get maximum overlap at the focus without being disturbed by high-current long-range beam-beam effects near the interaction point. This is part of the R&D discussions, and I am sure that CERN people will participate in your test somehow, because the machine people need this sort of solutions. And if it is done in view of the super beauty factory, as we were told last week, it's beautiful. We should combine all the know-how we have in order to make maximum use of it for LHC in ten years from now, or in whatever time we can. By the way, one thing I wanted to correct... We did leave a wrong impression of CLIC needing power equivalent of many nuclear power reactors. Wrong numbers were multiplied. Actually, the official CLIC power (AC power) for 3 TeV CLIC is 410 MW which is twice what CERN uses, anyway. So it is not a power completely out of range.

By the way, I also found a nice publication that says there was a beautiful talk at Erice on the CLIC scenario, so you will probably find it in the documents of the Erice schools of last year.

Another question that was coming up was, what do you gain by a factor of 10 in luminosity, well, now studies of ATLAS+CMS estimate about 25-30% mass range gain in the Higgs sector and SUSY sector. So I refer to the publications you find, especially, in the CERN summer schools of this August. It is maybe easier to increase the luminosity of a machine, rather than for the detector, to handle it. So once more, I think you can observe on the LHC schedule the planned time for the cool-down and the powering of an octant. The successful processing of this will show one has resolved every problem, which was not yet seen, maybe, which arises when powering the whole octant of the machine. But this is a crucial test. Of course Fermilab's, luminosity is limited by missing antiprotons. For a proton-proton machine 10^33 is a nice luminosity, and that's what one will start with .10^34 is the first challenge upgrade for many reasons. But especially because the injector chains are not optimized for this sort of LHC energies, and that's why many people feel that you should have pulsed magnets, pulsed PS, you should have a new linac, and then you can optimize the injector parameters to the LHC energy. This will be probably the first major improvement on the luminosity and long stability of the machine.

- *Preghenella:*

Is it possible to build an electron machine in order to be able to collide both e+ and e- with the LHC protons?

- *Wenninger:*

Well there are some studies and you also can see a miniCLIC collide with LHC.

- *Preghenella:*

I am thinking about CLIC energies.

- *Wenninger:*

Everything is possible, but for the moment I think there are ways how you could do low-energy electron collisions in parallel with collisions in the LHC. You see when HERA was approved, we said we keep the electron machine of LEP and LHC in the same tunnel and then physicists have the option of ep collisions at any moment, because you have two machines at the same time. Would HERA have had spectacular physics results, then pressure from the users would have forced this, maybe it would have happened. The electron machine was finally taken out, because it would anyway need new investment, as the energy was not optimized for polarized electrons. But you can always add this to an existing proton machine and some studies have been done.

- *Preghenella:*

From the plot of the integrated luminosity it seems that the LHC upgrade will take more or less one year. Is it a realistic prediction?

- *Wenninger:*

I think this is something one has to learn to live with, that any intervention is a few months for the detectors and for the machine, and if you completely redesign in the interaction region, it certainly is of the order of a year. You see, thc process as is planned now is that you have a proton run and at the end of each year you have a heavy ion run. Of course all this is discussed by the physics community. The scenario is that all the machine planning is based on physics requirements.

- *Lopez Mateos:*

While the commissioning of the machine is under way, will we be able to take data? If yes, then what will the challenges of understanding these data be? If not, how long are the commissioning steps going to take?

- *Wenninger:*

I think this proton machine needs interaction with the detectors at all stages, and the more you can participate even at the injection energy, first collisions, which is assumed that you do, the more one also learns about the machine and the luminosity. It is all interlinked. The proton machine is not a machine where the detector does something and the machine does something, it's is a combined effort and therefore I think all the experiments will try to see as much as they can, as long as it is not too dangerous for their tracker or something like that. In principle, I think it is planned, that the physics starts with 450 GeV collisions at the injection and then goes up to higher energies.

- *Lopez Mateos:*

Well, all your processes take a short time...

- *Wenninger:*

I think it will be tested, and then one knows that all the equipment is safe and you don't lose beam into the magnet, and then you will be on. You may be afraid that the machine will make a hole in your beam pipe. I think the real problem now is that we have a superb design of a collimation system, but we don't know how much collimation we really need to avoid that the machine magnet quenches etc. So once this is experienced the confidence will grow and, then, of course, the luminosity also. And it is this process which goes together with the experiment and all the monitors etc. But we would like to use much less collimation, than we now have planned for, but we have to plan, because we can know how much collimation we need, and the accident we had at the injection was just simply a failure of the power supply for microseconds, it didn't set the septum at the right moment. This sort of things we have to test out. This is how the time will be used in the first months.

- *Csanad:*
What happened to the TOTEM experiment?

- *Wenninger:*
It is a combined experiment, with CMS.

- *Sens:*
The connection between LHC and the GRID... At the startup, I imagine that there will be decisions to be made as how many and with what factors you will pre-trigger the incoming rate, assuming you will have 10^11 protons per bunch. And then there will be pre-scaling and then pre-triggering and then there will be triggering and then there will be an assembly onto the pipeline for these events, and then comes the question where this GRID comes in, assuming not the first days or the first weeks, but in the long run. Is there an organization now (the answer is undoubtedly yes. Could you detail it?) for sending events in large quantities overseas by satellites and by lines to different laboratories, participating in the analysis of the data.

- *Wenninger:*
I think the GRID challenges are ongoing activities, of course, on simulated data which have been transferred and which are really analyzed by many centers, so the GRID challenge, as I call it, on simulated data is an ongoing activity to learn the whole process.

- *Sens:*
Is this all already in place?

- *Wenninger:*
Well, it is in good advance, even though not all the machines are installed at CERN, which you want for full or higher luminosity, the system is under test, and as I understand between Italy, Scandinavia and CERN the GRID challenges are successfully ongoing activities in parallel with the machine construction.

- *Sens:*
Right away, at the startup, you'll have to decide where you pre-scale, pre-trigger this-that and other things. I myself have been working seven years on a beam line which had 10^11 protons on the target, and so I know what sort of problems you are going to run into, and one has to be very careful about....

- *Wenninger:*
I think you are completely right. It is clear that the GRID was developed, while the LHC was under construction and testing, and naively one assumed that the GRID means you do the computing in the member states and very little at CERN. In the end, I

think, that the first line 0 TIER at CERN is really what will be necessary. There is an organization behind, from each experiment and from the CERN computer center which certainly work together.

- *Sens:*

And a related little detail: you mentioned collimators, are these collimators remote controlled nowadays?

- *Wenninger:*

Oh, yes!

- *Sens:*

And the opening angles, are they under remote control?

- *Wenninger:*

Yes! Not all phases are complete in the design yet, because one knows that at full luminosity you have to do more, once you understand how it works.

- *Pinfold:*

Have there been any studies on the use of polarized beams at the LHC?

- *Wenninger:*

Yes, I think there is a study group, which is looking into this possibility.

- *Cifarelli:*

It would be wise to make very clear to the physics community what would be the power actually needed to run CLIC, especially since the R&D for CLIC will indeed focus on extremely powerful klystrons.

- *Wenninger:*

Obviously, it is clear that we understand the weaknesses of a warm technique over a cold technique, but we also understand the tricks how they want to be more smart, than just get 10 MW klystrons in big numbers from industry and do this. At an energy of 500 GeV CLIC the AC power amounts to170 MW that is more or less that for TESLA. I think it is not so far off. But, TESLA has an advantage: at low energies experimentation is easier. However, you cannot go beyond 40 MV/m and beyond 1 TeV, and this is really what physics should say. We have to exploit LHC and then see what comes out to decide what to do and at which energy we want to continue e+e- physics.

- *Bourjaily:*

I would like to make a quick comment about the upgrade. I did spend a little time on the GRID project. As Prof. t'Hooft mentioned earlier, we are waiting for most of the technology to be built, so most of the GRID does not exist, yet. It's in planning phase. There are the locations where it is to be established and there are a couple of data challenges where they have run GEANT simulations; which are complicated. That really pushed the limits of the communications and the storage and communication between different sites. So that has been tested and we're expecting to be ready, when LHC turns on.

- *Wenninger:*

I always hear positive statements after the challenges that have been resolved between the various centers.

- *Abramowicz:*

I want to make a comment on the issue of LHC, ILC and CLIC. We are making a big mistake by thinking sequentially. First of all, LHC will be a discovery machine, because otherwise we are really in trouble. But it is not a precision machine. So running the two together has the advantage of the synergy we discussed here. We can try to look at precision measurements and hopefully extract not only limits. As you saw, ILC is basically sensitive to regions of 5 TeV, although the CM energy will be 1 TeV. So scientifically it is still very much motivating, especially that we are a very large community, and it is unthinkable that we will all be sitting on the same experiment and at the same level of frustration. The other thing is that I don't see what is going on around ILC. It is very important in terms of detector development, whatever be our future. In LHC, the proton-proton machine, the level of background is so overwhelming, that we are all the time trying to design the detector to deal with the background. With ILC we certainly dream of an ideal detector, which basically beats all the limits, so I think it is really in the interest of the community to support both projects at the same time, whatever the costs, if we want to remain a dynamic community.

- *Wenninger:*

I think the support is probably there. It's a question of timing, you see. In the 90's you could still dream, you could make it really in parallel without waiting for results from LHC. I think that if LHC is a success, then in 2011, everybody will agree, that you have to try to move forward. But before LHC is running, you should prepare everything, and there is a lot of R&D going on, so we should not stop it. Nobody ever said that one should stop continuing the R&D to be ready at the right moment. The detector R&D is going on very strongly on this project, since many years already. I think, now, that we are so close to the running of LHC, so any plans not to wait for the results will not be understood. Just a couple of years of patience. Two years, three years...

How to Detect Extra-dimensions

I. Antoniadis[1]

Department of Physics, CERN
Theory Division
CH-1211 Geneva 23, Switzerland

Abstract

Lowering the string scale in the TeV region provides a theoretical framework for solving the mass hierarchy problem and unifying all interactions. The apparent weakness of gravity can then be accounted by the existence of large internal dimensions, in the submillimeter region, and transverse to a braneworld where we must be confined. I review the main properties of this scenario and its implications for observations at both particle colliders, and in non-accelerator gravity experiments.

1 Introduction

During the last few decades, physics beyond the Standard Model (SM) was guided from the problem of mass hierarchy. This can be formulated as the question of why gravity appears to us so weak compared to the other three known fundamental interactions corresponding to the electromagnetic, weak and strong nuclear forces. Indeed, gravitational interactions are suppressed by a very high energy scale, the Planck mass $M_P \sim 10^{19}$ GeV, associated to a length $l_P \sim 10^{-35}$ m, where they are expected to become important. In a quantum theory, the hierarchy implies a severe fine tuning of the fundamental parameters in more than 30 decimal places in order to keep the masses of elementary particles at their observed values. The reason is that quantum radiative corrections to all masses generated by the Higgs vacuum expectation value (VEV) are proportional to the ultraviolet cutoff which in the presence of gravity is fixed by the Planck mass. As a result, all masses are "attracted" to become about 10^{16} times heavier than their observed values.

Besides compositeness, there are two main ideas that have been proposed and studied extensively during the last years, corresponding to different approaches of explaining the mass hierarchy problem. (1) Low energy supersymmetry with all superparticle masses in the TeV region. Indeed, in the limit of exact supersymmetry, quadratically divergent corrections to the Higgs self-energy are exactly cancelled, while in the softly broken case, they are cutoff by the supersymmetry breaking mass splittings. (2) TeV scale strings, in which quadratic divergences are cutoff by the string scale and low energy supersymmetry is not needed. Both ideas are experimentally testable at high-energy particle colliders and in particular at LHC. Below, I discuss their implementation in string theory.

[1]On leave from CPHT (UMR 7644 of CNRS), Ecole Polytechnique, F-91128 Palaiseau.

The appropriate and most convenient framework for low energy supersymmetry and grand unification is the perturbative heterotic string. Indeed, in this theory, gravity and gauge interactions have the same origin, as massless modes of the closed heterotic string, and they are unified at the string scale M_s. As a result, the Planck mass M_P is predicted to be proportional to M_s:

$$M_P = M_s/g\,, \tag{1}$$

where g is the gauge coupling. In the simplest constructions all gauge couplings are the same at the string scale, given by the four-dimensional (4d) string coupling, and thus no grand unified group is needed for unification. In our conventions $\alpha_{\rm GUT} = g^2 \simeq 0.04$, leading to a discrepancy between the string and grand unification scale $M_{\rm GUT}$ by almost two orders of magnitude. Explaining this gap introduces in general new parameters or a new scale, and the predictive power is essentially lost. This is the main defect of this framework, which remains though an open and interesting possibility.

The other idea can be naturally realized in the framework of type I string theory with D-branes. Unlike in the heterotic string, gauge and gravitational interactions have now different origin. The latter are described again by closed strings, while the former emerge as excitations of open strings with endpoints confined on D-branes [1]. This leads to a braneworld description of our universe, which should be localized on a hypersurface, i.e. a membrane extended in p spatial dimensions, called p-brane (see Fig. 1). Closed strings propagate in all nine dimensions of string theory: in those extended along the p-brane, called parallel, as well as in the transverse ones. On the contrary, open strings are attached on the p-brane. Obviously, our p-brane world must have at least the three known dimensions of

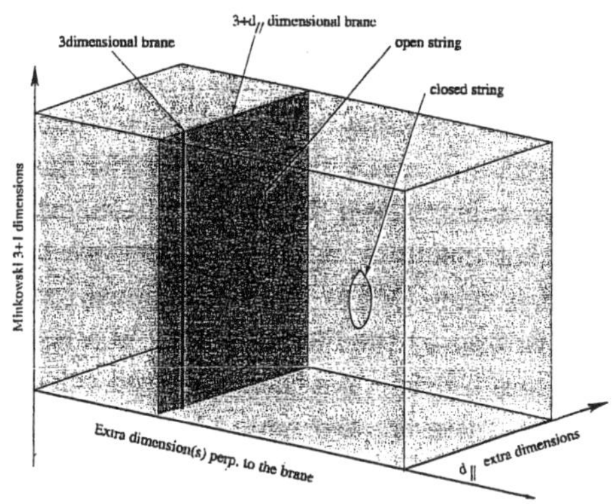

Figure 1: In the type I string framework, our Universe contains, besides the three known spatial dimensions (denoted by a single blue line), some extra dimensions ($d_\parallel = p-3$) parallel to our world p-brane (green plane) where endpoints of open strings are confined, and some transverse dimensions (yellow space) where only gravity described by closed strings can propagate.

space. But it may contain more: the extra $d_\parallel = p-3$ parallel dimensions must have a finite size, in order to be unobservable at present energies, and can be as large as $\text{TeV}^{-1} \sim 10^{-18}$ m [2]. On the other hand, transverse dimensions interact with us only gravitationally and experimental bounds are much weaker: their size should be less than about 0.1 mm [3]. In the following, I review the main properties and experimental signatures of low string scale [4, 5].

2 Low string scale: Framework

In type I theory, the different origin of gauge and gravitational interactions implies that the relation between the Planck and string scales is not linear as (1) of the heterotic string. The requirement that string theory should be weakly coupled, constrain the size of all parallel dimensions to be of order of the string length, while transverse dimensions remain unrestricted. Assuming an isotropic transverse space of $n = 9 - p$ compact dimensions of common radius $R_\perp$, one finds:

$$M_P^2 = \frac{1}{g^4} M_s^{2+n} R_\perp^n \,, \qquad g_s \simeq g^2 \,. \tag{2}$$

where g_s is the string coupling. It follows that the type I string scale can be chosen hierarchically smaller than the Planck mass [6, 4] at the expense of introducing extra large transverse dimensions felt only by gravity, while keeping the string coupling small [4]. The weakness of 4d gravity compared to gauge interactions (ratio M_W/M_P) is then attributed to the largeness of the transverse space $R_\perp$ compared to the string length $l_s = M_s^{-1}$.

An important property of these models is that gravity becomes effectively $(4+n)$-dimensional with a strength comparable to those of gauge interactions at the string scale. The first relation of Eq. (2) can be understood as a consequence of the $(4+n)$-dimensional Gauss law for gravity, with $M_*^{(4+n)} = M_s^{2+n}/g^4$ the effective scale of gravity in $4+n$ dimensions. Taking $M_s \simeq 1$ TeV, one finds a size for the extra dimensions $R_\perp$ varying from 10^8 km, .1 mm, down to a Fermi for $n = 1, 2$, or 6 large dimensions, respectively. This shows that while $n = 1$ is excluded, $n \geq 2$ is allowed by present experimental bounds on gravitational forces [3, 7]. Thus, in these models, gravity appears to us very weak at macroscopic scales because its intensity is spread in the "hidden" extra dimensions. At distances shorter than $R_\perp$, it should deviate from Newton's law, which may be possible to explore in laboratory experiments (see Fig. 2).

The main experimental implications of TeV scale strings in particle accelerators are of three types, in correspondence with the three different sectors that are generally present: (i) new compactified parallel dimensions, (ii) new extra large transverse dimensions and low scale quantum gravity, and (iii) genuine string and quantum gravity effects. On the other hand, there exist interesting implications in non accelerator table-top experiments due to the exchange of gravitons or other possible states living in the bulk.

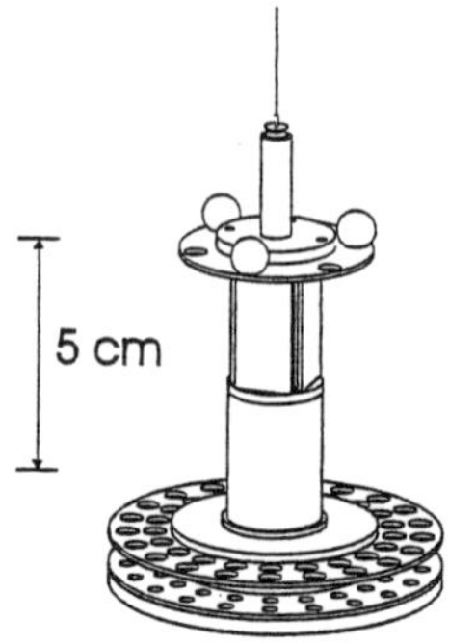

Figure 2: Torsion pendulum that tested Newton's law at 55 μm.

3 Experimental implications in accelerators

3.1 World-brane extra dimensions

In this case $RM_s \gtrsim 1$, and the associated compactification scale $R_\parallel^{-1}$ would be the first scale of new physics that should be found increasing the beam energy [2, 8]. There are several reasons for the existence of such dimensions. It is a logical possibility, since out of the six extra dimensions of string theory only two are needed for lowering the string scale, and thus the effective p-brane of our world has in general $d_\parallel \equiv p-3 \leq 4$. Moreover, they can be used to address several physical problems in braneworld models, such as obtaining different SM gauge couplings, explaining fermion mass hierarchies due to different localization points of quarks and leptons in the extra dimensions, providing calculable mechanisms of supersymmetry breaking, etc.

The main consequence is the existence of Kaluza-Klein (KK) excitations for all SM particles that propagate along the extra parallel dimensions. Their masses are given by:

$$M_m^2 = M_0^2 + \frac{m^2}{R_\parallel^2} \quad ; \quad m = 0, \pm 1, \pm 2, \ldots \tag{3}$$

where we used $d_\parallel = 1$, and M_0 is the higher dimensional mass. The zero-mode $m = 0$ is identified with the 4d state, while the higher modes have the same quantum numbers with the lowest one, except for their mass given in (3). There are two types of experimental signatures of such dimensions [8, 9, 10]: (i) virtual exchange of KK excitations, leading to deviations in cross-sections compared to the SM prediction, that can be used to extract bounds on the compactification scale; (ii) direct production of KK modes.

On general grounds, there can be two different kinds of models with qualitatively different signatures depending on the localization properties of matter fermion fields. If the latter are localized in 3d brane intersections, they do not have excitations and KK momentum is not conserved because of the breaking of translation invariance in the extra dimension(s). KK modes of gauge bosons are then singly produced giving rise to generally strong bounds on the compactification scale and new resonances that can be observed in experiments. Otherwise, they can be produced only in pairs due to the KK momentum conservation, making the bounds weaker but the resonances difficult to observe. When the internal momentum is conserved, the interaction vertex involving KK modes has the same 4d tree-level gauge coupling. On the other hand, their couplings to localized matter have an exponential form factor suppressing the interactions of heavy modes. This form factor can be viewed as the fact that the branes intersection has a finite thickness.

To simplify the analysis, let us consider first the case $d_{\parallel} = 1$ where some of the gauge fields arise from an effective 4-brane, while fermions are localized states on brane intersections. Since the corresponding gauge couplings are reduced by the size of the large dimension $R_{\parallel}M_s$ compared to the others, one can account for the ratio of the weak to strong interactions strengths if the $SU(2)$ brane extends along the extra dimension, while $SU(3)$ does not. As a result, there are 3 distinct cases to study [10], denoted by (t,l,l), (t,l,t) and (t,t,l), where the three positions in the brackets correspond to the three SM gauge group factors $SU(3) \times SU(2) \times U(1)$ and those with l (longitudinal) feel the extra dimension, while those with t (transverse) do not.

In the (t,l,l) case, there are KK excitations of $SU(2) \times U(1)$ gauge bosons: $W_{\pm}^{(m)}$, $\gamma^{(m)}$ and $Z^{(m)}$. Performing a χ^2 fit of the electroweak observables, one finds that if the Higgs is a bulk state (l), $R_{\parallel}^{-1} \gtrsim 3.5$ TeV [11]. This implies that LHC can produce at most the first KK mode. Different choices for localization of matter and Higgs fields lead to bounds, lying in the range $1-5$ TeV [11]. In addition to virtual effects, KK excitations can be produced on-shell at LHC as new resonances [9] (see Fig. 3). There are two different channels, neutral Drell–

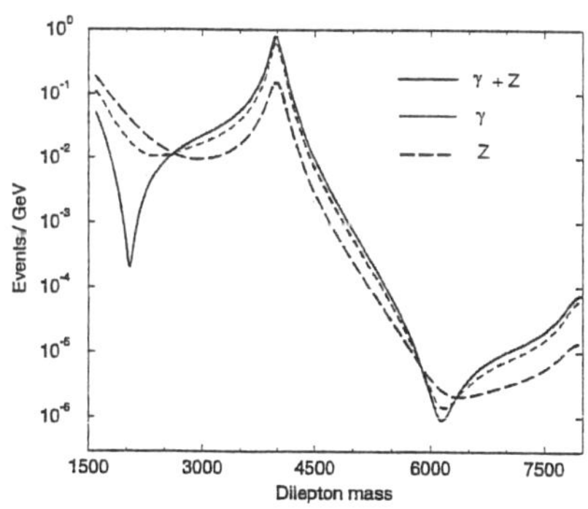

Figure 3: Production of the first KK modes of the photon and Z-boson at LHC, decaying to e^+e^- pairs. The number of expected events is plotted as a function of the energy of the pair.

Yan processes $pp \to l^+l^-X$ and the charged channel $l^{\pm}\nu$, corresponding to the production of the KK modes $\gamma^{(1)}, Z^{(1)}$ and $W_{\pm}^{(1)}$, respectively. The discovery limits are about 6 TeV, while the exclusion bounds 15 TeV. An interesting observation in the case of $\gamma^{(1)} + Z^{(1)}$ is that interferences can lead to a "dip" just before the resonance. There are some ways to distinguish the corresponding signals from other possible origin of new physics, such as models with new gauge bosons. In fact, in the (t,l,l) and (t,l,t) cases, one expects two resonances located practically at the same mass value. This property is not shared by most of other new gauge boson models. Moreover, the heights and widths of the resonances are directly related to those of SM gauge bosons in the corresponding channels.

In the (t, l, t) case, only the $SU(2)$ factor feels the extra dimension and the limits set by the KK states of $W^{\pm}$ remain the same. On the other hand, in the (t, t, l) case where only $U(1)_Y$ feels the extra dimension, the limits are weaker and the exclusion bound is around 8 TeV. In addition to these simple possibilities, brane constructions lead often to cases where part of $U(1)_Y$ is t and part is l. If $SU(2)$ is l the limits come again from $W^{\pm}$, while if it is t then it will be difficult to distinguish this case from a generic extra $U(1)'$. A good statistics is needed to see the deviation in the tail of the resonance as being due to effects additional to those of a generic $U(1)'$ resonance. Finally, in the case of two or more parallel dimensions, cross-sections become bigger yielding stronger bounds, while resonances are closer implying that more of them could be reached by LHC.

On the other hand, if all SM particles propagate in the extra dimension (called universal)[2], KK modes can only be produced in pairs and the lower bound on the compactification scale becomes weaker, of order of 300-500 GeV. Moreover, no resonances can be observed at LHC, so that this scenario appears very similar to low energy supersymmetry. In fact, KK parity can even play the role of R-parity, implying that the lightest KK mode is stable and can be a dark matter candidate in analogy to the LSP [12].

3.2 Extra large transverse dimensions

The main experimental signal is gravitational radiation in the bulk from any physical process on the world-brane. In fact, the very existence of branes breaks translation invariance in the transverse dimensions and gravitons can be emitted from the brane into the bulk. During a collision of center of mass energy $\sqrt{s}$, there are $\sim (\sqrt{s}R_\perp)^n$ KK excitations of gravitons with tiny masses, that can be emitted. Each of these states looks from the 4d point of view as a massive, quasi-stable, extremely weakly coupled (s/M_P^2 suppressed) particle that escapes from the detector. The total effect is a missing-energy cross-section roughly of order:

$$\frac{(\sqrt{s}R_\perp)^n}{M_P^2} \sim \frac{1}{s}\left(\frac{\sqrt{s}}{M_s}\right)^{n+2}. \tag{4}$$

Explicit computation of these effects leads to the bounds given in Table 1. However, larger radii are allowed if one relaxes the assumption of isotropy, by taking for instance two large dimensions with different radii.

Fig. 4 shows the cross-section for graviton emission in the bulk, corresponding to the process $pp \to jet + graviton$ at LHC, together with the SM background [13]. For a given value of M_s, the cross-section for graviton emission decreases with the number of large transverse dimensions, in contrast to the case of parallel dimensions. The reason is that gravity becomes weaker if there are more dimensions because there is more space for the gravitational field to escape. There is a particular energy and angular distribution of the

[2]Although interesting, this scenario seems difficult to be realized, since 4d chirality requires non-trivial action of orbifold twists with localized chiral states at the fixed points.

Table 1: Limits on $R_\perp$ in mm.

Experiment	$n = 2$	$n = 4$	$n = 6$
Collider bounds			
LEP 2	5×10^{-1}	2×10^{-8}	7×10^{-11}
Tevatron	5×10^{-1}	10^{-8}	4×10^{-11}
LHC	4×10^{-3}	6×10^{-10}	3×10^{-12}
NLC	10^{-2}	10^{-9}	6×10^{-12}
Present non-collider bounds			
SN1987A	3×10^{-4}	10^{-8}	6×10^{-10}
COMPTEL	5×10^{-5}	-	-

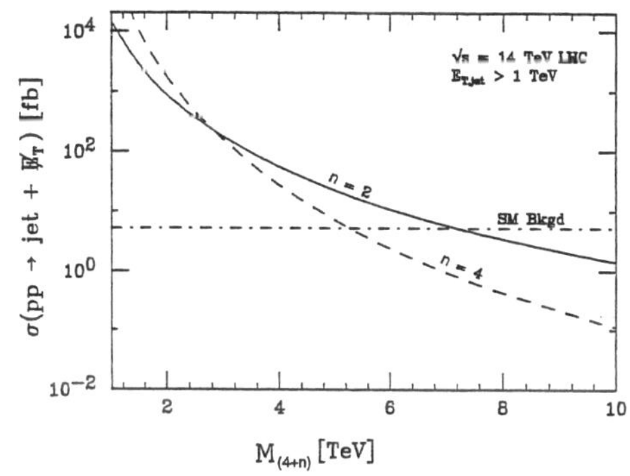

Figure 4: Missing energy due to graviton emission at LHC, as a function of the higher-dimensional gravity scale M_*, produced together with a hadronic jet. The expected cross-section is shown for $n = 2$ and $n = 4$ extra dimensions, together with the SM background.

produced gravitons that arise from the distribution in mass of KK states of spin-2. This can be contrasted to other sources of missing energy and might be a smoking gun for the extra dimensional nature of such a signal.

In Table 1, there are also included astrophysical and cosmological bounds. Astrophysical bounds [14, 15] arise from the requirement that the radiation of gravitons should not carry on too much of the gravitational binding energy released during core collapse of supernovae. In fact, the measurements of Kamiokande and IMB for SN1987A suggest that the main channel is neutrino fluxes. The best cosmological bound [16] is obtained from requiring that decay of bulk gravitons to photons do not generate a spike in the energy spectrum of the photon background measured by the COMPTEL instrument. Bulk gravitons are expected to be produced just before nucleosynthesis due to thermal radiation from the brane. The limits assume that the temperature was at most 1 MeV as nucleosynthesis begins, and become stronger if temperature is increased.

3.3 String effects

At low energies, the interaction of light (string) states is described by an effective field theory. Their exchange generates in particular four-fermion operators that can be used to extract independent bounds on the string scale. In analogy with the bounds on longitudinal extra dimensions, there are two cases depending on the localization properties of matter fermions. If they come from open strings with both ends on the same stack of branes, exchange of massive open string modes gives rise to dimension eight effective operators, involving four fermions and two space-time derivatives [17, 18]. The corresponding bounds on the string scale are then around 500 GeV. On the other hand, if matter fermions are localized on non-trivial brane intersections, one obtains dimension six four-fermion operators and the bounds become stronger: $M_s \gtrsim 2 - 3$ TeV [18, 5]. At energies higher than the string scale, new spectacular phenomena are expected to occur, related to string physics and quantum gravity effects, such as possible micro-black hole production [19]. Particle accelerators would then become the best tools for studying quantum gravity and string theory.

4 Supersymmetry in the bulk and short range forces

4.1 Sub-millimeter forces

Besides the spectacular predictions in accelerators, there are also modifications of gravitation in the sub-millimeter range, which can be tested in "table-top" experiments that measure gravity at short distances. There are three categories of such predictions:
(i) Deviations from the Newton's law $1/r^2$ behavior to $1/r^{2+n}$, which can be observable for $n = 2$ large transverse dimensions of sub-millimeter size. This case is particularly attractive on theoretical grounds because of the logarithmic sensitivity of SM couplings on the size of transverse space [20], that allows to determine the hierarchy [21].
(ii) New scalar forces in the sub-millimeter range, related to the mechanism of supersymmetry breaking, and mediated by light scalar fields φ with masses [22, 4]:

$$m_\varphi \simeq \frac{m_{susy}^2}{M_P} \simeq 10^{-4} - 10^{-6}\ \text{eV}\,, \tag{5}$$

for a supersymmetry breaking scale $m_{susy} \simeq 1 - 10$ TeV. They correspond to Compton wavelengths of 1 mm to 10 μm. m_{susy} can be either $1/R_\parallel$ if supersymmetry is broken by compactification [22], or the string scale if it is broken "maximally" on our world-brane [4]. A universal attractive scalar force is mediated by the radion modulus $\varphi \equiv M_P \ln R$, with R the radius of the longitudinal or transverse dimension(s). In the former case, the result (5) follows from the behavior of the vacuum energy density $\Lambda \sim 1/R_\parallel^4$ for large $R_\parallel$ (up to logarithmic corrections). In the latter, supersymmetry is broken primarily on the brane, and thus its transmission to the bulk is gravitationally suppressed, leading to (5). For $n = 2$, there may be an enhancement factor of the radion mass by $\ln R_\perp M_s \simeq 30$ decreasing its wavelength by an order of magnitude [21].

The coupling of the radius modulus to matter relative to gravity can be easily computed and is given by:

$$\sqrt{\alpha_\varphi} = \frac{1}{M}\frac{\partial M}{\partial \varphi}\ ; \quad \alpha_\varphi = \begin{cases} \frac{\partial \ln \Lambda_{QCD}}{\partial \ln R} \simeq \frac{1}{3} \ \text{for } R_\parallel \\ \\ \frac{2n}{n+2} = 1 - 1.5 \ \text{for } R_\perp \end{cases} \tag{6}$$

where M denotes a generic physical mass. In the longitudinal case, the coupling arises dominantly through the radius dependence of the QCD gauge coupling [22], while in the case of transverse dimension, it can be deduced from the rescaling of the metric which changes the string to the Einstein frame and depends slightly on the bulk dimensionality ($\alpha = 1 - 1.5$ for $n = 2 - 6$) [21]. Such a force can be tested in microgravity experiments and should be contrasted with the change of Newton's law due the presence of extra dimensions that is observable only for $n = 2$ [3, 7]. The resulting bounds from an analysis of the radion effects are [23]: $M_* \gtrsim 6\,\text{TeV}$. In principle there can be other light moduli which couple with even larger strengths. For example the dilaton, whose VEV determines the string coupling, if it does not acquire large mass from some dynamical mechanism, can lead to a force of strength 2000 times bigger than gravity [24].
(iii) Non universal repulsive forces much stronger than gravity, mediated by possible abelian gauge fields in the bulk [14, 25]. Such fields acquire tiny masses of the order of M_s^2/M_P, as in (5), due to brane localized anomalies [25]. Although their gauge coupling is infinitesimally small, $g_A \sim M_s/M_P \simeq 10^{-16}$, it is still bigger that the gravitational coupling E/M_P for typical energies $E \sim 1$ GeV, and the strength of the new force would be $10^6 - 10^8$ stronger than gravity. This is an interesting region which will be soon explored in micro-gravity experiments (see Fig. 5). Note that in this case supernova constraints impose that there should be at least four large extra dimensions in the bulk [14].

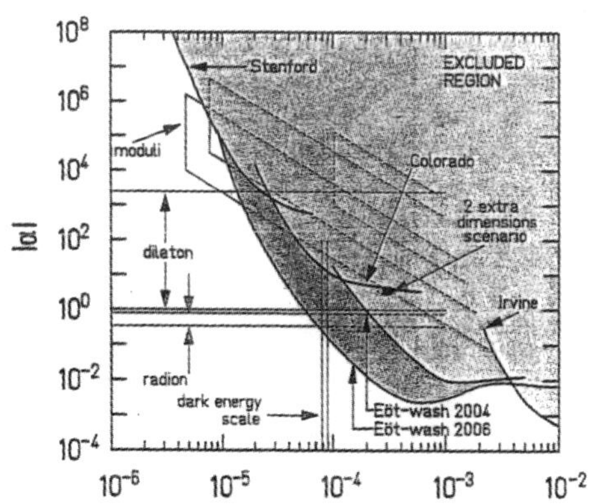

Figure 5: Present limits on new short-range forces (yellow regions), as a function of their range λ and their strength relative to gravity α. The limits are compared to the forces mediated by the graviton in the case of two large extra dimensions, and by the radion.

In Fig. 5 we depict the actual information from previous, present and upcoming experiments [7, 21]. The solid lines indicate the present limits from the experiments indicated. The excluded regions lie above these solid lines. Measuring gravitational strength forces at short distances is challenging. The dashed thick lines give the expected sensitivity of the various experiments, which will improve the actual limits by roughly two orders of magnitude, while the horizontal dashed lines correspond to the theoretical predictions for the graviton in the case $n = 2$ and for the radion in the transverse case. These limits are compared to those obtained from particle accelerator experiments in Table 1. Finally, in Figs. 6 and 7, we display recent improved bounds for new forces at very short distances by focusing on the right hand side of Fig. 5, near the origin [7].

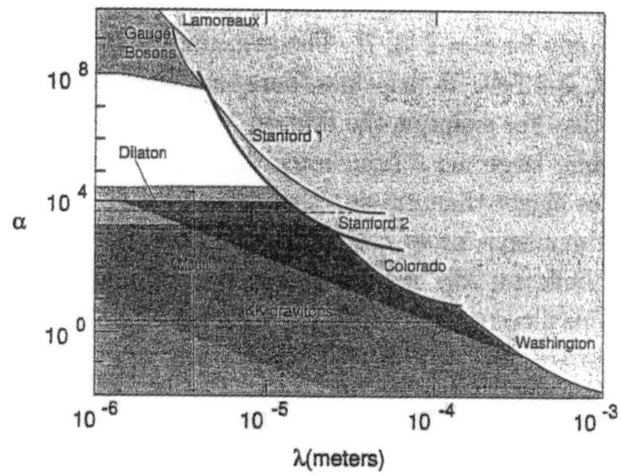

Figure 6: Bounds on non-Newtonian forces in the range 6-20 μm (S. J. Smullin et al. in [7]).

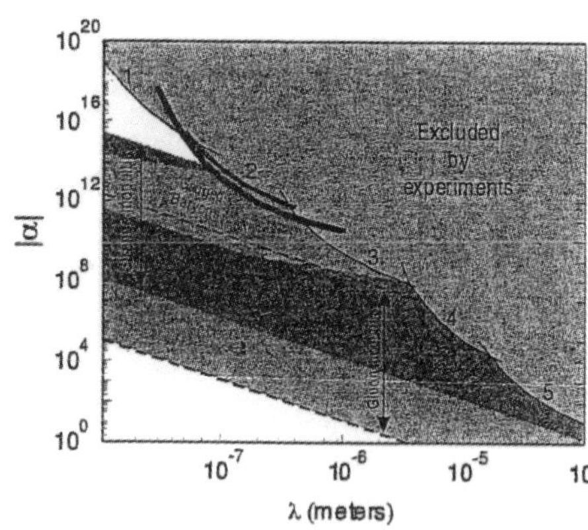

Figure 7: Bounds on non-Newtonian forces in the range around 200 nm (see R. S. Decca et al. in Ref. [7]). Curves 4 and 5 correspond to Stanford and Colorado experiments, respectively, of Fig. 6 (see also J C. Long and J. C. Price of Ref. [7]).

5 Standard Model on D-branes

The gauge group closest to the Standard Model one can easily obtain with D-branes is $U(3) \times U(2) \times U(1)$. The first factor arises from three coincident "color" D-branes. An open string with one end on them is a triplet under $SU(3)$ and carries the same $U(1)$ charge for all three components. Thus, the $U(1)$ factor of $U(3)$ has to be identified with *gauged* baryon number. Similarly, $U(2)$ arises from two coincident "weak" D-branes and the corresponding abelian factor is identified with *gauged* weak-doublet number. Finally, an extra $U(1)$ D-brane is necessary in order to accommodate the Standard Model without breaking the baryon number [26]. In principle this $U(1)$ brane can be chosen to be independent of the other two collections with its own gauge coupling. To improve the predictability of the model, we choose to put it on top of either the color or the weak D-branes [27]. In either case, the model has two independent gauge couplings g_3 and g_2 corresponding, respectively, to the gauge groups $U(3)$ and $U(2)$. The $U(1)$ gauge coupling g_1 is equal to either g_3 or g_2.

Let us denote by Q_3, Q_2 and Q_1 the three $U(1)$ charges of $U(3) \times U(2) \times U(1)$, in a self explanatory notation. Under $SU(3) \times SU(2) \times U(1)_3 \times U(1)_2 \times U(1)_1$, the members of a family of quarks and leptons have the following quantum numbers:

$$\begin{array}{ll} Q & (\mathbf{3}, \mathbf{2}; 1, w, 0)_{1/6} \\ u^c & (\bar{\mathbf{3}}, \mathbf{1}; -1, 0, x)_{-2/3} \\ d^c & (\bar{\mathbf{3}}, \mathbf{1}; -1, 0, y)_{1/3} \\ L & (\mathbf{1}, \mathbf{2}; 0, 1, z)_{-1/2} \\ l^c & (\mathbf{1}, \mathbf{1}; 0, 0, 1)_1 \end{array} \tag{7}$$

The values of the $U(1)$ charges x, y, z, w will be fixed below so that they lead to the right hypercharges, shown for completeness as subscripts.

It turns out that there are two possible ways of embedding the Standard Model particle spectrum on these stacks of branes [26], which are shown pictorially in Fig. 8. The quark

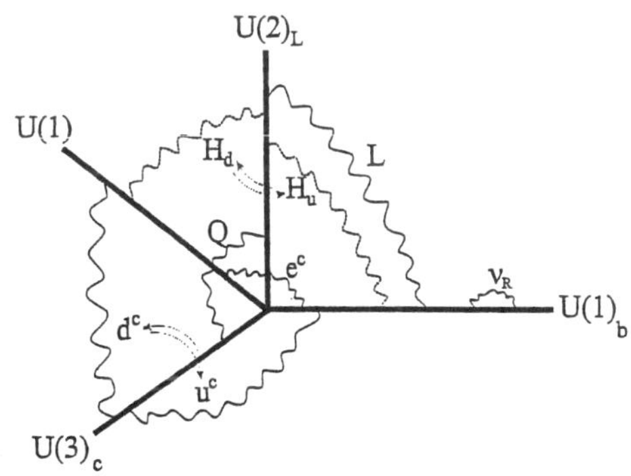

Figure 8: A minimal Standard Model embedding on D-branes.

doublet Q corresponds necessarily to a massless excitation of an open string with its two ends on the two different collections of branes (color and weak). As seen from the figure, a fourth brane stack is needed for a complete embedding, which is chosen to be a $U(1)_b$ extended in the bulk. This is welcome since one can accommodate right handed neutrinos as open string states on the bulk with sufficiently small Yukawa couplings suppressed by the large volume of the bulk [28]. Moreover, a linear combination of the four $U(1)$s turns out to be the lepton number, needed to protect small neutrino masses from dimension-five operators involving two lepton doublets and two higgses. The two models are obtained by an exchange of the up and down antiquarks, u^c and d^c, which correspond to open strings with one end on the color branes and the other either on the $U(1)$ brane, or on the $U(1)_b$ in the bulk. The lepton doublet L arises from an open string stretched between the weak branes and $U(1)_b$, while the antilepton l^c corresponds to a string with one end on the $U(1)$ brane and the other in the bulk. For completeness, we also show the two possible Higgs states H_u and H_d that are both necessary in order to give tree-level masses to all quarks and leptons of the heaviest generation.

The weak hypercharge Y is a linear combination of the three $U(1)$'s:

$$Y = Q_1 + \frac{1}{2}Q_2 + c_3Q_3 \quad ; \quad c_3 = -1/3 \text{ or } 2/3\,, \tag{8}$$

where Q_N denotes the $U(1)$ generator of $U(N)$ normalized so that the fundamental representation of $SU(N)$ has unit charge. The corresponding $U(1)$ charges appearing in eq. (7) are $x = -1$ or 0, $y = 0$ or 1, $z = -1$, and $w = 1$ or -1, for $c_3 = -1/3$ or $2/3$, respectively. The hypercharge coupling g_Y is given by [3]:

$$\frac{1}{g_Y^2} = \frac{2}{g_1^2} + \frac{4c_2^2}{g_2^2} + \frac{6c_3^2}{g_3^2}\,. \tag{9}$$

It follows that the weak angle $\sin^2\theta_W$, is given by:

$$\sin^2\theta_W \equiv \frac{g_Y^2}{g_2^2 + g_Y^2} = \frac{1}{2 + 2g_2^2/g_1^2 + 6c_3^2 g_2^2/g_3^2}\,, \tag{10}$$

where g_N is the gauge coupling of $SU(N)$ and $g_1 = g_2$ or $g_1 = g_3$ at the string scale. In order to compare the theoretical predictions with the experimental value of $\sin^2\theta_W$ at M_s, we plot in Fig. 9 the corresponding curves as functions of M_s. The solid line is the experimental curve. The dashed line is the plot of the function (10) for $g_1 = g_2$ with $c_3 = -1/3$ while the dotted-dashed line corresponds to $g_1 = g_3$ with $c_3 = 2/3$. The other two possibilities are not shown because they lead to a value of M_s which is too high to protect the hierarchy. Thus, the second case, where the $U(1)$ brane is on top of the color branes, is compatible with low energy data for $M_s \sim 6-8$ TeV.

From Eq. (10) and Fig. 9, we find the ratio of the $SU(2)$ and $SU(3)$ gauge couplings at the string scale to be $\alpha_2/\alpha_3 \sim 0.6$. This ratio can be arranged by an appropriate choice of the relevant moduli. For instance, one may choose the color and $U(1)$ branes to be D3 branes while the weak branes to be D7 branes. Then, the ratio of couplings above can be explained

[3]The gauge couplings $g_{2,3}$ are determined at the tree-level by the string coupling and other moduli, like radii of longitudinal dimensions. In higher orders, they also receive string threshold corrections.

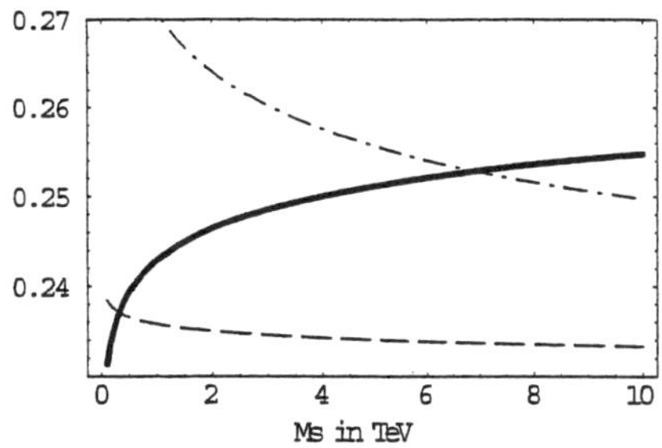

Figure 9: The value of $\sin^2\theta_W$ (thick curve), and the theoretical predictions (10).

by choosing the volume of the four compact dimensions of the seven branes to be $V_4 = 2.5$ in string units. This being larger than one is consistent with the picture above. Moreover it predicts an interesting spectrum of KK states for the Standard model, different from the naive choices that have appeared hitherto: the only Standard Model particles that have KK descendants are the W bosons as well as the hypercharge gauge boson. However, since the hypercharge is a linear combination of the three $U(1)$'s, the massive $U(1)$ KK gauge bosons do not couple to the hypercharge but to the weak doublet number.

Acknowledgments

This work was supported in part by the European Commission under the RTN contract MRTN-CT-2004-503369, and in part by the INTAS contract 03-51-6346.

References

[1] C. Angelantonj and A. Sagnotti, *Phys. Rept.* **371**, 1 (2002) [Erratum-ibid. **376**, 339 (2003)] [arXiv:hep-th/0204089].

[2] I. Antoniadis, *Phys. Lett.* B **246** (1990) 377.

[3] D. J. Kapner, T. S. Cook, E. G. Adelberger, J. H. Gundlach, B. R. Heckel, C. D. Hoyle and H. E. Swanson, Phys. Rev. Lett. **98** (2007) 021101.

[4] N. Arkani-Hamed, S. Dimopoulos and G. R. Dvali, *Phys. Lett.* B **429**, 263 (1998) [arXiv:hep-ph/9803315]; I. Antoniadis, N. Arkani-Hamed, S. Dimopoulos and G. R. Dvali, *Phys. Lett.* B **436**, 257 (1998) [arXiv:hep-ph/9804398].

[5] For a review see e.g. I. Antoniadis, *Prepared for International School of Subnuclear Physics: 41st Course: From Quarks to Black Holes: Progress in Understanding the Logic of Nature, Erice, Sicily, Italy, 29 Aug - 7 Sep 2003*; and references therein.

[6] J. D. Lykken, *Phys. Rev.* D **54**, 3693 (1996) [arXiv:hep-th/9603133].

[7] J. C. Long and J. C. Price, *Comptes Rendus Physique* **4**, 337 (2003); R. S. Decca, D. Lopez, H. B. Chan, E. Fischbach, D. E. Krause and C. R. Jamell, *Phys. Rev. Lett.* **94**, 240401 (2005); S. J. Smullin, A. A. Geraci, D. M. Weld, J. Chiaverini, S. Holmes and A. Kapitulnik, arXiv:hep-ph/0508204; H. Abele, S. Haeßler and A. Westphal, in 271th WE-Heraeus-Seminar, Bad Honnef (2002).

[8] I. Antoniadis and K. Benakli, *Phys. Lett.* B **326** (1994) 69.

[9] I. Antoniadis, K. Benakli and M. Quirós, *Phys. Lett.* B **331** (1994) 313 and *Phys. Lett.* B **460** (1999) 176; P. Nath, Y. Yamada and M. Yamaguchi, *Phys. Lett.* B **466** (1999) 100 T. G. Rizzo and J. D. Wells, *Phys. Rev.* D **61** (2000) 016007; T. G. Rizzo, *Phys. Rev.* D **61** (2000) 055005; A. De Rujula, A. Donini, M. B. Gavela and S. Rigolin, *Phys. Lett.* B **482** (2000) 195;

[10] E. Accomando, I. Antoniadis and K. Benakli, *Nucl. Phys.* B **579** (2000) 3.

[11] P. Nath and M. Yamaguchi, *Phys. Rev.* D **60** (1999) 116004; *Phys. Rev.* D **60** (1999) 116006; M. Masip and A. Pomarol, *Phys. Rev.* D **60** (1999) 096005; W. J. Marciano, *Phys. Rev.* D **60** (1999) 093006; A. Strumia, *Phys. Lett.* B **466** (1999) 107; R. Casalbuoni, S. De Curtis, D. Dominici and R. Gatto, *Phys. Lett.* B **462** (1999) 48; C. D. Carone, *Phys. Rev.* D **61** (2000) 015008; A. Delgado, A. Pomarol and M. Quirós, *JHEP* **1**, 30 (2000).

[12] G. Servant and T. M. P. Tait, *Nucl. Phys.* B **650**, 391 (2003).

[13] G. F. Giudice, R. Rattazzi and J. D. Wells, *Nucl. Phys.* B **544** (1999) 3; E. A. Mirabelli, M. Perelstein and M. E. Peskin, *Phys. Rev. Lett.* **82** (1999) 2236; T. Han, J. D. Lykken and R. Zhang, *Phys. Rev.* D **59** (1999) 105006; K. Cheung and W.-Y. Keung, *Phys. Rev.* D **60** (1999) 112003; C. Balázs *et al.*, *Phys. Rev. Lett.* **83** (1999) 2112; L3 Collaboration (M. Acciarri *et al.*), *Phys. Lett.* B **464** (1999) 135 and **470**, 281 (1999): J. L. Hewett, *Phys. Rev. Lett.* **82** (1999) 4765.

[14] N. Arkani-Hamed, S. Dimopoulos and G. Dvali, *Phys. Rev.* D **59** (1999) 086004.

[15] S. Cullen and M. Perelstein, *Phys. Rev. Lett.* **83** (1999) 268; V. Barger, T. Han, C. Kao and R. J. Zhang, *Phys. Lett.* B **461** (1999) 34.

[16] K. Benakli and S. Davidson, *Phys. Rev.* D **60** (1999) 025004; L. J. Hall and D. Smith, *Phys. Rev.* D **60** (1999) 085008.

[17] S. Cullen, M. Perelstein and M. E. Peskin, *Phys. Rev.* D **62** (2000) 055012; D. Bourilkov, *Phys. Rev.* D **62** (2000) 076005; L3 Collaboration (M. Acciarri *et al.*), *Phys. Lett.* B **489** (2000) 81.

[18] I. Antoniadis, K. Benakli and A. Laugier, *JHEP* **0105** (2001) 044.

[19] S. B. Giddings and S. Thomas, *Phys. Rev.* D **65** (2002) 056010; S. Dimopoulos and G. Landsberg, *Phys. Rev. Lett.* **87**, 161602 (2001).

[20] I. Antoniadis, C. Bachas, *Phys. Lett.* B **450** (1999) 83.

[21] I. Antoniadis, K. Benakli, A. Laugier and T. Maillard, *Nucl. Phys.* B **662** (2003) 40 [arXiv:hep-ph/0211409].

[22] I. Antoniadis, S. Dimopoulos and G. Dvali, *Nucl. Phys.* B **516** (1998) 70; S. Ferrara, C. Kounnas and F. Zwirner, *Nucl. Phys.* B **429** (1994) 589.

[23] E. G. Adelberger, B. R. Heckel, S. Hoedl, C. D. Hoyle, D. J. Kapner and A. Upadhye, Phys. Rev. Lett. **98** (2007) 131104.

[24] T. R. Taylor and G. Veneziano, *Phys. Lett.* B **213** (1988) 450.

[25] I. Antoniadis, E. Kiritsis and J. Rizos, *Nucl. Phys.* B **637** (2002) 92.

[26] I. Antoniadis, E. Kiritsis and T. N. Tomaras, Phys. Lett. B **486** (2000) 186 [arXiv:hep-ph/0004214]; I. Antoniadis, E. Kiritsis, J. Rizos and T. N. Tomaras, Nucl. Phys. B **660** (2003) 81 [arXiv:hep-th/0210263]; R. Blumenhagen, B. Kors, D. Lust and T. Ott, Nucl. Phys. B **616**, 3 (2001) [arXiv:hep-th/0107138]; I. Antoniadis and J. Rizos, 2003 unpublished work.

[27] G. Shiu and S.-H. H. Tye, *Phys. Rev.* D **58** (1998) 106007; Z. Kakushadze and S.-H. H. Tye, *Nucl. Phys.* B **548** (1999) 180; L. E. Ibáñez, C. Muñoz and S. Rigolin, *Nucl. Phys.* B **553** (1999) 43.

[28] K. R. Dienes, E. Dudas and T. Gherghetta, *Nucl. Phys.* B **557** (1999) 25 [arXiv:hep-ph/9811428]; N. Arkani-Hamed, S. Dimopoulos, G. R. Dvali and J. March-Russell, *Phys. Rev.* D **65** (2002) 024032 [arXiv:hep-ph/9811448]; G. R. Dvali and A. Y. Smirnov, *Nucl. Phys.* B **563** (1999) 63.

CHAIRMAN: I. ANTONIADIS

Scientific Secretaries: V.Alexandrov, A.Sviatkovsky

DISCUSSION

- Krotov:

You mentioned the possibility to extract the experimental limit on the constants alpha and lambda in the gravity modification law from the Casimir Force. Are these experiments realistic? Because measuring the Casimir Effect itself is rather subtle. And you want to extract from it even more subtle effects of gravity modification.

- Antoniadis:

Actually the Casimir Force is measured by one of the experimental groups I mentioned very well. Moreover, one has to know very well the theoretical prediction for the experimental apparatus, so that one is able to measure deviations. This experiment is based on the Casimir Force between a sphere and a plane at a distance of 200 nm.

- Naumenko:

Can you compare your model to the Randall-Sundrum scenario?

- Antoniadis:

The Randall-Sundrum Scenario is very different. In the model I considered, extra dimensions are flat, while in the RS model extra dimensions are curved, there is a warp factor in the metric. The hierarchy is not due to the large volume of the internal space, but due to the distance of the observable world from the place where gravity is localized.

- Neri:

What do you consider as the most striking experimental evidence of your theory?

- Antoniadis:

The most spectacular effect is the possibility that the string scale can be at low energies. Accelerators can then be used to study string theory and quantum gravity, since gravitation becomes strong at energies of the order of the string scale. If such an idea is realized, there will be a revolution in particle physics.

- Aschieri:

Therefore the models you discussed here do not fit to grand unification?

- Antoniadis:

Usually grand unification has baryon number violation and is not appropriate when the string scale is low. Moreover, there is no particular compelling explanation that gauge couplings should unify at a high-energy scale when the string scale is low. In other constructions, in which the string scale is high, there are GUT models like the one you see here using this alternative brane configuration.

- *Pacetti:*

Assuming the Heisenberg uncertainty principle, if the graviton is produced in the final state at LHC and it escapes in some compactified dimensions of a size of about 1 fm, it should have an uncertainty of about 100 MeV on its momentum which should manifest itself on the total energy balance. Is this reasonable?

- *Antoniadis:*

This is correct. It implies simply that you cannot measure the internal momentum and the position in the extra dimensions to this accuracy.

-*Marrani:*

I would like to know if the feature of missing energy attributed to a graviton escaping into the transverse extra dimensions is a peculiar feature of your model. It seems to me that this is not the case, since many other theories could produce such a feature?

- *Antoniadis:*

Actually, this is what I said in my talk. If one observes missing energy it does not imply necessarily that it is due to extra dimensions, but one has to do more work in order to differentiate this source of missing energy from other sources. One difference is due to the spin of the particle that escapes detection, in this case the spin is 2, showing up in the angular distribution. Another difference is that in the case of SUSY or another 4-dimensional theory, there is just one particle that produces missing energy, leading to a pick in the missing p_T-distribution, while for the higher-dimensional graviton this is not the case.

-*Pinfold:*

Can you tell us about limits on extra dimensions from astrophysical/cosmological observations – for example from reheating?

- *Antoniadis:*

From reheating, there is only a constraint on the reheating temperature to be sufficiently low, but not on the size of extra dimensions. Moreover, one has to assume that their size is stabilized early enough to avoid cosmological problems such as with nucleosynthesis.

-*Bourjaily:*

How is the mass scale of neutrino fitted into a model with TeV string scale?

- *Antoniadis:*

To explain the smallness of neutrino masses when the string scale is low, one possibility is to introduce right-handed neutrino in the bulk, so that neutrino masses are suppressed due to the wave function. For instance, in this construction of the Standard Model I presented, one has a brane in the bulk that gives a right-handed neutrino which has a localized coupling with the Higgs and lepton doublets. From the Higgs expectation value, one then finds a neutrino mass which is suppressed by the volume of the bulk.

-Zichichi:

How do you get small neutrino mass?

- Antoniadis:

You have a right-handed neutrino as a bulk state that lives in extra dimensions. You decompose it into Kaluza-Klein excitations and the zero mode has a volume factor in its kinetic term. Then, there is a coupling of this right-handed state with the lepton doublet and the Higgs. When the Higgs gets an expectation value you obtain a Dirac mass between the left-handed and right-handed states, the first coming from the brane and the second from the bulk. This mass is inversely proportional to the volume of the bulk from the normalization of the kinetic term of the bulk state.

- Zichichi:

But you are copying the see-saw mechanism?

- Antoniadis:

I would not call it see-saw mechanism.

- Zichichi:

But it is the same!

- Antoniadis:

One difference is that the see-saw mechanism generates usually Majorana neutrino masses, while in this case one obtains Dirac masses. These lead also to neutrino oscillations compatible with the present experimental data.

- Zichichi:

But why the denominator in your formula cannot be very small?

- Antoniadis:

Because this is the size of the extra dimensions that are taken to be large in this model.

- Zichichi:

But it brings you back to the standard formulations. You obtain the same formula, roughly speaking.

- Antoniadis:

No, this is another mechanism that produces small neutrino masses in the framework of extra dimensions.

- t'Hooft:

When you mentioned the graviton going away as missing energy, you were talking about the spin of the graviton. Of course you have the spin 2 particle running away. But can you see the lower spins, because of extra dimensions?

- *Antoniadis:*

As I said, you study the angular distribution of the missing transverse momentum. This distribution is dependent on the particle that produces the missing p_T. For instance, in the case of SUSY the spin is ½ and you obtain a particular distribution. In the case of graviton we have a different distribution. This is not only due to one mode of spin 2, but also due to the whole tower of KK excitations.

-*Hisamatsu:*

I am stimulated by your lecture, and have started to think about other possibilities to measure the extra dimensions. With photon beams, there is a possibility to probe extra dimensions via photino and Kaluza–Klein gravitino production. With my naïve thought, this reaction may happen. So if we prepare a lot of photons and wait for gravitinos, we should be able to measure the effect of extra dimensions. Having a lot of photons is possible I think because we have a laser. Can I have your comment on this possibility?

- *Antoniadis:*

I think that this may be indeed a possibility. However, you do an assumption on the SUSY realization. These models do not need to be globally (at least) supersymmetric, so the photino may not exist. On the other hand, in linear colliders, one can still study the signatures of extra dimensions due to missing energy from graviton emission.

Supercomputing: general purpose and custom architectures

Roberto Petronzio

Università di Roma "Tor Vergata", I-00133 Rome, Italy
INFN sezione di Roma "Tor Vergata", I-00133 Rome, Italy

Current status of general purpose and custom supercomputers for basic science is discussed with a couple of examples of recent applications.

I. INTRODUCTION

For many physical phenomena the knowledge of the equations rarely leads to analytical solutions and many interesting problems can be only addressed by numerical techniques. Many body problems can be only solved on a computer and complex systems require a huge amount of computing power (measured in Flops, i.e. the number of floating point operations performed per second; the notations GF, MF, TF, PF refer to GigaFlops, MegaFlops, TeraFlops and PetaFlops respectively). In order to fulfill these requirements there are mainly two possibilities. The first consists in assembling together a sufficiently large number of general purpose computing units, essentially desk-top computers, and connecting them through a fast and scalable network: the resulting architecture is a so-called "PC-Cluster". In some case it is more convenient and/or more efficient to build computer architectures dedicated to the solution of a restricted class of problems; the computing units and the interconnecting network are customized to some peculiar aspects of the systems: these are the so-called "custom machines". A few examples of research fields where supercomputers are commonly employed are given in table I were the different entries show the computer power available nowadays, the one that will be available in the near future and the name of the custom architectures currently running.

Different supercomputers are usually ranked in terms of their overall peak performances but the choice of the architecture to be used in order to solve a given problem is related to the efficiency of that machine on the actual calculation (sustained performances). A measure of the efficiency of a computer architecture is the so-called "balance-ratio" (R). This is a structural parameter of the architecture and is defined as the number of floating point operations to be performed divided by the number of accesses to the data memory required to fetch data and store results (example: in order to compute $c = a + b$ one has to fetch from the memory 2 numbers and perform one floating point operation; another access to the memory is required in the end to store the result, i.e. $R = 1/3$). Once the balance ratio is known for a class of problems the corresponding optimized architecture can be devised.

II. THE APE EXPERIENCE

The core routine in the study of non-perturbative phenomena in quantum chromodynamics is the inversion of the discretized version of the Dirac operator, i.e. the inversion of a large sparse complex matrix. The inversion is carried out by applying iteratively the lattice Dirac operator and converging toward an approximate solution within the requested accuracy. The balance-ratio for the application of the Wilson discretization of the Dirac operator is $R = 4$. At the beginning of the 80's people realized that a boost to lattice QCD calculations can be given through the construction of dedicated computers. The first supercomputer of the APE family (Array Processor Experiment) [1] was made of 16 computing nodes delivering 1 GF of peak performances. The second generation machine, APE100, has been the first custom machine for lattice QCD made of 2048 floating point units and delivering 300 GF. The third generation of APE machines, APE1000, has been devised and built in collaboration between the INFN and DESY. The sum of all the installations in Europe amounted to 2 TF of peak performances. The last supercomputer of the APE family is apeNEXT. Also in this case the machine has been devised in a European collaboration between INFN, DESY and ORSAY. The installations of Rome and DESY Zeuthen amount to 12 TF of peak performances for 64 bit arithmetics.

The main innovations introduced by the APE supercomputers are the complex "normal" operation, the very long instruction word (VLIW) and the single instruction multiple data (SIMD) architecture with strong 3D connectivity. The normal operation performed by an APE machine is $(a+ib)(A+iB)+C+iD$, i.e. the basic operation occurring in lattice QCD calculations. The VLIW technique simplifies the decoding effort allowing to devise an efficient compiler and a simple chip. In addition the power consumption of the APE machines is very low with respect to other

	today	2009	custom architectures
Astrophysics and Plasma Physics	70-100 TF	> 500 TF	Grape (JP/EU)
High Energy Physics (LQCD)	10-100 TF	500-1000 TF	APE (EU), QCDOC (USA/UK)
Climatology, Earth Sciences	10-30 TF	200-300 TF	Earth Simulator (JP)
Life Sciences (protein folding, drug design)	100-300 TF	$N * \text{PF}$	IBM BlueGene (USA)

TABLE I: MegaFlops (MF), TeraFlops (TF), PetaFlops (PF).

	n. processors	peak perf. (TF)	sustained perf. (TF)	GF/(proc. perf. peak)	GF/(proc. perf. sustained)
BlueGene/L	65536	367	73	5.6	1.12
apeNEXT	8192	13.2	8.0	1.6	0.97

TABLE II: Comparison of BlueGene/L and apeNEXT performances.

architectures: as an example one can compare the GF delivered per Watt by apeNEXT, 46 MF/W, with the ones delivered by the Earth Simulator [18] (this is a custom machine by NEC in Japan to handle with Climatology and Earth Science problems, with 5120 processors, 36 TF sustained and with 20 MF/W). The SIMD architecture means that the same computer code runs in parallel on all the nodes by processing different data. Data exchange between the memories of neighbouring nodes is made possible at the hardware level by the 3D network and it is handled at the software level by a few simple parallel instructions available in the high level programming language called TAO.

The apeNEXT machine is a 3D mesh of computing nodes. Each node hosts a custom processor (VLSI 200 MHz) performing the normal operation in 64 bit arithmetics, 256 MB of resident memory with access band of 3.2 GB/s and delivers 1.6 GF. Furthermore, each node is equipped with 6 network channels to allow 3D first neighbours communications at 200 MB/s per channel. The apeNEXT is a scalable system, from the 26 GF of a single processing board (node topology $4 \times 2 \times 2$) to the 12 TF of a multi-rack (node topology, rack $8 \times 8 \times 8$, multi rack $12 \times 8 \times 8 \times 8$).

According to the top500 ranking, the best supercomputer currently running is the IBM BlueGene/L (n.1 in current top500 ranking, 9 systems in the first 20 entries). The project started in 2000 based on the custom machine for QCD calculations designed by the Columbia University, QCDOC. The cost of the development amounted to 100 million dollars: the outcome is a machine made of 131072 processors delivering 280 TF sustained (100 times smaller footprint and 75 times smaller power consumption with respect to the Earth Simulator). Architectural and assembly parameters of the BlueGene are very similar to the historical APE project as can be seen in fig. 1. Both machines have computing units with low power consumption and powerful networks integrated at the chip level. Both architectures are modular and total peak performances depend upon the number of racks per installation. A comparison of the performances of the two machines is shown in table II: it emerges that BlueGene/L is only a factor 3.5 faster if peak performances are considered while it is almost as fast as apeNEXT if one considers the sustained performances in the case of QCD applications.

III. HIGHLIGHTS OF HIGH ENERGY PHYSICS ON SUPERCOMPUTERS

A central goal of high energy physics is the determination of the Cabibbo–Kobayashi–Maskawa [2, 3] quark mixing matrix. A set of redundant and precise measurements can also provide informations about possible new physics [4]. In turn, precise measurements need an adequate theoretical determination of hadron matrix elements of the weak currents (table III). Non perturbative tools and in particular lattice QCD can eventually provide the required precision. The accuracy of lattice calculations goes with the reduction of discretization effects and, on a finite volume, this requires a number of lattice points as big as possible, i.e. more and more computer power and memory. Chiral symmetry is broken at finite lattice spacing and can be recovered in the continuum limit that, in order to be taken, requires simulations for different values of the lattice spacing. Alternatively, one can resort to lattice formulations that retain exact chiral symmetry at finite lattice spacing paying, again, a huge cost in terms of CPU and memory.

The APE supercomputers had a key role in reducing the errors of the unitarity triangle analysis. On the APE1 where performed [5] the first calculations of weak matrix elements on lattices, see upper top plot in fig. 2, with a number of points of the order of 20×10^3 in the quenched approximation. On the supercomputers of the APE100 family

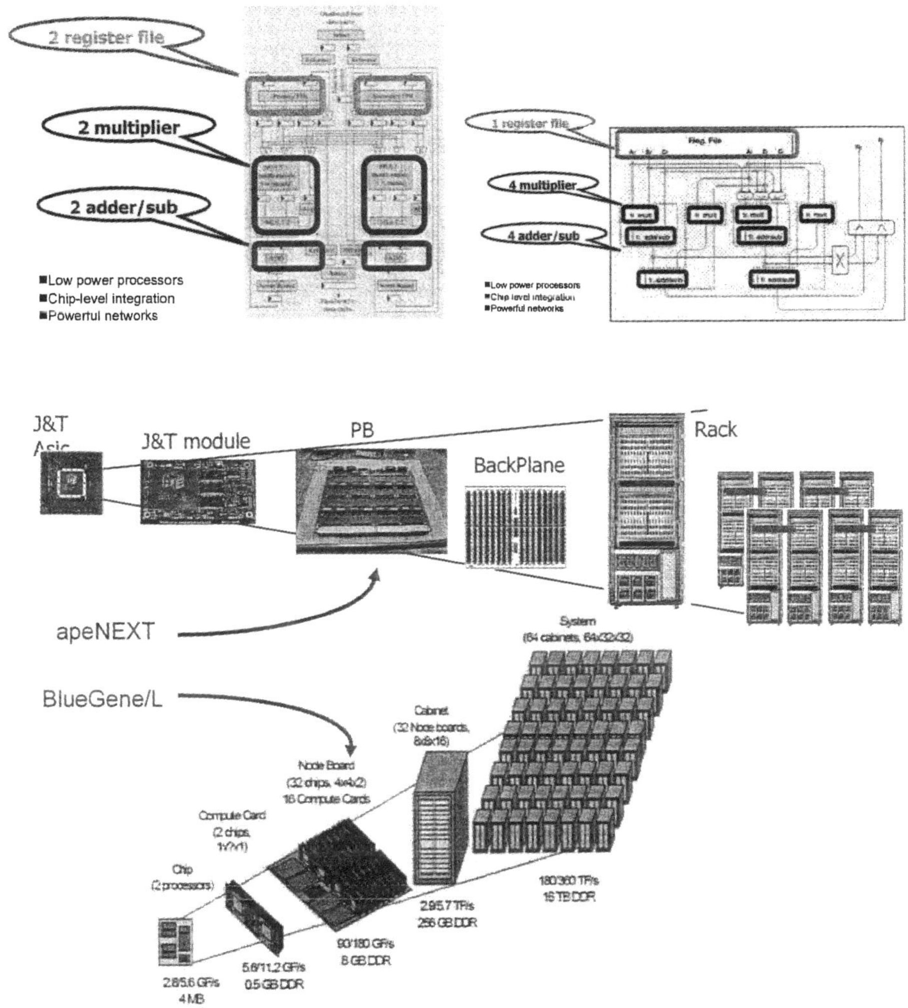

FIG. 1: Computing engines of BlueGeen/L (upper left) and of apeNEXT (upper right). System assembly of BlueGeen/L and apeNEXT (lower).

have been developed and applied the "improvement" techniques aimed to ameliorate the approach to the continuum limit; typical lattices had a number of points of the order of 48×24^3 in the quenched approximation (upper right plot in fig. 2). With the APE1000 and the Japanese CP-PACS supercomputers quenched lattice calculations entered the precision era and physical volumes of the order of 2 fm have been simulated with lattice spacings as small as 0.05 fm; some of the matrix elements of table III have been calculated in the continuum limit of quenched lattice QCD (lower plots in fig. 2).

	$\lvert\varepsilon_K\rvert$	ΔM_{B_s}	$\Delta M_{B_s}/\Delta M_B$	$B \to (\pi,\rho)\ell\nu$	$B \to (D,D^*)\ell\nu$
CKM	$\Im\{V_{td}\}$	$\lvert V_{ts}\rvert^2$	$\lvert V_{ts}\rvert^2/\lvert V_{td}\rvert^2$	$\lvert V_{ub}\rvert^2$	$\lvert V_{cb}\rvert^2$
LQCD	$\hat{B}_K$	$f_{B_s}^2 B_{B_s}$	$f_{B_s}^2 B_B/f_{B_s}^2 B_B$	$\lvert\langle(\pi,\rho)\lvert J_L^{ub}\rvert B\rangle\rvert^2$	$\lvert\langle(D,D^*)\lvert J_L^{cb}\rvert B\rangle\rvert^2$

TABLE III: Non perturbative inputs needed in the unitarity triangle analysis.

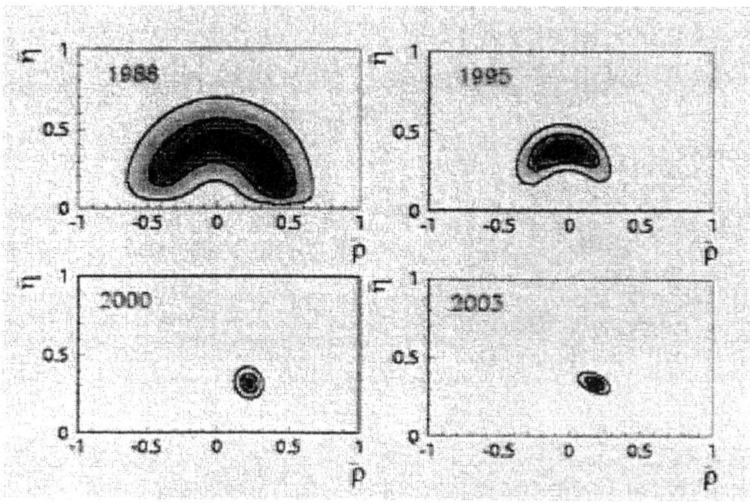

FIG. 2: Evolution during the last 15 years of the allowed region in the (ρ-η) plane. Plot taken from the UTfit collaboration website.

The calculations that are currently carried out on the supercomputers of the last generation, including the apeNEXT machines, account for the dynamical generation of q-$\bar{q}$ pairs (unquenching). The inclusion of the quark determinants in the Monte Carlo calculations requires the evaluation of a non-local effective gauge action and standard algorithms scale badly with the volume and light quark masses. Having more computing power and memory may not be enough to solve such kind of problems. The cost of the standard Hybrid Monte Carlo (HMC) algorithm for $N_f = 2$ Wilson Fermions is approximated by the so called "Berlin Wall" formula,

$$C_{cost} = C\left[\frac{\#confs}{1000}\right]\cdot\left[\frac{0.6}{M_P/M_V}\right]^6\cdot\left[\frac{L}{3\text{fm}}\right]^5\cdot\left[\frac{a^{-1}}{2GeV}\right]^7 Tflops/year \qquad C \simeq 2.8 \tag{1}$$

as discussed by A. Ukawa at the 2001 annual lattice conference in Berlin [6]. The bad scaling of the HMC with the volume and light quark masses is related to the fact that the inclusion of sea quark loops needs the evaluation of the determinant of an operator with a large hierarchy of scales, from the mass to the chiral cutoff. All the solutions that have been found to such a problem consists in different methods to implement a separation of the scales. In ref. [7] M. Lüscher proposed an hybrid algorithm for lattice QCD calculations based on domain decomposition techniques. The idea consists in dividing the lattice in a series of sub-domains with sizes in each direction of the order of 0.5 fm (see fig. 3). The Wilson-Dirac operator, and its determinant, is written as the product of a term that acts on the restriction of the quark fields to the blocks and a term that accounts for the interaction between the different blocks. The small physical extent of the block's volume is such that the lowest modes of the block operator experience a cutoff of the order $\lambda_{min} \simeq 1/L_{block}$. At the same time the block interaction operator is proportional to the square of the quark propagator and it is suppressed at least as the inverse sixth power of the distance among the blocks. Each factor in the quark determinant produces a force in the molecular dynamic equations that have to be solved in order to update the gauge fields. The domain decompositions translate in different force terms that are calculated a number of times inversely proportional to their magnitude in a single molecular dynamics trajectory, according to

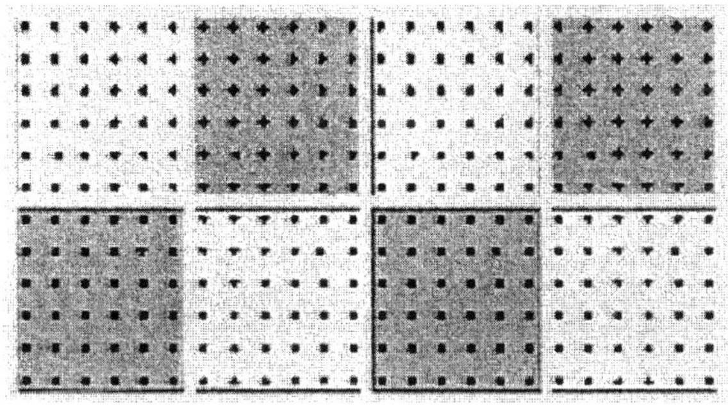

FIG. 3: Domain decomposition of a two dimensional lattice.

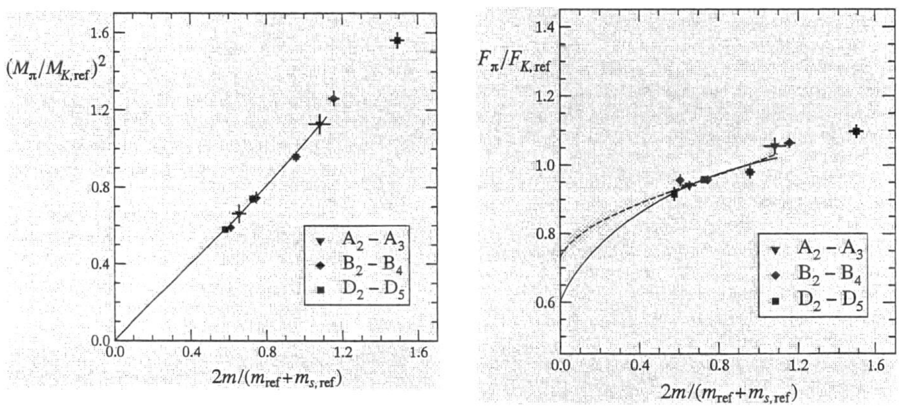

FIG. 4: Chiral fits to lattice data for M_π and F_π.

the multiple time steps integration techniques [8]. The net result is that forces that require expensive calculation are evaluated a number of times smaller with respect to the ones that are easier to evaluate. The cost formula for domain decomposed HMC lattice QCD calculations is [9]

$$C_{cost} = k\left[\frac{\#confs}{1000}\right]\cdot\left[\frac{20MeV}{m_q}\right]\cdot\left[\frac{L}{3\mathrm{fm}}\right]^5\cdot\left[\frac{0.1fm}{a}\right]^6 \; Tflops/year \qquad k \simeq 0.5 \tag{2}$$

i.e. the algorithm scales linearly with the inverse running light quark mass. Recent developments can further improve the performances of the Lüscher's algorithm by reaching an almost flat scaling with respect to the light quark masses [10]. A first impact on phenomenology due to this improvement has been exploited in ref. [9, 11] where for the first time dynamical lattice calculations with Wilson fermions have been performed at values of the running light quark masses in the range of validity of the chiral effective theory. In fig. 5 are shown the chiral fits to the lattice data from witch in ref. [9] have been extracted constrains on the low energy constants of the chiral lagrangian at the next to leading order (see quoted references for further details). On the theoretical side, the simulations of dynamical quark masses as low as about a fourth of the strange quark mass revealed unambiguously the effect of sea quark loops on correlation functions (see fig. 5).

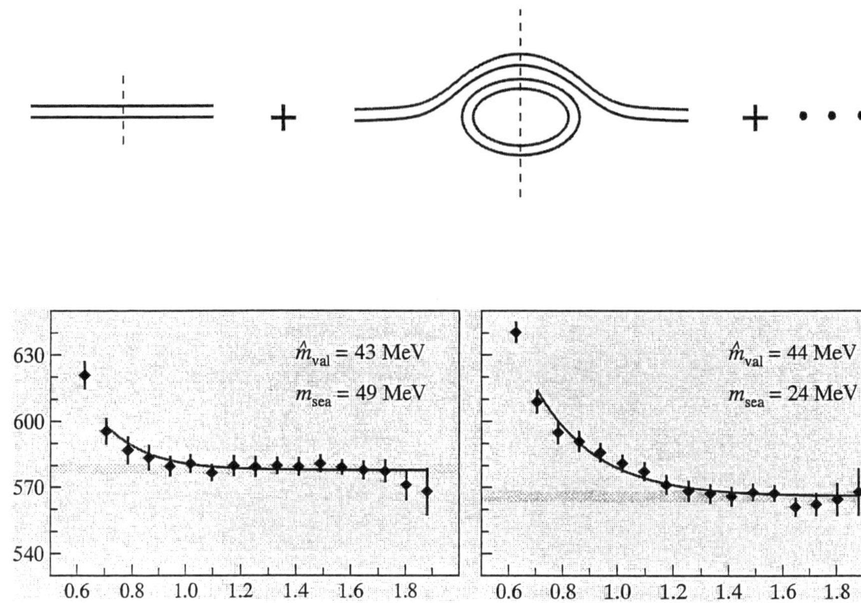

FIG. 5: Upper graphs: diagrams contributing to the two-point correlation functions of a pseudoscalar meson interpolating operator in the unquenched case. Lower plots: the lower is the mass of the valence quarks the bigger is the contamination of the dynamical quark loops to the correlation functions.

IV. HIGHLIGHTS OF BIOLOGICAL SIMULATIONS ON SUPERCOMPUTERS

Classical molecular dynamics biological simulations are characterized by less degree of freedom per point (particle) with respect to lattice QCD calculations and, consequently, are much less memory demanding. Currently one is able to propagate the Newton's equations of motions for a system made up of about 10^5 atoms and a simulation time of about 100 ns. The folding time of a protein is of the order of the second and conformational changes of biologically relevant systems cannot be studied by standard techniques.

In the case of ab-initio calculations one has to solve the Schrödinger equation for the electronic degrees of freedom and the size of the systems is limited to a number of atoms of the order of 10-100 for a time evolution of about 10-100 ps. Quantum mechanics and molecular mechanics can be combined together by isolating in a big system, 10^4-10^5 atoms, an active site of "quantum atoms" (10-100), and by treating the remaining particles as classical. In such kind of calculations the problem is that of finding the "interface" between the quantum and classical parts of the lagrangian. This problem has been successfully addressed in the recent years [13, 14].

In any case the goal of statistical simulations of biological system is the calculation of the free energy landscape of the system. Complicated systems like proteins, membranes, etc. present an extremely corrugated free energy landscape characterized by many different minima separated by high barriers. The crossing of such barriers is a rare event and happens on time scales of the order of seconds or more (like in the case of (un)folding of proteins). Typical simulations start with a thermalized system behaving in one of these minima and end, after a few hundred of ns, in the same minimum. Efficient sampling techniques have been devised over the years in order to calculate the free energy of the system through a smarter dynamics. Among these method there is the so called "metadynamics" [15]. Given a system whose standard dynamics can be described in terms of a few collective variables (order parameters that univocally characterize the states of the system at the macroscopic level), its free energy is given by

$$F(\underline{s}) = V(\underline{s}) - \frac{1}{\beta} \ln P(\underline{s}) \tag{3}$$

where the collective variables are functions of the microscopic atomic coordinates, $\underline{s} = \underline{s}(x_1, \ldots, x_N)$, $V(\underline{s})$ is the potential expressed in terms of the s's and $P(\underline{s})$ is the entropy of the system. For each point $\underline{s}'$ visited by the system during its evolution, a small gaussian centered around $\underline{s}'$ it is added to the potential, thus preventing the system to stay around this point. In such a way the system behaves an history dependent trajectory with an effective potential given by

$$F^{eff}(\underline{s};t) = F(\underline{s}) + \sum_{t'<t} w\, e^{-\frac{[\underline{s}-\underline{s}(t')]^2}{2\sigma^2}} \tag{4}$$

The minima of the original free energy of the system are "filled" during the evolution that stops when the process becomes purely diffusive, i.e. when $F^{eff}(\underline{s};t^*)$ is constant. In the end the true free energy of the system is obtained by

$$F(\underline{s}) = -\sum_{t'<t*} w\, e^{-\frac{[\underline{s}-\underline{s}(t')]^2}{2\sigma^2}} \tag{5}$$

Note that the evolution of the system is performed at the microscopic level. Indeed the force acting on each atomic coordinate is obtained from

$$F_i = \frac{\partial V^{eff}(s)}{\partial x_i} = \frac{\partial V^{eff}(s)}{\partial s} \frac{\partial s(x_1, \ldots, x_N)}{\partial x_i} \tag{6}$$

The method can fail if the collective variables chosen to characterize the states of the system are inappropriate. In this case the original free energy profile also changes during the evolution and the diffusive regime is never reached.

Among the biological systems that are suitable to be studied by using metadynamics there are the amyloids. Amyloid fibrils are big molecular aggregates responsible of diverse neurodegenerative diseases like Alzheimer, Parkinson, Type II diabetes etc. The aggregates are non crystalline and non soluble in water (see fig. 7) and experimental determinations of their atomic structures pose severe challenges. The issues related to the stability and the aggregation mechanism can only be studied through numerical simulations. In ref. [16] it has been considered a short fragment, $A\beta_{16-22}$, of the protein responsible of the Alzheimer disease and it has been carried out a systematic study of the stability of aggregates as a function of the number of peptides involved. It has been found that the minimal number of peptides required in order to form a stable aggregate is of the order of 16 with the system oscillating between different peptide's stackings (see fig. 8). The inter and intra-sheets distances of the aggregates have been extracted from simulation data by calculating the diffraction pattern that the system would have show in an X-ray experiment (see fig. 9) and have been found in very well agreement with available experimental data. Once the stability issues have been addressed one can start metadynamics simulations and study the aggregation mechanism. In fig. 10 are shown the free energy profiles for the process of attachment of a single $A\beta_{16-22}$ peptide to a system already made of 5 $A\beta_{16-22}$ peptides. The collective variables used in the metadynamics simulations are the number of hydrogen bonds and the number of hydrophobic contacts. These preliminary results show a peculiar L-shape pattern indicating that the process of aggregation is initially driven by hydrophobic interactions to be subsequently stabilized by the formation of hydrogen bonds.

V. TRENDS

Future supercomputers will reach multi-petaflops performances. From the INFN point of view there are mainly two ways to take part in this challenge: assembling commercial CPU's or going to custom technology again.

Instead of keep pushing on work frequency, present commercial processors are becoming smaller and less power demanding, leading to the possibility of putting more computing cores inside the same silicon package. A famous example of this technology is represented by the IBM Cell processor [19], which implements a general purpose PowerPC processor with 8 specialized floating point engines. This processor, which can reach 256 GF of peak performance in single precision on the top speed model and a power dissipation up to 250 W, is sold in the latest PlayStation game consoles and actually represents an extreme approach to vector SIMD processing, even if the context where it's used is quite different. Cell processor may become the building block of a next-generation SIMD supercomputer. On the other hand, its memory interface and balancing doesn't seem to fit scientific application purposes, the I/O interface needs to be reworked to build scalable systems and a power dissipation of 250 W avoid a very dense integration. Furthermore, double precision peak performance is declared to be 25 GF, much less than the single precision value.

On the custom side there is still room to build a N*100000 processors computational system in the PF range, which must be reliable, fine grained and low cost. Using the state of the art VLSI process, which can actually be 65nm, large memory buffers can be integrated on-chip in order to reduce bandwidth requirements towards external memory and to reduce the number of memory chips per computing element. A multi-core approach can even be taken, considering the FP nodes as a simple tile of a scalable architecture, with a strong embedded serial technology interconnection system; this would require two scales of connections (intra- and extra-chip), coping with LQCD algorithms developments. A conservative system clock frequency could be adopted (400-500 MHz), keeping power dissipation low and allowing to build very dense systems with high reliability. An estimate value for power consumption of this next generation floating point engine is approximately 3 GF/W. A new system engineering for heat removal and 5-10000 computing elements per cubic meter integration is also needed: a possibility is a 3D stacking technology for multi-board system integration, allowing up to 512 FP nodes per board, 32 boards per rack, giving an amount of 16k nodes per rack. At this scale of integration, it would be possible to assembly racks with 100-300 TF peak performance, thus reaching the PF range with a dozen of racks. Finally, the Digital Signal Processing (DSP) technology, especially in embedded systems, and its key features, optimized architectures for particular algorithms together with low power consumption, are clearly appealing to the supercomputers world.

Acknowledgments

I thank the organizers of the Erice School for the warm hospitality and R. Ammendola and N. Tantalo for their advise on current computer architectures.

---

[1] P. Bacilieri *et al.*, "The Ape Project: A Computer For Lattice QCD" IFUP-TH 84/40, Dec 1984
[2] N. Cabibbo, Phys. Rev. Lett. **10** (1963) 531.
[3] M. Kobayashi and T. Maskawa, Prog. Theor. Phys. **49** (1973) 652.
[4] M. Bona *et al.* [UTfit Collaboration], JHEP **0610** (2006) 081 [arXiv:hep-ph/0606167].
[5] N. Cabibbo, G. Martinelli and R. Petronzio, Nucl. Phys. B **244** (1984) 381.
[6] A. Ukawa [CP-PACS and JLQCD Collaborations], Nucl. Phys. Proc. Suppl. **106** (2002) 195.
[7] M. Luscher, Comput. Phys. Commun. **165** (2005) 199 [arXiv:hep-lat/0409106].
[8] J. C. Sexton and D. H. Weingarten, Nucl. Phys. B **380** (1992) 665.
[9] L. Del Debbio, L. Giusti, M. Luscher, R. Petronzio and N. Tantalo, JHEP **0702** (2007) 056 [arXiv:hep-lat/0610059].
[10] M. Luscher, "Local coherence and deflation of the low quark modes in lattice QCD," arXiv:0706.2298 [hep-lat].
[11] L. Del Debbio, L. Giusti, M. Luscher, R. Petronzio and N. Tantalo, JHEP **0702** (2007) 082 [arXiv:hep-lat/0701009].
[12] L. Guidoni, contribution at the workshop SEDS06: Perspectives on Science and Engineering Driven Supercomputing, May 14-16, 2006.
[13] A. Laio *et al.*, J. Chem. Phys. **116**, 6941 (2002)
[14] M. Colombo *et al.*, CHIMIA **56**, 11 (2002)
[15] A. Laio and M. Parrinello, PNAS USA **99**, 12562 (2002)
[16] U.F. Röhrig, A. Laio, N. Tantalo, M. Parrinello and R. Petronzio, BIOPHYS. J. **91**, 3217 (2006)
[17] `www.top500.org`
[18] Earth Simluator, `http://www.es.jamstec.go.jp/index.en.html`
[19] IBM Cell Processor `http://www.research.ibm.com/cell/home.html`

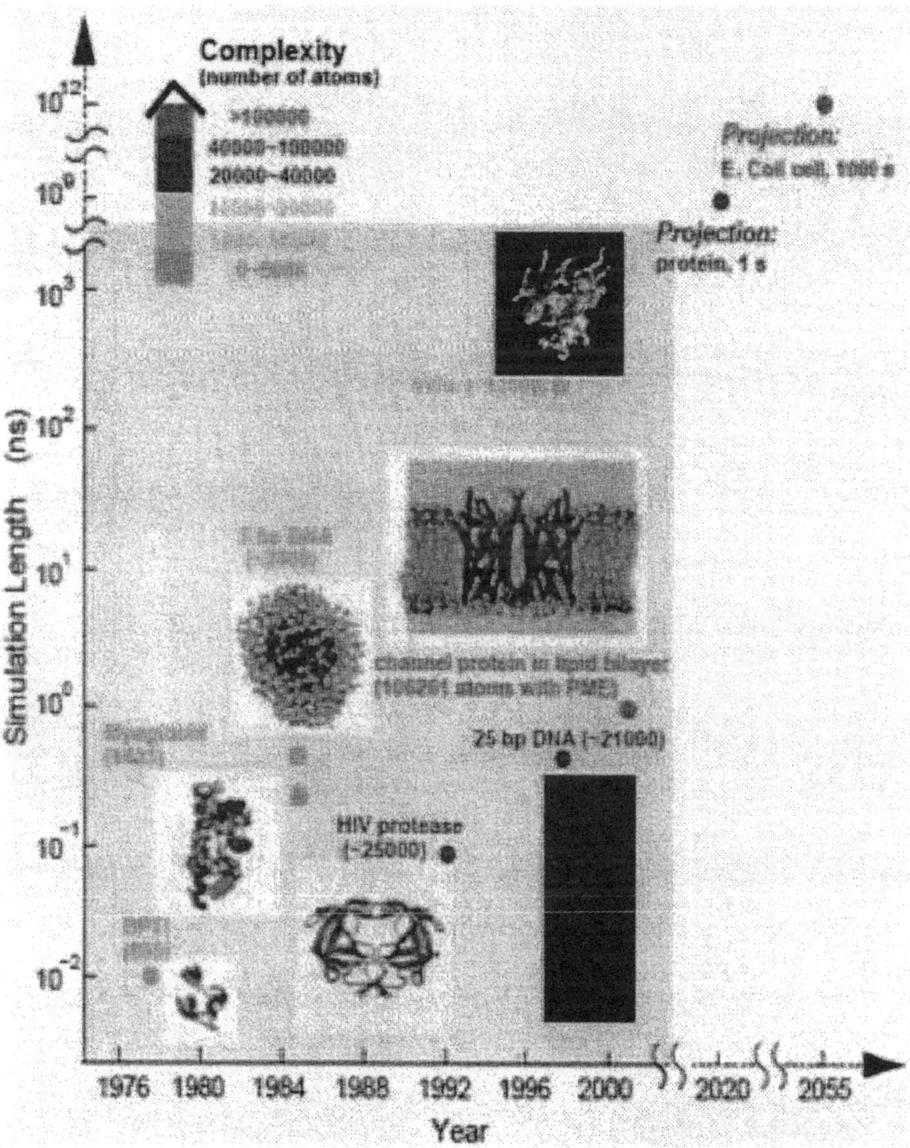

FIG. 6: Complexity of classical biological simulations [12].

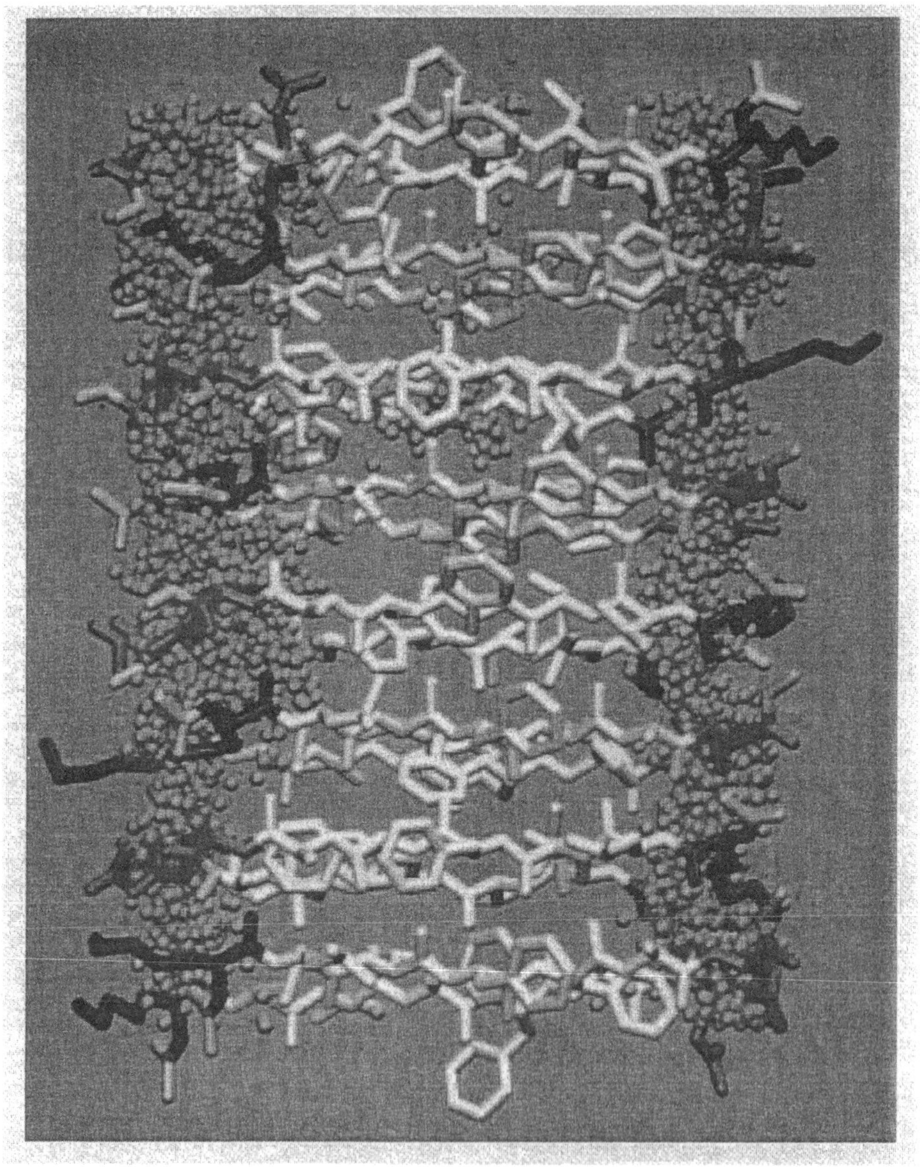

FIG. 7: β-sheet aggregate of a fragment, $A\beta_{16-22}$, of the human protein responsible of the Alzheimer disease.

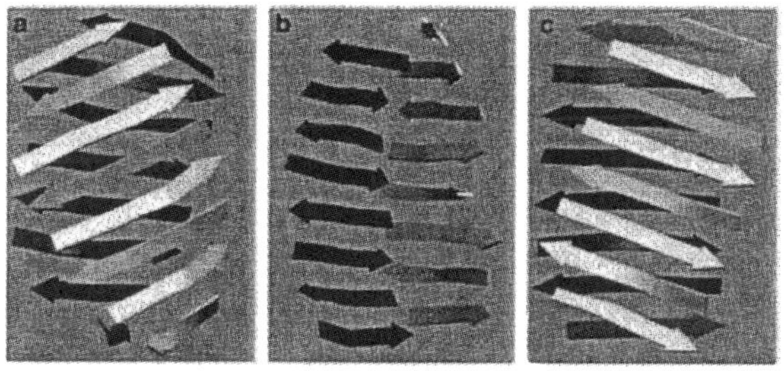

FIG. 8: Stable β-sheet aggregates of a 16-mer of $A\beta_{16-22}$ peptides in different stackings.

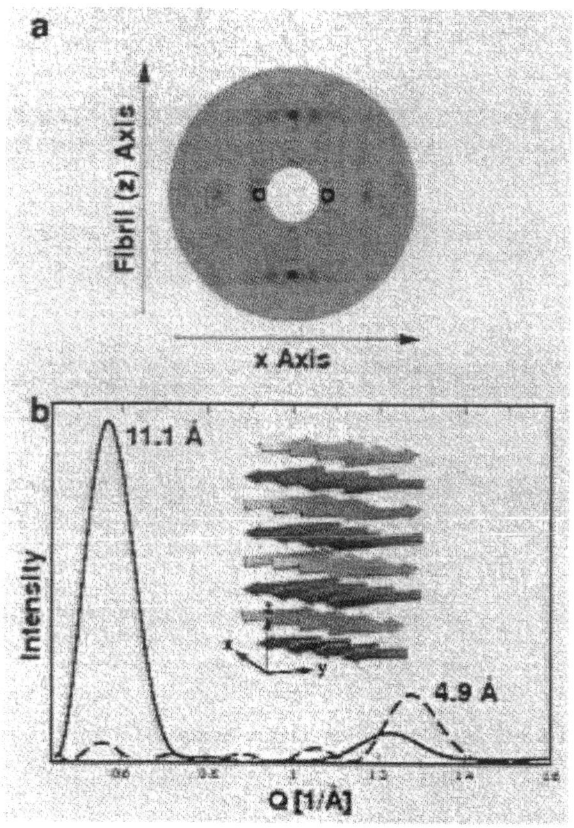

FIG. 9: X-ray diffraction pattern of a 32-mer of $A\beta_{16-22}$ peptides as calculated from numerical simulations.

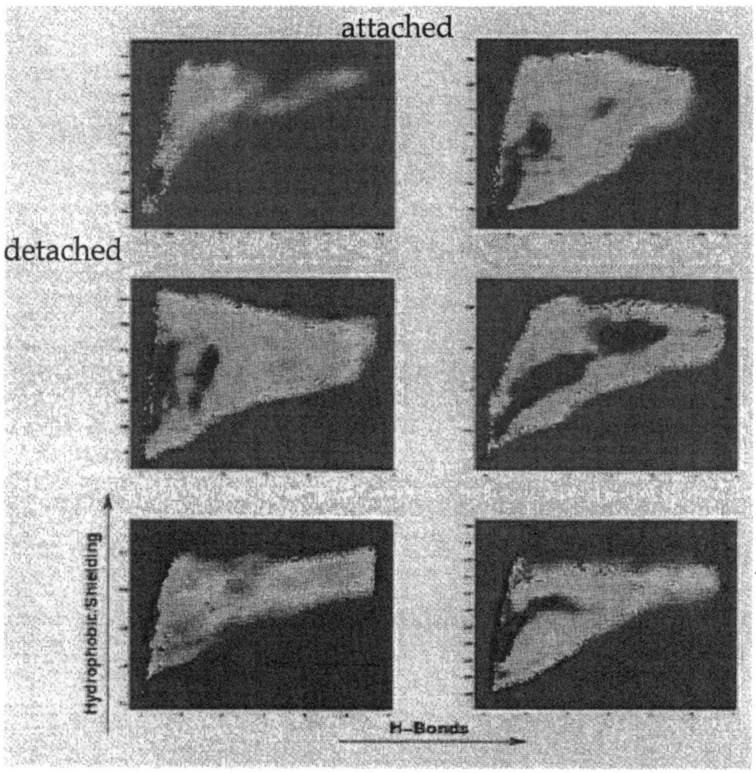

FIG. 10: Free energy for the process of attachment of a single $A\beta_{16-22}$ peptide to a system already made of 5 $A\beta_{16-22}$ peptide. The collective variables used in the metadynamics simulations are the number of hydrogen bonds (x-axis) and the number of hydrophobic contacts (y-axis).

CHAIRMAN: R. PETRONZIO

Scientific secretaries: P. Aschieri, F. Noferini

DISCUSSION

-Naumenko:

Can you compare the computing power of traditional supercomputer with the computing power of quantum computers?

-Petronzio:

Quantum computers are still in their infancy. They are still much more an interesting research item than a reliable tool to support reseach in computational physics. I do not see any real machine, there are only a few examples of toy quantum computers. None of them has been extensively used for scientific computing. I am not an expert in the field. Basically, if you succeed in representing a bit of information through some physical object that has true quantum behaviour (e.g., the two independent polarization states of a photon) suitable superpositions of the states associated to input data will be all "processed" at the same time, a novel and potentially effective approach to parallel computing. If your system is described by some quantum type entanglement of the wave function, you can easily take shortcuts to your computations. Once again, we are still at the level of development and demonstration that some algorithm may work on some quantum architecture, but close to nothing is done at the commercial level or for intensive use.

-'t Hooft:

They are still in Euclidean space.

- Petronzio:

Yes, of course.

-'t Hooft:

In principle they could do better than any of your computers.

-Petronzio:

Yes, in principle but we will have to wait several more years before they become useful computing tools.

-Brandt:

Can you point out the most important scientific discoveries made with supercomputers? In particular, can you point out the most important results coming from the Earth Simulator?

-Petronzio:

I am not able to comment on the Earth Simulator since I am not a climatologist. I can try to pin down what has been done with supercomputers in QCD which is my own field. You may also want to refer to the lecture by Richard Kenway who has covered similar material.

Let me first stress that only today - after twenty years since we entered the game - we are starting to approach the real domain, that is the complete theory. At the unquenced level, the computation of the hadron spectrum is a major result. Another point which I think is very important is the evidence for confinement, which we got from numerical computations. We still lack somehow a proof, an analytic proof of that, but the fact that the so called strong coupling regime, where confinement goes by default, is related to the weak coupling regime without phase transitions is a very solid argument that QCD does explain confinement as we see it in nature…
At the moment of course there are many numbers which are being quoted for this calculation. Numerical calculations no matter what you do, they always offer a number. So the danger is to play with numbers and provide numerical results without understanding exactly their physical meaning or the context in which they are a good approximation to the real ones. This should be criticized. What I want to stress is that we have heard many times that lattice QCD has been solved. I think that a few more years are still needed before we will see real tests of the full QCD case. Real test is something you can bring to an accuracy which only depends on your computer power, but nothing depends on systematics. So, once again, I think we are entering this era just now, in spite of the fact that we have been in the field for twenty years.
Of course, results on the quenched theory have been very useful for phenomenology. They proved to be accurate enough for the spectrum so people used them very often for weak interaction matrix elements of hadronic states. I showed you the CKM triangle (unitarity triangle). No doubt corrections will show up, probably at the level of 5% or 10%. Also in this case, the real result has to come, as we start to explore real life that includes truly real light quarks.

-Csanad:

You have described the operation scheme of multiplying two complex numbers and adding a third to them. Why is this exactly the basic operation?

-Petronzio:

Well, in lattice QCD you work with complex numbers, in most cases. Rarely with real numbers. The are two issues. One is the R parameter which tells how well you match the speed at which you fetch something from the memory on one side, and the speed at which you do complex multiplications and sums on the other side. The ratio of these two speeds must be optimally balanced. The other point is that APE has several operation units in parallel; they do exactly what you need for a complex multiplication and addition. Of course you could also do it in series, you have eight real operations that you could do one after the other. However they naturally go in parallel on this stack. So the only thing which is parallelized is the basic complex operation. That is the only thing. Then you have a different style of parallelism which means that you consider different portions of space-time in parallel (that is, many processors perform the same computation on different portions of space-time in parallel). However, from the point of view of the CPU the things which you parallelize are the operations which are required for complex arithmetics.

-Csanad:

So basically you first figure out the mathematical operations specific to the problem which you need to solve and then try to optimize the processor for this task...

-Petronzio:

Well, you can keep track with paper and pencil of the operations the computer does. For instance, you can estimate how many times a data element will be reused in different operations, before you need to fetch a new one. That you do by hand, and then you try to devise the computer that does this in an optimised way. Of course this accurate book-keeping is viable only for those problems where computer time is spent moslty in just a few very specific places. There are codes where your computer time is distributed evenly on all places in the code, and they are much more difficult to optimize. Luckily enough, QCD has only a few subroutines where you crunch numbers and you do always the same operations, basically matrix multiplications of unitary matrices or a matrix times a vector. It is just plain linear algebra. So if you identify these operations, and you optimize carefully you may expect a large gain. In other cases, for example in biology, it is very difficult to optimize, because you have all sort of operations in classical molecular dynamics, you have real operations, you have complex patterns of data transfers, and, of course, a biological object has much less simmetries you can leverage on. Another point is that the tradition in QCD, in our field, is that you build your own code, you write it. In other fields the codes are industrially developed and then bought by users. So, in most cases the people doing research only use existing codes, whose details they do not understand in details and they are not interested in. In this case, you can hardly modify the code, even if it might potentially have a major impact on performance. In QCD we can afford the luxury of optimizing both the computers and the programs.

-Lopez Mateos:

You have spoken of the history, challenges and progress in computing applied to theoretical high energy physics. Could you comment on all these aspects in experimental high energy physics?

-Petronzio:

I am not an expert but I followed a bit the area. One key challenge in experimental physics is the development of Monte Carlo simulations of something that controls the showers of particles. Hadronization is a typical problem that you have to deal when analyzing experimental data, in order to bridge the gap betwenn theory and measurement. From the computational point of view, the key point is that events are independent object. One single CPU unit with its own memory is enough to deal with one event. So the key problem there is to distribute events to independent processors, each processor handling all details of that specific event. This is quite at variance with lattice QCD where you distribute the space-time among different processors.

This difference explains why the GRID is a very effective technique in experimental physics. You need to have a smart way of distributing the code and the single problems to processors. After this is done, processing goes on independently and asynchronously for each event. It is really a different approach. Some proposals have been set forth to use GRID techniques for theoretical physics. This may be possible in some cases, but probably not in situations in which very large amounts of data have to be repeatedly moved back and forth among processors.

-Aschieri:

I just have this question about what policy you have about the patenting of what you develop?

-Petronzio:

The policy that I try to push forward is what is now usually known open source. It is very difficult to bind a scientific development to an industrial application from the beginning. So basically what you do, is to develop everything in open source mode. This approach has been taken sofar for some very succesfull software developments, such as Linux. Everything is available to everybody, is published. Of course, at some point it may happen (and in fact it did happen) that an industry is interested in taking that open source and constructing on top of it a specific application. Then comes a question of a contract that you do with the industry and you try to have a deal for the specific application and this deal gives back some royalties to your institute and so on. But at the moment we are really going with the open source idea which I think is the only one that allows any scientist to contribute. If you start to protect too early, you basically drain the exchange of ideas and the opportunities for new and unexpected collaboration that are typical of our way of doing things.

-Shebalin:

You discuss advances in hardware. What advances in software do you see? For example, automatic parallelization, language extension, advanced operating system, etc, all of which may make computer usage easier for researchers.

-Petronzio:

In our case most of this is done not in a dramatic way, is done by constructing the code. Of course there are objects like MPI for standard clusters which are very useful. MPI is a standard way to move data among different processors, through the exchange of messages. Also in APE the language has some specific instructions which describe where your data has to come from, e.g., from a first neighbour or beyond first neighbour. In general there is not much you do, all the idea of customization was to have a computer, for example the 3D-mesh, which was suitable for parallelization from the very beginning. So I can hardly see that you can do by software something useful if the underlying architecture is bad for your problem. If you had just one very powerful processor you could do everything in serial mode, but this is not the way you can hope to go: basic laws of physics tell us that in principle we have much more to gain by making our computers more and more parallel. Of course, our approach to parallelism changes with time. Thirty years ago, the first CRAY computer had what is called vector processing, that exploits the idea of the pipeline: if the time needed to perform an operation is too long, you split it in smaller tasks, each simple enough that it can be performed in one clock period. So you pipeline many operations in a stack, and then you get them out one every clock. There you have help in programming because, of course, you could declare that you were forming a vector stack so there was something in the software development. The vector approach has been pushed aggressively by some computer companies, especially in Japan, as a way to 'hide' to the user the fact that the computer does things in parallel, but now I think that there is much more to gain if you accept that things have to be done explicitely the parallel way.

-'t Hooft:

You have explained how the next generation supercomputers, "super hardware", might require 12 nuclear reactors to power them. As the previous questioner, I wonder whether not a similar effort should be spent on "super software". In QCD of course we know that improved programming can reduce the finite lattice size effects as well as the finite volume effects, but should such improvements not be further intensified? An extreme possibility is the development of genuine "artificial intelligence", the art of having the computer program itself. I know that such attempts in the past had only very limited success. Do you know of any attempts to use supercomputers to do artificial intelligence?

-Petronzio:

No, no that I know. The tricks you have mentioned about finite volume are algorithmic improvements that physicists have invented with a combination of physics intuition and sound mathematics. Then they were converted to better and more efficient programming, a simple, even if tedious task, for the physicists, but something that machines are not able to do efficiently. Nevertheless, I would say that some software that helps in programming in an efficient way would be very useful, for fields where people are not used to write their own code. So that could help in getting good performances. This is already done, to some admittedly limited extent, by dedicated libraries of routines for specific applications. This is a sort of higher-level language, higher-level software which does not give you the basic building blocks but some more sophisticated blocks of your computation. These blocks can then be carefully optimized by the computer experts, following the evolution of computer architectures.

-Kenway:

Do you think the particle physics community is spending the right amount of money on computing?

-Petronzio:

You mean computer for theoretical physics or general purpose computers?

-Kenway:

Whichever you want to answer.

-Petronzio:

I think that if you try to estimate how much we have been spending on computers, especially in theoretical physics, you will acknowledge it is not big business. Let me try to give a figure: It is probably not beyond one hundred million dollars in the last 10 years, all over the world; I would say no more than that. And that is very little money with respect to the type of investment in accelerators and experimental apparatuses for particle physics. I think we do not spend enough. I think computing is still considered not as an instrument, a vital instrument, but just as an aside object that if you have it, of course it helps, but it is not clearly crucial. Money is not the only issue, however. There are situations in which just pumping money will not solve your problem. Take the grid as an example: the whole story started becauses our experimental colleagues found out they could not solve their problems just making their farms bigger and bigger. So they had to invent something. We face a similar problem also in our field, which is somehow at the frontier in the use of supercomputing, and, even worse, we are not generally rewarded for the amount of effort we put in pushing the frontier forward.

-'t Hooft:

We saw in the presentation on LHC how its design is limited by computer hardware. Present visitors are shown empty spots for the computer with the explanation that the engineers are waiting for the very latest version.

-Petronzio:

They want to buy later to have the best possible technology and performance per unit price, at the old price, this is a typical (and quite reasonable, for that matter) argument that you hear at LHC. Generally speaking, I believe that LHC results will depend comparatively more from analysis than from detectors. What I mean is that the detectors that LHC is going to use are not widely different from those used by other experiments. The real challenge is the really huge rate of events that you have to trigger, store and analyze. That is order of magnitude more than ever attempted so far, and it all depends on computing.

-'t Hooft:

They want to decide on the spot whether an event is physically important or not...

-Petronzio:

This is the key point. The risk with these huge event rates is that you only see (as was said many years ago) what you want to see because you trigger on what you expect to see, and you do not trigger for events your present understanding tells you they should not be there. So, computer power is again a key point; if you can relax your trigger criteria and perform more analysis then you are able to leave some doors open for the unexpected.

-Brandt:

Do you think we are not spending enough on GRID?

-Petronzio:

GRID spending at this time is just what is barely needed. If we tried to cut more, we would not be able to see the data that is going to come from LHC and analyze them. So, in a sense, I think that GRID is not pushed by itself but as an instrument for something else, which is data processing. Of course GRID costs seem so large because they come at the end of the development phase of the experiments, when one has almost forgotten the big money that went into constructions, and because the GRID concept was almost unknown when LHC started, so there were no funds allocated for this from the very beginning. But now the GRID is a substantial part of the game, so I think that what people are spending for GRID is not unjustified.

-Haidt:

What is eventually limiting (in principle) further progress?

-Petronzio:

As I was trying to say earlier, in my opinion there has already been considerable progress.

-Haidt:

What is limiting at the end?

-Petronzio:

Well, the silicon. How much you can put on a silicon chip. There is a quantum limit, I am not an expert but they cannot shrink indefinitely the size of a transistor.

-Haidt:

Is that of a factor 100?

-Petronzio:

We can already capitalize about a factor 100, of course you have to engineer it, so it means a three year project but it is visible from the fact that everybody is aiming to that, so that probably can be done. However, one must understand that this factor is going to come essentially from architectural improvements, that invariably boil down to using more parallelism, Beyond that probably there are more fancy solutions which are not yet commercial.

-'t Hooft:

Are there special principles regarding how far can one (in principle) go with miniaturizing the chip, and its power consumption?

-Petronzio:

I do not have a clear cut answer to that. Today, we routinely build transistors with a channel length of 90 nm. Experts predict that in the next ten years we can go to 20 or 30 nm. This is more a factor 3 than a factor 10. However, a factor 3 means one order of magnitude more transistor density, each switching 3 times faster (basic solid state physics tells you that the switching time of a MOS transistor scales with the channel length). So one or two orders of magnitudes are possible and not too far away in time. Howerer I think there is not much more space for improvements beyond that: as you come close to the atomic scale, things become wildly different and radically new ideas are needed.

- Korthals-Altes:

Nino has been talking about the UEEC (Unexpected Event of Enormous Consequences), with this minimalistic GRID we risk to miss things.

-Petronzio:

I agree with you. I hope that will come. Of course, there will be time to adjust because the luminosity of LHC at the beginning will be low and slowly increasing in time. We had some details on this in the beautiful talk by Wenninger, but I am not sure we will get easily to 10^34. So there will be hopefully time to learn to cope with high luminosities. I fully agree that we should avoid having prejudices which may really hide the new physics. LHC is really a superposition of machines, a nightmare we have to face. It is not like e+ e- machines where you can accurately tune the beam energies and scan in c.m. energy. Again this somehow follows the fashion so if we only look for missing energy events may be we run the risk to miss the big thing. So I agree the minimum bias event should be recorded as much as possible. Maybe experimentalists have a different opinion on this point, so it would be interesting to hear their opinion.

Dick Dalitz: Examples of His Contributions to Particle Physics

Gary R. Goldstein*
Department of Physics and Astronomy
Tufts University
Medford, MA 02155 USA

I had been asked to speak about the contributions of Dick Dalitz to Particle Physics. I was very pleased to be able to do this as an expression of my good fortune to have worked with him. This was an opportunity to present a synopsis of his contributions to an audience that may not have known about some of this history. In choosing a sampling of Dalitz's work, I relied on the talks that were given at the Memorial Meeting in Oxford, June 1, 2006, the extensive bibliography compiled by Aitchison, Close, Gal, and Millener [1] and my own recollections during many years of collaboration.

Richard Henry Dalitz was born in 1925 in rural Australia, Dimboola, Victoria. His mother was from Scottish forebears. His father was descended from a small, distinctive ethnic group, the Wends or Sorbs. A large contingent of that group had emigrated from Germany to Australia in the mid-19th century. Later in life Dick discovered that he was not of German descent when he uncovered this geneaology. At age two his family moved near to Melbourne, at his mother's urging. He had speculated that perhaps she was conscious of his precocious abilities, even that early, and wanted to be near better schools. This was apparently not an easy change for his father, who managed to earn a modest income as a laborer. In any case, Dick distinguished

*Supported in part by U.S. Department of Energy grant DE-FG02-92ER40702.

himself throughout his schooling, culminating in his undergraduate work at University of Melbourne. He was then awarded a traveling scholarship that allowed him to pursue graduate studies at Cambridge University.

In Cambridge Dick worked with the theorist, Nicholas Kemmer, on nuclear questions, receiving his D. Phil. in 1950. His dissertation was on 0→0 transitions in ^{16}O. The gamma transition is forbidden by angular momentum conservation. But the conversion of a virtual longitudinal γ into an e^+e^- pair is allowed and observed. So he encountered pair production early in his career. He soon applied this idea to particle decays, surmising that in electromagnetic decays like $\pi^0 \to \gamma\gamma$ one or both of the γ's could appear as an e^+e^- pair, since known as Dalitz pairs. Such pairs were seen in nuclear emulsions soon thereafter. The $\pi^0 \to e^+e^-e^+e^-$ decay was used to determine the parity of the π^0 [2] through the azimuthal correlations of the pairs. A similar analysis of Σ^0 decays into $\Lambda^0 + e^+e^-$ showed that the two hyperons had the same parity [3].

It is interesting to speculate about how outstanding scientists came to their insights; where there creative ideas originated; what habits of mind were instrumental in their important work. With Dick Dalitz there is evidence from this earliest work on nuclear transitions and Dalitz pairs that thorough, detailed studies of the phenomena in question open up the mind for creative leaps. In much of his work he began with experimenters showing him puzzling phenomena or asking difficult questions about poorly understood data. The physics was in the observable phenomena. In this sense he was always a phenomenologist, although he could get quite involved in the subtleties of field theory in many instances.

While still working on his thesis he spent part of a year visiting the cosmic ray group of C. Powell in Bristol where strange particles were being discovered and studied. He became very interested in hyperfragments and hypernuclei, nuclei with a hyperon replacing a nucleon. He continued to revisit this subject throughout his career, seeing it as a rich source for understanding the many manifestations of the strong, nuclear interaction. I will not discuss his extensive work on hypernuclei here (a summary and references are presented in the aforementioned ref. [1] and by his collaborator, A. Gal in ref. [4]). In this period he was drawn into the puzzling properties of strange particles, especially strange mesons that decayed into pions. Two seemingly different strange mesons, the θ^+ and the τ^+, had about the same mass but decayed into two and three π's, respectively. For the two π decay

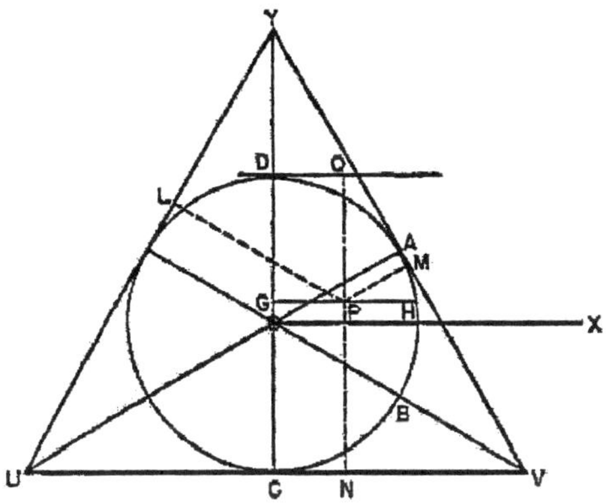

Figure 1: Triangular diagram for the representation of τ-meson decay events, from Ref. [5].

the angular momentum of the θ had to be $0^+, 1^-, 2^+$, etc. To determine the angular momentum of the τ he developed a geometrical analysis. Three particles in the rest frame of the decay have to have their energy add to the mass of the τ and their 3-momenta add to zero. With each π's energy related to its 3-momentum, there are only two independent variables left to characterize any particular decay. The distribution of events in a 2-dimensional plot will reveal the allowed mutual orbital angular momenta. He called this a "phase space plot", but its remarkable usefulness warranted the eponymous designation as the "Dalitz plot". When applied to the τ data it showed that the particle had to have even spin and odd parity, $0^-, 2^-$, etc. This pair of nearly equal mass strange particles have opposite parity, a circumstance known then as the $\tau - \theta$ puzzle. He speculated that perhaps there was something amiss with parity conservation, but found little sympathy for this idea among his colleagues. In 1956 T.D. Lee and C.N. Yang wrote a paper in which they concluded that there was no evidence that parity was conserved in weak processes, although it was clearly conserved in strong and electromagnetic processes. This would solve the puzzle - the τ and θ are the same particle and parity is not conserved in their decays. By 1957, C-S. Wu and collaborators showed that parity was not conserved in nuclear beta decay, ushering in a new era in particle physics.

The plots that were displayed in Dalitz's first Physical Review article on the subject [5] are shown here in Figs 1,2,3. Here was a major step forward in analysis and it is applied with great care. It is particularly noteworthy that

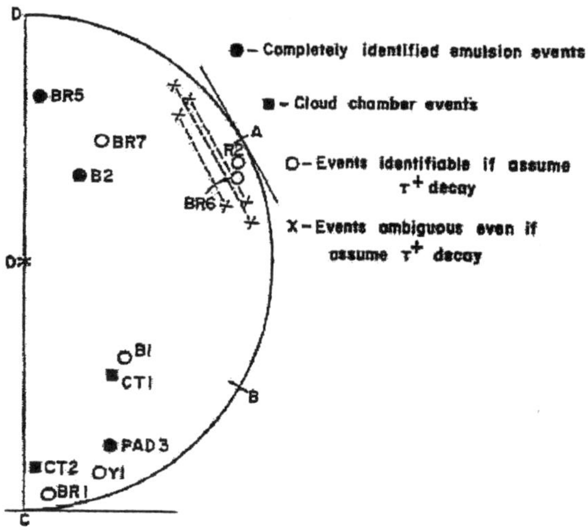

Figure 2: Data from τ-meson events in which the π-meson charges are established, from Ref. [5].

his hand drawn figures have such precision. He respected the importance of data. Throughout his career he valued working closely with experimentalists. There are many instances in which discussions with experimenters and study of current data inspired his theoretical breakthroughs. The Physical Review article also demonstrates the care and precision of his writing, in general. His text (and his physics) is never carried away by hyperbole. Clarity of expression was central to his approach to writing and speaking. In the article he states his conclusion cautiously, "At the present stage, the number of τ-decay events giving a slow unlike π meson rather suggests that the least L is $L = 0$, which would imply that the τ meson belongs to a class of even j and odd w, a class for which the 2π decay is forbidden." He awaits further data to confirm this assignment, but sees that it is quite likely, given the distribution of this small number of events. Soon thereafter more data did confirm his observation, τ was 0^-.

After leaving Cambridge and then Bristol, where he came to know the cosmic ray data that inspired his interest in strange particles, he became a Lecturer in Birmingham at the invitation of Rudolph Peierls. Peierls became a lifelong colleague and friend. In Birmingham and in visits to several US facilities, he pursued the analyses that expanded on discoveries of strange par-

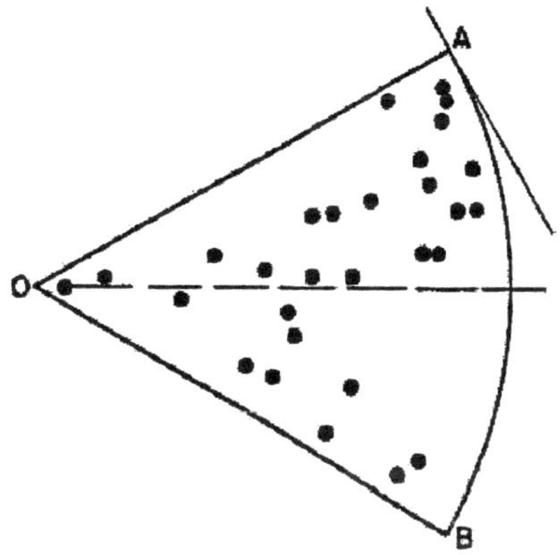

Figure 3: 29 τ-meson decay events from Ref. [5].

ticle resonances. The Dalitz plot was instrumental in finding the $\Sigma^*(1385)$ in 1960 [6] and later the $\Xi^*(1530)$. Both had spin 3/2 and are strange partners of the first hadronic resonance, the $\Delta(1236)$ discovered by Fermi and Anderson in 1952 [7]. With various collaborators, Dalitz studied strong production processes of strange particles - mesons and hyperons. He pursued this research through multichannel partial wave analyses, supplemented by a growing understanding of the analyticity of the scattering amplitudes. Much of this is thoroughly summarized in several review articles and a nicely written 1962 book for workers in the field and graduate students, Strange Particles and Strong Interactions [8].

Of special note in the middle 1950's is Dalitz's prediction, with S-F. Tuan, of a $\Lambda(1405)$ hyperon below the $K^- + p$ threshold [9], which appeared in the partial wave analysis extended below threshold via a multichannel K-matrix method. This state was discovered subsequently, but its complicated interactions have remained an interesting area of study itself. In this period Dalitz probed deeply into the subtleties of partial wave analyticity and, along with Castillejo and Dyson in 1956, found an unnoticed phenomenon, since called the CDD pole [10]. The dispersion relations for amplitudes that arise from strong interaction dynamics in meson-baryon scattering (for example the Low equation) contain resonance poles and threshold cuts. But the implementation of analyticity and unitarity also allows a set of extra poles with arbitrary strengths and positions - the CDD poles. These can be associated with fundamental, elementary particles or with overall ambiguity in the dynamical equations. These poles continue to be of use in understanding many two-body scattering processes.

Dick settled into the University of Chicago from 1956 to 1963, during which there were many experimental discoveries of new meson and baryon resonances. The Dalitz plot had application to many of those discoveries. It was clear, though, that the plethora of particles could not all be fundamental. Theorists looked for patterns that would allow grouping these particles. Dalitz's study of strange states primed him to appreciate efforts to generalize from isospin to more encompassing symmetries. Gell-Mann's and Ne'eman's SU(3) was proposed in 1961 and Dalitz soon studied its impact on the spectrum of strange states. A new notion about universality emerged and to address strangeness changing weak decays, Cabibbo, in 1963, proposed a mixing angle between the strange and non-strange sector. Dick was interested in comparing the mixing angles for pseudoscalar mesons to vector mesons. He also saw tests of universality as experimentally accessible and urged experimenters to test its ramifications [11]. When Dick first heard about Gell-Mann's proposal at a 1961 Summer School in India, he was struck by the notion that there could be more fundamental particles, the quarks, even though they were initially only considered to be a mathematical device (triplet representation of flavor SU(3)) for forming group representations. In the next few years, during which he moved to Oxford as the Royal Society Professor (at Peierls' urging), he thought about how to form the observed hadrons out of actual bound physical quarks. By the time of the Les Houches Summer School in 1965 [12] he presented a thoroughly developed quark model of the baryons.

Dalitz's "symmetric quark model" is a non-relativistic construction in which 3 quarks are bound by unspecified two-body forces in various orbital angular momentum configurations with spatially symmetric, antisymmetric or mixed symmetry wavefunctions. The overall space-spin-flavor symmetry is symmetric under pair interchanges, thereby violating Fermi statistics. He knew that this was a problem to be solved somehow, but, nevertheless saw the model as incorporating most of the known baryons into well-defined multiplets. The ground state spin $\frac{1}{2}$ octet and spin $\frac{3}{2}$ decuplet were s-states with masses separated by hyperfine splitting. The first excited states were $L = 1$ and $L = 2$. Fig. 4 and Fig. 5 show these states along with Frank Close's [13] additions of the currently known masses for these states. The connection to current data is remarkable, albeit that 40 years have elapsed. A glance at

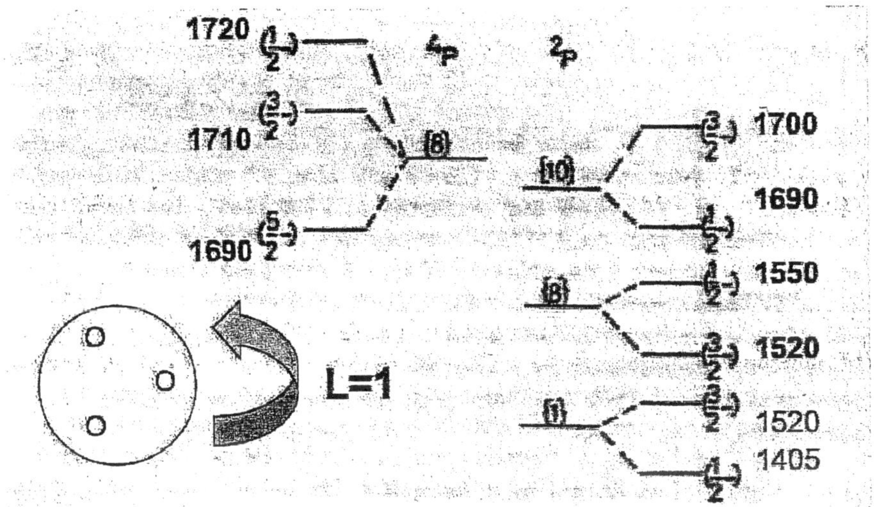

Figure 4: L=1 multiplet from Ref. [12] with masses from Ref. [13].

the current Particle Data Group Listings of baryons shows how remarkable is the reduction of the spectrum to three quark states. This model holds up, given the addition of the color SU(3) symmetry introduced several years later that allows the "symmetric" quark model to become the antisymmetric quark model in compliance with Fermi statistics. While Dick worked on this model, in 1964 the unification of spin and SU(3) had been suggested via SU(6) by Sakita and by Gürsey and Radicati [14]. The Dalitz model ground state baryons then fit nicely into a **56** representation of SU(6), while the next excited states are in **70** and **20** representations.

In the late 1960's through the 1980's Dick continued to work on hadron spectroscopy from the quark model QCD perspective, hadron scattering as described via analyticity and multichannel unitarity, hypernuclei, CP violation and other symmetries in weak decays. Dick even worked on the formulation of potential models in Lattice QCD. He worked with many colleagues, post-docs and students, and built the Theoretical Physics Department at Oxford after R. Peierls stepped down. He also became interested in the biographies of his notable forerunners and peers, writing several scientific biographies, including those of Peierls, Dirac, Skyrme, Tamm and Sakharov.

It was in 1987 that I began collaborating with Dick. R. Marshall, an

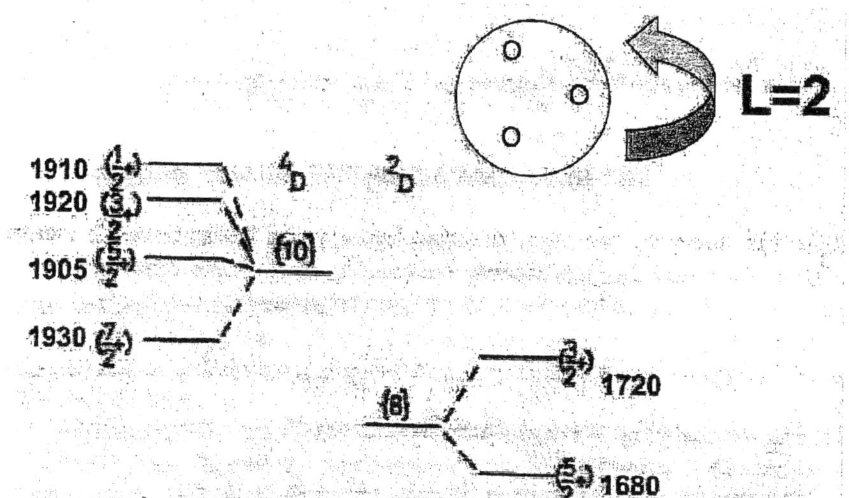

Figure 5: L=2 multiplet from Ref. [12] with masses from Ref. [13].

experimenter at Rutherford Lab and member of a DESY collaboration, had raised a question to Dick. How does a quark spin polarization get transferred to the jet of hadronic fragments that the quark produces? Dick invited me to spend part of my sabbatical in Oxford to address this question and whatever else might arise. We decided that this could be done in principle by considering three of the most energetic mesons in the jet. To identify the particular quark flavor would be difficult unless it were a heavy flavor. At that time there were many charmed quark processes, but few bottom quark mesons and the top quark had not been discovered. So we focused on c-quarks and $D\pi\pi$ fragments. The correlation represented by $\langle\vec{S}\cdot\vec{p}_D\rangle\langle\vec{p}_1\cdot\vec{p}_2\rangle$ can be non-zero for non-zero c-quark polarization $\vec{S}$, either helicity or transversity. Careful examination shows that this correlation arises from an interference between a tree-level amplitude and the imaginary part of an amplitude for final state interactions, a so-called T-odd effect (although there is no violation of Time Reversal invariance). We performed a model calculation involving the interference between independent amplitudes for c-quark $\rightarrow$ s-wave $D\pi\pi$ and c-quark $\rightarrow D^*\pi$ [15]. Disappointingly, the predicted effect was small, which was confirmed at SLAC. Subsequently fragmentation of transversely polarized light quarks into mesons with transverse momentum was reexamined by J. Collins [16] and the prediction and measurement of the "Collins function"

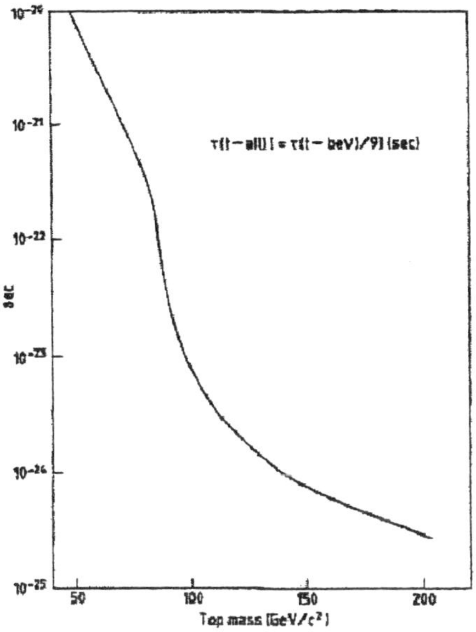

Figure 6: Top quark lifetime vs. mass from Ref. [17].

has become a very active pursuit.

In 1989 five quark flavors were well established and the sixth, the top quark was being sought at LEP and Fermilab. Non-discovery at LEP put a lower limit on the top mass of 100 GeV. Dick and I realized that for a mass considerably larger than that, the top would decay very rapidly into a b-quark and an on-shell W boson. The Standard Model calculation showed that the lifetime would go as $1/m_t^5$ and would be comparable to strong decays. Hence, the t-quark would decay as fast or faster than the time it would take to form color singlet, top flavored hadrons, as shown in Fig. 6 and 7. The possibility of producing tops at LEP inspired our first paper [17] on e^+e^- production characteristics. In the clean reaction $e^+e^- \to t+\bar{t}$, the incoming energy would fix the top mass. Since the mass was too high for LEP, however, it would fall to the Tevatron to search for it. The collider at that time provided near 1 TeV for $\bar{p}p$ energies, but the top mass is no longer tied to the incoming energy. Furthermore, the most striking signature for the production would involve two missing neutrinos, which would seem to make kinematic reconstruction difficult and full of ambiguity. But this was the kind of challenge that inspired Dick to push forward. We thought about how particular events would have to appear in momentum space, given the chains of subsequent decays. A very attractive geometrical picture emerged [18]. We used the Standard Model

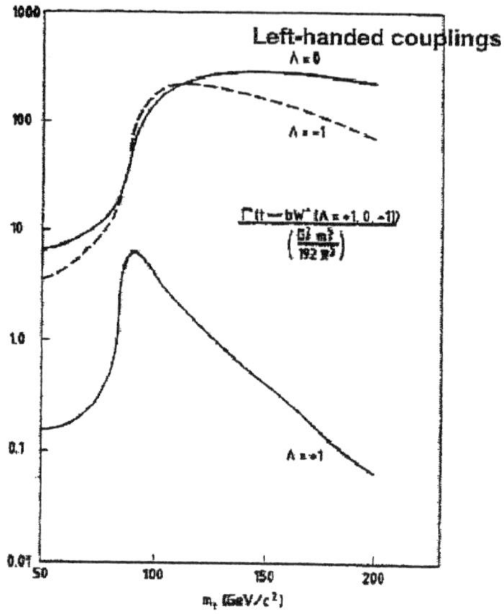

Figure 7: Partial rates for different helicities Ref. [17].

to determine the probability $P_i(m_t)$ that in an event i a measured set of jet and lepton momenta corresponds to top quark pair production and decay for top quark mass m_t. Underlying this probability is the interpretation that an observed event is consistent with the hypothesis

$$\bar{p} + p \to \bar{t} + t + \text{other hadrons}$$

$$\text{followed by} \quad t \to W^+ b$$

$$\text{where} \quad W^+ \to l^+\nu_l \text{ or } u\bar{d} \text{ or } c\bar{s}$$

$$\text{and} \quad \bar{t} \to W^- \bar{b}$$

$$\text{where} \quad W^- \to l^-\bar{\nu}_l \text{ or } d\bar{u} \text{ or } s\bar{c}.$$

Given that the W decay into quarks involves pairs of colored quarks, the quark jets are about 9 times more likely than the lepton mode. Nevertheless, the pair of leptonic decays produces very energetic, large angle leptons that signal a large mass origin. In this dilepton channel the probability depends critically on the assumed mass of the top quark, since there are two unobserved neutrinos. In the unilepton channel [19] the probability is more constrained, since the purely hadronic decay of $t \rightarrow b + q + \bar{q}'$ should give an invariant mass for the decaying top. The invariant mass is smeared, however, by the hadronization process as well as experimental uncertainties in momentum assignments for jets.

The geometrical construction that yields a probability distribution for the top mass is based on 4-momentum conservation, as follows. Consider one hypothetically produced top, with unknown mass m_t, energy E_t and 3-momentum $\vec{t}$. The possible $\vec{t}$ values are plotted in 3-dimensional momentum space. With decay products' energy and momenta labeled by b, l, ν, the 3-momenta and energies are constrained via

$\left(\vec{t} - \vec{b}\right)^2 = \vec{W}^2$, a sphere in $\vec{t}$-space centered on $\vec{b}$,

$\left(\vec{t} - \vec{b} - \vec{l}\right)^2 = \vec{\nu}^2$, a sphere centered on $\vec{b} + \vec{l}$,

$\vec{W}^2 = (E_t - E_b)^2 - M_W^2$, fixing the radius of the first sphere,

$E_\nu^2 = (E_t - E_b - E_l)^2$, fixing the radius of the second sphere.

For fixed t 4-momentum, the $\vec{t}$ lies on the circle of intersection of the two spheres. The circle lies in a plane perpendicular to the $\vec{l}$. The set of all E_t's forms a parabaloid with axis along $\vec{l}$ as shown in Fig. 8. For fixed m_t, $E_t^2 - \vec{t}^2$ is constrained, cutting the paraboloid in an ellipse. The collection of ellipses for all allowed m_t values covers the paraboloid. A corresponding set of ellipses for the anti-top can be constructed also. The quark+antiquark pair in each event that annihilate to form the top+antitop pair have limited transverse momenta, so that $\vec{t}_T + \vec{\bar{t}}_T \simeq 0$. Hence the intersections of each pair of fixed m_t ellipses determine possible values of the top and antitop momenta, as in the simulated example of Fig. 9. At each intersecting pair of momenta for a possible m_t there is a Standard Model probability assigned. So each event gets a distribution of probabilities as a function of top mass.

As candidate events became available our method was applied to determine the relative probability that an event involved top production. For likely cases the probability distribution as a function of possible mass yielded peak values. Collecting the first sizable set of likely top events and their peak mass probabilities gave us an estimate of the actual top mass. This value, around 160 GeV with a big spread, was on the low side of the currently reported

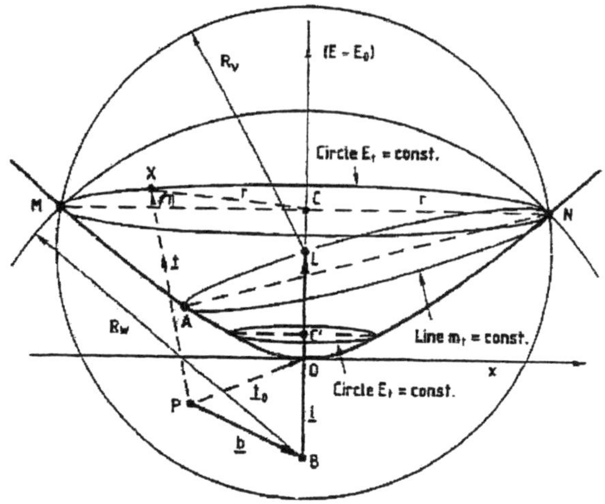

Figure 8: Construction of allowed top momenta $\vec{t}$ for given b and l momenta, see Ref. [18].

mass nearer to 174 GeV [20]. More recently a variant of this method was used by the D0 group to further constrain the top mass and to set limits on the Higgs mass [21].

Dick Dalitz continued to work on hypernuclei and scientific biographies into the last years. His pace slowed, but his enthusiasm for interesting problems never waned, nor did the seriousness with which he approached his work. He was an active participant in Particle Physics during its most exciting years of discovery. He contributed significantly to those discoveries. Dick's deep, sincere, overarching appreciation of physics as a pursuit of knowledge and understanding, without cant or boasting, was genuine and rare. He never expressed doubt that this pursuit was of unquestionable value and importance. Those who knew Dick Dalitz and those who were affected by his work will miss him. His dedication to physics was a beacon for us all.

Acknowledgement

It is a pleasure to thank Prof. Zichichi for inviting me to the Erice meeting and to thank the staff that makes this Summer School so successful. This is a truly remarkable institution whose longevity and reputation are well deserved.

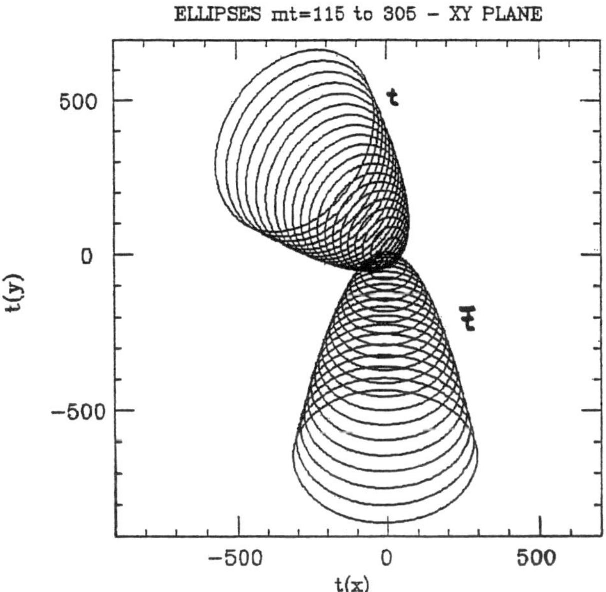

Figure 9: Pair of transverse plane projection of allowed top and antitop momenta.

References

[1] I.J.R. Aitchison, F.E. Close, A. Gal, and D.J. Millener (arXiv:physics/0603219).

[2] Plano, *et al.*, Phys. Rev. Lett. **3** (1959) 525.

[3] H. Courant, et al., Phys. Rev. Lett. **10**(1963) 409.

[4] A. Gal, arXiv:physics/0701019v2 (2007).

[5] R.H. Dalitz, Phys. Rev.**94** (1954) 1046.

[6] M. Alston, *et al.*, Phys. Rev. Lett. **5** (1960) 520; *ibid*, Phys. Rev. Lett. **6** (1961) 698.

[7] H.L. Anderson, *et al.*, Phys. Rev. **85** (1952) 936.

[8] R.H. Dalitz, "Strange Particles and Strong Interactions" (Oxford University Press, London, 1962).

[9] R.H. Dalitz, S.F. Tuan, Phys. Rev. Lett. **2** (1959) 425.

[10] L. Castillejo, R.H. Dalitz, F.J. Dyson, Phys. Rev. **101** (1956) 453.

[11] A. Zichichi, Proc. Roy. Soc. A **285** (1965) 175.

[12] R.H. Dalitz, in*High Energy Physics, Les Houches 1965*, eds. C DeWitt, M. Jacob (Gordon and Breach, New York, 1965) pp. 251-323.

[13] F. Close, Talk presented at "Dalitz Memorial Meeting", Oxford, June 1, 2006.

[14] B. Sakita, Phys. Rev. **136** (1964) 1756; F. Gürsey and L. Radicati, Phys. Rev. Letters, **13** (1964) 173.

[15] R.H. Dalitz, Gary R. Goldstein, R. Marshall, Z. Phys. C **42** (1989) 441; *ibid* Phys. Lett. B **215** (1988) 783.

[16] J. Collins, Nucl. Phys. B **396** (1993) 161.

[17] R.H. Dalitz and Gary R. Goldstein, Phys. Rev. D **45** (1992) 1531.

[18] R.H. Dalitz and Gary R. Goldstein, Phys. Lett. B **287** (1992) 225.

[19] Gary R. Goldstein, K. Sliwa, R.H. Dalitz, Phys. Rev. D **47** (1993) 967.

[20] R.H. Dalitz and Gary R. Goldstein, Proc. Roy. Soc. A **455** (1999) 2803, and references cited therein.

[21] D0 Collaboration, Nature **429** (2004) 638.

CHAIRMAN: G.R. GOLDSTEIN

Scientific Secretaries: Y. Petukhin, V. Konchakovskyi

DISCUSSION

- *Cifarelli:*

Could you illustrate the details of the top search method due to Dalitz?

- *Goldstein:*

We were puzzled, as were many people, about why the top quark was not the phenomenon to be seen at CDF. So we started to think about how it could be detected first in an e^+e^--machine and then we thought about detection at a collider.

The first thing we did was to look at how the decays of the top depended on its mass. As others had realized before, the top's decay time is extremely short because it has a high mass. The heavy top quark's weak decay is as fast as the strong decays, so it is unlikely we would see its decay as any time delay process. At first we thought that we could take advantage of the spin properties and here you see that in the decay, as a function of the mass, the helicity zero W in that decay begins to dominate the helicity minus one state and so the spin method for using the spin analysis to facilitate the top were not going to be very truthful. Yet, we thought about how the CDF people were talking about a golden event being one in which there were two very energetic leptons. So, we did a tree level QCD-model of the top production from p anti-p. We knew this and everyone heard that there are two missing neutrinos. So several people told us that this made doing a kinematic analysis quite difficult. However, I think that realizing that it was difficult made it more interesting to think about, so we came up with geometric construction. It showed us that even with two neutrinos' four-momentum missing for a given mass of top the QCD tree level picture constrained things considerably.

There are two spheres connected with three-momentuma for each top and anti-top. For a given energy those two spheres intersect in a circle and that constrains the top momentum. So, using those kinematic constraints we were able to come out with a construction that would give us a range of probability for a given kinematic configuration of decay products that corresponds to a top of given mass. Then we applied that to the publicly presented event that suggested very low mass: something greater than 130 MeV. Then, as real events became available, we put them through our method. I show examples from CDF in my lecture notes which are probability distributions versus mass.

In 1992 Dick Dalitz gave a talk here presenting our scheme for disentangling top quarks from the background, and at that time we were beginning to look at actual events.

So finally if you collect together all the probability distributions for the published single leptons (the work of Dalitz, Goldstein and Sliwa) and for the dileptons, they intersect in the region around 170 MeV. This was done several years ago to estimate the mass. Although we worked on this thinking about CDF data, it was the D0 group who applied these methods first. Two years ago the D0 group used this scheme to further narrow the top mass in order to constrain the Higgs mass.

Noncommutative Gravity and the $\star$-Lie algebra of diffeomorphisms *

Paolo Aschieri

Centro Studi e Ricerche Enrico Fermi
Compendio Viminale, I-00184, Roma, Italy

Dipartimento di Scienze e Tecnologie Avanzate
Universitá del Piemonte Orientale, and INFN - Sezione di Torino
Via Bellini 25/G 15100 Alessandria, Italy.
E-mail: aschieri@to.infn.it

Abstract

We construct functions and tensors on noncommutative spacetime by systematically twisting the corresponding commutative structures. The study of the deformed diffeomorphisms (and Poincaré) Lie algebra allows to construct a noncomutative theory of gravity.

PACS: 02.40.Gh, 02.20.Uw, 04.20.-q, 11.10.Nx, 04.60.-m. 2000 MSC: 83C65, 53D55, 81R60, 58B32

This article is based on common work with Christian Blohmann, Marija Dimitrijević, Frank Meyer, Peter Schupp, Julius Wess [1, 2] and on [3].

1 Introduction

We present a differential geometry theory on noncommutative spacetime. The simplest and most discussed example is canonical noncommutative spacetime. There we have $x^\mu \star x^\nu - x^\nu \star x^\mu = i\theta^{\mu\nu}$. After introducing noncommutative functions and tensors we study their infinitesimal transformations laws and the corresponding deformed Lie algebra of infinitesimal diffeomorphisms (and as a particular case Poincaré transformations). Starting from these basic notions we can then define covariant derivatives (so that we can implement the principle of general covariance on noncommutative spacetime), and torsion and curvature tensors. With these geometric tools we formulate noncommutative Einstein gravity. We here for simplicity consider noncommutative spacetime to be $\mathbb{R}^4$ with canonical commutation relations [1, 3]. Gravity on more general noncommutative manifolds, and its coordinate independent formulation is presented in [2].

Among the motivations for the study of noncommutative field theories and gravity I would like to recall that in the passage from classical mechanics to quantum mechanics classical observables become noncommutative. Similarly we expect that in the passage from classical gravity to quantum gravity, gravity observables, i.e. spacetime itself, with its coordinates and metric structure, will become noncommutative. Thus by formulating Einstein gravity on noncommutative spacetime we may learn about quantum gravity.

Planck scale noncommutativity is further supported by Gedanken experiments that aim at probing spacetime structure at very small distances. They show that due to gravitational backreaction one cannot test spacetime at those distances. For example, in relativistic quantum mechanics the position of a particle can be detected with a precision at most of the order of its Compton wave length $\lambda_C = \hbar/mc$. Probing spacetime at infinitesimal distances implies an extremely heavy particle that in turn curves spacetime itself. When λ_C is of the order of the Planck length, the spacetime curvature radius due to the particle has the same order of magnitude and the attempt to measure spacetime structure beyond Planck scale fails.

This Gedanken experiment supports finite reductionism. It shows that the description of spacetime as a continuum of points (a smooth manifold) is an assumption no more justified at Planck scale. It is then natural to relax this assumption and conceive a noncommutative spacetime, where uncertainty relations and discretization naturally arise. In this way the dynamical feature of spacetime that prevents from testing sub-Plankian scales is explained by incorporating it at a deeper kinematical level. A similar mechanism happens for example in the passage from Galilean to special relativity. Contraction of distances and time dilatation can be explained in Galilean relativity: they are a consequence of the interaction between ether and the body in motion. In special relativity they have become a kinematical feature.

2 ⋆-Products and Twists

The star product that implements the $x^\mu \star x^\nu - x^\nu \star x^\mu = i\theta^{\mu\nu}$ noncommutativity is given by

$$(h \star g)(x) = e^{\frac{i}{2}\theta^{\mu\nu}\frac{\partial}{\partial x^\mu}\frac{\partial}{\partial y^\nu}} h(x)g(y)|_{x=y} \tag{2.1}$$

where h and g are arbitrary functions. This star-product between functions can be obtained from the usual pointwise product $(hg)(x) = h(x)g(x)$ via the action of a twist operator $\mathcal{F}$ [4, 5]

$$f \star g := \mu \circ \mathcal{F}^{-1}(f \otimes g) , \tag{2.2}$$

where μ is the usual pointwise product between functions, $\mu(f \otimes g) = fg$, and the twist operator and its inverse are

$$\mathcal{F} = e^{-\frac{i}{2}\theta^{\mu\nu}\frac{\partial}{\partial x^\mu}\otimes\frac{\partial}{\partial x^\nu}} \quad , \quad \mathcal{F}^{-1} = e^{\frac{i}{2}\theta^{\mu\nu}\frac{\partial}{\partial x^\mu}\otimes\frac{\partial}{\partial x^\nu}} . \tag{2.3}$$

Despite the indices ${}^{\mu\ \nu}$ notation, we will consistently consider the entries $\theta^{\mu\nu}$ of the antisymmetric matrix θ as fundamenal dimensionful constants, like c or $\hbar$. In particular the deformed spacetime symmetries we consider will leave invariant the θ matrix. The point is that the exponent of $\mathcal{F}$,

$$\theta^{\mu\nu}\frac{\partial}{\partial x^\mu} \otimes \frac{\partial}{\partial x^\nu}$$

is not the Poisson tensor associated to the ⋆-product. The difference lies in the tensor-product $\otimes$. The Poisson tensor is

$$\theta^{\mu\nu}\frac{\partial}{\partial x^\mu} \otimes_A \frac{\partial}{\partial x^\nu} \tag{2.4}$$

where we have explicitly written that the tensorproduct is over the algebra $A = Fun(\mathbb{R}^4)$ of functions on spacetime. On the other hand the tensorproduct in $\mathcal{F}$ is over the complex numbers, we should write

$$\mathcal{F} = e^{-\frac{i}{2}\theta^{\mu\nu}\frac{\partial}{\partial x^\mu}\otimes_{\mathbb{C}}\frac{\partial}{\partial x^\nu}} .$$

That is why $\theta^{\mu\nu}$ in $\mathcal{F}$ is not a tensor but a set of constants. In this respect, a better notation for $\mathcal{F}$ is

$$\mathcal{F} = e^{\frac{-i}{2}\theta^{ab}X_a \otimes X_b} . \tag{2.5}$$

where $a, b = 1, ...4$ and $X_1 = \frac{\partial}{\partial x^1}, \ldots X_4 = \frac{\partial}{\partial x^4}$, are globally defined vectorfields on spacetime.

We shall frequently write (sum over α understood)

$$\mathcal{F} = \mathrm{f}^\alpha \otimes \mathrm{f}_\alpha \quad , \quad \mathcal{F}^{-1} = \overline{\mathrm{f}}^\alpha \otimes \overline{\mathrm{f}}_\alpha , \tag{2.6}$$

so that

$$f \star g := \overline{\mathrm{f}}^{\alpha}(f)\overline{\mathrm{f}}_{\alpha}(g) \ . \tag{2.7}$$

Explicitly we have

$$\mathcal{F}^{-1} = e^{\frac{i}{2}\theta^{\mu\nu}\frac{\partial}{\partial x^{\mu}}\otimes\frac{\partial}{\partial x^{\nu}}} = \sum \frac{1}{n!}\left(\frac{i}{2}\right)^{n} \theta^{\mu_1\nu_1}\ldots\theta^{\mu_n\nu_n}\partial_{\mu_1}\ldots\partial_{\mu_n}\otimes\partial_{\nu_1}\ldots\partial_{\nu_n} = \overline{\mathrm{f}}^{\alpha}\otimes\overline{\mathrm{f}}_{\alpha} \ , \tag{2.8}$$

so that α is a multi-index. We also introduce the universal R-matrix

$$\mathcal{R} := \mathcal{F}_{21}\mathcal{F}^{-1} \tag{2.9}$$

where by definition $\mathcal{F}_{21} = \mathrm{f}_{\alpha}\otimes\mathrm{f}^{\alpha}$. In the sequel we use the notation

$$\mathcal{R} = R^{\alpha}\otimes R_{\alpha} \quad , \qquad \mathcal{R}^{-1} = \overline{R}^{\alpha}\otimes\overline{R}_{\alpha} \ . \tag{2.10}$$

In the present case we simply have $\mathcal{R} = \mathcal{F}^{-2}$ but for more general twists this is no more the case. The R-matrix measures the noncommutativity of the $\star$-product. Indeed it is easy to see that

$$h \star g = \overline{R}^{\alpha}(g) \star \overline{R}_{\alpha}(h) \ . \tag{2.11}$$

The permutation group in noncommutative space is naturally represented by $\mathcal{R}$. Formula (2.11) says that the $\star$-product is $\mathcal{R}$-commutative in the sense that if we permute (exchange) two functions using the R-matrix action then the result does not change.

Note: The class of $\star$-products that can be obtained from a twist $\mathcal{F}$ is quite rich, (for example we can obtain star products that give the commutation relations $x{\star}y = qy{\star}x$ in two or more dimensions). Moreover we can consider twists and $\star$-products on arbitrary manifolds not just on $\mathbb{R}^4$. For example, given a set of mutually commuting vectorfields $\{X_a\}$ $(a = 1, 2, \ldots n)$ on a d-dimensional manifold M, we can consider the twist

$$\mathcal{F} = e^{\frac{-i}{2}\theta^{ab}X_a\otimes X_b} \ . \tag{2.12}$$

Another example is $\mathcal{F} = e^{\frac{1}{2}H\otimes ln(1+\lambda E)}$ where the vectorfields H and E satisfy $[H, E] = 2E$. In these cases too the $\star$-product defined via (2.2) is associative and properly normalized in the sense that for any function h we have $h \star 1 = 1 \star h = h$. In general an element $\mathcal{F}$ is a twist if it is invertible, if it satisfies a cocycle condition and if it is properly normalized [6]. The cocycle and the normalization conditions imply associativity of the $\star$-product and the normalization $h \star 1 = 1 \star h = h$. A deformed gravity theory based on this class of $\star$-products can also be constructed [2].

3 Vectorfields and Tensorfields

We now use the twist to deform the commutative geometry on spacetime into the twisted noncommutative one. The guiding principle is the one used to deform the product of functions into the $\star$-product of functions. Every time we have a bilinear map

$$\mu : X \times Y \to Z$$

where X, Y, Z are vectorspaces, and where there is an action of $\mathcal{F}^{-1}$ on X and Y we can combine this map with the action of the twist. In this way we obtain a deformed version $\mu_\star$ of the initial bilinear map μ:

$$\mu_\star := \mu \circ \mathcal{F}^{-1} , \tag{3.1}$$

$$\begin{aligned} \mu_\star : X \times Y &\to Z \\ (\mathsf{x}, \mathsf{y}) &\mapsto \mu_\star(\mathsf{x}, \mathsf{y}) = \mu(\bar{\mathrm{f}}^\alpha(\mathsf{x}), \bar{\mathrm{f}}_\alpha(\mathsf{y})) . \end{aligned}$$

The $\star$-product on the space of functions is recovered setting $X = Y = A = Fun(M)$. We now study the case of vectorfields and tensorfields.

Vectorfields $\Xi_\star$. We deform the product $\mu : A \otimes \Xi \to \Xi$ between the space $A = Fun(M)$ of functions on spacetime M and vectorfields. A generic vectorfield is $v = v^\nu \partial_\nu$. Partial derivatives acts on vectorfields via the Lie derivative action

$$\partial_\mu(v) = [\partial_\mu, v] = \partial_\mu(v^\nu)\partial_\nu . \tag{3.2}$$

According to (3.1) the product $\mu : A \otimes \Xi \to \Xi$ is deformed into the product

$$h \star v = \bar{\mathrm{f}}^\alpha(h)\bar{\mathrm{f}}_\alpha(v) . \tag{3.3}$$

Since $\mathcal{F}^{-1} = e^{\frac{i}{2}\theta^{\mu\nu}\partial_\mu \otimes \partial_\nu}$, iterated use of (3.2) gives

$$h \star v = \bar{\mathrm{f}}^\alpha(h)\bar{\mathrm{f}}_\alpha(v) = \bar{\mathrm{f}}^\alpha(h)\bar{\mathrm{f}}_\alpha(v^\nu)\partial_\nu = (h \star v^\nu)\partial_\nu . \tag{3.4}$$

It is then easy to see that $h \star (g \star v) = (h \star g) \star v$, i.e. that the $\star$-multiplication between functions and vectorfields is consistent with the $\star$-product of functions. We denote the space of vectorfields with this $\star$-multiplication by $\Xi_\star$. As vectorspaces $\Xi = \Xi_\star$, but Ξ is an A-module while $\Xi_\star$ is an $A_\star$-module.

Tensorfields $\mathcal{T}_\star$. Tensorfields form an algebra with the tensorproduct $\otimes$ (over the algebra of functions). We define $\mathcal{T}_\star$ to be the noncommutative algebra of tensorfields. As vectorspaces $\mathcal{T} = \mathcal{T}_\star$; the noncommutative and associative tensorproduct is obtained by applying (3.1):

$$\tau \otimes_\star \tau' := \bar{\mathrm{f}}^\alpha(\tau) \otimes \bar{\mathrm{f}}_\alpha(\tau') . \tag{3.5}$$

There is a natural action of the permutation group on undeformed tensorfields:

$$\tau \otimes \tau' \xrightarrow{\sigma} \tau' \otimes \tau \ .$$

In the deformed case it is the R-matrix that provides a representation of the permutation group on $\star$-tensorfields:

$$\tau \otimes_\star \tau' \xrightarrow{\sigma_{\mathcal{R}}} \overline{R}^\alpha(\tau') \otimes_\star \overline{R}_\alpha(\tau) \ .$$

It is easy to check that, consistently with $\sigma_{\mathcal{R}}$ being a representation of the permutation group, we have $(\sigma_{\mathcal{R}})^2 = id$.

If we consider the local coordinate expression of two tensorfields, for example of the type

$$\tau = \tau^{\mu_1,\dots\mu_m} \partial_{\mu_1} \otimes_\star \dots \partial_{\mu_m}$$
$$\tau' = \tau'^{\nu_1,\dots\nu_n} \partial_{\nu_1} \otimes_\star \dots \partial_{\nu_n}$$

then their $\star$-tensor product is

$$\tau \otimes_\star \tau' = \tau^{\mu_1,\dots\mu_m} \star \tau'^{\nu_1,\dots\nu_n} \partial_{\mu_1} \otimes_\star \dots \partial_{\mu_m} \otimes_\star \partial_{\nu_1} \otimes_\star \dots \partial_{\nu_n} \ . \tag{3.6}$$

4 $\star$-Lie algebra of Diffeomorphisms

In the commutative case we have two derivations that map tensors to tensors (of the same type): the Lie derivative $\mathcal{L}_v$ and the covariant derivative ∇_v along the vectorfield v. If we act on functions (tensors of type 0) Lie derivative and covariant derivative coincide and are the usual action of the vectorfield v on a function h. The result of this action is the variation of the function h under the infinitesimal transformation generated by v,

$$\delta_v h = \mathcal{L}_v h = v(h) \ .$$

In this section we study the $\star$-action of a vectorfield on a function, i.e. we deform the notion of Lie derivative (infinitesimal diffeomorphism), and study the deformed Lie algebra of these infinitesimal diffeomorphisms. Once this deformed derivation is understood then the notion of deformed covariant derivative will naturally emerge, and the construction of the Rimemann curvature tensor, the Ricci tensor and the curvature scalar will follow. These tensors will transform covariantly under infinitesimal diffeomorphisms.

The $\star$-Lie derivative on the space of functions $A_\star$ is obtained following the general rule (3.1). We combine the usual Lie derivative on functions $\mathcal{L}_u h = u(h)$ with the twist $\mathcal{F}$

$$\mathcal{L}^\star_u(h) := \overline{\mathrm{f}}^\alpha(u)(\overline{\mathrm{f}}_\alpha(h)) \ . \tag{4.1}$$

By recalling that every vectorfield can be written as $u = u^\mu \star \partial_\mu = u^\mu \partial_\mu$ we have

$$\begin{aligned}\mathcal{L}^\star_u(h) &= \overline{\mathrm{f}}^\alpha(u^\mu\partial_\mu)(\overline{\mathrm{f}}_\alpha(h)) = \overline{\mathrm{f}}^\alpha(u^\mu)\,\partial_\mu(\overline{\mathrm{f}}_\alpha(h)) \\ &= u^\mu \star \partial_\mu(h)\ , \end{aligned} \tag{4.2}$$

where in the second equality we have considered the explicit expression (2.8) of $\overline{\mathrm{f}}^\alpha$ in terms of partial derivatives, and we have iteratively used the property $[\partial_\nu, u^\mu\partial_\mu] = \partial_\nu(u)\,\partial_\mu$. In the last equality we have used that the partial derivatives contained in $\overline{\mathrm{f}}_\alpha$ commute with the partial derivative ∂_μ.

The differential operator $\mathcal{L}^\star_u$ satisfies the deformed Leibniz rule

$$\mathcal{L}^\star_u(h \star g) = \mathcal{L}^\star_u(h) \star g + \overline{R}^\alpha(h) \star \mathcal{L}^\star_{\overline{R}_\alpha(u)}(g)\ . \tag{4.3}$$

This deformed Leibniz rule is intuitive: in the second addend we have exchanged the order of u and h, and this is achieved by the action of the R-matrix that as observed in Section 2 and 3 provides a representation of the permutation group. A proof of (4.3) is not difficult

$$\begin{aligned}\mathcal{L}^\star_u(h \star g) = u^\mu \star \partial_\mu(h \star g) &= u^\mu \star \partial_\mu(h) \star g + u^\mu \star h \star \partial_\mu(g) \\ &= \mathcal{L}^\star_u(h) \star g + \overline{R}^\alpha(h) \star \overline{R}_\alpha(u^\mu) \star \partial_\mu(g) \\ &= \mathcal{L}^\star_u(h) \star g + \overline{R}^\alpha(h) \star \mathcal{L}^\star_{\overline{R}_\alpha(u)}(g)\ . \end{aligned} \tag{4.4}$$

From (4.2) it is also immediate to check the compatibility condition

$$\mathcal{L}^\star_{f\star u}(h) = f \star \mathcal{L}^\star_u(h)\ , \tag{4.5}$$

that shows that the action $\mathcal{L}^\star$ on functions is the one compatible with the $A_\star$ module structure of vectorfields.

In the commutative case the commutator of two vectorfields is again a vectorfield, we have the Lie algebra of vectorfields. In this $\star$-deformed case we have a similar situation. We first calculate

$$\mathcal{L}^\star_u\mathcal{L}^\star_v(h) = \mathcal{L}^\star_u(\mathcal{L}^\star_v(h)) = u^\mu \star \partial_\mu(v^\nu) \star \partial_\nu(h) + u^\mu \star v^\nu \star \partial_\nu\partial_\mu(h)$$

Then instead of considering the composition $\mathcal{L}^\star_v\mathcal{L}^\star_u$ we consider $\mathcal{L}^\star_{\overline{R}^\alpha(v)}\mathcal{L}^\star_{\overline{R}_\alpha(u)}$, indeed the usual commutator is constructed permuting (transposing) the two vectorfields, and we have just remarked that the action of the permutation group in the noncommutative case is obtained using the R-matrix. We have

$$\begin{aligned}\mathcal{L}^\star_{\overline{R}^\alpha(v)}\mathcal{L}^\star_{\overline{R}_\alpha(u)}(h) &= \mathcal{L}^\star_{\overline{R}^\alpha(v)}(\overline{R}_\alpha(u^\mu) \star \partial_\mu h) = \overline{R}^\alpha(v^\nu) \star \partial_\nu(\overline{R}_\alpha(u^\mu) \star \partial_\mu h) \\ &= \overline{R}^\alpha(v^\nu) \star \overline{R}_\alpha(\partial_\nu u^\mu) \star \partial_\mu h + \overline{R}^\alpha(v^\nu) \star \overline{R}_\alpha(u^\mu) \star \partial_\nu\partial_\mu h \end{aligned}$$

In conclusion

$$\mathcal{L}^\star_u \mathcal{L}^\star_v - \mathcal{L}^\star_{\overline{R}^\alpha(v)} \mathcal{L}^\star_{\overline{R}_\alpha(u)} = \mathcal{L}_{[u,v]_\star} \tag{4.6}$$

where we have defined the new vectorfield

$$[u, v]_\star := (u^\mu \star \partial_\mu v^\nu)\partial_\nu - (\partial_\nu u^\mu \star v^\nu)\partial_\mu \ . \tag{4.7}$$

The bracket $[\ ,\ ]_\star$ is a bilinear map

$$\begin{aligned} [\ ,\ ]_\star \ : \ \Xi_\star \times \Xi_\star &\to \Xi_\star \\ (u, v) &\mapsto [u, v]_\star \end{aligned} \tag{4.8}$$

and the space of vectorfields equipped with this bracket becomes the $\star$-Lie algebra of vectorfields $\Xi_\star$. It is easy to see that the bracket $[\ ,\ ]_\star$ has the $\star$-antisymmetry property

$$[u, v]_\star = -[\overline{R}^\alpha(v), \overline{R}_\alpha(u)]_\star \ . \tag{4.9}$$

$\star$-Jacoby identities can be proven as well

$$[u, [v, z]_\star]_\star = [[u, v]_\star, z]_\star + [\overline{R}^\alpha(v), [\overline{R}_\alpha(u), z]_\star]_\star \ . \tag{4.10}$$

In conclusion we have constructed the deformed Lie algebra of vectorfields $\Xi_\star$. As vectorspaces $\Xi = \Xi_\star$, but $\Xi_\star$ is a $\star$-Lie algebra. We previously constructed $\Xi_\star$ as an $A_\star$-module. The compatibility between these two structures reads

$$[u, h \star v]_\star = \mathcal{L}^\star_u(h) \star v + \overline{R}^\alpha(h) \star [\overline{R}_\alpha(u), v]_\star \ .$$

We stress that a $\star$-Lie algebra is not a generic name for a deformation of a Lie algebra. Rather it is a quantum Lie algebra in the sense of [10], see also [11, 12], [3] and [13]. In this respect the deformed Leibnitz rule (4.3), that states that only vectorfields (or the identity) can act on the second argument g in $h \star g$ (no higher order differential operators are allowed on g) is of fundamental importance, and later on it will be a key ingredient for the definition of a covariant derivative along a generic vectorfield. Another main property of quantum Lie algebras, and of the quantum Lie algebra of infinitesimal diffeomorphisms we have presented, is that it can be shown that the bracket $[u\ v]_\star$ is the $\star$-adjoint action of u on v. Had we chosen an exactly antisymmetric bracket of vectorfields, we would have lost this geometric property.

A general comment on the approach adopted is now in order. Given the deformation $Fun_\star(M)$ of the algebra of functions $Fun(M)$, we can

- consider the derivations of $Fun_\star(M)$, i.e. the infinitesimal transformations of $Fun_\star(M)$ that satisfy the usual Leibnitz rule, $v(h \star g) = v(h) \star g + h \star v(g)$. As is easily seen expanding in power series of $\theta^{\mu\nu}$, these maps are only the vectorfields that leave invariant the Poisson tensor (2.4). Thus while in the commutative case any vectorfield is a derivation, in the deformed case the space of derivations is smaller. This viewpoint for our pourposes is too restrictive, for example infinitesimal Poincaré transformations are not derivations. In this approach we have that Poincaré invariance is spontaneously broken by the presence of $\theta^{\mu\nu}$.

- consistently deform the notion of derivation so that to any infinitesimal transformation of $Fun(M)$ there correspond one and only one deformed infinitesimal derivation. This is what we have achieved with the map $\mathcal{L}_v \to \mathcal{L}^\star_v$, where $\mathcal{L}^\star_v$ satisfies the deformed Leibnitz rule (4.3). This is the quantum groups and quantum spaces approach [10–12, 14]. The bonus of this approach is that instead of dealing with a spontaneously broken diffeomorphisms (or Poincaré) symmetry we have an unbroken quantum diffeomorphisms (or Poincaré) symmetry. In this way we retain a symmetry property that is as strong as the one of commutative spacetime.

Deformed Poincaré Lie algebra

The quantum or twisted approach in the case of Poincaré symmetry of canonical noncommutative spacetime was considered in [4, 5, 7], where the twisted Poincaré Hopf algebra is presented (see also [8, 9]). The description of this twised symmetry in terms of $\star$-Poincaré Lie algebra, i.e. the quantum Lie algebra of the Poincaré Hopf algebra, is in [3]. Using the bracket (4.7), we have that the infinitesimal generators

$$P_\mu = i\partial_\mu \ , \ \ M_{\mu\nu} = i(x_\mu\partial_\nu - x_\nu\partial_\mu) \ , \tag{4.11}$$

close the $\star$-Poincaré Lie algebra

$$\begin{aligned} [P_\mu, P_\nu]_\star &= 0 \ , \\ [P_\rho, M_{\mu\nu}]_\star &= i(\eta_{\rho\mu}P_\nu - \eta_{\rho\nu}P_\mu) \ , \\ [M_{\mu\nu}, M_{\rho\sigma}]_\star &= -i(\eta_{\mu\rho}M_{\nu\sigma} - \eta_{\mu\sigma}M_{\nu\rho} - \eta_{\nu\rho}M_{\mu\sigma} + \eta_{\nu\sigma}M_{\mu\rho}) \ . \end{aligned} \tag{4.12}$$

In this particularly simple case of canonical noncommutativity it happens that the structure constants are the same as in the undeformed case, however the $\star$-bracket differs from the commutator, $\mathcal{L}^\star_{[M_{\mu\nu},M_{\rho\sigma}]_\star} \neq \mathcal{L}^\star_{M_{\mu\nu}}\mathcal{L}^\star_{M_{\rho\sigma}} - \mathcal{L}^\star_{M_{\rho\sigma}}\mathcal{L}^\star_{M_{\mu\nu}}$. The deformed Leibnitz rule of these derivations, according to (4.3), reads

$$\mathcal{L}^\star_{M_{\mu\nu}}(h \star g) = \mathcal{L}^\star_{M_{\mu\nu}}(h) \star g + h \star \mathcal{L}^\star_{M_{\mu\nu}}(g) - i\theta^{\alpha\beta}\eta_{\beta\mu}\partial_\alpha(h) \star \partial_\nu(g) + i\theta^{\alpha\beta}\eta_{\beta\nu}\partial_\alpha(h) \star \partial_\mu(g) \ .$$

From (4.10) or immediately from (4.12) we obtain the $\star$-Jacoby identities:

$$[t\,, [t', t'']_\star]_\star + [t'\,, [t'', t]_\star]_\star + [t''\,, [t, t']_\star]_\star = 0 \ , \tag{4.13}$$

for all elements t, t', t'' of the $\star$-Poincaré Lie algebra.

5 Covariant Derivative, Torsion and Curvature

The noncommutative differential geometry set up in the previous section allows to develop the formalism of covariant derivative, torsion, curvature and Ricci tensors just by following the usual classical formalism.

We define a $\star$-covariant derivative $\nabla^\star_u$ along the vector field $u \in \Xi$ to be a linear map $\nabla^\star_u : \Xi_\star \to \Xi_\star$ such that for all $u, v, z \in \Xi_\star$, $h \in A_\star$:

$$\nabla^\star_{u+v} z = \nabla^\star_u z + \nabla^\star_v z \ , \tag{5.1}$$

$$\nabla^\star_{h \star u} v = h \star \nabla^\star_u v \ , \tag{5.2}$$

$$\nabla^\star_u (h \star v) \ = \ \mathcal{L}^\star_u(h) \star v + \overline{R}^\alpha(h) \star \nabla^\star_{\overline{R}_\alpha(u)} v \tag{5.3}$$

where in the last line we have used the same deformed Leibnitz rule that appears in (4.3). Epression (5.3) is well defined because $\overline{R}_\alpha(u)$ is again a vectorfield.

The covariant derivative is extended to tensorfields using again the deformed Leibniz rule

$$\nabla^\star_u (v \otimes_\star z) = \nabla^\star_u(v) \otimes_\star z + \overline{R}^\alpha(v) \otimes_\star \nabla^\star_{\overline{R}_\alpha(u)}(z) \ .$$

The (noncommutative) connection coefficients ${\Gamma_{\mu\nu}}^\sigma$ are given by

$$\nabla^\star_\mu \partial_\nu = {\Gamma_{\mu\nu}}^\sigma \star \partial_\sigma = {\Gamma_{\mu\nu}}^\sigma \, \partial_\sigma \ , \tag{5.4}$$

where $\nabla^\star_\mu = \nabla^\star_{\partial_\mu}$. Let $z = z^\mu \star \partial_\mu$, $u = u^\nu \star \partial_\nu$, then (5.2) and (5.3) imply

$$\nabla^\star_z u = z^\mu \star \partial_\mu(u^\nu) \, \partial_\nu + z^\mu \star u^\nu \star {\Gamma_{\mu\nu}}^\sigma \, \partial_\sigma \ . \tag{5.5}$$

Similarly the covariant derivative on 1-forms is given by

$$\nabla^\star_\mu (\omega_\rho dx^\rho) = \partial_\mu(\omega_\rho) \, dx^\rho - {\Gamma_{\mu\rho}}^\nu \star \omega_\nu \, dx^\rho \tag{5.6}$$

and $\nabla^\star_z = z^\mu \star \nabla^\star_\mu$.

The torsion T and the curvature R associated to a connection $\nabla^\star$ are the linear maps $\mathsf{T} : \Xi_\star \times \Xi_\star \to \Xi_\star$, and $\mathsf{R}^\star : \Xi_\star \times \Xi_\star \times \Xi_\star \to \Xi_\star$ defined by

$$\mathsf{T}(u, v) := \nabla^\star_u v - \nabla^\star_{\overline{R}^\alpha(v)} \overline{R}_\alpha(u) - [u, v]_\star \ , \tag{5.7}$$

$$\mathsf{R}(u, v, z) := \nabla^\star_u \nabla^\star_v z - \nabla^\star_{\overline{R}^\alpha(v)} \nabla^\star_{\overline{R}_\alpha(u)} z - \nabla^\star_{[u,v]_\star} z \ , \tag{5.8}$$

for all $u, v, z \in \Xi_\star$. From the $\star$-antisymmetry property of the bracket $[\ ,\ ]_\star$, see (4.9), it easily follows that the torsion T and the curvature R have the following $\star$-antisymmetry property

$$\mathsf{T}(u, v) = -\mathsf{T}(\overline{R}^\alpha(v), \overline{R}_\alpha(u)) \ ,$$

$$\mathsf{R}(u, v, z) = -\mathsf{R}(\overline{R}^\alpha(v), \overline{R}_\alpha(u), z) \ .$$

The presence of the R-matrix in the definition of torsion and curvature insures that T and R are left $A_\star$-linear maps [2], i.e.

$$\mathsf{T}(f \star u, v) = f \star \mathsf{T}(u,v) \quad , \quad \mathsf{T}(\partial_\mu, f \star v) = f \star \mathsf{T}(\partial_\mu, v)$$

(for any index μ), and similarly for the curvature. The $A_\star$-linearity of T and R insures that we have a well defined torsion tensor and curvature tensor.

One can also prove (twisted) first and second Bianchi identities [2].

The coefficients $\mathsf{T}_{\mu\nu}{}^\rho$ and $\mathsf{R}_{\mu\nu\rho}{}^\sigma$ with respect to the partial derivatives basis $\{\partial_\mu\}$ are defined by

$$\mathsf{T}(\partial_\mu, \partial_\nu) = \mathsf{T}_{\mu\nu}{}^\rho \partial_\rho \quad , \qquad \mathsf{R}(\partial_\mu, \partial_\nu, \partial_\rho) = \mathsf{R}_{\mu\nu\rho}{}^\sigma \partial_\sigma \tag{5.9}$$

and they explicitly read

$$\begin{aligned} \mathsf{T}_{\mu\nu}{}^\rho &= \Gamma_{\mu\nu}{}^\rho - \Gamma_{\nu\mu}{}^\rho \ , \\ \mathsf{R}_{\mu\nu\rho}{}^\sigma &= \partial_\mu \Gamma_{\nu\rho}{}^\sigma - \partial_\nu \Gamma_{\mu\rho}{}^\sigma + \Gamma_{\nu\rho}{}^\beta \star \Gamma_{\mu\beta}{}^\sigma - \Gamma_{\mu\rho}{}^\beta \star \Gamma_{\nu\beta}{}^\sigma \ . \end{aligned} \tag{5.10}$$

As in the commutative case the Ricci tensor is a contraction of the curvature tensor,

$$\mathsf{Ric}_{\mu\nu} = \mathsf{R}_{\rho\mu\nu}{}^\rho . \tag{5.11}$$

A definition of the Ricci tensor that is independent from the $\{\partial_\mu\}$ basis is also possible.

6 Metric and Einstein Equations

In order to define a $\star$-metric we need to define $\star$-symmetric elements in $\Omega_\star \otimes_\star \Omega_\star$ where $\Omega_\star$ is the space of 1-forms. Recalling that permutations are implemented with the R-matrix we see that $\star$-symmetric elements are of the form

$$\omega \otimes_\star \omega' + \overline{R}^\alpha(\omega') \otimes_\star \overline{R}_\alpha(\omega) \ . \tag{6.1}$$

In particular any symmetric tensor in $\Omega \otimes \Omega$,

$$g = g_{\mu\nu} dx^\mu \otimes dx^\nu \ , \tag{6.2}$$

$g_{\mu\nu} = g_{\nu\mu}$, is also a $\star$-symmetric tensor in $\Omega_\star \otimes_\star \Omega_\star$ because

$$g = g_{\mu\nu} dx^\mu \otimes dx^\nu = g_{\mu\nu} \star dx^\mu \otimes_\star dx^\nu \tag{6.3}$$

and the action of the R-matrix is the trivial one on dx^ν. We denote by $g^{\star\mu\nu}$ the star inverse of $g_{\mu\nu}$,

$$g^{\star\mu\rho} \star g_{\rho\nu} = g_{\nu\rho} \star g^{\star\rho\mu} = \delta^\mu_\nu \ . \tag{6.4}$$

The metric $g_{\mu\nu}$ can be expanded order by order in the noncommutative parameter $\theta^{\rho\sigma}$. Any commutative metric is also a noncommutative metric, indeed the $\star$-inverse metric can be constructed order by order in the noncommutativity parameter. Contrary to [15], we see that in our approach there are infinitely many metrics compatible with a given noncommutative differential geometry, noncommutativity does not single out a preferred metric.

A connection that is metric compatible is a connection that for any vectorfield u satisfies, $\nabla^\star_u g = 0$, this is equivalent to the equation

$$\nabla^\star_\mu g_{\rho\sigma} - \Gamma_{\mu\rho}{}^{\nu} \star g_{\nu\sigma} - \Gamma_{\mu\sigma}{}^{\nu} \star g_{\rho\nu} = 0 \ . \tag{6.5}$$

Proceeding as in the commutative case we obtain that there is a unique torsion free metric compatible connection [1]. It is given by

$$\Gamma_{\mu\nu}{}^{\rho} = \frac{1}{2}(\partial_\mu g_{\nu\sigma} + \partial_\nu g_{\sigma\mu} - \partial_\sigma g_{\mu\nu}) \star g^{\star\sigma\rho} \tag{6.6}$$

We now construct the curvature tensor and the Ricci tensor using this uniquely defined connection. Finally the noncommutative Einstein equations (in vacuum) are

$$\mathsf{Ric}_{\mu\nu} = 0 \tag{6.7}$$

where the dynamical field is the metric g.

Acknowledgements

I acknowledge the stimulating and fruitful atmospheres of the Erice International school of Subnuclear Physics, 44t course (2006). Partial support from Ettore Majorana Foundation and Centre for Scientific Studies, Erice, is kindly acknowledged. Partial support from the European Community's Human Potential Program under contract MRTN-CT-2004-005104 and from the Italian MIUR under contract PRIN-2003023852 is also acknowledged.

References

[1] P. Aschieri, C. Blohmann, M. Dimitrijevic, F. Meyer, P. Schupp and J. Wess, *A gravity theory on noncommutative spaces,* Class. Quant. Grav. **22** (2005) 3511 hep-th/0504183.

[2] P. Aschieri, M. Dimitrijevic, F. Meyer and J. Wess, *Noncommutative geometry and gravity,* Class. Quant. Grav. **23** (2006) 1883, hep-th/0510059.

[3] P. Aschieri, *Noncommutative Symmetries and Gravity,* J. Phys. Conf. Ser. **53** (2006) 799 [arXiv:hep-th/0608172].

[4] R. Oeckl, *Untwisting noncommutative R^d and the equivalence of quantum field theories*, Nucl. Phys. **B581**, 559 (2000), hep-th/0003018.

[5] M. Chaichian, P. Kulish, K. Nishijima, and A. Tureanu, *On a Lorentz-invariant interpretation of noncommutative space-time and its implications on noncommutative QFT*, Phys. Lett. **B604**, 98 (2004), hep-th/0408069.

[6] V. G. Drinfeld, *On constant quasiclassical solutions of the Yang-Baxter equations,* Soviet Math. Dokl. **28** (1983) 667-671. *Quasi-Hopf Algebras,* Lengingrad Math. J. **1** (1990) 1419 [Alg. Anal. **1N6** (1989) 114].

[7] J. Wess, *Deformed coordinate spaces: Derivatives*, Lecture given at the Balkan workshop BW2003, August 2003, Vrnjacka Banja, Serbia , hep-th/0408080.

[8] F. Koch and E. Tsouchnika, *Construction of θ-Poincare algebras and their invariants on M_θ*, Nucl. Phys. B **717** (2005) 387 hep-th/0409012.

[9] C. Gonera, P. Kosinski, P. Maslanka and S. Giller, *Space-time symmetry of noncommutative field theory*, Phys. Lett. B **622** (2005) 192 hep-th/0504132,

[10] S. L. Woronowicz, *Differential Calculus On Compact Matrix Pseudogroups (Quantum Groups),* Commun. Math. Phys. **122** (1989) 125.

[11] P. Aschieri and L. Castellani, *An Introduction to noncommutative differential geometry on quantum groups,* Int. J. Mod. Phys. A **8** (1993) 1667 hep-th/9207084.

[12] P. Schupp, P. Watts and B. Zumino, *Bicovariant quantum algebras and quantum Lie algebras,* Commun. Math. Phys. **157** (1993) 305 hep-th/9210150.

[13] P. Aschieri, *Lectures on Hopf Algebras, Quantum Groups and Twists,* to appear in the proceeding of the second Modave Summer School in Mathematical Physics, August, 6-12, 2006. hep-th/0703013

[14] J. Wess and B. Zumino, *Covariant differential calculus on the quantum hyperplane.* Nucl. Phys. Proc. Suppl. **18B** (1991) 302.

[15] M. Buric, T. Grammatikopoulos, J. Madore and G. Zoupanos, *Gravity and the structure of noncommutative algebras,* JHEP **0604** (2006) 054, hep-th/0603044.

Events with Isolated Leptons and Missing Transverse Momentum in *ep* Collisions at HERA

*Gerhard Brandt**

Physical Institute, Philosophenweg 12, 69120 Heidelberg

Abstract

The analysis of events with isolated leptons and missing transverse momentum in the H1 experiment is discussed for the electron, muon and tau channels. In the Standard Model (SM) framework, production of real W-bosons gives rise to such topologies. Contributions to the background are dominated by QCD processes. An excess of observed signal over background presents a chance of the discovery of new physics. The results using the HERA 1994-2006 data set corresponding to 341 pb^{-1} are presented While the e^-p sample shows good agreement between data and SM expectation, in e^+p collisions an excess over the SM expectation with 3.4σ significance is observed at high hadronic transverse momentum.

1 Introduction

Events with high-P_T isolated leptons and missing transverse momentum are sensitive to physics beyond the Standard Model (BSM). Both H1 and ZEUS have previously published searches for events with isolated electrons and muons [1, 2, 4, 3, 5] and τ−leptons [6, 7]. The observation of such events in *ep* collisions at HERA has created considerable interest in the high-energy physics community (many references can be found in [2]) and is continously updated as new data become available [8, 9, 10]. Section 2 reviews the experimental conditions (collider and detector) of the search, section 3 presents the most important physics processes involved, section 4 presents the search

*email: gbrandt@mail.desy.de

for $e/\mu + P_T^{\mathrm{miss}}$ events, section 5 the search for $\tau + P_T^{\mathrm{miss}}$ and section 6 discusses the results and possible interpretations.

2 Experimental Conditions

HERA operating at DESY in Hamburg is the worlds only *ep* collider. It collides electrons or positrons[1] with protons at a center-of-mass energy $\sqrt{s} =$ 320 GeV for most of its operating time. HERA data taking is divided in two main running periods, HERA-1 in the years 1994 to 2000, and HERA-2 in the years 2000 to 2007. During HERA-1, H1 collected a data set corresponding to 118 pb^{-1} of data, dominated by about 90% by e^+p data. For the HERA-2 period, the luminosity has been upgraded and spin rotators installed to provide a longitudinally polarised electron beam. For the results discussed here about 250 pb^{-1} of HERA-2 data have been available to H1. This includes the full e^-p data sample which corresponds to a total integrated luminosity of 182 pb^{-1}. Data taking continues with e^+p running until the end of HERA operation, which is scheduled for June 2007.

Two collider experiments, H1 and ZEUS, operate at HERA. This article focusses on H1, described in detail in [11]. A drawing of the H1 detector is shown in figure 1a. It has almost 4π angular coverage to detect missing transverse momentum and the subdetectors are arranged for excellent lepton identification. Track information is provided by central jet chambers, a forward tracker and silicon vertexing detectors. The trackers in a solenoid magnet field allow momentum measurement of charged particles. Energy measurement for electromagnetic and hadronic showers is provided by a LAr (liquid argon) calorimeter in the central and forward (proton beam direction) region, and a lead-fibre spaghetti calorimeter ("SpaCal") in the backward direction. For muon detection, the iron flux return yoke of the magnet is instrumented with limited streamer tubes to measure escaping muons. In the forward direction, there is an additional forward muon detector installed.

To compare recorded data to SM predictions, the relevant physics processes are modelled by Monte Carlo generators. The detector response is simulated by H1 software based on GEANT 3 [12]. The simulated events are then processed through the complete reconstruction and analysis chain, identical to the data treatment.

[1] In this article, the term *electron* refers to both electrons and positrons unless explicitly stated.

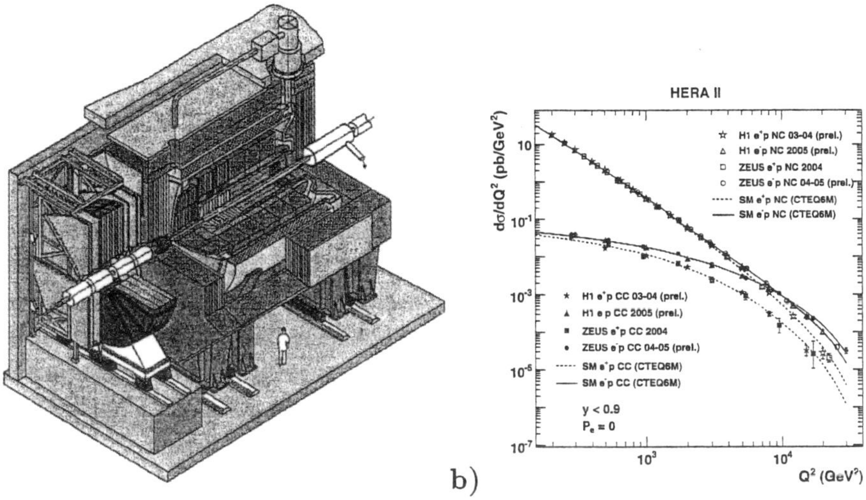

a) b)

Fig. 1: **a)** The H1 detector at HERA [11]. Electrons enter from the left side, and protons from the right side; **b)** A HERA "textbook" measurement: Inclusive unpolarised cross sections of NC and CC DIS processes in *ep* collisions measured by the H1 and ZEUS experiments (taken from [13]). The unification of electroweak forces is visible for $Q^2 > 1000$ GeV.

3 Physics in *ep* Collisions

In the Standard Model events with isolated leptons and missing energy are mainly expected from real W production with subsequent leptonic decay,

$$ep \to eW^{\pm}(\hookrightarrow l\nu)X,$$

where X denotes the hadronic final state (HFS). This is composed of the scattered quark jet and the proton remnant. The main contribution to this process is radiation of a W boson off the scattered quark line, shown in figure 2a. This and other contributing processes are simulated by the EPVEC generator [14]. The total cross section is about 1 pb. With a typical detection efficiency of 40% for the total leptonic branching ratio this means about 100 events are expected in an ep data sample of 250 pb^{-1}. For SM W production, the transverse momentum of the HFS P_T^X is predicted to be well below 25 GeV. At high P_T^X the SM prediction is low. Looking at its P_T^X provides a measure for how untypical, or interesting, an event is. An event with a P_T^X of 29 GeV is shown in figure 3.

Several physics processes contribute to the background. The most prominent physics processes measured at HERA are neutral current (NC) $ep \to eX$

and charged current (CC) $ep \to \nu X$ in deep inelastic scattering (DIS), shown in figures 2b and c respectively. These processes are simulated by the RAPGAP [15] and DJANGO [16] generators. The regime of DIS is defined by the negative momentum transfer square Q^2 of the boson exchanged between the indicent electron and the quark scattered from the proton, at $Q^2 > 1$ GeV. The regime $Q^2 < 1$ GeV known as photoproduction is simulated in the PYTHIA [17] framework. Here, two jets are produced by boson-gluon-fusion as shown in figure 2d. For the analysis presented in this paper NC and CC in DIS are important background processes due to their high cross sections and event topologies. They are also tools, because they provide high statistics samples of final state objects such as leptons and QCD jets whose properties need to be well understood to make accurate predictions. Figure 1b shows measurements of NC and CC cross sections done at HERA.

As a template for physics beyond the Standard Model producing isolated lepton and missing transverse momentum topologies, the anomalous production of top quarks via flavor changing neutral currents (FCNC) is used (shown in Fig. 2e). This process was generated by the ANOTOP [18] programme and used to optimise acceptance for a possible BSM signal.

4 Search for Events with Isolated Electrons and Muons

Electrons are detected as isolated, compact electromagnetic clusters in the calorimeter and a high quality track pointing to that cluster. Muons are detected by signals in the muon detectors, matching tracks in the trackers and energy deposits in the LAr calorimeter typical for minimally ionising particles. Electrons and Muons are detected in the phase space of lepton transverse momentum $P_T^l > 10$ GeV and polar angular range $5°\theta_l < 140°$ [19, 21]. Leptons are required to be isolated against other tracks and jets in the event. This reduces leptons in or close to QCD jets which are expected to be at lower P_T^l. In DIS the scattered electron is mostly seen in the backward region $\theta_l > 155°$ of the detector. The missing transverse momentum P_T^{miss} is reconstructed from the four-vector sum of all reconstructed final state particles (hadrons and leptons) and required to be greater than 12 GeV. This greatly reduces NC and photoproduction contributions to the background, where P_T^{miss} only appears due to fluctuations in the energy measurement. Additionally SM processes are suppressed topologically, for example by requiring acoplanarity of the lepton and the hadronic final state four-vector in the transverse plane. For neutral current and photoproduction events a back-to-back topology of the final state objects is expected from momentum conservation.

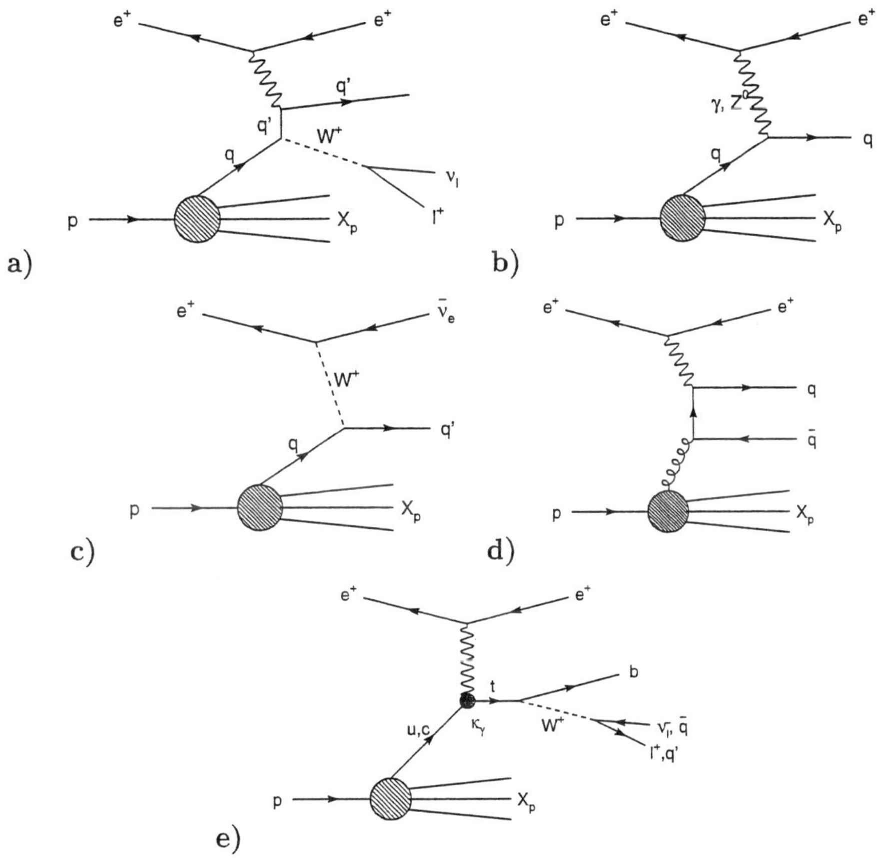

Fig. 2: Physics processes at HERA contributing to $l + P_T^{\mathrm{miss}}$ topologies as signal (a and e) or background due to mismeasurements (b,c,d): **a)** Production of real W-Bosons in the Standard Model; **b)** Neutral current; **c)** Charged current; **d)** Photoproduction of jets by boson-gluon-fusion; **e)** Anomalous production of single *top* quarks via FCNC, the template used for possible BSM contributions.

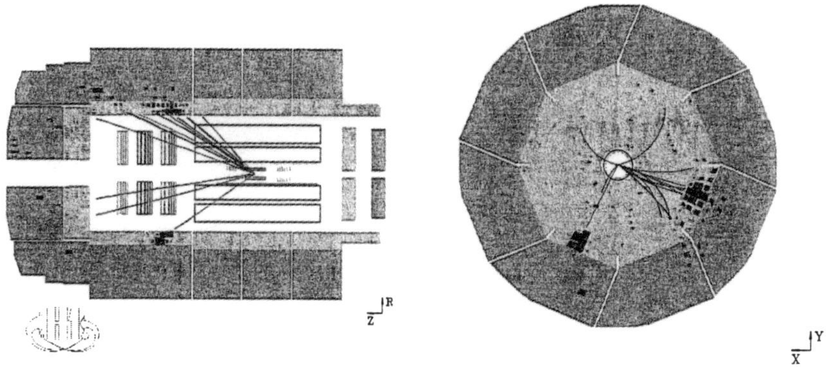

Fig. 3: Event display for an event with an isolated electron of $P_T^e = 37\text{GeV}$, missing transverse momentum of $P_T^{\text{miss}} = 44\text{GeV}$ and a hadronic system of $P_T^X = 29$ GeV, observed in HERA-2 e^+p data. The Standard Model expectation for events with this topology is low.

Figure 4 shows the results of the search for the combined electron and muon channels for the e^+p data sample corresponding to 158 pb^{-1} and the e^-p data sample corresponding to 184 pb^{-1}. A total of 46 events are observed in the complete data, with 43.0 ± 6.0 expected from the SM. At large hadronic transverse momentum $P_T^X > 25$ GeV a total of 18 events are observed, but only 11.5 ± 1.8 predicted. 15 of these events are observed in the e^+p sample, where only 4.6 ± 0.8 are expected. Only 3 events are seen in the e^-p data sample, where 6.9 ± 1.0 are expected. The significance of the observed excess in the e^+p data sample is 3.4σ. No excess is observed in the e^-p data sample. This observation represents a chance for discovery of new physics at HERA.

5 Search for Events with Isolated Tau Leptons

Due to lepton universality in the Standard Model, the tau channel is expected to behave identical to the electron and muon channels. Leptonic τ-decays are included in the electron and muon channels. This search adds hadronic τ-decays with one charged hadron, leading to a 1-prong signature, which covers about 50% of the total τ-lepton branching ratio [20, 21]. 3-prong decays are not included because the background is too large to yield a significant signal-to-background ratio.

Tau-jets are identified based on hadronic jets in the phase space $P_T^{jet} > 7$ GeV and $20° < \theta^{jet} < 120°$. The candidate jets are expected to be collimated due to the boost of the τ-lepton. This is ensured by requiring a jet radius

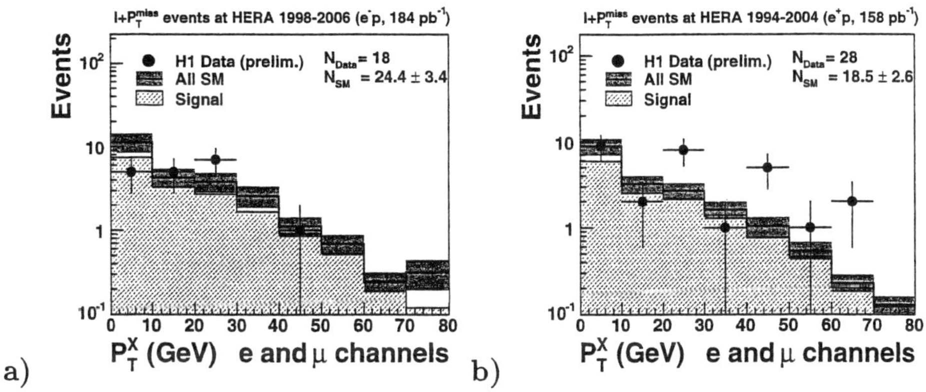

Fig. 4: Transverse momentum of the hadronic system P_T^X for the combined e- and μ–channels in the $e/\mu + P_T^{\text{miss}}$ search. The points show the data with statistical errors. The open histogram shows the SM expectation. Systematic and statistical errors added in quadrature are shown as the shaded band. The signal contribution of the SM expectation dominated by real W production is shown as a hatched histogram and amounts to about 70%. While the e^-p data (**a**) agree well with the SM expectation , the e^+p data (**b**) show an excess of events over the SM prediction amounting to 3.4σ significance.

$R_{jet} < 0.12$ in the $\eta - \phi$–plane. Exactly one track is required in a cone $D = 1.0$ in the $\eta - \phi$–plane around the jet axis. Neutrinos from the decay chain $W \rightarrow \nu\tau \hookrightarrow l\nu_l\bar{\nu}$ are taken into account by requiring $P_T^{\text{miss}} > 12$ GeV. Because of the similarity of τ-jets to narrow QCD jets with a low track mulitplicity, the signal to background ratio is less favorable than in the electron and muon channels. Jets in photoproduction ($Q^2 < 1$ GeV) may enter the sample due to fake P_T^{miss} from fluctuations in the energy measurement and QCD jets faking τ-jets. CC background is much higher than in the electron and muon channel. The correct description of dijet CC events becomes important since their signature is very similar to the events in the interesting region at high PTX. The second jet, which could fake a τ-jet, may come from next to leading order processes or final state radiation.

Figure 5 shows the preliminary results for the τ analysis. In e^+p data, 8 candidates are observed for 10.8 $\pm$ 2.5 expected. All candidates are in the lowest P_T^X bin where most of the background is expected. In e^-p data 17 events are observed, with 13.5 $\pm$ 2.8 expected. Three events are observed at $P_T^X > 25$ GeV, while only 0.74 $\pm$0.18 are expected. At first glance this seems to suggest a trend opposite to the electron and muon channels, but

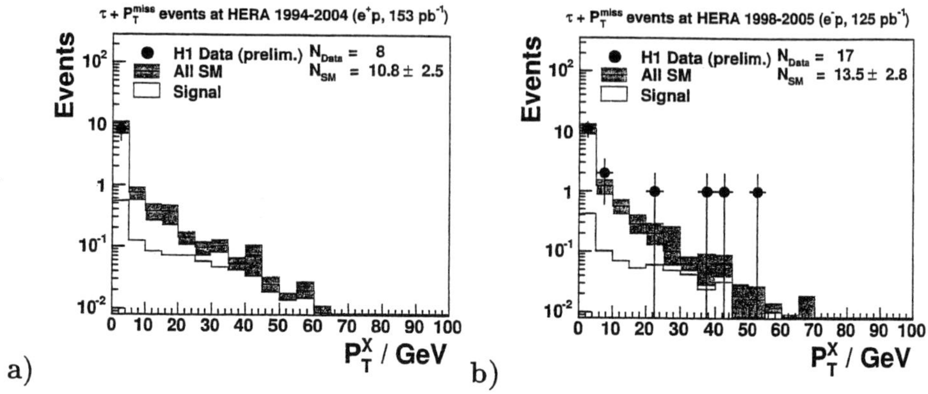

Fig. 5: Transverse momentum of the hadronic system P_T^X in $\tau + P_T^{\rm miss}$ events. While there are no candidates at high P_T^X observed in the e^+p sample (a), there are three candidates at high P_T^X observed in the e^-p sample.

this conclusion is premature. The statistical significance of the result is low. Also at closer inspection, some of the three observed candidates seem quite untypical for τ-jets and are more likely background that can possibly be removed in an improved analysis.

6 Interpretation

While the low number of observed events in these analyses does not yet show a clear pattern pointing to a strong hypothesis for a BSM process to explain the excess, it is already intriguing to think about possible mechanisms.

Upon the first observation of these events at HERA-1, one of the most exciting interpretations was the production of single *top*-quarks. While this is allowed kinematically and possible in CC processes at HERA, the cross section is too low to see a signal in the available data samples. If observed, *top*-quark events could therefore only be produced anomalously via flavor changing neutral currents (FCNC). The process is illustrated in figure 2e where the scattered quark couples to the exchanged photon (or Z^0, not shown) via a FCNC coupling and produces a *top*-quark. The subsequent decay $t \to bW$ followed by $W \to l\nu$ or $W \to q\bar{q}$ produces either the searched $l + P_T^{Miss}$ signature or a 3 jet topology. The b-jet gives rise to a highly energetic hadronic system X, giving rise to the high P_T^X. These leptonic and hadronic signatures have been analysed and limits have been set which are competitive to limits on anomalous top production set at other colliders [22].

Another possibility to explain the observation is to explicitly construct a model. This could be based on a new particle N with fermion number F=0 coupling to $e - q$. If such a particle had a high mass m_N it would have to be produced at large $x_{Bj} = M_N^2/s$. In this region the parton density of valence quarks in the proton is much larger than the density of sea quarks, hence the production cross section for such a particle is much larger in e^+p running than in e^-p running. In the framework of R-parity violating supersymmetry, such particles could be resonantly produced *stop*-squarks, or *sbottom*-squarks produced in the t-channel [23]. These preliminary ideas have yet to be intensively studied.

A major issue with the observation is that ZEUS cannot confirm the excess observed by H1, although the analysis is performed in a more restricted phase space. Studies to properly compare the results and test the compatibility of both experiments are ongoing [8].

For the remainder of HERA data taking e^+p data will be recorded. If the data taking continues at the current rate, an additional integrated luminosity of $\mathcal{L} \sim 150\ \text{pb}^{-1}$ can be expected. At the time of writing the analysis of events with isolated leptons and missing transverse momentum is followed with full effort to extract the maximum of information from this puzzling observation.

References

[1] T. Ahmed *et al.* [H1 Collaboration], DESY 94-248 (1994).

[2] C. Adloff *et al.* [H1 Collaboration], Eur. Phys. J. C **5** (1998) 575 [hep-ex/9806009]

[3] V. Andreev *et al.* [H1 Collaboration], Phys. Lett. B **561** (2003) 241 [hep-ex/0301030].

[4] J. Breitweg *et al.* [ZEUS Collaboration], Phys. Lett. B **471** (2000) 411 [hep-ex/9907023].

[5] S. Chekanov *et al.* [ZEUS Collaboration], Phys. Lett. B **559** (2003) 153 [hep-ex/0302010].

[6] S. Chekanov *et al.* [ZEUS Collaboration], Phys. Lett. B **583** (2004) 41 [hep-ex/0311028].

[7] A. Aktas *et al.* [H1 Collaboration], submitted to Eur. Phys. J. C [hep-ex/0604022].

[8] H1 Collaboration, H1prelim-06-162 [3].

[9] ZEUS Collaboration, ZEUS-prel-06-012 [4].

[10] H1Collaboration, H1prelim-06-064 [3].

[11] H1 Collaboration, *The H1-Detector at HERA*, Nucl. Instr. and Meth. A **386** (1997) 310 and 348.

[12] GEANT 3, R. Brun *et al.*, CERN-DD/EE/84-1.

[13] H1 Collaboration, H1prelim-06-041 [3];
B. Portheault, Dissertation, Université Paris-Sud XI (2005) [5].

[14] U. Baur, J. A. M. Vermaseren and D. Zeppenfeld, Nucl. Phys. B **375** (1992) 3.

[15] H. Jung, Comput. Phys. Commun. **86** (1995) 147, RAPGAP program manual (1998) http://www-h1.desy.de/~jung/RAPGAP.html.

[16] G. A. Schüler and H. Spiesberger, DJANGO 2.1, Proc. of the Workshop "Physics at HERA" (1991), Eds. W. Buchmüller and G. Ingelman., Vol. 3, p. 1419.

[17] T. Sjöstrand, PYTHIA 5.7, Comput. Phys. Commun. 82 (1994) 74.

[18] E. Perez, ANOTOP, unpublished.

[19] D. South, Dissertation, University of Manchester (2003) [5].

[20] C. Veelken, Dissertation, University of Liverpool (2006) [5].

[21] G. Brandt, Dissertation, University of Heidelberg (*2007, in preparation*) [5].

[22] A. Aktas *et al.* [H1Collaboration], Eur. Phys. J C **33** (2004) 9 [hep-ex/0310032].

[23] E. Perez, private communication
*For more information on (*R*-parity violating) SUSY try:*
J. Haller, Dissertation, University of Heidelberg (2003) [5]
S. Martin, *A Supersymmetry Primer*, [hep-ph/9709356].

[3] H1 preliminary results available at:
http://www-h1.desy.de/publications/H1preliminary.short_list.html
[4] ZEUS preliminary results available at:
http://www-zeus.desy.de
[5] H1 theses available at:
http://www-h1.desy.de/publications/theses_list.html

From Quark Gluon Plasma to a Perfect Fluid of Quarks and Beyond

M. Csanád

Dept. Atomic Phys., ELTE, H-1117 Budapest, Pázmány P. 1/a, Hungary

Abstract

With high energy heavy ion collisions one tries to create a new forms of matter that is similar to the one present at the birth of our Universe. Recent development on flow pattern, initial energy-density and freeze-out temperature shows that most likely this new form of matter is in a deconfined state, has colored degrees of freedom and is more fluid-like than gas-like. In present paper we calculate estimations on the physical properties of this new-old matter.

"We simply do not yet know enough about the physics of elementary particles to be able to calculate the properties of such a melange with any confidence. ... Thus our ignorance of microscopic physics stands as a veil, obscuring our view of the very beginning."

S. Weinberg, about the first hundredth of a second [1]

1 Introduction

Ultra-relativistic collisions, so called "Little Bangs" of almost fully ionized Au atoms are observed in four major experiments at the RHIC accelerator at the highest currently available colliding energies of $\sqrt{s_{NN}} = 200$ GeV. The aim of these experiments is to create new forms of matter that existed in Nature a few microseconds after the Big Bang, the creation of our Universe.

Quantum Chromodynamics (QCD), the theory of quarks and gluons, the strong force interacting between them and their color degree of freedom was formulated and established soon after Weinberg's famous book [1] about the early Universe had been published. Confinement is an important (though mathematically never proven) property of QCD, its consequence is that quarks are bound into hadrons in a matter of normal temperature and pressure.

But in the early Universe, energy density was many orders of magnitude higher than today, and at that high energy densities, deconfined phases of colored matter might have existed. Quark-gluon plasma (QGP) is such a phase, that might have existed during the first few microseconds after the Universe came into existence. This type of matter was searched for at the SPS, and experiments at RHIC are continuing this effort. Evidence for formation of a hot and dense medium in gold-gold collisions was found based on a phenomenon called jet quenching, and confirmed by its disappearance in deuteron-gold collisions [2].

A consistent picture emerged after the first three years of running the RHIC experiment: quarks and gluons indeed become deconfined, but also behave collectively, hence this hot matter acts like a liquid [3], not like an ideal gas theorists had anticipated when defining the term QGP. The situation is similar to as if prisoners (quarks and gluons confined in hadrons) have broken out of their cells at nearly the same time, but they find themselves on the crowded jail-yard coupled with all the other escapees. This strong coupling is exactly what happens in a liquid [4].

1.1 A sign for hydrodynamic behavior: elliptic flow

Azimuthal asymmetry of single particle spectra measured in relativistic heavy ion collisions is called elliptic flow (v_2). It is an indication of liquid-like behavior [5], and can be explained by hydrodynamics [6, 7, 8]. In the hydrodynamic picture it turns out, that elliptic flow can result from the initial spatial asymmetry but also from momentum-space asymmetry. Important is, that in contrast to a uniform distribution of particles expected in a gas-like system, this liquid behavior means that the interaction in the medium of these copiously produced particles is rather strong, as one expects from a fluid. Detailed investigation of these phenomena suggests that this liquid flows with almost no viscosity [9].

1.2 Relativistic perfect fluids

Perfect hydrodynamics is based only on local conservation of charge and energy-momentum and on the assumption of local thermal equilibrium, and this is the tool that we use to describe and calculate the properties of the matter created in relativistic heavy ion collisions at RHIC. While there are accelerating non-relativistic solutions in the literature, recent development shows also relativistic solutions that can be compared to the data [10, 11].

Local conservation of charge and four-momentum reads as

$$\partial_\nu(nu^\nu) = 0, \tag{1}$$
$$\partial_\nu T^{\mu\nu} = 0, \tag{2}$$
$$T^{\mu\nu} = (\varepsilon + p)u^\mu u^\nu - pg^{\mu\nu}. \tag{3}$$

We find the following solution (for arbitrary λ in $d = 1$, $\kappa = 1$ and for $\lambda = 2$ in arbitrary d with $\kappa = d$) [10]:

$$v = \tanh \lambda\eta \tag{4}$$
$$n = n_0 \left(\frac{\tau_0}{\tau}\right)^{\lambda d} \nu(s), \tag{5}$$
$$T = T_0 \left(\frac{\tau_0}{\tau}\right)^{\lambda d/\kappa} \frac{1}{\nu(s)}. \tag{6}$$

As an illustration, fluid trajectories of this solution are shown on fig. 1.

2 Results

2.1 An advanced estimate on the initial energy density

Based on the above solution of eqs. 4-6 let us estimate the initial energy density of relativistic heavy ion or p+p reactions. As our solution is an accelerating one, and we do not neglect the initial acceleration period, we improve the renowned Bjorken estimate both quantitatively and qualitatively. Similarly to Bjorken's method [12], we can estimate the initial energy density (see ref. [10] for details). Finally we get a correction to the widely used Bjorken formula, depending on the acceleration parameter (λ):

$$\frac{\varepsilon_0}{\varepsilon_{Bj}} = \frac{\alpha}{\alpha - 2}\left(\frac{\tau_f}{\tau_0}\right)^{1/(\alpha-2)} = (2\lambda - 1)\left(\frac{\tau_f}{\tau_0}\right)^{\lambda-1}, \tag{7}$$

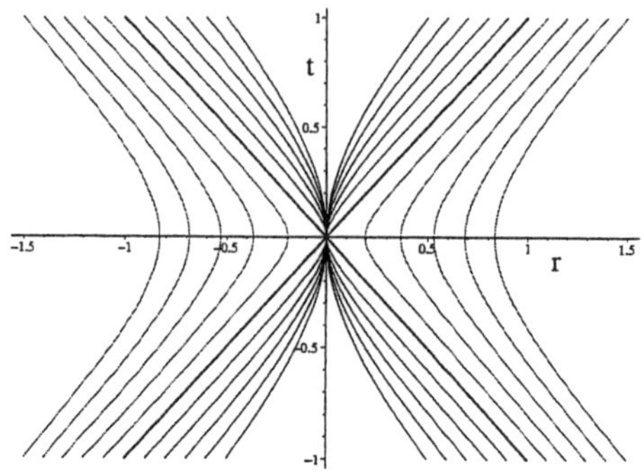

Figure 1: Fluid trajectories of the new exact solution of perfect fluid hydro, corresponding to and $\lambda = 2$. The trajectories are shown both inside and outside the lightcone.

The acceleration parameter can be extracted from the measured rapidity distribution [10]. For flat rapidity distributions, $\alpha \to \infty$ ($\lambda \to 1$, i.e. no acceleration) and the Bjorken estimate is recovered. For $\lambda > 1$, the correction factor is bigger than 1. Hence we conclude that the initial energy densities are under-estimated by the Bjorken formula. For realistic RHIC data from BRAHMS [13], the correction factor can be as big as $\varepsilon/\varepsilon_{Bj} \approx 2.2$ [10]. Thus smaller initial bombarding (or colliding) energies are needed to reach the critical energy density in high energy heavy ion collisions, than thought previously using Bjorken's renowned formula.

2.2 Estimating the freeze-out temperature

We estimated the freeze-out temperature of these Little Bangs, fitting data to the Buda-Lund hydro model [6, 14]. Recently, Fodor and Katz calculated the phase diagram of lattice QCD at finite net baryon density: their results indicate that the transition from confined to deconfined matter is a cross-over with a nearly constant critical temperature, $T_c = 175 \pm 2$ MeV [15]. The result of the Buda-Lund fits to RHIC Au+Au data of refs. [13, 16, 17, 18, 19, 20, 21, 22] (shown on fig. 3), in particular the value of the fit parameter T_0 (central freeze-out temperature, see details in ref. [23]), indicates the existence of a region several standard deviations hotter than the critical temperature. This is an indication on quark deconfinement in Au + Au collisions with

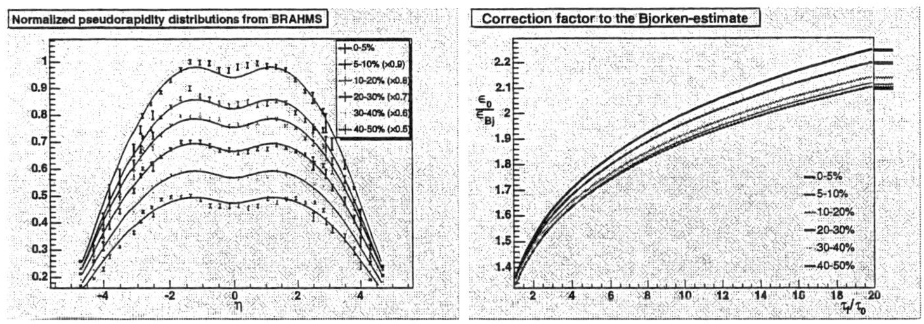

Figure 2: Left: charged particle $\frac{dn}{d\eta}$ distributions of ref. [13] fitted with the result of the relativistic hydro solution of ref. [10]. Right: the correction parameter (obtained from fits show on the left panel) as a function of freeze-out time versus thermalization time (τ_f/τ_0). At a reasonable value of 15, the correction to the Bjorken estimate of energy density is a factor of ~2.2.

$\sqrt{s_{\rm NN}}$ = 130 and 200 GeV at RHIC [23, 24, 25], confirmed by the analysis of p_t and η dependence of the elliptic flow [6]. A similar analysis of Pb+Pb collisions at CERN SPS energies yields central temperatures lower than the critical value, $T_0 < T_c$ [26, 27].

2.3 Universal scaling of the elliptic flow

The Buda-Lund calculation of the elliptic flow results (under certain conditions detailed in ref. [6]) in the following simple universal scaling law:

$$v_2 = \frac{I_1(w)}{I_0(w)}, \tag{8}$$

thus the model predicts a *universal scaling:* every v_2 measurement is predicted to fall on the same *universal* scaling curve I_1/I_0 when plotted against the scaling variable w (see details in ref. [28]).

This means, that v_2 depends on any physical parameter (transverse or longitudinal momentum, center of mass energy, centrality, type of the colliding nucleus etc.) only through the (universal) scaling parameter w.

In ref. [28] we have shown that the excitation function of the transverse momentum and pseudorapidity dependence of the elliptic flow in Au+Au collisions (RHIC data from refs. [5, 29, 30]) is well described with the formulas

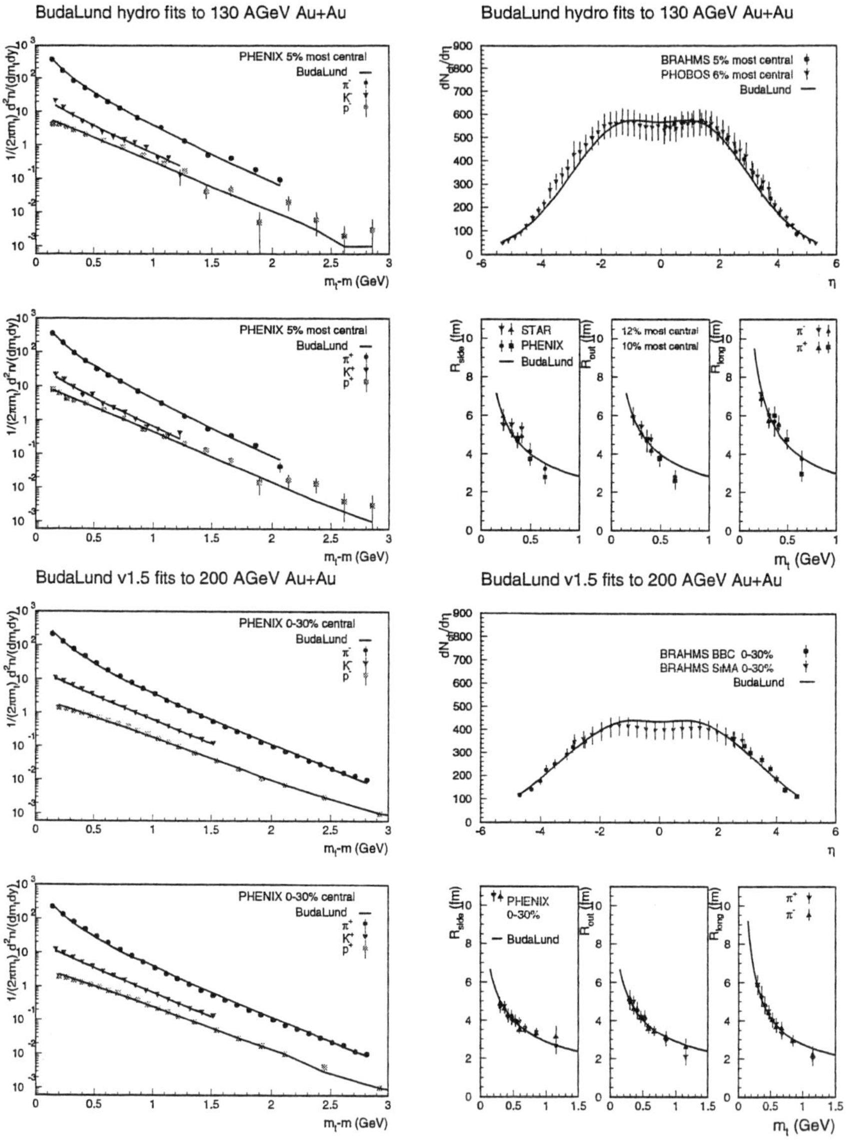

Figure 3: The upper four panels show a simultaneous Buda-Lund fit to 0-5(6) % central Au+Au data on p_t and η spectra and HBT radii at $\sqrt{s_{\rm NN}} = 130$ GeV [16, 17, 18, 19, 20]. The lower four panels show similar fits to 0-30 % central Au+Au data at $\sqrt{s_{\rm NN}} = 200$ GeV [13, 21, 22]. Fit parameters are summarized in ref. [23].

that are predicted by the Buda-Lund type of hydrodynamical calculations. We have provided a quantitative evidence of the validity of the perfect fluid picture of soft particle production in Au+Au collisions at RHIC up to 1-1.5 GeV but also show here that this perfect fluid extends far away from mid-rapidity, up to a pseudorapidity of $\eta_{\rm beam} - 0.5$. The universal scaling of PHOBOS $v_2(\eta)$ [29], PHENIX $v_2(p_t)$ [5] and STAR $v_2(p_t)$ [30], expressed by Eq. (8) and illustrated by Fig. 4.e provides a successful quantitative as well as qualitative test for the appearance of a perfect fluid in Au+Au collisions at various colliding energies at RHIC.

2.4 Chiral symmetry restoration

Correlation functions are important to see the collective properties of particles and the space-time structure of the emitting source, e.g. the observed size of a system can be measured by two-particle Bose-Einstein correlations [31]. The m_t dependent strength of two-pion correlations, the so-called λ_* parameter, which is related to the extrapolated value of the correlation function at zero relative momentum, can be used to extract information on the mass-reduction of the η' meson, a signal of $U_A(1)$ symmetry restoration in the source [32, 33, 34, 35].

PHENIX analyzed [36] $\lambda_*(m_t)$ with fits to two-pion correlation functions using different shapes, Gauss, Levy and Edgeworth, described in refs. [27, 36, 37]. A comparison of the measurements with model calculations of ref. [32] using FRITIOF results for the composition of the long-lived resonances and a variation of the η' mass is presented in fig. 5. If we re-norm the $\lambda_*(m_t)$ curves with their maximum on the investigated m_t interval, they confirm the existence and characteristics of the hole in the $\lambda_*(m_t)$ distribution.

Gauss fit results agree with former PHENIX measurements (see ref. [22]). Regarding $U_A(1)$ symmetry restoration, conclusion is that at present, results are critically dependent on the understanding of statistical and systematic errors, and additional analysis is required to make a definitive statement.

3 Summary and conclusions

In summary, we can make the definitive statement, based on elliptic flow measurements and the broad range success of analytic hydro models, that in relativistic Au+Au collisions observed at RHIC we see a perfect fluid. Based

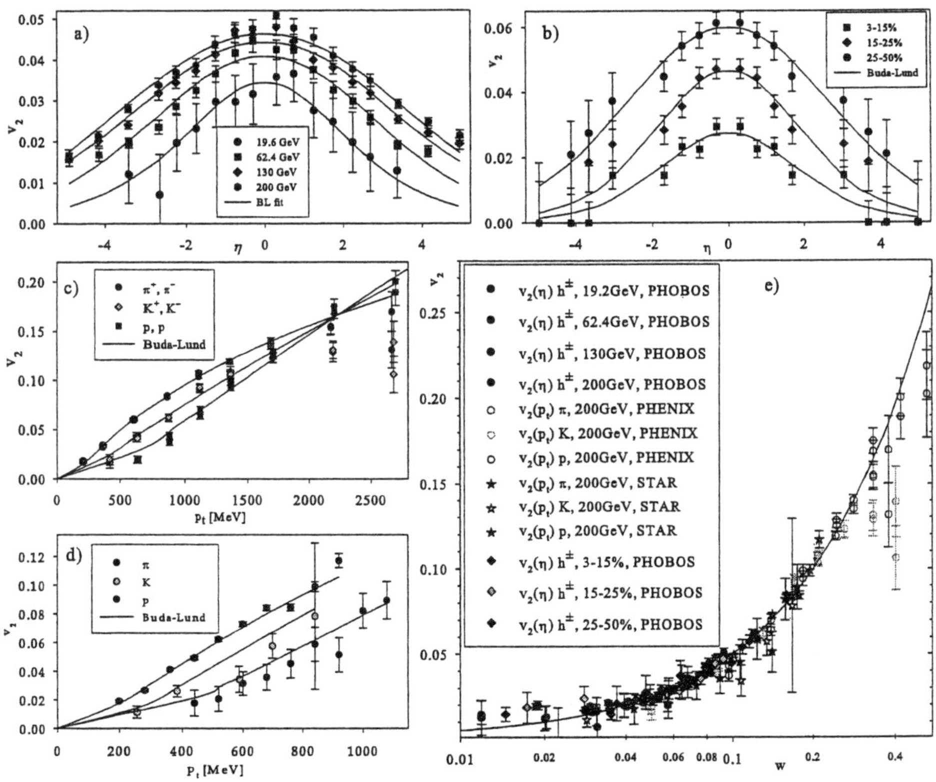

Figure 4: PHOBOS [29] (a-b), PHENIX [5] (c) and STAR [30] (d) data on elliptic flow, v_2, plotted versus p_t and η and fitted with Buda-Lund model. Elliptic flow versus variable w is plotted in panel (e): data points of plots (a-d) show the predicted [6] universal scaling. See fit parameters in ref. [28]

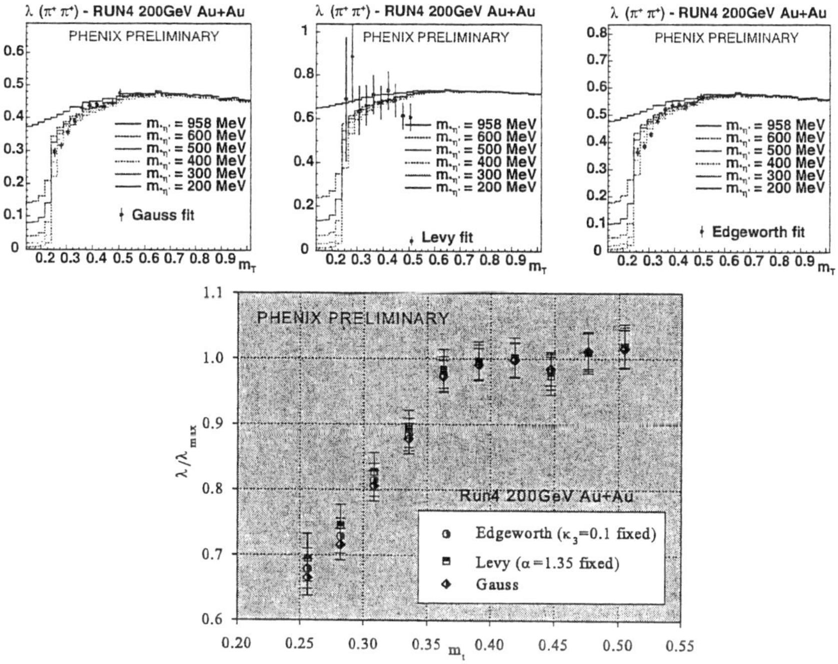

Figure 5: Top three figures: measured $\lambda_*(m_t)$ of ref. [36] compared to calculations using the model of ref. [32] with various η' mass values. Bottom figure: $\lambda_*(m_t)$ curves with equal number of fit parameters and re-normed with their maximal value on the interval of 0.20GeV< m_t <0.55GeV all show the same shape. See details of the fits in ref. [36].

on our estimates on the temperature and energy density we also conclude that the observed matter is in a deconfined state. We also see a possible signal of partial symmetry restoration in the mass reduction of η' bosons. Future plan is to explore all properties of the Quark Matter, by analyzing more data and using higher luminosity. We are after the full map of the QCD phase diagram, and in order to explore it, we also have to go to higher energies and compare them to lower energy data. If the Quark Matter is the New World, then Columbus just realized he is not in India, but on a new continent.

"It does not make any difference how beautiful your guess is. It does not make any difference how smart you are, who made the guess, or

what his name is — if it disagrees with experiment it is wrong."

R. P. Feynman, about discovering new laws [38]

Acknowledgements

The author would like to thank Professor Antonio Zichichi for the invitation to the fascinating International School of Subnuclear Physics; and for the kind hospitality of the organizers, especially Silvia Arcelli and Chiara Zampolli. The author also would like to express his gratitude to his colleagues Tamás Csörgő, Bengt Lörstad, Márton Nagy and András Ster for discussions, help and all the work done to acquire the results of this paper.

References

[1] S. Weinberg, The First Three Minutes Basic Books, New York

[2] S. S. Adler *et al.*, Phys. Rev. Lett. **91**, 072303 (2003)

[3] K. Adcox *et al.*, Nucl. Phys. A **757**, 184 (2005)

[4] M. Riordan and W. A. Zajc, Sci. Am. **294N5**, 24 (2006).

[5] S. S. Adler *et al.*, Phys. Rev. Lett. **91**, 182301 (2003)

[6] M. Csanád, T. Csörgő and B. Lörstad, nucl-th/0310040.

[7] Y. Hama *et al.*, Nucl. Phys. A **774**, 169 (2006)

[8] W. Broniowski, A. Baran and W. Florkowski, AIP CP **660**, 185 (2003)

[9] A. Adare *et al.*, nucl-ex/0608033.

[10] T. Csörgő, M. I. Nagy and M. Csanád, nucl-th/0605070.

[11] S. Pratt, nucl-th/0612010.

[12] J. D. Bjorken, Phys. Rev. D **27** (1983) 140.

[13] I. G. Bearden *et al.*, Phys. Rev. Lett. **88** (2002) 202301

[14] T. Csörgő and B. Lörstad, Phys. Rev. C **54** (1996) 1390

[15] Y. Aoki, Z. Fodor, S. D. Katz, K. K. Szabó, P. L. B **643**, 46 (2006)

[16] I. G. Bearden *et al.*, Phys. Lett. B **523** (2001) 227

[17] K. Adcox *et al.*, Phys. Rev. Lett. **88** (2002) 242301

[18] K. Adcox *et al.*, Phys. Rev. Lett. **88** (2002) 192302

[19] B. B. Back *et al.*, Phys. Rev. Lett. **87** (2001) 102303

[20] C. Adler *et al.*, Phys. Rev. Lett. **87** (2001) 082301

[21] S. S. Adler *et al.*, nucl-ex/0307022.

[22] S. S. Adler *et al.*, nucl-ex/0401003.

[23] M. Csanád, T. Csörgő, B. Lörstad and A. Ster, J. P. G **30**, S1079 (2004)

[24] M. Csanád, T. Csörgő, B. Lörstad, A. Ster, Act. P. Pol. B**35**, 191 (2004),

[25] M. Csanád, T. Csörgő, B. Lörstad, A. Ster, Nukleonika **49**, S49 (2004)

[26] A. Ster, T. Csörgő and B. Lörstad, Nucl. Phys. A **661**, 419 (1999)

[27] T. Csörgő, Heavy Ion Phys. **15**, 1 (2002)

[28] M. Csanád *et al.*, nucl-th/0512078.

[29] B. B. Back *et al.*, Phys. Rev. Lett. **94**, 122303 (2005)

[30] J. Adams *et al.* nucl-ex/0409033.

[31] R. Hanbury Brown and R. Q. Twiss, Nature **178**, 1046 (1956)

[32] S. E. Vance, T. Csörgő and D. Kharzeev, P. R. L. **81**, 2205 (1998)

[33] J. I. Kapusta, D. Kharzeev and L. D. McLerran, P. R. D **53** (1996) 5028

[34] Z. Huang and X. N. Wang, Phys. Rev. D **53** (1996) 5034

[35] T. Hatsuda and T. Kunihiro, Phys. Rept. **247**, 221 (1994)

[36] M. Csanád for the PHENIX Coll., Nucl. Phys. A **774**, 611 (2006)

[37] T. Csörgő, S. Hegyi and W. A. Zajc, Eur. Phys. J. C **36**, 67 (2004)

[38] Feynman videos, `http://heelspurs.com/zpics/feynman9.rm`

Analog Models beyond Kinematics

November 17, 2006

Serena Fagnocchi
Centro Studi e Ricerche "Enrico Fermi", Compendio Viminale, 00184 Roma, Italy
INFN sezione di Bologna, via Irnerio 46, 40126 Bologna, Italy
e-mail: fagnocchi@bo.infn.it

Abstract

In this paper I discuss the extension of the analogy between gravitation and some systems of condensed matter physics from kinematics to dynamics. I will focus my attention on two applications of the analogy to the dynamics of fluids that have been recently proposed: the study of backreaction effects and the calculation of the depletion in Bose-Einstein condensates, showing how this extension is possible and stressing the main differences with respect to the gravitational context. I will conclude with some remarks about the actual reliability of the proposed scheme, pointing out the basis issues that have still to be addressed.

One of the most peculiar features of actual theoretical physics is that some of its branches suffer an incredible lack of experimental data that seriously risks to compromise their development. Among them is Quantum Field Theory (QFT) in curved space, with its most famous result – the Hawking effect [1], widely considered a milestone of theoretical physics – still unobserved. In contrast with what previously believed about black holes, Hawking, studying the propagation of quantum fields on the background of a collapsing body forming a black hole[1], showed that a thermal flux of particles emerges after

[1]This effect has been obtained within the semiclassical theory of Gravity, where the spacetime is described by the classical Einstein theory of General Relativity (GR), while the fields propagating on this background are quantized.

the horizon forms, and therefore the black hole radiates. It is of fundamental importance that it is a pure kinematical effect, related only to the geometrical propagation of the field modes next to the horizon and completely independent of the details of the collapse itself. Therefore it will occur whenever the same spacetime structure emerges.
As already pointed out, for the extreme weakness of the produced effect, the Hawking radiation has so far no experimental observation confirming it: in fact for astrophysical black holes the temperature associated to the Hawking emission is of order $10^{-8}\ K$ or lower, far beyond the CMB temperature ($\sim 3\ K$) which hoplessly covers it. A possible test in next generation accelerators collisions has been also proposed, provided an ad hoc extra-dimensions theory lowers the scale of Quantum Gravity at $\sim TeV$ [2].
A more promising way out can be in the so called *Analog Models*, i.e. some systems of condensed matter which, under some circumstances, behave as gravitational ones. As shown by Unruh in his seminal work [3], the fluctuations in an Eulerian fluid propagate *exactly* as a massless scalar field on a fictitious curved background, whose metric tensor depends on the characteristics of the underlying fluid. In fact an inviscid irrotational fluid is described by the hydrodynamical equations

$$\partial_t \rho + \vec{\nabla} \cdot (\rho \vec{v}) = 0 \,, \qquad \partial_t \theta + \frac{1}{2}\vec{v}^2 + \mu(\rho) = 0 \,, \tag{1}$$

with ρ the density, $\vec{v} = \vec{\nabla}\theta$ the velocity, $\mu(\rho) \equiv \frac{du}{d\rho}$ and u the internal energy. Expanding eqs.(1) at first order in the linear fluctuations ρ_1 and θ_1 around a mean field solution of eqs.(1) one gets

$$\Box\theta_1 = 0 \,, \qquad \rho_1 = -\frac{\rho}{c^2}\left(\partial_t\theta_1 + \vec{\nabla}\theta \cdot \vec{\nabla}\theta_1\right) \,, \tag{2}$$

where $\Box \equiv \frac{1}{\sqrt{-g}}\partial_\mu(\sqrt{-g}g^{\mu\nu}\partial_\nu)$ is the covariant d'Alembertian calculated from the so called "acoustic metric"

$$g_{\mu\nu} \equiv \frac{\rho}{c}\begin{pmatrix} -(c^2 - v^2) & -\vec{v}^T \\ -\vec{v} & \mathbf{1} \end{pmatrix} \,, \tag{3}$$

and c is the sound speed $c^2 = \rho \left.\frac{d\mu}{d\rho}\right|_{\rho_0}$.
Thus the field θ_1 satisfies the equation of motion for a massless scalar field in the curved background described by the tensor metric (3). For a homogeneous fluid the acoustic metric reduces to the flat minkowskian one, while it

can show a black hole form if $|\vec{v}| = c$: there g_{00} vanishes and a sonic horizon forms [4]. The supersonic region is the acoustic analog of the black hole, where sound is trapped as light is in the gravitational field of a black hole. Once the hydrodynamical fluctuations are quantized, following Hawking's arguments, a flux of phonons (the sonic quanta) at a temperature $T_H = \frac{\hbar k}{2\pi \kappa_B c}$ is expected, where $k = \frac{1}{2}\nabla(c^2 - v^2)|_{hor}$ is the acoustic analog of the surface gravity of the horizon [3]. This is the sonic analog of the Hawking effect. For this reason hydrodynamical systems can in principle be used as "laboratories" to observe this effect (among other effects otherwise unfathomable). One of the most promising setting to this aim are Bose Einstein condensates (BEC), for which $T_H \approx 10\ nK$ with respect to a condensate temperature $T_c \approx 100\ nK$ [5]. Unfortunately, the analog Hawking radiation is still difficult to point out directly, both for the fact that T_H is actually very low and for the noise produced by other effects that risks to cover it. Yet the recent technological improvement in handling BEC and the possibility to study analog models more in depth make us confident that it can be soon observed.

Dynamics

As shown in the previous section, the analogy works perfectly at the kinematical level: it occurs for the amazing fact that sound waves in a fluid "feel" inhomogeneities exactly as a quantum field on a gravitational background "feels" curvature and it is therefore possible to use the GR formalism to describe their motion. But the features of the acoustic metric strictly depend on the underlying medium and its flow, which can be described in no way using the GR framework: they are indeed non-relativistic ($c_{sound} \ll c_{light}$) and newtonian. The equations governing the dynamics in the two setting evidently differ: on one side are Einstein equations, on the other hydrodynamical equations.

Does this mean a full stop in the application of analogy beyond the kinematical issues? Or is there still the possibility to go further addressing also dynamical ones?

I will show how is still possible, with special care, to use analog models to show up also behaviors depending on the dynamics. Of course the analogy will be not "perfect" any more and the outcomes in the condensed matter context could have nothing to do with what known in GR. Yet the formalism developed in QFT in curved space turns out to be a powerful tool to eval-

uate analytically quantities otherwise nasty to evaluate with the standard techniques.
As examples, I will show in the following two seemingly very different applications of this formalism: the backreaction of the analog Hawking radiation and the depletion in BEC.

Backreaction in sonic black holes

As already mentioned, Hawking radiation is still difficult to be directly observed even in the acoustic setting. It could be easier to try and see it indirectly through the modifications its presence will induce on the fluid: this beck-effect is the so called backreaction. Of course it will depend on dynamics and there is no reason to expect that what found for gravitational black holes still holds for these completely different objects.
In GR backreaction is governed by the semiclassical Einstein equations, namely $G_{\mu\nu}(g_{\alpha\beta}) = \hbar\langle T_{\mu\nu}(g_{\alpha\beta})\rangle$, where $\langle T_{\mu\nu}\rangle$ is the stress energy tensor of the quantum field which acts as a source in the semiclassical equations and modifies the background spacetime. The first issue is to find the analog of the semiclassical equations for the hydrodynamical systems. In [6] it is shown how to solve this problem going further in the expansions of eqs.(1) up to quadratic order in the fluctuations, yielding

$$\partial_t\rho_B+\vec{\nabla}\cdot(\rho_B\vec{v}_B) = -\partial_i(\sqrt{-g^B}\langle T^{0i}\rangle)\,, \qquad \dot{\psi}_B+\frac{1}{2}\vec{v}_1^2+\mu(\rho_B) = \frac{1}{2}\frac{\rho_B}{c}\langle T\rangle\,, \quad (4)$$

with the subscript $_B$ meaning "corrected by backreaction".
In eqs.(4) the pseudo-stress energy tensor for the fluctuations (namely for a massless scalar field θ_1) has been introduced: $T_{\mu\nu} = \partial_\mu\theta_1\partial_\nu\theta_1 - \frac{1}{2}g_{\mu\nu}\partial_\alpha\theta_1\partial^\alpha\theta_1$ and T is its trace. This procedure is equivalent to what one usually do in QFT in curved space to write down the backreaction equations for gravitational waves. It is worth noting that eqs.(4) are the most general equations governing backreaction effects in the hydrodynamical setting for $c = const$: they are not linked to the presence of a sonic black hole and describe classical as well as quantum fluctuations backreaction. All the informations relative to the different situations are enclosed in $\langle T_{\mu\nu}\rangle$: the brackets are stochastic mean values in describing classical fluctuations or expectation values in the quantum case. The quantum state which properly describes black hole formations and Hawking radiation is the so called Unruh state [7].
Under some simplifying hypothesis, eqs.(4) has been solved analytically [6]

and numerically [8] for a sonic black hole in the Unruh state. It has been shown that, due to the presence of the Hawking radiation, next to the horizon the fluid slows down and the black hole size shrinks. Moreover the emission temperature gets lower and lower as the evaporation goes on, reaching zero temperature emission in an infinite amount of time. This is not a surprising behavior: the evaporation proceeds at the expenses of the kinetic energy of the fluid, which in fact slows down, and the temperature (related to the velocity on the horizon) also decreases. Yet it is completely different from what is known about Schwarzschild black holes, that get hotter and hotter as the evaporation goes on. The sonic black hole looks rather like a near-extremal charged black hole: this kind of black hole reaches zero temperature emission in an infinite time, but with different power law with respect to the sonic ones [9].

Depletion in Bose-Einstein condensates

As well as other ultra-cold atoms systems, BEC can be described in some regimes by hydrodynamical equations [10]. In the mean field limit the bosonic operator $\hat{\Psi}$ describing the system can be split as $\hat{\Psi} = \psi + \hat{\varphi}$: $\psi = \langle\hat{\Psi}\rangle$ describes the condensed part, while $\hat{\varphi}$ is a quantum fluctuation. Under some hypothesis and writing $\psi = \sqrt{\rho}e^{i\theta/\hbar}$, the condensed part satisfies the classical equation of motion analog to eqs.(1), where now ρ is the density of particles and $\vec{v} = m^{-1}\vec{\nabla}\theta$ the velocity of the fluid. In practice this description is possible if the scale of variation of the inhomogeneities L is much bigger than the healing length $\xi = \hbar/(\sqrt{2}mc)$, that is related to the mean distance between the constituents of the fluid. Moving to the fluctuation part, without loss of generality, $\hat{\varphi}$ can be rewritten in the acoustic representation as

$$\hat{\varphi} = e^{i\theta/\hbar}\left(\frac{1}{2\sqrt{\rho}}\hat{\rho}_1 + i\frac{\sqrt{\rho}}{\hbar}\hat{\theta}_1\right), \tag{5}$$

with $\hat{\rho}_1$ and $\hat{\theta}_1$ two real fields. The equations describing them assume the hydrodynamical form and can therefore be rewritten using the gravitational formalism as in eqs.(2) [4]. It is worth stressing that this description is correct only for modes with wavelength $\lambda \gg \xi$: the modes with very high frequency in fact do not obey the relativistic dispersion relation $\omega = ck$ implicit in the GR framework, but the full Bogoliubov dispersion relation for BEC, namely $\omega = ck\sqrt{1+\xi^2k^2}$.

In the hydrodynamical regime it is therefore possible to use BEC as labs for testing GR effects, but also to handle the analogy to give analytical estimations of quantities that are hardly evaluable with the standard tools. In this spirit, QFT in curved space has been used to work out the depletion for a non-homogeneous BEC [11]. The total particles density can be written as a sum of condensed and non condensed part

$$\rho_{tot} = \langle \hat{\Psi}^\dagger \hat{\Psi} \rangle = \psi^* \psi + \langle \hat{\varphi}^\dagger \hat{\varphi} \rangle = \rho + \tilde{\rho} \,. \tag{6}$$

The last term describes the so called depletion: because of quantum and temperature fluctuations, a non vanishing fraction of the total particles number is not in the condensed phase. The depletion is explicitely known only for homogeneous BEC where $\rho = \rho_0 = const.$ At low temperature ($\kappa_B T \ll \rho_0 g$), it reads [10]

$$\tilde{\rho}_0 = \frac{8}{3\pi^{1/2}} (\rho_0 a)^{3/2} \left[1 + \left(\frac{\pi \kappa_B T}{2\rho_0 g} \right)^2 \right] . \tag{7}$$

Moving to non homogeneous condensates with vanishing velocity, the depletion is usually approximated by eq.(7) with now ρ_0 replaced by the actual space-time varying density $\rho(\vec{x}, t)$, in the so called *local density approximation* (LDA) [12].

It is clear that this approximation is not able to take the effects of spatial derivatives into account. It is worth noting that the derivation of the zero temperature quantum depletion, which does not vanish in the $T \to 0$ limit, is based on the full Bogoliubov dispersion relation for the modes. Unlike this, the low temperature correction comes from the phononic part of the spectrum only, i.e. that one assumed in the GR analogy. The integral giving the thermal depletion is dominated by modes with wavelength $\lambda \gg \xi$, so the result is insensitive to the high k behavior of the dispersion relation. Therefore, while the correct description of the ground state requires the knowledge of the full "microscopic" theory[2], deviations from the ground state value can be evaluated using the low energy "macroscopic" phonon theory. In other words, we assume that the corrections coming from the inhomogeneities of the condensate to the ground state result, sensitive to the whole quantum theory describing the system, are determined by those modes with wavelength comparable with the scale of the inhomogeneities (as curvature is seen by

[2]The linear phonon dispersion relation, characteristic of a massless relativistic free field would lead (after renormalization) to a vanishing quantum depletion.

modes with λ of the same order of the curvature itself). Those modes therefore satisfy the hydrodynamical approximation request ($\lambda \sim L \gg \xi$) and for them the analogy works. [3]
Using techniques developed within QFT in curved space, it is possible to overtake the difficulties in handling analytical calculations for non-homogeneous BEC [11]. Now, noting that

$$\tilde{\rho}_{in} = \langle \hat{\varphi}^\dagger \hat{\varphi} \rangle = \frac{\langle \hat{\rho}_1^2 \rangle}{4\rho} + \frac{\rho \langle \hat{\theta}_1^2 \rangle}{\hbar^2} \approx \frac{\rho \langle \hat{\theta}_1^2 \rangle}{\hbar^2} \left(1 + O\left(\frac{\xi^2}{L^2} \right) \right) , \tag{8}$$

the calculation of the depletion has been reduced to the calculation of the expectation value of the squared of a massless scalar field minimally coupled to the acoustic spacetime (3). For a spherically symmetric static BEC configuration, using renormalization techniques peculiar of QFT in curved spacc, it is possible to give an explicit description for the thermal and quantum depletion for an inhomogeneous BEC in term of a general density $\rho(r)$:

$$\tilde{\rho}_{in} = \frac{\kappa_B^2 T^2}{12\hbar^3} \left(\frac{m^3}{g\rho} \right)^{1/2} - \frac{(mg\rho)^{1/2}}{96\pi^2\hbar} \left[-\frac{7}{8}\frac{\rho'^2}{\rho^2} + \frac{5}{2}\frac{\rho''}{\rho} + \right.$$
$$\left. + \frac{5}{r}\frac{\rho'}{\rho} \right] \ln \mu^2 \frac{g^{1/2}\rho^{3/2}}{m^{3/2}} + \frac{(mg\rho)^{1/2}}{48\pi^2\hbar} \left[-\frac{39}{16}\frac{\rho'^2}{\rho^2} + \frac{3}{4}\frac{\rho''}{\rho} + \frac{3}{2r}\frac{\rho'}{\rho} \right] , \tag{9}$$

where m is the mass of the atoms and g is a constant proportional to the scattering length a: $g = 4\pi\hbar^2 a/m$. Note that the thermal term exactly reproduces the thermal depletion in LDA. The other terms instead generalize the zero-temperature depletion in LDA taking the spatial derivatives of the condensate density explicitely into account. In fact all of them vanish for a homogeneous BEC.
Moving to configurations with a radial stationary velocity ($\vec{v} = v(r)\vec{r}/|\vec{r}|$), the concept of thermal equilibrium loses its meaning, but it is still possible to evaluate the effects of this velocity on the quantum depletion, yielding [11]

$$\tilde{n}_{in}^q = -\frac{m}{96\pi^2\hbar c^2} \left[\frac{3}{2} cc'^2 + 5c^2c'' + -3v^2c'' - 2cv'^2 - 2cvv'' + \frac{9}{2c}v^2c'^2 \right.$$
$$\left. -4vc'v' + \frac{2}{r}(5c^2c' - 4cvv' - v^2c') - \frac{2}{r^2}v^2c \right] \ln \mu^2 \frac{c(c^2 - v^2)}{g} +$$

[3] This is more than a guess, since it is partially confermed by the results in this direction coming from the Casimir effect [13], even if so far there is not a conclusive answer.

$$+\frac{m}{48\pi^2\hbar c^2}\left\{-\frac{1}{2c(c^2-v^2)}(3c^2c'-v^2c'-2cvv')(3c^2c'-2v^2c'-cvv')\right.$$
$$-\frac{15}{4}cc'^2-\frac{1}{4c}v^2c'^2-\frac{5}{2}vc'v'++\frac{3}{2}c^2c''-\frac{1}{2}v^2c''-cv'^2-cvv''+$$
$$\left.+\frac{1}{r}(3c^2c'-v^2c'-2cvv')\right\}, \tag{10}$$

where we have introduced the local sound speed c, that for BEC is related to the density by the equation of state $mc^2 = gn$. This expression generalizes eq.(9) considering also contributions due to the velocity of the condensate. It reduces to eq.(9) only for $v(r) = 0$: for $v = const$ in fact a curvature is induced since a radial velocity can not be trivially eliminated by a Galileo transformation.

In the case of an acoustic black hole, the expression in eq.(10) diverges on the acoustic horizon ($v^2 = c^2$): this is a familiar divergence in GR, related to the fact that the quantum state over which the expectation value of $\langle\hat{\theta}_1^2\rangle$ has been calculated is not the right one to describe the region next to the horizon. In GR it has been shown that for a Schwarzschild black hole the same calculation for the Unruh state leads to a regular result: the same is expected also for acoustic black holes.

Conclusions

In this paper I have illustrated the first step in extending the analogy between gravitation and condensed matter systems also to dynamical issues. It has been stressed how this extension has to be handled with particular care, without expecting in the acoustic setting an exact counterpart of the results already obtained in GR (unlike what happens with the kinematical ones).

However there is still a crucial issue to be addressed: how big are the errors in neglecting high momenta effects?

This question involves both QFT in curved space and condensed matter physics. As already mentioned, for BEC as in general for fluids, very high k modes see the coarse grain of the medium, and, as a consequence, they do not obey the relativistic dispersion relation, but a short-distance modified one.

In the gravitational context instead the question is related to the so called *transplanckian problem*: the frequency of the modes which give the Hawking radiation is indeed transplanckian, i.e. beyond the actual reliability limit of

the theory used to study the process, namely QFT in curved space. This fact does raise doubts on the reliability of the Hawking effect itself, as a consequence of the breaking of QFT in curved space beyond Planck energy scale, where Quantum Gravity effects should become important. Thought a full Quantum Gravity theory is not known and therefore it is not possible to safely say how these behaviors will emerge, it is widely believed that a first hint of such effects should show up in a short-distance modification of the dispersion relation [14] (like what happens in the acoustic setting). Recent investigations on Hawking effect within a QFT with modified dispersion relation have shown that Hawking radiation is only slightly affected by these modifications, turning out to be unaltered in its fundamental features [15, 9]. This is a good starting point to carry out deeper investigations involving also the calculation of expectation values in presence of such modified dispersion relations. Moreover also the first result concerning the calculation of renormalized expectation value goes in this direction: in fact the Casimir energy density between two conducting plates can be evaluated with great accuracy using the phonon dispersion relation, provided the plate separation is much larger than the interatomic distance of the constituents of the plates [13]. Similar analysis has been performed also in the cosmological context, yielding so far only partial results [16]. Nothing has still been obtained for a black hole background.
A deeper knowledge of these effects will be of crucial importance in several field: cosmology, black hole physics and cold-atoms systems. A more accurate use of QFT in curved space will turn out to be very useful both in studying gravitation directly and in leading a more reliable use of the analogy in the dynamics of the condensed matter systems.

Acknowledgments: I would like to thank R. Balbinot for helpful suggestions and contributions to this paper. I also thank "Enrico Fermi" Center for supporting this research.

References

[1] S.W. Hawking, *Nature* **248**, 30 (1974)

[2] See for ex.: S.B. Giddings, S. Thomas, *Phys. Rev.* **D65**, 056010 (2002)

[3] W.G. Unruh, *Phys. Rev. Lett.* **46**, 1351 (1981)

[4] C. Barcelo, S. Liberati, M. Visser, *Living Rev. Rel.* **8**, 12 (2005)

[5] M. Visser, C. Barcelo, S. Liberati, *Gen. Rel. Grav.* **34**, 1719 (2002)

[6] R. Balbinot, S. Fagnocchi, A. Fabbri, G.P. Procopio, *Phys. Rev. Lett.* **94**, 161302 (2005);
R. Balbinot, S. Fagnocchi, A. Fabbri, *Phys. Rev.* **D71**, 064019 (2005)

[7] See for ex. : *Black Hole Physics: basic concepts and new developments*, V.P. Frolov and I.D. Novikov, Kluwer Ed., Dordrecht, (1997)

[8] R. Balbinot, A. Fabbri, S. Fagnocchi, A. Nagar, *Nuovo Cimento* **B 121**, 201 (2006)

[9] R. Balbinot, A. Fabbri, S. Fagnocchi, R. Parentani, *La Rivista del Nuovo Cimento* **28**, 1 (2005)

[10] See for ex.: *Bose-Einstein condensation*, L.P. Pitaevskii, S. Stringari, Clarendon Press, Oxford (2003)

[11] All the details can be found in: R. Balbinot, S. Fagnocchi, A. Fabbri, *The depletion in Bose Einstein condensates using Quantum Field Theory in curved space*, cond-mat/0610367 (2006)

[12] E. Timmermans, P. Tommasini, K. Huang, *Phys. Rev.* **A55**, 3645 (1997)

[13] S. Bachmann, A. Kempf, gr-qc/0504076 (2005)

[14] D. Mattingly, *Living Rev. Rel.* **8**, 5 (2005)

[15] W.G. Unruh, R. Schützhold, *Phys. Rev.* **D71**, 0240028 (2005)

[16] R.H. Brandemberger, J. Martin, *Phys. Rev.* **D71**, 023504 (2005);
D. Lopez Nacir, F.D. Mazzitelli, C. Simeone, *Phys. Rev.* **D72**,124013 (2005)

Complexity in cosmic structures

Francesco Sylos Labini

Enrico Fermi Center, Via Panisperna 89 A, Compendio del Viminale, 00184 - Rome, Italy and

ISC-CNR Via dei Taurini, 19 00185 Rome Italy

Abstract

Although dark matter is supposed to provide with more than 0.9 of the total fraction of the mass-energy in universe, its amount and properties can only be defined a posteriori. In this context, a crucial point concerns the identification of a possible clear feature of dark matter fields which is not arbitrary, i.e. a property which has to be satisfied by dark matter fluctuations under some very general theoretical conditions. We discuss the fact that this property, in standard cosmological models, is represented by super-homogeneity, i.e. a very fine tuned balance between negative and positive correlations of density fluctuations, which must be imprinted both in the anisotropies of the CMBR and in the large scale distribution of galaxies. We review the main aspects of this property, considering examples of super-homogeneous systems well-studied in statistical physics, and discuss its possible observational evidences.

1 Introduction

More than twenty years ago it has been surprisingly discovered that galaxy velocity rotation curves remain flat at large distances from the galaxy center while the density profile of luminous matters rapidly decays (e.g. [1]). This is one of the strongest indications of the need from dynamically dominant dark matter in the universe. Most attention has been focused on the fact that these bound gravitational systems contain large quantities of unseen matter and an intricate paradigm has been developed in which non-baryonic dark matter plays a central role not only in accounting for the dynamical mass of galaxies and galaxy clusters [2] but also for providing the initial seeds which have given rise to the formation of structure via gravitational collapse [3].

In current standard cosmological models, different forms of dark matter are needed to explain a number of different phenomena. In fact, the results of several observations, such as the scale size of fluctuations of the Cosmic Microwave Background Radiation (CMBR) (e.g., [4]), the measurements of clustering mass on large scales (e.g., [5]), the magnitude-redshift relation of type Ia supernovae (SNe Ia) (e.g., [6]), are interpreted to give consistent measurements of the amount of dark matter. In this framework, baryons (Ω_B), which can be detected in the form of, for example, luminous objects such as stars and galaxies, would only be the 5% of the total mass in the universe; the rest is made of entities about which very little is understood: dark matter and dark energy. More specifically dark matter, in form of non-baryonic elementary particles, would contribute to the $\sim$ 30% of the total mass of universe ($\Omega_m \sim 0.3$). It is worth noticing that its direct detection in laboratory experiments is still lacking and that the standard model of particle physics does not predict the existence of candidate dark matter particles with the necessary properties from a cosmological point of view.

Several evidences, from supernovae and other observations, shows that the expansion of the Universe, rather than slowing because of gravity, is increasingly rapid. Within the standard cosmological framework, this must be due to a substance, which has been termed dark energy, that behaves as if it has negative pressure. This is a mysterious form of energy which would cause the accelerating expansion of the universe and it should account for about 70% (i.e. $\Omega_\Lambda \sim 0.7$) of the mass-energy in the Universe. It is thus not surprising that great observational and theoretical effort is devoted to the understanding of the nature and properties of dark matter and dark energy which, giving the main contribution to the mass-energy density of the universe, play a crucial role, for example, in the problem of structure formation.

The previous discussion enlightens the fact that we know very little about the nature of cosmological dark matter both from a fundamental and observational points of view. Although dark matter is so central in modern cosmology its amount and properties can only be defined a posteriori. In this context a crucial question concerns a possible clear property of dark matter density fields which is not arbitrary, i.e. a property which has to be satisfied by dark matter fluctuations under some very general theoretical conditions. In fact, from the above discussion it seems that much freedom is left for the choice of dark matter, its physical properties and its statistical distribution. However there is an important constraint which must be valid for any kind of initial matter density fluctuation field in the framework of Friedmann-Robertson-Wlaker (FRW) models and which represents a consistency condition to be satisfied by any fluctuation field compatible with the FRW metric. As we discuss below this must be imprinted both in the fluctuations of the CMBR and in the large scale distribution of galaxies.

2 The signature of dark matter

The most prominent feature of the initial conditions in the early universe, in standard theoretical models, derived from inflationary mechanisms, is that matter density field presents on large scale super-homogeneous features [7]. This means the following. If one considers the paradigm of uniform distributions, the Poisson process where particles are placed completely randomly in space, the mass fluctuations in a sphere of radius R growths as R^3, i.e. like the volume of the sphere. A super-homogeneous distribution is a system where the average density is well defined (i.e. it is uniform) and where fluctuations in a sphere grow slower than in the Poisson case, e.g. like R^2: in this case there are the so-called surface fluctuations to differentiate them from Poisson-like volume fluctuations.

Note that a uniform system with positive correlations present fluctuations which grow faster than Poisson. For example a perfect cubic lattice of particle is a super-homogeneous system. A well known system in statistical physics systems of this kind is the one component plasma [8] (OCP) which is characterized by a dynamics which at thermal equilibrium gives rise to such configurations. The OCP is simply a system of charged point particles interacting through a repulsive $1/r$ potential, in a uniform background which gives overall charge neutrality. Simple modifications of the OCP can produce equilibrium correlations of the kind assumed in the cosmological context [8]. In the cosmological context inflationary models predict a spectrum of fluctuations of this type.

In terms of the normalized mass variance

$$\sigma^2(R) = \frac{\langle M(R)^2 \rangle - \langle M(R) \rangle^2}{\langle M(R) \rangle^2} , \tag{1}$$

where $\langle M(R) \rangle$ is the average mass in a sphere of radius R and $\langle M(R)^2 \rangle$ is the average of the square mass in the same volume. Thus for a Poisson distribution, where there are no correlation between particles (or density fluctuations) at all, one simply has

$$\sigma^2(R) \sim R^{-3} . \tag{2}$$

For an ordered system characterized by small-scale anti-correlation the variance behaves as

$$\sigma^2(R) \sim R^{-4} , \tag{3}$$

which is the fastest possible decay for discrete or continuous distributions [7]. By comparing Eq.2 with Eq.3 one may notice that the long-range order of the distribution corresponds to the presence of mass fluctuations which are depressed with respect to the uncorrelated Poisson case.

The reason for this peculiar behavior of primordial density fluctuations is the following. In a FRW cosmology there is a fundamental characteristic length scale, the horizon scale $R_H(t)$. It is simply the distance light can travel from the Big Bang singularity $t = 0$ until any given time t in the evolution of the Universe, and it grows linearly with time. The Harrison-Zeldovich (H-Z) criterion can be written as

$$\sigma^2_M(R = R_H(t)) = \text{constant}. \tag{4}$$

This conditions states that the mass variance at the horizon scale is constant: this can be expressed more conveniently in terms of the power spectrum (PS) of density fluctuations [7]

$$P(\vec{k}) = \left\langle |\delta_\rho(\vec{k})|^2 \right\rangle \tag{5}$$

where $\delta_\rho(\vec{k})$ is the Fourier Transform of the normalized fluctuation field $(\rho(\vec{r}) - \rho_0)/\rho_0$, being ρ_0 the average density. It is possible to show that Eq.4 is equivalent to assume $P(k) \sim k$: in this situation matter distribution present fluctuations of the type given by Eq.3 [7].

In order to illustrate more clearly the physical implications of this condition, one may consider gravitational potential fluctuations $\delta\phi(\vec{r})$ which are linked to the density fluctuations $\delta\rho(\vec{r})$ via the gravitational Poisson equation: $\nabla^2 \delta\phi(\vec{r}) = -4\pi G \delta\rho(\vec{r})$. From this, transformed to Fourier space, it follows that the PS of the potential $P_\phi(k) = \left\langle |\delta\hat{\phi}(\vec{k})|^2 \right\rangle$ is related to the density PS $P(k)$ as: $P_\phi(k) \sim \frac{P(k)}{k^4}$.

The H-Z condition corresponds therefore to $P_\phi(k) \propto k^{-3}$ so that gravitational potential fluctuations become constant as a function of scale. In fact, the variance in real space spheres of the gravitational potential fluctuations can be written as [7] $\sigma^2_\phi(R) \approx \frac{1}{2} P_\phi(k) k^3|_{k=R^{-1}}$. The H-Z condition fixes this variance to be constant as a function of R. This is a consistency constraint in the framework of FRW cosmology. In fact the FRW is a cosmological solution for a homogeneous Universe, about which fluctuations represent an inhomogeneous perturbation: if density fluctuations obey to a different condition than Eq.4, then the FRW description will always break down in the past or future, as the amplitude of the perturbations become arbitrarily large or small. For this reason the super-homogeneous nature of primordial density field is a fundamental property independently on the nature of dark matter. This is a very strong condition to impose, and it excludes even Poisson processes ($P(k)$ =const. for small k) [7].

Various models of primordial density fields differ for the behavior of the power spectrum at large wave-lengths, i.e. at relatively small scales. For example, for the case the Cold Dark Matter scenario (CDM), where elementary non-baryonic dark matter particles have a small velocity dispersion, the power spectrum decays as a power law $P(k) \sim k^{-2}$ at large k. For Hot Dark Matter (HDM) models, where the velocity dispersion is large, the power spectrum present an exponential decay at large k. However at small k they both exhibit the H-Z tail $P(k) \sim k$ which is in fact the common feature of all density fluctuations compatible with FRW models.

3 Super-homogeneity in large scale galaxy structures

In terms of correlation function $\xi(r)$ (the Fourier conjugate of the PS) CDM/HDM models present the following behavior: it is positive at small scales, it crosses zero at a certain scale and then it is negative approaching zero with a tail which goes as r^{-4} (in the region corresponding to $P(k) \sim k$) [9]. The super-homogeneity (or H-Z) condition says that the volume integral over all space of the correlation function is zero

$$\int_0^\infty d^3r\xi(r) = 0 \,. \tag{6}$$

This means that there is a fine tuned balance between small-scale positive correlations and large-scale negative anti-correlations [7, 9]. This is the behavior that one would like to detect in the data in order to confirm inflationary models. Up to now this search has been done through the analysis of the galaxy PS which has to go correspondingly as $P(k) \sim k$ at small k (large scales). However one should consider an additional complication.

In standard models of structure formation galaxies result from a *sampling* of the underlying CDM density field: for instance one selects (observationally) only the highest fluctuations of the field which would represent the locations where galaxy will eventually form. It has been shown that sampling a super-homogeneous fluctuation field changes the nature of correlations [10]. The reason can be found in the property of super-homogeneity of such a distribution: the sampling necessarily destroys the surface nature of the fluctuations, as it introduces a volume (Poisson-like) term in the mass fluctuations, giving rise to a Poisson-like PS on large scales $P(k) \sim$ constant. The "primordial" form of the PS is thus not apparent in that which one would expect to measure from objects selected in this way. This conclusion should hold for any generic model of bias and its quantitative importance has to established in any given model [10]. On the other hand one may show [10] that the negative

r^{-4} tail in the correlation function does not change under sampling: on large enough scales, where in these models (anti) correlations are small enough, the biased fluctuation field has a correlation function which is linearly amplified with respect to the underlying dark matter correlation function. For this reason the detection of such a negative tail would be the main confirmation of the super-homogeneous character of primordial density field [9].

Up to now, no quantitative evidences in this respect have been reported as measurements of galaxy data are not extended enough to reach the region where $\xi(r) \sim -r^{-4}$. Future surveys, like the complete SDSS catalog [11], may sample this range of scales, but a precise study of the crossover to homogeneity, discretization effects, sampling effects and statistical noise is still needed. At smaller scales instead there have been found evidences for the presence of highly clustered structures which exhibit power law correlation, i.e. a situation which is extremely different from a super-homogeneous distribution. Let us spend some few comments on this issue.

In the past twenty years observations have provided several three dimensional maps of galaxy distribution, from which there is a growing evidence of large scale structures. This important discovery has been possible thanks to the advent of large redshift surveys: angular galaxy catalogs, considered in the past, are in fact essentially smooth and structure-less. In the CfA2 catalog[12], which was one of the first maps surveying the local universe, it has been surprisingly observed the giant "Great Wall" a filament linking several groups and clusters of galaxies of extension of about 200 Mpc/h [1] and whose size is limited by the sample boundaries. Recently the SDSS project [11] has reveled the existence of structures larger than the Great Wall, as for example the so-called "Sloan Great Wall" which is almost double longer than the Great Wall. Nowadays this is the most extended structure ever observed, covering about 400 Mpc/h, and whose size is again limited by the boundaries of the sample [13].

The search for the "maximum" size of galaxy structures and voids, beyond which the distribution becomes essentially smooth, is still an open problem. Instead the fact that galaxy structures are strongly irregular and form complex patterns has become a well-established fact. From the theoretical point

[1]The typical mean separation between nearest galaxies is of about 0.1 Mpc. By local universe one means scales in the range $[1, 100]$ Mpc/h, where space geometry is basically Euclidean and dynamics is Newtonian, i.e. effects of General Relativity are negligible. The size of the universe, according to standard cosmological models is about 5000 Mpc/h, where 1 Mpc $\simeq 3\times10^{22}$ m; distances are given in units of h, a parameter which is in the range [0.5,0.75] reflecting the incertitude in the value of the Hubble constant (H=100 h km/sec/Mpc) used to convert redshift z into distances $d \approx c/Hz$ (where c is the velocity of light).

of view the understating of the statistical characterization of these structures represents the key element to be considered by a physical theory dealing with their formation. The primary questions that such a situation rises are therefore: (i) which is the nature of galaxy structures and (ii) which is the maximum size of structures ? A number of statistical concepts can be used to answer to these questions: in general one wants to characterize n-point correlation properties which are able to capture the main elements of points distributions [9]. For example, [14] have recently measured the conditional average density in a sample of Luminous Red Galaxies (LRG) from a data release of the SDSS. Such a statistics is very useful to determine correlation properties in the regime of strong clustering and the spatial extension of strong fluctuations in a given sample. This was firstly introduced by [15] and then measured in many samples by [16]. The conditional density gives the average density of points in a spherical volume (or a spherical shell) centered around a galaxy (see e.g. [9]). The results obtained by [14] can be summarized as follows:

(i) A simple power-law scaling corresponding to a correlation exponent $\gamma \approx 1$ gives a very good fit to the data up to at least 20 Mpc/h, over approximately a decade in scale. We note that these results are in good agreement with those obtained by[16] through the analyses of many smaller samples and more recently by [17] in another galaxy survey, the 2dFGRS.

(ii) The second important result of [14] is that at larger scales (i.e. $r > 30$ Mpc/h) the conditional density continues to decrease, but less rapidly, until about $\sim$ 70 Mpc/h, above which it seems to flatten up to the largest scale probed by the sample (100 Mpc/h). The transition between the two regimes is slow, in the sense that the conditional density at $\sim$ 20 Mpc/h is about twice the asymptotic mean density. Note that [16] found evidences for a continuation of the small scale power-law to distances of order hundreds of Mpc/h, although with a weaker statistics, which seems to be not confirmed by [14]. A discussion of systematic unaveraged fluctuations and of problems induced by finite volume effects on the determination of the conditional density in the new data release of the SDSS is now in progress [18]. Thus the precise scale of homogeneity, if any, is not yet determined and a detailed analysis of the new coming data will clarify the situation in few years from now.

4 Super-homogeneity in the Cosmic Microwave Background

Primordial density fluctuations have imprinted themselves on the patterns of radiation also, and those variations should be detectable in the CMBR. Three decades of observations have revealed fluctuations in the CMBR of amplitude of order 10^{-5} [4] [2]. It is in fact to make these measurements compatible with observed structures that it is necessary to introduce non-baryonic dark matter which interact with photons only gravitationally, and thus in a much weaker manner than ordinary baryonic matter. Thus in standard models of structure formation dark matter plays the dominant role of providing density fluctuation seeds which, from the one hand are compatible with observations of the CMBR and from the other hand they are large enough to allow the formation, through a complex non linear dynamics, of galaxy structures we observe today.

In standard cosmological theories the CMBR represents a bridge between the very early universe and the universe as we observe today and in particular the galaxy structures. On the one hand the CMBR probes the very early hot universe at extreme energies through the theories proposed — notably "inflation" — to explain the origin of these perturbations. On the other hand the anisotropies reflect the local very small amplitude perturbations which give the initial conditions for the gravitational dynamics which should subsequently generate the galaxy structures observed today.

In the CMBR one measures fluctuations in temperature on the sky i.e., on the celestial sphere. We will not enter here into the detail of the physical theory in standard models which link these temperature fluctuations to the mass density field [21]. It is useful however for what follows to give the precise relation between the two quantities. The temperature fluctuation field $\frac{T(\theta,\phi)-\langle T\rangle}{\langle T\rangle} \equiv \frac{\delta T}{T}(\theta,\phi)$, where θ , ϕ are the two angular coordinates, is conventionally decomposed in spherical harmonics on the sphere:

$$\frac{\delta T}{T}(\theta,\phi) = \sum_{l=0}^{\infty} \sum_{m=-l}^{+l} a_{lm} Y_{lm}(\theta,\phi) \ . \tag{7}$$

The variance of these coefficients a_{lm} is then related to the matter power spectrum through

$$C_l \equiv \langle |a_{lm}|^2 \rangle = \frac{H_0^4}{2\pi} \int_0^\infty dk \frac{P(k)}{k^2} |j_l(k\eta)|^2 \tag{8}$$

where j_l is the spherical Bessel function and $\eta \simeq 2H_0^{-1}$ is a constant at fixed time (H_0 is the Hubble constant today). Note that the ensemble average contains no dependence on m because of the assumption of statistical isotropy.

Taking $P(k) \sim k$ in (8) we get that the $l > 2$ multi-poles are given by $C_l \sim (l(l+1))^{-1}$, so that the H-Z condition for the power spectrum $n = 1$ corresponds to a constant value of the quantity $l(l+1)C_l$. For this reason it is

[2]Recently evidences for an important non-cosmic contamination of the latest WMAP data [4] have been reported by [19].

usually in terms of this combination of l and C_l that the data from the CMBR are represented.

The WMAP team [4] has found that the two point correlation function, the Fourier transform of the angular power spectrum C_l, nearly vanishes on scales greater than about 60 degrees, contrary to what the standard theories, where the super-homogeneous $P(k) \sim k$ is present, predict, and in agreement with the same finding obtained from COBE data about a decade earlier [20]. Note that the vanishing of power is much more apparent in real space than in multiple space (i.e in C_l). In addition, particularly puzzling are the alignments of low multi-poles with the solar system features as this implies that CMBR anisotropy should be correlated with our local habitat: the observed correlations seem to hint that there is contamination by a foreground [19]. Thus the observational situation for the CMBR is not able at the moment to determine whether fluctuations in the radiation and matter density fields really show the crucial super-homogeneous feature.

5 Conclusions

Statistical properties of primordial density fields show interesting analogies with systems in statistical physics, like the one-component plasma, whose main characteristic is the ordered, or super-homogeneous, nature. In the FRW models the super-homogeneous (or Harrison-Zeldovich) condition arises as a kind of consistency constraint: other, more inhomogeneous, stochastic fluctuations, like the uncorrelated Poisson case, will always break down in the FRW models in the past or future as the amplitude of perturbations in the gravitational potential may become arbitrarily large. We discussed that the observational detection of the super-homogeneous character of the matter density field, through the observation of galaxy distribution or of the CMBR anisotropies, is still lacking. However the main feature of galaxy two-point correlation function is represented by its power-law character in the strongly non-linear region. We stressed that a clear crossover to homogeneity is also not well established in the data, and thus the transition from the highly clustered phase to the highly uniform (super-homogeneous) one is the main observational tests for theories of the early universe. For galaxies this should be evidenced as a negative correlation function behaving as $-r^{-4}$ at large scales. Future galaxy surveys, like the SDSS project, will provide with samples large enough to clarify this issue.

It is a pleasure to thank A Gabrielli, M. Joyce, L. Pietronero and B. Marcos for useful collaborations and discussions.

References

[1] V. C. Rubin, N. Thonnard, W.K. Ford, ApJ, **238** 471 (1980)

[2] S.M. Faber, J.S. Gallagher, Astron.Astrophys.Ann.Rev., **17**, 135, (1979)

[3] P.J.E. Peebles, *Large Scale Structure of the Universe*, (Princeton University Press, Princeton, New Jersey, 1980)

[4] D.N. Spergel, et al. `astro-ph/0603449`

[5] D.J. Eisenstein, et al., ApJ **633**, 560 (2005)

[6] S. Perlmutter, et al., Astrophys. J., **517**, 565, (1999)

[7] A. Gabrielli, M. Joyce, F. Sylos Labini, Phys. Rev. **D65**, 083523 (2002)

[8] A. Gabrielli, B. Jancovici, M. Joyce, J. Lebowitz, L. Pietronero, F. Sylos Labini Phys.Rev., **D67**, 043406,(2003)

[9] A. Gabrielli, F. Sylos Labini, M. Joyce, L. Pietronero, *Statistical physics for cosmic structures*, (Springer Verlag, Berlin, 2004)

[10] R. Durrer,A. Gabrielli, M. Joyce, F. Sylos Labini, Ap.J. **585**, L1, (2003)

[11] D. York, et al., Astron.J., **120**, 1579 (2000)

[12] V. De Lapparent, M. Geller, J. Huchra, Ap.J. **343**, 1 (1989)

[13] J.R. Gott, J.R. III, et al., Ap.J., **624**, 463 (2005)

[14] D.W. Hogg, et al., Ap.J., **624**, 54 (2005)

[15] L. Pietronero, Physica A, **144**, 257 (1987)

[16] F. Sylos Labini, M. Montuori, L. Pietronero, Phys.Rep., **293**, 66, (1998)

[17] N.L. Vasilyev, Yu.V. Baryshev, F. Sylos Labini, A&A **447**, 431 (2006)

[18] F. Sylos Labini, N.L. Vasilyev, Yu.V. Baryshev, A&A in the press (2006)

[19] C. J. Copi, D. Huterer, and G. D. Starkman Phys. Rev. D **70** 043515 (2004)

[20] C.L. Bennett, et al., Ap.J., **436**, 423, (1994)

[21] T. Padmanabhan, *Structure formation in the universe*, (Cambridge University Press, Cambridge, 1993)

Inclusive Measurements as an mSUGRA Signal with ATLAS

David López Mateos

Supersymmetry (SUSY) is a baryon-fermion symmetry postulated to be underlying the symmetries of the Standard Model of Particle Physics (SM). This symmetry was first proposed in the context of String Theory in 1971[1], and soon after the first SUSY theories in 3+1 dimensions were built by Wess and Zumino[2] and others. Ever since, SUSY searches have been part of the Physics program of every major collider experiment. Theoretically, SUSY is a very attractive idea, since it was proved in [3] to be the only space-time extension to the Poincaré algebra and it is required to exist at some energy scale by String Theories. In addition, SUSY at the TeV scale is especially attractive because it solves the Hierarchy Problem, protects the Higgs mass against radiative corrections and easily provides a candidate for Dark Matter. However, so far, no experiment has been able to discover TeV-scale SUSY. Nevertheless, with the advent of the LHC, a discovery machine designed precisely to scan an energy scale of up to a few TeV, the prospects of either discovering or ruling out TeV SUSY have become very plausible in the vast majority of the existing models.

In this context, it has become compulsory for the LHC experiments to be ready to make the measurements that will lead to the discovery or non-discovery of SUSY. In the following sections we summarize the experimental approach under preparation by the ATLAS collaboration in order to perform the so-called *inclusive* measurements, which will soon after the LHC startup shed light into the questions of the existence and nature of Supersymmetry.

1 Realizations of SUSY: the MSSM and SUGRA

Obviously, in order to give a phenomenological value to the theoretical idea of Supersymmetry, it is necessary to put it in a model which implements the concept and agrees at the same time with all current experimental data. The simplest of these models is the so-called Minimal Supersymmetric extension to the Standard Model (MSSM). In this model, each of the degrees of freedom of the fields in the SM has a supersymmetric degree of freedom counterpart, corresponding to one of the superpartner fields. In addition, the analyticity of the superpotential in the Lagrangian forces the existence of two Higgs isodoublets, thus increasing the number of parameters in the Higgs sector with respect to the most common SM Higgs sector. Calculations of the radiative corrections to the Higgs mass have been performed in the MSSM and for a large region of its parameter space the Hierarchy Problem is solved [4]. However, even in such a minimal extension, there are still 105 parameters in the supersymmetric sector which cannot be all constrained by the currently available experimental data. Therefore, in order to perform systematic studies on the MSSM parameter space, it is necessary to further constrain the model. It is worth noticing at this point that even if the true theory does not correspond to any of the points in the parameter space of the MSSM, the MSSM should be a good first approximation; and some of its experimental signatures are characteristic of many other supersymmetric theories as well. Studies illustrating this can be found in [5] and elsewhere. Similarly, some of these experimental signatures are preserved when further simplifying the MSSM. Such a simplification, thus, does not compromise the validity of the general results, while allowing systematic studies to be carried out. In Section 1.1 we describe one such simplifications used in the studies shown in Section 5 in order to visualize the exclusion power of a given measurement. For now, in order to motivate inclusive measurements in a more general way, it suffices to say that if the Lightest Supersymmetric Partner (LSP) is stable (a hypothesis which would make it a very good candidate for Dark Matter) then, when supersymmetric particles are created at the LHC, they will decay in model-dependent decay chains which will leave as by-products jets, maybe leptons and, most significantly, two undetected LSPs giving rise to missing transverse energy (missing E_T). Since this reasoning is very stable against changes of parameters and is even valid in many models in which the LSP is unstable, we can scan large areas of parameter space by making inclusive measurements, in which we look at all events with these by-products and search for an excess of events with respect to the SM background. In this way, if a discovery is made, the parameter space is greatly constrained, and studies of specific superpartner properties can be performed through guided measurements.

1.1 SUGRA

One of the most daunting questions arising when one accepts supersymmetry as a symmetry of nature is: Why is supersymmetry so strongly broken? The answer is not known, but the impossibility of the MSSM to be broken spontaneously compels the postulation of an explicit breaking at a high scale which is inaccessible to current experiments. Physically motivated assumptions can be made at this scale to reduce the number of free parameters in the model. Renormalization group evolution (RGE) of these parameters to the weak scale maintains the diversity of the SUSY parameters at currently accessible scales. A possible assumption which does not violate current precision measurements is that the explicit symmetry breaking is communicated by gravitational-strength flavor-blind interactions. This is what is known in this context as Supergravity or SUGRA and it introduces new symmetries (and, thus, simplifications) to the MSSM Lagrangian. Further assumptions are made to complete a model usable to perform search studies known as minimal or mSUGRA. These assumptions are made at some scale where gauge couplings unify and which RGE in the MSSM suggests to be at $M_U = 2 \times 10^{16}$ GeV. This simplification leaves a theory with only five parameters (4 continuous and 1 discrete):

- Masses of the scalars at M_U (m_0).
- Masses of the gauge boson superpartners (gauginos) at M_U ($m_{1/2}$).
- Trilinear couplings between the scalars at M_U (A_0).
- The ratio of the expectation values of the two Higgs isodoublets (u and d which give mass to the up- and down-like quarks respectively): $\tan\beta = \frac{v_u}{v_d}$.
- Sign of μ, a bilinear coupling between the two Higgs supermultiplets related to the higgsino mass.

2 The ATLAS Detector

The ATLAS detector is one of the two general-purpose detectors of the LHC. Its electronics and read-out system are designed to withstand the high radiation doses from the high-energy proton-proton collisions and the high rates from the small bunch spacing required to achieve the high luminosity which will allow to push the statistical limits of its Physics program. ATLAS is the largest collider detector designed to date with the muon chambers spanning over 20 meters in diameter and over 40 meters in length. A precision in the measurement of the bending coordinate down to 30 μm allows obtaining a resolution of down to 2% in muon transverse momentum. The bending power for the muon spectrometer measurement is provided by toroidal magnets withstanding currents of up to 21 kAmps. Inside the volume of the muon chambers, the calorimeters provide hermetic coverage for missing E_T measurements. The calorimeter active media are chosen to match the radiation environments of the different regions of the detector: liquid argon (LAr) is used in the regions with a higher radiation exposure due to its radiation hardness, while scintillating tile is used in the outer part of the barrel, where radiation exposure is smaller. Finally, an inner tracker for precise momentum measurement of charged particles is located inside the calorimeter volumes, providing up to over 50 tracking points distributed between a minimum radial distance to the interaction point of 5.5 cm provided by the b-layer of the Pixel Detector up to a maximum radial distance of approximately 120 cm provided by the last straws of the Transition Radiation Tracker. The geometry of the detector as described above is shown in Figure 1. With its over 6,000 tons of material, the

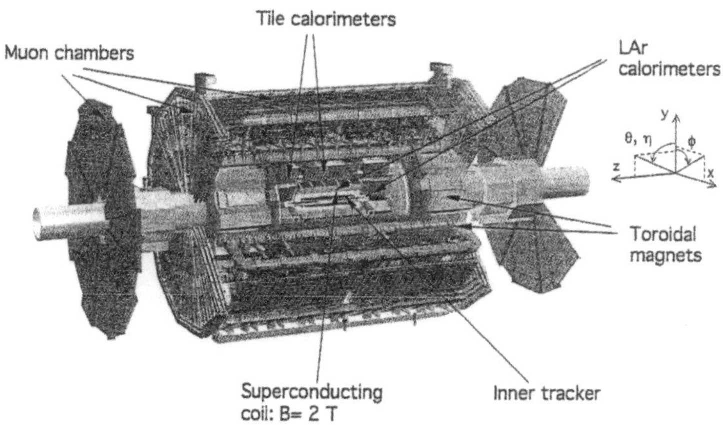

Figure 1: Schematic visualization of the ATLAS detector with its different subdetectors and components. For future reference, the detector coordinate system is also shown, where the pseudorapidity is defined as $\eta = -\ln\left(\tan\frac{1}{2}\theta\right)$.

ATLAS detector remains an extremely complex system; and its understanding will be one of the most challenging tasks of any analysis done on its data. In the particular case of inclusive measurements, all subdetectors should be well understood. However, since the main discriminator between signal and background events is the missing E_T, the energy imbalances in the detector and, in particular, in the calorimeter need to be extremely well understood. The set of all the activities required to understand the energy measurements in the calorimeters are referred to as the *determination of the energy scale*, and to them we devote the next section.

3 Determination of the Energy Scale

In order to understand how particles deposit their energy in the calorimeters, the ATLAS collaboration has carried out and will continue to carry out several efforts. At this stage in the detector construction, these efforts comprise from the use of test beam data to the combination of information from the muon chambers and the calorimeters during the cosmic runs with the magnetic field on. However, the final resolution will only be achieved by a combination of the information coming from all the subdetectors with p-p collision data and the use of known processes. Regardless, the missing-E_T measurement is limited by the resolution achievable in the determination of the energy scale of hadronic jets, which is itself limited by theoretical knowledge about the fragmentation and hadronization of the original partons. The efforts from the experimental side must then focus on pushing the experimental errors below the theoretical errors. This is done through a deep understanding of the detector effects contributing to the determination of the energy scale, which is being addressed by activities in progress, and a comprehensive on-line calibration program whose design has already been studied and continues to be studied within the collaboration.

When we refer to detector effects, we refer to the detector limitations, inherent to the detector design. These detector limitations have been studied in the test beam with real data and in simulations. The result is a parameterization of the error in the energy measurement on each cell of the calorimeter which includes, as it is common for this kind of detector, a noise, a sampling and a constant term. These terms are purely controlled by detector design and particle type. Figure 2 shows these effects on the jet energy resolution at some fixed η. Ultimately, we are interested in the effect of this jet-energy resolution on the missing-E_T resolution. Studies addressing this question[6] have shown that the missing-E_T resolution can be parameterized by $\sigma\left(E_T^{miss}\right) = 0.46 \times \sqrt{E_T^{depo}}$, all variables in GeV. With a typical transverse energy deposition of $\sim$ 8000 GeV and missing E_T of 1000 GeV in SUSY events, we see that the resolution becomes $\sim 10\%$ for typical events.

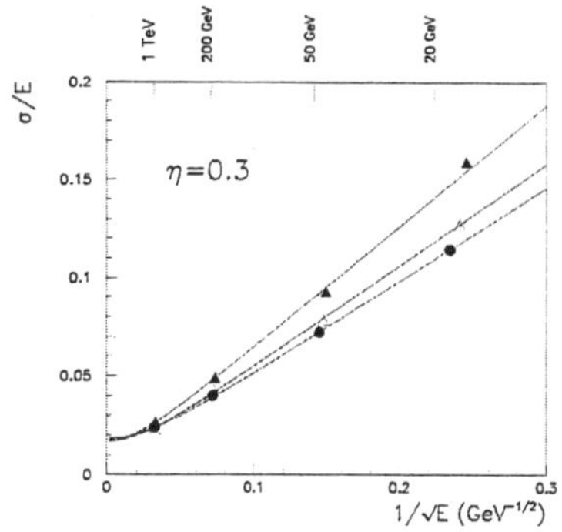

Figure 2: Plot showing the jet energy resolution of ATLAS for different cone sizes in $\eta \times \phi$ space (solid circles: $\Delta R = 1.5$, open triangles: $\Delta R = 0.7$ and black triangles: $\Delta R = 0.4$) at fixed η[6]. Studies on these and other specific detector effects can be found in [6].

However, to obtain these ultimate resolution tables it is necessary to perform calibrations so that voltages and currents coming from the detector front-end electronics can be appropriately transformed into energy measurements. These calibrations have already been performed in the test beam, but obviously need to be redone online with the first data samples when the whole detector is fully assembled and running. Since real data is used for these calibrations, further reduction of calibration errors is achieved with increasing amount of data collected. After 10 fb^{-1} of data, statistics allow reducing the error in the jet-energy measurements due to calibration to 1%[6], and the error becomes dominated by the detector effects.

4 Understanding the Background

As in all measurements looking at the tails of distributions, in performing the inclusive searches, it is crucial to understand whether there are other SM events that can create a similar signal and what their cross-sections might be. Beyond the inaccuracy in the determination of the energy scale discussed in the previous section, there is only one type of SM particle which can create missing E_T: the neutrino (ν). Therefore, the backgrounds that need to be under control are those coming from processes which can in principle produce high-energy νs, that is, Z/W events and QCD events. The efforts to understand these backgrounds are many in the ATLAS collaboration and cannot be all surveyed in this review. Here we just briefly outline the challenges of eliminating the $t\bar{t}$ background, which is the one with the largest theoretical uncertainty and from which we would expect the neutrinos with the highest energy. An illustration of such uncertainty is shown in Figure 3 where the use of different shower algorithms in the initial state shower yields very significant differences in the cross-sections for associated jet production. Since large missing E_T can come from the top decay and the jet hardness is the other most important discriminating variable in inclusive measurements, the large uncertainty shown in this figure could easily cause an underestimation of background and the identification of background events as signal events in the ulterior analysis. This type of argument has rendered the use of data,

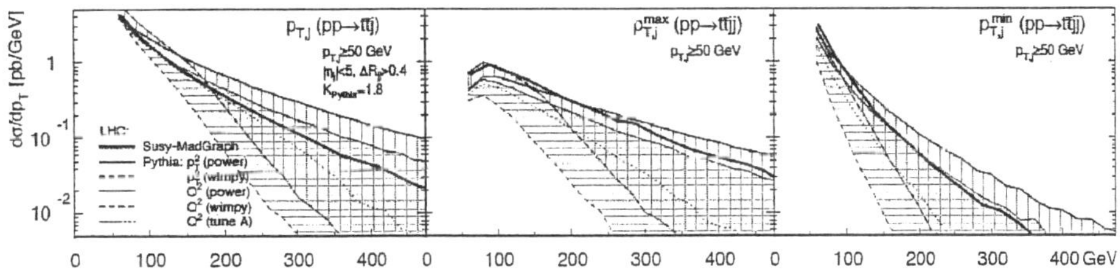

Figure 3: This plot illustrates the theoretical uncertainty on the cross-section for associated top-jet production processes at the LHC by showing the cross-section obtained by different Monte Carlos and different methods for calculating the initial state shower. Further discussion of these studies can be found at the source of this figure, at [7].

and not Monte Carlo, compulsory in order to correctly perform the calculation of $t\bar{t}$ background to SUSY inclusive events. Using data might be also desirable for the other backgrounds, however, the error derived from the use of Monte Carlo in those cases is much smaller than in the case of the top background as the figure hints.

At this point in time, the collaboration has identified the $t\bar{t} \to b\bar{b}l\nu q\bar{q}$ and $t\bar{t} \to l\nu l\nu q\bar{q}$ samples as some of the most important control samples for understanding the top background, since they may both fake missing E_T signals when one lepton is misidentified. As with all other data-driven calibrations, the statistical features of this background estimation will affect the error on the final SUSY measurement. However, there are indications[8] that the data obtained on these control samples after less than 1 fb^{-1} integrated luminosity will be enough to match the errors coming from detector-calibration effects. Similar errors are achievable for the other backgrounds, whose contribution is, however, significantly smaller than the $t\bar{t}$ background. This gives us confidence that the backgrounds can be under control to perform the inclusive measurement program described in more detail on the next section.

5 Discovery Reach of ATLAS in mSUGRA

Ever since the potential of inclusive measurements to discover SUSY at the LHC was noticed[9] the LHC experiments have tried to improve the studies on these so-called "golden-channels", using updated information from detector studies on the energy scale and theoretical developments on the understanding of the background cross-sections. In what follows, we go through the experimental procedure that will lead to performing these inclusive measurements with the help of the latest studies from the ATLAS collaboration on the discovery reach of ATLAS for an mSUGRA signal[5].

If after a systematic search for the optimal set of cuts and after acquiring a good understanding of the background, no significant signal is found over the background in events triggering on missing E_T, then, exclusion limits can be obtained by performing analyses guided by specific models like mSUGRA. This was done in the studies in [5] and some of the parameter space investigated there is shown in Figure 4. For

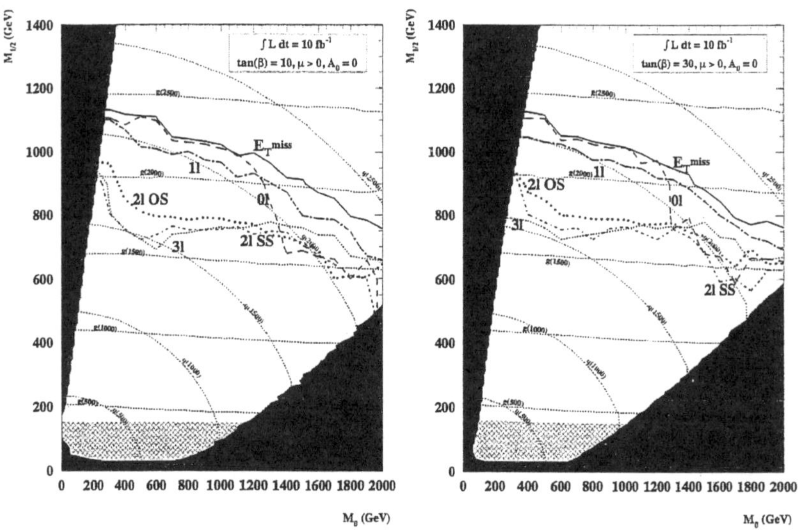

Figure 4: Discovery reach in mSUGRA parameter space with a luminosity of 10 fb^{-1}. The trilinear scalar coupling A_0 is fixed to 0, and the sign of μ is taken positive, as favored by $b \to s\gamma$ measurements [4]. The hatched area is the area already ruled out by LEP or the Tevatron. The solid area is not allowed, because the theory would either give a charged LSP (not allowed by cosmological constraints) or does not match electroweak data. The doted lines show the masses of the gluinos and the lightest squark in different points in the parameter space. The lines labeled 0l, 1l, 2l OS, 2l SS, 3l and E_T^{miss} correspond to the exclusion regions derived from the lepton-veto, 1-lepton, 2-lepton-opposite-sign, 2-lepton-same-sign, 3-lepton and combined channels, respectively.

these analyses, the backgrounds considered were W+jet, Z+jet and QCD events ($t\bar{t}$ and lighter-quark production). Then, the mSUGRA parameter space was scanned with different data samples and cuts on up to 10 significant variables (among them, obviously, missing E_T and jet p_T) were selected optimally to reject background. Significance was quantified by $S_f = S/\sqrt{B}$ with $S \geq 10$. The exclusion regions in Figure 4 correspond to regions where $S_f > 5$. It can be seen from these plots that the discovery reach of the ATLAS detector after running for a year at initial luminosity expands greatly the current limits on the masses of supersymmetric particles in mSUGRA. While 10 fb^{-1} might be an overestimation of the amount of data that will be delivered in the first Physics run of the LHC in 2008, the discovery reach is still very significant with respect to the limits from current data even with 0.1 fb^{-1} (worth less than a week of run in the low luminosity regime of the LHC) as Figure 5 illustrates. In order to see how excluding this area would affect naturalness arguments on

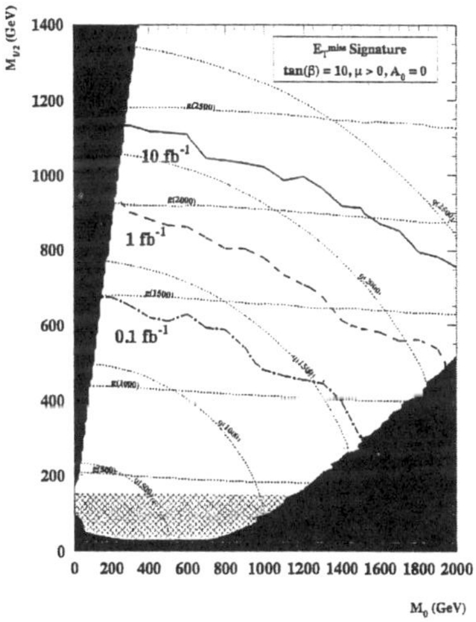

Figure 5: Variation of the exclusion limits for the combined inclusive measurement for $\tan\beta = 10$ as a function of integrated luminosity. For a description of other features please refer to the caption of Figure 4.

the Higgs mass we refer the reader to [4]. For the purposes of this review, it suffices to say that after very few fb^{-1} of data, if SUSY is not discovered, its most favored theoretical realizations will no longer succeed in solving the Hierarchy Problem.

However, if these same set of cuts give a significant signal over the SM background, then, these same inclusive measurements will provide essential information about the validity of mSUGRA or other SUSY models and will significantly constrain the allowed regions of parameter space in these theories. More details on how this might happen can be found in [5].

6 Conclusions

It is impossible to predict what Physics awaits to be discovered as we embark ourselves in unveiling the secrets of the TeV scale. However, it is clear that SUSY will be one of the great discoveries or disappointments that the Particle Physics community will face during the beginning of this century. It is our duty as experimentalists to make sure that the measurements leading to these events are carried out rigorously; and thousands of scientists at CERN and all over the world will continue to spend their efforts until this is made a reality. Whatever the discoveries at the LHC are, it is worthwhile remembering that SUSY is well justified theoretically, and it can solve many of the current problems of Particle Physics if realized in some form. But most importantly, if inclusive measurements indicate that SUSY exists, and this is confirmed by other measurements, a new era of Particle Physics will start, where many new measurements will be needed and new fundamental questions will arise.

References

[1] A. Neveu and J.H. Schwarz, Nucl. Phys. **B31** (1971) 86.

[2] J. Wess and B. Zumino, Nucl. Phys. B70 (1974) 39.

[3] R. Haag, J. Łopusznski and M. F. Sohnius, Nucl. Phys. **B188** (1975) 61.

[4] J. Ellis, K. A. Olive and Y. Santoso, New Jour. Phys. **4** (2002) 32.

[5] D. R. Tovey, Eur. Phys. Jour. Direct. **4** (2002) 1.

[6] ATLAS Detector and Physics Performance TDR. Volume I. CERN/LHCC 99-14.

[7] T. Plehn, D. Rainwater and P. Skands, hep-ph/0510144.

[8] S. Asai, Talk delivered at ATLAS Physics and Trigger Week, 1st Nov. 2006.

[9] H. Baer, C. Chen, F. Paige and X. Tata, Phys. Rev. **D52** (1995) 5.

Unraveling the f_0 nature by connecting *KLOE* and *BABAR* data through analyticity

S. Pacetti

Laboratori Nazionali di Frascati dell'INFN, Frascati, Italy

Abstract

We define a general procedure, based on analyticity and dispersion relations, to estimate low energy amplitudes for processes like: $\phi \to e^+e^-M$ and $\phi \to \gamma M$, starting from cross section data on $e^+e^- \to \phi M$, where M is a generic light scalar or pseudoscalar meson. In particular this procedure is constructed to obtain predictions on the radiative decay rate which are crucially linked on the assumed quark structure for the meson M under consideration.
Three cases are analyzed: $M = \eta$, $M = f_0(q\overline{q})$ and $M = f_0(qq\overline{q}\overline{q})$. While in the η case the estimate of the branching fraction for the radiative decay $\phi \to \eta\gamma$ is in agreement with the data, in the case of f_0, such agreement is obtained only under the hypothesis of a tetraquark scalar meson.

1 Introduction

1.1 The known properties of the $f_0(980)$ scalar meson

The $f_0(980)$ is the more extensively studied isospin zero scalar meson [1]. The properties of such a meson are very peculiar and, in some cases, in spite of the deep experimental investigation, they are still poorly known. In fact, while the mass is rather well measured $M_{f_0} = 980 \pm 10\ MeV$ [1], the width is known with a large uncertainty: $\Gamma_{f_0} = 40 \div 100\ MeV$ [1]. This is a consequence of the scalar nature of this meson, which does not allow unambiguous interpretations when the resonance is studied by means of the usual Breit-Wigner analysis. The f_0 has the quantum number of the vacuum, the Higgs boson and also of the expected lowest-lying glueball [2].
Usually, this scalar meson is considered as a "bridge" between light (u and d) and strange quarks because it can be produced by both strange and non-strange particles.

1.2 Long outstanding questions

By following the standard quark model [1], where mesons are described as colorless $q\overline{q}$ bound states and classified in J^{PC} multiplets, we expect there should be the conventional flavor $SU(3)$ scalar nonet (0^{++}):

$$q^3 \otimes \overline{q}^{\overline{3}} = [\overline{q}q]^1 \oplus [\overline{q}q]^8. \tag{1}$$

In such a scenario the f_0 represents the singlet and, due to the large $s\overline{s}$ component, it should be the heaviest meson of this multiplet, in contradiction with the experimental data which provide: $M_{f_0} \simeq M_{a_0}$ and no degeneracy between a_0 and σ_0 ($M_{a_0} > M_{\sigma_0} \sim 450\ MeV$). Also the absolute values of the masses are not in agreement with the expectation, in fact, being $q\overline{q}$ bound in P-wave, these masses should lie in the $> 1\ GeV$ energy region of other P-wave states. This means that the $q\overline{q}$ bound state is only a small component of the f_0 Fock space.
Since some years, the interpretations in terms of tetraquark states (in different combinations) [4–10] is object of deep theoretical and experimental investigation and also in this work we will focus on this possibility.

2 Probing the intrinsic nature of f_0

2.1 Common practice

At present, the main source of information on the f_0 structure is the radiative decay $\phi \to f_0\gamma$ [10], whose surprisingly high branching fraction is interpreted in terms of an abnormally strong affinity between the ϕ

vector meson and the f_0 itself. Such an affinity has been described in different ways by different authors, even though, each of them put the cue on the key role played by the coupling $g^{f_0}_{KK}$. In particular, there are two schools of thought, the one describes this coupling by introducing in the $\phi \to f_0\gamma$ amplitude a $K\overline{K}$-loop [4], while the second, in order to avoid any model dependent description, adopts a usual parameterization in terms of meson propagators with free couplings [7]. In both these descriptions, with different languages, the message is the same: the coupling $(g^{f_0}_{KK})^2$ is more than twice the $(g^{f_0}_{\pi\pi})^2$ [11].

2.2 The strategy

The idea of this work consists in probing the intrinsic nature of the f_0 by considering different processes to investigate the transition form factor (tff) which describes the coupling $\phi f_0 \gamma$. All the information on the tff are then used under the constraint of an assumed f_0 quark structure, which strongly affects every step of the procedure.

3 First step: parameterization in the resonance region $[3M_\pi, \sim 3\ GeV]$

The amplitude of $e^+e^- \to \phi M$ (fig. 1) is parameterized in terms of a tff, which describes the coupling between the virtual photon and the intermediate vector mesons V_j, i.e.:

$$\mathcal{M}_{\text{Annihi}} = \sum_j^{N_V} \frac{g^{V_j}_{\phi M}}{M_j^2 - q^2 - i\Gamma_j M_j} \frac{M_j^2}{F_{V_j}} \epsilon_\nu \, \mathcal{T}^{\mu\nu}_M \, \frac{1}{q^2} J_\mu, \tag{2}$$

where $J_\mu = -ie\overline{v}(p_+)\gamma_\mu u(p_-)$, ϵ_ν is the ϕ polarization vector and, F_{V_j} and $g^{V_j}_{\phi M}$ are couplings with the initial and the final state. The second-rank tensors $\mathcal{T}^{\mu\nu}_M$ accounts for the dynamics of the meson M.

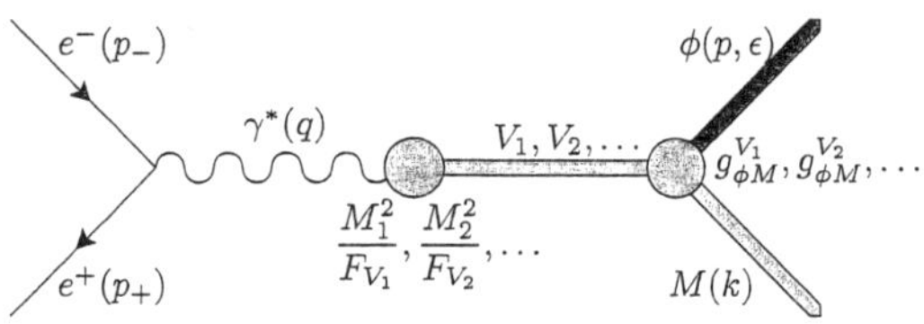

Figure 1: Feynman diagram for the $e^+e^- \to \phi M$ in Born approximation, with intermediate vector mesons V_j.

The formal expression of the ϕM tff, in the resonance region, as extracted from the amplitude of eq. (2), is:

$$F^{\text{Res}}_{\phi M}(q^2) = \sum_j^{N_V} \frac{M_j \, g^{V_j}_{\phi M}}{eF_{V_j} \, \Gamma_j} \frac{\Gamma_j M_j}{M_j^2 - q^2 - i\Gamma_j M_j}. \tag{3}$$

This parameterization integrates all the information about the meson M structure[1]. In case of radiative decay, by using the previous arguments, the amplitude is:

$$\mathcal{M}_{\text{Rad}} = F_{\phi M}(0) \, \mathcal{T}^{\mu\nu}_M \, \epsilon_\mu(p)\epsilon^*_\nu(q), \tag{4}$$

it is constant and proportional to the value of the tff at $q^2 = 0$. To give the explicit expressions of the second-rank tensors $\mathcal{T}^{\mu\nu}_M$ we have to specify the nature of the mesons under consideration. We will study two cases:

$$\mathcal{T}^{\mu\nu}_\eta = \varepsilon^{\mu\nu\rho\sigma} p_\rho q_\sigma, \qquad \mathcal{T}^{\mu\nu}_{f_0} = p^\mu q^\nu - g^{\mu\nu}(pq), \tag{5}$$

[1] E.g., for $M \equiv f_0$, a strong affinity of the f_0 with the $K\overline{K}$ intermediate state (kaon loop) should manifest itself in an enhancement of the coupling $g^{\phi}_{\phi f_0}$, since in the s-channel the $K\overline{K}$ state is almost completely resonant in $\phi(1020)$.

where p and q are 4-momenta as labeled in fig. 1.
The formal expressions for total annihilation cross sections and decay rates are [eqs. (2,5)]:

$$\begin{array}{ll} \sigma_{\phi\eta}(s) = \rho^{\rm kin}_{\phi\eta}(s)|F_{\phi\eta}(s)|^2 & \Gamma_{\phi\eta} = \tau^{\rm kin}_{\phi\eta} F_{\phi\eta}(0)^2 \\ \sigma_{\phi f_0}(s) = \rho^{\rm kin}_{\phi f_0}(s)|F_{\phi f_0}(s)|^2 & \Gamma_{\phi f_0} = \tau^{\rm kin}_{\phi f_0} F_{\phi f_0}(0)^2, \end{array} \tag{6}$$

where $s = q^2$. These quantities are products between kinematic factors ($\rho^{\rm kin}(s)$ and $\tau^{\rm kin}$), which, depending on the nature of the meson M, are different in the two cases ($M \equiv \eta$ and $M \equiv f_0$), and the modulus squared of the corresponding tff's.

3.1 Step two: selection of the contributions

In this so-called resonance region, i.e., from the theoretical threshold, which is $s_0 = (3M_\pi)^2$ since the isoscalar final state, up to $\sim (3\ GeV)^2$, we adopt the parameterization of eq. (3). The first criterion used to select the vector mesons which contribute to the tff's is the quantum number conservation. For both $\phi\eta$ and ϕf_0 final state we have: $I^G(J^{PC}) = 0^-(1^{--})$, hence only contributions from the ω- and ϕ-family are expected. However, if we consider explicitly the structure of the mesons in terms of valence quarks, assuming the f_0 as $q\overline{q}$ bound state, we have.

$$|\phi\rangle = s\overline{s} \qquad \begin{array}{l} |\eta\rangle = X_\eta|u,d;-\rangle + Y_\eta|s;-\rangle \\ |f_0\rangle = X_{f_0}|u,d;+\rangle + Y_{f_0}|s;+\rangle, \end{array} \tag{7}$$

with the normalization: $X^2_{\eta,f_0} + Y^2_{\eta,f_0} = 1$ and where the state $|q_1, q_2, \ldots; P\rangle$ indicates a normalized combination of $q_1\overline{q}_1$, $q_2\overline{q}_2$, etc., with parity P and $J^C = 0^+$.

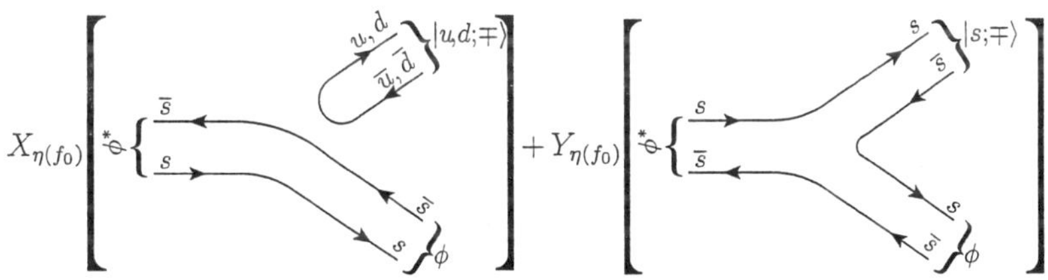

Figure 2: Feynman diagram for $\phi^* \to \phi\eta(f_0)$ with: $|\phi\rangle = s\overline{s}$ and $|\eta(f_0)\rangle = X_{\eta,f_0}|u,d;\mp\rangle + Y_{\eta,f_0}|s;\mp\rangle$. The two components $|u,d;\mp\rangle$ and $|s;\mp\rangle$ are separately shown.

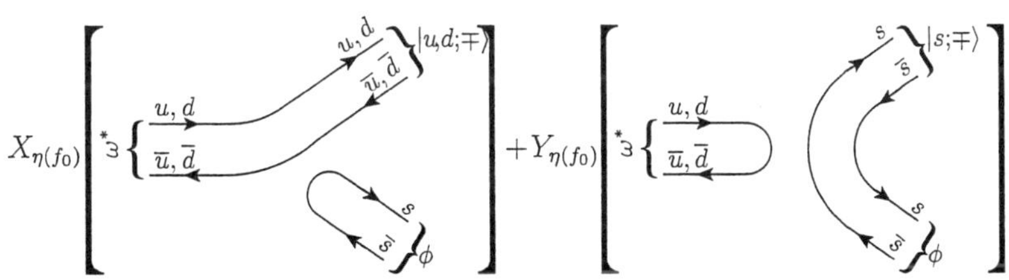

Figure 3: Feynman diagram for $\omega^* \to \phi\eta(f_0)$ with: $|\phi\rangle = s\overline{s}$ and $|\eta(f_0)\rangle = X_{\eta,f_0}|u,d;\mp\rangle + Y_{\eta,f_0}|s;\mp\rangle$. The two components $|u,d;\mp\rangle$ and $|s;\mp\rangle$ are separately shown.

While the ϕ-family contributions are allowed (fig. 2), the ω-family contributions are instead OZI-forbidden [12]. In fact, as it is shown in fig. 3, the corresponding Feynman diagrams have disconnected flavor lines in the vector sector.
It follows that, in light of the quantum number conservation and the OZI rule, only ϕ-family contributions are expected, not only for the $\phi\eta$, but also for ϕf_0 tff under the hypothesis of a $q\overline{q}$ f_0. In addition, the

uncertainty due to the unknown $\omega-\phi$ mixing, allows only to reduce the $\phi\eta\gamma$ coupling with respect to $\phi f_0\gamma$. In fact, by putting $Y_\eta = 0$, it should remain the only $|u,d;-\rangle$ component, which, being OZI forbidden (see first diagram in fig. 2), should give $g^{\phi}_{\eta\gamma}=0$. Contrary, by putting $Y_{f_0}=0$, the remaining component $|u,d;+\rangle$ is a combination of u and d quarks with scalar quantum numbers ($J^{PC}=0^{++}$) and then it is not OZI suppressed [13].

3.2 The asymptotic region

Another crucial point of this procedure is the knowledge of the time-like asymptotic behaviour of a hadronic form factor. To this purpose, the nominal power law, as predicted by the perturbative QCD (pQCD) [14], is used. The power to be considered linearly depends on the number of elementary constituents, i.e. valence quarks, even though further sub-structures are present (e.g.: deuteron [15]).
In addition, in order to describe the asymptotic behaviour of the tff's under consideration, we have also to account for the flip of the hadronic helicity[3] [16], which passes from zero, in the leptonic initial state, to one in the $\phi\eta(f_0)$ final state ($\lambda_{\eta(f_0)}=0$, $|\lambda_\phi|=1$). This implies an additional suppression, and then the asymptotic behaviour to be considered, starting from a certain energy $s_{\rm asy}^{\eta(f_0)}$, is [16,17]:

$$F^{\rm asy}_{\phi\eta(f_0)}(s) \propto \left(\frac{1}{s}\right)^{n_H+\frac{n_\lambda+l_q-1}{2}=2\left(\frac{5}{2}\right)} \qquad \text{as: } s > s_{\rm asy}^{\eta(f_0)}, \tag{8}$$

where: $n_\lambda = |\lambda_{\eta(f_0)}+\lambda_\phi| = 1$, $n_H=2$, $l_q=0(1)$. The power is a function of the number of hadronic fields n_H, the modulus of the total helicity n_λ and the relative angular momentum of the quarks which constitute the meson η ($l_q=0$) or f_0 ($l_q=1$). The parameterization of eq. (8) provides different asymptotic powers for $\phi\eta$ and $\phi f_0(q\bar{q})$ tff.

3.3 The region below the theoretical threshold s_0

Unitarity assures that a generic hadronic tff $F_H(s)$ is an analytic function in the s-complex plane with a cut, along the real axis, starting from the theoretical threshold s_0 up to infinity. This property, together with a power-law vanishing asymptotic behaviour allows to use spectral representations for the tff's as, for instance, the dispersion relation (DR) for the imaginary part [18]. However, to use directly the experimental data and the asymptotic behaviour [eq. (8)] in the time-like region, we use the DR for the logarithm [19]:

$$\ln[F_H(t)] = \frac{\sqrt{s_0-t}}{\pi}\int_{s_0}^{\infty}\frac{\ln|F_H(s)|}{(s-t)\sqrt{s-s_0}}ds, \tag{10}$$

which relates the real values of the tff below the theoretical threshold ($t<s_0$) to its modulus over the cut. The DR for the logarithm can be used in the form of eq. (10) only if $F_H(s)$ has neither zeros nor poles on the physical sheet. The parameterization used in the resonance region [eq. (3)] guarantees the absence of singularities for the tff's in the physical sheet.
Since we know the modulus of the tff $F_{\phi\eta(f_0)}(s)$ over the cut [eqs. (3,8)], the DR for the logarithm allows to perform an analytic continuation of the parameterization below the theoretical threshold and it gives:

$$\begin{aligned} F^{\rm an}_{\phi\eta(f_0)}(t) &= \exp\left[\frac{\sqrt{s_0-t}}{\pi}\left(\int_{s_0}^{s_{\rm asy}^{\eta(f_0)}}\frac{\ln|F^{\rm Res}_{\phi\eta(f_0)}(s)|}{(s-t)\sqrt{s-s_0}}ds\, + \right.\right. \\ &\left.\left. + \int_{s_{\rm asy}^{\eta(f_0)}}^{\infty}\frac{\ln|F^{\rm asy}_{\phi\eta(f_0)}(s)|}{(s-t)\sqrt{s-s_0}}ds\right)\right], \qquad t<s_0. \end{aligned} \tag{11}$$

[3] The virtual photon of the annihilation $e^+e^- \to \gamma^* \to \phi M$, at high energies, always has spin ±1 along the beam axis. Angular momentum conservation implies that the total angular momentum is 1. In the center of mass frame: $\vec{p}_\phi = -\vec{p}_M \equiv \vec{p}$ it follows that

$$\left|\frac{\vec{p}_\phi\cdot\vec{s}_\phi}{|\vec{p}_\phi|} - \frac{\vec{p}_M\cdot\vec{s}_M}{|\vec{p}_M|}\right| = |\lambda_\phi-\lambda_M| = \left|\frac{\vec{p}\cdot\vec{s}_{\rm tot}}{|\vec{p}|}\right| = 0,1, \tag{9}$$

where λ_ϕ and λ_M are the helicities of the final mesons. But, since hadronic helicity conservation requires: $\lambda_\phi+\lambda_M=0$, or $\lambda_\phi-\lambda_M=2\lambda_\phi=-2\lambda_M$, eq. (9) holds, in case of mesons, only if $|\lambda_\phi|=|\lambda_M|=0$. In the cases under consideration, the coupling $\gamma^*\phi M$ ($M=\eta, f_0$) is described in terms of only one tff $F_{\phi M}(s)$ and this requires $|\lambda_\phi|=1$. Helicity zero would mean that the ϕ spin lies in the plane orthogonal to its 3-momentum, in this case we would need an additional degree of freedom (for instance: the azimuthal angle) and then an additional tff to describe this process. But since the tff is only one it follows that the ϕ spin lies in 3-momentum direction, i.e. $|\lambda_\phi|=1$.

3.4 The overall parameterization

We are now ready to write down the parameterization which covers the whole s-real axis (in principle the whole s-complex plane), it is a three-fold definition:

$$F_{\phi\eta(f_0)}(s) = \begin{cases} F^{\rm an}_{\phi\eta(f_0)}(s) & s \le s_0 & \text{[eq. (11)]} \\ F^{\rm Res(\phi)}_{\phi\eta(f_0)}(s) & s_0 < s \le s^{\eta(f_0)}_{\rm asy} & \text{[eq. (3)]} \\ F^{\rm asy}_{\phi\eta(f_0)}(s) & s > s^{\eta(f_0)}_{\rm asy} & \text{[eq. (8)]}. \end{cases} \qquad (12)$$

The real axis has been divided in three intervals to parameterize the tff. From higher energies: the asymptotic region, where the pQCD power law is used with power 2 for the η and 5/2 for the $f_0(q\overline{q})$; the resonance region, where the tff is written in terms of vector propagators and coupling constants; the analytic region, where the tff is reconstructed by means of a DR.

4 The $\phi\eta$ transition form factor

In fig. 4 the reconstructed $\phi\eta$ tff is shown. This result has been obtained by considering only ϕ-family contributions in the resonance region, namely the $\phi(1020) \equiv \phi$ and $\phi(1680) \equiv \phi'$, and a power law asymptotic behaviour $\propto (1/s)^2$. Higher mass resonances are not included because their effect in the low-energy region under consideration is negligible.
By using only the *BABAR* data (red points in fig. 4) and the analyticity to fix the free parameters of the procedure, the function defined in eq. (12) reproduces the low-energy behaviour of the tff in good agreement with the data. The obtained parameters are shown in tab. 1.

Res	$g^V_{\phi\eta}(GeV^{-1})$	$M_V(MeV)$	$\Gamma_V(MeV)$
ϕ	13 ± 1.4	1019.4	4.3
ϕ'	56 ± 6	1622 ± 15	203 ± 12

Table 1: Parameters obtained for the $\phi\eta$ tff. The values without error are fixed.

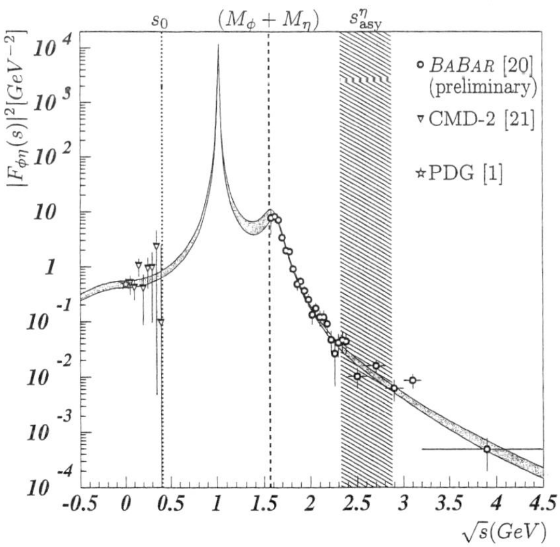

Figure 4: Reconstructed $|F_{\phi\eta}(s)|^2$ tff. The grey band represents the error. The circles are the preliminary *BABAR* data used as input, while the triangles (CMD-2: $\phi \to \eta e^+e^-$) and the star (PDG: $\phi \to \eta\gamma$) are not included in the computation. The dotted and the dashed vertical lines represent the theoretical and the physical thresholds, and the lined region in the obtained interval for $s^\eta_{\rm asy}$.

By using eq. (6) we get: $BR(\phi \to \eta\gamma) = (1.3 \pm 0.2)\%$, to be compared with [1]: $BR_{\rm PDG}(\phi \to \eta\gamma) = (1.301 \pm 0.024)\%$ (see fig. 4). This good agreement provides an important positive check for two key points of the procedure i.e., the description in terms of resonances in the low-energy region and the power-law asymptotic behaviour.

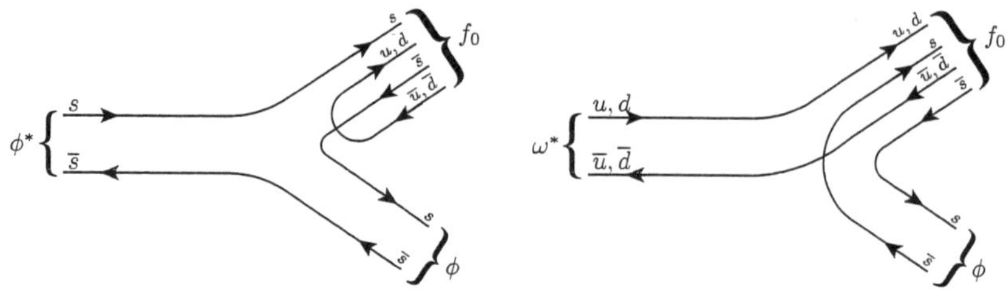

Figure 6: Feynman diagrams for the ϕ^* and ω^* contribution to the ϕf_0 final state with a tetraquark f_0.

6.1 Results for $f_0(qq\overline{qq})$

In fig. 7 is shown the ϕf_0 tff obtained under the hypothesis of a tetraquark f_0. In this scenario our prediction for the tff, which gives: $BR(\phi \to f_0\gamma) = (3.9 \pm 0.9) \cdot 10^{-4}$ is in good agreement with the PDG value (mainly due to *KLOE* [23]), which is: $BR_{\text{PDG}}(\phi \to f_0\gamma) = (4.40 \pm 0.21) \cdot 10^{-4}$.
This drastic change in the low-energy result, with respect to the previous case without the ω, is the consequence of a combined effect of the two modifications; the faster vanishing asymptotic behaviour enhances the resonance region and the additional ω contribution moves, down in energy, its mean value. The sum of these two effects is just a large enhancement of the tff at $s = 0$.

Res	$g^V_{\phi f_0}(GeV^{-1})$	$M_V(MeV)$	$\Gamma_V(MeV)$
ω	112 ± 13	783.7	8.5
ϕ	52 ± 7	1019.4	4.3
ϕ'	12 ± 7	1622	203
ϕ''	$F_{\phi''}(1.9 \pm 0.4) \cdot 10^{-3}/M_{\phi''}$	2176	51

Table 3: Parameters obtained in the case of the $\phi f_0(qq\overline{qq})$ tff. The values without error are fixed.

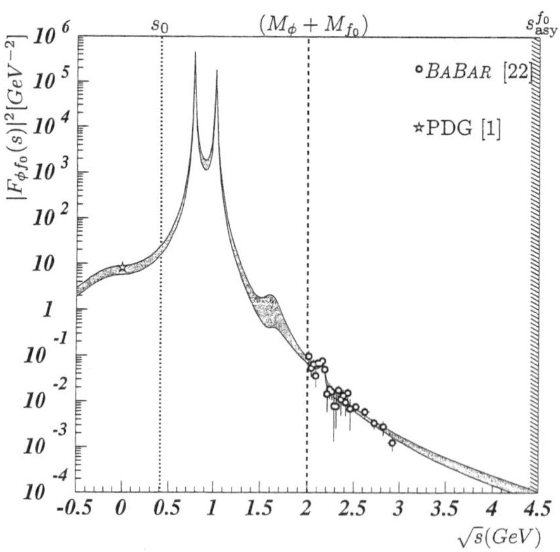

Figure 7: Reconstructed $\phi f_0(qq\overline{qq})$ tff. The grey band represents the error. The circles, on the right, are the *BABAR* data used as input, while the star (PDG: $\phi \to f_0\gamma$) is not included in the computation. The dotted and the dashed vertical lines represent the theoretical and the physical thresholds, and the lined region in the obtained interval for $s^{f_0}_{\text{asy}}$.

7 Duality

The pQCD nominal power law of eq. (8) behaves like a counter of valence quarks involved in a certain process. Nevertheless, there is an objective difficulty in exploiting experimentally this power counting. In fact, the high energy values of the tff's, due to their power law vanishing, are by definition very small quantities. In the low-energy region, where the tff's attain their higher values and then the experimental observation should be more feasible, the power law behaviour is covered by the resonances.
However, following the so-called quark-hadron duality [24], the asymptotic behaviour may be "restored" also at low energy by averaging the tff's and then by canceling the local effects of the resonances.
Fig. 8 shows the asymptotic powers of the $\phi\eta$, ϕf_0 tff's and also of the pion form factor $F_{\pi\pi}(s)$, which is used as a check of the averaging procedure. The results are in agreement with the expectations and they confirm the further suppression due to the helicity flip in case of $F_{\phi\eta}(s)$ (first experimental observation) and the additional hadronic filed in case of $F_{\phi f_0}(s)$.

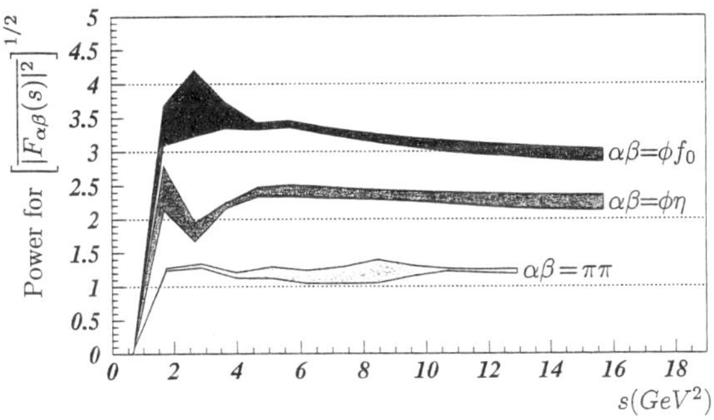

Figure 8: Asymptotic powers for the mean values of $\pi\pi$ form factor (light grey band) and, the $\phi\eta$ (medium grey band) and ϕf_0 (dark grey band) tff's.

8 Conclusions

We have defined a dispersive technique to construct, for a generic ϕM tff, an analytic parameterization which is defined in the whole s-complex plane (M is a pseudoscalar or scalar light meson). The main ingredients of this technique are: the pQCD counting rule [14] (and helicity rule [16]) to describe the asymptotic behaviour; data on the annihilation cross section $\sigma(\phi M)$; Breit-Wigner parameterization in the resonance region and DR's for the logarithm.
In the case of the $\phi\eta$ tff, by considering only the ϕ-family contributions, assuming the asymptotic behaviour ($1/s^2$), and using the *BABAR* data on $\sigma(\phi\eta)$, we achieved a parameterization $F_{\phi\eta}(s)$ (fig. 4), which is in good agreement with the low energy data, not used in the analysis.
In the case of the ϕf_0 tff, the same procedure, i.e., an analysis under the hypothesis of a $q\bar{q}$ f_0 with the opportune ($1/s^{5/2}$) asymptotic behaviour, gives a prediction in huge disagreement with the low-energy data (fig. 5).
We modified the procedure in order to account for a tetraquark f_0. The modification concerns two steps: a new contribution coming from the ω-family is included; the asymptotic behaviour becomes ($1/s^3$) because of the additional quarks in the final state and the disappearing of the angular momentum attenuation. These two upgrades, if individually considered, give opposite effects: the first reduces, while the second enhances the low-energy values of the tff. However, the combined effect gives an important enhancement of the tff, which reconciles the prediction with the data (fig. 7).
In addition, we have performed an analysis in terms of quark-hadron duality, which, by restoring the power-law behaviour at low energy, allows to extract the asymptotic power from the data and the parameters of the resonances. The results are in good agreement with a tetraquark f_0 and with the helicity rule, which is for the first time experimentally confirmed.
Since, by means of this procedure we have tested two extreme hypotheses concerning the f_0 structure, i.e. the two- and the four-quark state, our conclusion is that: the available data, the analyticity requirement

and the pQCD asymptotic behaviour prove that the tetraquark component in the f_0 is the dominant one. A still in progress analysis, where both the components are included, gives for the $q\bar{q}$ content an upper limit of 12%. It follows that the four-quark state represents at least the 88% of the f_0 structure.

Acknowledgments

I warmly acknowledge R. Baldini, G. Isidori, L. Maiani, G. Pancheri, A. Polosa and A. Zallo for precious discussions on the subject of this talk and, *BABAR* and *KLOE* collaboration for support and cooperation.

References

[1] W.-M. Yao *et al.*, J. Phys. G **33** (2006) 1.

[2] E. Klempt, arXiv:hep-ph/0404270.

[3] R.L. Jaffe, Phys. Rev. D **15** (1977) 267.

[4] N.N. Achasov and V.V. Gubin, Phys. Rev. D **63** (2001) 094007; V.E. Markushin, Eur. Phys. J. A **8** (2000) 389 [arXiv:hep-ph/0005164]; J.A. Oller, Phys. Lett. B **426** (1998) 7 [arXiv:hep-ph/9803214]; E. Marco, S. Hirenzaki, E. Oset and H. Toki, Phys. Lett. B **470** (1999) 20 [arXiv:hep-ph/9903217]; J.A. Oller, Nucl. Phys. A **714** (2003) 161 [arXiv:hep-ph/0205121].

[5] M. Boglione and M.R. Pennington, Eur. Phys. J. C **30** (2003) 503 [arXiv:hep-ph/0303200].

[6] L. Maiani, F. Piccinini, A.D. Polosa and V. Riquer, Phys. Rev. Lett. **93** (2004) 212002 [arXiv:hep-ph/0407017].

[7] G. Isidori, L. Maiani, M. Nicolaci and S. Pacetti, JHEP **0605**, 049 (2006) [arXiv:hep-ph/0603241].

[8] R.L. Jaffe, Phys. Rept. **409** (2005) 1 [Nucl. Phys. Proc. Suppl. **142** (2005) 343] [arXiv:hep-ph/0409065].

[9] F.E. Close, AIP Conf. Proc. **717** (2004) 919 [arXiv:hep-ph/0311087].

[10] N.N. Achasov and V.N. Ivanchenko, Nucl. Phys. B **315** (1989) 465; S. Nussinov and T.N. Truong, Phys. Rev. Lett. **63** (1989) 1349 [Erratum-ibid. **63** (1989) 2002]; J.L. Lucio Martinez and J. Pestieau, Phys. Rev. D **42** (1990) 3253; F.E. Close, N. Isgur and S. Kumano, Nucl. Phys. B **389** (1993) 513 [arXiv:hep-ph/9301253]; N. Brown and F.E. Close, The DNE Physics Handbook, (ed. L. Maiani, G. Pancheri and N. Paver), (INFN, Frascati, 1995) pp. 447-464.

[11] F. Ambrosino *et al.* [KLOE Collaboration], Phys. Lett. B **634** (2006) 148 [arXiv:hep-ex/0511031].

[12] G. Zweig, CERN report S419/TH412 (1964), unpublished; S. Okubo, Phys. Lett, **5** (1963) 165; I. Iizuka, K. Okuda, O. Shito, Prog. Theor. Phys. **35** (1966) 1061.

[13] P. Geiger and N. Isgur, Phys. Rev. D **47** (1993) 5050; S. Descotes-Genon, L. Girlanda and J. Stern, JHEP **0001** (2000) 041 [arXiv:hep-ph/9910537]; N. Isgur and H. B. Thacker, Phys. Rev. D **64** (2001) 094507 [arXiv:hep-lat/0005006].

[14] S.J. Brodsky, G.R. Farrar, Phys. Rev. D **11** (1975) 1309.

[15] S.J. Brodsky and B.T. Chertok, Phys. Rev. D **14** (1976) 3003.

[16] S.J. Brodsky, G.P. Lepage, Phys. Rev. D **24** (1981) 2848. S.J. Brodsky and G.F. de Teramond, Phys. Lett. B **582** (2004) 211 [arXiv:hep-th/0310227].

[17] V.L. Chernyak and A.R. Zhitnitsky, Phys. Rept. **112** (1984) 173; V. Chernyak, arXiv:hep-ph/9906387; G.R. Farrar and D.R. Jackson, Phys. Rev. Lett. **35** (1975) 1416.

[18] See for instance: E.C. Tichmarsch, *The theory of functions*, London, Oxford University Press, 1939.

[19] B.V. Geshkenbein, Yad. Fiz. **9** (1969) 1232.

[20] A. Zallo [BABAR Collaboration] private comunication.

[21] M.N. Achasov, V.M. Aulchenko, K.I. Beloborodov and A.V. Berdyugin, Phys. Lett. B **504** (2001) 275.

[22] G. Solodov [BABAR Collaboration], "Initial state radiation study at *BABAR* and the application to the R measurement and hadron spectroscopy " Presented at ICHEP 06, Moscow, Russia.

[23] A. Aloisio *et al.* [KLOE Collaboration], Phys. Lett. B **537** (2002) 21 [arXiv:hep-ex/0204013].

[24] J.J. Sakurai, Phys. Lett. B **46** (1973) 207.

Dynamic Time Scales in Colored Glass Nuclear Matter

Vivek Parihar
Physics Department, Northeastern University, Boston MA USA

1 Introduction

Ultra high energy collisions between heavy nuclei can be employed to experimentally probe the QCD properties of dense quark-gluon systems [1, 2, 3, 4, 5]. The properties of the high density quark gluon system have also been compared to a color glass[6, 7, 8, 9, 10, 11, 12, 13]. High energy proton-proton scattering also probes the QCD vacuum. The QCD vacuum differs from the QED vacuum via properties largely due to the nature of the electrically charged and color charged dielectric screening response functions. In the QCD case, there is a quark-quark potential containing both a Coulomb-like piece as well as a linear *confining* potential. The confining *force* may be viewed in terms of the tension σ in a "string" connecting two quarks having the experimental value

$$\sigma \approx \frac{0.18 \text{ GeV}^2}{\hbar c} \approx 1.45 \times 10^5 \text{ Newton.} \tag{1}$$

The QCD string may be visualized as a color electric field flux tube. The string vibrational and rotational states describe the mesons contributing to nuclear forces. The QCD string has a much lower tension than some theoretical strings said to describe gravity; e.g.

$$\sigma_{\text{gravity}} = \frac{c^4}{8\pi G} \approx 4.8 \times 10^{42} \text{ Newton.} \tag{2}$$

In what follows only properties of the low tension QCD string will be invoked.

As the string excitation energies grow ever larger, the enormous number of possible string excitations can be described by the entropy as a function of energy.

In computing quantum transitions, the number Ω of possible quantum states is determined by the entropy S via

$$S = k_B \ln \Omega. \tag{3}$$

Since one must average over initial states and sum over the final states, transition rates have the "detailed balance" form

$$\Gamma_{i \to f} = \nu_\infty \left\{ \frac{\Omega_f}{\Omega_i} \right\} = \nu_\infty \exp\left[\frac{S_f - S_i}{k_B} \right]. \tag{4}$$

Eq.(4) will be applied to the time scales of nuclear matter motions which occur shortly after an ultra high energy collision between two heavy nuclei. The time scales are familiar from the the theory of glasses[15, 16, 17, 18]. The time scales are

$$\tau = \tilde{\tau} \exp\left[\frac{\Phi}{k_B (T - T_H)} \right], \quad \text{for} \quad T > T_H, \tag{5}$$

wherein T_H is the Hagedorn temperature[19] and Φ is the activation energy. Experimentally,

$$k_B T_H \approx 207 \text{ MeV} \quad \text{and} \quad \Phi \equiv \frac{7 k_B T_H}{2} \approx 725 \text{ MeV}. \tag{6}$$

The notion of final state particle product entropy is also of use in discussing the total cross section as a function of center of mass energy.

2 Vacuum Stability and Dielectric Screening

The Vacuum law of force between two charges $z_1 e$ and $z_2 e$ may be written as

$$V(r) = e^2 z_1 z_2 \int \left[\frac{e^{i\mathbf{Q}\cdot\mathbf{r}}}{Q^2 \varepsilon(Q^2)} \right] \frac{d^3\mathbf{Q}}{(2\pi)^3} = \left[\frac{z_1 z_2 e^2}{4\pi r} \right] \chi(r), \tag{7}$$

wherein the vacuum screening function $\chi(r)$ is determined by the vacuum dielectric response function $\varepsilon(Q^2)$ via

$$\chi(r) = \frac{2}{\pi} \int_0^\infty \sin(Qr) \left[\frac{dQ}{Q \varepsilon(Q^2)} \right]. \tag{8}$$

In the more usual notation of QED, one defines the dielectric response of the vacuum in terms of a running coupling strength

$$\alpha = \frac{e^2}{4\pi\hbar c} \quad \Rightarrow \quad \alpha(Q^2) = \frac{e^2}{4\pi\hbar c\varepsilon(Q^2)} \, . \tag{9}$$

In the limit of large distances, $\hbar/mc \ll r$, we have

$$\lim_{Q^2\to 0^+} \varepsilon(Q^2) = \lim_{r\to\infty} \chi(r) = 1, \tag{10}$$

which implies a simple Coulomb law. The subtracted dispersion relation then reads

$$\varepsilon(Q^2) = 1 - \frac{Q^2}{\pi}\int_0^\infty \left[\frac{\Im m\, \varepsilon(-\nu - i\,0^+)}{\nu + Q^2}\right]\frac{d\nu}{\nu} \, . \tag{11}$$

One has a dissipative QED vacuum with positive electrical conductivity for time-like wave vectors,

$$\Im m\, \varepsilon(-\nu - i\,0^+) \geq 0, \tag{12}$$

at the expense of a Landau $\alpha(Q^2)$ singularity for some space-like wave vector implicit in Eqs.(9) and (12). Such a singularity has been referred to as a Landau ghost[20, 21, 22, 23].

For the QCD case the potential between color charges analogous to Eq.(7) may be written as

$$V_s(r) = \eta^{ab} t_{1a} t_{2b} \left(\frac{g^2}{4\pi r}\right) \chi_s(r),$$
$$\chi_s(r) = \frac{2}{\pi}\int_0^\infty \sin(Qr)\left[\frac{dQ}{Q\varepsilon_s(Q^2)}\right], \tag{13}$$

wherein the strong screening dielectric response arises from the strong running coupling constant

$$\alpha_s = \frac{g^2}{4\pi\hbar c} \quad \Rightarrow \quad \alpha_s(Q^2) = \frac{g^2}{4\pi\hbar c\varepsilon_s(Q^2)} \, . \tag{14}$$

The subtracted dispersion relation for the strong color screening response function is governed by

$$\varepsilon_s(Q^2) = -\frac{Q^2}{\pi}\int_0^\infty \left[\frac{\Im m\, \varepsilon_s(-\nu - i\,0^+)}{\nu + Q^2}\right]\frac{d\nu}{\nu},$$
$$-\frac{\Im m\, \varepsilon_s(-\nu - i\,0^+)}{\pi} = \frac{\alpha_s}{4\pi}\left[\frac{11}{3}N_c - \frac{2}{3}n_f\right] + \ldots \, . \tag{15}$$

For QCD[24] it then follows that $\varepsilon_s(Q^2) \geq 0$ so there is no QCD Landau ghost. On the other hand, the color conductivity is *not* dissipative. In detail,

$$\Im m\, \varepsilon_s(-\nu - i0^+) \leq 0 \qquad \text{(color amplifier)}. \tag{16}$$

The *perturbative* QCD vacuum as described by Eqs.(15) and (16) is thereby unstable. The length scale L associated with strong color screening $\varepsilon_s(Q^2)$, i.e.

$$L^2 = -2 \lim_{Q^2 \to 0^+} \frac{\varepsilon_s(Q^2)}{Q^2} . \tag{17}$$

The two body color potential (as $r \to \infty$) may then be written as a QCD string linear potential

$$V_s(r) = -\eta^{ab} t_{1a} t_{1b} \sigma\, r \quad \text{where} \quad \sigma = \frac{g^2}{4\pi L^2} . \tag{18}$$

is the string tension.

3 Nuclear Matter and String Models

The liquid droplet model of a nucleus, assumes a *non-relativistic dilute fluid* of atomic mass A and radius $R = aA^{1/3} \approx 1.2 \times 10^{-13}\text{cm}\ A^{1/3}$ filling up a Fermi sphere of momentum radius $p_F = \hbar k_F$,

$$A = \left[\frac{4\pi R^3}{3}\right] \left[\frac{4}{(2\pi)^2}\right] \left[\frac{4\pi k_F^3}{3}\right] = \frac{8}{9\pi}(k_F R)^3. \tag{19}$$

Eq.(19) allows for the computation of the Fermi velocity $v_F = \hbar k_F / M$; i.e.

$$\frac{v_F}{c} = \left(\frac{9\pi}{8}\right)^{1/3} \frac{\hbar}{Mca} \approx \frac{1}{4}. \tag{20}$$

The liquid drop thereby appears to be internally relativistic. Another notion of nuclear liquids being relativistic arises in shell models. A strong $\mathbf{L} \cdot \mathbf{S}$ coupling exists characteristic of the Dirac relativistic spinor model of fermions. The QCD string constituents of nuclear matter starts from a relativistic viewpoint.

In the string model, the atomic number $Z = 2N_u + N_d$ and the atomic mass $A = 3(N_u + N_d)$ can be expressed in terms of the net numbers of up and down quarks. The mesons internal to the nucleus are made of quark anti-quark pairs connected by strings. The strings consist of color electric flux tubes. The electric field colorless condensate $\mathcal{E} = \sqrt{\eta_{ab}\langle \mathbf{E}^a \cdot \mathbf{E}^b - \mathbf{B}^a \cdot \mathbf{B}^b \rangle}$ within the string determines the the magnitude of the string tension $\sigma = g\mathcal{E}$.

To understand the spectrum of excited states of a QCD string, one may consider a length Λ of a rotating string along with an eigenvalue angular velocity

$$\omega = \frac{\pi c}{\Lambda} = \frac{\pi c \sigma}{E} = \frac{\pi \sigma}{Mc} \tag{21}$$

wherein $E = Mc^2 = \sigma\Lambda$ is the energy of the string. Since the angular velocity of the string with angular momentum J obeys

$$\omega = \frac{dE}{dJ} = \frac{\pi c \sigma}{E} \quad \Rightarrow \quad \frac{dJ}{d(E^2)} = \frac{1}{2\pi c \sigma}, \tag{22}$$

it follows that the angular momentum energy-squared variation is a linear classical Regge trajectory. The quantum mechanical version of Eq.(22) is usually written

$$J = J_0 + \left(\frac{E^2}{2\pi c \sigma}\right) = \hbar(\alpha_0 + \alpha' M^2) = \hbar\alpha_0 + \left(\frac{c^3}{2\pi\sigma}\right) M^2. \tag{23}$$

From the experimental slope of the meson spectrum, the measured value of the QCD string tension is given in Eq.(1). As the energy grows ever larger, the degeneracy of the excited states grows exponentially. The resulting increasing entropy and the large string energy degeneracy leads to notion of a Hagedorn temperature.

4 String Entropy

For a fixed angular momentum $J = \hbar N$ and for two polarizations

$$N = \sum_{p=1,2} \sum_{k=1}^{\infty} k N_{kj}. \tag{24}$$

The number of partitions of the integer N into the sum of other integers equal or less than N yields the degeneracy number Ω[25, 26]. For large N,

$$\begin{aligned} \ln\Omega(N) &= 2\pi\sqrt{\frac{2N}{3}} + \ln\left(\frac{3^{1/2}}{(6N)^{7/4}}\right) + \dots, \\ N &= \frac{E^2}{2\pi\hbar c\sigma} + \dots, \\ S(E) &= k_B \ln\Omega(N). \end{aligned} \tag{25}$$

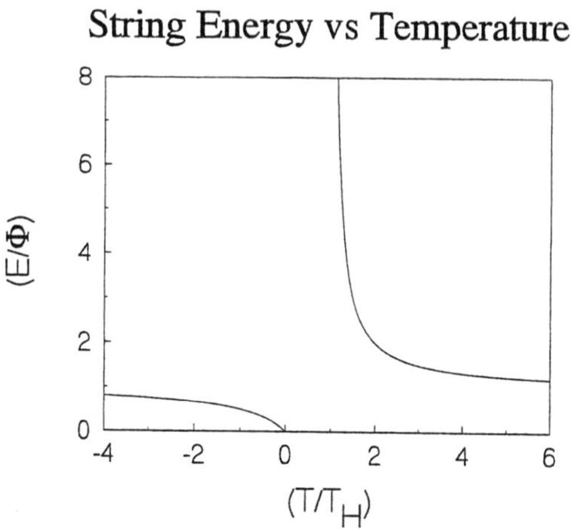

Figure 1: The temperature in the interval $0 < T < T_H$ is forbidden. For a small amount of energy $E < \Phi$ fed into a string from a high energy nuclear collision, the temperature will be negative. In the high energy transfer regime, $E > \Phi$, the temperature will be positive. The temperature passes from negative to positive through the points $T = \pm\infty$ as E passes through Φ. For the very highest collision energy passed to a string, the temperature will "cool" down from above to T_H. The Hagedorn temperature is the lowest possible positive temperature of the string. In this regard, note the negative heat capacity $C = (dE/dT) < 0$.

The microcanonical string entropy determines the temperature via

$$\frac{1}{T} = \frac{dS}{dE} \quad \text{and} \quad k_B T_H = \sqrt{\frac{3\hbar c\sigma}{4\pi}},$$
$$E = \left(\frac{T}{T - T_H}\right)\frac{7k_B T_H}{2} = \left(\frac{T}{T - T_H}\right)\Phi. \tag{26}$$

Eq.(26) describes the thermodynamic excitation energy above the Hagedorn temperature $k_B T_H \approx 207$ MeV; i.e. $\lim_{T\to\pm\infty} E = \Phi \approx 725$ MeV. The string energy is plotted as a function of temperature in Figure1.

The following is a simple physical picture for describing how QCD strings behave during a high energy nuclear collisions. During a collision the meson (quark anti quark) or nucleon (three quark) string connections are excited. If a small amount of collision energy is transferred into a string, then the string temperature will be negative $0 < E < \Phi \;\Rightarrow\; 0 > T > -\infty$. If a large amount of collision energy is transferred into a string, then the string temperature will be positive $\Phi < E < \infty \;\; \Rightarrow \;\; T > T_H$ over and above the Hagedorn temperature. The string heat capacity

$$C = \frac{dE}{dT} = -\frac{\Phi T_H}{(T - T_H)^2} < 0 \qquad \text{(QCD string)}. \tag{27}$$

The *color glass-like* time scales of string motions in the high energy limit are given in Eq.(5). As $T \to T_H + 0^+$, the time scales grow ever larger consistent with the VFT law often applied to polymer and other glasses. The QCD strings in nuclear matter thereby play the role of polymer (or other network) chains in melted glass with viscosity

$$\eta = \rho c^2 \tau = \rho c^2 \tilde{\tau} \exp\left[\frac{\Phi}{k_B(T - T_H)}\right], \quad (T > T_H), \tag{28}$$

wherein ρ is the fluid mass density. A color glass near the Hagedorn temperature is similar to a very viscous fluid. If the duration of a collision is sufficiently short, then there is little time to excite highly energetic string states leaving one with an intermediate temperature $T \gg T_H$ which implies a short time scale and a low viscosity. The resulting low viscosity in an almost *ideal fluid* is sometimes called a "perfect fluid"[27, 28].

Finally, we note in passing that a black hole has thermodynamic analogies to a QCD string; Black holes, as well as some observed *normal stars*, also have negative heat capacity; c.g. for the black hole

$$\begin{aligned} \frac{S}{k_B} &= 4\pi\left(\frac{GM^2}{\hbar c}\right) = 4\pi\left(\frac{GE^2}{\hbar c^5}\right), \\ \frac{dE}{dS} = T \quad \Rightarrow \quad E &= \left(\frac{\hbar c^5}{8\pi G}\right)\frac{1}{k_B T} = \left(\frac{\hbar c \sigma_{\text{gravity}}}{k_B T}\right), \\ C &= \frac{dE}{dT} = -\left(\frac{\hbar c \sigma_{\text{gravity}}}{k_B T^2}\right) < 0. \end{aligned} \tag{29}$$

5 String Fragmentation

What happens to a highly excited state QCD string when it decays into final product particles in a nuclear collision? If the string is highly excited, then what appears in the final state products are jets of the many parts of the strings which fragment into lower mass elementary particles. To see what is involved, consider a meson made of a $q\bar{q}$ quark pair each with mass m which sit on the ends of a QCD string. The quark pair Hamiltonian matrix may be written as

$$H = \begin{pmatrix} cp & mc^2 \\ mc^2 & -cp \end{pmatrix} \quad \text{with eigenvalues} \quad E_{\pm}(p) = \pm\sqrt{c^2p^2 + m^2c^4}. \tag{30}$$

The transition rate per unit time to fragment a point on a string cereating a $q\bar{q}$ pair follows by applying Fermi's golden rule to the Hamiltonian matrix Eq.(30); i.e.

$$\gamma = \left(\frac{2\pi}{\hbar}\right) |mc^2|^2 \delta(2cp). \tag{31}$$

The probability of finding the fragmentation phase point $\in (dpdx)/2\pi\hbar$ is

$$d^2P = \left(\frac{dpdx}{2\pi\hbar}\right) e^{-\int \gamma dt}. \tag{32}$$

Since the rate of change of momentum of a quark in a color condensate electric field is given by

$$\frac{dp}{dt} = g\mathcal{E} = \sigma, \tag{33}$$

it follows that the transition per unit time per unit length for a QCD string to fragment producing a $q\bar{q}$ pair is given by

$$\frac{d^2P}{dxdt} = \left(\frac{\sigma}{2\pi\hbar}\right) \exp\left[-\frac{\pi m^2 c^3}{\hbar\sigma}\right], \tag{34}$$

which is exponentially suppressed if the quark mass m is high.

6 Conclusions

It was shown how the perturbation theory QED vacuum is stable to decay but has a short space-like distance Landau ghost. The perturbation theory QCD vacuum has no Landau ghost, but is in an excited energetic state of negative temperature. From the viewpoint of QCD, nuclear matter consists of relativistic nucleons each with three quarks and relativistic mesons each with a quark anti quark pair. The quarks are tied together by gluons in the form of strings having tension σ. The excitation of QCD strings at low energy has a negative temperature and at high energy has a positive temperature always higher than T_H. Very high energy strings $T \to T_H + 0^+$ move very slowly as a viscous melted glass with very high viscosity. However, in a very short collision time it is difficult to transfer the initial collision kinetic energy into the internal energy of a few strings. The low energy strings at high $T \gg T_H$ have a low viscosity giving rise to an almost perfect fluid. The role of strings in a nucleus is closely analogous to the role of polymer chains in some viscous glass beads.

7 Acknowledgements

V. P. thanks the organisers of the International School of Subnuclear Physics for the very illuminating meeting and for the oppurtunity to present this talk. We are grateful to A. Widom and Y. N. Srivastava for a great deal of guidance and discussions. This talk is based on our previous work [14].

References

[1] L. Maiani, F. Piccinini, A.D. Polosa and V. Riquer, *Nucl.Phys.* **A748**, 209 (2005).

[2] E.V. Shuryak, *Phys.Rev.* **D70**, 054507 (2004).

[3] E.V. Shuryak, *Nucl.Phys.* **A702**, 83 (2002).

[4] S. Soff, S. A. Bass and A. Dumitru, *Phys. Rev. Lett.* **86**, 3981 (2001).

[5] D. Teaney, J. Lauret, E.V. Shuryak, *Phys.Rev.Lett.* **86**, 4783 (2001).

[6] M. Gyulassy and L. McLerran, *Nucl. Phys.* **A750**, 30 (2005).

[7] Dmitri Kharzeev and Kirill Tuchin, *Nucl.Phys.* **A753**, 316 (2005).

[8] E. Iancu and R. Venugopalan in *QGP 3*, R. Hwa and X.-N. Wang Editors, World Scientific, Singapore (2004).

[9] J. Jalilian-Marian, *Mod. Phys. Lett.* **A19**, 1251 (2004).

[10] Raju Venugopalan, hep-ph/0412396 (2004).

[11] E. Iancu, A. Leonidov, and L. McLerran, hep-ph/0202270 (2002).

[12] Raju Venugopalan, *Comments Nucl. Part.Phys.* **22**, 113 (1997).

[13] A. Ayala, J. Jalilian-Marian, L. McLerran and R. Venugopalan, *Phys. Rev.* **D52**, 2935 (1995) .

[14] V. Parihar, A. Widom and Y.N. Srivastava, *Phys. Rev.* **C73**, 017901 (2006) .

[15] H. Vogel, *Phys. Z.* **22**, 645 (1921).

[16] G.S. Fulcher, *J. Am. Chem. Soc.* **8**, 339 (1925); *J. Am. Chem. Soc.* **8**, 789, (1925).

[17] G. Tammann and W. Hesse, *Z. Anorg. Allg. Chem.* **156**, 245 (1926).

[18] H. Tanaka, *Phys. Rev. Lett.* **90**, 055701 (2003).

[19] R. Hagedorn, *Nuovo Cim. Suppl.* **3**, 147 (1965)

[20] L.D. Landau, in *Neils Bohr and the Development of Physics*, W. Pauli Editor, Pergamon Press, London (1955).

[21] S. Sivasubramanian, Y. Srivastava and A. Widom, *Mod. Phys. Lett.* **B16** 1201 (2002).

[22] V.B. Berestetskii, E.M. Lifshitz and L.P. Pitaevskii, *Quantum Electrodynamics*, §114, Butterworth-Heinemann, Oxford (1997).

[23] Y. Srivastava and A. Widom, *Phys. Rev.* **D63** 077502 (2001).

[24] Y. Srivastava, S. Pacetti, G. Pancheri and A. Widom, "Dispersive techniques for α_s, R_{had} and instability of the perturbative vacuum", Invited talk at the conference "e^+e^- *physics at intermediate energies workshop*", *eConf C010430* **T19** (2001).

[25] G.H. Hardy and S. Ramanujan *Proc. London Math. Soc.* **17**, 75 (1918)

[26] K. Huang and S. Weinberg, *Phys. Rev. Lett.* **25**, 895 (1970).

[27] K. Adcox, et. el., *Nucl. Phys.* **A757**, 184 (2005).

[28] John Adams, et. al. ,*Nucl. Phys.* **A757**, 102 (2005).

Mapping the Transverse Size of the Proton

Ori Smith
School of Physics and Astronomy
Raymond and Beverly Sackler Faculty of Exact Sciences
Tel Aviv University, Tel Aviv, Israel.

November 2006

Abstract

We discuss how the exclusive production of ρ^0 in deep inelastic ep scattering at HERA can be used to map the transverse distribution of gluons in the proton. We then present the experimental challenge in a precise determination of the slope of the four-momentum transfer squared distribution, which can be directly linked to the size of the proton.

1 Introduction

Prof. Zichichi opened this Ettore Majorana Summer School in Sicily with a lecture entitled "Complexity vs. Simplicity." It is interesting to think of his lecture when discussing the composition of the proton. Not so long ago the proton was regarded as one of the fundamental particles of physics, and the Hydrogen atom is still considered one of the simplest examples of quantum mechanics calculations. However, since the discovery of partons, the complex nature of the proton system started to emerge. Even today, we do not have an exact knowledge of the distribution of partons inside the proton. The subject of this talk will be the use of exclusive ρ^0 electroproduction in the ZEUS detector [1] at the Hadron Electron Ring Accelerator (HERA) to advance our understanding of the transverse map of the proton.

To define the problem, we have to ask what exactly is the transverse size of the proton? Much research has been undertaken to determine the electromagnetic radius of the proton [2], but these experiments are not sensitive to the gluon distribution. In the research described here, along with other studies conducted on diffractively produced particles, the cross section is sensitive to the gluon distribution [3]. If we combine the two pictures, we get a full description of what the transverse extension of the proton looks like. The information obtained in this manner will be important for the LHC experiments and theoretical predictions in the future.

The process investigated in this analysis is exclusive diffractive ρ^0 electroproduction, namely $ep \rightarrow e\rho^0 p$. The electron emits a virtual photon that undergoes the interaction with the proton. Hence, the effective process is $\gamma^* p \rightarrow \rho^0 p$. Diffraction means that the proton stays intact after the interaction. For a clear description of diffraction, the reader is referred to the lecture given by B. Loehr at this school.

2 HERA Accelerator

The Hadron Electron Ring Accelerator (HERA) is designed to accelerate electrons/positrons and protons in a 6.3 km ring. In 1998-2000 the 920 GeV protons collided with 27.5 GeV electrons/positrons. The ZEUS detector, weighing 3600 tons and standing 12 meters in height, is located 30 meters underground in the south experimental hall of the main HERA ring.

The photons emitted by the electrons have virtualities in the range of zero to $O(10000)$ GeV^2. This means that unlike previous proton-proton experiments, the size of one of the interacting particles in HERA can be varied (the transverse size changes as $1/Q^2$). This can be used, as will be explained later, to determine a region where hard scale dominates and where pQCD calculations can be applied.

3 Theoretical Introduction

The leading order diagram for the exclusive production of vector mesons in pQCD is shown in Fig. 1 with the notation used for the incoming and outgoing particles momenta. Also shown in the figure are the variables that are used to describe the interaction. They are the center of mass energy of the photon-proton system

$$W^2 = (p+q)^2 \quad , \tag{1}$$

the four-momentum squared of the virtual photon

$$-Q^2 = (k' - k)^2 \quad , \tag{2}$$

and the four-momentum transfer at the proton vertex

$$t^2 = (p' - p)^2 \quad . \tag{3}$$

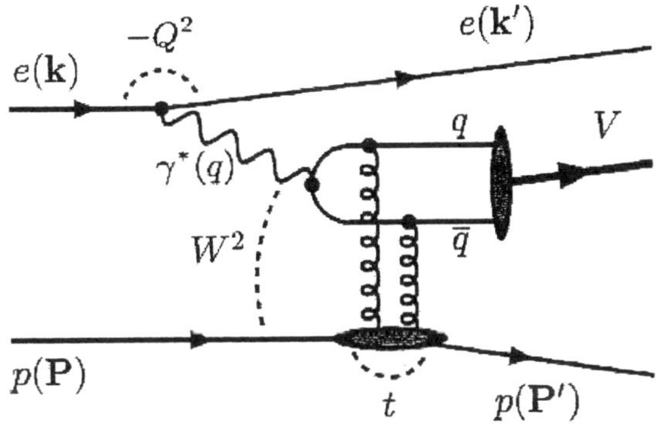

Figure 1: Leading order diagram for the process $ep \to eVp$, where V stands for vector meson, and the notation for the particle momenta and the variables used to describe the interaction.

In this picture, the photon fluctuates into a $q\bar{q}$ pair. The $q\bar{q}$ interacts elastically with the proton and combines again to form the vector meson (ρ^0 in our case). In this description factorization is used [4] which means that the total amplitude is divided into the $q\bar{q}$ fluctuation, given by the photon wave function, the interaction amplitude with the proton and the wave function of the vector meson. Justification for factorization comes from the typical lifetime of the $q\bar{q}$ fluctuation, as well as the time for the ρ^0 formation being much longer than the time of the interaction with the proton. Another important point is to make sure we are in the pQCD regime. For this, a large scale must exist in the interaction. Since our experiment is performed at low t and the ρ^0 is a low mass vector meson, large scale can only enter through the photon virtuality, Q^2.

Thus, at the first stage of the analysis, it is necessary to find the threshold of Q^2 for pQCD calculations to dominate. I will demonstrate how we do it below. In the case that the two above conditions are satisfied, in first order approximation the cross section is proportional to the square of the gluon distribution in the proton [3, 5]:

$$\frac{d\sigma}{dt}(\gamma^* p \to \rho p) \propto [xg(x, Q^2)]^2, \tag{4}$$

where x is the part of the longitudinal momentum of the proton carried by the interacting parton and $xg(x, Q^2)$ is the gluon density distribution. For the gluon distribution, the results extracted from the F_2 measurements can be used [6]. It should be stressed that the identification with the ordinary diagonal parton distribution is only a good approximation for very small values of x and t. Otherwise the process depends on off diagonal parton distributions [7].

4 t Distribution

When discussing the transverse projection of the proton, an important quantity is the "impact parameter", b. As an insight to the meaning of b, we write the Fourier transform of the scattering amplitude T,

$$T(W,t) = 4W^2 \int d^2 b e^{i\vec{k}\vec{b}} f(W,\vec{b}) \ , \tag{5}$$

where $\vec{k}$ is the two-dimensional transverse momentum vector ($|\vec{k}^2| = -t$). Using then our knowledge of the behaviour of the scattering amplitude we obtain

$$f(W,\vec{b}) \propto \frac{1}{B(W)} e^{-\vec{b}^2/2B(W)} \ , \tag{6}$$

where $2B(W) =< \vec{b}^2 >= R^2_{int}$ describes the mean square transverse extension of the scattering system or the size of the interaction radius. The cross section dependence on t is then expected to be a decaying exponent with a slope $B(W)$,

$$\frac{d\sigma}{dt} \propto e^{-B(W)|t|} \ . \tag{7}$$

As can be seen from the notation, B may dependent on the energy and its behavior is usually parameterized as

$$B(W) = B_0 + 4\alpha' \ln \frac{W}{W_0} \ . \tag{8}$$

Here B_0 is the value of the slope at some arbitrary value of $W = W_0$. At high Q^2 the size of the $q\bar{q}$ fluctuation from the virtual photon is small compared to the size of the proton. In this high Q^2 regime we can therefore expect that B will measure the size of the proton. The parameter α' is a measure of how fast the interaction region increases with energy. QCD calculations predict a small α', much smaller than its value at low Q^2.

What exactly happens when we increase the energy? The connection between W and x is $W^2/Q^2 = 1/x - 1$, i.e. at constant Q^2 increasing W results in decreasing x. As we increase the energy, gluons with lower and lower x contribute to the cross section. Therefore, by measuring the value of B as a function of W we can infer the transverse distribution of the gluons of a given x in the proton.

device and the hadron electron separator. The main contribution to the t uncertainty comes from the electron position reconstruction.

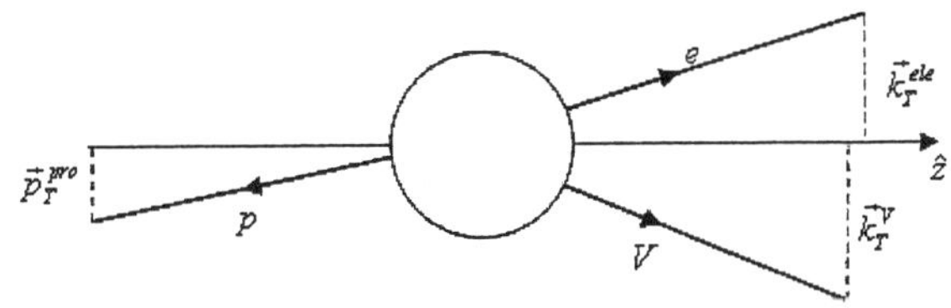

Figure 2: A sketch demonstrating momentum conservation for the particles exiting from the interaction

It is interesting to understand how the uncertainty on t affects the determination of the slope, $B(W)$. Since work is still in progress and no final results have been published yet, I will describe the analysis procedure done on the Monte Carlo simulation (MC) and present some preliminary results in the concluding section.

In section 4 I mentioned that the dependence of the cross section on t can be parameterised as a decaying exponent. This is a good approximation for values of $|t| < 1\text{GeV}^2$ as in this analysis. Hence, much more events are produced with t value close to zero. As the error on the different values of t does not change considerably, more events will be mistakenly reconstructed at higher t than at lower t, resulting in smearing of the distribution. As the value of B increases, the distortion of the original distribution amplifies. In order to correctly extract the true value of the data slope, unfolding is needed. In practice, the input B value is adjusted in the MC so that the ensuing t distribution reproduces well that of the data. In this process, we heavily rely on our understanding of the detector components and therefore it is of utmost importance to make sure that the detector resolutions in the MC and data agree. Otherwise the extracted value of B will be incorrect. An example of the effect of modifying the resolution appears in Fig. 3. On the left, the MC t distribution obtained at the detector level is shown and compared to the generated distribution. At the generation level a value $B = 7$ GeV^{-2} is assumed. The slope of the reconstructed distribution comes out to be $B_{det} \backsim 5.5$ GeV^{-2}. If an increase of 50% is assumed for the detector resolution, the slope of the reconstructed distribution is $B_{det} \backsim 4.7$ GeV^{-2}. For higher values of generated B this difference would increase. The goal of the analysis is thus to develop an algorithm leading to the best possible resolution on position reconstruction and to devise a method to validate the performance of the detectors in the MC simulation. A better t resolution will result in a smaller dependence on our level of understanding of detector

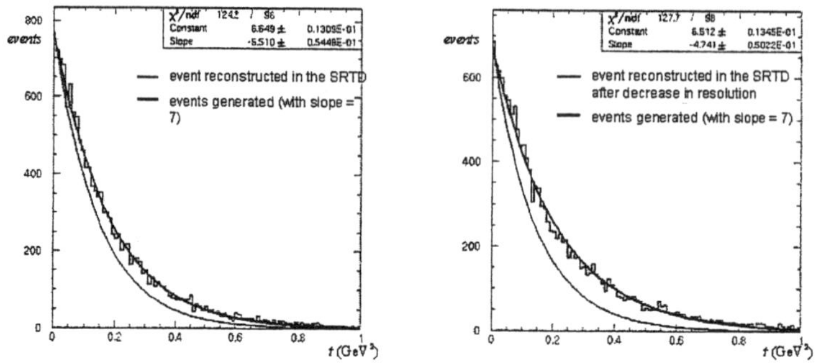

Figure 3: The effect of smearing caused by the detector reconstruction. The MC is generated with a slope of $B = 7$ GeV^{-2}. On the left, the nominal resolution is assumed. On the right, the resolution is increased by factor 1.5.

performance. The validation will result in reliable systematic uncertainties. To achieve both goals, we determine the position of the scattered electron by using the straight line fit method as described below.

7 Straight Line Fit Method

Three different detectors are positioned in the region of the outgoing electron. As already mentioned, these are the rear calorimeter (RCAL), the hadron electron separator (HES), and the small-angle rear tracking detector (SRTD). The mean resolution of the RCAL is $\backsim$0.75 cm while the SRTD and the HES have similar resolutions of $\backsim$0.45 cm. In the region of interest, the trajectory of the electron is little affected by the presence of the solenoidal magnetic field and therefore the electron should follow a straight line from the vertex through all the detectors. The parameters of the straight line are determined through a fit (SLF) in which each point is weighted according to the appropriate detector resolution. This is schematically shown in Fig. 4. So far the procedure is relatively straight forward. A complication arises from the need to select the appropriate fiducial volume for each detector, which in turn requires a good understanding of all edge effects.

The immediate benefit of the SLF method is the improvement in the t resolution. A further advantage can be achieved by examining the residuals, δ, between the detector position and the one projected from the line fit (shown also in Fig. 4). A plot of the residuals in the MC is presented in Fig. 5. The width of the distributions is a direct measure of the detectors

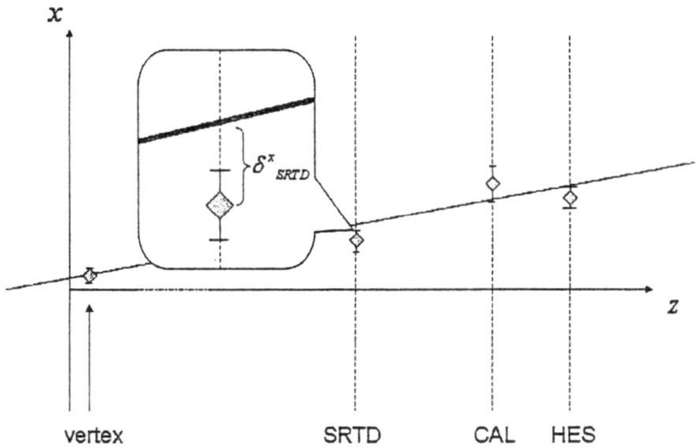

Figure 4: Description of the Straight Line Fit method as applied to the outgoing electron.

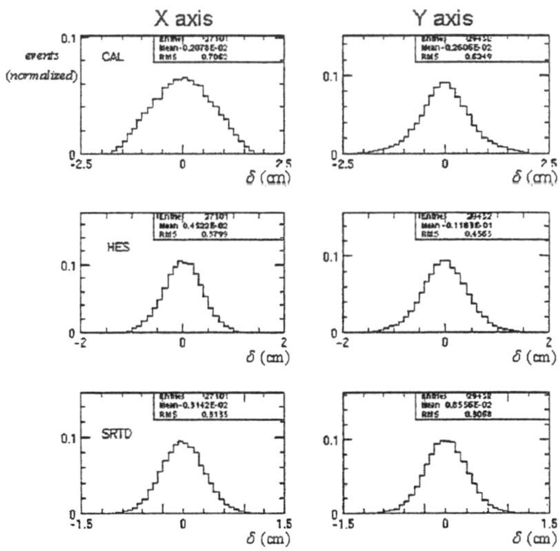

Figure 5: Distributions of events around the position determined by the line fit. The width of the distribution corresponds to the resolution of the detector.

resolution. A similar plot can be drawn for the data and the resolutions in the MC can be tuned accordingly. In addition, the distribution of residuals in the data is sensitive to detector shifts which can then be corrected for by requiring the distribution to be centered around zero.

8 Preliminary Results

Fig. 6 displays the slope of the different vector mesons produced in diffraction. For vector mesons with higher mass than the ρ, large scale can enter through a combination of mass and Q^2. The J/ψ, for example, is produced by hard interaction even at very low Q^2. This is reflected in the lack of Q^2 dependence. To account for the mass dependence, the slope is drawn as a function of $Q^2_{eff} = Q^2 + M^2_V$. We can see that indeed the values of the slope converge into a single value of around 4.5GeV$^-$2. For the ρ meson the slope becomes almost independent of Q^2_{eff} from $Q^2_{eff} \backsim 15\text{GeV}^2$. This is the region where hard interaction dominates and we can expect pQCD calculations to be valid. The description given here is only a qualitative one and for the exact numbers it is necessary to wait until the final results will be published.

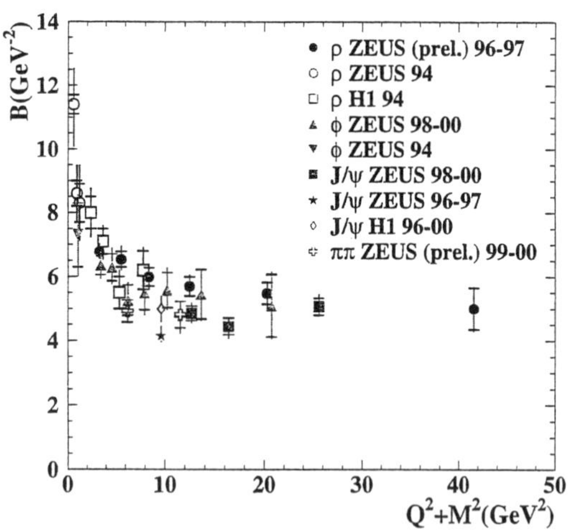

Figure 6: The slope B as a function of Q^2_{eff}. Preliminary results of ρ production are combined with the other mesons produced by diffraction.

9 Conclusions

We argue that exclusive vector meson electroproduction at high energy and high Q^2 can be used to map the transverse size of the proton. In the region of x explored by HERA ($10^{-4} < x < 10^{-2}$), the exclusive production of vector mesons is dominated by the gluon contribution in the proton. From the electromagnetic form factor measurements the slope of the t distribution comes out to be $B \backsim 3\text{GeV}^{-2}$ while in this analysis $B \backsim 4.5\text{GeV}^{-2}$ is obtained.

The electromagnetic form is only sensitive to the quark content of the proton. Combining this information with the results from vector meson electroproduction a picture for the transverse profile of the proton emerges: the proton has a disc shape that is filled with a dense cloud of gluons of an area approximately the same size as that filled by quarks.

10 Acknowledgements

I would like to thank Prof. H. Abramowicz and Prof. A. Levy for their close supervision and constant support. I would also like to warmly thank Dr. S. Kananov for his help and advice.

References

[1] ZEUS Coll. U. Holm (ed.), *The ZEUS Detector*, Status Report (unpublished), DESY (1993), available on `http://www-zeus.desy.de/bluebook/bluebook.html`.

[2] see for example H. Kowalski and J. Bartels, Eur. Phys. J. C (2001).

[3] S.J. Brodsky et al., Phys. Rev. D **50**, 3134 (1994).

[4] J. C. Collins, L. Frankfurt and M. Strikman, Phys. Rev. D **56**, 2982 (1997).

[5] M.G. Ryskin: Z. Phys. C **57**, 89 (1993).

[6] ZEUS Collaboration, S. Chekanov et al., Phys. Rev. D **67**, 012007 (2003).

[7] D. Mueller et al., Fortschr. Phys. **42**, 347 (1994); X. Ji, Phys. Rev. Lett. **78**, 610 (1997)

Scalar Higher Dimensional Theories in 1/N-expansion

G.S.Vartanov

Museo Storico della Fisica e Centro Studi e Ricerche "Enrico Fermi", Rome, Italy
Bogoliubov Laboratory of Theoretical Physics, Joint Institute for Nuclear Research, Dubna, Russia
e-mail: Vartanov@theor.jinr.ru

Abstract

This talk is based on the paper hep-th/0607177 where we demonstrate how one can construct renormalizable perturbative expansion in formally nonrenormalizable higher dimensional scalar theories. It is based on $1/N$-expansion and results in a logarithmically divergent perturbation theory in arbitrary high odd space-time dimension. The resulting effective coupling is dimensionless and is running in accordance with the usual RG equations. The corresponding beta function is calculated in the leading order and is nonpolynomial in effective coupling. It exhibits either UV asymptotically free or IR free behaviour depending on the dimension of space-time. In this talk we mainly discussed the problems concerning the Renormalization Group.

1 Introduction

In the paper [1] we constructed a renormalizable expansion in a formally nonrenormalizable higher dimensional scalar theory which was based on the $1/N$-expansion(for review of the $1/N$-expansion in the space-time with the number of dimensions not greater than four see [2]). The aim of this paper is checking of the renormalization group(RG) equations so the main part of this paper will be devoted to the calculation of the diagrams using the $1/N$-expansion. In the first part of the paper we will introduce our model and remind some main features of the $1/N$ expansion. In the second part the renormalization coefficients will be calculated up to the second order. Contrary to the usual expansion in the coupling constant we will see that in the leading order we have not only one-loop diagrams but there is also a two-loop diagram which does give contribution in this order. And in the second order of the $1/N$ expansion we will have diagrams with two, three, four and even five loops. In the third part of our paper we will write the so-called pole equations [4] that express the coefficients of the higher order poles via the coefficients of a simple pole. For checking the pole equations but we do not have to calculate the diagram completely we need only the leading divergence of it. We will see that the coefficients obtained from this equations coincide with the coefficients calculated from the diagrams.

2 1/N-expansion

Before going to the calculation of the diagrams up to the second order in $1/N$ expansion and checking of the RG equations in the higher dimensional scalar theories let us just remain the main features of the $1/N$-expansion [1]. For the beginning let us start with the usual N component scalar field theory in D dimensions, where D takes an arbitrary odd value, with ϕ^4 self-interaction. The Lagrangian looks like

$$\mathcal{L} = \frac{1}{2}(\partial_\mu \vec{\phi})^2 - \frac{1}{2}m^2\vec{\phi}^2 - \frac{\lambda}{8N}(\vec{\phi}^2)^2, \tag{1}$$

where N is the number of components of ϕ. After rewriting the initial Lagrangian introducing an auxiliary field σ [3] we have

$$\mathcal{L} = \frac{1}{2}(\partial_\mu \vec{\phi})^2 - \frac{1}{2}m^2\vec{\phi}^2 - \frac{1}{2\sqrt{N}}\sigma(\vec{\phi}^2) + \frac{1}{2\lambda}\sigma^2. \tag{2}$$

After summing up the zeroth order diagrams for the σ field propagator we obtain the effective Lagrangian

$$\mathcal{L}_{eff} = \frac{1}{2}(\partial_\mu \vec{\phi})^2 - \frac{1}{2\sqrt{N}}\sigma(\vec{\phi}^2) + \frac{1}{2\lambda}\sigma^2 + \frac{1}{2}f(D)\sigma(\partial^2)^{D/2-2}\sigma. \tag{3}$$

One can now construct the diagrams using the propagators from the Lagrangian(3) and the triple vertex having in mind that any closed cycle of ϕ gives an additional factor of N and any vertex gives $1/\sqrt{N}$. In [1] it was found that there are only two types of divergent diagrams in the dimensional regularization, they are the ϕ propagator and the triple vertex. All other diagrams do not contain the global divergencies what means that all other divergencies are those of the subgraphs eliminated by renormalization of the ϕ field and the vertex.

We see from Lagrangian(3) that $1/\sqrt{\lambda}$ plays the role of mass for σ field what means that the only parameter in the term for interaction is the number of components N which we should renormalize while introducing the renormalization constants. But the number of components N cannot be changed. In order to solve this problem we introduce the new coupling constant h which is associated *only with the triple vertex* (and not with the σ propagator) as

$$\mathcal{L}_{int} = -\frac{\sqrt{h}}{2\sqrt{N}}\sigma\vec{\phi}^2. \tag{4}$$

After introducing the coupling constant h we again resum the zeroth order diagrams since they will have the additional factor $\sqrt{h}$ in the vertex resulting in the following UV effective Lagrangian

$$\mathcal{L}_{eff} = \frac{1}{2}(\partial_\mu \vec{\phi})^2 - \frac{\sqrt{h}}{2\sqrt{N}}\sigma(\vec{\phi}^2) + \frac{1}{2\lambda}\sigma^2 + \frac{1}{2}f(D)\sigma(\partial^2)^{D/2-2}\sigma(1+h), \tag{5}$$

[1]more details can be found in the paper [1]

From this effective Lagrangian we have the modified Feynman rules: the ϕ propagator which is the usual one while the σ propagator is given by the following expression

$$\frac{i}{1/\lambda + f(D)(-k^2)^{D/2-2}(1+h)}, \tag{6}$$

where

$$f(D) = \frac{\Gamma^2(D/2-1)\Gamma(2-D/2)}{2^{D+1}\Gamma(D-2)\pi^{D/2}}$$

and we put $m = 0$ for simplicity.

3 Leading and the second order in 1/N expansion

As it was mentioned above this chapter will be devoted to the calculation of the coefficients for the renormalization constants. The first step will be the calculation of the poles in the leading order. We start with the $1/N$ terms for the propagator of ϕ and the triple vertex. One has the diagrams shown in Fig.1. Notice that besides the one-loop diagrams in the same order of $1/N$ expansion one has the two-loop diagram for the vertex which we do not have in the usual perturbative expansion.

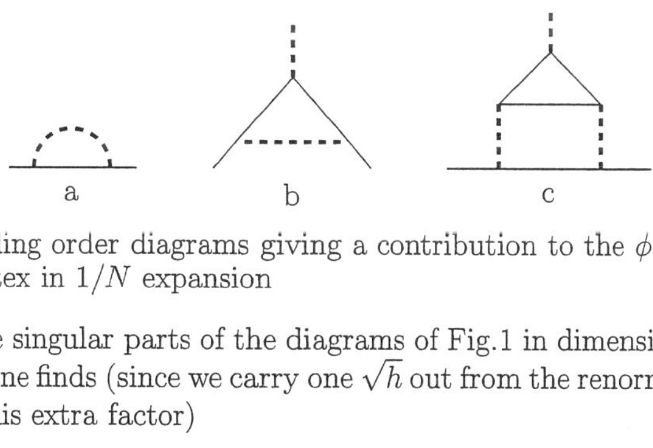

Figure 1: The leading order diagrams giving a contribution to the ϕ field propagator and the triple vertex in $1/N$ expansion

Calculating the singular parts of the diagrams of Fig.1 in dimensional regularization with $D' = D - 2\varepsilon$ one finds (since we carry one $\sqrt{h}$ out from the renormalization constant we do not write this extra factor)

$$Diag.a \;\Rightarrow\; \frac{h}{\varepsilon N(1+h)}A, \quad Diag.b \;\Rightarrow\; \frac{1}{\varepsilon N}\frac{hB}{1+h}, \quad Diag.c \;\Rightarrow\; \frac{1}{\varepsilon N}\frac{h^2 C}{(1+h)^2}, \tag{7}$$

$$A = \frac{2\Gamma(D-2)}{\Gamma(D/2-2)\Gamma(D/2-1)\Gamma(D/2+1)\Gamma(2-D/2)}, \quad B = \frac{D}{4-D}A, \quad C = \frac{D(D-3)}{4-D}A.$$

The diagram a in Fig.1 is multiplied by the term $\frac{h}{1+h}$ because there are two vertexes which give h and there is one propagator of the σ field giving the doniminator $\frac{1}{1+h}$. The same argument works for the diagram b. The diagram c has more vertexes and so we have h^2 in the nominator and since it has two σ propagators there is a square of $(1+h)$ in the denominator (it will be different in the gauge case [5]).

The corresponding renormalization constants in the $\overline{MS}$ scheme are then

$$Z_2^{-1} = 1 - \frac{1}{N\varepsilon}\frac{Ah}{1+h}, \tag{8}$$

$$Z_1 = 1 - \frac{h}{N\varepsilon}\frac{Bh}{1+h} - \frac{Ch^2}{N\varepsilon\,(1+h)^2}, \tag{9}$$

$$h_B = (\mu^2)^\varepsilon h Z_1^2 Z_2^{-2} = h\left(1 - \frac{1}{N\varepsilon}\frac{2(A+B)h}{1+h} - \frac{1}{N\varepsilon}\frac{2Ch^2}{(1+h)^2}\right). \tag{10}$$

We consider the next order of $1/N$ expansion. The corresponding diagrams for the ϕ propagator are shown in Fig.2. Again one can see that the $1/N^2$ terms contain not only the two-loop diagrams but also the three- and even four-loop ones.

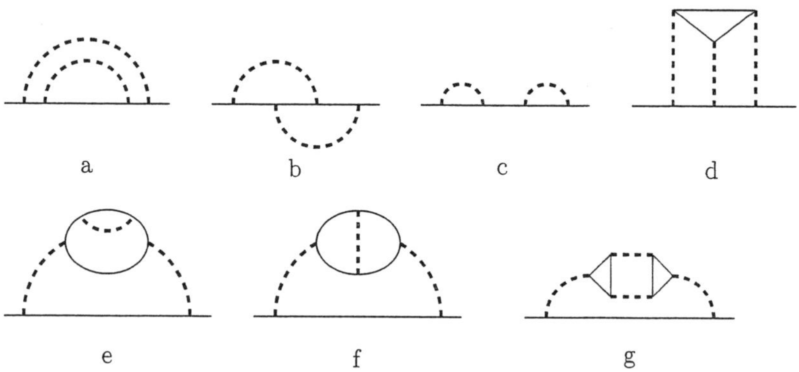

Figure 2: The second order diagrams giving a contribution to the ϕ field propagator in $1/N$ expansion

All these diagrams are double logarithmically divergent, i.e., contain both single and double poles in dimensional regularization. We calculate the leading double pole after subtraction of the divergent subgraphs, i.e. perform the R'-operation. The answer is:

$$Diag.a \Rightarrow -\frac{1}{\varepsilon^2 N^2}\frac{A^2h^2}{2(1+h)^2},\quad Diag.b \Rightarrow -\frac{1}{\varepsilon^2 N^2}\frac{ABh^2}{(1+h)^2},\quad Diag.c \Rightarrow -\frac{1}{\varepsilon^2 N^2}\frac{A^2h^2}{(1+h)^2},$$

$$Diag.d \Rightarrow -\frac{1}{\varepsilon^2 N^2}\frac{4ACh^3}{3(1+h)^3},\quad Diag.e \Rightarrow \frac{1}{\varepsilon^2 N^2}\frac{2A^2h^3}{3(1+h)^3},\quad Diag.f \Rightarrow \frac{1}{\varepsilon^2 N^2}\frac{2ABh^3}{3(1+h)^3},$$

$$Diag.g \Rightarrow \frac{1}{\varepsilon^2 N^2}\frac{ACh^4}{(1+h)^4}. \tag{11}$$

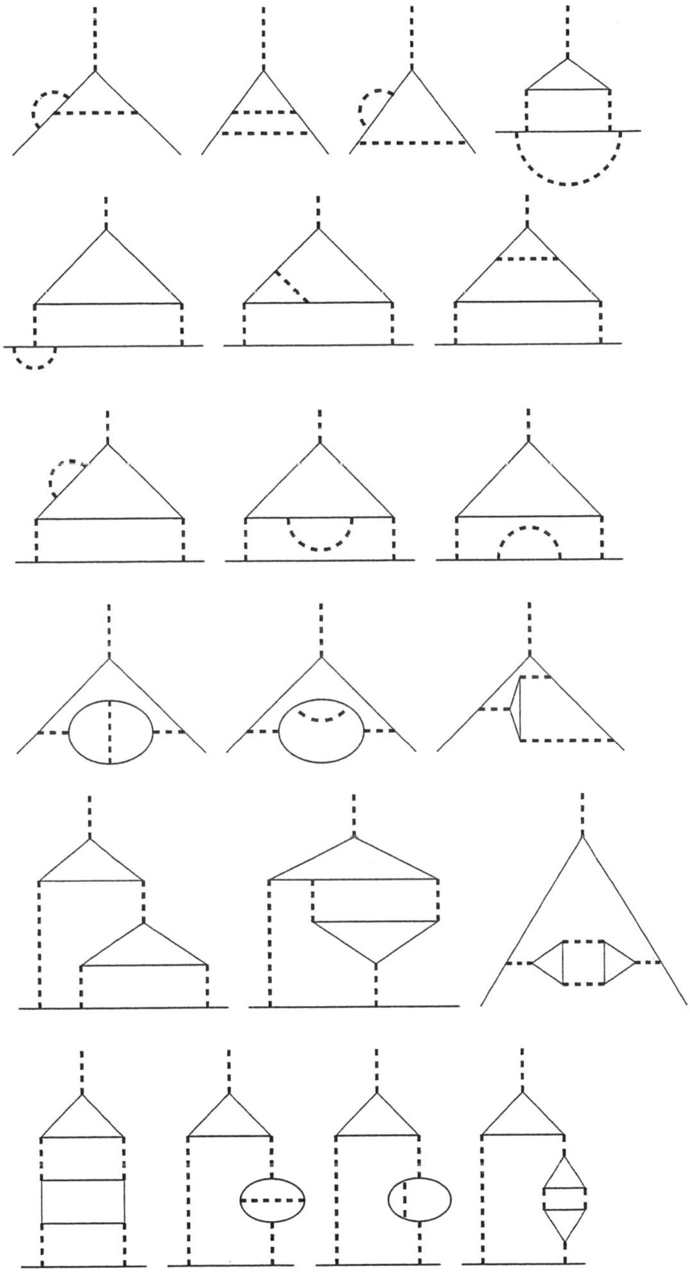

Figure 3. All vertex diagrams giving contribution to the leading logarithm in the second order of the 1/N expansion

The next our step is to find all vertex diagrams which give contribution to the leading logarithm in the second order in the $1/N$ expansion. They are shown in Fig.3, from this figure we see that here we have even a five-loop diagram which is certainly absent in the expansion in the initial coupling constant.

For example let us find the contribution to the double pole of the last diagram from the Fig.3. Writing the R' operation to this diagram we will have

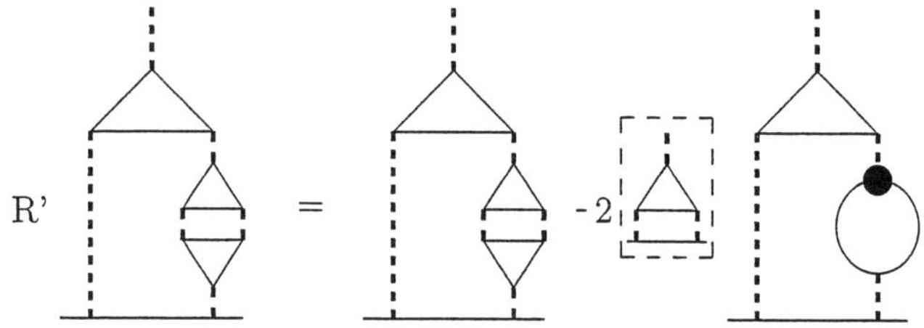

Figure 4. R' operation of the last diagram from the Fig.3

Fig.4 shows us the procedure of the R' operation. For calculation of the leading pole for the given diagram we do not need to calculate the initial diagram. Due to the locality of the theory which means that after the R' operation we have the local answer we need only to calculate the contertearm which is a lower loop diagram. The result of the R' operation for this diagram will give us

$$\frac{8}{5N^2\varepsilon^2}\frac{C^2h^5}{(1+h)^5}$$

Having all this in mind we come to the final expressions for the Z factors in $1/N$ expansion:

$$\begin{aligned} Z_1 &= 1-\frac{1}{\varepsilon N}\left(\frac{Bh}{1+h}+\frac{Ch^2}{(1+h)^2}\right)+\frac{1}{\varepsilon^2N^2}\left(\frac{3}{2}\frac{B^2h^2}{(1+h)^2}+\frac{ABh^2}{(1+h)^2}\right. \\ &+\frac{11}{3}\frac{BCh^3}{(1+h)^3}+\frac{4}{3}\frac{ACh^3}{(1+h)^3}-\frac{2}{3}\frac{ABh^3}{(1+h)^3}-\frac{2}{3}\frac{B^2h^3}{(1+h)^3}-2\frac{BCh^4}{(1+h)^4} \\ &\left.-\frac{ACh^4}{(1+h)^4}+\frac{5}{2}\frac{C^2h^4}{(1+h)^4}-\frac{8}{5}\frac{C^2h^5}{(1+h)^5}\right)+O(\frac{1}{\varepsilon N^2}), \end{aligned} \tag{12}$$

$$\begin{aligned} Z_2^{-1} &= 1-\frac{1}{\varepsilon N}\frac{Ah}{1+h}+\frac{1}{\varepsilon^2N^2}\left(\frac{3}{2}\frac{A^2h^2}{(1+h)^2}+\frac{ABh^2}{(1+h)^2}-\frac{2}{3}\frac{A^2h^3}{(1+h)^3}\right. \\ &\left.-\frac{2}{3}\frac{ABh^3}{(1+h)^3}+\frac{4}{3}\frac{ACh^3}{(1+h)^3}-\frac{ACh^4}{(1+h)^4}\right)+O(\frac{1}{\varepsilon N^2}). \end{aligned} \tag{13}$$

As we mentioned above we do not have renormalization of the σ field what means that

$$Z_3^{-1} = 1. \tag{14}$$

Having Z_1, Z_2^{-1} and Z_3^{-1} we can obtain the renormalization of the coupling h

$$\begin{aligned} Z_h &= 1-\frac{1}{\varepsilon N}\left(\frac{2(A+B)h}{1+h}+\frac{2Ch^2}{(1+h)^2}\right)+\frac{1}{\varepsilon^2N^2}\left(4\frac{(A+B)^2h^2}{(1+h)^2}-\frac{4}{3}\frac{(A+B)^2h^3}{(1+h)^3}\right. \\ &\left.+\frac{28}{3}\frac{(A+B)Ch^3}{(1+h)^3}-4\frac{(A+B)Ch^4}{(1+h)^4}+6\frac{C^2h^4}{(1+h)^4}-\frac{16}{5}\frac{C^2h^5}{(1+h)^5}\right). \end{aligned} \tag{15}$$

4 Renormalization group(RG) in 1/N expansion

Having these expressions for the Z factors one can construct the coupling constant renormalization and the corresponding RG functions. One has as usual in the dimensional regularization (let us remind that $Z_3^{-1} = 1$)

$$h_B = (\mu^2)^\varepsilon h Z_1^2 Z_2^{-2} = (\mu^2)^\varepsilon \left(h + \sum_{n=1}^{\infty} \frac{a_n(h,N)}{\varepsilon^n} \right), \tag{16}$$

$$Z_i = 1 + \sum_{n=1}^{\infty} \frac{c_n^i(h,N)}{\varepsilon^n}, \tag{17}$$

where the first coefficients a_n and c_n^i can be deduced from eqs.(12,13).

This allows one to get the anomalous dimensions and the beta function defined as

$$\gamma(h,N) = -\mu^2 \frac{d}{d\mu^2} \log Z = h \frac{d}{dh} c_1, \tag{18}$$

$$\beta(h,N) = 2h(\gamma_1 + \gamma_2) = (h\frac{d}{dh} - 1) a_1. \tag{19}$$

With the help of eqs.(12,13) one gets in the leading order of $1/N$ expansion[2]

$$\gamma_2(h,N) = -\frac{1}{N} \frac{Ah}{(1+h)^2}, \quad \gamma_1(h,N) = -\frac{1}{N} \left(\frac{Bh}{(1+h)^2} + \frac{2Ch^2}{(1+h)^3} \right), \tag{20}$$

$$\beta(h,N) = -\frac{1}{N} \left(\frac{2(A+B)h^2}{(1+h)^2} + \frac{4Ch^3}{(1+h)^3} \right). \tag{21}$$

It is instructive to check the so-called pole equations [4] that express the coefficients of the higher order poles in ε of the Z factors via the coefficients of a simple pole. For Z_2^{-1} one has, according to (13),

$$c_1(h,N) = -\frac{1}{N} \frac{Ah}{1+h}, \tag{22}$$

$$c_2(h,N) = \frac{1}{N^2} \left(\frac{3}{2} \frac{A^2h^2}{(1+h)^2} + \frac{ABh^2}{(1+h)^2} - \frac{2}{3} \frac{A^2h^3}{(1+h)^3} - \frac{2}{3} \frac{ABh^3}{(1+h)^3} + \frac{4}{3} \frac{ACh^3}{(1+h)^3} - \frac{ACh^4}{(1+h)^4} \right). \tag{23}$$

At the same time the coefficient c_2 can be expressed through c_1 via the pole equations as

$$h \frac{dc_2}{dh} = \gamma_2 c_1 + \beta \frac{dc_1}{dh}, \tag{24}$$

which gives

$$h \frac{dc_2}{dh} = \frac{1}{N^2} \frac{Ah}{(1+h)^2} \frac{Ah}{1+h} + \frac{1}{N^2} \left(\frac{2(A+B)h^2}{(1+h)^2} + \frac{4Ch^3}{(1+h)^3} \right) \frac{A}{(1+h)^2}.$$

[2]Note that the anomalous dimension of a field γ_2, is defined with respect to Z_2^{-1}.

Integrating this equation one gets for c_2 the expression coinciding with (23) which was obtained by direct diagram evaluation.

We have also checked the pole equations for the renormalized coupling. Calculating the diagrams we obtain the renormalization constant $Z_h = Z_1^2 Z_2^{-2}$ of the renormalized coupling

$$a_1(h,N) = -\frac{1}{N}\left(\frac{2(A+B)h^2}{1+h} + \frac{2Ch^3}{(1+h)^2}\right), \tag{25}$$

$$\begin{aligned} a_2(h,N) &= \frac{1}{N^2}\left(4\frac{(A+B)^2h^3}{(1+h)^2} - \frac{4}{3}\frac{(A+B)^2h^4}{(1+h)^3}\right.\\ &\left.+\frac{28}{3}\frac{(A+B)Ch^4}{(1+h)^3} - 4\frac{(A+B)Ch^5}{(1+h)^4} + 6\frac{C^2h^5}{(1+h)^4} - \frac{16}{5}\frac{C^2h^6}{(1+h)^5}\right). \end{aligned} \tag{26}$$

From the pole equations we can also get the coefficient a_2 using

$$(h\frac{d}{dh} - 1)a_2 = \beta\frac{da_1}{dh}. \tag{27}$$

Again integrating this equation one gets for a_2 the expression coinciding with (26) which was obtained by direct diagram evaluation. In the leading order in h when

$$a_1(h,N) \simeq -\frac{2(A+B)h^2}{N} \quad \text{and} \quad \beta(h,N) \simeq -\frac{2(A+B)h^2}{N}$$

one should have a geometric progression

$$a_n(h,N) = a_1(h,N)^n.$$

We have checked this relation up to third order in $1/N$ expansion and confirmed its validity.

Having expression for the β function one may wonder how the coupling is running. The crucial point here is the sign of the β function. One has

$$\beta(h,N) = -\frac{1}{N}\frac{4\Gamma(D-2)\left(\frac{2h^2}{(1+h)^2} + \frac{D(D-3)h^3}{(1+h)^3}\right)}{\Gamma(D/2-2)\Gamma(D/2-1)\Gamma(D/2+1)\Gamma(3-D/2)}. \tag{28}$$

It can also be rewritten as

$$\beta(h,N) = -\frac{1}{N}\frac{2^{D-1}\Gamma(D/2-1/2)(-)^{(D-1)/2}\left(\frac{2h^2}{(1+h)^2} + \frac{D(D-3)h^3}{(1+h)^3}\right)}{\Gamma(1/2)\pi\Gamma(D/2+1)}, \tag{29}$$

that clearly indicates the the theory is UV asymptotically free for $D = 2k+1$, k - even and IR free for k-odd. Solution of the RG equation looks somewhat complicated, but for the small coupling in the leading order it simply equals the usual leading log approximation

$$h(t,h) \simeq \frac{h}{1-\beta_0 ht}, \quad \beta_0 = -\frac{1}{N}\frac{2^D\Gamma(D/2-1/2)(-)^{(D-1)/2}}{\Gamma(1/2)\pi\Gamma(D/2+1)}. \tag{30}$$

For example, for $D = 5, 7$ the beta function equals $\beta_0 = -256/15\pi^2 N$ and $2^{12}/105\pi^2 N$, respectively. Thus we have that for $D = 5$ our theory has UV asymptotically free behaviour and for $D = 7$ it exhibits IR free behaviour.

5 Conclusion

We conclude that even in a formally non-renormalizable theory it is possible to construct renormalizable $1/N$ expansion which obeys all the rules of a usual perturbation theory. In the usual perturbation theory the expansion in coupling constant coincides with the expansion in number of loops and here we do not have such a feature. We found that even in the leading order we have the diagram with two loops and in the second order of the $1/N$ expansion we have even five-loop diagram. The expansion parameter is dimensionless, the coupling is running logarithmically, all divergencies are absorbed into the renormalization of the wave function and the coupling. The propagator of the spurion field σ behaves as $1/(p^2)^{D/2-2}$ which provides a better convergence of the diagrams. The original dimensionful coupling plays a role of a mass. Expansion over this coupling is singular and creates the usual nonrenormalizable terms.

Here we checked the renormalization group(RG) equations up to the second order of the $1/N$ expansion. We found that the beta-function is not polynomial in coupling constant as usual perturbative theory but exhibits nonpolynomial behaviour. Moreover the UV behaviour of the theory depends on the dimension D of the space-time and we can have either the UV asymptotically free behaviour for some dimensions or the IR free behaviour for other dimensions.

Acknowledgements

The author is grateful to D.I.Kazakov for valuable discussions. Also the author would like to thank the organizers of the school ISSP 2006 for a given opportunity to take part in this school and to give a talk. G.V. would like to thank "Enrico Fermi" center and Dynasty Foundation for support. Financial support from RFBR grant # 05-02-17603 and grant of the Ministry of Education and Science of the Russian Federation # 5362.2006.2 is kindly acknowledged.

References

[1] D.I.Kazakov and G.S.Vartanov, arXiv:hep-th/0607177 and references therein.

[2] M.Moshe, J.Zinn-Justin, Phys.Rept., **385** (2003) 69-228 (arXiv:hep-th/0306133).

[3] I.Ya.Aref'eva, Theor.Math.Phys., **29** (1976) 147; *ibid* **31** (1977) 3.

[4] G.t'Hooft, Nucl. Phys. **B61** (1973) 455.

[5] D.I.Kazakov and G.S.Vartanov, the paper is in preparation.

Eur. Phys. J. C (2008)
DOI 10.1140/epjc/s10052-008-0538-1

THE EUROPEAN PHYSICAL JOURNAL C

Special Article – Scientific Note

On the precision of a length measurement

X. Calmet[a, b]

Université Libre de Bruxelles, Service de Physique Théorique, CP 225, Boulevard du Triomphe, 1050 Bruxelles, Belgium

Received: 19 July 2007 / Revised version: 2 January 2008 /

Abstract. We show that quantum mechanics and general relativity imply the existence of a minimal length. To be more precise, we show that no operational device subject to quantum mechanics, general relativity and causality could exclude the discreteness of spacetime on lengths shorter than the Planck length. We then consider the fundamental limit coming from quantum mechanics, general relativity and causality on the precision of length measurement.

PACS. 04.20.-q; 03.65.-w

1 Introduction

Twentieth century physics has been a quest for unification. The unification of quantum mechanics and special relativity required the introduction of quantum field theory. The unification of magnetism and electricity led to electrodynamics, which was unified with the weak interactions into the electroweak interactions. There are good reasons to believe that the electroweak interactions and the strong interactions originate from the same underlying gauge theory: the grand unified theory. If general relativity is to be unified with a gauge theory, one first needs to understand how to unify general relativity and quantum mechanics, just as it was first necessary to understand how to unify quantum mechanics and special relativity before three of the forces of nature could be unified. The aim of this paper is much more modest – we want to understand some of the features of a quantum mechanical description of general relativity using some simple tools from quantum mechanics and general relativity. In particular, we shall show that if quantum mechanics and general relativity are valid theories of nature up to the Planck scale, they imply the existence of a minimal length in nature.

We shall address two questions: Is there a minimal length in nature, and is there a fundamental limit on the precision of distance measurement? The first question will be addressed in the second section while the second will be considered in the third section.

The usual approach to address the question of a minimal length is to do a scattering thought experiment [1–5], i.e. one studies the high energy regime of the scattering and finds that one cannot measure a length shorter than the Planck length. Here we shall argue that this is not enough to exclude a discreteness of spacetime with a lattice spacing shorter than the Planck length. The key new idea concerns how to define distance measurement more precisely and, in particular, the fact that such measurement actually involves two measurements.

We then apply our framework to the old thought experiment of Salecker and Wigner [6] and show that contractive states cannot beat the uncertainty due to quantu m mechanics for the measurement of a length. We then conclude.

2 Minimal length from quantum mechanics and general relativity

In this section we review the results obtained in [7, 8]. We show that quantum mechanics and classical general relativity considered simultaneously imply the existence of a minimal length, i.e. no operational procedure exists that can measure a distance less than this fundamental length. The key ingredients used to reach this conclusion are the uncertainty principle from quantum mechanics and gravitational collapse from classical general relativity.

A dynamical condition for gravitational collapse is given by the hoop conjecture [9]: if an amount of energy E is confined at any instant to a ball of size R, where $R < E$, then that region will eventually evolve into a black hole[1]. Although the hoop conjecture is, as its name says, a conjecture, it rests on firm footing. The least favorable case, i.e. as asymmetric as possible, is the one of two particles colliding

[a] e-mail: xcalmet@gmail.com

[b] "New Talent" Presentation Award at the 44th Course of the International School of Subnuclear Physics, Erice, Italy, 29 Aug–7 Sep 2006

[1] We use natural units where $\hbar$, c and Newton's constant (or l_P) are unity. We also neglect numerical factors of order one.

head on. It has been shown that even in that case, when the hoop conjecture is fulfilled, a black hole is formed [10, 11].

From the hoop conjecture and the uncertainty principle, we immediately deduce the existence of a minimum ball of size l_{P}. Consider a particle of energy E which is not already a black hole. Its size r must satisfy

$$r \gtrsim \max[1/E, E] \,, \tag{1}$$

where $\lambda_{\mathrm{C}} \sim 1/E$ is its Compton wavelength and E arises from the hoop conjecture. Minimization with respect to E results in r of order unity in Planck units or $r \sim l_{\mathrm{P}}$. If the particle *is* a black hole, then its radius grows with mass: $r \sim E \sim 1/\lambda_{\mathrm{C}}$. This relationship suggests that an experiment designed (in the absence of gravity) to measure a short distance $l \ll l_{\mathrm{P}}$ will (in the presence of gravity) only be sensitive to scales $1/l$.

Let us give a concrete model of minimum length. Let the position operator $\hat{x}$ have discrete eigenvalues $\{x_i\}$, with the separation between eigenvalues either of order l_{P} or smaller. (For regularly distributed eigenvalues with a constant separation, this would be equivalent to a spatial lattice, as seen in Fig. 1.) We do not mean to imply that in Nature a minimum length is realized in this particular fashion – most likely, the physical mechanism is more complicated and may involve, for example, spacetime foam or strings. However, our concrete formulation lends itself to detailed analysis. We show below that this formulation cannot be excluded by any gedanken experiment, which is strong evidence for the existence of a minimum length.

Quantization of position does not by itself imply quantization of momentum. Conversely, a continuous spectrum of momentum does not imply a continuous spectrum of position. In a formulation of quantum mechanics on a regular spatial lattice, with spacing a and size L, the momentum operator has eigenvalues which are spaced by $1/L$. In the infinite volume limit the momentum operator can have continuous eigenvalues even if the spatial lattice spacing is kept fixed. This means that the displacement operator

$$\hat{x}(t) - \hat{x}(0) = \hat{p}(0)\frac{t}{M} \,, \tag{2}$$

(where t is the time of the measurement and M the mass of the system under consideration) does not necessarily have discrete eigenvalues (the right hand side of (2) assumes free evolution; we use the Heisenberg picture throughout). Since the time evolution operator is unitary, the eigenvalues of $\hat{x}(t)$ are the same as $\hat{x}(0)$. Importantly, though, the spectrum of $\hat{x}(0)$ (or $\hat{x}(t)$) is completely unrelated to the spectrum of the $\hat{p}(0)$, even though they are related by (2). A measurement of arbitrarily small displacement (2) does not exclude our model of minimum length. To exclude it, one would have to measure a position eigenvalue x *and* a nearby eigenvalue x', with $|x - x'| \ll l_{\mathrm{P}}$.

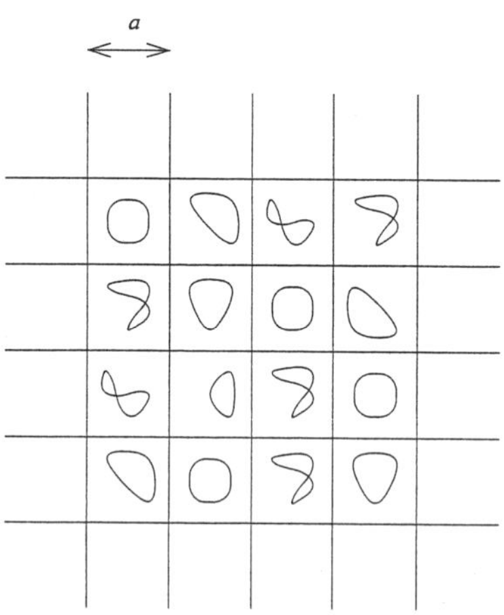

Fig. 1. We choose a spacetime lattice of spacing a of the order of the Planck length or smaller. This formulation does not depend on the details of quantum gravity

Many minimum length arguments are obviated by the simple observation of the minimum ball. However, the existence of a minimum ball does not by itself preclude the localization of a macroscopic object to very high precision. Hence, one might attempt to measure the spectrum of $\hat{x}(0)$ through a time of flight experiment in which wavepackets of primitive probes are bounced off of well-localised macroscopic objects. Disregarding gravitational effects, the discrete spectrum of $\hat{x}(0)$ is in principle obtainable this way. But detecting the discreteness of $\hat{x}(0)$ requires wavelengths comparable to the eigenvalue spacing. For eigenvalue spacing comparable or smaller than l_{P}, gravitational effects cannot be ignored because the process produces minimal balls (black holes) of size l_{P} or larger. This suggests that a direct measurement of the position spectrum to accuracy better than l_{P} is not possible. The failure here is due to the use of probes with very short wavelength.

A different class of instrument, the interferometer, is capable of measuring distances much smaller than the size of any of its sub-components. Nevertheless, the uncertainty principle and gravitational collapse prevent an arbitrarily accurate measurement of eigenvalue spacing. First, the limit from quantum mechanics – consider the Heisenberg operators for position $\hat{x}(t)$ and momentum $\hat{p}(t)$ and recall the standard inequality

$$(\Delta A)^2 (\Delta B)^2 \geq -\frac{1}{4}(\langle[\hat{A}, \hat{B}]\rangle)^2 \,. \tag{3}$$

Suppose that the position of a *free* test mass is measured at time $t = 0$ and *again* at a later time. The position operator at a later time t is

$$\hat{x}(t) = \hat{x}(0) + \hat{p}(0)\frac{t}{M} \,. \tag{4}$$

We assume a free particle Hamiltonian here for simplicity, but the argument can be generalized [7, 8]. The commutator between the position operators at $t = 0$ and t is

$$[\hat{x}(0), \hat{x}(t)] = \mathrm{i}\frac{t}{M} \,, \tag{5}$$

so using (3) we have

$$|\Delta x(0)||\Delta x(t)| \geq \frac{t}{2M}\,. \tag{6}$$

We see that at least one of the uncertainties $\Delta x(0)$ or $\Delta x(t)$ must be larger than of order $\sqrt{t/M}$. As a measurement of the discreteness of $\hat{x}(0)$ requires *two* position measurements, it is limited by the greater of $\Delta x(0)$ or $\Delta x(t)$, that is, by $\sqrt{t/M}$,

$$\Delta x \equiv \max[\Delta x(0), \Delta x(t)] \geq \sqrt{\frac{t}{2M}}\,, \tag{7}$$

where t is the time over which the measurement occurs and M the mass of the object whose position is measured. In order to push Δx below l_{P}, we take M to be large. In order to avoid gravitational collapse, the size R of our measuring device must also grow such that $R > M$. By causality, however, R cannot exceed t. Any component of the device a distance greater than t away cannot affect the measurement, hence we should not consider it part of the device. These considerations can be summarized in the inequalities

$$t > R > M\,. \tag{8}$$

Combined with (7), they require $\Delta x > 1$ in Planck units, or

$$\Delta x > l_{\mathrm{P}}\,. \tag{9}$$

Notice that the considerations leading to (7), (8) and (9) were in no way specific to an interferometer, and hence are *device independent*. We repeat: no device subject to quantum mechanics, gravity and causality can exclude the quantization of position on distances less than the Planck length.

It is important to emphasize that we are deducing a minimum length which is parametrically of order l_{P}, but may be larger or smaller by a numerical factor. This point is relevant to the question of whether an experimenter might be able to transmit the result of the measurement before the formation of a closed trapped surface, which prevents the escape of any signal. If we decrease the minimum length by a numerical factor, the inequality (7) requires $M \gg R$, so we force the experimenter to work from deep inside an apparatus which has far exceeded the criteria for gravitational collapse (i.e. it is much denser than a black hole of the same size R as the apparatus). For such an apparatus a horizon will already exist before the measurement begins. The radius of the horizon, which is of order M, is very large compared to R, so that no signal can escape.

An implication of our result is that there may only be a finite number of degrees of freedom per unit volume in our universe – no true continuum of space or time. Equivalently, there is only a finite amount of information or entropy in any finite region of our universe.

One of the main problems encountered in the quantization of gravity is a proliferation of divergences coming from short distance fluctuations of the metric (or graviton). However, these divergences might only be artifacts of perturbation theory: minimum length, which is itself a non-perturbative effect, might provide a cutoff which removes the infinities. This conjecture could be verified by lattice simulations of quantum gravity (for example, in the Euclidean path integral formulation), by checking to see if they yield finite results even in the continuum limit.

3 Limits on the measurement of large distances from fundamental physics

In the section, we study whether quantum mechanics and general relativity can limit the precision of a distance measurement. In order to address this question we shall reconsider the thought experiment first proposed by Salecker and Wigner almost 50 years ago. In order to measure the distance l we shall consider a clock which emits a light ray at a time $t = 0$. The clock will suffer a recoil from the emission of the light ray which induces a position uncertainty in the position of the clock $x(0)$. The mirror which is at a distance l from the clock will reflect the light ray, which is reabsorbed at a time t at $x(t)$ by the clock, and again there is a recoil effect and the position of the clock will have some uncertainty (see Fig. 2).

Consider the Heisenberg operators for position $\hat{x}(t)$ and momentum $\hat{p}(t)$ and recall the standard inequality

$$(\Delta A)^2(\Delta B)^2 \geq -\frac{1}{4}(\langle[\hat{A}, \hat{B}]\rangle)^2\,. \tag{10}$$

Suppose that the position of a *free* test mass is measured at time $t = 0$ and *again* at a later time. The position operator at a later time t is

$$\hat{x}(t) = \hat{x}(0) + \hat{p}(0)\frac{t}{M}\,. \tag{11}$$

The commutator between the position operators at $t = 0$ and t is

$$[\hat{x}(0), \hat{x}(t)] = \mathrm{i}\frac{t}{M}\,, \tag{12}$$

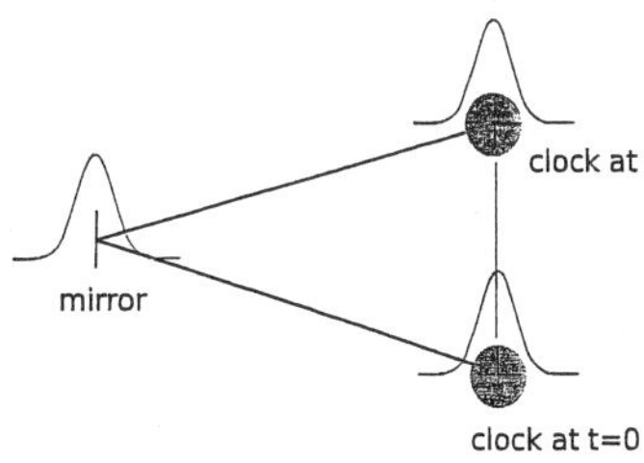

Fig. 2. Salecker and Wigner thought experiment to measure a length. A clock emits a light beam at a time $t = 0$ which is reflected by a mirror and reabsorbed by the clock at a later time $t = \tau$. Quantum mechanics implies a spread of the wave function of the clock and of the mirror

so using (10) we have

$$|\Delta x(0)||\Delta x(t)| \geq \frac{t}{2M}. \tag{13}$$

Since the total uncertainty for the measurement of the distance l is given by the sum of the uncertainties of $x(0)$ and $x(t)$, we find:

$$\delta l \sim \sqrt{\frac{t}{2M}}. \tag{14}$$

Note that we are not forced to take the mass of the clock to be large, as in the previous section. There are actually two options, one of which is to allow the mass of the clock to grow at the same rate as t, the time necessary for the measurement, in which case we have

$$\delta l \sim 1, \tag{15}$$

or

$$\delta l \sim l_{\mathrm{p}}. \tag{16}$$

The other option is to consider a fixed, finite, mass. This case applies to, e.g., the measurement of a distance performed with an interferometer such as LIGO [12]. The mass is at most the mass of the region of spacetime which feels one wavelength of the gravitational wave. In that case, the standard quantum limit [13–15] applies, and this is the well-known statement that LIGO operates at the standard quantum limit. Note that contractive states [16–18] cannot help to beat the standard quantum limit in a parametric manner. Again, as in [7, 8], the reason is that we need two measurements. Contractive states allow to make $\delta x(t)$ very small at the price of losing all the information about the uncertainty of $x(0)$ (see Appendix).

4 Conclusions

In this work we have shown that quantum mechanics and classical general relativity considered simultaneously imply the existence of a minimal length, i.e. no operational procedure exists that can measure a distance less than this fundamental length. The key ingredients used to reach this conclusion are the uncertainty principle from quantum mechanics and gravitational collapse from classical general relativity. Furthermore, we have shown that contractive states cannot be used to beat the limit obtained by Salecker and Wigner on the precision of a length measurement. Note that in that case we are not forced to consider very massive objects, and thus the gravitational collapse condition does not necessarily provide a bound. If we are forced to consider very massive objects, then the best precision that can be achieved for the measurement of a length is the minimal length itself. Our results have deep consequences for the detectability of quantum foam using astrophysical sources [19–26]. This, however, goes beyond the scope of this paper and shall be considered elsewhere.

Acknowledgements. I would like to thank G. 't Hooft and A. Zichichi for the wonderful Erice Summer School on Subnuclear Physics they have organized and where part of this work was done. It is with great pleasure that I acknowledge that the results of the third section were worked out during this summer school with S. Hsu. I am very grateful to A. Zichichi for the financial support that made my participation to this school possible as well as to H. Fritzsch who had nominated me for this school. Finally, I would like to thank F.R. Klinkhamer for enlightening discussions and for drawing my attention to the work of Salecker and Wigner. I would like to thank M. Ozawa for a helpful communication and for sending me a copy of his work on contractive states. This work was supported in part by the IISN and the Belgian science policy office (IAP V/27).

Appendix: Contractive states

Here we briefly review contractive states, following Ozawa's original work [18]. One introduces the operator $\hat{a}$ defined by

$$\hat{a} = \sqrt{\frac{m\omega}{2\hbar}}\hat{x} + \sqrt{\frac{1}{2\hbar m\omega}}\mathrm{i}\hat{p}. \tag{A.1}$$

The quantization of $\hat{x}$ and $\hat{p}$ implies $[\hat{a}, \hat{a}^\dagger] = 1$. The parameter ω is free. The twisted coherent state $|\mu\nu\alpha\omega\rangle$ is the eigenstate of $\mu\hat{a} + \nu\hat{a}^\dagger$ with eigenvalue $\mu\hat{\alpha} + \nu\hat{\alpha}^*$. The normalization of the wave function implies $|\mu|^2 - |\nu|^2 = 1$. The free Hamiltonian is given by

$$\hat{H} = \hat{p}/2m = \hbar\omega/2\left(\hat{a}^\dagger\hat{a} + \frac{1}{2} - \frac{1}{2}\hat{a}^2 - \frac{1}{2}\hat{a}^{\dagger 2}\right) \tag{A.2}$$

and the wave function of this state is given by

$$\langle x|\mu\nu\alpha\omega\rangle = \left(\frac{m\omega}{\pi\hbar|\mu-\nu|^2}\right)^{1/4} \times \exp\left(-\frac{m\omega}{\pi\hbar}\frac{1+2\xi\mathrm{i}}{|\mu-\nu|^2}(x-x_0)^2 + \mathrm{i}p_0(x-x_0)\right), \tag{A.3}$$

with $\xi = \mathrm{Im}(\mu^*\nu)$, $\alpha = (m\omega/2\hbar)^{1/2}x_0 + 1/(2\hbar m\omega)^{1/2}\mathrm{i}p_0$ where x_0 and p_0 are real. The position fluctuation for a free-mass is given by:

$$\Delta x(t)^2 = \frac{1}{4\xi}\frac{\hbar\tau}{m} + \frac{2\hbar}{m\omega}\left(|\mu+\nu|\frac{\omega}{2}\right)^2(t-\tau)^2, \tag{A.4}$$

with

$$\tau = \xi\hbar m/\Delta p(0)^2. \tag{A.5}$$

When $\xi > 0$ the x-dependent phase leads to a narrowing of $\Delta x(t)$ compared to $\Delta x(0)$. States with the property $\xi > 0$ are called contractive states. The absolute minimum is achieved for a time τ given by

$$\tau = \frac{2\xi}{\omega|\mu+\nu|^2)} = \frac{\xi\hbar m}{\Delta p(0)^2} \tag{A.6}$$

and one obtains

$$\Delta x(\tau) = \frac{\hbar}{2\Delta p(0)} = \frac{\Delta x(0)}{\sqrt{1+4\xi^2}}\,. \qquad \text{(A.7)}$$

The price to pay to make $\Delta x(\tau)$ very small, i.e. smaller than the Planck length, for example, is to pick ξ very large, which implies that τ is very large and thus $\Delta x(0)^2$ is very large as well. Keeping in mind that the measurement of a distance implies two measurements, we see that it is not possible to parametrically make the uncertainty in the measurement of a distance arbitrarily small. This is equivalent to the statement that LIGO and similar interferometers operate at the quantum limit – one can beat by some small factor the standard quantum limit, but it cannot be beaten parametrically.

References

1. L.J. Garay, Int. J. Mod. Phys. A **10**, 145 (1995) [arXiv:gr-qc/9403008]
2. C.A. Mead, Phys. Rev. B **135**, 849 (1964)
3. T. Padmanabhan, Class. Quantum Grav. **4**, L107 (1987)
4. K. Fredenhagen et al., Commun. Math. Phys. **172**, 187 (1995)
5. K. Fredenhagen et al., Phys. Lett. B **331**, 33 (1994)
6. H. Salecker, E.P. Wigner, Phys. Rev. **109**, 571 (1958)
7. X. Calmet, M. Graesser, S.D.H. Hsu, Phys. Rev. Lett. **93**, 211101 (2004) [arXiv:hep-th/0405033]
8. X. Calmet, M. Graesser, S.D.H. Hsu, Int. J. Mod. Phys. D **14**, 2195 (2005) [arXiv:hep-th/0505144]
9. K.S. Thorne, Nonspherical gravitational collapse: A short review. In: Magic Without Magic, ed. by J.R. Klauder (W.H. Freemann, San Francisco, 1972) pp. 231–258
10. D.M. Eardley, S.B. Giddings, Phys. Rev. D **66**, 044011 (2002) [arXiv:gr-qc/0201034]
11. S.D.H. Hsu, Phys. Lett. B **555**, 92 (2003) [arXiv:hep-ph/0203154].
12. http://www.ligo.caltech.edu/
13. V.B. Braginsky, Y.I. Vorontsov, Usp. Fiz. Nauk. **114**, 41 (1974)
14. V.B. Braginsky, Y.I. Vorontsov, Sov. Phys. Uspekhi **17**, 644 (1975)
15. C.M. Caves, Phys. Rev. Lett. **54**, 2465 (1985)
16. H.P. Yuen, Phys. Rev. Lett. **51**, 719 (1983)
17. H.P. Yuen, Phys. Rev. Lett. **52**, 788 (1984)
18. M. Ozawa, Realization of Measurement and the Standard Quantum Limit, Squeezed and Nonclassical Light. In: Squeezed and Nonclassical Light, ed. by P. Tombesi, E.R. Pike (Plenum, New York, 1989), pp. 263–286
19. Y.J. Ng, H. van Dam, Phys. Lett. B **477**, 429 (2000) [arXiv:gr-qc/9911054]
20. Y.J. Ng, H. van Dam, Found. Phys. **30**, 795 (2000) [arXiv:gr-qc/9906003]
21. G. Amelino-Camelia, T. Piran, Phys. Lett. B **497**, 265 (2001) [arXiv:hep-ph/0006210]
22. R. Lieu, L.W. Hillman, arXiv:astro-ph/0211402
23. R. Aloisio, P. Blasi, A. Galante, P.L. Ghia, A.F. Grillo, Astropart. Phys. **19**, 127 (2003) [arXiv:astro-ph/0205271]
24. Y.J. Ng, H. van Dam, W.A. Christiansen, Astrophys. J. **591**, L87 (2003) [arXiv:astro-ph/0302372]
25. R. Lieu, L.W. Hillman, Astrophys. J. **585**, L77 (2003) [arXiv:astro-ph/0301184]
26. W.A. Christiansen, Y.J. Ng, H. van Dam, Phys. Rev. Lett. **96**, 051301 (2006) [arXiv:gr-qc/0508121]

Eur. Phys. J. C 52, 477–485 (2007)
DOI 10.1140/epjc/s10052-007-0383-7

THE EUROPEAN
PHYSICAL JOURNAL C

Special Article – Young Scientists Paper

Search for the lepton flavour-violating decay $\mu \to e\gamma$

The MEG experiment probes GUT scale physics with MeV particles

Y. Hisamatsu[a]

Department of Physics, Graduate School of Science, The University of Tokyo, 7-3-1 Hongo, Bunkyo-ku, Tokyo, Japan

Received: 25 June 2007 /
Published online: 17 August 2007 – © Springer-Verlag / Società Italiana di Fisica 2007

Abstract. I introduce the upcoming MEG experiment, which will search for the rare decay $\mu \to e\gamma$ down to the branching ratio of 10^{-13}. In order to suppress the background and achieve this unprecedented sensitivity, a great deal of thought went into designing this experiment. Here, I describe the essential components of this experiment, the beam line, the positron spectrometer, and the liquid xenon γ-ray detector.

PACS. 29.40.Gx; 29.40.Mc

1 Introduction

Studies on flavour-violating processes have provided us with a detailed picture of the flavour structure of quarks and leptons. The flavour mixing among quarks arises when quarks couple to the Higgs field by Yukawa interactions. This is described by the off-diagonal terms in the Cabibbo–Kobayashi–Maskawa matrix within the framework of the Standard Model (SM), and has been studied intensively in many experiments. The mass matrix of charged leptons, on the other hand, is fully diagonal because of the absence of right-handed neutrinos. Therefore, flavour mixing in the lepton sector is explicitly forbidden in the SM.

The recent discovery of neutrino mixing has, however, shed light on the physics beyond the SM. Neutrino masses and mixing can be explained by introducing the Yukawa coupling of right-handed neutrinos, and the natural explanation of small neutrino masses is provided by the see-saw mechanism which predicts that neutrinos are Majorana particles. This extended-SM scenario implies the existence of flavour violation among charged leptons, and one such process, $\mu \to e\gamma$, which is the focus of the MEG experiment, is shown in its SM-form in Fig. 1a. However, the predicted rate of flavour mixing for charged leptons is still infinitesimal. The branching ratio of $\mu \to e\gamma$ is predicted to be $\mathrm{Br}(\mu \to e\gamma) \leq O(10^{-40})$ [1].

In the SUSY-GUT framework [2–4], quarks and leptons have superpartners and are unified into larger multiplets. When such theories are brought down to the SUSY-breaking scale, the Yukawa interaction induces the radiative corrections, which result in the off-diagonal term in the slepton mass matrix and an observable rate of flavour mixing for charged leptons. Taking again $\mu \to e\gamma$, which so far has never been observed, Fig. 1b shows one of the possible processes as predicted by SUSY–GUT scenarios. Such models, together with the neutrino parameters obtained from the observation of neutrino oscillations, predict the rate of $\mu \to e\gamma$ to be around $10^{-11} - 10^{-14}$.

In the following sections I introduce the upcoming MEG experiment [5, 6], a dedicated search for the $\mu \to e\gamma$ decay.

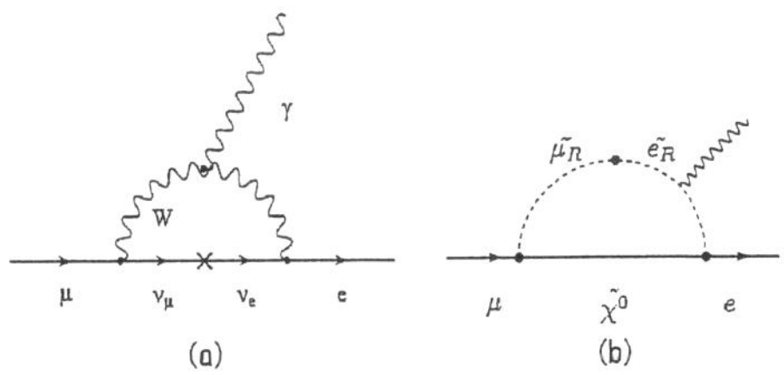

Fig. 1. Possible processes of $\mu \to e\gamma$ in (a) the SM and (b) SUSY–GUT

2 The MEG experiment

The MEG experiment plans to search for $\mu \to e\gamma$ down to a branching ratio of 10^{-13}, two orders of magnitude below the present limit of 1.2×10^{-11} [7]. Therefore, with the above-mentioned current predictions, MEG has a real chance for probing GUT-scale physics.

In order to achieve this, however, we need a large number of low energy muons and a precision detector. The MEG experiment will measure the decays of up to 10^8 stopped muons

[a] e-mail: pekochan@icepp.s.u-tokyo.ac.jp

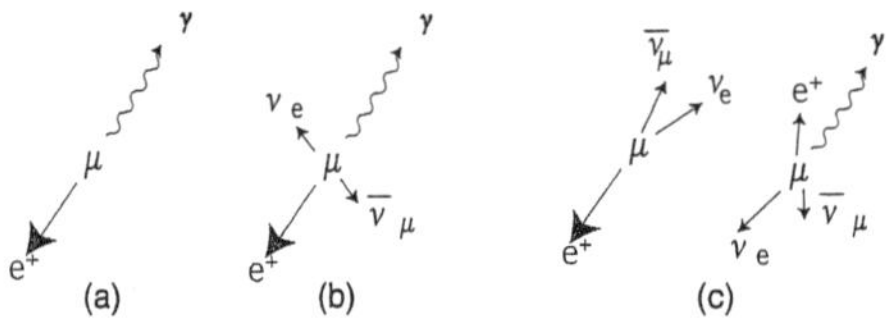

Fig. 2. Kinematics of (1) $\mu \to e\gamma$, (2) radiative muon decay and (3) e.g. accidental overlap event between e^+ from Michel decay and a γ-ray from radiative decay

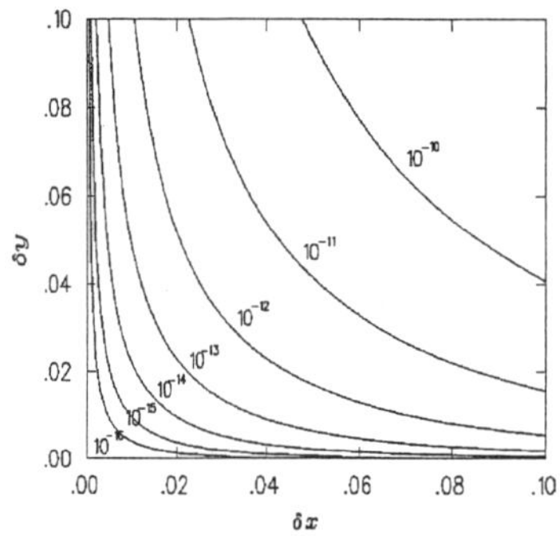

Fig. 3. Effective branching ratio of the physics background from muon radiative decay as a function of the e^+ energy resolution δx and photon energy resolution δy. δx and δy are defined as the half-widths of $\mu \to e\gamma$ signal region for $x = 2\,E_e/m_\mu$ and $y = 2\,E_\gamma/m_\mu$ [8]

per second with unprecedented sensitivity. The resolution of the detectors will be introduced in Sect. 4.

The kinematics of $\mu \to e\gamma$ is quite simple (Fig. 2); a two-body decay with a positron and a γ-ray emitted back-to-back, at the same momentum (52.8 MeV/c), and coincident in time. Concerning the background of the $\mu \to e\gamma$ decay, there are two major sources; 1) physics background from radiative muon decay $\mu^+ \to e^+ v_e \overline{v}_\mu \gamma$, and 2) accidental background, the coincidence between a positron from Michel decay ($\mu^+ \to e^+ v_e \overline{v}_\mu$) and a γ-ray coming from other sources (Fig. 2). Taking into account our detector performance, the accidental background is the more serious of the two backgrounds since the phase space of the radiative decay is strongly limited when the positron and photon energies are almost half of the muon mass, thus mimicking a $\mu \to e\gamma$ decay as shown in Fig. 3 for good energy resolutions.

The key to suppressing such accidental background is a good quality muon beam and precision detectors with unprecedented sensitivity. A great deal of very careful consideration went into designing this challenging experiment; the beam line provides a high-rate continuous muon beam in order to reduce accidental background. A tremendous effort went into making the drift chambers out of low mass material so that they could measure low energy positrons precisely. Great care was taken in selecting the scintillating material for the timing counter to achieve the best possible timing resolution, 100 psec (FWHM). An innovative gamma detector based on liquid xenon was developed for the precise measurement of position, energy and timing of the γ-ray.

In the following sections, these essential components of the experiment are described.

3 Beam line

3.1 π E5 beam line at the Paul Scherrer Institut

The MEG experiment will be conducted at the π E5 beam line at the Paul Scherrer Institut (PSI) Switzerland, where the world's most intense DC muon beam is available.

PSI has a long tradition of pion and muon physics with its isochronous ring cyclotron capable of accelerating protons to 590 MeV at a current close to 2 mA. The extracted 1.1 MW beam with 50 MHz time structure is delivered to two graphite target stations which serve several secondary beam lines.

The π E5 beam line extracts so-called "surface muons", 100% polarized and initially monoenergetically produced muons of 29.79 MeV/c from stopped pion decay, close to the surface of the production target [9]. These are transmitted in a narrow momentum band ($\delta p/p \sim 8\%$ FWHM), centered around 28 MeV/c, and originate from muons born at somewhat greater depths in the target. Due to the muon lifetime and the radio-frequency structure of the proton beam, a DC muon beam is produced, which has distinct advantages for a coincidence experiment such as $\mu \to e\gamma$. For example, the instantaneous rate is lower than a pulsed beam and offers less accidental background, which increases quadratically with beam rate. The narrow momentum band also makes it easier to stop the particles in a very thin target, which also helps to minimise the production of background particles which can enter the detector.

3.2 MEG beam transport system

The purpose of the MEG beam transport system is 4-fold: (1) to transmit the high intensity surface muon beam with minimal losses, (2) stop the maximum number of muons in the thinnest and smallest possible target at the centre of the MEG detector, (3) produce a source of maximally depolarised muons at the target, (4) minimise the transmission of unwanted beam positrons and background.

In order to achieve these aims, a beam transport system comprising two sets of quadrupole magnets arranged as "Triplets" is used (see Fig. 4). The first triplet enables the optimal injection of the beam into the crossed field separator (WIEN-filter), which runs at 200 kV. The second triplet focuses the well-separated (7.5σ) muon and beam-positron spots onto a collimator system placed between it and the beam transport solenoid BTS, allowing the transmission of a pure muon beam from the initial e/μ ratio of 8 : 1.

The BTS, a 2.8 m long, 38 cm diameter warm-bore superconducting solenoid, runs at a maximum magnetic

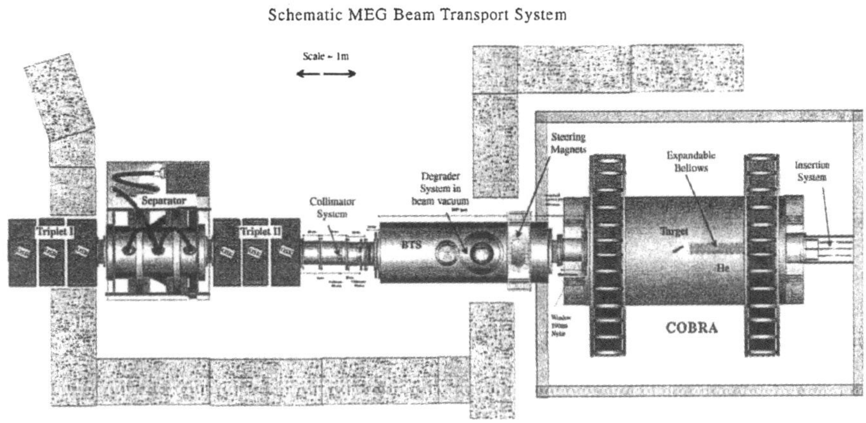

Fig. 4. Schematic of the π E5 MEG beam line at PSI. Muons enter from the left through a quadrupole triplet magnet, and then pass through a Wien filter, and are focused by a second quadrupole triplet onto a collimator, which transmits muons to the BTS and degrader system before being focused onto a target at the centre of the MEG detector (COBRA)

field of 0.5 T. A CH_2 degrader system placed at the central focus of the BTS reduces the momentum of the muons, which must traverse 1.5 m of He-gas to reach the centre of the COBRA spectrometer, where they are stopped in a 150 μm thick CH_2 target that is inclined at about 20° relative to the incoming beam.

The MEG beam transport system has so far been able to deliver a beam of 1.1×10^8 μ^+/s at 1.8 mA proton beam current, with a beam spot size of $\sigma \sim 11$ mm to the target.

4 The MEG detector

The MEG detector consists of a positron spectrometer called COBRA (constant-bending-radius) and a liquid xenon gamma detector. The spectrometer is comprised of 3 main components: 1) the COBRA magnet, which is a special gradient-field superconducting magnet, as described in Sect. 4.1.1; 2) a set of low-mass radial drift chambers as tracking devices, detailed in Sect. 4.1.2; and lastly, 3) a set of fast double-layered scintillation timing-counter arrays, placed at either end of COBRA to give both timing and trigger information. These are described in Sect. 4.1.3. The γ-ray, originating from the target, traverses the thin walls of COBRA and is converted into an electromagnetic shower in the liquid-xenon gamma detector, described in Sect. 4.2.

4.1 COBRA positron spectrometer

4.1.1 COBRA magnet

The superconducting COBRA magnet [10] forms a special gradient field (1.27 T at the centre to 0.49 T at both ends), as shown in Fig. 6. The magnet consists of five coils of three different radii, with the field gradient produced by this step-structure in the coil layout as well as by the winding density of the conductor in each coil. A high-strength aluminum-stabilised conductor was developed for this magnet to minimise the thickness of the support structure. The conductor itself, a copper matrix with a NbTi multi-filamentary core, is embedded in an aluminium stabiliser mechanically reinforced by "micro-alloying" and "cold work hardening" technologies [11, 12]. The aluminum stabiliser can be reinforced by adding a small amount of metal, such as nickel, magnesium or copper, while keeping electrical resistivity as low as possible; in this case, 5000 ppm of nickel was added. The total thickness of the central part of the magnet, where the γ-ray traverses, is in this way reduced down to 0.193 X_0.

The magnet is equipped with a pair of resistive compensation coils in order to cancel the stray field from the superconducting magnet in the vicinity of the LXe gamma detector. Any external magnetic field would deteriorate the performance of the photomultiplier tubes used in the LXe gamma detector. A residual field of less than 50 Gauss is achieved over the whole region of the LXe gamma detector.

In the gradient field, positrons with the same absolute momenta follow trajectories with a constant projected bending radius, independent of their emission angle as shown in Fig. 7a. This allows only the higher momentum positrons from the high-rate Michel decays to enter the drift chambers, thus helping to prevent accidental overlap events, which cause unwanted dead-time. This also helps to sweep away positrons quickly and prevents them from making many turns inside the tracking volume, as shown in Fig. 7b.

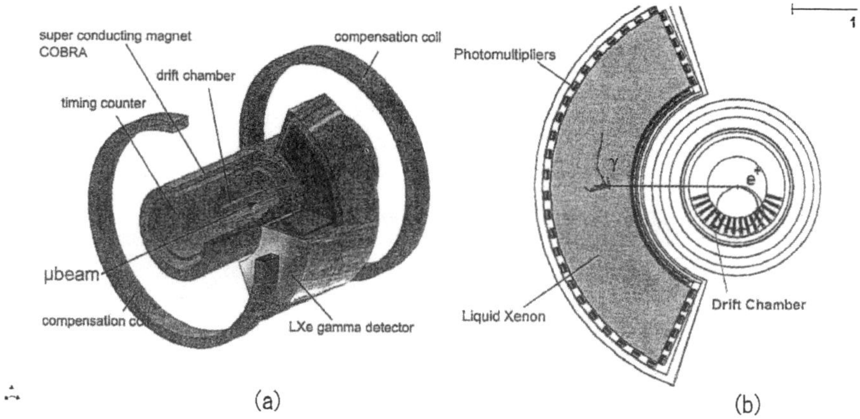

Fig. 5. (a) Schematic view of the MEG detector, and (b) cross-sectional view of the MEG detector

4.1.2 Drift chamber

The MEG drift chamber system [13] is placed radially around the target region, as indicated by the yellow trapezoid in Fig. 5.

It is composed of 16 modules, each with 2 layers of anode planes of nine drift cells each. The drift-cell of each layer is characterised by a central anode wire and two potential wires on either side, spaced 9 mm apart. Orthogonal to and 3.5 mm on either side of the wire plane are two Vernier Pad cathode planes. The two layers are staggered by a half-cell to resolve the left–right ambiguity. The sense wires are read out at both ends, as shown in Fig. 8. The ratio of charges measured at both ends provides the approximate position of the particle along the wire (z-coordinate) with $\sigma \sim 1$ cm resolution. From this information, the correct cycle of the repeating Vernier pad pattern is chosen, and a more precise z-coordinate is determined from the induced charge on the Vernier pads of the two associated parallel cathode planes on either side of the wire. The Vernier pattern of each cathode is shifted by 1/4 of a period to obtain a homogeneous resolution along the wire. For the z-direction, $\sigma_z = 300$–$500\,\mu$m, and by meas-

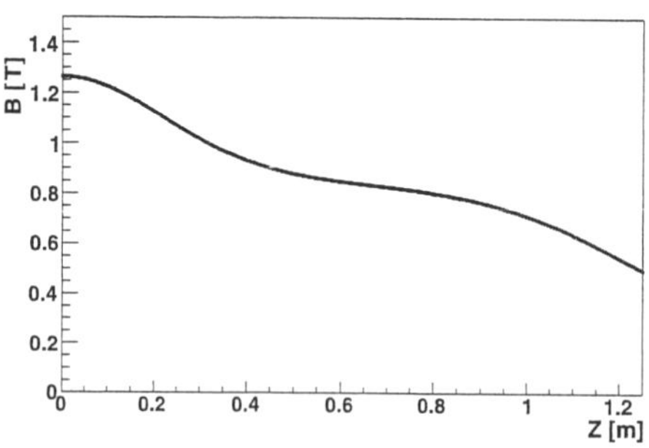

Fig. 6. Measured magnetic field of the COBRA magnet along the beam axis

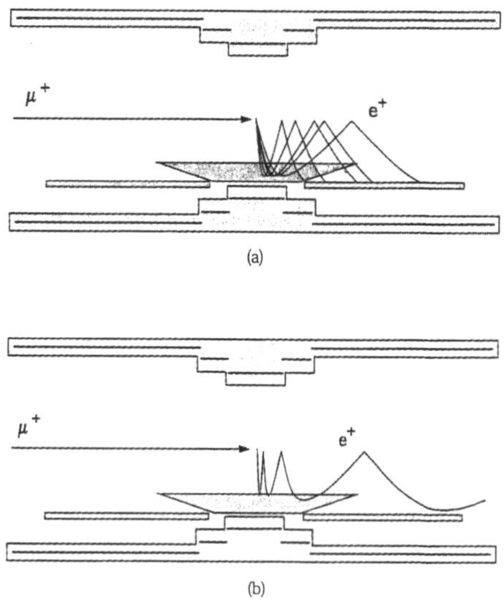

Fig. 7. Principle of the COBRA spectrometer; advantages of a gradient magnetic field

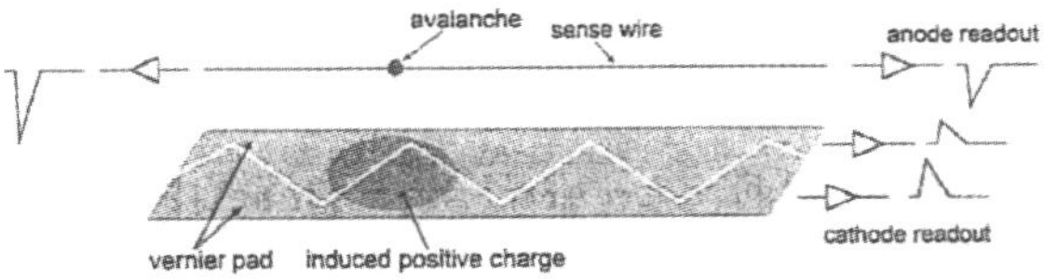

Fig. 8. Read-out scheme of the drift chamber. The sense wire is read out from both ends for a rough position reconstruction. Vernier pads are used for a more precise reconstruction

Fig. 9. Anode frame with wires (*front*); cathode frame with the Vernier foil (*middle*); completed drift chamber (*back*)

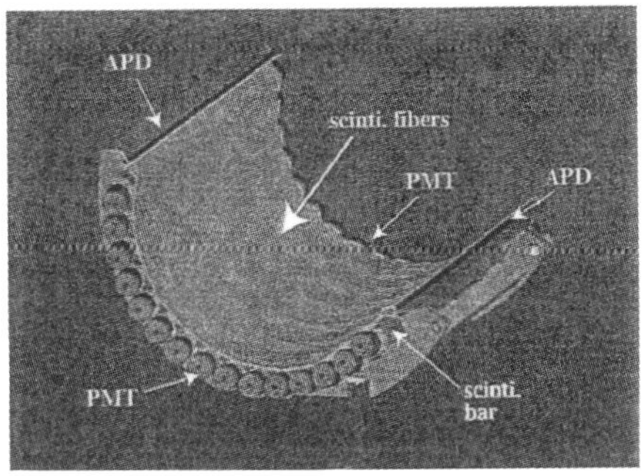

Fig. 10. Schematic view of a timing counter array. PMTs are connected to both ends of the scintillator bar. Scintillation fibres cover the scintillator bars and are read out by APDs

uring the drift time for the distance of the track to the wire, $\sigma_r = 100$–$200\,\mu$m are resolutions that have been obtained from tests with a prototype chamber. This good position resolution of the drift chamber is crucial for the precise determination of the opening angle between the positron and the gamma as well as for determining the positron momentum.

The MEG drift chambers are designed as a low-mass chamber system in order to minimise multiple scattering and γ-ray generation inside the chamber. The chamber gas is a helium-based mixture (He:$C_2H_6 = 1:1$). The chamber walls are made of a thin foil (12.5 μm thick polyimide with 2500 Å aluminium deposition), and the frames of the chamber are made of carbon-fibre. Moreover, material could further be reduced by making the chamber frames in a truncated "V-shaped" structure, open on the target side (Fig. 9). Thanks to these exhaustive efforts, the overall material along the positron track is reduced to 1.5×10^{-3} of a radiation length.

The drift chambers successfully experienced their first engineering run at the π E5 beam line of PSI in December 2006, where they were exposed to Michel positrons generated by $10^7 \sim 10^8$ muons/s. Currently, the data from the chamber is being analysed with respect to the chamber calibration and resolutions.

4.1.3 Timing counter

After leaving the drift chambers, positrons enter a set of plastic scintillation counters and scintillating fibre bundles called timing counters, which are placed axially on either side of the drift chambers [14]. The counters are intended to provide the precise timing of the positrons with 100 ps FWHM resolution, whereas the fibre bundles are used for the first-level trigger to determine the positron direction with respect to the γ-ray. The schematic view of a timing counter array is shown in Fig. 10.

The counters are made up of 15 plastic scintillator bars whose dimensions are $4 \times 4 \times 90$ cm, and their output is read by photomultipliers connected at both ends. The bars are made of Saint-Gobain Crystals – BC404 scintillator with a fast rise time of 700 ps. They are rotated by 20° with respect to the radial axis, as shown in Fig. 11a, in order to obtain a uniform track length for all positrons, which helps to make the pulse height distribution narrow. The PMTs are directly attached to each bar without using light guides to avoid a timing spread. Since one of the PMTs is placed close to the centre of the COBRA magnet, where the magnetic field is 1 T, fine-mesh PMTs of type R5924 from Hamamatsu Photonics are adopted. The PMTs are also tilted by $\sim 20°$ with respect to the magnetic field in order to create a "zero-field" environment inside them, which helps to reduce the loss of their gain (Fig. 11).

The timing resolution of the counters was measured with an electron beam at the Test Beam Facility in the Laboratori Nazionali di Frascati. The counters were irradiated with 25–550 MeV e^+ and 25–800 MeV e^- with a spot-size of $\sigma_z = 5$ mm. Here, z is the coordinate along the counters. The timing resolution was measured at 6 points along the bar at intervals of 10 cm, and the excellent resolution of ~ 100 ps FWHM was achieved at all measured points.

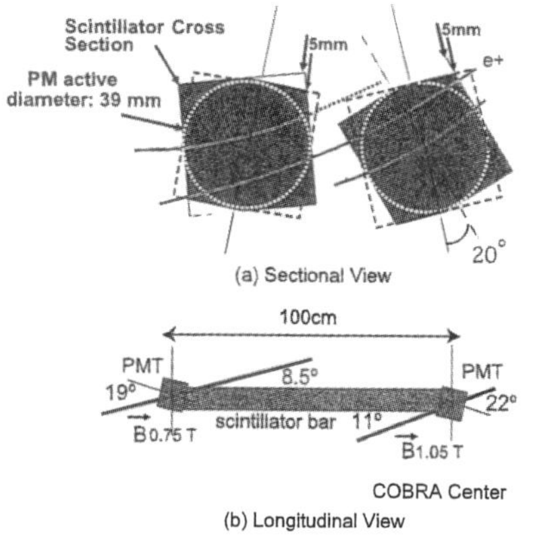

Fig. 11. (**a**) Cross-sectional view of a TC bar. The scintillator bar is slightly tilted to give a uniform track length of e^+. (**b**) Longitudinal view of a TC bar. The PMT is attached to the scintillator bar and is also slightly rotated

4.2 Liquid xenon gamma ray detector

An innovative liquid xenon (LXe) scintillation detector was devised for this experiment in order to make very precise measurements of the energy, position and timing of the γ-rays. The detector consists of 900 litres of LXe as a scintillator and 846 photomultipliers (PMTs) for the collection of scintillation light. LXe is viewed from all sides by PMTs which are directly immersed in LXe (Fig. 5). This characteristic design, which is similar to that of the Kamiokande detector, has distinct advantages: a high light yield from the scintillation light of LXe, fast scintillation signals and a homogeneous, non-segmented medium viewed via direct light, which gives a very good uniformity and allows a large active volume.

The performance of such a detector depends largely on the amount of scintillation light collected by the PMTs, but also on the transparency of the medium as well as a homogeneous PMT coverage. The energy of a γ-ray is determined by the total amount of scintillation light, and the incident position is calculated from the distribution of scintillation light on the detector walls. The arrival time of scintillation light at each PMT determines the timing of the γ-ray. This led us to conduct an intensive study on the optical properties of LXe and the development of PMTs to improve the collection of scintillation light, as described in Sect. 4.2.1 and Sect. 4.2.3.

Some of the novelties of this detector include how its special C-shaped form helps to minimise the quantity of LXe needed, and that the PMTs are operated inside the LXe, which is challenging from the point of view of both γ-ray detectors and cryogenics. The cryogenic detector operation was also optimised by the introduction of a pulse-tube refrigerator, suitable for LXe, which significantly reduces the vibrational noise at the cold-head. This is introduced in Sect. 4.2.2.

4.2.1 Liquid xenon as scintillator

LXe has a high light yield and a fast decay time, which are advantageous for precise energy measurements and the identification of pile-up γ-rays. The optical properties of LXe are listed in Table 1. The conversion from energy to scintillation photons is done by the formation of an excited state of Xe; $Xe_2^* \to 2Xe + h\nu$. Here, $h\nu$ is a vacuum-ultraviolet photon, which is not absorbed by the xenon itself since the inverse reaction, $2Xe + h\nu \to Xe_2^*$, does not occur. In principle, LXe is transparent to its scintillation light. However, impurities like water (even at the ppm-level) or oxygen (to a lesser extent) absorb scintillation light considerably, which leads to a deterioration of the detector performance. Hence, the evaluation of the LXe absorption length [15] is essential for understanding the LXe detector response. This is briefly described in Sect. 4.2.4.

Our group worked intensively to develop a purification system for LXe [16]. The initial efforts were focused on gas-phase purification, from which it was found that the main absorption comes from water [17]. This led to the current liquid-phase purification system, dedicated to the removal of water. This system consists of molecular sieves (MS13A) and a cryogenic centrifugal pump from BarBer-Nichols Inc., and it aims at removing water, the main impurity responsible for the absorption of scintillation light in our detector. Its performance was tested with 100 litres of LXe, and the impurity concentration was successfully reduced from 250 ppb to 50 ppb within 5 hours. With this purification system, LXe is free from the problem of the degradation of its optical properties during the detector operation. This is one great advantage that the LXe detector has over other crystal detectors, which are difficult to purify during their operation.

4.2.2 Cryogenics for liquid xenon

One of the main requirements of the LXe Cryogenic System is that of temperature stability, which is necessary to ensure the stable operation of the PMTs in the cryo-fluid. This was the reason for choosing a refrigeration system over conventional cooling. Furthermore, by using a pulse-tube refrigerator [19], which has no moving parts in the cold-head, a minimum of intervention is needed inside the

Table 1. Properties of LXe

Emission peak	178 nm
Spectral width	14 nm
Refractive index	1.62 ± 0.05 [18]
Decay time (recombination)	45 ns
Decay time (fast components)	4.2 ns
Decay time (slow components)	22 ns

system and no vibrational noise is transmitted to the detector. To allow for efficient liquefaction, retention and re-condensation of the LXe in our detector, we employed a compact system (developed by KEK and the Iwatani Gas Co.) with a heat capacity of 200 W, which was sufficient for our needs. During the operation of the prototype detector, a pressure and temperature stability of ± 0.001 MPa and ± 0.1 K, respectively, were maintained.

4.2.3 Photomultiplier for liquid xenon

In order to view the 178 nm scintillation light, the PMTs are directly immersed in 167 K LXe, a severe environment for their operation. This environment poses constraints on the types of materials that can be used, as is apparent from the perspectives of contamination and thermal shock. Heat dissipation was carefully considered during the design of the base-circuit of the PMTs. Finally, the fact that the PMT housing would have to withstand pressures of up to 3 bar also required careful R&D. A photocathode sensitive to the VUV range was required, with added difficulty coming from the high electrical resistivity of the photocathode at low temperatures, which decreases quantum efficiency and degrades photoelectron emission in high-rate counting environments.

To overcome these difficulties, a new type of PMT (Fig. 12) was developed together with Hamamatsu Photonics Inc. It has a special low surface resistivity K−Cs−Sb photocathode, made by forming an aluminium strip pattern on the silica window first, and then evaporating the photocathode material. In order to make a compact type of device, a metal channel dynode structure that could function in magnetic fields up to 50 Gauss was used.

The characteristics of the base circuit (shown in Fig. 13) feature a high resistance voltage divider to minimise heat dissipation and Zener diodes placed between the last two dynodes as a voltage regulator to stabilise the PMT gain in high-rate environments. Noise filtering resistors are also placed in series with the diodes that filter at low temperature. The PMT output linearity has been confirmed above a dynode current of 4 μA, which is sufficient for our operational conditions.

So far, over 1000 PMTs have been manufactured by Hamamatsu Photonics Inc. in Japan, and tested for performance prior to their installation in the LXe γ-ray

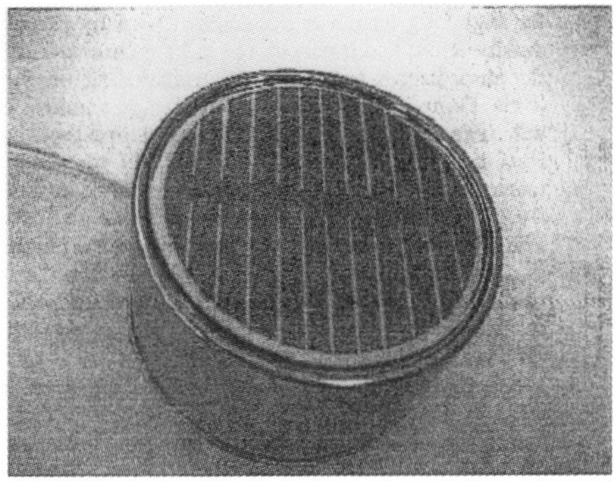

Fig. 12. Photomultiplier for the MEG LXe detector

detector. The tests were conducted using a large prototype detector, which will be described in Sect. 4.2.4, and a PMT test facility at INFN Pisa [20]. The quantum efficiency (QE) was measured to be 16% on average, and their good stability in high-rate counting environments was confirmed [21].

4.2.4 Monitoring and calibration of the detector

The good performance of the LXe detector is assured by stability monitoring and precise calibration. The stability is monitored using special α-sources mounted in the detector. Several point-like ^{241}Am α-sources (200–300 Bq each) are deposited on a 100 μm diameter wire, several of which are stretched inside the detector [22]. α-particles deposit a constant energy in the liquid, resulting in the emission of a constant amount of scintillation light from all points, as if the sources were floating in the liquid. The PMT long-term stability is monitored using this method. In addition, the PMT response linearity can be investigated by observing α-source events under different conditions of beam intensity.

The transparency of the liquid is another important parameter to be monitored. The scintillation light in the liquid suffers from Rayleigh scattering and absorption. The

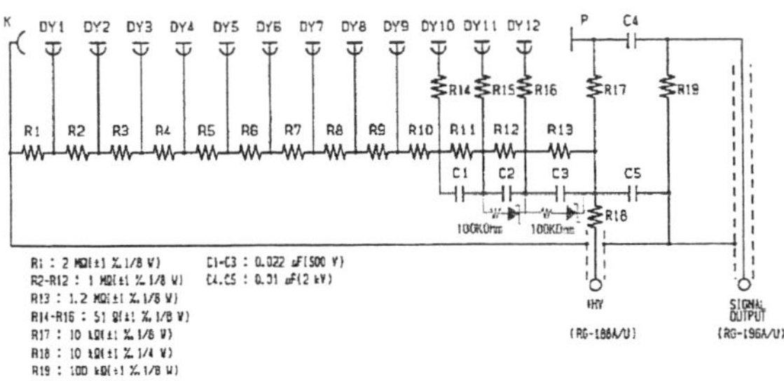

Fig. 13. PMT Base circuit: a Zener diode is placed between the last two dynodes, as indicated in *red*

Rayleigh scattering length is determined by the electron density of the liquid and is considered to be constant, while the absorption length can differ depending on impurity levels. Point-like α-sources are utilized to monitor this as well. The amount of scintillation light observed by each PMT is plotted as a function of the PMT to α-source distance and compared to a simulation, taking into account the effect of detector geometry and Rayleigh scattering. The resulting plot is then fitted with an exponential function to evaluate the absorption length, which in the simulation is assumed to be infinite. This method has been tested in the detector development phase using a prototype, as explained in [22].

Calibration of the detector is performed in several steps. First, the PMT gains are adjusted to within 3% of each other by using blue LEDs mounted in the detector. The gain of each PMT is evaluated using photo–electron statistics, as described in [15], and then is periodically measured during detector operation. The quantum efficiency (QE) of the PMTs is evaluated by comparing Monte Carlo (MC) simulation data with α-source data taken in cold ($< 170\,\mathrm{K}$) gaseous xenon. Gaseous xenon is used since it contains fewer impurities, so that the QE can be obtained from the PMT outputs without having to take into account the absorption. All of these calibration data are stored in a database and used in offline analysis.

The detector is both monitored and calibrated periodically with γ-rays of various energies. Four methods are employed, over a wide energy range: 1) 4.43 MeV γ-rays produced by an Am/Be radioactive source, 2) 9 MeV γ-rays produced by neutron capture on nickel, from a polyethylene-moderated Californium neutron source, 3) 17.6 MeV γ-rays produced by the $^7\mathrm{Li}(p,\gamma)^8\mathrm{Be}$ resonance excitation in a LiF target, bombarded by a 500 keV proton beam from a Cockcroft–Walton accelerator, 4) 55 MeV, 83 MeV and 129 MeV γ-rays obtained from dedicated runs with a negative pion beam, stopped in a liquid hydrogen target placed at the centre of our detector. The lower two energies originate from the pion charge-exchange reaction $\pi^- p \to \pi^0 n$, selecting back-to-back γs from π^0-decay, while the higher energy γ-ray comes from the radiative capture reaction $\pi^- p \to \gamma n$.

4.2.5 Performance evaluation

We constructed a prototype detector to evaluate its performance and to gain experience in operating such a detector and handling xenon liquefaction and purification. The active volume of the prototype is $372 \times 372 \times 496\,\mathrm{mm}^3$ (69 liters) and it is surrounded by 228 PMTs assembled in a rectangular formation in a cryostat, as shown in Fig. 14. Performance evaluation was carried out by using γ-rays from π^0 decay in the charge-exchange reaction and from radiative capture. The PMT gain and QE calibrations were conducted as described above. The xenon was purified before starting data acquisition by using the method described in [17] to obtain a sufficiently long ($\sim 3\,\mathrm{m}$) absorption length relative to the detector size. An array of NaI crystals with a $\sim 5\,\mathrm{cm}$ diameter collimator for tagging γs was located on the opposite side of the LXe detector.

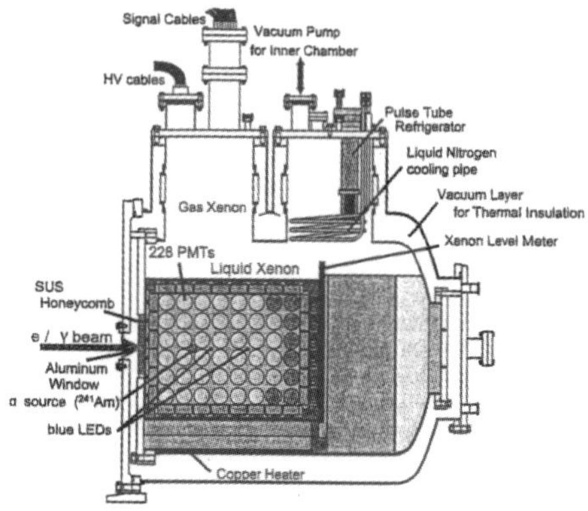

Fig. 14. Schematic view of the large prototype LXe detector

The γ-ray in the LXe detector is reconstructed by using the charge information collected by the PMTs. Its energy is reconstructed by summing all of the PMT outputs after calibration. In reconstructing the position and timing of the γ-ray interaction, we select only those PMTs that observed a relatively large charge, on the entrance face, in order not to suffer from shower shape fluctuations. The γ-ray interaction position is reconstructed by taking the weighted mean of PMT locations according to the observed charge as expressed below:

$$X = \frac{\sum_{\mathrm{selected}} q_i X_i}{\sum_{\mathrm{selected}} q_i} . \tag{1}$$

The depth of the interaction point is evaluated by using the spread of the light distribution on the entrance face. If a γ-ray deposits energy near the entrance face, the entrance face receives scintillation light with a small spread, while if it occurs deep in the detector, the spread becomes broader, as shown in Fig. 15.

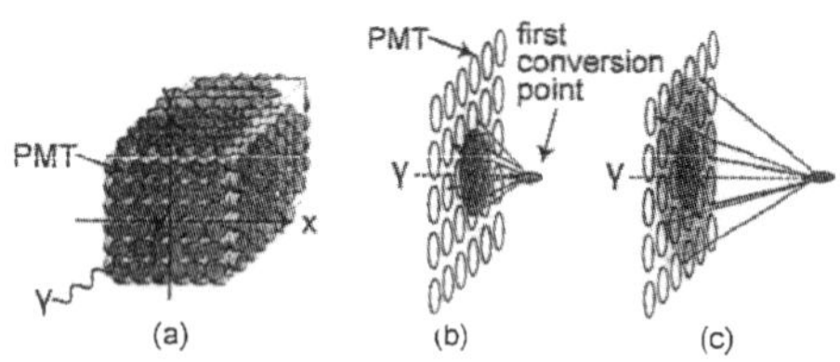

Fig. 15. (a) γ-ray impinging at the centre of the large prototype "front face". The "depth" is the distance along the z-axis. (b) Light distribution on front face with a shallow conversion point. (c) Light distribution on front face with a deep conversion point

Performance evaluation tests were carried out twice at PSI with different sets of PMTs, resulting in an energy resolution of 4.8% (FWHM), a position resolution of 4.9 mm (FWHM) and a timing resolution of 45 ps (sigma) for 55 MeV γ-rays [21]. This performance has convinced us of the future success of the LXe detector. Construction of the final LXe detector is in progress now and its engineering run will be performed in 2007.

5 Summary

Lepton flavour violation is a clear sign of physics beyond the Standard Model. The upcoming $\mu \to e\gamma$ experiment, MEG, is very attractive in the sense that we can probe GUT-scale physics by measuring MeV-scale particles. However, this requires the careful design and operation of novel detectors. The MEG is designed to search for the flavour violating decay of muons with an unprecedented sensitivity, 10^{-13}. For the start of the physics run in 2007, we are intensively preparing the beam line, the positron spectrometer and the LXe γ-ray detector. The MEG beam line has succeeded in delivering more than 10^8 muons/s to the centre of the MEG detector. The positron spectrometer has recently measured its first high-rate muon decays in the special gradient magnetic field created by COBRA. The LXe γ-ray detector will measure the energy, incident position, and timing of incoming γ-rays using newly developed cryogenics and photomultipliers. The prototype detector has confirmed the feasibility of obtaining the required resolutions with the final detector. After years of developing our novel detectors, we are now getting ready to probe GUT-scale physics.

Acknowledgements. I would like to thank the International School of Subnuclear Physics (ISSP) Committee of the Ettore Majorana Foundation and the director of this school, Prof. A. Zichichi and Prof. G. 'tHooft for giving me the great opportunity to broaden my view of physics. So many wonderful lectures, from the quark–gluon plasma world to the landscape of string theory vacua, stimulated and exited me. I also appreciate the great encouragement I received from the lecturers and scientists of this school, and this great opportunity to submit a paper as a young scientist.

I would like to express my deep gratitude to Prof. M. Koshiba, Prof. S. Komamiya, and Prof. T. Mori, who sent me to ISSP, thus giving me the opportunity to attend wonderful lectures and stimulating discussions with great scientists.

I would also like to send many thanks to my MEG colleagues, especially, Dr. J. Egger, Dr. P.-R. Kettle, Dr. S. Mihara, Prof. T. Mori and Dr. W. Ootani, who gave me much advice on this paper. Many thanks also go to my two senior colleagues, H. Nishiguchi and R. Sawada. Our discussions on the drift chamber and the calorimetry of the LXe detector taught me many things.

Finally, I would like to express my appreciation to Prof. Luisa Cifarelli and Dr. Silvia Arcelli of the University of Bologna for their great encouragement and patience with the writing of this article.

This work was supported by a Japan MEXT Grant-in-Aid for Scientific Research on Priority Areas 441.

References

1. T.P. Cheng, L.F. Li, Phys. Rev. Lett. **45**, 1908 (1980)
2. J. Hisano, D. Nomura, Phys. Rev. D **59**, 116005 (1999)
3. L.J. Hall, V.A. Kostelechy, S. Raby, Nucl. Phys. B **267**, 415 (1986)
4. F. Borzumati, A. Masiero, Phys. Rev. Lett. **57**, 961 (1986)
5. T. Mori et al., Research Proposal to Paul Scherrer Institut **R-99-5** (1999)
6. A. Baldini et al., Research Proposal to INFN, (2002)
7. M.L. Brooks et al., Phys. Rev. Lett. **83**, 1521 (1999)
8. Y. Kuno, Y. Okada, Phys. Rev. Lett. **77**, 434 (1996)
9. A.E. Pifer, T. Bowen, K.R. Kendall, Nucl. Instrum. Methods **135**, 39 (1976)
10. W. Ootani et al., IEEE Trans. Appl. Supercond. **14**, 568 (2005)
11. A. Yamamoto, Y. Makida, Y. Doi, T. Kondo, K. Wada, S. Meguro, Nucl. Phys. B **78**, 565 (1999)
12. A. Yamamoto, Y. Makida et al., IEEE Trans. Appl. Supercond. **12**, 438 (2002)
13. H. Nishiguchi, Nuc. Phys. B (2006)
14. S. Dssoniu, PhD thesis (University of Genova, 2006)
15. A. Baldini et al., Nucl. Instrum. Methods A **545**, 753 (2005)
16. S. Mihara et al., Cryogenics **46**, 688 (2006)
17. S. Mihara et al., Cryogenics **44**, 223 (2004)
18. private communication with S. Nakamura from Yokohama National Univ. and with T. Haruyama from KEK
19. T. Haruyama et al., AIP Conf. Proc. **823**, 1695 (2006)
20. A. Baldini et al., Nucl. Instrum. Methods A **566**, 294 (2006)
21. T. Iwamoto, Nucl. Phys. B (2006)
22. A. Baldini et al., Nucl. Instrum. Methods A **565**, 589 (2006)

Reprinted with permission from the European Physical Journal C

Eur. Phys. J. C (2008)
DOI 10.1140/epjc/s10052-008-0535-4

THE EUROPEAN
PHYSICAL JOURNAL C

Special Article – Scientific Note

The descent of off-shell supersymmetry to the mass shell

short review

D. Krotov[1,2,3,a]

[1] Institute for Nuclear Research of the Russian Academy of Sciences, Moscow, Russia
[2] Institute of Theoretical and Experimental Physics, Moscow, Russia
[3] Moscow State University, Department of Physics, Moscow, Russia

Received: 19 July 2007 / Revised version: 19 December 2007 /

Abstract. It is shown that classical actions for some "physically interesting" quantum field theories can be obtained as effective actions from the single "fundamental" theory of the Chern–Simons form. The physical degrees of freedom are encoded in the space of cohomologies of a certain differential operator. This observation suggests a different perspective on some of the supersymmetric properties of these effective theories. Namely, it is possible to construct a superfield formalism which allows to find off-shell SUSY actions for the on-shell supersymmetric theories, where conventional superfield formalism does not work. This formalism contains even auxiliary variables λ^α in addition to conventional odd variables θ^α. This idea is similar to the Pure Spinor construction. This paper is a short review of papers [11, 12]. Original results discussed below were obtained in collaboration with V. Alexandrov, A. Gorodentsev, A. Losev and V. Lysov.

1 Introduction

The central statement of the paper is that different classical field theories can be written in the form of action (2) by the addition of auxiliary fields. The fact that the theory (2) can be written in such a compact form is due to the introduction of the pure spinor fields $\mathcal{A}$ and $\mathcal{P}$ (the definition is given in Sect. 2). The construction of these fields was suggested in [1–3], [4]. What is claimed in this paper is that these fields can be considered as superfields and are needed to restore the off-shell SUSY invariance for the various on-shell supersymmetric theories. Another result is that the descent from off-shell supersymmetric action (2) down to its on-shell version is realized through the conventional procedure of quantum field theory – evaluation of the Feynman path integral. It happens that the perturbation series for this integral is finite, which means that it can be summed up explicitly.

As a universal framework for the calculation of "effective action" (elimination of auxiliary fields), we use the Batalin–Vilkovisky [6–10] formalism. This formalism allows us to control the descent of a symmetry of the action (2) down to its effective action. This technique is applied to the case of supersymmetry.

Another message of the present paper is that, in contrast to N. Berkovits' construction, the matrices $C^\mu_{\alpha\beta}$ in the definition of the pure spinor constraints (see Sect. 2 for details) are *not* necessarily Dirac $\gamma^\mu_{\alpha\beta}$-matrices. We consider them as arbitrary constant matrices. This "extension" allows us to obtain a plethora of different theories described by the construction.

In this short review we avoid rigorous proofs and technical details. Its aim is to explain our main ideas, presented in [11] and [12].

2 A pure spinor superfield formalism

The construction starts from the set of odd Grassmann variables θ^α and an equal number of even variables λ^α (the index α runs in the range $\alpha = 1...k$, where k is arbitrary). These variables λ^α are restricted by the set of quadratic constraints $f^\mu = \lambda^\alpha C^\mu_{\alpha\beta}\lambda^\beta = 0$. Here, $C^\mu_{\alpha\beta}$ are arbitrary constant matrices, symmetric in α, β (allowing only those combinations of λ that are not proportional to $f^\mu(\lambda)$). In the end, the variables θ^α will play the role of conventional supercoordinates. In a way that is similar to the "standard" approach, we introduce two superfields $\mathcal{A}$ and $\mathcal{P}$. $\mathcal{A}$ is a generic element in the space of θ^α and λ^α with the component fields taking values in a (adjoint) representation of a given gauge group. The superfield $\mathcal{P}$ takes value in the space that is dual to that of θ^α and λ^α. The generating elements of this superfield are denoted by $\underline{\lambda}^\alpha$ and $\underline{\theta}^\alpha$. There is canonical pairing between the elements of the space of $\mathcal{A}$ and $\mathcal{P}$ denoted by $<,>$. It is defined as $<\underline{e}^a, e_b> = \delta^a_b$. Here, e_b denotes basis elements in the space

[a] "New Talent" Presentation Award at the 44th Course of the International School of Subnuclear Physics, Erice,Italy, 29 Aug-7 Sep 2006
e-mail: krotov@inr.ac.ru, krotov@itep.ru

of $\mathcal{A}$ and $\underline{e}^a$ denotes basis elements in the space of $\mathcal{P}$. The number of basis elements e_b is equal to the number of independent component fields in the superfield $\mathcal{A}$. The same is true for the superfield $\mathcal{P}$. The component fields in $\mathcal{A}$ and $\mathcal{P}$ are different. Introduction of the dual superfield is necessary to achieve the invariance of the construction under linear transformations of the basis elements e_a.

The next ingredient is the operator [1–3, 5]

$$Q_B = Q + \Phi = \lambda^\alpha \frac{\partial}{\partial \theta^\alpha} + \frac{1}{2}\theta^\alpha \frac{\partial f^\mu}{\partial \lambda^\alpha}\partial_\mu . \qquad (1)$$

Here $f^\mu(\lambda)$ is the set of quadrics defined above, and ∂_μ is the derivative with respect to the space-time coordinates. At this level we suppose that the number of quadrics is equal to the number of space-time coordinates. In Sect. 4 we will show (as described also in [11]) that one can make different reductions assuming that, for example, some derivatives ∂_μ act as zero on the component fields.

The central object of our consideration is the action

$$S^{\mathrm{Fund}} = \int \mathrm{Tr}\left(<\mathcal{P}, Q_B\mathcal{A}> + g <\mathcal{P}, \mathcal{A}^2>\right) . \qquad (2)$$

In this action, $\mathcal{A}$ and $\mathcal{P}$ are the superfields defined above and parameter g is the gauge coupling constant. One can rewrite S^{Fund} in terms of component fields. To find the contribution of the first term, one should apply the operator Q_B to the generic superfield $\mathcal{A}$ and calculate the canonical pairing of the result with the superfield $\mathcal{P}$. In the second term, one should multiply the two $\mathcal{A}$ superfields, expand the result into the basis elements in the space of $\mathcal{A}$ and then evaluate the pairing $<,>$. The result of this calculation is some combination of component fields and space-time derivatives. Each component field is an element of a representation of the gauge group and depends on the space-time coordinates. To find the scalar action, one should calculate the trace over the gauge group indices and integrate over the space-time coordinates. These operations are presented in (2).

After the calculation of the canonical pairing $<,>$ the action (2) contains many fields. Some of these fields are physical and some of them are auxiliary. We would like to emphasize that the number of these fields and the way they contribute to the action is determined by the number k of λ^α and θ^α and the structure of the matrices $C^\mu_{\alpha\beta}$ in the definition of the quadrics. The crucial element of this construction is the assumption that among these fields the physical degrees of freedom are those which have a $\lambda - \theta$ structure that is proportional to the cohomologies of the operator $Q = \lambda^\alpha \frac{\partial}{\partial\theta^\alpha}$.

We would like to emphasize the difference between operators Q_B and Q. The latter represents only the first term in Q_B. The physical spectrum of the theory coincides with the cohomologies of operator Q.

3 Calculation of effective action

Our strategy is to integrate out all the fields in the action (2) that are not cohomologies of the operator Q. The effective action on cohomologies is a classical Batalin–Vilkovisky action for a "physically interesting" theory. To avoid confusion, we would like to stress that the fields which are integrated out are auxiliary. Hence the terminology "effective action" may seem inappropriate for this object. However, as we will show below, the calculation of this "effective action" is the standard procedure in the Batalin–Vilkovisky formalism. In accordance with this observation, we retain this somewhat unconventional terminology.

3.1 Batalin–Vilkovisky formalism

In the approach to supersymmetric gauge theories that is the subject of the present paper, the Batalin–Vilkovisky formalism plays the role of universal framework for the calculation of effective action [16, 17]. What is important for us is that the BV formalism is a convenient tool that allows us to control the descent of the arbitrary symmetry (global or local) of the "fundamental action" down to the effective action.

The central object in the formalism is the classical BV master equation

$$0 = \frac{\delta S^{\mathrm{BV}}}{\delta \chi^n}\frac{\delta S^{\mathrm{BV}}}{\delta \chi^*_n} \qquad (3)$$

Here S^{BV} is called BV action and depends on the fields, the ghosts and their BV antifields (χ^n stands for all the fields of the theory, χ^*_n denotes the corresponding antifields). This equation has many solutions, but the most important one for us is called classical BV action

$$S^{\mathrm{BV}} = S^{\mathrm{cl}}[A] + \delta A^n A^*_n + \frac{1}{2} f^c_{ab} c^a\, c^b c^*_c \qquad (4)$$

Here $S^{\mathrm{cl}}[A]$ is classical action, which depends only on the physical fields A^n and is invariant under the symmetry transformations δA^n of the second term. The parameter of these transformations is c^a. The fields A^*_n and c^*_a are BV antifields for A^n and c^a. Parameters f^c_{ab} are the structure constants for the algebra of symmetry. In the case of ordinary gauge symmetry, δA^n is a BRST transformation of A^n and c^a is a Faddeev–Popov ghost.

It is important to note that if an action satisfies (3) and can be written in the form of classical BV action (4), this is equivalent to the following conditions:

- $S^{\mathrm{cl}}[A]$ is invariant under the algebra of symmetry of the second term.
- This algebra of symmetry is closed off-shell and the structure constants are given by f^c_{ab}.

The important point is that there exists a procedure for the calculation of the effective action that preserves the BV invariance. Namely, if the initial action satisfies the BV equation, then the effective action (which is obtained by integrating out some fields) satisfies the BV equation as well. The procedure involves integrating over a lagrangian submanifold in the space of fields [16, 17] (for the simple explanation see Sect. 2 of [11]). This fact makes it possible to analyse the symmetries of the effective action.

We apply this technique to our calculation, starting from the action (2). It is possible to show that this action satisfies the BV equation (3). The BV fields are the component fields of $\mathcal{A}$, and the BV antifields are represented in the superfield $\mathcal{P}$. We are going to decompose the superfields $\mathcal{A}$ and $\mathcal{P}$ into two parts: $\mathcal{A} = A + a$ and $\mathcal{P} = P + p$. A is the superfield that takes values in the space of Q-cohomologies, and a lies in the orthogonal complement to the Q-cohomologies. P and p are the elements of spaces dual to those of A and a. The effective action is defined as

$$e^{-S^{\text{eff}}(P,A)} = \int_{\substack{\text{lagrangian}\\ \text{submanifold}}} Dp\, Da\, e^{-S^{\text{Fund}}(P,p,A,a)} \qquad (5)$$

The fact that S^{Fund} satisfies equation (3) is inherited in the gauge invariance of $S^{\text{eff}}(P, A)$.

Up to this point we have discussed the descent of the gauge invariance from the action (2) to S^{eff}. This descent is "trivial" in the sense that both actions (2) and S^{eff} are gauge invariant. In Sect. 5 we analyse the descent of supersymmetry from the same perspective.

3.2 Feynman diagram technique

In this subsection we explain how the calculation of $S^{\text{eff}}(P, A)$ in (5) is done by summation of all possible connected Feynman diagrams, with external legs being the cohomologies of the Q-operator. In spite of the presence of the non-standard variables λ and θ in the definition of S^{Fund}, this calculation is simply the calculation of a Feynman integral in conventional quantum field theory. The remarkable fact is that its perturbative expansion is finite and contains only a quite limited number of diagrams. These diagrams can be summed up explicitly into a rather simple and local effective action. Thus, the expansion of the integral (5) is exact.

Our suggestion is to consider the operator $< \mathcal{P}, Q\mathcal{A} >$ (the first part of Q_B) as a kinetic term and the rest of S^{Fund}, namely $< \mathcal{P}, \Phi\mathcal{A} >$ and $g < \mathcal{P}, \mathcal{A}^2 >$, as perturbation. With the decomposition $\mathcal{A} = A + a$ and $\mathcal{P} = P + p$, where A and P are defined by Q-cohomologies, it becomes apparent that the kinetic term reduces to $< p, Qa >$. The terms QA and PQ vanish. There are many contributions into the vertex part of the action. Some of them are $< P, \Phi A >$, $< p, \Phi A >$, $< P, \Phi a >$, $< p, \Phi a >$, $g < P, a^2 >$, ..., 12 contributions in total. Some of these vertices are drawn in Fig. 1 (only two vertices are illustrated here, but the remainder can be derived in an obvious way).

In this Figure, the solid in-coming line denotes the external field A, the solid out-going line denotes the field P, and the wavy lines denote the propagating fields a and p.

Fig. 1. Examples of vertices

Fig. 2. Possible diagrams

We would like to emphasize that the kinetic term $< p, Qa >$ has a large number of zero modes (for example $\delta a = Q\varepsilon$). To define the propagator "Q^{-1}" this gauge ambiguity should be fixed. This is done in (5) by restricting the domain of integration to the lagrangian submanifold.

Eventually, one could apply conventional Feynman technique to draw all possible diagrams. There are two rules to be followed.

- All quantum loops are absent in the theory (2).
- Even at the tree level, the length of the diagram is limited by θ structure. Indeed, propagator "Q^{-1}" increases the power of θ. Since the number of θ^α is equal to k, the number of propagators in each branch of the diagram can not be bigger than k. Usually this limit is even stronger, for instance, due to the λ structure.

In Fig. 2 we present two possible diagrams as an example. The first one contributes to the kinetic term of the effective action, while the second one contributes to the interaction.

4 The space of effective actions

As already mentioned above, all information about degrees of freedom and the structure of effective action is encoded in the number k of λ and θ and the structure of quadrics $f^\mu(\lambda)$. It happens that there are two large classes of quadrics, which are called **regular** and **singular** systems. We do not give precise definitions [11] of these systems here. What is important for us is that for the case in which the number of quadrics is smaller than the number k, almost all possible systems are regular. In the bulk of regular systems there is a small but still infinite number of singular quadrics.

The point is that in the case of regular quadrics there is a certain universality in the spectrum of Q-cohomologies and hence in the structure of effective action. For an arbitrary regular system of quadrics, the effective theory is the AKSZ (Alexandrov, Kontsevich, Schwarz, Zaboronsky) version of the Chern–Simons theory (BF-theory) [13]. For the system of N quadrics $f^\mu(\lambda)$ (index μ runs $\mu = 1...N$) this theory is defined in N dimensions. Therefore, in the regular case the effective action is not sensitive to the particular structure of matrices $C^\mu_{\alpha\beta}$ in the definition of quadrics. Only the number N is important.

The opposite is true for singular systems. In this case the structure and degrees of freedom of effective action are highly sensitive to the structure of $C^\mu_{\alpha\beta}$. For each system of quadrics its own effective theory can be obtained. Though in the regular case effective action is a "trivial" topological theory, in the case of singular systems the effective action has a quite non-trivial spectrum and dynamics. Thus, we

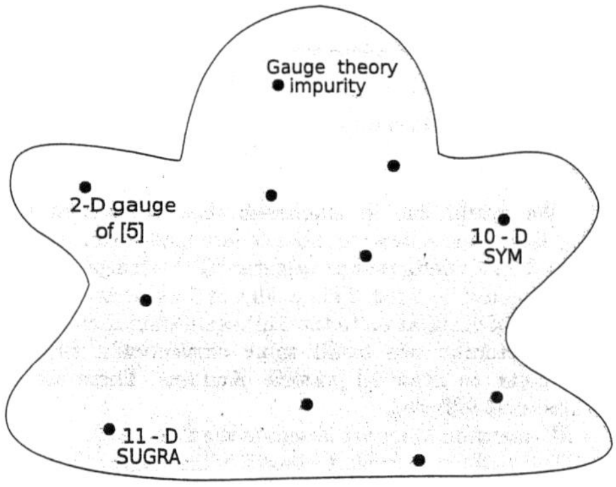

Fig. 3. The space of effective theories. The *gray background* represents the bulk of AKSZ theories. We emphasize that this picture is a schematic illustration only (see the comments below and footnote 1)

come to the conclusion that "physically interesting" theories are singularities in the bulk of trivial "topological" theories. This idea is schematically depicted in Fig. 3.

Until now, only a quite small number of singular systems has been investigated in detail. Some of them are depicted in the figure. It would be of great interest to search for such singular systems with different numbers k of λ and θ and different numbers N and structures of quadrics.

It should be emphasized that this picture is only a schematic illustration. For example, the points "$2-d$ gauge model of [5]" and $10-d$ super Yang–Mills have different statuses. The action of $2-d$ gauge model is obtained from the fundamental action (2) exactly as outlined in the discussions of the previous Section. To obtain the action of $10-d$ SYM, on the other hand, an additional Z_2 projection must be made onto the space of fields (see Sect. 6). The same seems to be true for $11-d$ supergravity[1].

We would also like to mention the possibility that there may be connections between these ideas and recent discussions regarding the Landscape program [14, 15].

5 The descent of supersymmetry

In the previous Sections we explained that by integrating out the orthogonal complement to Q-cohomologies in the action (2) one can obtain "physically interesting" effective action. In the present Section we show that all the fields that are integrated out are auxiliary. These fields are needed to restore the off-shell SUSY invariance for the on-shell supersymmetric effective theory. The central statement is that the action S^{Fund} is the off-shell supersymmetric version of S^{eff} for a given system of quadrics. We also show that for an arbitrary set of quadrics, the action S^{eff} is at least on-shell supersymmetric.

To prove these statements one uses the general ideology of BV formalism, as follows. To study the descent of supersymmetry from the action (2) to S^{eff}, one introduces the sources for each generator of the symmetry into the action. The algebra of SUSY is generated by Q^s_α and the translations ∂_μ.

$$Q^s_\alpha = \frac{\partial}{\partial\theta^\alpha} - C^\mu_{\alpha\beta}\theta^\beta\partial_\mu \,. \qquad (6)$$

The commutation relations are $\{Q_\alpha, Q_\beta\} = -2C^\mu_{\alpha\beta}\partial_\mu$. This algebra is closed. We introduce the ghosts for SUSY ε^α and the ghosts for translations η^μ. Then one can write the new action.

$$S_s = S^{\text{Fund}} + \int \text{Tr}\left(< \mathcal{P}, (\varepsilon Q_s + \eta^\mu\partial_\mu)\mathcal{A} > -\eta^*_\mu(\varepsilon C^\mu\varepsilon)\right) \qquad (7)$$

Since we are discussing global symmetries, the ghosts ε and η do not depend on space-time coordinates. It is straightforward to check that the action (7) satisfies the BV equation (3) over the fields $\mathcal{A}, \mathcal{P}, \varepsilon^\alpha, \varepsilon^*_\alpha, \eta^\mu, \eta^*_\mu$ and has the structure of the classical BV action. Hence, according to the discussion in Sect. 3.1, the action S^{Fund} is invariant under the SUSY transformations and the SUSY algebra is closed off-shell. Then we integrate out the complement to Q-cohomologies from the action (7) to obtain S^{eff}_s. We do not integrate over parameters ε^α and η^μ. Since the integration is done over a lagrangian submanifold, S^{eff}_s also satisfies the classical BV equation. However, this action does not have the structure of (4). In addition to the standard BV structures, some "new" terms appear in S^{eff}_s. These terms can be divided into two classes. We tried to determine the physical meaning of these terms [18–23]. Their definitions and interpretations are summarized in the Table 1. Taking into account these facts, we conclude that if S^{eff}_s satisfies the BV equation, then it follows that S^{eff} (which is the first part of S^{eff}_s, independent of ε and η) is on-shell supersymmetric. The action S^{Fund} is its off-shell version.

Table 1. Interpretation of non-standard terms in the master action

Term	Interpretation
$(\varepsilon\psi^*)^2$	The terms are quadratic in antifields. The appearance of such terms in BV action signals that the algebra of symmetry is closed only on-shell.
$c^*(\varepsilon\gamma^\mu\varepsilon)A_\mu$	Such terms are responsible for the gauge fixing. For example, they appear in the BV action when one fixes the Wess–Zumino gauge in $\mathcal{N}=1, D=4$ SYM theory. The appearance of such terms signals that the algebra of SUSY is closed only up to some gauge transformation.

[1] To obtain the interaction terms for SUGRA, one should modify the interaction in the action (2), see [24].

As a concluding remark, we would like to emphasize that the introduction of ghosts ε and η is only a technical trick, which allows us to control the SUSY properties of S^{Fund} and S^{eff}. On the other hand, the fields $\mathcal{A}$ and $\mathcal{P}$ can be considered as superfields. All this is true for the theories, similar to [11], which do not require Z_2 projection. For the Z_2 projected theories the situation is a little bit different.

6 The descent of SUSY in Z_2 projected theories

We start this Section with the simplest example of Z_2 projected theories – 10-dimensional SYM. This system is described [1–3] by the set of quadrics $f^\mu(\lambda) = \lambda^\alpha \gamma^\mu_{\alpha\beta} \lambda^\beta$, where $\gamma^\mu_{\alpha\beta}$ are 10 dimensional Dirac γ-matrices (index α runs in the range $\alpha = 1...16$, index μ runs over $\mu = 1...10$). The spectrum of cohomologies is given in the Table 2.

The first column presents all the fields of effective theory, and the second column contains their BV antifields. From this table one can see that in addition to the fields (c, A_μ, ψ^α) (which are F–P ghost, gauge field and gluino) and their antifields $(c^*, A^*_\mu, \psi^*_\alpha)$, which are needed for the BV action of SYM, there are additional fields with the tilde sign. These additional fields are not simply decoupled from effective action, but actually contribute to it in a rather non-trivial way. To obtain the action of SYM, one should carry out Z_2 reduction, by factorizing out the tilde-fields according to the rule $\widetilde{c}^* = -c$, $\widetilde{A}^*_\mu = A_\mu, \widetilde{\psi}^* = -\psi$, $\widetilde{\psi} = \psi^*, \widetilde{A}_\mu = -A^*_\mu$, $\widetilde{c} = c^*$. After this Z_2 reduction, the effective action calculated for (2) reduces to BV action for SYM.

We would like to emphasize that although the calculation of effective action before Z_2 projection for the theory (2) is a well-defined procedure in quantum field theory, Z_2 projection itself is done "by hand". We do not yet know how it can be realized through the Feynman path integral. A quite similar situation holds for the case of 11-d supergravity. It would be interesting to understand how this Z_2 projection can be realized within the framework of BV formalism (evaluation of path integral). This would shed light on the structure of off-shell SUSY invariant actions for these Z_2 projected theories (including 10-d SYM, 11-d SUGRA, etc.).

Table 2. The spectrum of cohomologies

Polarization	A	P
1	c	c^*
$(\lambda\gamma^\mu\theta)$	A_μ	A^*_μ
$(\lambda\gamma^\mu\theta)\ (\theta\gamma_\mu)_\alpha$	ψ^α	$\psi^{*\alpha}$
$(\lambda\gamma^\mu\theta)(\lambda\gamma^\nu\theta)(\theta\gamma_{\mu\nu})^\alpha$	$\widetilde{\psi}_\alpha$	$\widetilde{\psi}^*_\alpha$
$(\lambda\gamma^\mu\theta)(\lambda\gamma^\nu\theta)(\theta\gamma_{\mu\nu\rho}\theta)$	$\widetilde{A}_\rho$	$\widetilde{A}^*_\rho$
$(\lambda\gamma^\mu\theta)(\lambda\gamma^\nu\theta)(\lambda\gamma^\rho\theta)(\theta\gamma_{\mu\nu\rho}\theta)$	$\widetilde{c}$	$\widetilde{c^*}$

Acknowledgements. It is a pleasure to thank the organizers of the International School of Subnuclear Physics (Erice, 2006) for suggesting the idea to write this review, and for the excellent atmosphere of the School. I am very indebted to V. Alexandrov, A. Losev, V. Lysov and V. Rubakov for useful critical discussions. This work was supported by the grant RFBR-05-02-17363- a and the grant of the President of the Russian Federation NS-7293.2006.2 (government contract 02.445.11.7370) and Dynasty foundation in 2007.

References

1. N. Berkovits, JHEP **0109**, 016 (2001) [hep-th/0105050]
2. N. Berkovits, JHEP **0004**, 018 (2000) [hep-th/0001035]
3. N. Berkovits, hep-th/0209059
4. B.E.W. Nilsson, Class. Quantum Grav. **3**, L41 (1986)
5. E. Witten, Nucl. Phys. B **266**, 245 (1986)
6. I.A. Batalin, G.A. Vilkovisky, Phys. Lett. B **102**, 27 (1981)
7. I.A. Batalin, G.A. Vilkovisky, Phys. Rev. D **28**, 2567 (1983)
8. I.A. Batalin, Phys. Rev. D **30**, 508 (1984)
9. B.L. Voronov, I.V. Tyutin, Theor. Math. Phys. **50**, 218 (1982)
10. B.L. Voronov, Teor. Mat. Fiz. **50**, 333 (1982)
11. D. Krotov, A. Losev, [hep-th/0603201]
12. V. Alexandrov, D. Krotov, A. Losev, V. Lysov, arXiv: hep-th/0705.2191
13. M. Alexandrov, M. Kontsevich, A. Schwartz, O. Zaboronsky, Int. J. Mod. Phys. A **12**, 1405 (1997) [hep-th/9502010]
14. L. Susskind, hep-th/0302219
15. J. Polchinski, hep-th/0603249
16. A. Schwarz, Commun. Math. Phys. **155**, 249 (1993) [hep-th/9205088]
17. A. Schwarz, Commun. Math. Phys. **158**, 373 (1993) [hep-th/9210115]
18. E.S. Fradkin, M.A. Vasiliev, Phys. Lett. B **72**, 70 (1977)
19. M.B. Green, C.M. Hull, QMC/PH/89-7. Presented at Texas A and M Mtg. on String Theory, College Station, TX, Mar 13-18, 1989
20. M.B. Green, C.M. Hull, Nucl. Phys. B **344**, 115 (1990)
21. M.B. Green, C.M. Hull, Mod. Phys. Lett. A **5**, 1399 (1990)
22. E. Bergshoeff, R. Kallosh, A. Van Proeyen, Class. Quantum Grav. **9**, 321 (1992)
23. R.E. Kallosh, Phys. Lett. B **225**, 49 (1989)
24. L. Anguelova, P.A. Grassi, P. Vanhove, Nucl. Phys. B **702**, 269 (2004) [hep-th/0408171]

Reprinted with permission from the European Physical Journal C

Eur. Phys. J. C 52, 487–494 (2007)
DOI 10.1140/epjc/s10052-007-0387-3

THE EUROPEAN PHYSICAL JOURNAL C

Special Article – Young Scientists Paper

Measurement of the CKM angle γ in $B^{\pm} \to DK^{\pm}$ decays with the BABAR detector: status and prospects

N. Neri[a]

INFN Pisa, Largo B. Pontecorvo 3, 56127 Pisa, Italy

Received: 25 June 2007 /
Published online: 15 August 2007 – © Springer-Verlag / Società Italiana di Fisica 2007

Abstract. We report on the measurement of the CKM angle γ in $B^{\pm} \to DK^{\pm}$ decays with the BABAR detector. A general overview of different methods of analysis and a critical discussion of the most sensitive methods are presented here.

1 Introduction

CP violation (CPV) was first established in $K_L \to \pi^+\pi^-$ decays in 1964 [1]. It has been accomodated in the standard model (SM) by a CP-violating phase in the matrix that describes the mixing of the quarks under the weak interaction, known as the CKM matrix [2, 3]. The unitarity constraints of the CKM matrix gives us $V_{ud}V_{ub}^* + V_{cd}V_{cb}^* + V_{td}V_{tb}^* = 0$, the so-called Unitarity Triangle relation, represented in Fig. 1. CPV is proportional to the area of the triangle and requires that the angles and sides are different from zero. The primary goal of the B-factories is the study of CP violation in the B_d and B_u meson system. Overcontraining of the Unitarity Triangle parameters, measuring the sides and the angles of the triangle, represents one of the most stringent tests of the SM. The precise measurement of the angle $\gamma \equiv \arg(-V_{ud}V_{ub}^*/V_{cd}V_{cb}^*)$ is a crucial goal for this scientific program, yet it is also one of the most difficult to achieve.

2 General overview of the methods

There are several decay modes that can be used to measure the angle γ, each with its own merits and drawbacks.

In $B^{\pm} \to DK^{\pm}$ decays[1], if we consider the decay modes of the neutral D meson that are accessible to both D^0 and $\overline{D^0}$, we can reach the final state through two different quark-level processes, as shown in Fig. 2. The interference between the two quark-level processes $b \to u\bar{c}s$ and $b \to c\bar{u}s$ (respectively $B^- \to D^0K^-$ and $B^- \to \overline{D^0}K^-$) introduces a relative phase γ in the decay amplitude.

By neglecting the $D^0 - \overline{D^0}$ mixing [4], it is possible to determine the angle γ without hadronic uncertainties, since the main contributions to the decay amplitude come from tree-level transitions. Several decay modes can be studied, including $B^{\pm} \to DK^{\pm}$, $B^- \to D^*K^-$ and $B^- \to DK^{*-}$, which have the same quark-level process in common. In the following, whenever we write $B^{\pm} \to D^{(*)}K^{(*)\pm}$ we intend all the above-mentioned decay modes, unless explicitely stated. Three different analysis methods have been used so far:

- GLW method [5, 6]: where the D is recontructed in CP eigenstates (D^0_{CP}) decay modes.
- ADS method [7, 8]: with D reconstructed in doubly Cabibbo suppressed decay modes.

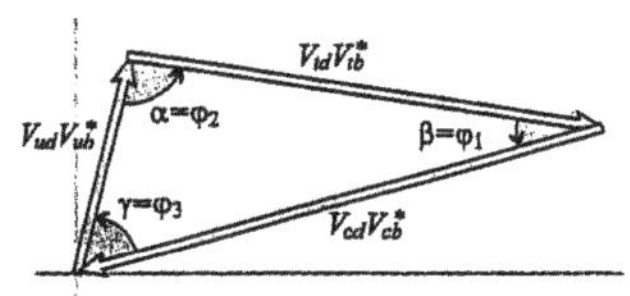

Fig. 1. Graphical representation of the unitarity constraint $V_{ud}V_{ub}^* + V_{cd}V_{cb}^* + V_{td}V_{tb}^* = 0$ as a triangle in the complex plane

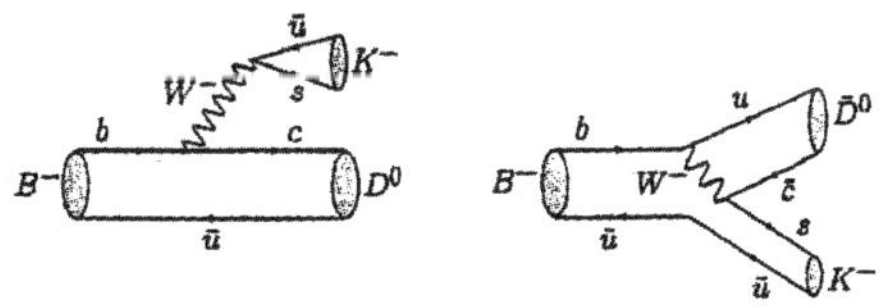

Fig. 2. Main diagrams contributing to $B^{\pm} \to DK^{\pm}$ decay. The *left diagram* proceeds via V_{cb} transition, while the *right diagram* proceeds via V_{ub} transition and is color suppressed

[a] email: nicola.neri@pi.infn.it

[1] In what follows, the symbol D refers to either D^0 or $\overline{D^0}$.

– Dalitz method [9]: where the D is reconstructed in 3-body final states and the angle γ is extracted through an analysis of the distribution of the events in the D Dalitz plane [10].

The sensitivity of the different methods to γ depends on the magnitude of the ratio

$$r_B = \left| \frac{A(b \to u\bar{c}s)}{A(b \to c\bar{u}s)} \right|$$

of the $b \to u\bar{c}s$ amplitude with respect to the $b \to c\bar{u}s$ one. The value of r_B is a key quantity which has a significant impact on the ability to measure the CKM angle γ at the B-factories and beyond. Note that r_B takes different values for different B decays. Theoretical expectations for r_B are in the range ≈ 0.1–0.2 [5, 6, 11], in agreement with the 90% C.L. upper limits on r_B set by BABAR ($r_B < 0.23$) [12, 13] and Belle ($r_B < 0.18$) [14] through the study of $B^- \to DK^-, D \to K^+\pi^-$ decays.

3 The GLW method: $B^\pm \to DK^\pm$ with D^0_{CP} decays

This method considers the $B^\pm \to DK^\pm$ decays , where the D decays to a CP eigenstate. The CP observables are:

$$\begin{aligned} R_{CP\pm} &\equiv \frac{2\left[\mathcal{B}\left(B^- \to D^0_{CP\pm}K^-\right) + \mathcal{B}\left(B^+ \to D^0_{CP\pm}K^+\right)\right]}{\mathcal{B}\left(B^- \to D^0K^-\right) + \mathcal{B}\left(B^+ \to \overline{D^0}K^+\right)} \\ &\equiv 1 + r_B^2 \pm 2r_B \cos\delta_B \cos\gamma\,, \\ A_{CP\pm} &\equiv \frac{\mathcal{B}\left(B^- \to D^0_{CP\pm}K^-\right) - \mathcal{B}\left(B^+ \to D^0_{CP\pm}K^+\right)}{\mathcal{B}\left(B^- \to D^0_{CP\pm}K^-\right) + \mathcal{B}\left(B^+ \to D^0_{CP\pm}K^+\right)} \\ &\equiv \frac{\pm 2 r_B \sin\delta_B \sin\gamma}{R_{CP\pm}}\,, \end{aligned}$$

where δ_B is the strong phase difference between the V_{ub} and the V_{cb} mediated amplitudes. Here, $D^0_{CP\pm} = (D^0 \pm \overline{D^0})/\sqrt{2}$ are the CP eigenstates of the neutral D meson system. The main advantage of this method is that γ can be extracted in a theoretically-clean manner if one reconstructs D^0_{CP}-even and D^0_{CP}-odd decays. In fact, the number of unknowns is three (r_B, γ, δ_B) and we have three linear independent observables. However, an 8-fold ambiguity on the value of γ is not resolved since the ambiguities on $(\gamma, \delta_B) \to (\delta_B, \gamma)$ and on the sign of $\sin\gamma$, which admits four different solutions, are indistinguishable. In principle, carrying out analyses for the different decay modes of $B^\pm \to D_{CP}X^\pm$, where $X^\pm = K^\pm, K^\pm\pi^0, K^0_S\pi^\pm, (K\pi\pi)^\pm$, makes it possible to solve the ambiguity on the magnitude of $\sin\gamma$, since each of the decay modes has the same weak phase γ but a different final-state phase difference δ_B. The event yield is similar for the CP-even and the CP-odd decay modes – almost 150 signal events with the present statistics in $B^\pm \to DK^\pm$ decay modes. Figure 3 shows the ΔE distribution for the signal and background events of the reconstructed modes: $D^0_{CP+} \to \pi^+\pi^-, K^+K^-$, and $D^0_{CP-} \to K^0_S\pi^0, K^0_S\omega, K^0_S\phi$. Here, ΔE is the difference between the measured B meson energy and the energy of the beam in the center of mass system, and peaks near zero for signal events.

The total reconstruction efficiencies, based on simulated signal events, are 30%–40% for the D^0_{CP+} modes and 10%–20% for the D^0_{CP-}. Experimentally, the $R_{CP\pm}$ ratios are computed using the $R_{CP\pm} \simeq R_\pm/R$ relations, where the quantities R and $R_\pm$ are defined as:

$$\begin{aligned} R &= \frac{\mathcal{B}(B^- \to D^0K^-) + \mathcal{B}(B^+ \to \overline{D^0}K^+)}{\mathcal{B}(B^- \to D^0\pi^-) + \mathcal{B}(B^+ \to \overline{D^0}\pi^+)}\,, \\ R_\pm &= \frac{\mathcal{B}(B^- \to D^0_{CP\pm}K^-) + \mathcal{B}(B^+ \to D^0_{CP\pm}K^+)}{\mathcal{B}(B^- \to D^0_{CP\pm}\pi^-) + \mathcal{B}(B^+ \to D^0_{CP\pm}\pi^+)}\,. \end{aligned}$$

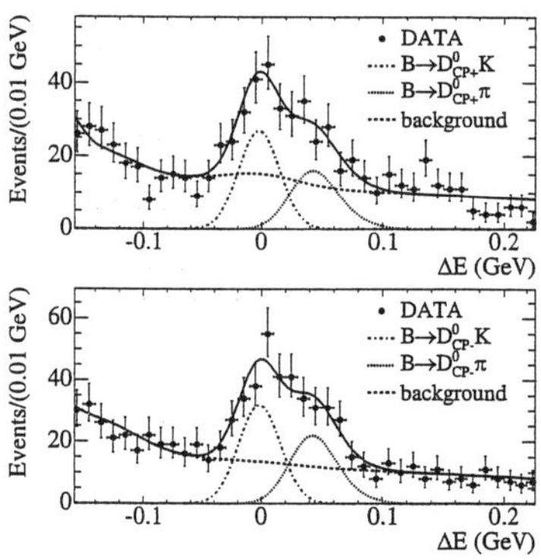

Fig. 3. Distributions of ΔE for events enhanced in $B^- \to DK^-$ signal. *Top:* $B^- \to D_{CP+}K^-$; *bottom:* $B^- \to D_{CP-}K^-$. *Solid curves* represent projections of the maximum likelihood fit; *dashed*, *dashed-dotted* and *dotted curves* represent the $B^- \to DK^-$, $B^- \to D\pi^-$ and background contributions

Table 1. Measured ratios $R_{CP\pm}$ and $A_{CP\pm}$ for CP-even and CP-odd D decay modes. The first error is statistical, the second is systematic

B mode	$N(B\bar{B}) \times 10^6$	R_{CP+}	A_{CP+}	R_{CP-}	A_{CP-}
$B^\pm \to DK^\pm$	232	$0.90 \pm 0.12 \pm 0.04$	$0.35 \pm 0.13 \pm 0.04$	$0.86 \pm 0.10 \pm 0.05$	$-0.06 \pm 0.13 \pm 0.04$
$B^\pm \to D^*K^\pm$	123	$1.06 \pm 0.26^{+0.10}_{-0.09}$	$-0.10 \pm 0.23^{+0.03}_{-0.03}$	–	–
$B^\pm \to DK^{*\pm}$	232	$1.96 \pm 0.40 \pm 0.11$	$-0.08 \pm 0.19 \pm 0.08$	$0.65 \pm 0.26 \pm 0.08$	$-0.26 \pm 0.40 \pm 0.12$

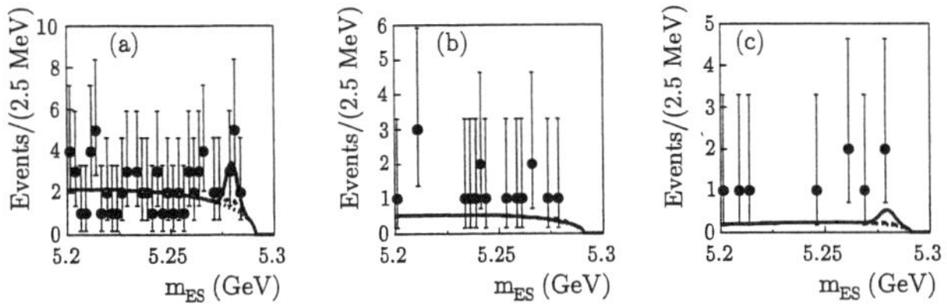

Fig. 4. m_{ES} distributions for candidate signal events with the fit model overlaid. **a** $\bar{D}K$ events; **b** $\overline{D^*}K$ events with $D^* \to D\pi^0$; **c** $\overline{D^*}K$ events with $D^* \to D\gamma$

Systematic uncertainties are canceled out in the measurement of these double ratios. The results for the R_{CP} and A_{CP} observables using BABAR data [15–17] are reported in Table 1. The precision of these measurements does not significantly constrain the value of γ, but when combined with the existing measurements of the $B^{\pm} \to D^{(*)}K^{(*)\pm}$ decays, it will improve the knowledge of the angle γ and of the parameter r_B.

4 The ADS method: $B^{\pm} \to DK^{\pm}$with Double-Cabibbo-Suppressed D decays

In the ADS method, the favored B decay ($B^- \to D^0K^-$) followed by the doubly CKM-suppressed D decay ($D^0 \to K^+\pi^-$) interferes with the suppressed B decay ($B^- \to \bar{D}^0K^-$) followed by the CKM-favored D decay ($\overline{D^0} \to K^+\pi^-$). As a result, the two interfering amplitudes become comparable. The CP asymmetry is potentially larger in these modes than in the GLW method, however the decays have a smaller branching ratio, on the order of 10^{-7}.

The observables sensitive to the CP parameters are

$$R_{\mathrm{ADS}} \equiv \frac{\mathcal{B}([K^+\pi^-]_D K^-) + \mathcal{B}([K^-\pi^+]_D K^+)}{\mathcal{B}([K^-\pi^+]_D K^-) + \mathcal{B}([K^+\pi^-]_D K^+)}$$
$$\equiv r_B^2 + r_D^2 + 2r_B r_D \cos\gamma \cos(\delta_B + \delta_D)$$
$$A_{\mathrm{ADS}} \equiv \frac{\mathcal{B}([K^+\pi^-]_D K^-) - \mathcal{B}([K^+\pi^-]_D K^+)}{\mathcal{B}([K^+\pi^-]_D K^-) + \mathcal{B}([K^+\pi^-]_D K^+)}$$
$$\equiv \frac{2r_B r_D \sin\gamma \sin(\delta_B + \delta_D)}{r_B^2 + r_D^2 + 2r_B r_D \cos\gamma \cos(\delta_B + \delta_D)},$$

where δ_D is the relative strong phase in the D^0 decay and $r_D \equiv \left|\frac{A(D^0 \to K^+\pi^-)}{A(D^0 \to K^-\pi^+)}\right|$ is the magnitude of the ratio of the amplitude of the doubly-Cabibbo-suppressed D^0 decay to that of the Cabibbo-allowed one. The value of r_D has been measured to be $r_D = 0.060 \pm 0.002$ [18].

In this method, each B decay mode has two indepedent equations (R_{ADS}, A_{ADS}) that cannot be solved for three unknowns (r_B, γ, $\delta_B + \delta_D$). In order to determine the value of γ, for a given B decay mode it is necessary to reconstruct at least two different D^0 decay modes, such as $B^- \to [K^+\pi^-]_D K^-$, $B^- \to [K^{*+}\pi^-]_D K^-$ or $B^- \to [K^+\rho^-]_D K^-$. In the case of two D^0 decay modes, it is possible to extract the value of γ up to a 16-fold ambiguity, while in the case of three D^0 decay modes there remains a 4-fold ambiguity [7, 8].

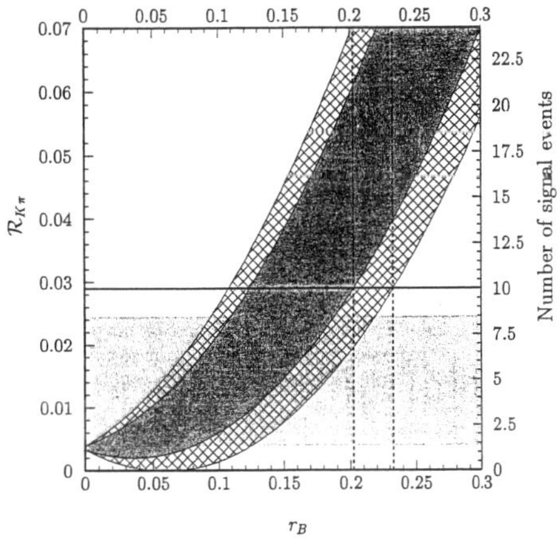

Fig. 5. Expectations for $\mathcal{R}_{K\pi}$ and the number of signal events vs. r_B. *Dark filled-in area*: allowed region for any value of δ, with $a \pm 1\sigma$ variation on r_D, and $51° < \gamma < 66°$. *Hatched area*: additional allowed region with no constraint on γ. Note that the uncertainty on r_D has a very small effect on the size of the allowed regions. The *horizontal line* represents the 90% C.L. limit $\mathcal{R}_{K\pi} < 0.029$. The *vertical dashed lines* are drawn at $r_B = 0.203$ and $r_B = 0.233$. They represent the 90% C.L. upper limits on r_B with and without the constraint on γ. The *light filled-in area* represents the 68% C.L. region corresponding to $\mathcal{R}_{K\pi} = 0.013\pm^{0.011}_{0.009}$

Table 2. Measured charge-independent ratios R_{ADS} for $B^{\pm} \to D^{(*)}K^{(*)\pm}$ decay modes. Where a single term for the error is specified, it includes the statistical and the systematic contribution, otherwise the first error is statistical, the second is systematic. The 90% C.L. limits reported are evaluated without any assumptions for the values of γ and $\delta_B + \delta_D$. The result for $B^{\pm} \to DK^{*\pm}$ is obtained combining the ADS and GLW measurements

B mode	R_{ADS}	r_B
$B^{\pm} \to DK^{\pm}$	< 0.029 90% C.L.	$r_B < 0.23$
$B^{\pm} \to D^*K^{\pm}$	< 0.023 90% C.L.	$r_B^{*\,2} < (0.16)^2$
$B^{\pm} \to DK^{*\pm}$	$0.046 \pm 0.031 \pm 0.08$	$0.28^{+0.006}_{-0.010}$

The addition of different B decay modes is helpful to constrain the value of γ. Particularly interesting are $B^- \to D^* K^-$ decays, where the D^* is reconstructed in $D^0\pi^0$ and $D^0\gamma$. In fact, there is an effective strong phase shift of π between the two cases [19], leading to two different R^*_{ADS} expressions:[2]

$$R^*_{\mathrm{ADS},D\pi^0} = r_B^{*2} + r_D^2 + 2r_B^* r_D \cos\gamma \cos(\delta_B^* + \delta_D),$$
$$R^*_{\mathrm{ADS},D\gamma} = r_B^{*2} + r_D^2 - 2r_B^* r_D \cos\gamma \cos(\delta_B^* + \delta_D),$$

where $R^*_{\mathrm{ADS},D\pi^0}$ ($R^*_{\mathrm{ADS},D\gamma}$) is the charge-independent ratio for the $B^- \to D^*K^-$ with $D^{*0} \to D^0\pi^0$ ($D^{*0} \to D^0\gamma$). Hence in the case of $B^- \to D^*K^-$, it is straightforward to determine the value for r_B^* through the relation

$$\frac{R^*_{\mathrm{ADS},D\pi^0} + R^*_{\mathrm{ADS},D\gamma}}{2} = {r_B^*}^2 + r_D^2 .$$

In Fig. 4 it is shown the m_{ES} distribution with fit model overload for candidate signal events. With the present statistics there is no evident signal in $B^{\pm} \to D^{(*)}K^{(*)\pm}$ decay modes at the B-factories. The experimental observables R_{ADS} have been measured in $B^{\pm} \to D^{(*)}K^{(*)\pm}$ decays and they were found to be consistent with zero. However, it is possible to set an upper limit to the value of r_B as shown in Fig. 5. The summary of the results is reported in Table 2.

5 The Dalitz method: $B^{\pm} \to DK^{\pm}$ with a Dalitz analysis of the $D^0 \to K^0_S\pi^+\pi^-$ decay

In the previously described methods, if the relative strong phases δ_B vanish, the sensitivity to γ is significantly reduced. In general, having large interfering amplitudes with relatively strong phases enhances the sensitivity to the phase γ. The main advantage of the method [9] is that it involves the entire resonant structure of the $D^0 \to K^0_S\pi^+\pi^-$ three-body decay, with interference between doubly-Cabibbo-suppressed, Cabibbo-allowed and CP-eigenstate amplitudes all providing the sensitivity to γ. No branching ratio measurements are needed and only charged particles are involved in the final states, which results in a higher reconstruction efficiency and low background. The price to pay is that it requires a detailed study of the resonances and their interference through a *Dalitz plot* technique [20].

Unless otherwise stated, we use the term "Dalitz plot" to refer to the allowed kinematic region in the two-dimensional squared space m_-^2 and m_+^2, where $K^0_S\pi^-$ and $K^0_S\pi^+$ m_-^2 and m_+^2 are the invariant masses of $K^0_S\pi^-$ and $K^0_S\pi^+$ respectively.

[2] Here and in the following the '*' symbol indicates that the specified value refers to the $B^- \to D^*K^-$ decay mode.

Let us focus on the following cascade decay[3]

$$B^- \to DK^- \to (K_S\pi^-\pi^+)_D K^- , \quad (1)$$

using the notation of Giri et al. [9] to define the amplitudes

$$A(B^- \to D^0K^-) \equiv A_B ,$$
$$A(B^- \to \overline{D^0}K^-) \equiv A_B r_B \mathrm{e}^{\mathrm{i}(\delta_B - \gamma)} . \quad (2)$$

The same definitions apply to the amplitudes for the CP conjugate cascade $B^+ \to DK^+ \to (K_S\pi^+\pi^-)_D K^+$ with the change of weak phase sign $\gamma \to -\gamma$ in (2). We have set the strong phase of A_B to zero by convention, so that δ_B is the difference of strong phases between the two amplitudes. The value of $|A_B|$ is known from the measurement of the $B^- \to D^0K^-$ decay width using flavor specific decays of D^0. The amplitude $A(B^- \to \overline{D^0}K^-)$ is color suppressed and cannot be determined from experiment in this way [7, 8].

Assuming CP is conserved in $D^0 \to K^0_S\pi^+\pi^-$ decay [21], we define the decay amplitude of the $A(B^- \to DK^-)$ decays, with $D^0 \to K^0_S\pi^+\pi^-$ as

$$\mathcal{A}_D(m_-^2, m_+^2) + \kappa r_B \mathrm{e}^{\mathrm{i}(\delta_B - \gamma)} \mathcal{A}_D(m_+^2, m_-^2) , \quad (3)$$

where $\mathcal{A}_D(m_-^2, m_+^2)$ is the $D^0 \to K^0_S\pi^+\pi^-$ decay amplitude. As a consequence of parity and angular momentum conservation in the $B^- \to D^*K^-$ decay, the factor κ takes the value $+1$ for $B^- \to D^*K^-$ ($D^* \to D^0\pi^0$), and -1 for $B^- \to D^*K^-$ ($D^* \to D^0\gamma$) [19].

A model dependent parameterization of the Dalitz structure can be introduced to reduce the number of unknown parameters to extract from the data. If the functional dependence of both the moduli and the phases of the D^0 meson decay amplitudes $\mathcal{A}_D(m_-^2, m_+^2)$ were known, then the analysis would be simplified. There would be only three variables, r_B, δ_B, and γ, that need to be fitted. A plausible assumption, confirmed by data, is that a significant part of the three-body $D^0 \to K_S\pi^-\pi^+$ decay proceeds via two-body resonances. The $D^0 \to K^0_S\pi^-\pi^+$ decay amplitude $\mathcal{A}_D(m_-^2, m_+^2)$ hence can be determined from an unbinned maximum-likelihood fit to the Dalitz plot distribution of a D^0 sample from $D^{*+} \to D^0\pi^+$ decays reconstructed on data.

A phenomenological model to describe $\mathcal{A}_D(m_-^2, m_+^2)$, based on Breit–Wigner (BW) parameterizations of a set of resonances, can be used. The decay amplitude of the model is then expressed as a sum of two-body decay-matrix elements (subscript r) and a non-resonant (subscript NR) contribution,

$$\mathcal{A}_D(m_-^2, m_+^2) = \Sigma_r a_r \mathrm{e}^{\mathrm{i}\phi_r} \mathcal{A}_r(m_-^2, m_+^2) + a_{\mathrm{NR}} \mathrm{e}^{\mathrm{i}\phi_{\mathrm{NR}}} , \quad (4)$$

where each term is parameterized with an amplitude a_r (a_{NR}) and a phase ϕ_r (ϕ_{NR}). The function $\mathcal{A}_r(m_-^2, m_+^2)$ is the Lorentz-invariant expression for the matrix element

[3] In the following discussion we neglect $D^0 - \overline{D^0}$ mixing, which is a good approximation in the context of the standard model [9].

Table 3. Complex amplitudes $a_r e^{i\phi_r}$ and fit fractions of the different components ($K_S\pi^-$, $K_S\pi^+$, and $\pi^+\pi^-$ resonances) obtained from the fit of the $D^0 \to K_S\pi^-\pi^+$ Dalitz distribution from $D^{*+} \to D^0\pi^+$ events. Errors are statistical only. Masses and widths of all resonances are taken from [18] with the exception of $K_0^*(1430)^+$ taken from [31]. The fit fraction is defined for the resonance terms as the integral of $a_r^2|\mathcal{A}_r(m_-^2, m_+^2)|^2$ over the Dalitz plane divided by the integral of $|\mathcal{A}_D(m_-^2, m_+^2)|^2$. The sum of fit fractions is 119.5%. A value different from 100% is a consequence of the interference among the amplitudes

Component	$\mathrm{Re}\{a_r e^{i\phi_r}\}$	$\mathrm{Im}\{a_r e^{i\phi_r}\}$	Fit fraction (%)
$K^*(892)^-$	-1.223 ± 0.011	1.3461 ± 0.0096	58.1
$K_0^*(1430)^-$	-1.698 ± 0.022	-0.576 ± 0.024	6.7
$K_2^*(1430)^-$	-0.834 ± 0.021	0.931 ± 0.022	3.6
$K^*(1410)^-$	-0.248 ± 0.038	-0.108 ± 0.031	0.1
$K^*(1680)^-$	-1.285 ± 0.014	0.205 ± 0.013	0.6
$K^*(892)^+$	0.0997 ± 0.0036	-0.1271 ± 0.0034	0.5
$K_0^*(1430)^+$	-0.027 ± 0.016	-0.076 ± 0.017	0.0
$K_2^*(1430)^+$	0.019 ± 0.017	0.177 ± 0.018	0.1
$\rho(770)$	1	0	21.6
$\omega(782)$	-0.02194 ± 0.00099	0.03942 ± 0.00066	0.7
$f_2(1270)$	-0.699 ± 0.018	0.387 ± 0.018	2.1
$\rho(1450)$	0.253 ± 0.038	0.036 ± 0.055	0.1
Non-resonant	-0.99 ± 0.19	3.82 ± 0.13	8.5
$f_0(980)$	0.4465 ± 0.0057	0.2572 ± 0.0081	6.4
$f_0(1370)$	0.95 ± 0.11	-1.619 ± 0.011	2.0
σ	1.28 ± 0.02	0.273 ± 0.024	7.6
σ'	0.290 ± 0.010	-0.0655 ± 0.0098	0.9

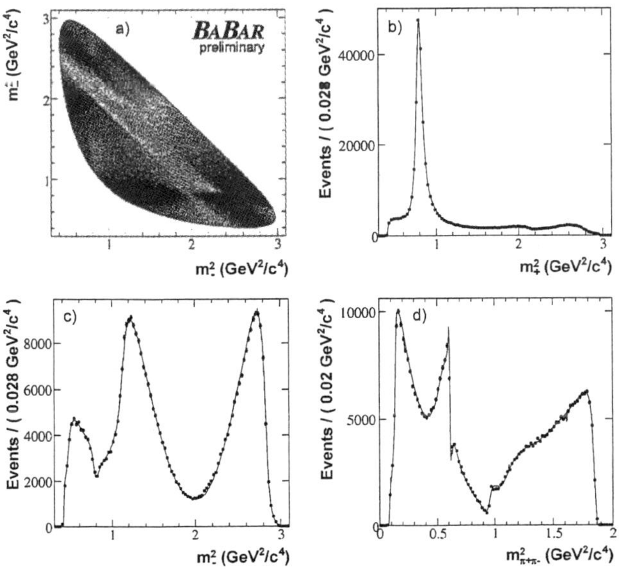

Fig. 6. (a) The $\bar{D}^0 \to K_S^0\pi^-\pi^+$ Dalitz distribution from $D^{*-} \to \bar{D}^0\pi^-$ events, and projections on (b) $m_+^2 = m_{K_S^0\pi^+}^2$, (c) $m_-^2 = m_{K_S^0\pi^-}^2$, and (d) $m_{\pi^+\pi^-}^2$. $D^0 \to K_S^0\pi^+\pi^-$ from $D^{*+} \to D^0\pi^+$ events are also included. The *curves* are the reference model fit projections

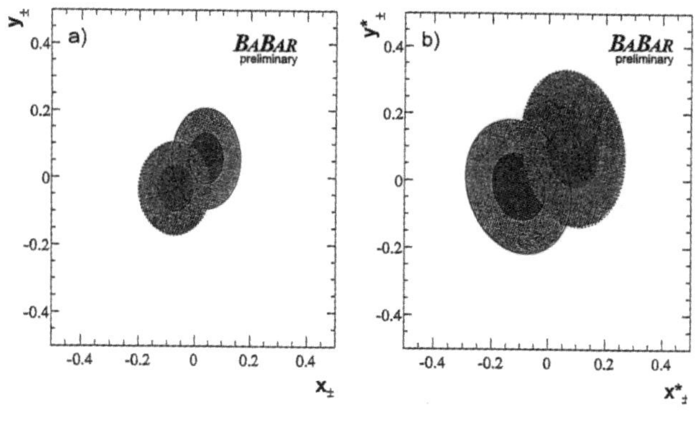

Fig. 7. Contours at 39.3% (*dark*) and 86.5% (*light*) confidence level (corresponding to two-dimensional one- and two-standard deviation regions), including statistical and systematic uncertainties, for the $(x_{\mp}^{(*)}, y_{\mp}^{(*)})$ parameters for B^- (*thick* and *solid lines*) and B^+ (*thin* and *dotted lines*) decays

of a D^0meson decaying into $K_S^0\pi^-\pi^+$ through an intermediate resonance r, parameterized as a function of position in the Dalitz plane. For $r = \rho(770)$ and $\rho(1450)$, we use the functional form suggested in [22], while the remaining resonances can be parameterized by a spin-dependent relativistic Breit–Wigner distribution [18]. An analogous phenomenological approach is represented by the K-matrix formalism [23, 24], which provides a direct way of imposing the unitarity constraint that is not guaranteed in the case of the BW model and is suited to the study of broad and overlapping resonances in multi-channel decays. In the $D^0 \to K_S^0\pi^+\pi^-$ decay, the K-matrix method is suited to solve the main limitation of the Breit–Wigner model to parameterize the $\pi\pi$ S-wave states [25], thus avoiding the need to introduce the σ scalars.

The Dalitz amplitude $\mathcal{A}_D(m_-^2, m_+^2)$ can be written in this case as a sum of two-body decay matrix elements for the spin-1, spin-2 and $K\pi$ spin-0 resonances (as in the Breit–Wigner model), and the $\pi\pi$ spin-0 piece denoted as F_1 is written in terms of the K-matrix. We have

$$\mathcal{A}_D(m_-^2, m_+^2) = F_1(s) + \sum_{r \neq \pi\pi \text{ S-wave}} a_r \mathrm{e}^{\mathrm{i}\phi_r} \mathcal{A}_r(m_-^2, m_+^2)\,,$$

where $F_1(s)$ is the contribution of $\pi\pi$ S-wave states,

$$F_1(s) = \sum_j [I - \mathrm{i}K(s)\rho(s)]_{1j}^{-1}\, P_j(s)\,.$$

Here, s is the squared mass of the $\pi\pi$ system $(m_{\pi^+\pi^-}^2)$, I is the identity matrix, K is the matrix describing the S-wave scattering process, ρ is the phase-space matrix, and P is the initial production vector [26]. The index j represents the jth channel ($1 = \pi\pi$, $2 = K\overline{K}$, 3 =multi-meson[4], $4 = \eta\eta$, $5 = \eta\eta'$ [27]). The K-matrix parameters can be obtained from [27] from a global fit of the available $\pi\pi$ scattering data from threshold up to 1900 MeV$/c^2$.

The BABAR analysis uses, as a "nominal" model, the isobar model, which consists of 13 resonances leading to 16 two-body decay amplitudes and phases (see Table 3), plus the non-resonant contribution.

All the resonances considered in this model are well established except for the two scalar $\pi\pi$ resonances, σ and σ', whose masses and widths are obtained from our sample.[5] Their addition to the model is motivated by an improvement in the description of the data. The $D^0 \to K_S^0\pi^+\pi^-$ Dalitz distribution and the fit projections are shown in Fig. 6. The possible absence of the σ and σ' resonances is considered in the evaluation of the systematic errors, fitting the data using the alternative K-matrix model. Once the $\mathcal{A}_D(m_-^2, m_+^2)$ amplitude is obtained from the fit on the $D^{*+} \to D^0\pi^+$ sample, it can be fed into (3). The extraction of the angle γ is then performed through a fit to the Dalitz distribution of the D^0 in the $B^{\pm} \to DK^{\pm}$ decays. The value of the CP-odd phase γ changes sign for B^+ and B^- in (3), leading to different rates in corresponding regions of the D^0 Dalitz plane for B^+ and B^- decays. We introduce here the CP *parameters* $x_{\mp}$ and $y_{\mp}$ defined respectively as the real and imaginary parts of $r_B \mathrm{e}^{\mathrm{i}(\delta_B \mp \gamma)}$, for which the constraint $r_B^2 = x_{\mp}^2 + y_{\mp}^2$ holds. Experimentally, it was demonstrated that $x_{\mp}$ and $y_{\mp}$ are well-behaving fitting parameters that are unbiased with Gaussian errors [28].

The results for the CP variables, using 347 million of $B\bar{B}$ events recorded with the BABAR detector, are reported in Table 4 [29]. The results of the fit are represented showing the 1σ and 2σ two-dimensional counters in Fig. 7.

A frequentist (Neyman) procedure [18, 30] has been adopted to interpret the measurement of the CP parameters $(x_{\mp}^{(*)}, y_{\mp}^{(*)})$ reported in Table 4 in terms of confidence regions on $\mathbf{p} = (\gamma, r_B, \delta_B, r_B^*, \delta_B^*)$. For a given $\mathbf{p}$, the five-dimensional confidence level $\mathcal{C}$ is calculated by integrating over all points in the fit parameter space closer (larger PDF) to $\mathbf{p}$ than the fitted data values. The one- (two-) standard deviation region of the CP parameters is defined as the set of $\mathbf{p}$ values for which confidence level $\mathcal{C}$ is smaller than 3.7% (45.1%). Figure 8 shows the

[4] Multi-meson channel refers to a final state with four pions.

[5] The σ and σ' masses and widths are determined from the data. We find (in MeV$/c^2$) $M_\sigma = 490 \pm 6$, $\Gamma_\sigma = 406 \pm 11$, $M_{\sigma'} = 1024 \pm 4$, and $\Gamma_{\sigma'} = 89 \pm 7$, Errors are statistical.

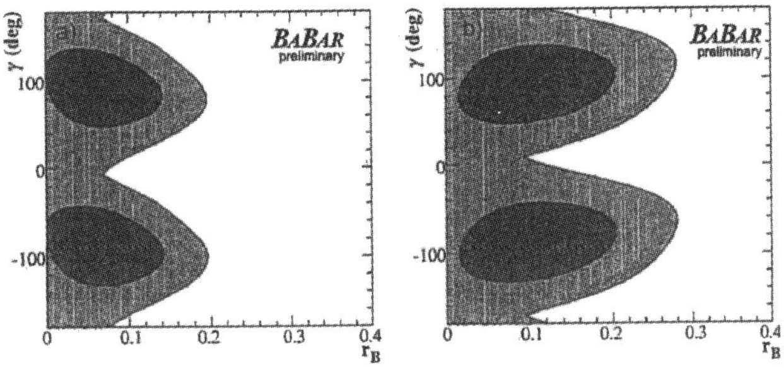

Fig. 8. Projections in the (a) $r_B - \gamma$ and (b) $r_B^* - \gamma$ planes of the five-dimensional one- (*dark*) and two- (*light*) standard deviation regions

Table 4. CP-violating parameters $x_{\mp}^{(*)}, y_{\mp}^{(*)}$ obtained from the CP fit to the $B^{\pm} \to D^{(*)}K^{\pm}$ samples. The first error is statistical, the second is experimental systematic uncertainty and the third is the systematic uncertainty associated with the Dalitz model

CP parameter	$B^{\pm} \to D^{(*)}K^{\pm}$
x_-	$0.041 \pm 0.059 \pm 0.018 \pm 0.011$
y_-	$0.056 \pm 0.071 \pm 0.007 \pm 0.023$
x_+	$-0.072 \pm 0.056 \pm 0.014 \pm 0.029$
y_+	$-0.033 \pm 0.066 \pm 0.007 \pm 0.018$
x_-^*	$-0.106 \pm 0.091 \pm 0.020 \pm 0.009$
y_-^*	$-0.019 \pm 0.096 \pm 0.022 \pm 0.016$
x_+^*	$0.084 \pm 0.088 \pm 0.015 \pm 0.018$
y_+^*	$0.096 \pm 0.111 \pm 0.032 \pm 0.017$

two-dimensional projections onto the $r_B - \gamma$ and $r_B^* - \gamma$ planes, including statistical and systematic uncertainties. The figure shows that this Dalitz analysis has a two-fold ambiguity, $(\gamma, \delta_B^{(*)}) \to (\gamma + 180°, \delta_B^{(*)} + 180°)$, as expected from (3). From the one-dimensional projections we obtain for the weak phase $\gamma = (92 \pm 41 \pm 11 \pm 12)°$, and for the strong phase differences $\delta_B = (118 \pm 63 \pm 19 \pm 36)°$ and $\delta_B^* = (-62 \pm 59 \pm 18 \pm 10)°$. No constraints on the phases are achieved at the level of two standard deviations and beyond. Similarly, for the magnitude of the ratio of decay amplitudes r_B and r_B^* we obtain the one (two) standard deviation constraints $r_B < 0.140$ ($r_B < 0.195$) and $0.017 < r_B^* < 0.203$ ($r_B^* < 0.279$). In all cases, the first error is statistical, the second systematic and the third is due to the parametrization of the $D^0 \to K_S^0 \pi^+ \pi^-$ decay amplitude.

6 Combined measurements of γ and projections for the future

The Dalitz method has the best sensitivity to γ with the current statistics, but it is still not possible to precisely determine the value. Combining the results of several methods and different $B^{\pm} \to D^{(*)}K^{(*)\pm}$ decay modes enlarges the sensitivity to the angle γ. The measurement is dominated by statistical error, but more data will improve

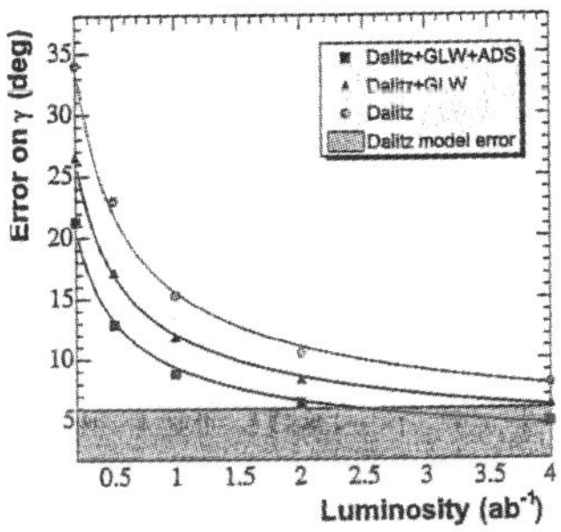

Fig. 9. Projection of the sensitivity to γ, assuming $r_B = 0.1$, for the Dalitz method (○), Dalitz + GLW combined (△) and Dalitz + GLW + ADS combined (□).The *horizontal band* represent the error projection due to the phenomenological parameterization of the D decay amplitude

the precision. The projections for the measurement are highly dependent on the value of r_B, hence, it is difficult at this point to make predictions for large statistics. However, it is possible to make predictions by choosing a specific value for r_B. In Fig. 9 we show the projection for the Dalitz method and for the combined measurement of the Dalitz method, GLW and ADS, assuming $r_B = 0.1$. In this scenario, by combining different methods it will be possible to measure the angle γ with 10° error with a 1 ab^{-1} data sample, which is within the reach of the BABAR experiment.

Acknowledgements. I wish to thank the EMFCSC of Erice for the invitation to the Summer School and the Istituto Nazionale di Fisica Nucleare (INFN) for their support. I would also like to thank M.A. Giorgi and F. Martinez-Vidal for very useful discussions.

References

1. J.H. Christenson, J.W. Cronin, V.L. Fitch, R. Turlay, Phys. Rev. Lett. **13**, 138 (1964)
2. N. Cabibbo, Phys. Rev. Lett. **10**, 531 (1963)

3. M. Kobayashi, T. Maskawa, Prog. Theor. Phys. **49**, 652 (1973)
4. Y. Grossman, A. Soffer, J. Zupan, Phys. Rev. D **72**, 031 501 (2005)
5. M. Gronau, D. London, Phys. Lett. B **253**, 483 (1991)
6. M. Gronau, D. Wyler, Phys. Lett. B **265**, 172 (1991)
7. D. Atwood, I. Dunietz, A. Soni, Phys. Rev. Lett. **78**, 3257 (1997)
8. D. Atwood, I. Dunietz, A. Soni, Phys. Rev. D **63**, 036 005 (2001)
9. A. Giri, Y. Grossman, A. Soffer, J. Zupan, Phys. Rev. D **68**, 054 018 (2003)
10. R.H. Dalitz, Philos. Mag. **44**, 1068 (1953)
11. M. Gronau, Phys. Lett. B **557**, 198 (2003)
12. B. Aubert et al., Phys. Rev. D **72**, 032 004 (2005)
13. B. Aubert et al., Phys. Rev. D **72**, 071 104 (2005)
14. M. Saigo et al., Phys. Rev. Lett. **94**, 091 601 (2005)
15. B. Aubert et al., Phys. Rev. Lett. **73**, 051 105 (2006)
16. B. Aubert et al., Phys. Rev. Lett. **72**, 071 103 (2005)
17. B. Aubert et al., Phys. Rev. D **71**, 031 102 (2005)
18. Particle Data Group, W.-M. Yao et al., J. Phys. G **33**, 1 (2006)
19. A. Bondar, T. Gershon, Phys. Rev. D **70**, 091 503 (2004)
20. Review on Dalitz plot analysis formalism in [18]
21. CLEO Collaboration, D.M. Asner et al., Phys. Rev. D **70**, 091 101 (2004)
22. G.J. Gounaris, J.J. Sakurai, Phys. Rev. Lett. **21**, 244 (1968)
23. E.P. Wigner, Phys. Rev. **70**, 15 (1946)
24. S.U. Chung et al., Ann. Phys. **4**, 404 (1995)
25. Review on Scalar Mesons in [18]
26. I.J.R. Aitchison, Nucl. Phys. A **189**, 417 (1972)
27. V.V. Anisovich, A.V. Sarantev, Eur. Phys. J. A **16**, 229 (2003)
28. BABAR Collaboration, B. Aubert et al., Phys. Rev. Lett. **95**, 121 802 (2005)
29. BABAR Collaboration, B. Aubert, arXiv:hep-ex/0607104
30. J. Neyman, Phil. Trans. Royal Soc. London, Series A **236**, 333 (1937), reprinted in: A selection of Early Statistical Papers on J. Neyman (University of California Press, Berkeley, 1967)
31. E791 Collaboration, E.M. Aitala et. al., Phys. Rev. Lett. **89**, 121 801 (2002)

Reprinted with permission from the European Physical Journal C

Eur. Phys. J. C 52, 247–251 (2007)
DOI 10.1140/epjc/s10052-007-0386-4

THE EUROPEAN PHYSICAL JOURNAL C

Special Article – Young Scientists Paper

Two-particle correlations from RHIC to LHC, a Monte Carlo approach [a]

F. Noferini[1,2,3,b]

[1] Dipartimento di Fisica dell'Università, Via Irnerio 46, 40126 Bologna, Italy
[2] Museo Storico della Fisica e Centro Studi e Ricerche "Enrico Fermi", Roma, Italy
[3] Sezione INFN, Bologna, Italy

Received: 25 June 2007 /
Published online: 3 August 2007 – © Springer-Verlag / Società Italiana di Fisica 2007

Abstract. The aim of this work is to extend to LHC the results observed for two-particle correlations at RHIC, especially in terms of jet quenching effects. In this study a parton quenching model developed in the BDMPS-Z-SW framework is considered and implemented as an afterburner for PYTHIA and HIJING. A simplified parametrization of the quenching mechanism at the parton level is included in one of the most popular Monte Carlo event generators for *AA* collisions, HIJING. The simulation method, tuned on the RHIC data, is then used to make predictions for the LHC energy regime in order to probe the scenario we will study in the ALICE experiment.

1 Introduction

The well-known results on two-particle correlations from RHIC have generated a substantial interest in jet studies. They provide a way to study the properties of the dense medium produced in heavy ion collisions. This work moves in this direction, first considering some of the jet properties observed at RHIC, and then studying the kinematic region that is more adequate for extending the same analysis to LHC in the parton quenching model scenario [1]. In Fig. 1 the results from the STAR collaboration are reported, showing the back-to-back suppression of the jet correlations in very central Au + Au collisions at intermediate p_T [2]. The absence of this suppression in $d+$Au collisions strongly suggests that the effect is due to the interaction between the jet and the medium, occurring only in *AA* collisions.

At RHIC it was observed [3] that the back-to-back correlation shows up again when the p_T value of the trigger particle is increased. Therefore, from RHIC we learn that the strength of the signal effect changes with the p_T of the trigger particle.

In Sect. 2, the quenching model with the Glauber geometry (parton quenching model) is described, together with some relevant results concerning jet physics [1,4,5]. The implementation of this model in a Monte Carlo simulation is described in Sect. 3, where the PYTHIA [6] and HIJING [7,8] generators are considered (see [9]). The results obtained with this Monte Carlo simulation at RHIC energy are then presented. In Sect. 4, the extrapolation at the LHC energy is performed in order to make predictions in the new energy regime. Finally, in Sect. 5 the conclusions are reported.

2 Quenching model description and general results

The quenching model developed in the BDMPS-Z-SW framework [4] is based on the idea that a fast parton interacts strongly with the medium formed in the collision, losing energy via gluonic bremsstrahlung. In particular,

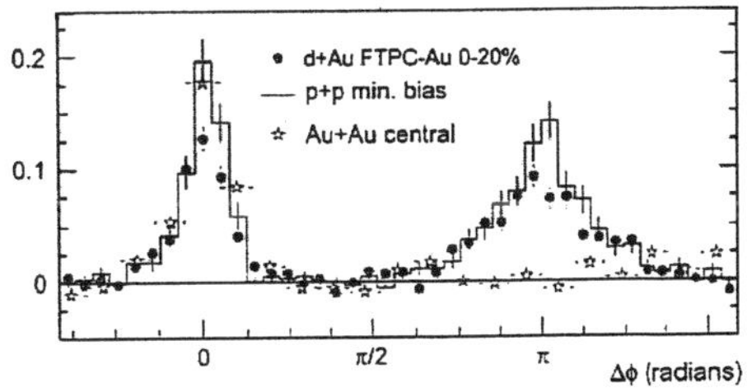

Fig. 1. Two-particle azimuthal distributions for central $d+$Au collisions compared with those seen in *pp* and central Au + Au collisions $\left(4 < p_T^{trig} < 6\ \text{GeV}/c, p_T^{assoc} > 2\ \text{GeV}/c\right)$. The corresponding pedestals have been subtracted. Figure taken from [2]

[a] "New Talent" Presentation Award at the 44th Course of the International School of Subnuclear Physics, Erice, Italy, 29 Aug–7 Sep 2006

[b] e-mail: noferini@bo.infn.it

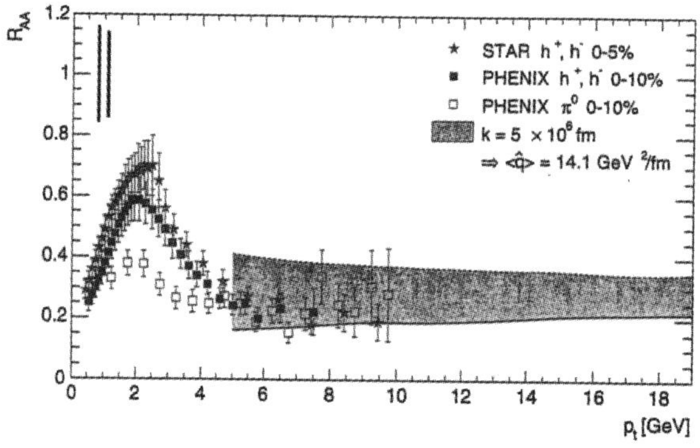

Fig. 2. $R_{AA}(p_T)$ for central Au+Au collisions at $\sqrt{s_{NN}} = 200$ GeV. The model band is obtained with a parton-by-parton calculation of ω_c and R. The average transport coefficient is 14 GeV^2/fm. The plot is taken from [1]

the formation of a deconfined medium (the so-called quark gluon plasma) would result in a very different kind of interaction compared to the case of purely hadronic matter. The main difference in the type of interaction is due to the fact that, in the former case, the interaction of the radiated gluons with the medium also has to be considered. In particular, the probability for a parton to lose a given amount of energy scales with the square of the path length instead of linearly. A simple explanation for this behaviour is that the amount of the energy lost is proportional to the number of scatterings ($\propto L$) and to the formation probability ($\propto L$), which is approximately an L^2-dependence. The details of the approximation made in this model can be found in [4], where a method to calculate the probabilities for the quenching processes (quenching weights) is also reported. Here, only the main parameters which characterize the model are described in order to help the reader understand the assumptions and the procedure used.

The quenching mechanism states that a parton that has moved in the medium on a path of length L will lose a given amount of energy that is independent of its initial energy. On average, its energy loss is proportional to L^2 (path length squared) [10].

If the two quantities ω_c and R (defined as $\omega_c = \hat{q}L^2/2$ and $R = \hat{q}L^3/2$, where $\hat{q}$ is the mean squared momentum per unit length and L the path length in the medium) are considered, the energy loss is fully determined by these two parameters.

If the Glauber geometry is also considered, the previous quantity can be defined as in the parton quenching model (PQM)[1], as follows:

$$\begin{aligned}(\hat{q}L)_{\text{eff}} &= \int_0^\infty k \times T_A(\chi;b)T_B(\chi;b)\,\mathrm{d}\chi\,,\\ I_n &= \int_0^\infty \chi^n \hat{q}(\chi;b)\,\mathrm{d}\chi\,,\\ \omega_c &= I_1\,,\\ R &= I_0^2/(2I_1)\,,\end{aligned} \tag{1}$$

where T_A and T_B are the impact parameter dependent transverse nuclear densities (b being the impact parameter) of nuclei A and B, and k is a parameter linked to the parton density in the nuclei.

It was also proven [4] that in the case of medium expansion the quenching mechanism can be described, as in the static case, over a large parameter range by replacing $\hat{q}$ with

$$\bar{\hat{q}} = \frac{2}{L^2}\int_{\chi_0}^{L+\chi_0} \mathrm{d}\chi(\chi-\chi_0)\hat{q}(\chi)\,,$$

where $\hat{q}(\chi)$ is the time-dependent transport coefficient.

The following figures present some of the results obtained with this model. In particular, Fig. 2 shows how the model describes the R_{AA} ratio suppression in central collisions [1] for a $\langle\hat{q}\rangle \sim 14$ GeV^2/fm. The quantity R_{AA} is defined as

$$R_{AA}(p_T) = \frac{1}{\langle N_{\text{coll}}\rangle}\left[\left(\frac{\mathrm{d}^2N_{AA}}{\mathrm{d}p_T\,\mathrm{d}\eta}\right)\Big/\left(\frac{\mathrm{d}^2N_{pp}}{\mathrm{d}p_T\,\mathrm{d}\eta}\right)\right],$$

which is the ratio of the charged multiplicity per unit rapidity in AA collisions over the pp one normalized with the average number of collisions in a given centrality class, as a function of p_T.

3 Quenching in the Monte Carlo

In the work presented in this paper, the quenching weight probabilities (see Sect. 2) have been implemented and tested in the PYTHIA and HIJING Monte Carlos. This choice has been driven by the fact that other existing quenching models do not provide a satisfactory description of AA collisions. Moreover, the quenching mechanism provided by the HIJING Monte Carlo generator used in this work is not compatible with our aims, since it simulates quenching effects in hadronic matter, where the energy loss is proportional to L and not L^2 (as predicted in the BDMPS-Z-SW framework).

It must be specified that the quenching procedure used here is not a Monte Carlo simulation, but a simplified parametrization which describes the model in a way that accounts for the nuclear geometry.

The possibility to test the same method in the two different generators has the advantage of making it possible to separate the effects due to the single jet modification from the background effects.

The following simulated results were obtained with the modified versions of PYTHIA and HIJING integrated into the ROOT and ALIROOT framework [11, 12].

3.1 PYTHIA

In this subsection the effect of jet quenching in single NN collisions that assume the Glauber geometry of AA collisions is taken into account. This makes it possible to probe jet modification in a way that excludes the background of the rest of the AA collision.

The method used here is based on the parton quenching model [1] and the radiation products are also taken into account.

The simulation is implemented in two steps [5].

1. High-p_T parton production was simulated using the PYTHIA event generator in pp mode with CTEQ5L [13] parton distribution functions.
2. Depending on whether the jet quenching effect is simulated or not, after the final state gluon radiation, the partonic system is either subsequently hadronised or first passed through an afterburner, as described in the previous section.

The probabilities coming from the quenching weights define the fraction z of total energy lost by the initial parton. The number of radiated gluons is given by the expression

$$1/(1-z)\,, \tag{2}$$

in order to limit their energy to be less than that of the outgoing leading parton. Moreover, the leading parton gets an additional transverse momentum (with respect to its initial direction) given by the expression $\sqrt{\hat{q}L}$.

Figure 3 reports the integrated number of correlated charged particles per trigger particle in the near and opposite side as a function of the centrality of the event. The dashed lines represent the PYTHIA prediction for pp events without quenching effects. It can be observed that the near side correlation (squares) is almost constant when varying the centrality, while the opposite side correlation (triangles) goes down with decreasing impact parameter. Although a residual correlation in the opposite side survives also in central collisions, it is a very small effect. In fact, the suppression factor for $\langle\hat{q}\rangle = 15\ \mathrm{GeV}^2/\mathrm{fm}$ is about 10.

3.2 HIJING

While the PYTHIA generator is a good tool to test the jet quenching model in the absence of background, the HIJING Monte Carlo provides a more realistic description

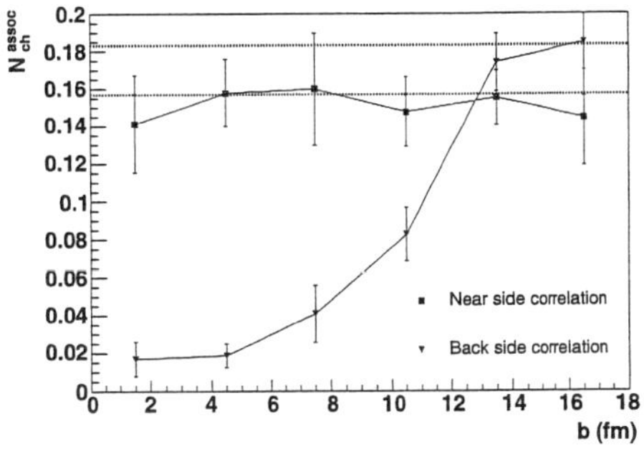

Fig. 3. Behaviour of near and away side correlation as a function of the centrality of the collisions, as explained in the text, when quenching is switched on (modified PYTHIA simulation)

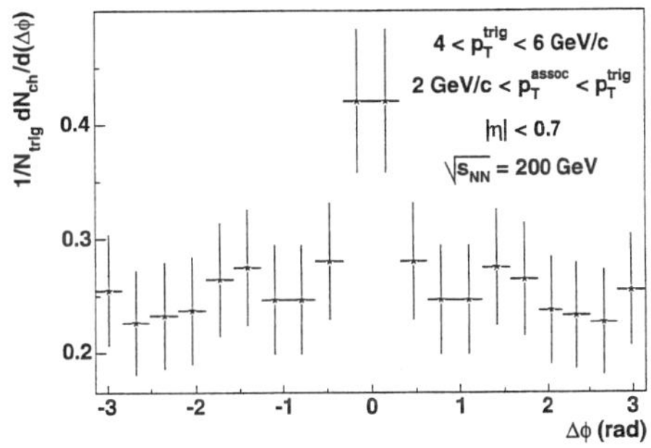

Fig. 4. HIJING simulation in the parton quenching model. The choice of p_T-cuts used here is reported as well

of the environment of AA collisions. For this reason the same jet quenching mechanism has been introduced also in the HIJING code.

Although the code structure is slightly different when the generation of hard processes is performed, the implementation of the jet quenching mechanism is organized in the same two steps as before. The results for central simulated Au + Au collisions at $\sqrt{s_{NN}} = 200$ GeV are shown in Fig. 4.

It is necessary to note that in the current version of HIJING the simulation of the elliptic flow contribution, as it was observed at RHIC, is not included.

The preliminary results obtained with the HIJING generator are consistent with the PYTHIA ones. Even if the statistics are still low, because of the background fluctuations and the small number of events ($N_{\text{trig}} = 1500$), the back-to-back total suppression for this choice of p_T-cuts (Fig. 4) agrees with what is seen in the RHIC data [2].

The modified version of HIJING used in this work also allows for the simulation of single NN collisions, as in

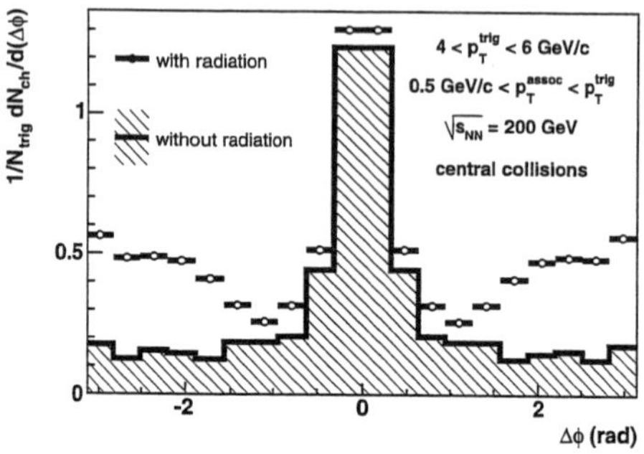

Fig. 5. HIJING simulations (without background) focused on the contribution of radiated gluons at low p_T for Au + Au collisions at $\sqrt{s_{NN}} = 200$ GeV. The *open* and *dashed histograms* refer to simulations with and without radiative contribution. The comparison shows that the contribution of radiated gluons is important and dominant at low p_T if formula (2) is assumed

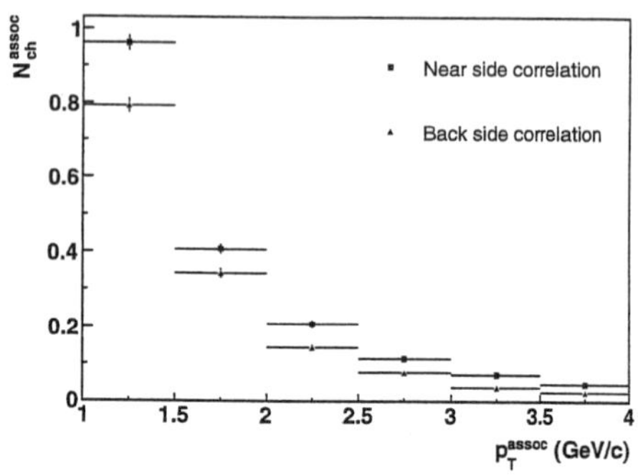

Fig. 6. The behaviour of the near side and away side correlation as a function of the p_T of the associated particles with $4 < p_T^{trig} < 6$ GeV/c, assuming central collisions

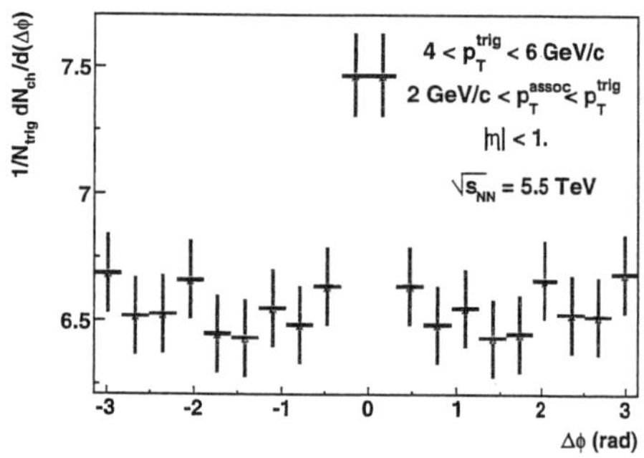

Fig. 7. Two-particle correlation at the LHC energy predicted from the HIJING generator when parton quenching effects are considered (5% most central collisions). The p_T-cut values are reported

PYTHIA, with jet quenching effects and with the Glauber geometry relative to Au + Au collisions taken into account. This capability was used to check directly the effect of the inclusion of the radiation used in the model presented here (see Sect. 3.1). Figure 5 shows the two-particle correlation for low p_T in the cases of radiation and no-radiation. This plot indicates that the radiation introduced in the generator gives a non-negligible contribution at low p_T. It must be noted that the comparison with data is not straightforward, so that the above should be regarded only as a checking of the procedure.

4 Predictions for LHC

As already mentioned, the main aim of this work is to make predictions for LHC. In this new regime, the range of p_T that can be investigated with the ALICE experiment extends to higher values than those at RHIC. Therefore, the PYTHIA and HIJING generators, which have been used to study the results obtained at RHIC, were used to predict two-particle correlations at the LHC energy.

Figure 6 shows PYTHIA simulations, with the quenching effects described in Sect. 3, at $\sqrt{s_{NN}} = 5.5$ TeV with the same choice of p_T for trigger particles used before ($4 < p_T^{trig} < 6$ GeV/c). The behaviours of the near side and away side correlation as a function of the p_T of the associated particles are considered, assuming central collisions.

The effect of the background is visible in Fig. 7, where HIJING simulations for Pb + Pb central collisions, at $\sqrt{s_{NN}} = 5.5$ TeV, are considered. In this case the increase of the charged particle multiplicity at intermediate p_T is clearly visible (both for the background and the signal). Moreover, it has been found that the rate of particles with $p_T > 4$ GeV/c changes from 1 every 40 events (at the RHIC energy) to two particles per event (at the LHC energy). However the definition of "high" p_T for this regime is not clear because the signal is not well separated from the background sources.

The recent results from RHIC [3] have shown that a higher p_T choice provides a signal that is background free. The background reduction with increasing p_T-cuts is expected to be stronger at LHC. To check this, an analysis at higher p_T was performed, and the results are shown in Fig. 8. Also in this case, the HIJING generator with parton quenching model has been used to simulate Pb + Pb collisions. As one can see, the background is reduced by a factor of 13 with respect to the previous case, and the jet-correlations dominate.

The behaviour of the near and away side correlation for $8 < p_T^{trig} < 15$ GeV/c, as a function of p_T^{assoc}, is shown in Fig. 9.

With this choice of cuts, the simulations at the LHC energy with the same quenching strength used for $\sqrt{s_{NN}} =$ 200 GeV show a very evident signal, demonstrating that

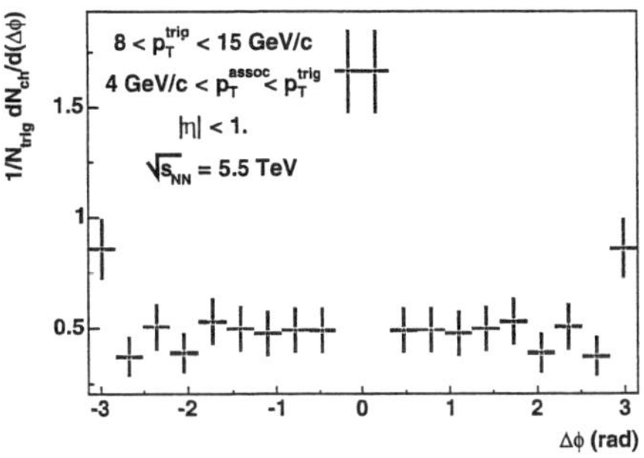

Fig. 8. Two-particle correlation at the LHC energy predicted from the HIJING generator when parton quenching effects are considered (5% most central collisions) for a different choice of p_T-cuts. The p_T-cut values are reported

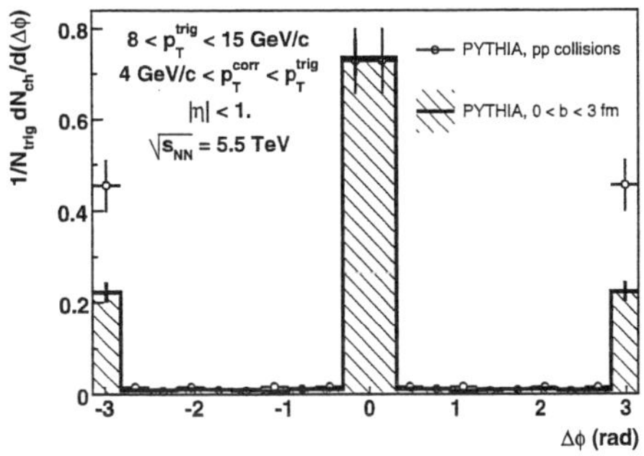

Fig. 10. A comparison between *pp* collisions at LHC energy performed using the PYTHIA generator with and without quenching effects (quenched events are for central collisions with $0 < b < 3\,\mathrm{fm}$)

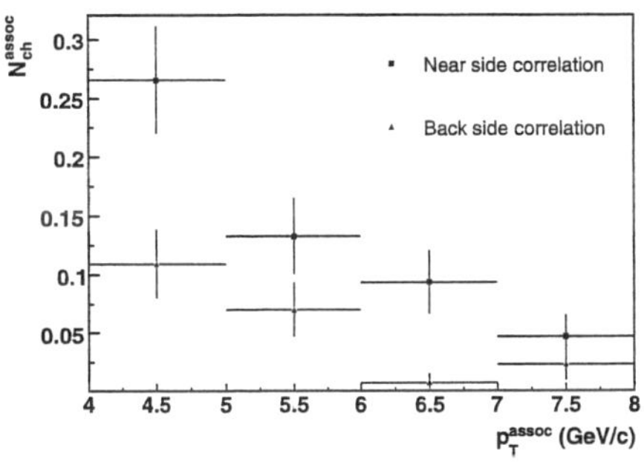

Fig. 9. The behaviour of the near side and away side correlation as a function of the p_T of the associated particles with $8 < p_T^{\mathrm{trig}} < 15\,\mathrm{GeV}/c$

this kind of selection may be better suited for extending the study to the jet-medium interaction at high-p_T. This is also suggested by the comparison of *pp* collisions obtained with the PYTHIA generator with and without quenching, as shown in Fig. 10. Here it can be noted that the back-to-back correlation is meaningfully suppressed when quenching effects are taken into account.

The results presented for the last choice of p_T-cuts reveal that two-particle correlations at LHC will be particularly sensitive to quenching effects, provided that the model employed and the values used here for the quenching strength are still valid.

5 Conclusions

A parton quenching model was implemented in Monte Carlo generators. This approach described the two-particle correlations observed in the RHIC data in a satisfactory way. The same model was used to determine the sensitivity of this kind of analysis at the LHC. In particular, it was shown that an increase of the p_T-cuts is necessary to separate the signal from the background and that in this case, quenching effects, when occurring, could be identified through this analysis.

One characteristic feature of the model is the inclusion of gluon radiation in the jet quenching mechanism, to which the analysis has shown itself to be sensitive.

The Monte Carlo approach appears to be a very useful tool for probing and predicting the behaviour of jet observables. In order to have a more realistic description of nuclear collisions, in the future we plan to add the elliptic flow contribution into this study.

References

1. A. Dainese, C. Loizides, G. Paic, Eur. Phys. J. C **38**, 461 (2005)
2. STAR Collaboration, J. Adams et al., Phys. Rev. Lett. **91**, 072304 (2003)
3. STAR Collaboration, J. Adams et al., arXiv:nucl-ex/0604018 (2006)
4. C.A. Salgado, U.A. Wiedemann, Phys. Rev. D **68**, 014008 (2003)
5. ALICE Collaboration, F. Carminati et al., ALICE: Physics performance report, Vol. II, CERN-LHCC-2005-030
6. T. Sjöstrand, Comput. Phys. Commun. **82**, 74 (1994)
7. X.N. Wang, M. Gyulassy, Phys. Rev. D **44**, 3501 (1991)
8. X.N. Wang, M. Gyulassy, Phys. Rev. Lett. **68**, 1480 (1992)
9. ALICE Collaboration, A. Morsch, J. Phys. G **31**, S597 (2005)
10. U.A. Wiedemann, Nucl. Phys. A **690**, 731 (2001)
11. http://aliceinfo.cern.ch/Offline
12. http://root.cern.ch
13. H.L. Lai et al., Eur. Phys. J. C **12**, 375 (2000)

DIPLOMAS

Twenty one Diplomas were open for competition among the participants. They have been awarded as follows:

- John S. BELL Diploma to:

  Xavier CALMET
  University of Brussels, Belgium

- Patrick M.S. BLACKETT Diploma to:

  Paolo ASCHIERI
  University of Piemonte Orientale, Alessandria, Italy

- James CHADWICK Diploma to:

  Nicola NERI
  INFN, Pisa, Italy

- Richard DALITZ Diploma to:

  Simone PACETTI
  INFN, Frascati-LNF, Italy

- Paul A.M. DIRAC Diploma to:

  Màtè CSANÀD
  Department of Atomic Physics, Budapest, Hungary

- Vladimir N. GRIBOV Diploma to:

  Dmitry KROTOV
  ITEP, Moscow, Russia

- Robert M.S. HOFSTADTER Diploma to:

  David LOPEZ MATEOS
  CALTECH, Pasadena, USA

- Gunnar KALLEN Diploma to:

  Grigory VARTANOV
  JINR, Dubna, Russia

- Yuval NE'EMAN Diploma to:

  Pedro MACHADO
  University of Utrecht, The Netherlands

- Giuseppe P.S. OCCHIALINI Diploma to:

  Francesco NOFERINI
  University of Bologna, Italy

- Oreste PICCIONI Diploma to:

  Ori SMITH
  University of Tel Aviv, Israel

- Bruno PONTECORVO Diploma to:

  Viktor BEGUN
  Bogolyubov Institute for Theoretical Physics, Kiev, Ukraine

- Isidor I. RABI Diploma to:

  Vyacheslav DIDENKO
  Lebedev Physical Institute, Moscow, Russia

- Giulio RACAH Diploma to:

  Boaz KEREN-ZUR
  University of Tel Aviv, Israel

- Bruno ROSSI Diploma to:

  Serena FAGNOCCHI
  Centro Fermi, Rome, Italy

- Andreij D. SAKHAROV Diploma to:

  Evgeny SKVORTSOV
  Lebedev Physical Institute, Moscow, Russia

- Alexandre SMIRNITSKIY Diploma to:

  Roberto PREGHENELLA
  University of Bologna, Italy

- Victor F. WEISSKOPF Diploma to:

  David ANTONIO
  University of Edinburgh, UK

- Eugene P. WIGNER Diploma to:

  Vivek Singh PARIHAR
  Northeastern University, Boston, USA

- Bjorn WIIK Diploma to:

  Gerhard BRANDT
  DESY, Hamburg, Germany

- Chien Shiung WU Diploma to:

  Yasuko HISAMATSU
  University of Tokyo, Japan

AWARDS

- The Award for the best "New Talents-Speaker" is divided into five parts:

i) for an original work in theoretical physics:

Dmitry KROTOV
ITEP, Moscow, Russia

ii) for valuable work in experimental physics:

Francesco NOFERINI
University of Bologna and INFN, Italy

iii) for an original work in data analysis:

Nicola NERI
INFN, Sezione di Pisa, Italy

iv) for the best theoretical presentation:

Xavier Calmet
University of Brussels, Belgium

v) for the best experimental presentation:

Yasuko Hisamatsu
University of Tokyo, Japan

- Award for BEST STUDENT given to:

  Yasuko HISAMATSU
  University of Tokyo, Japan

- Awards for BEST SCIENTIFIC SECRETARY given to:

  David ANTONIO
  University of Edinburgh, United Kingdom

PARTICIPANTS

Halina ABRAMOWICZ
Israel

Department of HEP
School of Physics and Astronomy
Tel Aviv University
Ramat Aviv, 69978, TEL AVIV, Israel

Victor A. ALEXANDROV
Russia

ITEP
B. Cheremushkinskaya 25
RU-117259 MOSCOW, Russia

Cristian ANDREI
Romania

International Institute for Nuclear Physics Engineering
Centre of Excellence
R-76900 BUCHAREST-MAGURELE, Romania

Laura ANDRIANOPOLI
Italy

CERN
CH-1211 GENEVA 23, Switzerland

Ignatios ANTONIADIS
Greece

CERN - Theory Division
CH-1211 GENEVA 23, Switzerland

Centre de Physique Théorique
Ecole Polytechnique
PALAISEAU, France

David ANTONIO
UK

JCMB Building
The University of Edinburgh
King's Buildings
EDINBURGH, EH9 3JZ, UK

Silvia ARCELLI
Italy

Dipartimento di Fisica
Università degli Studi
Via Irnerio 46 - 40100 BOLOGNA, Italy

Paolo ASCHIERI
Italy

Dipartimento di Scienze e Tecnologie Avanzate
Università degli Studi
Via Bellini 25/G - 15100 ALESSANDRIA, Italy

Yakov AZIMOV
Russia

Theory Division
St. Petersburg Nuclear Physics Institute
RU-188350 GATCHINA, St. Petersburg, Russia

Roberto BALBINOT
Italy

Dipartimento di Fisica
Università degli Studi
Via Irnerio 46 – 40126 BOLOGNA, Italy

Rinaldo BALDINI FERROLI
Italy

Enrico Fermi Centre
Via Panisperna, 89/A - 00184 ROMA, Italy

Christian BECK
UK

Queen Mary, University of London
Mile End Road
LONDON, E1 4NS, UK

Viktor BEGUN
Ukraine

Bogolyubov Institute for Theoretical Physics
14 b Metrologicheskaya Str.
UA-03143 KIEV, Ukraine

Ernesto Pinheiro BORGES
Brazil

Escola Politécnica, Departamento de Engenharia Química
Universidade Federal da Bahia
Rua Aristides Novis 2, Federacao
SALVADOR, BA 40210-630, Brazil

Jacob BOURJAILY
USA

CALTECH
1201 East California Boulevard
PASADENA, CA 91125, USA

Gerhard BRANDT
Germany

DESY
Notkestrasse 85
D-22607 HAMBURG, Germany

Stefania BUFALINO
Italy

INFN e Dipartimento di Fisica Sperimentale
Università degli Studi
Via Pietro Giuria 1 - 10125 TORINO, Italy

Xavier CALMET
France

Department of Physics Theorique
Université Libre de Bruxelles
B-1050 BRUSSELS, Belgium

Leonardo CASTELLANI
Italy

Facoltà di Scienze MFN
Università del Piemonte Orientale
Via Bellini 25/G - 15100 ALESSANDRIA, Italy

Dipartimento di Fisica Teorica
Università degli Studi
Via P. Giuria 1 - 10125 TORINO, Italy

Ciro CATTUTO
Italy

Enrico Fermi Centre
Via Panisperna, 89/A – 00184 ROMA, Italy

Oana CATU
Romania

AW Wright Nuclear Structure Laboratory
Yale University
NEW HAVEN, CT 06520-8124, USA

Luisa CIFARELLI
Italy

Dipartimento di Fisica
Università degli Studi
Via Irnerio 46 – 40126 BOLOGNA, Italy

Fabrizio COCCETTI
Italy

Enrico Fermi Centre
Via Panisperna, 89/A – 00184 ROMA, Italy

Eugenio COCCIA
Italy

Laboratori Nazionali del Gran Sasso – LNGS
S.S. 17 Bis km 18+910 - 67010 ASSERGI, L'Aquila, Italy

Gilberto COLANGELO
Italy

Berna University
Institute for Theoretical Physics
Sidlerstrasse 5
CH-3012 BERN, Switzerland

Mátè CSANÁD
Hungary

FLTE TTK
Department of Atomic Physics
H-1117 BUDAPEST XI, Hungary

Frederik DENEF
Belgium

NHETC - Department of Physics and Astronomy
Rutgers, The State University of NJ
126 Frelinghuysen Rd.
PISCATAWAY, NJ 08854-8019, USA

Denis DERKACH
Russia

V. Fock Institute for Physics
St. Petersburg State University
RU-198504 ST PETERSBURG, Russia

Vyacheslav DIDENKO
Russia

I.E. Tamm Department of Theoretical Physics
Lebedev Physical Institute
RU-119991, MOSCOW, Russia

Michael R. DOUGLAS
USA

NHETC - Department of Physics and Astronomy
Rutgers, The State University of NJ
126 Frelinghuysen Rd.
PISCATAWAY, NJ 08854-8019, USA

Michael James DUFF
UK

Blackett Laboratory
Imperial College London
South Kensington Campus
LONDON SW7 2AZ, UK

Vladimir DUDCHENKO
Germany

ITEP
B. Cheremushkinskaya 25
RU-117259 MOSCOW, Russia

Callum DURNFORD
UK

Durham University
Science Site, South Road
DURHAM, DH1 4FF, UK

Serena FAGNOCCHI
Italy

Dipartimento di Fisica
Università degli Studi
Via Irnerio 46 - 40126 BOLOGNA, Italy

Marco GARBINI
Italy

Enrico Fermi Centre
Via Panisperna, 89/A – 00184 ROMA, Italy

Istituto di Fisica, Università degli Studi
Via Irnerio 46 - 40126 BOLOGNA, Italy

Marcello GIORGI
Italy

INFN e Università degli Studi
Viale Buonarroti 2 – 56100 PISA, Italy

Ivan GNESI
Italy

Dipartimento Fisica Generale "A. Avogrado"
Università degli Studi
Via P. Giuria 1 - 10125 TORINO, Italy

Wlodzimierz GODLOWSKI
Poland

Astronomical Observatory
Jagellonian University
Ul. Oria 171
30-244 KRAKÓW, Poland

Jacek GOLBIAK
Poland

Department of Theoretical Physics
Catholic University of Lublin
PL-20-039 LUBLIN, Poland

Pavel GOLDIN
Russia

Theoretical Department
St. Petersburg Nuclear Physics Institute
RU-288300 GATCHINA, ST. PETERSBURG, Russia

Gary R. GOLDSTEIN
USA

Department of Physics and Astronomy
Tufts University
MEDFORD, MA 02155, USA

Alexey GOLOVNEV
Russia

Institute of Physics
St. Petersburg State University
198904 Ulianovskaya ul. 1, Stary Petergof
ST. PETERSBURG, 288300, Russia

Dieter HAIDT
Germany

EPJC – DESY
Notkestrasse 85
D-22603 HAMBURG, Germany

Stephen D.H. HSU
USA

Department of Physics
University of Oregon
EUGENE, OR 97403, USA

Haim HARARI Israel	Department of Particle Physics Weizmann Institute 76100 REHOVOT, Israel
John W. HARRIS USA	Physics Department - Yale University 272 Whitney Avenue NEW HAVEN, CT 06520-6124, USA
Yasuko HISAMATSU Japan	International Centre for Elementary Particle University of Tokyo 7-3-1 Hongo, Bunkyo ku, TOKYO, Japan
Jay N. IHRY USA	Department of Physics and Astronomy University of North Carolina at Chapel Hill CHAPEL HILL, NC 27599-3255, USA
Pratima JINDAL India	Department of Physics Panjar University 160014 CHANDIGARTH, India
Marek KARLINER Israel	Department of Physics Faculty of Exact Sciences Tel Aviv University, RAMAT AVIV, Israel
Manjit KAUR India	Department of Physics Centre of Advanced Study in Physics Panjab University 160 014 CHANDIGARH, India
Richard D. KENWAY UK	Department of Physics and Astronomy University of Edinburgh The King's Buildings EDINBURGH, EH9 3JZ, Scotland, UK
Boaz KEREN-ZUR Israel	Department of Physics Tel Aviv University 69978 TEL AVIV, Israel
Volodymyr KONCHAKOVSKYI Ukraine	Bogolyubov Institute for Theoretical Physics 14b Metrologicheskaya Str. UA-03143 KIEV, The Ukraine
Christian P. KORTHALS-ALTES The Netherlands	Centre de Physique Théorique CNRS-Luminy F-13288 MARSEILLE Cedex 9, France
Dmitry KROTOV Russia	ITEP B. Cheremushkinskaya 25 RU-117259 MOSCOW, Russia
Alexander KUREPIN Russia	Moscow Engineering Physical Institute and the Institute for Nuclear Research prospekt 60-letiya Oktyabrya 7a RU-117312 MOSCOW, Russia
Aharon LEVY Israel	School of Physics and Astronomy Tel Aviv University Ramat Aviv, 69978, TEL AVIV, Israel
Nigel LOCKYER USA	Department of Physics and Astronomy University of Pennsylvania 209 South 33rd PHILADELPHIA, PA 19104-3696, USA

Bernd LOEHR
Germany

DESY
D-22607 HAMBURG, Germany

David LOPEZ MATEOS
Spain

CALTECH
1200 E California Boulevard
PASADENA, 910125 CA, USA

Pedro MACHADO
Brazil

Institute for Theoretical Physics
Leuvenlaan 4
NL-3584 CE UTRECHT, The Netherlands

Alessio MARRANI
Italy

INFN - Laboratori Nazionali di Frascati
Via Enrico Fermi 40 - 00044 FRASCATI, Italy

Alexey MATVEEV
Russia

I.E. Tamm Department of Theoretical Physics
Lebedev Physical Institute
RU-119991, MOSCOW, Russia

Anton NAUMENKO
Ukraine

ITEP
B. Cheremushkinskaja, 25
RU-117259 MOSCOW, Russia

Nicola NERI
Italy

INFN Sezione di Pisa
Largo Bruno Pontecorvo 3 - 56127 PISA, Italy

Francesco NOFERINI
Italy

Dipartimento di Fisica
Università degli Studi
Via Irnerio 46 - 40100 BOLOGNA, Italy

Simone PACETTI
Italy

Sezione INFN - LNF
Via Enrico Fermi 40 - 00044 FRASCATI, Italy

Vivek Singh PARIHAR
India

Department of Physics
Northeastern University
BOSTON, MA-02115, USA

Roberto PETRONZIO
Italy

Istituto Nazionale Fisica Nucleare
Piazza de' Capretari 70 - 00186 ROMA, Italy

Yury PETUKHIN
Russia

Theoretical Department
St. Petersburg Nuclear Physics Institute
RU-288300 GATCHINA, ST. PETERSBURG, Russia

James PINFOLD
Canada

Department of Physics
Centre for Subatomic Research
University of Alberta
EDMONTON, Alberta T6G 2N5, Canada

Guido PIRAGINO
Italy

Dipartimento di Fisica Generale "Amedeo Avogadro"
Università degli Studi
Via P. Giuria 1 - 10125 TORINO, Italy

Gil PONTECORVO
Italy

INFN e Dipartimento Fisica Generale "A. Avogrado"
Università degli Studi
Via P. Giuria 1 - 10125 TORINO, Italy
Enrico Fermi Centre
Via Panisperna, 89/A – 00184 ROMA, Italy

Roberto PREGHENELLA
Italy

Dipartimento di Fisica
Università degli Studi
Via Irnerio 46 – 40126 BOLOGNA, Italy

Paul RESKA
Germany

Institute for Theoretical Physics
Leuvenlaan 4
NL-3584 CE UTRECHT, The Netherlands

Igor RUBINSKIY
Russia

DESY/ZEUS
Notkestrasse 85
D-22607 HAMBURG, Germany

Gabriella SARTORELLI
Italy

Istituto Nazionale di Fisica Nucleare - INFN
Sezione di Bologna
Via Irnerio 46 - 40126 BOLOGNA, Italy

Cristina SBARRA
Italy

Dipartimento di Fisica
Università degli Studi
Via Irnerio 46 – 40126 BOLOGNA, Italy

Johannes C. SENS
France

Institut Non-Linéaire de Nice
Université de Nice, Sophia Antipolis
1361 Route des Lucioles
F-06560 VALBONNE, France

Maksym SERBYN
Ukraine

ITEP
B. Cheremushkinskaya 25
RU-117259 MOSCOW, Russia

John V. SHEBALIN
USA

NASA - Lyndon B. Johnson Space Center
HOUSTON, TX,

Alexey SIZANOV
Russia

Theoretical Department
St. Petersburg Nuclear Physics Institute
RU-288300 GATCHINA, ST. PETERSBURG, Russia

Evgeny SKVORTSOV
Russia

I.E. Tamm Department of Theoretical Physics
Lebedev Physical Institute
RU-119991, MOSCOW, Russia

Ori SMITH
Israel

Particle Physics Department
Tel Aviv University
TEL AVIV, Israel

Jacob SONNENSCHEIN
Israel

Department of the Particle Physics
Faculty of Exact Sciences - Tel Aviv University
Kaplun 405 - Raymond and Beverly Sackler
69978 TEL AVIV, Israel

Yogendra N. SRIVASTAVA
India

INFN e Dipartimento di Fisica
Università degli Studi
Via A. Pascoli – 06100 PERUGIA, Italy

Leonard SUSSKIND
USA

Department of Physics
Stanford University
STANFORD, CA 94305-4060, USA

Alexey SVIATKOVSKY
Russia

Theoretical Department
St. Petersburg Nuclear Physics Institute
RU-288300 GATCHINA, ST. PETERSBURG, Russia

Francesco SYLOS-LABINI
Italy

Enrico Fermi Centre
Via Panisperna, 89/A – 00184 ROMA, Italy

Institute for Complex Systems CNR
Via dei Taurini, 19 - 00100 ROMA, Italy

Constantino TSALLIS
Greece

Centro Brasileiro de Pesquisas Fisicas
Rua Xavier Sigaud 150
22290-180 RIO DE JANEIRO - RJ, Brazil

Gerardus 't HOOFT
The Netherlands

Department of Theoretical Physics
Spinoza Institute - Utrecht University
NL-3508 TD UTRECHT, The Netherlands

Grigory VARTANOV
Russia

Bogoliubov Theoretical Physics Laboratory (BLTP)
JINR - Joint Institute for Nuclear Research
DUBNA, Russia

Horst WENNINGER
Germany

CERN
CH-1211 GENEVA 23, Switzerland

Chiara ZAMPOLLI
Italy

Dipartimento di Fisica
Università degli Studi
Via Irnerio 46 – 40126 BOLOGNA, Italy

Antonino ZICHICHI
Italy

CERN
CH-1211 GENEVA 23, Switzerland

Dipartimento di Fisica - Università degli Studi
Via Irnerio 46 – 40126 BOLOGNA, Italy